MISCELLANEOUS NON-SI TO SI CONVERSIONS

Name	Symbol	SI Value
inch	in	54×10^{-2} m
mile	m	09.344 m
nautical mile		52 m
angstrom	Å	$^{-10}$ m
pound	lb	5359237 kg
metric ton	t	1000 kg
atomic mass unit	amu	$1.6605402 \times 10^{-23}$ kg
atmosphere	atm	101325 N m^{-1} (Pa)
liter	l, L	10^{-3}m$^3 = 1$ dm^3
liter-atmosphere	L atm	101.325 J
bar	bar	10^5 Pa
liter-bar	L bar	10^2 J
torr	torr	$101325/760$ N m^{-1} (Pa)
British thermal unit	BTU	1055.056 J
electron volt	eV	$1.60217733 \times 10^{-19}$ J
erg	erg	10^{-7} J
dyne	dyne	10^{-5} N
poise	P	0.1 Pa s
oersted	Oe	$(1000/4\pi)$ A/m
maxwell	Mx	10^{-8} Wb
kilowatt-hour	kWh	3.6×10^6 J
calorie (thermochemical)	cal$_{th}$	4.184 J
calorie (at 15°C)	cal$_{15}$	4.185 J
calorie (international)	cal$_{IT}$	4.186 J
debye (dipole moment)	D	3.33564×10^{-30} C m
esu (of charge)	esu	$3.3356410 \times 10^{-10}$ C
curie	Ci	3.7×10^{10} Bq (s^{-1})
roentgen	R	2.58×10^{-4} C kg^{-1}
rad	rad	10^{-2} Gy
rem	rem	10^2 Sv
micron	μ	10^{-6} m
barn	b	10^{-28} m^2
maxwell	Mx	10^{-8} W
gauss	G	10^{-4} T
minute	min	60 s
hour	h	3600 s
day	d	86400 s
degree	°	$(\pi/180)$ rad
minute	'	$(\pi/10800)$ rad
second	"	$(\pi/64000)$ rad

Sources: National Bureau of Standards, "Policy for NBS Usage of SI Units," *J. Chem.;* National Bureau of Standards Special Publication 330, U.S. Government Printing Office, Washington, D.C., 1981; and Cohen, E. R., and B. N. Taylor, *The 1986 Adjustment of the Fundamental Physical Constants,* CODATA Bulletin 63, Pergamon Press, Elmsford, New York, 1986.

PHYSICAL CHEMISTRY

Methods, Techniques, and Experiments

Rodney J. Sime

California State University,
Sacramento

 **SAUNDERS GOLDEN SUNBURST SERIES**

SAUNDERS COLLEGE PUBLISHING

Philadelphia   Ft. Worth   Chicago
San Francisco   Montreal   Toronto
London   Sydney   Tokyo

Text Typeface: Times Roman
Compositor: TAPSCO, Inc.
Acquisitions Editor: John Vondeling
Managing Editor: Carol Field
Project Editor: Mary Patton
Copy Editor: Ruth Melnick
Manager of Art and Design: Carol Bleistine
Art Director: Christine Schueler
Art and Design Coordinator: Doris Bruey
Text Designer: Dorothy Chattin
Cover Designer: Lawrence R. Didona
Text Artwork: Rolin Graphics, Inc.
Director of EDP: Tim Frelick
Production Manager: Charlene Squibb

Cover Credit: Samuels Studio © 1988   Al Micciche, photographer

Printed in the United States of America

PHYSICAL CHEMISTRY: METHODS, TECHNIQUES, AND EXPERIMENTS

ISBN 0-03-009499-2

Library of Congress Catalog Card Number: 89-043114

4567   045   98765432

To my family—Ruth, Karl,
Erik, Karen, and Jennifer—
and to my students in Chemistry 141,
the Physical Chemistry Laboratory

PREFACE

Traditionally, work in the physical chemistry laboratory has offered students the opportunity to reinforce their understanding of general principles by executing selected experiments and reporting the data.

In addition, the laboratory has inherited the responsibility for acquainting students with a number of skills that are required of undergraduates but often are not systematically part of the chemistry curriculum. Foremost among the professional skills students must master informally are reading, writing, and programming. By "reading" I mean the ability to search the literature of chemistry systematically. By "writing" I mean the ability to analyze one's work critically and to communicate clearly its meaning, its significance, and its limitations to the scientific community. Finally, it is axiomatic that chemistry undergraduates should at least be computer literate. In this experimental physical chemistry laboratory course, computer literacy includes the ability to write reports with a typical word processor and to write data reduction programs in the Pascal language.

The approach to physical chemistry laboratory work described in this book has evolved from a one-semester course I have taught for many years at California State University, Sacramento. Students work in the laboratory for about two and one-half hours twice a week, with an accompanying lecture once a week. The laboratory course is entirely independent of the physical chemistry lecture course, which consists of three lectures per week for two semesters. The physical chemistry laboratory may be taken concurrently with the second semester of the lecture course, an option that most students choose. It is expected that students complete all physical chemistry laboratory and lecture work in the third, or junior, year.

Completion of physical chemistry in the junior year allows time in the senior year for a variety of special projects or faculty-directed research. Consequently, I find it appropriate to give students in the physical chemistry laboratory an introduction to the methods of research. There is little difference between the mechanics of executing a predesigned physical chemistry experiment and the data collection phase of a research problem. Students are encouraged to select and treat each of their experiments as if they were the original investigator. With this slight shift in attitude, students eagerly accept guidance in the methods of research described in

Parts One and Two of this book, while they are experimentally testing the hypotheses described in Part Three.

Part One, The Elements of Research, consists of eight chapters on the methodology of research. From choosing a problem to writing the report, the actions of student and researcher are nearly identical except for designing the experiment. Even here, students are encouraged to criticize the design of experiments and offer alternatives. Even though the physical chemistry laboratory is not a research laboratory, it is possible for students to do their experimental work with an awareness of how their activities mesh with the larger framework of research. Chapters 1 through 8 break up the process of research into some of its components, some of which are immediately useful to the undergraduate experimental physical chemist.

It was once common practice to offer a chemical literature course to chemistry students, but as the chemistry curriculum became more congested, this course has almost disappeared. Yet the continuously increasing torrent of scientific literature makes it even more important to be able to cope with the flood of scientific knowledge. In a physical chemistry laboratory modeled after research methodology, it is appropriate to acquaint students with some of the main features of the literature of chemistry. Besides the practical value of this approach, it is pedagogically useful for students to compare their results with literature values. Students are invariably thrilled, sometimes amazed, to discover that their results agree, within experimental error, with numbers on a printed page.

In the natural sequence of research, the choice of the problem and a search of the literature are followed by the design and execution of experiments. Although students are not routinely involved in the design of physical chemistry laboratory experiments, two short chapters on the design and execution of experiments serve as a brief introduction. Indeed, some experiments have evolved over the years as a result of student special projects that gave them an opportunity to design as well as execute an experiment.

Experiments in physical chemistry often generate fairly long lists of data—that is, parallel arrays of dependent and independent variables. It is usually possible to find a linear relationship between the two or carry out an appropriate linear transformation. Consequently, in Chapter 6 the method of least squares is described in detail, as well as the transformations of polynomials to linear form.

Analyzing the errors in an experiment implies more than the calculation of the standard error or propagated error. It suggests a conscientious, self-critical perspective. Students nearly always find it difficult to be simultaneously self-critical and totally objective. In Chapter 7 they learn how to evaluate their data and the methods of communicating their evaluation to other scientists in the language of science and statistics.

Most students of experimental physical chemistry find report writing even more challenging than error analysis. It takes a certain tenacity on the part of both instructor and student to maintain consistently high standards of writing. Although my own students have been known to grumble about report writing, when I meet them years later they express the most appreciation for what they learned in this area. I stress that high-quality laboratory work is of little or no value unless properly communicated to the scientific world. Writing physical chemistry laboratory reports in a typical scientific journal style prepares students to write a report, thesis, or journal article. Some of the elements of scientific writing are presented in the last chapter of Part One.

I believe that the physical chemistry laboratory is the place for a student to begin to make the transition from student to scientist. We accomplish this by treating each experiment as a small research problem, and by thinking about it in terms of choosing a problem, searching the literature, designing the experiment, executing the experiment, reducing the data, analyzing the errors, and writing the report. These are the elements that make up Part One of this book.

Part Two covers some of the tools and techniques used in physical chemistry, beginning with microcomputers. Aficionados like to tick off long lists of uses for computers by scientists. In resisting that temptation, I have tried to limit computer use in the physical chemistry laboratory to word processing and number crunching. More than any other course in chemistry, physical chemistry laboratory work tends to generate large data sets requiring considerable numerical calculation. The power of a computer to alleviate the burden motivates students to use computers and to write programs of their own for computers. I have found that students who want to learn are easy to teach.

It is important, however, to resist the temptation to computerize the laboratory and turn an experimental chemistry course into a computer science course. Computers are just a tool, not an end in themselves. This book presents a compromise that has proved workable. Many of the experiments are accompanied by computer programs, since it would take far too much time for students to regularly write programs for most of their own experimental work in the physical chemistry laboratory. In the student's report, however, a sample calculation of a data pair should be included that verifies the agreement between a single hand calculation and the program output. In addition, students learn sufficient programming to understand the programs furnished for them and also to write their own simple programs if the occasion arises.

Pascal was chosen as the programming language for the physical chemistry laboratory for a number of reasons. It is a modern language that lends itself to structured programming in general and good programming in particular. Even to someone not skilled in the language, a Pascal program is quite readable—much more so than other languages. Consequently, it is relatively simple for beginning students to relate Pascal algorithms to the physical chemistry from which they originate. Even more important, students *like* Pascal. Students experienced in BASIC or FORTRAN nearly always come to prefer Pascal, and students who have had absolutely no contact with computers of any kind have no problem with Pascal as their first language. Finally, an inexpensive yet superb type of Pascal software is available that runs on the IBM-PC or compatibles (PC-DOS or MS-DOS): Turbo Pascal.[1] Turbo Pascal is widely available and has become the standard against which other microcomputer Pascals are compared.[2,3]

Most important, writing programs in Pascal encourages good programming and problem-solving techniques. It requires the student to break the problem up into simply understood units analogous to the solution of a physical chemistry problem. It compels a student to identify constants and variables. It forces a student to recognize which variable assignments correspond to experimental observation and input and which variable assignments correspond to output derived from some physical chemical algorithm.

Because the Turbo Pascal Editor has the "look and feel" of WordStar, the use of WordStar[4] as a word processor eliminates the need to give instruction on the Turbo Pascal Editor. I begin the physical chemistry laboratory course with two 50-minute lectures describing the course (one-half lecture), MS-DOS (one-half lecture), and WordStar (one lecture). This enables students to write their first report with WordStar on a PC-compatible computer. It is not necessary to *master* WordStar in order to use it effectively for report writing. A relatively small subset of commands, especially those that overlap the Turbo Pascal Editor commands, suffices to write a physical chemistry laboratory report, or even a journal article for that matter. Interestingly, students who cannot type at all are as appreciative of the utility of word processing as are those who type well—perhaps more so, since they are more apt to make typing errors, which are so easily corrected with a word-processing program.

The next two lectures cover error analysis, followed by about five lectures serving as an introduction to Pascal with an emphasis on processing numbers. The material in Chapter 10, Pascal for Physical Chemistry, is a tutorial in elementary Pascal programming with an emphasis on handling numbers as opposed to characters and strings. At first, students just run some of the programs furnished in Chapter 10 and in selected experiments. This gets them to a skill level at which they can write quite powerful programs on their own. For physical chemistry laboratory students this introduction works better than the traditional Pascal instruction given by computer science departments, which place much more emphasis on text and character processing.

Chapter 11 offers an introduction to glass blowing the way physical chemists blow glass, not the way glass blowers blow glass. No specific

---

[1] Borland International, Inc., 4585 Scotts Valley Drive, Scotts Valley, California 95066. Turbo Pascal is also implemented for the Macintosh.

[2] Bridger, M., "Turbo Pascal 3.0," *Byte Magazine,* February 1986.

[3] Shammas, N. C., "Pascal for the IBM PC," *Byte Magazine,* December 1986.

[4] The manuscript for this book was typed not with WordStar but with WordPerfect, which is by far the most popular MS-DOS word-processing program. However, the Turbo Pascal Editor does emulate WordStar. The use of WordStar eliminates the need to learn two word-processing programs.

experiments in glass blowing are given, but a collection of selected exercises could be used as a physical chemistry experiment in glass blowing. Most of our students use this chapter as they move on to special projects and undergraduate research.

Pressure and temperature are so important in physical chemistry that a chapter on each is included. The emphasis is on vacuum technique in Chapter 12 and on the definition and measurement of temperature in Chapter 13. Chapter 14, the final chapter of Part Two, treats instruments with an emphasis on spectrometry.

Part Three contains 40 experiments in physical chemistry in the areas of thermodynamics, structure, and kinetics. About one fourth of the experiments include a Pascal computer program to assist with data reduction. More than half the experiments generate sufficient linearly related data to justify a linear-least-squares treatment of data.

Since a full-year physical chemistry lecture course has not been completed by most students enrolled in the physical chemistry laboratory, each experiment includes sufficient background theory to execute the experiment with understanding. References for further reading appear at the end of each experiment.

ACKNOWLEDGMENTS

I am grateful to the Literary Executor of the late Sir Ronald A. Fisher, F.R.S.; Dr. Frank Yates, F.R.S.; and the Longman Group Ltd., London, for permission to reprint Table 7–8 from their book *Statistical Tables for Biological, Agricultural and Medical Research* (Sixth Edition, 1974). I wish to thank the following reviewers for their helpful comments:

Salim Banna, Vanderbilt University
Gary Bertrand, University of Missouri at Rolla
Alan Campion, University of Texas at Austin
Joseph Cantrell, Miami University of Ohio
Joseph Chaiken, Syracuse University
Edward Grant, Purdue University
Colin Hubbard, University of New Hampshire
Jack Morgan, University of Nevada at Reno
P. L. Polavarapu, Vanderbilt University
Susanne Rayner-Kipnis, Rutgers University
Emil Slowinski, Macalester College, St. Paul
David Waldeck, University of Pittsburgh
Richard Wilde, Texas Tech University

It has been a great pleasure to work with the staff at Saunders College Publishing. I wish to thank John Vondeling, Associate Publisher, for his support from the very beginning. Copy editor Ruth Melnick and proofreader Bob Griffin made numerous invaluable contributions for which I am grateful. Most of all I wish to thank Mary Patton, project editor, for her gentle but constant and effective leadership.

Over the years, many students enrolled in Chemistry 141, the physical chemistry laboratory, have contributed to this book; special thanks are due Paul Aldrich, Patrick Derochers, Hans Deuel, John Freitas, Andrei Tokmakoff, and John Wooley.

I thank my wife Ruth for her encouragement and inspiration; my daughter Jennifer for forgiving my absence during some of her important moments; and my daughter Karen for her photographic contributions to this book and for the joy of working with her.

FOREWORD TO THE STUDENT

At many universities the physical chemistry laboratory course has a reputation as a "tough course," but that sobriquet is applied to many courses in which a lot of learning is going on. P. Chem. lab, as this course is most often called, offers a combination of learning experiences that do indeed require hard work but that are just as often fun and rewarding.

Just as a young medical resident is not quite practicing medicine on her own in a hospital, you are not quite doing scientific research on your own in the P. Chem. lab, but like the resident you are in a special environment. Here you can learn the *elements of research*. Look at the table of contents for the eight chapters that make up Part One of this book, The Elements of Research.

During this semester you will execute seven or eight experiments in physical chemistry and make some kind of sense out of your observations with appropriate calculations. You will also critically evaluate your work, compare your results with the scientific literature, and report all of this in writing. If your semester is 16 weeks long, this corresponds approximately to a report every 2 weeks, so you will need to continually organize your preparation, work, and writing.

Writing is hard work, but sometime during your undergraduate education you need to learn how to write in the language of your chosen profession. Science writing, report writing, and writing for scientific journals are slightly different from the traditional expository writing to which you have already been exposed. Writing your P. Chem. lab reports will give you a better understanding of how scientists communicate the results of their work to other scientists. At the end of the semester, when you finally hold in your hand all of your completed P. Chem. lab reports, I hope you will look back on the experience as rewarding and sometimes even fun.

Just as there is much more to becoming a professional scientist than mastering the principles and theories, executing a P. Chem. lab experiment is much more than figuring out which buttons to push and which knobs to turn in what order. Your experience in the P. Chem. lab is a kind of apprenticeship during which you learn the operations practiced by experienced scientists. As you work on an experiment, try to think of yourself as the original researcher who first chose the problem, designed the experiment, executed and evaluated it, and finally wrote it up and published it.

Rodney J. Sime
Sacramento, California

CONTENTS

APPENDICES

Those who have handled sciences have been either men of experiment or men of dogmas. The men of experiment are like the ant: they only collect and use; the reasoners resemble the spiders, who make cobwebs out of their own substance. But the bee takes a middle course, it gathers its materials from the flowers of the garden and the field, but transforms and digests it by a power of its own. Not unlike this is the true business of philosophy; for it neither relies solely or chiefly on the powers of the mind, nor does it take the matter which it gathers from natural history and mechanical experiments and lay it up in memory whole, as it finds it; but lays it up in the understanding altered and digested.

Francis Bacon (1561–1626), "Novum Organum," Aphorism 95.

Matthew Thompson McClure, ed., *Bacon Selections,* Charles Scribner's Sons, New York, 1928.

THE ELEMENTS
OF RESEARCH

CHAPTER 1

Introduction

Before going on to the elements of research, it is useful to review the overall nature of science and to see how laboratory experimentation fits into the scheme of things. In Part One, following a review of the methodology of science in Chapter 1, Chapters 2 through 8 present a brief introduction to research for undergraduates. In Part Two, Chapters 9 through 14 describe some of the important tools of physical chemistry, especially microcomputers. Part Three contains about 40 physical chemistry lab experiments. As you will see in Part One, executing an experiment is equivalent to testing a hypothesis.

The scientific method is the process by which science grows; science differs from other intellectual disciplines, not only in its content, but in its procedures. The scientific method may be divided into a series of consecutive steps:

1. Observation
2. *Facts*
3. Induction
4. *Hypothesis*
5. Deduction
6. *Experimentation*
7. Observation

It all begins by observing, seeing, and reflecting—in nature, in the laboratory, or even in books, where the observations of other scientists are reported. Through the magic of induction an idea is born: a hypothesis, a trial guess concerning some suspected regularity exhibited by nature. If the hypothesis is fruitful, it gives rise to further predictions concerning the behavior of nature, which in turn are testable by experimentation.

The experiments in the physical chemistry laboratory are examples of the testing of hypotheses. Besides illustrating and clarifying well-established hypotheses, theories, and laws, they provide a forum for practicing the scientific method and imitating the research methods of scientists.

OBSERVATION: THE SELECTION OF FACTS

Observation is included in our list twice because the scientific method is cyclic. One could argue that the list really should consist of items 1, 3, and 5, so that the scientific method involves the collection of facts through observation; the construction of a hypothesis, an inductive process; and the testing of a hypothesis by experimentation, a deductive process. When a hypothesis has withstood sufficient testing, it may be promoted to the status of a law or theory.

Notice that it is possible to write a detailed description of the various components, hardware, and apparatus that make up an *experiment*. We can list the chemicals, the measuring instruments, the spectrometers, and the temperature- and pressure-sensing devices. It is certainly possible to describe a *hypothesis* in words by listing its postulates. For example, one of the postulates of the Bohr Theory is that the angular momentum of an electron can take on only integral multiples of $h/2\pi$, and the First Law of Thermodynamics states that energy is conserved in a cyclic process. The *facts* of science first are written in laboratory notebooks, then appear in reports, and eventually are preserved in libraries. It is a straightforward matter to write a description of an experiment, a hypothesis, or a fact.

On the other hand, the remaining three components of the scientific method are ephemeral, human actions. Observation is a very personal thing, dependent on sensations including seeing, hearing, and touching. It involves selection and rejection. It requires processing sensory output with the brain and interpreting the brain's signals, making decisions on their meaning. Induction and deduction are specific ways of thinking, of using one's brain to arrive at a conclusion. Observation, induction, and deduction are the procedures by which we advance among fact, hypothesis, and experiment.

Laws and theories are hypotheses that have been sufficiently tested so that they become part of the navigation system of science, providing guidance and direction for further explorations into nature. The relationship between laws and theories and hypotheses is shown in Figure 1–1. What is the difference, then, between a law and a theory?

In chemistry, laws tend to be generalizations based upon macroscopic properties of matter, while theories are generalizations or models based upon microscopic properties. For example, the three laws of thermodynamics are based on the bulk properties of matter, and their validity is completely independent of the existence of atoms. On the other hand, statistical thermodynamics is a theory, formulated in terms of behavior on the atomic or molecular scale.

Not every problem resolution in science achieves the status of a law or theory. More commonly, as problems in science arise, scientists respond, and as problems are resolved they simply become part of scientific knowledge without ever becoming a law or theory.

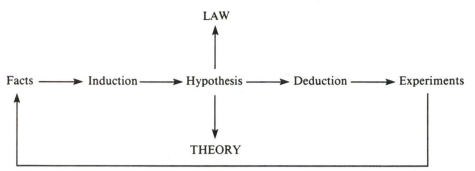

Figure 1–1 The elevation of a hypothesis into a theory or law.

Observation includes all knowledge currently available: that which a scientist may draw upon from his own mind, from his own observations on his own experimental inquiries, and from the literature of science, that is, from authority in science. Scientists are skeptical of authority and reserve the right to test it with their own observations. Nevertheless, what stands in the literature is, by consensus and peer review, what constitutes present-day scientific knowledge. That knowledge is dynamic, changing from year to year as new theories replace old theories. The facts of science, if gathered by careful observation and reported with clarity, may be more permanent.

Many scientists seldom stray far from simply observing. Some botanists spend a lifetime observing, collecting, classifying, and adding to currently available botanical knowledge. Large numbers of physical chemists are engaged in the careful measurement of thermodynamic properties, an important activity that adds to our available knowledge. Crystallographers determine the structures of crystals furnished to them by synthetic organic and inorganic chemists, an undeniably important contribution to our scientific knowledge.

Strictly speaking, observation isolated from the whole research cycle is not in itself the scientific method, yet it is an important activity of virtually all practicing scientists. Nevertheless, an investigation into a new area is likely to lead to surprising results demanding further explanation. For example, in the 1920s Alfred Stock systematically investigated compounds between hydrogen and boron, a virtually untapped area. Eventually these observations led to the construction of new hypotheses on the nature of bonding in boron hydrides and other electron-deficient compounds. In 1980, Lipscomb was awarded the Nobel Prize for extending the concept of the three center–two electron bond to the topologies of complex boron hydrides.

INDUCTION: CONSTRUCTING HYPOTHESES

Induction is generalization. It is the thought process of drawing inferences valid for a whole set from observations on a limited number of members of the set. For example, although I have not observed all living crows, every crow that I have seen has been black. From that large but incomplete set, I construct a hypothesis: all crows are black. In science, induction is the thought process by which a hypothesis is constructed from a limited number of observations or experiments.

Deduction, on the other hand, is the process of reasoning in which a conclusion follows necessarily from the premise, or hypothesis. If the premise is true, then the deduced conclusion must be true. If the bird is a crow, I deduce that it is black. Induction means to lead in. Deduction means to lead from. The relationship between induction and deduction is shown schematically in Figure 1–2.

Science has its share of dramatic moments when an intuitive leap occurs in a scientist's mind, allowing him to formulate a hypothesis that ties together a number of isolated facts. Let us examine some familiar examples.

Archimedes' Principle (a hypothesis) is as follows: if a solid body is immersed in a fluid it loses weight, and the weight it loses equals the weight of the displaced fluid. It is said that Archimedes was a philosopher and not an experimentalist. That may be true, but he was a very talented observer. If he never experimented with floating bodies, he must have spent many an hour watching floating objects, perhaps his own body in his bath. When Archimedes thought of the connection between weight loss and the weight of displaced fluid, he was, so the story goes, so excited that he left his bath and ran naked down the streets of Syracuse shouting "Eureka! Eureka!" (I have found it! I have found it!).

What conditions are favorable for forming hypotheses, for making the connections and generalizations? Reflective, controlled thinking is sometimes fruitful. Once in a while it is necessary to stop experimental work, change perspective, and turn over the ideas in one's mind in an ordered and consecutive manner, as opposed to daydreaming. Taking stock periodically is a powerful tactic.

On the other hand, controlled thinking in isolation can get one

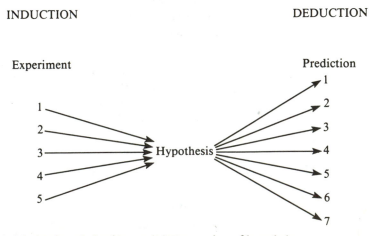

Figure 1–2 Induction, deduction, and the expansion of knowledge.

caught in an established pattern that is not productive. Discussing a problem with colleagues helps one escape from dead ends. Colleagues can sometimes furnish new ideas and even identify errors. Regular research meetings are useful; so are short morning and afternoon coffee breaks with colleagues.

The main idea is that it is as important to think as to work. Even daydreaming is sometimes productive. The story of F. A. Kekulè's solution to the structure of benzene is worth retelling. In 1865, most scientists felt confident that carbon was tetravalent and that they had correctly worked out the constitution of saturated hydrocarbons. However, the relatively low ratio of hydrogen to carbon in unsaturated hydrocarbons, including benzene, presented a perplexing problem. Kekulè described his thoughts vividly:

> I was sitting, writing at my text-book; but the work did not progress; my thoughts were elsewhere. I turned my chair to the fire and dozed. Again the atoms were gambolling before my eyes. This time the smaller groups kept modestly in the background. My mental eye, rendered more acute by repeated visions of the kind, could now distinguish larger structures, of manifold conformation: long rows, sometimes more closely fitted together; all twining and twisting in snake-like motion. But look! What was that? One of the snakes had seized hold of its own tail, and the form whirled mockingly before my eyes. As if by a flash of lightning I awoke; and this time also I spent the rest of the night in working out the consequences of the hypothesis.[1]

The "consequences," well known today, are the hexagonal ring structure of benzene and its alternating single and double bonds.

Usually, however, science does not advance in such dramatic leaps. It is more likely that small sets of observations lead to a less sweeping, more tentative, lesser hypothesis. Upon testing, if the tentative hypothesis holds up, more observations may permit the construction of a more robust hypothesis. The whole process repeats itself over and over. As reports of substantial progress appear in the literature, other workers may join in; gradually a new area of research may be identified, and out of the work of many, a significant new hypothesis may emerge that may eventually become an accepted theory. "Science has its cathedrals, built by the efforts of a few architects and of many workers."[2] Thus, in the cycle of the scientific method (Fig. 1–3), different workers join in at different points, as knowledge accumulates and science progresses.

CAUSE AND EFFECT

Another way of looking at the construction of hypotheses is to consider the connection between cause and effect. First, we need to define these

---

[1] Ihde, A. J., *The Development of Modern Chemistry,* Harper and Row, New York, 1964.
[2] Lewis, G. N., and M. Randall, *Thermodynamics,* McGraw-Hill, New York, 1926.

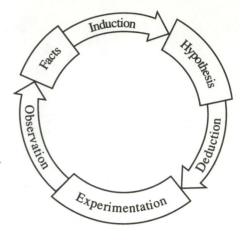

Figure 1–3 The scientific method: a cyclic process in which experimentation, facts, and hypothesis are connected by the human actions of observation, induction, and deduction.

terms. If, when event A is allowed to occur, event B always follows subsequently, and if, when event A is not allowed to occur, event B never occurs subsequently, then event A is said to be the cause of event B. The idea of cause and effect is a common one and is often self-evident. Whenever we use the word "because," we are asserting a cause-and-effect relationship. Our definition suggests a certain caution, which is well advised in searching for causes (constructing hypotheses). A few systematic methods are recognized as being of value in searching for causes: the method of agreement, the method of differences, and the method of concomitant variations. Let us examine these one at a time by comparing some definitions and looking at some examples.

The Method of Agreement

"If several instances of a phenomenon have one antecedent or circumstance in common, then that antecedent or circumstance in which alone all the instances agree (occur together) is the cause or effect of the phenomenon."[3] If that definition is not perfectly clear, let us compare it with the following: "If the circumstances leading up to a given event have in all cases had one factor in common, that factor may be the cause sought."[4] We can illustrate the method of agreement symbolically in a quite simple manner:

> If factors A, B, C, . . .
> and factors A, D, E, . . .
> and factors A, F, G, . . .
> all lead to event Z, then A is the cause of Z.

[3] Peterson, M. S., *Scientific Thinking and Scientific Writing,* Reinhold, New York, 1961.
[4] Wilson, E. B., Jr., *An Introduction to Scientific Research,* McGraw-Hill, New York, 1952.

As an example, consider the study of the Dutch scientist Eijkman on the cause of beriberi, a tropical disease.[5,6] Beriberi is a disease of the nervous system that causes general malaise and stiffness of the long muscles. In the Dutch East Indies, rice is an important dietary staple available in two forms, polished and unpolished. Eijkman observed that there seemed to be a disease like beriberi contracted by laboratory chickens, which ate primarily polished rice. It also appeared to him that the disease was more prevalent in humans who ate polished rice than in those who ate unpolished rice. Eijkman then proposed the hypothesis that beriberi is a disease caused not by a bacterium or a living agent, but rather by a dietary deficiency; and furthermore, that a dietary factor essential to good health is present in the hulls of unpolished rice and absent in polished rice. Subsequent experimental tests verified Eijkman's hypothesis and eventually led to the identification of thiamine, or vitamin B_1, as the essential dietary factor in unpolished rice hulls (and other foodstuffs).

As already pointed out, it is necessary to test a hypothesis with experiments that are quite independent of the observations that led to the construction of the hypothesis. "There is, for example, the story of the scientist who was overliberal in enjoying Scotch and soda at a party. The next morning he felt very poorly; so that night he tried rye and soda, again rather too freely. The following day, he again noted the same distressing symptoms. The third night he switched to bourbon and soda, but the morning after was no more pleasant than the others. Analyzing the evidence, he concluded that thereafter he would omit the soda from his drinks, since it was the common ingredient in the three observed cases."[7]

The Method of Differences

"If an instance where a phenomenon occurs and an instance where it does not occur have every antecedent or circumstance in common except one, then that one circumstance is the cause of the occurrence of the phenomenon."[8] Perhaps another point of view will help clarify: "If two sets of circumstances differ in only one factor and the one containing the factor leads to the event and the other does not, then this factor can be considered the cause of the event."[9] Again, let us represent this method symbolically:

> If factors A, B, C, D, E, . . . lead to event Z,
> and factors A, C, D, E, . . . do not lead to event Z,
> then B is the cause of event Z.

---

[5] Peterson, *Scientific Thinking.*
[6] Baker, J. J., and G. E. Allen, *Hypothesis, Prediction and Implication in Biology,* Addison-Wesley, New York, 1968.
[7] Wilson, *An Introduction to Scientific Research.*
[8] Peterson, *Scientific Thinking.*
[9] Wilson, *An Introduction to Scientific Research.*

Perhaps you have already noticed that this is the method usually referred to as "experiment and control," a method that often uses laboratory rats as experimental subjects. If one group of rats (the experimental group) is treated in an identical manner to a second group of rats (the control group), except that the experimental group diet contains saccharin, and it is observed that the experimental group develops malignant tumors while the control group develops none, then one may hypothesize that saccharin causes cancer in rats.

The Method of Concomitant Variations

"If quantitative changes in an antecedent bring about quantitative changes in a consequent, then antecedent and consequent are interrelated."[10] Compare this with the following definition: "If the variation in the intensity of a factor results in a parallel variation of the effect, then this factor is a cause."[11] Whereas the method of agreement and the method of differences are used to find a simple cause-and-effect relationship, the method of concomitant variations asks for a quantitative relationship between cause and effect. In a way this is an extension of the method of agreement: if the common factor A is increased, is there an increase (in the effect)? The relationship may be mathematically complex, that is, nonlinear, but nevertheless convincing, as shown, for example, in Figure 1–4.

The determination of the rate law of a chemical reaction by the method of initial rates is an example of the Principle of Concomitant Variation. For a reaction

$$a\text{A} + b\text{B} = d\text{D} + e\text{E}$$

the initial rate R is given by

$$R = k[\text{A}]^{\alpha}[\text{B}]^{\beta}$$

	Cause	Effect
	[A] [B]	[R]
Experiment 1	[A] [B]	[R]
Experiment 2	[2A] [B]	[2R]
Experiment 3	[3A] [B]	[3R]

Because our chemical systems are so well defined, and because we can control our relevant variables so completely, we use general and mathematical approaches to determine the rate law for a reaction. In the social and behavioral sciences, however, factorial design experiments often use the Principle of Concomitant Variation to find convincing cause-and-effect relationships.

[10] Peterson, *Scientific Thinking.*
[11] Wilson, *An Introduction to Scientific Research.*

"THEY'RE NOT SCIENTIFIC — THEY'RE JUST A LOT OF FUNNY COINCIDENCES"

Figure 1–4 The search for causes: "*The principle of concomitant variation . . .* states that if the variation of the intensity of a factor results in a parallel variation of the effect, then this factor is a cause." E. Bright Wilson, Jr., *An Introduction to Scientific Research,* McGraw-Hill, New York, 1952. (Herblock cartoon courtesy of *The Washington Post.*)

DEDUCTION: TESTING A HYPOTHESIS

How good is a hypothesis? What is the nature of the truth it contains? Can a false hypothesis predict true results? Since a hypothesis is based on a limited number of observations, it would seem that a hypothesis based on a large number of observations would more likely be true than one based on a small number of observations: "It probably can be said that no generalization is ever completely true and that few generalizations based on very many data are completely false."[12]

[12] Wilson, *An Introduction to Scientific Research.*

Experimentation

To test a hypothesis, an experiment must be designed, executed, and interpreted. In the example just described the experiment is to measure the rate law for the reaction. If the hypothetical rate law agrees with the measured rate law, then the mechanism may be true. This is a necessary, but not sufficient, condition to prove that the mechanism is true. In complex situations a "truth table" can be of assistance in sorting out difficult relationships. Table 1–1 is a truth table for a hypothesis. If a hypothesis is true, it always gives true predictions. If a hypothesis is false, however, it may give either true or false predictions. Consequently, it is impossible to *prove* the validity of any hypothesis. Nevertheless, in science hypotheses are tested, and as long as they work, they are used to guide further progress in acquiring knowledge and understanding.

A good hypothesis must be both fruitful and testable, and it should be compatible with science in general. The best hypothesis is one that leads us to something we might not have expected, or at least puts us on the right track, so that we find or learn something that we might not have if the hypothesis were not available to us. A hypothesis must be testable by experiment. It must have some kind of consequence that can be deduced from it by the laws of logic. If the hypothesis is true, then some event should occur that can be observed or measured according to some experimental procedure. If the event does not occur as predicted, then the hypothesis is false. If the event does occur as predicted, then the hypothesis *may* be true.

Logical Deductions

The rules of symbolic logic are useful in complex situations, but if the hypothesis is a simple generalization, then the logical relationship is usually self-evident. Some examples of logical relationships are familiar:

If	A = B		If	A > B
and	B = C		and	B > C
then	A = C		then	A > C

In symbolic logic A and B are propositions; in science they could be the postulates of a hypothesis; in computer science they would be values of variables or constants. In symbolic logic, such words as "not," "and," "or," and "if" are called logical connectives; in computer science they

TABLE 1–1

Truth Table for Hypotheses

True hypothesis	→	True predictions
False hypothesis	→	True predictions and false predictions

TABLE 1-2

Truth Tables

And Proposition			Or Proposition		
P	Q	P and Q	P	Q	P or Q
T	T	T	T	T	T
T	F	F	T	F	T
F	T	F	F	T	T
F	F	F	F	F	F

are called logical operators, but their use is quite similar. In computer science, the symbols >, <, =, <=, >=, and *in* are called relational operators and have the same meaning as in symbolic logic.

The truth table for *and* is seen to be much more restrictive than that for *or*, since *P and Q* is true only if both *P* and *Q* are true, but *P or Q* is true unless *P* and *Q* are both false (Table 1–2).

As an example from chemistry, consider the mechanism of a chemical reaction. A reaction mechanism is a hypothesis. It cannot be deduced; on the contrary, like any hypothesis, a mechanism is formulated by induction. Scientists who formulate the elementary steps that make up a mechanism use their imagination, background, experience, intuition, and knowledge of thermodynamics and of the mechanisms of other reactions. *If* the mechanism (the hypothesis) is true, *then* the (hypothetical) rate law for the overall reaction can be deduced from the mechanism.

In fact, the truth table for hypotheses (Table 1–1) indicates that it is not possible to design an experiment such that the conditions are not only necessary but also sufficient to prove a hypothesis true. A chemical reaction has only one observed rate law, but many different mechanisms are consistent with the observed rate law.[13]

When an experiment does not agree with the predictions of a hypothesis, a research scientist must consider two reasons:

1. The hypothesis is not true
2. The experiment is
   a. Poorly designed
   b. Poorly executed
   c. Poorly interpreted

The design, execution, and interpretation of experiments are treated later. In the physical chemistry laboratory, you may be well assured that the hypotheses, theories, and laws upon which the experiments are based are well established. Because you may consider the execution of a physical chemistry laboratory experiment as the testing of a hypothesis, you should have the underlying law or theory clearly in mind.

---

[13] See the discussion of the mechanism of $2 NO + O_2 = 2 NO_2$ in Experiment 25.

Occasionally, bad deductions are made from good theories. For example, according to the theory of covalent bonding, it is assumed that the bond between two atoms is most stable if each atom can achieve a rare-gas type of configuration, that is, eight outer (valence) electrons. Since the inert gases already have eight outer electrons, it was generally believed (deduced) that the rare gases were inert and could form no bonds to other elements. All that changed when Neil Bartlett synthesized $XePtF_4$ in 1961. Subsequently, many rare gas compounds have been synthesized. Was the theory of covalent bonding wrong? No, just incomplete. Science and scientific theories are not infallible but are constantly adapting and evolving.

One of the most common misconceptions that the public has about science concerns fallibility. Science is fallible, it is provisional; it is not only testable, but *welcomes* testing. This last characteristic is one that distinguishes it from other human activities such as business, politics, and religion.

REFERENCES

1. Beck, L. W., *Philosophic Inquiry,* Prentice-Hall, New York, 1952.

2. Beveridge, W. I. B., *The Art of Scientific Investigation,* W. W. Norton & Co., New York, 1951.

3. Branscomb, L. M., "Integrity in Science," *Am. Sci.* 73:421, 1985.

4. Bunge, M., *Scientific Research,* vols. I, II, Springer-Verlag, New York, 1967.

5. DiMarzio, E. A., "What Is Basic Research," *Physics Today,* April 1980.

6. Freedman, P., *The Principles of Scientific Research,* Pergamon Press, New York, 1960.

7. Weatherall, M., *Scientific Method,* English Universities Press, London, 1968.

CHAPTER

2

Choosing a Problem

Life is a series of choices, some major, most minor. In this chapter we wish to examine two kinds of choices: the choice of a physical chemistry experiment and the choice of a research problem. While a physical chemistry experiment is smaller in scope than a research problem, making choices in either domain involves some common factors.

CHOOSING A PHYSICAL CHEMISTRY EXPERIMENT

Part Three of this book describes far more experiments than a student can carry out in the course of a one- or even two-semester course, especially if it is required that carefully and critically written reports be submitted for completed experiments. Consequently, it is necessary to select from the available experiments. The experiments have been grouped into several areas. It is recommended that each student work in as many areas as possible. Most students arrive at the physical chemistry laboratory with some quite well-formed chemical interests—and sometimes even prejudices! It is not unusual, however, for some simple physical chemistry experiment to quite unexpectedly excite interest in a topic that might otherwise have been overlooked.

At one time or another, each experiment in Part Three of this book was the chosen research problem of an experienced physical chemist. How do you choose a research problem and how does that differ from choosing an experiment in the physical chemistry laboratory? The group of problems from which selections are made grows as you accumulate experience and knowledge. Consider the following roles you may take on:

1. Physical chemistry laboratory student
2. Independent study or senior thesis author
3. First-year graduate student
4. Third-year graduate student
5. Postdoctoral fellow
6. Research director

In the physical chemistry laboratory, you are choosing to repeat a research problem that someone else has already completed. As a junior or senior you may work on a senior thesis research project, which generally involves becoming part of a project already chosen by the professor directing the research. At this stage of your professional development, you are choosing a professor with whom to work as much as you are choosing a research problem, which is still quite true if you go on to graduate school. As you mature professionally, you may select areas of research yourself. As your experience and knowledge increase, so does your autonomy in choosing a problem.

While choosing a physical chemistry experiment and working on it, try to adopt the same frame of mind that you imagine the original investigator had. Work on it with the same kind of objectivity you would give to your own chosen research problem. In short, treat each physical chemistry laboratory experiment as a small research problem in every respect. In the following chapters we will examine in more detail the art of research and the methodology of science.

CHOOSING A RESEARCH PROBLEM

Even at this early stage in your scientific career, you should realize the importance of choosing a problem. "Many scientists owe their greatness not to their skill in solving problems, but to their wisdom in choosing them."[1] Of foremost importance is compatibility between the researcher and his or her chosen problem.

Factors to Consider

You must ask yourself why you believe that you are uniquely qualified to work on a particular problem. Do you have some special technical skill outside of chemistry that increases your versatility? Perhaps you have a talent for computer programming that fits well with your chosen problem. Maybe you are interested in the history of science and are fortunate enough to be fluent in German, French, and Russian. Perhaps electronics has been a hobby of yours for years, and you are interested in chemical instrumentation. Try to put all your skills to work.

Do you really understand the scope of the problem? Be sure that

---

[1] Wilson, E. B., Jr., *An Introduction to Scientific Research,* McGraw-Hill, New York, 1952.

you can clearly pose the questions to be answered. Can you state the question so that other workers can easily understand what is to be done? Be sure that you have clearly defined the problem that you have chosen to solve.

Do you have a genuine interest in the problem? It is much more likely that your imagination will generate creative solutions if you are deeply curious about and interested in your problem. You will probably be less successful if you choose a problem that appears easy but in which you have little interest than if you choose a difficult problem that holds a certain fascination for you.

As an undergraduate you still have a great deal of flexibility in choosing an undergraduate research project or special problem. Occasionally, a student chooses an undergraduate research project not because she is particularly good at the research area, but on the contrary, because she feels she is weak in that area and needs extra work to improve her skills; she can always choose her best area in graduate school. Life is full of challenges and choices. Sometime we choose to meet challenges head on, and sometimes we are more cautious. You may have the same conflicts in choosing experiments in the physical chemistry laboratory. You may choose some because they look easy, others because they seem interesting. You may even choose an experiment because you recognize that it challenges one of your weaknesses and you are willing to meet the challenge. It is not just the skills, but some of the attitudes that you develop in the physical chemistry laboratory that you will carry with you throughout your professional life.

How does your chosen problem fit into the larger picture of what is going on in the contemporary scientific community? Are you working at the leading edge of a new area of scientific interest? Or is your research problem in a well-established area, sailing smoothly down the mainstream? Perhaps you have found an inconsistency in an old problem that needs new attention because new techniques are now available. Although any of these categories may provide an excellent choice of research problem, it may be easier to find institutional financial support for some than others. Unfortunately, it costs money to do research. To paraphrase Jesse Unrue, a California politician: "Money is the mother's milk of research."

As an undergraduate registered in the physical chemistry laboratory or working on a senior thesis, you may not find this consideration of great interest. It may play a role, however, in your selection of a particular university or research professor if you are in need of a research fellowship because you are not independently wealthy. If all things go well, your chosen research problem will be in an area that is currently funded. One must be realistic, however, for the allocation of funds to support research is influenced as much by state and federal politics as by academic altruism.

Assistance in Choosing

The *Directory of Graduate Research,* published by the American Chemical Society, is a sensible starting point for choosing an undergraduate or graduate research advisor. At this point in your research career, choosing a research problem is quite equivalent to choosing a research advisor. In the *Directory of Graduate Research* you will find, for most schools in the country that have a master's or doctor's program, a list of the chemistry faculty members, their research interests, and citations to some recent publications reflecting their recent research activities.

Another source of information on research activities is the *Bulletin of Chemical Thermodynamics.* Although not as general as the *Directory of Graduate Research,* it is more general than its name suggests. Some spectroscopists list their current research interests in this journal because spectroscopic data are used to calculate thermodynamic data. Crystallographers are interested in phase relationships. This journal is published once a year and has a fairly large section, of 100 pages or so, called "Reports of Current Research" that is worth looking at when you begin thinking about choosing your research.

REFERENCES

1. Beveridge, W. I. B., *The Art of Scientific Investigation,* W. W. Norton & Co., New York, 1951.

2. *Bulletin of Chemical Thermodynamics,* 1985.

3. DiMarzio, E. A., "What Is Basic Research," *Physics Today,* April 1980.

4. *Directory of Graduate Research,* American Chemical Society, Easton, Pennsylvania, 1989.

5. Freedman, P., *The Principles of Scientific Research,* Pergamon Press, New York, 1960.

Searching the Literature

Of all the tools available to the scientist, the literature is by far the most powerful. As a student of physical chemistry, the chemistry you will have mastered upon graduation is but a drop in the vast ocean of scientific knowledge. As a professional scientist, your scientific education *begins* upon graduation if you have developed the ability to use the literature of chemistry to your advantage. Textbooks are just one of the parts of the literature at your disposal.

Some of the components of the literature, from most general to most specific, are

> Encyclopedias
> Textbooks
> Monographs
> Journals
> Reviews

In addition, information is available through the use of

> Numerical data compilations
> Chemical abstracts
> Computer databases
> Handbooks

As a student in the physical chemistry laboratory, you will primarily be interested in obtaining information from numerical data compilations in order to compare them with your own experimental results. In that way you can judge how successful your work is: does it agree with the literature value within the calculated experimental error? As you begin to work more independently, perhaps on a research project, you will use more of the resources available in the literature. Usually it is advantageous to begin with the more general sources such as encyclopedias or major references and work toward the more specific. After you are more experienced and involved in research, you may still find it advantageous to use this approach, even though the journals contain the most recent information.

ENCYCLOPEDIC COLLECTIONS

Three encyclopedic collections of chemical and physical data tower over all others in importance. By their short names, they are referred to as (1) Landolt-Börnstein, (2) Gmelin, and (3) Beilstein. All are published in Germany in German, although some of the newer volumes are bilingual (English/German) or even published only in English. Perhaps a fourth collection should be included: the *International Critical Tables.* This was published in 1928 (McGraw-Hill, New York) and is still a useful source of classical thermodynamic data and physical properties. Although its data were critically evaluated and are reliable, *International Critical Tables* has never been updated in its original form.

Landolt-Börnstein

Landolt-Börnstein's *Zahlenwerke und Funktionen aus Physik, Chemie, Astronomie, Geophysik und Technik* (numerical data and functional relationships in physics, chemistry, astronomy, geophysics, and technology; 6th ed., H. Borchers et al., eds., Springer-Verlag, New York) is the most comprehensive compilation of numerical data in print. The book was begun in 1883 as a single 281-page volume; the current sixth edition, begun about 1950 and completed in 1979, consists of 11 volumes and nearly 9000 pages. Although written in German, it is simple to use because the data are almost exclusively numerical and in tabular form. Supplementary volumes on special topics are published irregularly in English or English and German side by side, so that the language problem is gradually disappearing. An extensive bilingual table of contents of the sixth edition is given in Appendix 1; a brief English table of contents follows:

Volume 1—Atomic and Molecular Physics
Part 1	Atoms and Ions
Part 2	Molecular Structure
Part 3	External Electron Rings
Part 4	Crystals
Part 5	Atomic Nucleus and Elementary Particles

Volume 2—Properties of Matter in Various States of Aggregation
Part 1	Thermal Mechanical State
Part 2	Equilibria (other than melting point)
Part 2a	Vapor–Liquid Equilibria and Osmotic Pressure
Parts 2b, 2c	Solution Equilibria

Part 3	Melting Point Equilibria and Interfacial Phenomena
Part 4	Calorimetry
Part 5a	Physical and Chemical Kinetics
Part 5b	Transport Phenomena, Kinetics, Homogeneous Gas Equilibria
Parts 6, 7	Electrical Characteristics, Conductivity of Electron Systems
Part 8	Optical Constants
Parts 9, 10	Magnetic Characteristics

Volume 3—Astronomy and Geophysics

Volume 4—Basic Techniques

Part 1	Mechanical Properties of Natural Materials
Part 2	Mechanical Properties of Metallic Materials:
	(a) Ferrous; (b) Heavy Metals; (c) Light Metals
Part 3	Electrical, Optical, and X-Ray Techniques
Part 4a	Thermodynamic Properties of Gases, Liquids, and Solids
Part 4b	Thermodynamic Properties of Mixtures

Now that the sixth edition is complete, all future volumes are to be published as part of the New Series. The majority of these volumes will be dedicated to currently expanding fields of research and applications, and will be supplemented at regular intervals as developments are reported. Other volumes will be concerned with the reappraisal and updating of material from the more classical fields (mostly covered by the sixth edition); supplements will be published at less frequent intervals in these areas, as rapid growth is not expected. The recently completed sixth edition and the New Series supplement each other, as demonstrated by the random selection of two topics, Properties of Solutions and Molecular Spectroscopy, shown in Table 3–1.

Landolt-Börnstein is widely used, not only by chemists but also by physicists, geologists, and engineers. Because the data contained in it have been critically evaluated, it is often the first choice of professional scientists. You would be well advised to become familiar with this important international work.

Gmelin

Gmelin's *Handbuch der anorganischen Chemie* was first published between 1817 and 1819 by Leopold Gmelin (1788–1853) under the title *Handbuch der theoretischen Chemie.* Originally it covered all areas of chemistry, including organic, but with the fifth edition it was restricted to inorganic chemistry. The organic section was removed and taken up by a new compilation appearing about that time (1883): *Beilsteins Handbuch der organischen Chemie.* The eighth and last edition of Gmelin (Springer-Verlag, New York) was begun in 1922 and finished around 1958. On the reverse side of the title page you can find the original publication date and the last date of literature coverage. If necessary, you can then search the literature elsewhere (e.g., in *Chemical Abstracts*) from that date on to complete your search.

No new editions are planned; instead, supplementary volumes ap-

TABLE 3–1

Sample Coverage of Landolt-Börnstein, Sixth Edition and New Series

Topic	New Series	6th Edition
Molecular Spectroscopy		
Molecular energy levels		I/2,3
Molecular constants derived from rotation vibration spectra	II/4,6	
Luminescence spectra	II/3	
Raman spectra		I/2
Spin resonance spectra	II/1,2,8, 9a–d,10,11	
Absorption and emission spectra (HF, IR, vis., UV, x-ray)		I/2,3
Properties of Solutions		
Heat capacities	IV/1b	IV/4a
Freezing and boiling points	IV/3	
Enthalpies of solution, mixing	IV/2	II/4
Phase and solution equilibria	IV/3	II/2a–c
Vapor and osmotic pressures	IV/3	II/2a
Viscosities and diffusion		II/5a
Densities	II/5; IV/1ab	II/1
Electrochemical properties		II/7
Optical constants		I/3; II/8

pear frequently on topics already included in the eighth edition. The eighth edition consists of 71 main volumes plus a few supplements that were published before the completion of the eighth edition. The goal of the Gmelin Institute is "exhaustive yet concise and critical presentation of our entire knowledge of inorganic chemistry and related sciences." The coverage is indeed extraordinary, as can be seen in Table 3–2.

TABLE 3–2

Topics Treated in Gmelin

Inorganic chemistry	Ore dressing
Physical chemistry	Chemical economics
Nuclear chemistry	Metallurgy
Analytical chemistry	Metallography
Colloidal chemistry	Iron and steel
Electrochemistry	Nonferrous metals
Corrosion	Light metals
Chemical equilibria	Experimental physics
Chemical technology	Nuclear physics
Mineralogy	Atomic physics
Crystallography	Radioactivity
Geology	Electric properties
Geochemistry	Magnetic properties
Historical information	Optical properties

Instead of a general index, each section has a very detailed table of contents. To illustrate, major subject headings for arsenic are listed in Table 3–3. Gmelin is not bilingual as shown in the table; nevertheless, because so many scientific words are clearly very similar in German and English, the English-speaking scientist can generally decipher the information sought. Beginning with the supplementary volumes published after 1957, English headings, subheadings, and indices have been included.

After a description of the properties of an element, the binary compounds of that element are discussed, then compounds of greater complexity. The properties listed in Table 3–3 for the element are also presented for the compounds. Bibliographic citations are given in the body of the text after each property.

Gmelin is organized by assigning a system number ranging from 1 to 71 to each element, with three exceptions. The inert gases are collected together under system number 1, the rare earth elements under system number 39, and the transuranium elements under system number 71. In the example given earlier, arsenic occupies system number 17, as shown in Table 3–4.

Notice that the nonmetals, the elements that tend to form anions, have the lowest numbers, roughly 16 and lower. Compounds of the elements are assigned the highest possible system number. Thus N_2O is found under system number 4, but NI_3 is found under system number 8. Similarly, $BaCrO_4$ is found under system number 52 (chromium), but Ag_2CrO_4 is found under system number 61 (silver).

TABLE 3–3

Subject Headings for Arsenic in Gmelin (German/English)

Inhaltsverzeichnis	Table of Contents
Geschichtliches	Historical
Vorkommen	Deposits
Verwendung	Application
Das Element	The Element
Bildung und Darstellung	Formation and Preparation
Physikalish Eigenschaften	Physical Properties
Atom-kern	Atomic nucleus
Atom	Atom
Molekel	Molecule
Krystallographische	Crystallographic
Mechanische	Mechanical
Thermische	Thermal
Optische	Optical
Magnetische	Magnetic
Electrische	Electrical
Elektrochemisches Verhalten	Electrochemical Behavior
Chemisches Verhalten	Chemical Behavior
Physiologisches Verhalten	Physiological Behavior
Die Verbindungen des Arsen	Compounds of Arsenic

TABLE 3–4

Gmelin System Numbers

1 Edelgase	25 Caesium—Cs	49 Niob—Nb
2 Wasserstoff—H	26 Beryllium—Be	50 Tantal—Ta
3 Sauerstoff—O	27 Magnesium—Mg	51 Protactinium—Pa
4 Stickstoff—N	28 Calcium—Ca	52 Chrom—Cr
5 Fluor—F	29 Strontium—Sr	53 Molybdän—Mo
6 Clor—Cl	30 Barium—Ba	54 Wolfram—W
7 Brom—Br	31 Radium—Ra	55 Uran—U
8 Jod—I	32 Zink—Zn	56 Mangan—Mn
9 Schwefel—S	33 Cadmium—Cd	57 Nickel—Ni
10 Selen—Se	34 Quecksilber—Hg	58 Kobalt—Co
11 Tellur—Te	35 Aluminum—Al	59 Eisen—Fe
12 Polonium—Po	36 Gallium—Ga	60 Kupfer—Cu
13 Bor—B	37 Indium—In	61 Silber—Ag
14 Kohlenstoff—C	38 Thallium—Tl	62 Gold—Au
15 Silicium—Si	39 Seltene Erder	63 Ruthenium—Ru
16 Phosphor—P	40 Actinium—Ac	64 Rhodium—Rh
17 Arsen—As	41 Titan—Ti	65 Palladium—Pd
18 Antimon—Sb	42 Zirkonium—Zr	66 Osmium—Os
19 Wismut—Bi	43 Hafnium—Hf	67 Iridium—Ir
20 Lithium—Li	44 Thorium—Th	68 Platin—Pt
21 Natrium—Na	45 Germanium—Ge	69 Tecnecium—Tc
22 Kalium—K	46 Zinn—Sn	70 Rhenium—Re
23 Ammonium—NH₃	47 Blei-Pb	71 Transurane
24 Rubidium—Rb	48 Vanadium—V	

Beilstein

Beilsteins Handbuch der organischen Chemie is a monumental reference work on all carbon compounds described in the literature whose constitutions are known. All material from the literature that eventually finds its way into Beilstein is critically reviewed, checked for internal consistency, and corrected for errors when appropriate. Begun by Friedrich Konrad Beilstein as his own personal notes on the literature of organic chemistry when he was Wohler's assistant in Göttingen, Beilstein is now published by the Beilstein Institute, Frankfurt, and is available from Springer-Verlag New York, Inc. More than 100 chemists and physicists sift through the literature for the continuing publication of this invaluable reference work.

THE COVERAGE OF BEILSTEIN

This is usually thought of as an organic chemist's reference work because it contains such detailed information as

- *Constitution and configuration*
- *Natural occurrence*
- *Preparation, formation, and purification*
- *Chemical properties*

■ *Characterization and analysis*
■ *Salts and addition compounds*

Beilstein is also useful to physical chemists because it contains many structural and thermodynamic properties:

■ *Bond lengths and angles*
■ *Electron distribution*
■ *Dipole moment*
■ *Quadrapole moment*
■ *Polarizability*
■ *Coupling phenomena*
■ *Molecular deformation*
■ *Molecular potentials*
■ *Dissociation energies*
■ *Ionization energies*
■ *Transport phenomena*
■ *Energy data*
■ *Optical data*
■ *Spectra*
■ *Magnetic properties*
■ *Electrochemical behavior*
■ *Phase equilibria*
■ *Conformer equilibrium*

In contrast to abstracts of the literature, which only name the information or property sought and its location in the literature, Beilstein contains the actual information sought, and in a quite complete and critically evaluated form.

GUIDES TO BEILSTEIN

Although the book is written in German, this should not deter anyone unfamiliar with German from using it. The vocabulary used has been carefully limited; in fact, the Beilstein Institute publishes a tiny (62-page) *German-English Dictionary for Users of the Beilstein Handbook of Organic Chemistry.* You may obtain your own copy free of charge by writing to:

Springer-Verlag New York, Inc.   Springer-Verlag KG
175 Fifth Avenue   Heidelberger Platz 3
New York, NY 10010   D-1000 Berlin 33
USA   West Germany

While you are writing, ask for The Beilstein Reference Chart and a guidebook, *How to Use Beilstein.* These are also free of charge. The dictionary contains about 2100 words commonly used in organic chemistry. In addition, an appendix contains about 28 German phrases commonly used in Beilstein and their English translations. For example,

Die früher unter dieser Konstitution beschriebene Verbindung ist
(wahrscheinlich) als . . . zu formulieren.
The compound formerly described by this constitution is to be
(probably) formulated as. . . .

The guidebook is brief (30 pages), but of great assistance in learning to
master the Beilstein system. The reference chart is about 60×63 cm
and should be posted in the library near the Beilstein collection. Speak
to your librarian if it is not. Its title states just what it is: "The Short Cut
to Locating a Compound in Beilstein's Handbook of Organic Chemistry
Using the Beilstein System."

You will find *Beilsteins Handbuch* in the reference section of the
library. If it is complete, it occupies many linear feet of shelving so you
cannot miss it. The original edition covered the literature up to 1909.
The periodic publication of supplements currently covers the literature
through 1979, as indicated in Table 3–5.

Each series is organized into 27 volumes. Within these volumes, the
individual compounds are arranged according to the Beilstein system as
shown in Table 3–6. The Beilstein system is a set of rules through which
each carbon compound is assigned one specific location within the overall
array of all carbon compounds. These are described in detail in the guidebook mentioned above. The total array is further subdivided into 4720
system numbers.

USING BEILSTEIN

So how do we find our way through this maze of information to
zero in on whatever it is we need to know? Two alternative approaches
are available:

1. Use the general indices, of which there are two for each completed
   series (H, E I, and E II, currently).
   a. The general formula index (Volume 28)
   b. The general subject index (Volume 29)

TABLE 3–5

The Series of Beilstein

Series	Abbreviation[a]	Period of Literature Covered	Color of Label
Basic Series	H	Up to 1909	Green
Supplementary Series I	E I	1910–1919	Red
Supplementary Series II	E II	1920–1929	White
Supplementary Series III	E III	1930–1949	Blue
Supplementary Series III/IV	E IV	1950–1959	Blue/black
Supplementary Series IV	E IV	1950–1959	Black
Supplementary Series V	E V	1960–1979	(published in English starting in 1984)

[a] H = *Hauptwerk* (main work); E = *Ergänzungswerke* (supplementary work).

TABLE 3-6

Organization of the 27 Volumes of Beilstein

Type of Registry Compound	Feature of the Functional Group	A (Acyclics)	B (Isocyclics)	C (Heterocyclics) — Type and Number of Ring Heteroatoms					
				$1\,O^a$	$2\,O^a\cdots$	$1\,N$	$2\,N$	$3\,N\cdots$	*Further Heteroatoms*
1. Compounds without functional groups		1	5	17	19	20	23	26	27
2. Hydroxy-compounds	—OH	1	6	17	19	21	23	26	27
3. Oxo-compounds	=O	1	7	17	19	21	24	26	27
	=O + —OH	1	8	18	19	21	25	26	27
4. Carboxylic acids	COOH; [COOH$_n$]	2	9	18	19	22	25	26	27
	COOH + —OH, + =O	3	10	18	19	22	25	26	27
	COOH + =O + —OH			18	19	22	25	26	27
5. Sulfinic acids	—SO$_2$H	4	11	18	19	22	25	26	27
6. Sulfonic acids	—SO$_3$H	4	11	18	19	22	25	26	27
7. Selenic, selonic, and telluric acids	—SeO$_2$H, —SeO$_3$H, and TeO$_2$H	4	11	18	19	22	25	26	27
8. Amines	—NH$_2$	4	12	18	19	22	25	26	27
	[NH$_2$]$_n$; —NH$_2$ + —OH		13	18	19	22	25	26	27
	—NH$_2$ + =O, etc.		14	18	19	22	25	26	27
9. Hydroxylamines and dihydroxyamines	—NH—OH	4	15	18	19	22	25	26	27
	—NH(OH)$_2$		15	18	19	22	25	26	27
10. Hydrazines	—NH—NH$_2$	4	16	18	19	22	25	26	27
11. Azo-compounds	—N=NH	4	16	18	19	22	25	26	27
12. Diazonium compounds	—N≡N$^+$	4	16	18	19	22	25	26	27
13. Compounds with groups of 3 or more N atoms	—NH—NH—NH$_2$, etc.	4	16	18	19	22	25	26	27
14. Compounds with C to P, As, Sb, or Bi	—PH$_2$, PH—OH, etc.	4	16	18	19	22	25	26	27
15. Compounds with C to Si, Ge, or Sn	—SiH$_3$, SiH$_2$(OH), etc.	4	16	18	19	22	25	26	27
16. Compounds with C to Groups IA, IIA, IIIA	—BH$_2$, —Mg$^+$, etc.	4	16	18	19	22	25	26	27
17. Compounds with C to Groups IB–VIIIB	—HgH, Hg$^+$, etc.	4	16	18	19	22	25	27	27

Column header span: **Beilstein Volume No. for Main Division**

^{a} Also S, Se, and Te instead of O.

2. Use the rules of the Beilstein system
   a. Detailed in the free publication *How to Use Beilstein* (see above)
   b. Summarized in the free reference chart (see above)

Whether you use approach 1 or approach 2, your chances of quick success are greatly increased if you cultivate the friendship and assistance of an organic chemist who is very good at organic nomenclature. Nevertheless, the use of the indices is quite straightforward, especially if we choose an organic compound so simple that the nomenclature problems disappear; let us try this approach first.

Method 1: Using the Indices. Suppose we wish to find the heat of fusion of *p*-dichlorobenzene, with the empirical formula $C_6H_4Cl_2$. This is simple enough that we should not need the services of our organic chemist friends. Shelved between Supplementary Series II and III we find the subject and formula indices, labeled *"Generalsachregister"* and *"Generalformelregister,"* respectively. In a German-English/English-German dictionary we find that heat of fusion in German is *Schmelzwarme* and benzene is *Benzol.* The arrangement in the index volume is more or less self-explanatory, and eventually we find on page 893 of the subject index a series of entries:

Dichlor-benzochinoxalin

—

—   [other derivatives in which we are not interested]

—

—benzol **5,** 201, 202, 203, I 111, II 153, 154

Just for the sake of comparison, let's check out the *Generalformelregister.* On page 147 we quickly locate $C_6H_4Cl_2$, even more easily than in the *Generalsachregister,* since no knowledge of nomenclature is necessary. In general, the formula index is preferred over the subject index because there are so many different ways of naming organic compounds. The entry on page 147 of the formula index appears as follows:

$C_6H_4Cl_2$ 1,2-Dichloro-benzol **5,** 201, I 111,
  II 153.
  1,3 Dichloro-benzol **5,** 202, I 111, II 154.
  1,4 Dichloro-benzol **5,** 202, I 111, II 154.

The **5** means that the entry is in Volume 5 of the main series, the *Hauptwerke.* On page 203 we find *1,4-Dichloro-benzol,* and considerable physical and chemical information, but no *Schmelzwarme.* At the top of the page, we note that the system number is 464; we jot it down because, as you'll see, it will come in handy later.

Going along in the index entry, we realize that the "I" refers to E I, Supplementary Series I (*Erstes Ergänzungswerke*). The volume and system numbers are the same as in the main series, namely 5 and 464. So we identify the first supplementary series by its red label, locate Volume 5, look for system 464, which covers only a few pages, and finally we locate *1,4-Dichloro-benzol,* but no information on *Schmelzwarme.*

This does not present a problem, even though the indices cover just the H, E I, and E II Series. The reason is very important: by using the indices to go to these series, we have discovered the Beilstein system number—in this case, 464. This is the system number for 1,4-dichlorobenzene not just in the completed series, but in *all* series, including those in preparation and future series for which no indices exist. Consequently, we can go directly to system number 464 of each of the Supplementary Series, where it is a simple matter to find our compound, 1,4-dichlorobenzene.

Doing exactly that, we find in Supplementary Series I, along with several other bits of information for 1,4-dichlorobenzene, the following entry in Volume 5, page 544, four lines down from the top:

Schmelzwarme: 29.5±1 cal/gm, Narbutt. Z. El. Chem *24* 339.

This we calculate to be 18 kJ/mole. In another two minutes, we look up system number 464 in Supplementary Series III, and in Volume 5, page 544, 16 lines down, we come across another value for the heat of fusion:

Schmelzwarme: 29.54 cal/gm, Roczniki Chem. *17* 140 (1937).

For this value, we calculate the heat of fusion to be 18.16 kJ/mole, which, it turns out, is the currently accepted value. In Landolt-Börnstein, Volume II, Part 4, the heat of fusion of 1,4-dichlorobenzene is given as 18.16 kJ/mole with the reference *Z. Naturforsch* 5a:101 (1950). Now you see why Landolt-Börnstein is so valuable: its workers have searched the literature for us, located all previous values, critically evaluated them, and selected the most reliable value.

To get a feel for Beilstein, take this simple example to the library and retrace the steps through the series. Then select a simple compound of your own and some particular physical property, and see if you can find the required information in Beilstein.

Method 2: Using the Rules of Beilstein. The second approach to using Beilstein is to use the rules of the Beilstein system, which requires a more complete understanding of the organization of Beilstein, as well as of the rules of organic nomenclature. We have already seen that each of the Beilstein series (H, E I, E II, E III, and E IV) is arranged into 27 volumes and 4720 systems. The arrangement is also chemical, as may be seen in Table 3–7. (A more detailed listing of the subdivisions of Beilstein is given in Table 3–6.)

TABLE 3–7

The Main Divisions of Beilstein

Main Division	Volume Numbers	System Numbers
A. Acyclic compounds	1–4	1–449
B. Isocyclic compounds	5–16	450–2358
C. Heterocyclic compounds	17–27	2359–4720

Some rules provide guidelines for finding the volume with the compound for which you are seeking information.

Rule 1. You must be able to write a structural formula for the compound you are seeking.

Rule 2. You assign your compound to one of the groups of the main division (Table 3–6).

Rule 3. Priority or rank is determined by the Principle of Latest Entry. Thus, compound A in Figure 3–1 is somewhere in Volumes 1 through 4, but compound B is located between Volumes 5 and 16. The Principle of Latest Entry is generally valid in ordering compounds in Beilstein.

The chemical classification naturally goes further than these three main divisions. The key to the Beilstein organization is the basic structure of the registry compounds.

Rule 4. A registry compound is a hydrocarbon (acyclic or isocyclic) with or without one or more functional groups (see Table 3–7).

Note 1: In the Beilstein system F, Cl, Br, I, NO, NO_2, and N_3 are *not* considered functional groups.
Note 2: Most other traditional functional groups are considered functional groups by Beilstein *if no derivative has been formed.*

Rule 5. Within a given system number, compounds are presented more or less as you would expect: C_6H_5Cl before $C_6H_4Cl_2$, $CH_3CH_2CH_2CH_3$ before $CH_3CH_2CH=CH_2$, etc.

Let us look at a few examples of registry compounds (Fig. 3–2), their rank, and derivatives (Fig. 3–3). Remember, the organization of Beilstein is based upon the registry compound organization described in Table 3–6. If a compound is a derivative of, say, two organic compounds A and B, then it is found with the organic compound of higher rank,

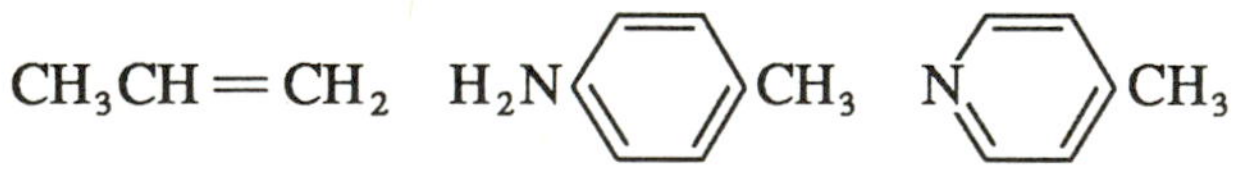

$$CH_3CH=CH_2 \quad H_2N\bigcirc CH_3 \quad N\bigcirc CH_3$$

A   B   C

Figure 3–1 Compounds to locate in Beilstein by the Principle of Latest Entry.

$$CH_3N_2 \quad \bigcirc\underset{O}{\overset{\|}{C}}-OH \quad Cl\bigcirc OH$$

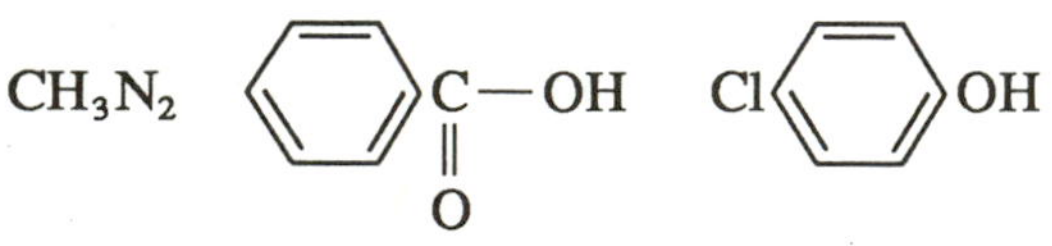

A   B   C

Figure 3–2 Some registry compounds in the Beilstein system.

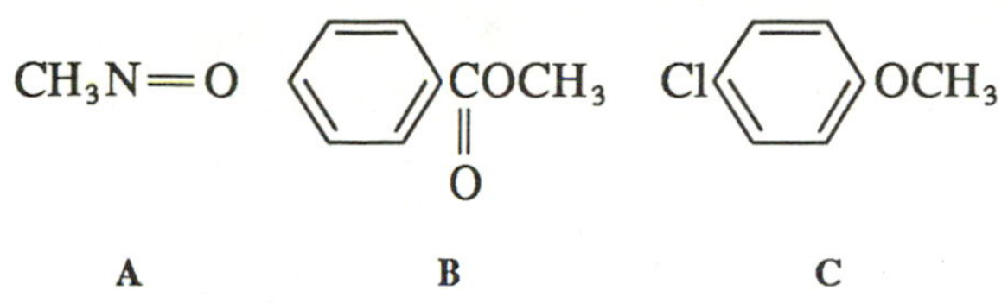

Figure 3–3 Some functional derivatives.

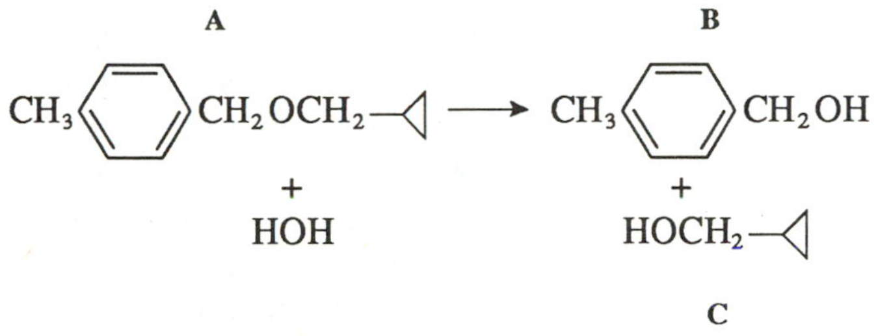

Figure 3–4 Formal hydrolysis.

according to the Principle of Latest Entry. The compounds A and B are found by carrying out a "formal hydrolysis" on the derivative (Fig. 3–4); the derivative (compound A) is found with the hydrolysis product with a free functional group. If formal hydrolysis results in two or more organic compounds, then attention is focused on the highest-ranking registry compound (Principle of Latest Entry). In this example, compound C in Figure 3–4 is the highest-ranking compound, and we would search initially for it, and then its derivatives.

REFERENCES TO THE USE OF BEILSTEIN

1. *Beilstein Handbook of Organic Chemistry,* Springer-Verlag, Berlin, 1981.
2. *Beilstein Dictionary,* Springer-Verlag, Berlin, 1979.
3. Huntdress, E. H., "1938: The One Hundredth Anniversary of the Birth of Friedrich Konrad Beilstein (1838–1906)," *J. Chem. Educ.* 15:303, 1938.
4. Richter, F., "How Beilstein is Made," *J. Chem. Educ.* 15:310, 1938.
5. Hancock, H. E. H., "An Introduction to the Literature of Organic Chemistry, Part III," *J. Chem. Educ.* 45:336, 1968.

CHEMICAL ABSTRACTS

Organization

Each week *Chemical Abstracts* (CA) publishes about 9000 short abstracts of original documents organized into 5 groups and 80 sections. Each abstract is assigned an abstract number. The groups are

1. Biochemistry
2. Organic chemistry

   3. Macromolecular chemistry
   4. Applied chemistry and chemical engineering
   5. Physical, inorganic, and analytical chemistry

Groups 1 and 2 are published together every other week; groups 3, 4, and 5 are also published every other week, alternating with the first two groups. The sections for physical chemistry are listed below:

   65. General physical chemistry
   66. Surface chemistry and colloids
   67. Catalysis and reaction kinetics
   68. Phase equilibriums, chemical equilibriums, and solutions
   69. Thermodynamics, thermochemistry, and thermal properties
   70. Nuclear phenomena
   71. Nuclear technology
   72. Electrochemistry
   73. Spectra by adsorption, emission, reflection, magnetic resonance, or other optical properties
   74. Radiation chemistry, photochemistry, and photographic processes
   75. Crystallization and crystal structure
   76. Electric phenomena
   77. Magnetic phenomena
   78. Inorganic chemicals and reactions
   79. Inorganic analytical chemistry
   80. Organic analytical chemistry

If you are using CA to keep up with the literature—for your personal "current awareness program"—you will generally consult one or two sections regularly as they appear in the library. Each biweekly issue is indexed by authors' names, patent numbers, and key word phrase chosen from the abstract text in the original article. The literature is so voluminous that keeping up with more than two sections in CA is too tedious. (See the section on online database searching.)

If you expect that your literature search for some specific datum will cover several years, then you will generally use one or more of the various volume or collective indices to CA. The use of the indices is not simple and you should surely consult the Index Guide, which is the key to the efficient use of the other indices. *Chemical Abstracts* makes available wall posters illustrating the use of its publications, and your library should have one displayed near the CA collection.

Volume Indices of *Chemical Abstracts*

A volume of CA consists of the weekly abstracts issued over a six-month period. For every volume, CA issues the following volume indices:

1. **The general subject index** connects subject terms, such as reaction, classes of substance, and procedure, with the corresponding CA abstract number.

2. **The chemical substance index** connects a CA chemical name with a corresponding CA abstract number. The index names are listed alphabetically, and Chemical Abstracts Service (CAS) Registry numbers (see below) are also given.

3. **The formula index** connects a molecular formula with a CA abstract number and also gives the CAS Registry number. The formulas are ordered according to the Hill system. If a compound contains carbon, C is listed first, followed by H, and then the symbols for the other atoms alphabetically. If no carbon is present, then the order of the elemental symbols is strictly alphabetical and numerical. For example, hydrogen cyanide is listed as CHN and potassium bromide as BrK. The information in the formula index is quite brief. It is usually advisable to take the CA name found there and go to the Chemical Substance Index to continue the search.

4. **The index of ring systems** lists the names of cyclic skeletons contained in organic compounds in the order determined by the number, size, and kind of atoms in the rings.

5. **The author index** connects the names of the author and coauthor to the CA abstract number. Coauthors are cross-referenced to the name of the first author, whose name is linked to the title of the original journal article.

6. **The patent index** lists the countries in which the patents were issued alphabetically. Within each country listing, the patents are listed in numerical order.

With six indices to choose from, it is not surprising that CA issues an index to the indices, which is called the Index Guide. First published in 1968, it is published at the beginning of each collective index period and updated every 18 months. It is particularly useful as an aid in using the first two indices listed above, the general subject index and the chemical substance index. It contains information on names, cross-references, synonyms, and general indexing policies. It is also a source for registry numbers. These are unique identifying numbers assigned to each substance mentioned in the abstracts. These are also called the CAS Registry numbers, as they are used in the CAS Chemical Registry System, a massive computer file of chemical information. Chemical Abstracts Service (CAS) is the division of the American Chemical Society (ACS) that publishes CA and maintains the computer database known as CAS ONLINE.

Using the Formula Index. Which of these indices you choose to use to begin with depends on what sort of information you have. Probably the most frequently used indices are the formula, subject, and author indices. As an example, let us suppose that you want to find some spectroscopic evidence on the existence of hypofluorous acid, the formula of which is usually written HOF. Formulas in the formula index are written in alphabetical and numerical order: FHO is where you find hypofluorous acid; similarly, FHO_2 is where you find fluorous acid. Searching formula indices, you find for 1967 to 1971 a volume marked

66–75
1967–71
FORMULAS
C_{34}–Z

Looking through the alphabetical listing of formulas, you find

FHO
Hypofluorous acid [14034-79-8], 66:14208y,
69123x; 69:31660u; 70:31781e; 72:13717r;
75:58037y

Notice that no clue is given as to the nature of the contents of the article abstracted. Fortunately, we need to check only six abstracts to determine if one deals with spectroscopic information.

Using the Subject Index. Let's compare this formula index option with the subject index by looking up hypofluorous acid. The volume found is labeled

66–75
1967–71
SUBJECTS
GLUCOPE–INDENA

Several more citations are listed in addition to those we recognize from the formula index; this time some identifying information is included and one abstract appears to be of interest:

Hypofluorous Acid . . .
Spectrum (ir) of, in nitrogen matrix,
from photolysis of fluorine with water,
69:31660u . . .

Leaving the indices, we look for the corresponding abstract and find on the abstract shelves:

Vol 69, 1968
30000–40000

On page 2957 we find the following abstract:

31660u Hypofluorous acid: infrared spectrum and vibrational potential function. Noble, Paul N., Pimentel, George C. (Univ. of California, Berkeley, Calif.). Spectrochim. Acta, Part A 1968,24(17),797–806(Eng). Photolysis of mixts. of R and water suspended in a N matrix at 20 and 14°K produces several new ir absorptions. Growth and diffusion studies, isotopic labeling, and normal coordinate anal. help to assign bands at 3483, 1393.0 and 884.0 cm^{-1} to HOF. A band at 3777 cm^{-1} is attributed to HF-F bonded in the same matrix cage to HOF. The HOF force field calcd. by normal coordinate anal. shows that the bonding is similar to that in H_2O and F_2O. 15 references.

Since infrared spectral information is sought, this looks promising enough that the entire journal article ought to be read in *Spectrochimica Acta.*

Using the Author Index. On the other hand, suppose that you need the data in a paper on the crystal structure of nalbuphine and you know it was published in the 1970s by Sime. If you looked in the author collective index for 1972 to 1976, you would find several publications by Sime, and among them the following:

> Sime, Rodney J.; Dobler, M.; Sime, R. L.
> The crystal structure of a narcotic agonist/antagonist: nalbuphine hydrochloride dihydrate. 84:158294t.

This citation in the index informs you that the abstract is in volume 84 and the number is 158294t. Moving from the shelves of indices to the shelves of abstracts, you find several volumes marked 84; one of them is labeled

vol 84, 1976
112821–167478

Looking through it you see that the abstracts are simply arranged in numerical order, and on page 555 you find the following entry:

> 84:158294t The crystal structure of a narcotic agonist/antagonist: nalbuphine hydrochloride dihydrate. Sime, Rodney; Dobler, Max; Sime, Ruth L. (Lab. Org. Chem., SwissFed. Inst. Technol., Zurich, Switz.). Acta Crystallogr., Sect. B 1976, B32(3), 809-12(Eng). Addnl. data considered in abstracting and indexing are available from a source cited in the original document. Nalbuphine (N-cyclobutylmethyl-7,8-dihydro-14-hydroxynormorphine) crystd. as the hydrochloridedihydrate in space group $P2_12_12_1$ with a 11.576, b 12.336, c 14.658 A, and Z = 4. Packing is largely detd. by extensive H bonding involving H_2O and Cl^-. Bond angles and distances were detd.

The information contained is self-explanatory and always appears in the above format, maybe a little shorter or a little longer. If more information is required, the complete article may be read in *Acta Crystallographica*.

It takes a little practice to become comfortable with using CA, but it is worth the effort. In fact, it is really mandatory, since there is no substitute. Currently, over 14,000 journals in 50 languages from 150 nations are abstracted; over 500,000 documents are abstracted in a single year.

ONLINE DATABASE SEARCHING

A database[1] is an organized collection of information that can be read by a computer. One of the earliest databases, the Chemical Titles (CT) database, was developed by CAS in 1961. Over the years it has evolved into the CAS ONLINE search system for finding information on the CA

[1] Carr, C., "Aids for Teaching Online Searching of the Chemical Literature," *J. Chem. Educ.* 66: 21–24, 1989.

file. The CA file database contains essentially the full contents of CA and its indices back to 1967. Consequently, searches can be made by inputting CAS Registry number, structure, chemical name, or subject name. The CA file is a bibliographic file; a successful search retrieves one or more CA bibliographic references (abstracts) published after 1962. It is possible to retrieve the complete CA abstract text for all abstracts published after 1975—and some back as far as 1967.

Besides bibliographic databases, full-text and numeric databases are in use. CAS ONLINE is a bibliographic database. On the other hand, a database that can retrieve for its user an entire journal article is called a full-text database. Various full-text databases contain the entire texts of newspaper articles, articles from encyclopedias, legal cases, and medical diagnoses. One of these, Bibliographic Retrieval Services (BRS), makes it possible to retrieve the full texts of 18 journals published by the American Chemical Society, several medical journals, and even textbooks.

A database containing mostly numbers along with a few necessary labels is called a numeric database. These also differ from bibliographic and full-text databases in that they can often carry out numerical calculations with the numerical information they contain. For example, a numeric database might store the Antoine constants for the Antoine vapor pressure equation. With such a database, you could request the boiling point of a compound or its vapor pressure at some desired temperature. The software of the database system would do the necessary calculations and report the results to you. One of the largest numeric databases is the Chemical Information System (CIS).

An example of a reference retrieved from a search with CAS ONLINE is shown in Figure 3–5. The structure of the molecule is also shown on the display monitor and is printed along with other relevant information on a dot-matrix printer. This example shows one reference, including abstract and index entries, retrieved from CAS ONLINE. See Table 3–8 for an explanation of the codes for the retrieved information. Suffix letters with Registry numbers indicate the following: P, a compound whose preparation is reported in the referenced document; D, a nonspecific derivative of the substance; DP, preparation of a nonspecific derivative.

Retrieving information[2] from a database with a computer is still expensive and complicated for the occasional user. Although online systems are becoming more "user friendly," they are still sufficiently complex that they are usually used by the professional information scientist or a librarian with special training who works with a scientist seeking specific

[2] Krumpolc, M., D. Trimakas, and C. Miller, "Searching Chemical Abstracts Online in Undergraduate Chemistry," Parts 1 and 2, *J. Chem. Educ.* 64:55–59 (1987) and 66:26–29 (1989).

Figure 3–5 An example of the kind of information that can be retrieved with CAS ONLINE. See Table 3–8 for an explanation of the codes. (Courtesy of Chemical Abstracts Service.)

CAS ONLINE®

SEARCH RESULTS -

REGISTRY NUMBER = 1695-77-8   ANSWER NUMBER = 1

INDEX NAME = 4*H*-Pyrano[2,3-*b*][1,4]benzodioxin-4-one,
              decahydro-4a,7,9-trihydroxy-2-methyl-6,8-bis(methylamino)-,
              [2*R*-(2α,4aβ,5aβ,6β,7β,8β,9α,9aα,10aβ)]- (9CI)

MATCH NAME = Spectinomycin

SYNONYM = Actinospectacin

MOLECULAR FORMULA = $C_{14}H_{24}N_2O_7$

SUBSTANCE CLASS = COM;

DELETED REGISTRY NUMBER(S) =   1404-65-5,   11002-81-6,   23559-31-1;

TEXT DESCRIPTOR = *

REFERENCE   4

AN   CA98(11):84380h

TI   Plasmids

AU   Katsumata, Ryoichi; Oka, Tetsuo; Furuya, Akira

CS   Kyowa Hakko Kogyo Co., Ltd.

LO   Japan

PI   Eur. Pat. Appl. EP 63763 A1, 3 Nov 1982, 25 pp.   Designated States: BE, CH, DE, FR, GB,
      IT, NL, SE

AI   Appl. 82/103222, 16 Apr 1982; JP Appl. 81/58186, 17 Apr 1981

CL   C12N15/00, C12R1/15, C12R1/13

SC   3-1 (Biochemical Genetics)

SX   10

DT   P

CO   EPXXDW

PY   1982

LA   Eng

AB   Several plasmids which bear the genes for resistance to the antibiotics *spectinomycin*
      [1695-77-8] and *streptomycin* [57-92-1] and which can replicate autonomously in bacteria
      of the genera *Corynebacterium* or *Brevibacterium* are isolated from *Corynebacterium*.
      One such plasmid was isolated from *Corynebacterium glutamicum* 225-250 by 1st treating
      the cells with sublethal doses of *penicillin* [61-33-6] (0.1-10 units/mL culture), then with
      egg white lysozyme, to disrupt the cell walls.   Plasmid DNA was extd. by SDS-NaCl
      treatment and pptd. with polyethylene glycol.   The ppt. was dissolved, and plasmid DNA was
      banded by centrifugation in an ethidium bromide-CsCl d. gradient.   The purified plasmid, pCG4,
      of 19 megadaltons, contained the drug resistance genes as well as multiple cleavage sites for
      restriction endonucleases.

KW   plasmid Corynebacterium Brevibacterium; drug resistance plasmid Corynebacterium Brevibacterium;
      spectinomycin streptomycin resistance Corynebacterium Brevibacterium

IT   Drug resistance
          (to streptomycin and spectinomycin, in *Corynebacterium* and *Brevibacterium*, plasmids for)

TABLE 3–8

**Key to Abbreviations Appearing on a
CAS ONLINE Search Result**

AB	Abstract text, if available
AI	Patent application information
AN	CA abstract number—volume and abstract number
AU	Authors
CL	Patent classification
CO	Document CODEN
CS	Corporate source (organization)
DT	Document type (J = journal, P = patent, etc.)
IN	CA index name
IS	ISSN (International Standard Serial Number)
IT	Index terms (from CA General Subject Index)
KW	Key word (indexing information)
LA	Language of original document
LO	Work location
MF	Molecular formula
PI	Patent information
PY	Publication year of original document
RN	CAS registry number
SC	CA section–subsection number and section title
SO	Bibliographic source (of the original document)
ST	Stereochemical information
SY	Synonyms
TI	Title

Source: Using CAS ONLINE, An Introduction, American Chemical Society, Columbus, Ohio, 1985.

information. To become comfortable with the command structure of the search software provided by the database vendor requires frequent use of the online system.

Vendors of online databases naturally charge for the use of their database systems. The main expense arises from the user's actual connect time to the database. Fortunately, vendors currently are giving academic users a substantial discount. For example, STN International gives a 90% discount for the use of the CAS files CA and REGISTRY, for which the nonacademic user pays $79 and $56, respectively, per hour of connect time. Slight additional costs result in an actual total cost to academic users of about $8 to $10 per hour. Recently, BEILSTEIN became available online at a connect-time rate of $49 per hour for nonacademic users or $24.50 per hour for academic users. These connect-time costs were effective January 1, 1989. If the search strategy is planned well in advance and carried out by a skilled user of the system, the entire search may require only 10 to 20 minutes. Other incidental fees plus telephone charges bring the cost of a single well-organized search to $25 to $50. If the planning is poor or carried out slowly online by an inexperienced user, the cost could be much higher. Nevertheless, to be able to search 20 years

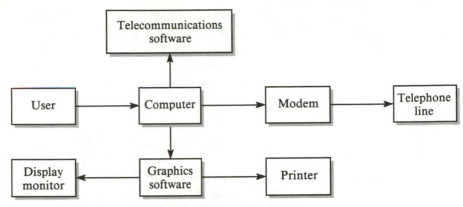

Figure 3–6 Capital equipment for online searching.

of CA in 20 minutes is impressive. It still may be necessary, however, to search in the old-fashioned way for abstracts published prior to 1967.

In addition to operational expenses, your librarian must consider the capital costs of the hardware and software shown in Figure 3–6.

A wide variety of ordinary microcomputers serve satisfactorily as terminals to an online searching system. CAS recommends a Tektronix PLOT 10 graphics terminal, but an Apple II or an IBM-PC will do the job with the help of graphics software to emulate the PLOT 10. A display monitor is always required, and if you wish to print the results of an online search, you will need a dot-matrix printer, such as an Epson FX-80 or similar model. A modem, Hayes 300B or 1200B or equivalent, is required to connect the computer to the telephone lines. Finally, a telecommunications software package is needed to interface the computer, modem, and telephone line. All this costs $2000 to $5000.

The online database is one more tool a scientist has in searching for information. Libraries are here to stay. Computer-assisted searches offer us a supplement, not an alternative.

SOURCES OF THERMODYNAMIC DATA: PRIMARILY INORGANIC

1. Wagman, D. D., et al., *The NBS Tables of Chemical Thermodynamic Properties: Selected Values for Inorganic and C_1 and C_2 Organic Substances in SI Units,* a National Bureau of Standards publication in cooperation with the American Chemical Society and the American Institute of Physics, New York, 1982. This volume replaces NBS Circular 500 and NBS Technical Note 270. (Beginning around 1965 NBS Circular 500 was gradually replaced by NBS Technical Note 270 over a period of 16 years.) The Wagman *NBS Tables* should be the starting point in a search for general thermodynamic data. The compilation primarily covers inorganic compounds, although carbon compounds containing up to two carbon atoms are also listed. When available, data are also given for aqueous solutions, often at a variety of concentrations. This volume and NBS Technical Note 270 list 26,000 values for the chemical thermo-dynamic properties of 14,300 substances, counting separately the same compound in different phases and at different concentrations. This volume is also referred to as *J. Phys. Chem. Ref. Data,* vol. II, suppl. 2, 1982.

2. Rossini, F. D., et al., *Selected Values of Chemical Thermody-namic Properties,* U.S. Government Printing Office, Washington, D.C., 1952. This was normally the first choice for an initial search for ther-modynamic data at 25°C, until the publication of the revised tables (see item 1 above). In Volume I the data are divided into two series; in the first are listed heats and free energies of formation, absolute entropies, and heat capacities, all at 25°C, for about 7000 substances. The energy unit is the calorie. In the second series one can find temperatures, heats and entropies of transition, fusion, and vaporization for about 3850 substances. Volume II contains literature references to the data in Volume I.

3. Chase, M. W., et al., *JANAF Thermochemical Tables,* 3rd ed., published by the American Chemical Society and the American Institute of Physics for the National Bureau of Standards, New York, 1986. This critically evaluated compilation of thermodynamic data for over 1100 inorganic substances is especially useful for calculating thermodynamic processes at high temperatures since the data are tabulated at 100-degree intervals from 0 to 6000 K. The functions tabulated are: C_P^0, S^0, $-(G^0 - H_{298}^0)/T$, $H^0 - H_{298}^0$, ΔH_f^0, ΔG_f^0, and log K_P.

4. Barin, I., and O. Knacke, *Thermochemical Properties of Inorganic Substances,* Springer-Verlag, New York, 1973. Over 900 inorganic sub-stances are listed in this collection of data, ranging from Ag to $ZrBr_2$. The functions tabulated include heat capacity, enthalpy of formation, absolute entropy, vapor pressure, and melting and boiling points.

5. Although Gmelin is usually thought of as a source of information on general inorganic chemistry, it also contains thermodynamic data for many inorganic and organometallic compounds. Similarly, do not over-look Beilstein as a source of thermodynamic data.

6. Quill, L. L., ed., *The Chemistry and Metallurgy of Miscellaneous Materials,* National Nuclear Energy Series IV-19B, McGraw-Hill, New York, 1950. This volume is part of a series reporting research work done during the Manhattan Project, from about 1942 to 1944. This particular volume is valuable for its extensive tabulations of thermodynamic prop-erties of transition-element halides, nitrides, carbides, sulfides, silicides, and phosphides between room temperature and high temperatures. Data include heats and free energies of formation, melting and boiling points, vapor pressures, heat-content functions, and free-energy functions.

7. Kubachewski, O., and E. L. Evans, *Metallurgical Thermochemistry,* 4th ed., Pergamon Press, New York, 1964. The second half of this book contains tables of thermodynamic data for inorganic substances, mostly binary compounds. Data include heats of formation, absolute entropy, melting and boiling temperatures, crystal structure type, lattice constants, heats of fusion and vaporization, heat-capacity equations of the form $C_p = a + bT + c/T^2$, and vapor-pressure equations of the form $\log P(\mathrm{mm}) = A/T + B \log T + CT + D$. The first half of the book has a quite readable treatment of some of the experimental methods used in obtaining many of the data tabulated within the book. Finally, Chapter 3 contains a useful treatment of the estimation of thermodynamic quantities.

8. Hultgren, R., et al., *Selected Values of Thermodynamic Properties of Binary Alloys,* American Society of Metals, Menlo Park, Ohio, 1973. In addition to phase diagrams, relative partial molar Gibb's free energy is listed, as well as other relevant thermodynamic functions.

9. Hultgren, R., et al., *Selected Values of Thermodynamic Properties of the Elements,* American Society of Metals, Menlo Park, Ohio (1973). This companion volume to item 8 above includes heat capacities, heat-content and free-energy functions, and absolute entropies.

SOURCES OF THERMODYNAMIC DATA: PRIMARILY ORGANIC

1. *TRC Tables, Selected Values of Properties of Chemical Compounds,* Thermodynamics Research Center, Texas A&M University, College Station, Texas, 1986. This collection contains data primarily for organic compounds. In addition to the usual thermodynamic properties such as heats of formation and combustion, the *TRC Tables* also include tables of physical properties such as refractive index, density, boiling point, melting point, viscosity, compressibility factors, and critical constants. The collection consists of eight three-ring binders so that current data may be added.

2. Zwolinski, B. J., and R. C. Wilhoit, *Comprehensive Index of API44-TRC Selected Data on Thermodynamics and Spectroscopy 1974,* Thermodynamics Research Center, Texas A&M University, College Station, Texas, 1974. This is an index to the *TRC Tables* listed above and to 11 other projects for the compilation of physicochemical, thermodynamic, and spectroscopic data undertaken by the TRC as of 1974. The thermodynamic data projects by the TRC and the American Petroleum Institute (API) are listed below. (The spectroscopic data projects are listed with the literature references to spectroscopy.)

Physicochemistry and Thermodynamics in API-TRC

Title	Vol. No.	Date
Selected Values of Properties of Hydrocarbons and Related Compounds		
API Research Project 44	A-68	12/31/73
Selected Values of Properties of Chemical Compounds		
TRC Data Project	A-33	12/31/73

For current information on TRC-API activities, write to

> Thermodynamics Research Center
> Data Distribution
> The Texas A&M University System
> College Station, TX 77843-3111

3. Selected Values of Properties of Hydrocarbons, National Bureau of Standards Circular 461, U.S. Government Printing Office, Washington, D.C., 1947. Among the thermodynamic properties you will find are vapor pressure, entropy, heat and free energy of formation, vaporization, and fusion, heat-content, and free-energy functions. In addition, refractive index, specific gravity, heat capacity, and heat of combustion are tabulated.

4. Timmermans, J., *Physico-Chemical Constants of Pure Organic Compounds,* Elsevier, Amsterdam, 1950, 1965. Most of the data in these two volumes are in tables of vapor pressure or specific heats; scattered throughout you will occasionally find values for density, dielectric constant, heats of combustion and vaporization, and surface tension. The data are for organic compounds containing C, H, O, N, S, and halogens. The index lists the entries by compound name; references are also listed.

5. Timmermans, J., *Physico-Chemical Constants of Binary Systems,* Wiley-Interscience, New York, 1959–1960. Four volumes.

6. Boublík, T., V. Fried, and E. Hála, *The Vapor Pressures of Pure Substances,* Elsevier, New York, 1973. This compact volume contains tables of vapor pressures between about 10 and 760 torr for about 900 elements and organic and inorganic compounds. For each entry, it also lists the three Antoine constants A, B, and C for $\log P = A - B/(t + C)$.

7. Wilhoit, R. C., and B. J. Zwolinski, *Handbook of Vapor Pressures and Heats of Vaporization of Hydrocarbons and Related Compounds,* Thermodynamics Research Center, Texas A&M University, College Station, Texas, 1971. The title says it all. In addition, the opening pages contain an excellent table of vapor-pressure data for water. A few sulfur-containing compounds are also included.

8. Horsley, L. H., ed., *Azeotropic Data,* Advances in Chemistry Series, American Chemical Society, vol. I (1952), vol. II (1962), and vol. III (1973).

SOURCES OF THERMODYNAMIC DATA: ORGANIC AND INORGANIC

1. Stephen, H., and T. Stephen, eds., *Solubilities of Inorganic and Organic Compounds,* vol. 1, Pergamon Press, Oxford, 1963.

2. Silcock, H. L., ed., *Solubilities of Inorganic and Organic Compounds,* Pergamon Press, Oxford (1964). Volume 1, published in two parts, contains not just solubility data, but also extensive data on two-component systems. Volume 2, published in three parts, lists data for ternary and multicomponent systems.

3. Landolt-Börnstein. A detailed description is given earlier in this chapter.

4. Washburn, et al., eds., *International Critical Tables,* McGraw-Hill, New York, 1926.

5. Touloukian, Y. S., J. K. Gerritsen, and N. Y. Moore, eds., *Thermophysical Properties Research Literature Retrieval Guide,* Plenum Press, New York, 1967. This guide, divided into three volumes, covers only seven properties, but for a very large number of substances (elements, compounds, and mixtures):

1. Thermal conductivity
2. Specific heat
3. Viscosity
4. Emissivity
5. Diffusion coefficient
6. Thermal diffusivity
7. Prandtl coefficient

The three volumes are used consecutively. For example, suppose we wish to find the viscosity of sulfuryl chloride. First, we locate the name in Volume I, where the name, formula, property number (1 through 7), and a seven-digit property code number are listed. We find the entry

Sulfuryl chloride SO_2Cl_2 1,3   106-0186

Since property 3 (viscosity) is listed, we jot down the property code number 106-0186. Then we open Volume II, and quickly find the property code number again, along with some unneeded information, and the serial number 5313, which we do need. Next, we turn to Volume III, and after the serial number 5313 we find the bibliographic citation to the information we need:

Thermal conductivity and diffusion coefficients in the gas phase. v. calculation of the coefficients of thermal conductivity. The relationship between the viscosity and thermal conductivity number.
Andrussow Leonid
Z. Elektrochem. 56:54–8, 1952

A quick look at *Zeitschrift für Elektrochemie* and we have our viscosity.

6. Bulletin of Chemical Thermodynamics, published once per year by Thermochemistry, Inc., Stillwater, Oklahoma. This bulletin of about 500 pages is useful for keeping up with the literature of thermodynamics. No data are contained in the *Bulletin of Chemical Thermodynamics*—it is a guide to the thermodynamics literature of the previous year. The major parts are the reports of current research and the substance property index, which are divided into four parts:

    I. Organic substances
   II. Organic mixtures
  III. Inorganic substances
  IV. Biochemical and macromolecular systems

As an example of its use, suppose we are following the molten salt chemistry of zinc halides. In the inorganic substances index, we find [under "Bull. Chem. Thermo. 25, 244 (1981)"] the following line:

$$ZnF_2\text{-}NaF(c/liq) \qquad Px \qquad 6025\text{-}81$$

The glossary on page 9 tells us that Px means the paper treats

Condensed phase equilibrium: solubility, freezing points, phase diagrams.

Turning to the Inorganic Substance Bibliography, which begins on page 435, we find the entry 6025-81 lies at the bottom of page 458 and reads as follows:

6025-81 Komlev, G. A., Lyazgin, B. I. and Nikitin, Yu. A., The NaF-ZnF_2 system, Zhur. Neorg. Khim 23, 2271-2272 (1978).

Another section, titled "Reports of Completed but Unpublished Research and Index to Contributions," is interesting for researchers who want to avoid unnecessarily duplicating the efforts of others. It is also of interest to undergraduates who wish to find out what is going on in the area of thermodynamics at various universities worldwide. A typical entry is this one from Iowa:

USA 079 1
Iowa State University, Chemistry Department and Ames Laboratory-DoE, Ames, Iowa 50011. We are currently working on the following systems using mass-loss, mass spectrometry and target collection analysis of Knudsen effusion rates: W-Se, Nb-Al, Pb-Yb, Ce-Sb[24/1981].

The final section of the *Bulletin* is called "News, Reviews and Comments," and is very useful for someone currently working in thermodynamics.

SOURCES OF THERMODYNAMIC DATA: ELECTROCHEMISTRY

1. Dobos, D., *Electrochemical Data,* Elsevier, New York, 1975. This book contains a wide range of tabulated electrochemical data. Extensive coverage is given to conductivities, transport numbers, and activity coefficients, in addition to electrode potentials.

2. Gibson, J. G., and J. L. Susworth, *Specific Energies of Galvanic Reactions and Related Thermodynamic Data,* Chapman and Hall, Ltd., London, 1973. This book contains values that were mostly calculated rather than directly measured for about 50,000 possible cell reactions and their temperature coefficients.

3. Meites, L., and P. Zuman, *Electrochemical Data,* Wiley-Interscience, New York, 1974. This compilation lists the electrochemical behavior of 2015 chemical compounds and the experimental techniques and conditions employed in studying them.

4. Latimer, W. M., *The Oxidation States of the Elements and their Potentials in Aqueous Solution,* Prentice-Hall, New York, 1938. Although dated, this classic is still readable and contains much thermodynamic data as well as electrode potentials.

5. Harned, H. S., and B. B. Owen, *The Physical Chemistry of Electrolytic Solutions,* Reinhold, New York, 1950. Besides being a treatise on the title subject, this volume contains many tables of useful data.

ESTIMATING THERMODYNAMIC DATA

Although libraries have thermodynamic data for thousands of organic and inorganic compounds, you will frequently need data for new compounds or other compounds for which data have not been measured or cannot be found. It is therefore wise to become familiar with some systematic methods for calculating or estimating thermodynamic data. The following monographs contain material especially useful for estimating thermodynamic parameters.

1. Benson, S. W., *Thermochemical Kinetics—Methods for the Estimation of Thermochemical Data and Rate Parameters,* 2nd ed., Wiley, New York, 1976. Both empirical methods and methods based on statistical thermodynamics are employed to estimate thermodynamic data.

2. Cox, J. D., and G. Pilcher, *Thermochemistry of Organic and Organometallic Compounds,* Academic Press, New York, 1970. This useful book contains a section on estimating heats of formation of organic compounds in addition to a critical compilation of thermodynamic data for about 3000 compounds.

3. Kubachewski, O., and E. L. Evans, *Metallurgical Thermochemistry,* 4th ed., Pergamon Press, New York, 1964. In Chapter 3 you will find some simple empirical methods for estimating heat capacities, heats and entropies of fusion, vaporization, and formation, and absolute entropies.

4. Janz, G. J., *Thermodynamic Properties of Organic Compounds—Estimation Methods, Principles, and Practice,* Academic Press, New York, 1967.

5. Lewis, G. N., and M. Randall, *Thermodynamics,* 2nd ed., revised by K. Pitzer and L. Brewer, McGraw-Hill, New York, 1961.

6. Margrave, J. L., in L. Brewer, N. K. Heister, and A. W. Searcy, eds., *High Temperature—A Tool for the Future,* Stanford Research Institute, Menlo Park, California, 1956.

7. Wenner, R. R., *Thermochemical Calculations,* McGraw-Hill, New York, 1941. This nice little book is out of print now, but it is worth looking for in your library because of Chapter 8 on estimating entropies, free energies, and heat capacities.

CRYSTALLOGRAPHY LITERATURE

1. For a general entry into crystal structure data, *Crystal Data: Determinative Tables* is most useful and direct. The first edition, published as Memoir 60 of the Geological Society of America (1954) consists of two parts: *Systematic Tables,* by W. N. Nowacki, and the *Determinative Tables* of J. D. H. Donnay. The second edition appeared in two parts as Monographs 5 and 6 of the American Crystallographic Association: *Determinative Tables,* by J. D. H. Donnay et al., and *Systematic Tables,* by W. N. Nowacki et al. The latest edition consists of six volumes:

Volume 1: Organic
Volume 2: Inorganic
Volume 3: Organic
Volume 4: Inorganic
Volume 5: Organic
Volume 6: Organic

The entries are listed, within each crystal system, according to increasing values of a determinative number: the a/b ratio in trimetric systems, the c/a ratio in dimetric systems, and the cubic edge a in the isometric system. Within each volume is a name and formula index. The important data listed are unit cell dimensions, space group, whether a complete structure is known with atomic coordinates, and references.

2. For powder diffraction data, a collection for inorganic compounds and another for organic compounds were published between 1960 and 1984—seven volumes for each collection, further divided into 26 sets.

The first five sets are published by the American Society for Testing Materials, as follows:

X-Ray Powder Diffraction File
Sets 1–5
[Inorganic/Organic]

The remaining sets are published by the Joint Committee on Powder Diffraction Standards,[3] divided as follows:

Sets 6–10, 11–15, 16–18, 19–20, 21–22,
23–24, 25–26, 27–28, 29–30, 31–32

In 1984 an alphabetical index was published, simplifying considerably the task of finding a desired compound.

As a simple example, let us look up cesium bromide in the alphabetical index. Under "cesium bromide," the only information given is the file number: 5-0588. This refers to entry 0588 in Set 5 of the inorganic volumes. In the volume labeled "Sets 1–5," we find entry 5-0588 on page 644. A facsimile of the data card shows the name of the compound, cesium bromide; the relative intensities of the three strongest lines; the diffraction conditions; a list of several d spacings and their Miller indices; and references (Fig. 3–7).

To find the name of an unknown compound starting with the powder diffraction pattern is more difficult, and two methods of searching are in general use: Fink and Hanalwalt. The *Fink Search Manual* was published in 1982 and the *Hanawalt Search Manual* in 1984. Brief directions for their use are in the manuals and in some of the monographs listed below.

3. The principal working handbooks of the crystallographer doing single-crystal x-ray diffraction studies are the *International Tables for X-Ray Crystallography:*

Volume I. Symmetry groups
Volume II. Mathematical tables
Volume III. Physical and chemical tables
Volume IV. Revised and supplementary tables

This is the second series; it was published between 1952 and 1974. A third series is in preparation, and to distinguish the four new volumes from the second series, they are to be called Volumes A, B, C, and D instead of Volumes I, II, III, and IV. Volume A was published in 1983.

4. Wyckoff, W. G., *Crystal Structures,* Interscience, New York, 1951–1969:

Volume 1. Elements and compounds of type RX, RX_2
Volume 2. R_nX_m to $R_n(MX_3)_m$

[3] Joint Committee on Powder Diffraction Standards, 1601 Park Lane, Swarthmore, Pennsylvania 19081.

5-0588 MINOR CORRECTION

d	3.04	1.75	1.15	4.29	CsBr
I/I_1	100	43	20	8	Cesium Bromide

Rad. $CuK\alpha_1$ λ 1.5405 Filter Ni
Dia. Cut off Coll.
I/I_1 G.C. Diffractometer d corr. abs.?
Ref. Swanson and Fuyat, NBS Circular 539, Vol. III, (1953)

Sys. Cubic (simple cubic) S.G. O_H^1 - Pm3m
a_0 4.296 b_0 c_0 A C
α β γ Z 1
Ref. Ibid.

δα nωβ1.703ξγ Sign
2V D_X4.456 mp Color
Ref.

Sample prepared at NBS. Spect. anal.: <0.01% Ca, K, Na; <0.001% Al, Ba, Cu, Fe, Mg, Si.
X-ray pattern at 25°C.

Replaces 1-0843

d Å	I/I_1	hkl	d Å	I/I_1	hkl
4.29	8	100	0.8424	9	510
3.039	100	110			
2.480	3	111			
2.148	18	200			
1.921	6	210			
1.754	43	211			
1.519	18	220			
1.432	3	300			
1.358	16	310			
1.2952	<1	311			
1.2397	6	222			
1.1919	1	320			
1.1482	20	321			
1.0741	1	400			
1.0125	9	411			
0.9856	1	331			
.9605	5	420			
.9157	3	332			
.8768	3	422			
.8590	<1	500			

Figure 3–7 Data card for x-ray powder diffraction.

Volume 3. Hydrates and ammoniates
Volume 4. Miscellaneous inorganic and silicate compounds
Volume 5. Aliphatic compounds
Volume 6. Benzene derivatives

This collection is a convenient source of complete structure determinations, usually with a sketch of the structure and a brief description of the interesting features.

5. Kennard, O., F. H. Allen, and D. G. Watson, eds., *Molecular Structure and Dimensions,* published for the Cambridge Crystallographic Data Centre and the International Union of Crystallography by Bohn, Scheltem and Holkema, Utrecht, 1977. This is the printed output of the Cambridge Crystallographic Data Center database. Volumes 1 through 8 contain formula, metal, and author indices. These volumes do not contain structural data, but rather bibliographies for each compound. Subsequent volumes contain permuted keyword-in-context compound name indices, and formula, permuted formula, and author indices.

6. Kennard, O., F. H. Allen, and D. G. Watson, eds., *Guide to the Literature 1935–1976—Organic and Inorganic Crystal Structures,* published as above (item 5) as an index to Volumes 1 through 8. This volume

is divided into compound name (11,116 names), molecular formula, permuted formula, author, and literature indices.

7. Trotter, J., ed., *Structure Reports,* vol. 8 (1940–41) to vol. 46A (1980); began as *Struktur Bericht,* vol. 1 (1913–28) to vol. 7 (1939). In recent years, the editor has divided each volume into Main Group Compounds and Transition Metal Compounds. Three indices (subject, formula, and author) for each volume facilitate finding the desired structural data. The entries include the name, formula, author, journal reference, unit cell data, abstract, and, if available, a drawing of the structure (ORTEP or line). Important bond angles and distances are included.

8. Structure Reports 1913–1973, 60-year index to above (item 7) *Structure Reports.*

SPECTROSCOPY

Specialized Collections

1. If a search of Gmelin or Beilstein fails to reveal the spectrum sought, then one of the more specialized collections may prove fruitful. Sadtler Laboratories of Philadelphia offers 3 one-volume handbooks, each containing a few thousand spectra:

1. *The Sadtler Handbook of NMR Spectra*
2. *The Sadtler Handbook to Infrared Spectra*
3. *The Sadtler Handbook of UV Spectra*

Although these three guides may be held in two hands, the complete *Sadtler Spectra Collection* occupies many linear feet of library shelves. The collection is divided into several groups:

1. Standard Infrared (30 volumes, 3 indices, 2 specfinders)
2. Standard IR Grating Spectra (75 volumes plus indices)
3. Inorganic IR Grating Spectra (5 volumes plus indices)
4. Standard Ultraviolet Spectra (86 volumes plus indices)
5. Standard Proton NMR Spectra (70 volumes plus indices)
6. Standard Carbon-13 NMR (90 volumes plus indices)
7. Organometallic Grating Spectra (1 volume)
8. Pharmaceutical Spectra IR (4 volumes plus index)
9. Steroids Grating Spectra (4 volumes)
10. Coblenz Society Spectra (11 volumes plus indices)

2. Zwolinski, B. J., and R. C. Wilhoit, *Comprehensive Index of API44-TRC Selected Data on Thermodynamics and Spectroscopy,* Thermodynamics Research Center, Texas A&M University, College Station, Texas, 1974. This index is also mentioned above for its value in finding thermodynamic data for organic compounds. It also is an index to five spectral data projects supported by TRC and the American Petroleum Institute (API). The complete set of thermodynamic and spectral projects is detailed below.

Physicochemistry and Thermodynamics in API44-TRC

Title	Vol. No.	Date
Selected Values of Properties of Hydrocarbons and Related Compounds		
API Research Project 44	A-68	12/31/73
Selected Values of Properties of Chemical Compounds		
TRC Data Project	A-33	12/31/73

Spectroscopy in API44-TRC

Title	Vol. No.	Date
Selected Infrared Spectral Data		
API Research Project 44	B-74	10/31/73
TRC Data Project	B-14	6/30/73
Selected Ultraviolet Spectral Data		
API Research Project 44	C-43	10/31/70
TRC Data Project	C-6	6/30/73
Selected Raman Spectral Data		
API Research Project 44	D-18	4/30/73
TRC Data Project	D-2	12/31/66
Selected Mass Spectral Data		
API Research Project 44	E-50	4/30/72
TRC Data Project	E-10	6/30/73
Selected NMR Spectral Data		
API Research Project 44	F-18	10/31/73
TRC Data Project	F-19	6/30/73

Not included in this index, but added later, is the following compilation: G. Selected ^{13}C NMR Spectral Data (Projects F-18 and F-19 handle only proton NMR spectra).

Handbooks of Spectroscopic Data

Four additional specialized handbooks worth knowing about are

1. *Varian Associates NMR Spectra Catalog*
2. *High Resolution NMR Spectra,* published by Japan Electron Optics Laboratories
3. *The Aldrich Library of Infrared Spectra,* published by the Aldrich Chemical Company
4. *Organic Electronic Spectral Data,* published by Interscience

These collections are generally qualitative; that is, they simply reproduce the spectra of various compounds. For quantitative numerical data derived from or related to spectroscopy some other sources are needed; these may be divided into data collections, which are primarily tabular, and into specialized monographs.

Tables of Spectroscopic Data

1. Bourcier, S., *Tables de Constantes et Données Numériques,* Pergamon Press, New York, 1970. This is an excellent source of spectroscopic and molecular parameters including electronic energies, fundamental vibration frequencies, and anharmonicity constants. Do not worry about the French table headings, because the symbols for the parameters are international and you will recognize them immediately.

2. Herzberg, G., and K. P. Huber, *Constants of Diatomic Molecules,* Van Nostrand, New York, 1979.

3. International Union of Pure and Applied Chemistry Commission on Molecular Spectroscopy, Tables of Provisional Wavenumber Standards for Calibration of Infrared Spectrometers, Buttersworth, London, 1961.

4. IUPAC Tables of Wavenumbers, Crane, Russak and Co., New York, 1961.

5. Harrison, G. R., ed., *MIT Wavelength Tables,* The M.I.T. Press, Cambridge, Massachusetts, 1969. This is a single table of wavelengths beginning at about 9300 Å and decreasing to 2000 Å for the wavelengths for emission spectra of the elements. It is useful for looking up a wavelength to determine which element could be the source.

6. Moore, C. E., *Atomic Energy Levels,* National Bureau of Standards Circular 467, U.S. Government Printing Office, vol. I (1949), vol. II (1952), vol. III (1958). These volumes contain wavelength tables for each element in the volume. Volume I covers H through V, Volume II covers Cr through Nb, and Volume III covers Mo through La and Hf through Ac. It also includes term symbols for many states so that these data are useful in calculating thermodynamic function by the methods of statistical thermodynamics.

7. Selected Constants: Spectroscopic Data Relative to Diatomic Molecules, Pergamon Press, New York, 1971.

8. Shimanouchi, T., et al., *Table of Molecular Vibration Frequencies,* National Standard Reference Data Series, U.S. Government Printing Office, Washington, D.C., 1972 (and later volumes).

9. Suchard, S. N., *Heteronuclear Diatomic Molecules,* Part A and Part B, Plenum Press, New York, 1976.

10. Suchard, S. N., *Homonuclear Diatomic Molecules,* Plenum Press, New York, 1976.

11. Sutton, L. E., ed., *Interatomic Distances and Configurations in Molecules and Ions,* Special Publication No. 11 (1958), Supplement for 1956–1959, Special Publication No. 18 (1965), The Chemical Society, London.

Monographs

1. Barrow, G. M., *Introduction to Molecular Spectroscopy,* McGraw-Hill, New York, 1962.

2. Brode, W. R., *Chemical Spectroscopy,* Wiley, New York, 1947.

3. King, G. W., *Spectroscopy and Molecular Structure,* Holt, Rinehart and Winston, New York, 1964.

4. Herzberg, G., *Atomic Spectra and Atomic Structure,* Dover, New York, 1944.

5. Herzberg, G., *Molecular Structure, I. Spectra of Diatomic Molecules,* Van Nostrand, New York, 1950.

6. Herzberg, G., *Molecular Structure, II. Infrared and Raman Spectra of Polyatomic Molecules,* Van Nostrand, New York, 1945.

7. Herzberg, G., *Molecular Structure, III. Electronic Spectra and Electronic Structure of Polyatomic Molecules,* Van Nostrand, New York, 1966.

8. Wilson, E. B., Jr., J. C. Decious, and P. C. Cross, *Molecular Vibrations,* McGraw-Hill, New York, 1955.

KINETICS

1. Bamford, C. H., and C. F. H. Tippes, eds., *Comprehensive Chemical Kinetics,* Elsevier, Amsterdam, New York, 1969.

2. Benson, S. W., and H. E. O'Neal, *Kinetic Data on Gas Phase Unimolecular Reactions,* NSRDS-NBS 21, National Bureau of Standards, Washington, D.C., 1970.

3. Trotman-Dickenson, A. F., and G. S. Milne, *Tables of Gas Phase Reactions,* NSRDS-NB59, NBS, U.S. Government Printing Office, Washington, D.C., 1967.

4. Amis, E. S., *Kinetics of Chemical Change in Solution,* Macmillan, New York, 1949.

5. Moelwyn-Hughes, E. A., *The Kinetics of Reactions in Solution,* Oxford University Press, London, 1947.

6. Benson, S. W., *The Foundations of Chemical Kinetics,* McGraw-Hill, New York, 1960.

7. Benson, S. W., *Thermochemical Kinetics—Methods for the Estimation of Thermochemical Data and Rate Parameters,* 2nd ed., New York, Wiley, 1976. Both empirical methods and methods based on statistical thermodynamics are employed to estimate thermodynamic data.

8. Tables of Chemical Kinetics, Homogeneous Reactions, National Bureau of Standards Circular 510, U.S. Government Printing Office, Washington, D.C., 1951. This volume gives rate constants for organic reactions for the most part. They are divided into seven types:

1. Rearrangement–isomerization
2. Condensation–solvolysis
3. Exchange–substitution
4. Elimination
5. Dissociation–decomposition
6. Association–addition
7. Oxidation–reduction

In addition, four supplements have been issued. The first two have the titles

Supplement 1 to Circular 510 (1956)
Supplement 2 to Circular 510 (1960)

The second two have the titles

Tables of Chemical Kinetics, Homogeneous Reactions (Supplement Tables), NBS Monograph 34, Volume 1, 1959

Tables of Chemical Kinetics, Homogeneous Reactions (Supplement Tables), NBS Monograph 34, Volume 2, 1964

ELECTRIC AND MAGNETIC PROPERTIES

1. Debye, P., *Polar Molecules,* Dover, New York, 1929.

2. Figgis, B. N., and J. Lewis, in J. Lewis and R. G. Wilkins, eds., *Modern Coordination Chemistry,* Interscience, New York, 1960.

3. McClellan, A. L., *Tables of Experimental Dipole Moments,* Freeman, San Francisco, 1963.

4. Smyth, C. P., *Dielectric Behavior and Structure,* McGraw-Hill, New York, 1955.

5. LeFevre, R. J. W., *Dipole Moments, Their Measurement and Applications in Chemistry,* Methuen, London, 1953.

6. Smyth, C. P., in A. Weissberger and B. W. Rossiter, eds., *Physical Methods of Chemistry,* vol. I, part IV, chap. VI, Interscience, New York, 1972.

7. Wesson, L. G., *Tables of Electric Dipole Moments,* The Technology Press of M.I.T., Cambridge, Massachusetts, 1948.

8. Maryott, A. A., and E. R. Smith, *Tables of Dielectric Constants of Pure Liquids,* National Bureau of Standards Circular 514, U.S. Government Printing Office, Washington, D.C., 1951.

GENERAL REFERENCES ON THE CHEMICAL LITERATURE

1. Gorin, G., *J. Chem. Educ.* 59:991, 1982.

2. Maizell, R. E., *How to Find Chemical Information,* Wiley, New York, 1979.

3. Mellon, M. G., *Chemical Publications, Their Nature and Use,* McGraw-Hill, New York, 1982.

4. Skolnik, H., *The Literature Matrix of Chemistry,* Wiley, New York, 1982.

5. Wiggins, G., *J. Chem. Educ.* 59:994, 1982.

6. Wilson, E. B., Jr., *An Introduction to Scientific Research,* McGraw-Hill, New York, 1952.

7. Wolman, Y., *Chemical Information,* John Wiley and Sons, New York, 1983.

8. Woodford, H. M., *Using the Chemical Literature,* Marcel Dekker, New York, 1974.

9. *CA Search for Beginners; An Introduction to Online Access to CA Search,* Chemical Abstract Services, Columbus, Ohio, 1980.

10. Sources of Adjuvant Data for Thermochemical Analyses, *Bull. Chem. Thermodynamics* 26:569, 1983.

Designing the Experiment

It is unusual for a physical chemistry laboratory student to participate in the design of the experiments. Still, if laboratory experience is to prepare a student for independent work, the physical chemistry laboratory is an appropriate place to become aware of the nature of experimental science and the factors entering into the design of an experiment.

Experiments are used to either construct or test hypotheses. In physical chemistry, we think of an experiment as a planned change on a carefully designed system that is allowed to take place at the will and under the control of an experimenter. This definition implies that we recognize and thoroughly understand the problem to be investigated, that we have a carefully thought out plan of attack, and that we have devised a scheme or apparatus to initiate some kind of change in such a way that the change can be controlled, observed, and recorded.

Experiments can be divided into two classes: statistical and nonstatistical. Experiments in the behavioral sciences, the life sciences, and applied research are often statistical. The variables that are controlled or allowed to vary tend to reflect large statistical fluctuations. On the other hand, experiments in the physical sciences are more often nonstatistical. It is easier to define the system, and the usual variables (e.g., pressure, temperature, and composition) fluctuate within relatively narrow limits. The physical design of the experiment controls the variables and defines the system. Even in nonstatistical experiments, however—such as chemical and physical experiments—where the relevant variables are individually and precisely controlled, statistics does indeed enter the final stages of interpreting the experiment and evaluating the random errors.

DESIGNING THE APPARATUS

While physical scientists occasionally use statistical design in the preliminary survey of an experiment that involves several variables, more often the design of a physical science experiment reduces to the design of the apparatus required to keep several variables constant and to control the variation of a single parameter.

Although the variety of apparatus used to solve problems in the physical sciences is nearly as great as the problems themselves, a number of general considerations are worth thinking about before running to the machine shop with a sketch on a paper towel. Naturally, it is important to have in mind a clear idea of the problem to be solved, and it is constructive to ask a few questions. What is it that we wish to accomplish? What do we want to observe? What kinds of things can be or need to be controlled if the change is chemical in nature (i.e., new substances are being created from old substances)? The laws of thermodynamics govern changes of state and should be kept firmly in mind. The rate of the change also affects the design or execution of an experiment.

Don't reinvent the wheel. In your literature search, you should take notes not only on the chemistry of the system you are about to study, but on the various experimental techniques and apparatus that have already been used. It is rare indeed that an entirely original apparatus is designed for a scientific investigation. More often, the actual apparatus that one designs and builds is a hybrid of several described in the literature, with some minor but highly original additions. The sources of these ideas should be acknowledged in appropriate references or footnotes in reports and journal articles.

Try to be realistic about the magnitude of the apparatus-building task you are to undertake. The kinds of apparatus that can be built differ enormously in complexity. Look at the total amount of time available and the fraction that can be allotted to apparatus design as opposed to the actual collection of data. The time spent on a special project for the physical chemistry laboratory may amount to only a few hours, on a senior thesis project several weeks, and on a doctoral dissertation a couple of years. Be sure some time is left over for *using* the apparatus. If the project you are getting into is construction-intensive, be sure that you realize this before you begin. If you are not particularly handy with machine tools, electronics, glass blowing, or computer programming, you should take these factors into consideration before choosing your research project.

The three laws of thermodynamics are the fundamental guiding principles in determining whether chemical change is to be observed or not, in particular the free energy:

$$\Delta G = f(P,\ T,\ n_1,\ n_2,\ \cdots)$$

Fortunately, only a limiting number of factors need be established, controlled, and measured. These generally reduce to pressure, temperature, and composition. From a more fundamental point of view, most measurements are eventually reduced to the actual measurement of length (area and volume), mass, and time. Ultimately these involve the measurement of length only: the deflection of a needle, the angle of a turning screw, the divisions on a scale. The measurement of temperature and pressure is dealt with later.

Composition is used in a very general sense here and is meant to convey the idea that all intensive and extensive properties of matter depend on the kind and/or amount of matter present in the system selected for study. In the beginning stages of planning an experiment it is a good idea to have clearly in mind what properties of matter are to be measured, whether they are extensive or intensive, and how they depend on other factors such as pressure or temperature or composition—the presence of other kinds of matter. Is the system open or closed? That is, does it matter if it leaks? Is it isolated? Does it matter if it exchanges heat and/or matter with its surroundings? Do the properties of any part of the system depend on time? If a one-component system is studied, what is its state? If a condensed phase is present, what is its vapor pressure? Is the presence of the vapor phase in addition to the condensed phase a problem? If so, should a lower temperature be selected? If the system is a solid phase, what is the state of subdivision? Does the phase consist of large single crystals or is it microcrystalline, or colloidal? Does surface area play a role? If the system consists of two or more phases, is there the possibility of solution? Solid solution? What are the activities of the components in each phase?

These kinds of questions may sound to you as if we are emphasizing thermodynamic studies. That is not the case at all. These questions need to be considered for any kind of chemical research or investigation. You need to know what the system really is that you are studying. You must establish that the spectrum you report is that of the substance you report and not of an addition product between the solvent and your substance. Instruments grind out data on poorly defined systems just as efficiently as they do on well-defined systems. Do not be deceived by the ease with which black boxes crank out lines and numbers.

In order to design an experiment we need to know what the characteristics of an experiment are.

1. An experiment has a clear cognitive purpose with the goal of seeking some kind of understanding. It should fit into the cycle of the scientific method and be part of either the construction or the testing of a hypothesis.
2. An experiment is selective; a portion of the universe is selected for study—a system is defined that has relatively small, tractable, artificial, noninterfering boundaries. The variables of the system must also be defined.

3. Experiments have a beginning, a middle, and an end, during which a change in the system is deliberately provoked.
4. The change must be capable of being
   a. observed by (Fig. 4–1)
      1. direct perception (seen with the eyes) or
      2. indirect measurement (with an instrument);
   b. interpreted by the observer; recognized as a significant change in a variable property of the system;
   c. described by the observer; recorded in such a fashion that it can be communicated to another scientist for testing or reproduction.
5. The system and the change subject to experimentation should be reproducible and testable by other scientists.

Experimentation requires the invention of a wide variety of techniques that depend on the specific nature of the experiment, that is, the variables to be controlled and the properties to be measured. Sometimes we observe experimental change with our eyes, but more often some kind of instrument intercedes on our behalf.

The measuring apparatus itself consists of several elements that emulate some of the functions of human sensation. A sensing element has some physical property that is proportional to the variable to be measured. The electromotive force of a thermocouple is, for example, proportional to the temperature difference between the system and a reference temperature. The conditioning element accepts the output of the sensing

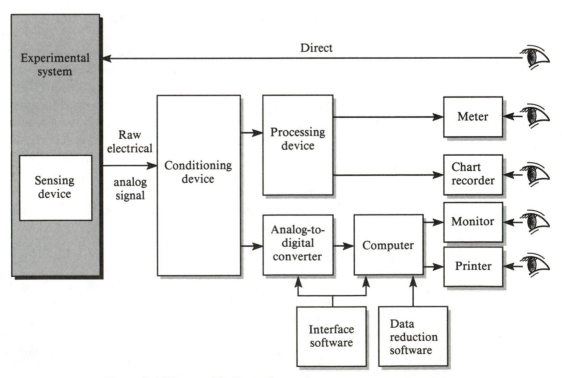

Figure 4–1 Direct and indirect observation.

TABLE 4–1

Components of Instrumental Observation

Sensing	Conditioning
Thermocouple	Wheatstone bridge
Scintillation detector	Amplifier
Bolometer	Oscillator
Photocell	Film developer
Capacitor	Operational amplifier
Inductor	Gates (AND, NAND, OR, NOR)
Film	
Thermister	

Processing	Displaying
Analog-to-digital converter	Meter
Correcting device	Chart recorder
Integrator	Digital readout
	Computer/CRT monitor
	Computer/printer

element and changes it into a form suitable for further processing. For example, an amplifier converts the feeble voltage of a thermocouple to a voltage capable of driving a chart recorder. The processing element refines the output of the conditioning element so that the output is simple to interpret. Since the output of a thermocouple is a nonlinear function of temperature, the processing element might correct for the nonlinearity. Finally, the data presentation element displays the conditioned output in some readable form, such as a line on a chart recorder, a series of printed digits, or the deflection of a meter with respect to a scale. Not every apparatus has all four of these elements, and some apparatus have a particular element more than once. Some possible components of a measuring apparatus are shown in Table 4–1.

CHOICE OF VARIABLES

In its simplest form, an experiment is an investigation of the relationship between a dependent variable and an independent variable. In the physical sciences, usually only one independent variable is allowed to change and the others are held constant by the design of the experiment. Only those variables that need to be controlled should be controlled; some are irrelevant. The phase of the moon is an example of an irrelevant variable—to a physical chemist, but perhaps not to an astronomer. While Friday the 13th is not relevant, the time of day may or may not be important.

After an independent variable has been chosen, a few more decisions affect the overall design of your apparatus. You must decide what the range of the independent variables will be. This includes the expected or needed maximum value, the minimum value, and the number of values in between. Very often some kind of linear relationship can be established between the dependent and the independent variables. For example, although the vapor pressure P is not a linear function of the temperature T, log P is a linear function of $1/T$, at least over short temperature ranges. Clearly, if the vapor pressures of interest range from 2 to 50 atm, the design of the apparatus is substantially different than if the vapor pressures range from 10^{-7} to 10^{-5} atm. Not only is the temperature range substantially different, but so is the method for sensing and measuring the pressure.

After you decide upon the overall range of the variables, you must determine how much error you can tolerate in the dependent variable that is the focus of your investigation. If you are measuring the vapor pressure from 2 to 50 atm, you must decide whether an uncertainty of 0.1 or 0.01 atm is tolerable. The design of the apparatus may be considerably more complex if a measurement of high accuracy is needed; the sensitivity of the pressure-measuring device would be much more sophisticated. It is important to realize at this point that decisions concerning the errors of measurement of the dependent variable affect the measurement and control of the independent variable—in this example, the temperature. Since the vapor pressure depends exponentially upon the temperature, small fluctuations in the temperature generate large fluctuations or uncertainties in the pressure. It does little good to design a pressure-measuring device good to 0.01 atm if you find it impossible to control or measure the temperature well enough to keep random pressure fluctuations less than 0.1 atm.

The choice of variables, their range, and their accuracy play a significant role in the design of an experiment. You should try to anticipate these factors and design the experiment with them in mind. Alternatively, you can play it by ear, improvising as you go. While this may work well for musicians, it is not recommended to neophyte scientists. A better procedure is to outline the components of your experiment, to list their functions and your expectations of their performance, and to judge the difficulty in achieving those expectations. Next, order your list according to your best guess as to the difficulty in fabricating each component. Begin your work with the most difficult component, because if you cannot get it to work, then there is no point in building the other components. If you do the easy part first, then all this effort is wasted if a crucial, difficult step proves to be impossible. It is better to change directions early in an investigation than later. This strategy is of general utility and is as valuable in such diverse activities as organic synthesis and computer programming as it is in physical chemistry.

If the toughest component of your experiment is really complex, you might consider building a rough prototype to test. If the prototype passes your tests, then you might consider expending the effort to create a full-fledged working model that might well be the core of your experiment. The easier components can then be added to the core until the experiment is complete.

HUMAN FACTORS

As the apparatus increases in complexity, it is particularly important to pay attention to human factors in addition to the usual physical details. All parts of the apparatus that are constantly used as the experiment is being executed must be conveniently reached by the experimenter. The arrangement of electrical switches and dials should be laid out like the words on a page; that is, they should be used and read from left to right and from top to bottom. It is easy to make a mistake if you have to throw one switch, skip one, throw another, and then back up. If several switches and meters must be read, they should be collected together in a control board for operating convenience. If it is absolutely essential that switch A be turned on before switch B, then some kind of physical or electrical method should be devised to deactivate switch B if it is inadvertently turned on before switch A. Meters should be located adjacent to the switch that activates them. Usually the switch should be below the meter, so that hand turning on the switch does not block the view of the meter. In designing an experiment, one should give close attention to the general problem of accessibility.

If the apparatus incorporates several electrical modules, these should be mounted in a vertical rack designed for that purpose. A vertical arrangement saves bench space, and if the rack is on wheels, it can be moved easily through a door without dismantling. An apparatus spread out horizontally on a table top is likely to be less portable than a vertically arranged apparatus. If some module of the apparatus is of general utility it should be easily portable and demountable. Repair work on an apparatus is much easier if the apparatus is modular in construction, easily accessible, and portable. In trouble shooting an apparatus, it is sometimes convenient to test by substitution. If one module can be pulled out and another substituted, the trouble can be quickly identified. If your apparatus can be calibrated, you should provide for convenient access to calibration points on the front panel.

The control points of the apparatus should be clearly labeled and the whole working area should be well illuminated. Do not rely on your memory if the operating sequence is fairly complex; instead, develop a checklist. If you have been working long hours on an experiment, fatigue can affect your performance. You are more apt to make blunders and mistakes. Ease and convenience of operation in a pleasant environment reduce errors caused by fatigue.

COMPUTERS

A computer program may play a role in the design of the experiment, especially if computerized data acquisition is involved. If you write a computer program, take care to use structured programming and document the program thoroughly. Structured programming follows naturally with languages such as Pascal (see Part Two) and C; less so with BASIC and FORTRAN.

The program should be documented internally. A brief explanation of what the program does should immediately follow the main program header. Every procedure header in Pascal should be followed by a comment explaining what action the procedure carries out. Every function header should be followed by a comment explaining exactly what value is being returned. All tricky little algorithms should have some kind of explanatory comment in the program. If the program has more than five or six procedures, they should be documented in comments at the very beginning of the program. Use self-documenting names for identifiers; for example, "Volts" (not "V") and "Time" (not "t"). Anyone—not just the programmer—should be able to read the source code and get a pretty good idea of what the program does and how it does it.

The program should be documented externally. This means a page or two—written in good English and as free as possible of computer jargon and acronyms—should be available to all users. The internal documentation should be repeated and expanded. When possible, the condition for a test run should be provided. Interactive input often acts as the visible interface between the program and the user. Interactive input should be user-friendly and self-documenting.

As you work in the physical chemistry laboratory, keep in mind the design factors relating to your experiment. Is there room for improvement or extension? When you begin research, either as a senior thesis student or as a graduate student, the design of an experiment will become more and more your responsibility.

REFERENCES

1. Baird, D. C., *Experimentation: An Introduction to Measurement Theory and Experiment Design,* Prentice-Hall, Englewood Cliffs, New Jersey, 1962.

2. Baker, J. W., and G. E. Allen, *Hypothesis, Prediction, and Implication in Biology,* Addison-Wesley, Reading, Massachusetts, 1968.

3. Bentley, J. P., *Principles of Measurement Systems,* Longman, London, 1983.

4. Beveridge, W. I. B., *The Art of Scientific Investigation,* Norton, New York, 1951.

5. Brinkworth, B. J., *An Introduction to Experimentation,* C. Tinling & Co. Ltd., Liverpool, 1968.

6. Bunge, M., *Scientific Research II: The Search for Truth,* Springer-Verlag, New York, 1967.

7. Cochran, W. G., and G. M. Cox, *Experimental Designs,* Wiley, New York, 1957.

8. De Beer, E. J., "The Place of Statistical Methods in Biological and Chemical Experimentation," in *Annals of the New York Academy of Science,* vol. 52 (March 10, 1950), art. 6, pp. 789–942.

9. Freedman, P., *The Principles of Scientific Research,* Pergamon Press, New York, 1960.

10. Hooke, R., *Scientific Inferences,* Holden-Day, San Francisco, 1963.

11. Montgomery, D. C., *Design and Analysis of Experiments,* Wiley, New York, 1984.

12. Peterson, M. S., *Scientific Thinking and Scientific Writing,* Reinhold, New York, 1961.

13. Weatherall, M., *Scientific Method,* English Universities Press, London, 1968.

14. Westmeyer, P., *A Guide for Use in Planning and Conducting Research Projects,* Thomas, Springfield, Illinois, 1981.

15. Wilson, E. B., Jr., *An Introduction to Scientific Research,* McGraw-Hill, New York, 1952.

Executing the Experiment

Most experiments are either performed more than once or, during the execution of an experiment, several values of the independent variable are measured. After an experiment has been carried out and the data collected, statistical methods are used to evaluate the quality of the measurements, that is, to estimate various measures of error attributed to the final value of the independent variable. The magnitude of such errors depends upon n, the number of observations made or the number of data collected. Consequently, while designing the experiment the experimenter must make some decision as to the number of runs to make.

The nature of experimental errors is treated in more detail in Chapter 7. Nevertheless, your intuition probably tells you that several measurements are better than one or two, and that increasing the number of observations leads to a reduction in error. If the errors are random, the error is proportional to the square root of the number of observations. In other words, the number of runs must be quadrupled to decrease the errors by a factor of one half. Unfortunately, increasing the number of observations decreases the error very slowly. A better method is to design the experiment more carefully; even random errors ultimately have their origin in the design of the experiment.

COLLECTING DATA

An experimenter may choose to collect data randomly or sequentially. Randomization is helpful in preventing human bias from affecting values obtained from direct visual observation or indirect observation with instruments. Suppose that some property of a substance is measured as a dial on a scale moves. If the dial moves exactly 5 units and a change is observed, then the next increase of exactly 5 units might prejudice the experimenter into expecting the same change. It would be safer to increase the dial by random increments to prevent the experimenter from having any preconceived idea of what he is about to observe.

Randomization can minimize such random effects on the experiment as ambient temperature, humidity, wear of parts, instrument performance, and the skill or patience of the experimenter. For example, if low-temperature measurements are always made in the morning and high-temperature measurements in the late afternoon, the high-temperature measurements may exhibit more experimental scatter because of the fatigue of the experimenter. Some experiments require that a sample represent the bulk sample or population from which it is drawn. A sample of water from a lake or a sample of ore from a mine must be obtained by randomized sampling.

Data collected sequentially are collected in numerical order from lowest to highest or vice versa. Consider three experiments: the dependence of pH on temperature in an equilibrium constant measurement, the dependence of pH on time in a kinetics experiment, and the dependence of pH on milliliters of base added in a titration experiment. After selecting a range of temperatures, the experimenter has the choice of beginning at the lowest temperature and measuring the pH at gradually increasing temperatures. Or she could begin at the highest selected temperature and work down to the lowest. On the other hand, she could make a series of measurements at random temperatures in between the extremes. In a reversible experiment like this, either approach is possible. It is probably more convenient experimentally to use the sequential method, however, since equilibrium is more quickly established after a small change in temperature than after a large change. Both the system and the surroundings (the thermostat bath) must come to thermal equilibrium after a change in the temperature. The sequential method also has the advantage that any hysteresis can be detected by a series of measurements taken at both increasing and decreasing temperatures. If the system and the experimental design are reversible, the same values for pH obtained with increasing temperatures should be obtained with decreasing temperatures.

When a property of a system is measured as a function of time, the data must be collected sequentially. We can cause the temperature to increase or decrease, but time increases relentlessly. For example, in a kinetics experiment, the system initially is not at equilibrium but is allowed to approach equilibrium as time goes on. We can study the change in composition at $t = 0, 5, 10, 20$, etc., but not in the order $t = 100, 90, 80$, etc.

The determination of a titration curve is a familiar example of the sequential collection of data. We measure the pH before adding any base ($ml = 0$) and then proceed to add the base in convenient increments, so that the milliliters of base added increases regularly. We measure the pH at base volumes of $0, 5, 10, 25$ ml, etc. The design of the experiment to

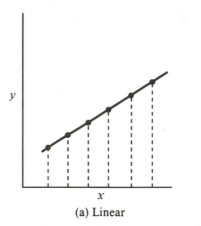

(a) Linear

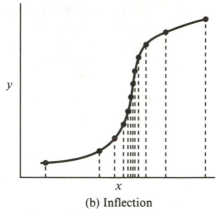

(b) Inflection

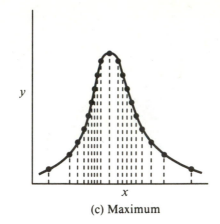

(c) Maximum

Figure 5–1 Spacing of experimental data.

measure the pH at decreasing volumes of base would be awkward, but possible.

The titration curve also suggests some features of spacing data when the data are nonlinear and exhibit points of inflection, maxima, or minima, as shown in Figure 5–1. When the data indicate a linear relationship between the dependent variable y and the independent variable x, the experimental design calls for more or less equal spacing between values of the independent variable, which results in equal spacing between values of the dependent variable y (Fig. 5–1a). When y is changing slowly with x, then the x values may be more widely spaced, for reasons of economy. This is true, for example, before and after the end point of a titration curve. Near the end point, however, y is changing very rapidly with x, and the x values must be chosen close together so that changes in y are kept more or less constant between all changes in x (Fig. 5–1b).

When the data show a maximum, as in Figure 5–1c, the data again can be more widely spread near the base of the peak, where y is changing slowly with x. They should be more closely spaced on the sides, where y is changing rapidly, and also near the maximum in order to locate the maximum precisely.

Starting Up

Learning to do an experiment is like learning to ride a bicycle: you get better at both with practice. Making up a check list is like making a dry run. Try it. Then go through the check list once or twice with the apparatus without actually doing the experiment. Make some notes where you anticipate difficulties or where you recognize crucial points. Plan ahead. Allow time to gather your supplies together before beginning the experiment: chemicals, standard solutions, indicators, liquid nitrogen, Dewar flasks, test instruments, standards, safety devices, cells, electrical leads, tools, etc.

Like a bicycle, an experimental apparatus must be brought under control. Perhaps the system must come to thermal equilibrium, a property must be brought to an initial value, or the vacuum must be reduced below a maximum value. The conditions are monitored and adjustments made. When a new apparatus does not respond as expected and trouble is found, the cause must be sought. The same methods for searching for causes in the construction of hypotheses can be used for trouble shooting an apparatus. Trouble shooting by substituting a good component for a suspected bad component is the same as searching for causes by the method of differences. The method of concomitant variation can also be used. If a suspected value can be changed, then one can see if the problem value changes simultaneously and in the suspected direction.

Disturbances affecting the success of an experiment can come from within the apparatus itself or from without. Overheating of a component can arise because of its location within the apparatus adjacent to a heat-generating component. On the other hand, failure of the air conditioning system on a record-setting hot day might result in an excessively high ambient room temperature. Insulation, isolation, and deliberate control are some of the means of eliminating disturbing factors such as those listed in Table 5–1.

A few years ago, the author's students were having a problem with the x-ray powder diffraction equipment located on the fifth story of the science building. Because exposures of several hours are required with small samples in a large-diameter camera, the machine was in constant use. It was common practice to begin a second exposure late in the morning and let it run through the noon hour so that a film could be developed after lunch. It invariably happened, however, that the machine turned itself off sometime during the lunch hour while everyone was gone. This never happened in the morning or the late afternoon. Finally, it was postulated that the water pressure dropped during the noon hour, resulting in insufficient cooling water for the x-ray tube, and safety devices built into the machine turned the power off to prevent damage to the tube.

TABLE 5–1

Disturbances

Heat, temperature	Friction
Pressure	Vibration
Magnetic fields	Voltage fluctuations
Electric field	Vacuum leaks
Light leaks	Backlash
Gamma or x rays	Drift
Humidity fluctuations	Hysteresis
Condensation	Noise
Corrosive vapors	Nonlinearity
Air currents	Oscillation
Static electricity	Electric shock

Evidently, the sudden surge of water used as toilets were flushed shortly after noon lowered the water pressure substantially. Fortunately, a self-contained water-cooling unit was installed and the unit has never shut itself down since.

Keeping Records: The Laboratory Notebook

The laboratory notebook is a permanent record of experimental data. Such a record is more permanent if the notebook is bound and the entries are in ink. Loose-leaf or spiral-bound notebooks are not satisfactory. The pages in the notebook should be numbered. Notebooks are available in plain, lined, or crosshatched (graph) paper. Since notebooks are personal and are not to be used by anyone except the owner, the choice of paper is a personal one. Crosshatching facilitates entering data neatly in tabular form. It is also useful in making rough graphs for calibration curves or in following the course of an experiment graphically.

Immediately after purchasing your notebook, you should enter your name, address, department, and home and office phone numbers on the inside of the front cover. It is surprising how many lost notebooks turn up in the chemistry office without any owner's information entered. It is impossible to return notebooks to an unknown owner. The first five or six pages of the notebook should be left blank for developing a table of contents as your experimental work progresses. Although an index is rarely necessary, the last few pages of the notebook may serve for Appendix A, B, etc., going backward from the end of the notebook. In Appendix A you might glue in a small periodic table with current atomic weights; in Appendix B you might attach a table of accurate physical constants; and Appendix C might hold a frequently used calibration curve, drawn on fine graph paper and glued to the notebook page. A list of the appendices should appear in the Table of Contents. Your notebook is now personalized and ready for use.

What should you enter into a notebook and how should you do it? All entries should be in ink and should be entered immediately after making an observation. Before recording anything, however, you should enter the date, perhaps even the hour. Your notebook should be at your side at all times. If you must walk across the laboratory to your notebook to record an observation, other people or events can distract you so that a weight of 28.4 g may be recorded as 24.8 g.

Anything worth writing down at all is worth writing in your notebook. In *no case* should you ever record data on pieces of scrap paper or paper towels. Scratch calculations should be done in the laboratory notebook even if you find this aesthetically unpleasing. Besides numerical data and text, you may also sketch certain observations that are easier to draw than to describe. For example, a crystallographer might sketch the appearance of a single crystal at two different settings in a goniometer head, as shown in Figure 5–2. Completeness, legibility, and neatness are

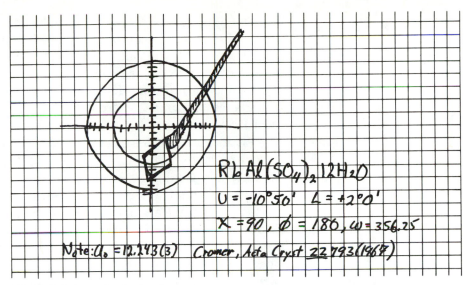

Figure 5–2 A notebook sketch.

important, in that order. Enter data directly; never transfer data from another sheet of paper to your notebook.

Some entries can be made in advance. Somewhere in your notebook, either in the main body or in an appendix, you should have a detailed sketch and description of your apparatus. This should be done carefully, since others may need to read it. Leave a blank page or two after the description of a complex apparatus to provide room for recording eventual modifications. Be sure to note the date of the original apparatus and the date of the modification. Include appropriate calibration curves and the dates during which they were used. At some later time it may be possible to correct certain observations if it is known when an auxiliary apparatus was used. If you have a work plan or a check list for the task at hand, record it in your notebook, not on a separate sheet of paper. If it is useful, you may want to use it again and you will know where to find it.

It is good practice to mark a permanent identification number on crucibles, pycnometers, sample bulbs, and similar small apparatus used for multiple measurements. If these numbers are then recorded in your laboratory notebook as they are used, the chances of inadvertently mixing them is virtually nil. Similarly, several sets of electrical reference cells, matched spectrophotometer cells, goniometer heads, etc., are often available. The serial number of the one used should be recorded. If no serial number is available, an appropriate identification number should be engraved and recorded. Spectrographic and x-ray film should be labeled with your name or initials and a notebook page; for example, RJS89S14. Here the initials are RJS, 89S stands for Spring 1989, and 14 is the page number. If several workers are developing the same kind of film, it is useful to cut or punch a simple code marking on one end of the film before placing it in the wash water.

It is a wise practice to make up your own solutions and not to trust those made up by others. Label your solutions with your name, the name of the solution, its concentration, and the date on which it was prepared. In your notebook you can then reference the solution by the name and date; if it is necessary, the solution can be checked if it is suspected of being faulty in any way. If the purity of the chemical is of particular importance, the grade or even lot number of the chemical should be recorded.

Besides creating a permanent record of your observations and data in your notebook, you will need to file film, recorder paper, and computer hard copy. A small box or accordion folder is useful for storing these organized in manila folders. Each spectrum recorded and each computer output should have an identification number referring to a notebook, as recommended above for film. You should not rely solely on microcomputer or mainframe computer disks for the storage of important data. These are not often lost, but when they are, the results can be personally catastrophic.

If you store data on floppy disks, be sure to back them up. Keep them away from strong magnetic fields, which might be found in the vicinity of an NMR spectrometer or magnetic susceptibility apparatus.

If you are using a microcomputer, always enter the correct date and time since this information is stored with saved disk files and can prove useful at a later time regarding the order in which data were collected. If you are using a hard disk, work out a sensible set of directories. Use extensions with your file names to further organize your files; for example, DICHLOR.RN1, DICHLOR.RN2. In your notebook, make a tree diagram of your hard disk directories and the classes of extensions you are using for file names.

Prejudice, Bias, and Fraud

Scientists must constantly be on guard against personal prejudice. Once a hypothesis has been formed, one must resist the temptation to perceive observations that tend to support the hypothesis, or to distort unsupportive observations to minimize their disagreement. Absolute objectivity is a hard discipline. Bias interferes with the testing of hypotheses at any time: in the designing of experiments and especially in the execution of experiments. Bias operates at all levels, including the way we perceive an observation, the way we record it in our notebook, the way we interpret it, and the way we choose to report it.

REFERENCES

1. Baird, D. C., *Experimentation: An Introduction to Measurement Theory and Experiment Design,* Prentice-Hall, Englewood Cliffs, New Jersey, 1962.
2. Beveridge, W. I. B., *The Art of Scientific Investigation,* Norton, New York, 1951.
3. Brinkworth, B. J., *An Introduction to Experimentation,* C. Tinling & Co. Ltd., Liverpool, 1968.

4. Freedman, P., *The Principles of Scientific Research,* Pergamon Press, New York, 1960.

5. Montgomery, D. C., *Design and Analysis of Experiments,* Wiley, New York, 1984.

6. Weatherall, M., *Scientific Method,* English Universities Press, London, 1968.

7. Westmeyer, P., *A Guide for Use in Planning and Conducting Research Projects,* Thomas, Springfield, Illinois, 1981.

8. Wilson, E. B., Jr., *An Introduction to Scientific Research,* McGraw-Hill, New York, 1952.

CHAPTER 6

Reducing the Data

Data collection in physical chemistry often generates large sets of numbers to which we must give meaning. To test a hypothesis we may develop a model that can be compactly represented by an equation. Can we fit our data to a linear equation? If not, can the data fit a nonlinear equation? In either case we desire an objective approach, one that minimizes subjectivity and prejudice.

Although curve fitting is one of the most important kinds of numerical analysis for the physical chemist, the solution to simultaneous linear equations and numerical integration also play an important role in reducing data to an understandable form.

LINEAR LEAST SQUARES

When the dependent variable y is expected to be linearly related to the independent variable x, the relationship is given by

$$y = mx + b \qquad (6\text{--}1)$$

The constant m is the slope of the line and the constant b is the intercept of the line. Because random experimental errors occur, a straight line cannot usually be drawn through all the points; instead, it is necessary to estimate visually the "best" straight line that can be drawn through the points. In Figure 6–1, for example, the scientist, now artist, has drawn a line through the points listed in Table 6–1 so that the line goes through or touches as many points as possible, and at the same time leaves about an equal number of points above and below the line, scattered about an equal distance from the line. This is clearly a pretty subjective task, and you can be quite sure that no two people will draw the line in identical positions.

Having drawn the line, we wish to obtain the slope m and intercept b so that the equation of the line can be written. The two-point equation of a line is most convenient for extracting the slope and intercept:

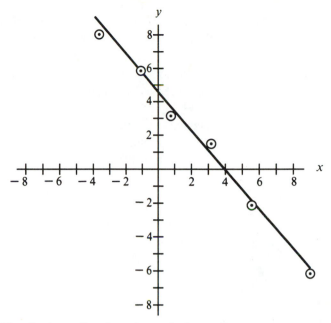

Figure 6–1 Visually drawn line through sample data points.

$$\frac{y - y_1}{x - x_1} = \frac{y_2 - y_1}{x_2 - x_1} \qquad\qquad \textbf{(6–2)}$$

Two arbitrary points (x_1, y_1) and (x_2, y_2) are selected—not from the experimental points, but from points that lie exactly on the line—that are separated from each other as far as possible. For example,

$$\frac{y - 8.8}{x + 4.0} = \frac{-5.2 - 8.8}{8.0 + 4.0} \qquad\qquad \textbf{(6–3)}$$

$$y = -1.17x + 4.1 \qquad\qquad \textbf{(6–4)}$$

The slope $m = -1.17$ and the intercept $b = 4.1$. Because both drawing the line and selecting the two points are very subjective operations, this method is not particularly satisfactory. A better method is the method of least squares.

TABLE 6–1

Sample Data Points

x	y
−3.80	8.2
−1.20	6.0
0.75	3.3
3.10	1.6
5.60	−2.0
8.00	−6.0

The Method of Least Squares

The method of least squares is a statistical method for determining the best line that represents a linear relationship between the dependent and independent variables. If there is a unique best line, then it must be characterized by a unique slope m and a unique intercept b, which are to be determined. In drawing a line by hand, we judge with our eye the relative position of the line with respect to the points, as shown for a few points in Figure 6–2. The residual r_1 for point (x_1, y_1) is the difference in the experimental value for y at x_1 and the calculated value for y at x_1, or

$$\text{residual} = y_{\text{experimental}} - y_{\text{calculated}} \tag{6-5}$$

We assume for the moment that we know the values of the slope m and the intercept b, so that we can calculate the value of y at each point and compare it with the measured value

$$r_i = y_i - (mx_i + b) \tag{6-6}$$

When we draw by hand the best line through a set of points, we use our eye to minimize the residuals. The Principle of Least Squares is quite similar. It can be shown statistically (Young, 1962) that the most probable values of x_i are those that minimize the sum of the squares of residuals. Let R be the sum of the squares of the residuals so that

$$R = \sum r_i^2 = \sum (y_i - mx_i - b)^2 \tag{6-7}$$

As usual for any function, the conditions for the minimum in the function are

$$\left(\frac{\partial R}{\partial m}\right)_b = 0 \qquad \text{and} \qquad \left(\frac{\partial R}{\partial b}\right)_m = 0 \tag{6-8}$$

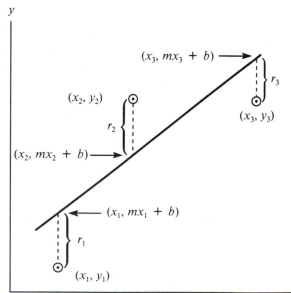

Figure 6–2 Residuals r_i.

When Equation 6–7 is expanded by squaring, the result is

$$R = \sum y_i^2 - 2m \sum x_i y_i - 2b \sum y_i + m^2 \sum x_i^2 \qquad (6\text{–}9)$$
$$+ 2m \sum x_i b + nb^2$$

$$\left(\frac{\partial R}{\partial m}\right)_b = 0 - 2 \sum x_i y_i - 0 + 2m \sum x_i^2 + 2b \sum x_i + 0 = 0 \qquad (6\text{–}10)$$

$$\left(\frac{\partial R}{\partial b}\right)_m = 0 - 0 - 2 \sum y_i + 0 + 2m \sum x_i + 2nb = 0 \qquad (6\text{–}11)$$

Equations 6–10 and 6–11 are two equations in two unknowns, m and b, which may be solved simultaneously to give the required slope and intercept of the statistically best line:

$$\text{slope} = m = \frac{n \sum y_i x_i - \sum x_i \sum y_i}{n \sum x_i^2 - (\sum x_i)^2} \qquad (6\text{–}12)$$

$$\text{intercept} = b = \frac{\sum x_i^2 \sum y_i - \sum x_i \sum x_i y_i}{n \sum x_i^2 - (\sum x_i)^2} \qquad (6\text{–}13)$$

This is the best method of determining the equation of a line, since it is completely objective. All that is required is the calculation of the four summations that arise in Equations 6–12 and 6–13.

Now let us use the data in Table 6–1 to determine by the method of least squares the equation of the line shown in Figure 6–1 (the appropriate summations are shown in Table 6–2):

$$m = \frac{n \sum y_i x_i - \sum x_i \sum y_i}{n \sum x_i^2 - (\sum x_i)^2} = \frac{6(90.125) - (12.45)(11.10)}{6(121.4125) - (12.45)^2} \qquad (6\text{–}14)$$

$$= -1.184$$

TABLE 6–2

Summations for Six Data Points ($n = 6$) from Table 6–1

	x	y	x^2	xy
	−3.80	8.2	14.4400	−31.1600
	−1.20	6.0	1.4400	−7.2000
	0.75	3.3	0.5625	2.4750
	3.10	1.6	9.6100	4.9600
	5.60	−2.0	31.3600	−11.2000
	8.00	−6.0	64.0000	−48.0000
$\sum =$	12.45	11.10	121.4125	−90.1250

$$b = \frac{\sum x_i^2 \sum y_i - \sum x_i \sum x_i y_i}{n \sum x_i^2 - (\sum x_i)^2} \tag{6-15}$$

$$= \frac{(121.4125)(11.10) - (12.45)(90.125)}{(6)(121.4125) - (12.45)^2}$$

$$= 4.307$$

The summations should not be rounded off, since the data are considered to be exact for the purposes of calculation. The number of significant figures, or more exactly the measures of the errors associated with the slope and intercept, are treated in the next chapter.

Least-Squares Fitting with a Computer

Although the least-squares calculation is straightforward, it can become a bit tedious as the number of points increases. Consequently, it is convenient to have a least-squares computer program available on a floppy disk for a microcomputer or stored on the disk library of a mainframe computer. Some of the graphics capabilities of currently available microcomputers such as the Apple IIe or IBM-PC also permit automatic graphing of the least-squares line and the points from which it was determined, as shown in Figure 6–3. The graph is equivalent to Figure 6–1 and the data are from Table 6–1.

A typical computer program for a least-squares calculation is depicted in Figure 6–4. It is written in Pascal and should run with little or no modification on any machine with a Pascal compiler. The Pascal language was chosen because the program is more readable than programs written in BASIC or FORTRAN. Even someone who has never programmed in Pascal can follow the general outline of the procedure and compare it with Equations 6–12 and 6–13. The main program (at the end of the listing) consists of just three procedures. The first, *GetData*, permits the user to input data interactively with the computer. The second, *DoLeastSquares*, does the least-squares calculation according to Equations 6–12 and 6–13. The third procedure, *WriteData*, writes the output to the screen. You will notice that the figures for the slope and intercept of the line differ considerably from those we calculated earlier—based on drawing a line through the plotted points by hand, and interpolating a pair of points in the two-point formula for calculating the slope and intercept. The least-squares method should always be used for calculating the equation of a line whenever it is determined by three or more points.

Graphics presentation of the least-squares line and data points adds complexity and decreases portability from a programming point of view, but it also increases the program's utility to the user. The program in Figure 6–4, extended with Turbo Pascal Graphics, is listed in Figure 6–18 at the end of this chapter. The graphics version has been popular with students, who find it easy to use. They have been encouraged to "borrow" procedures from it for their own programs. Input is interactive either

(Text continues on page 81.)

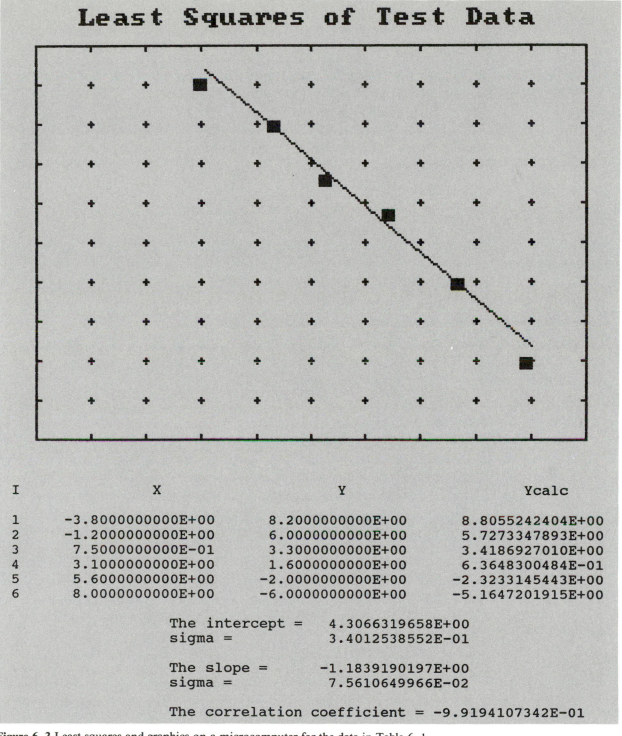

Figure 6-3 Least squares and graphics on a microcomputer for the data in Table 6-1. Compare with Figure 6-1.

```
Program LSPChem;

(*********************************************************************)
(*                                                                   *)
(* A program to fit N X,Y pairs to a straight line by least          *)
(* Squares.  The slope m, the intercept b and their standard         *)
(* deviations are calculated for a line of the form Y = mX + b.      *)
(*                                                                   *)
(*********************************************************************)

uses crt,printer;

Const Max = 50;

Type Ary = Array [1..Max] of Real;

VAR
  X,Y,Ycalc          :  Ary;
  N                  :  Integer;
  Answer             :  Char;
  A,b,m              :  Real;
  CorrelCoef, Sigmam, Sigmab:  Real;

(*********************************************************************)

PROCEDURE ReadFile(VAR Xry,Yry:  Ary;
                       VAR Num:  Integer);

   Var I:  Integer;
       F:  Text;

Begin (* ReadFile *)
   Assign(F,'LS.DAT');
   Reset(F);
   Readln(F,N);
   For I:= 1 to Num Do
      Readln(F,Xry[I],Yry[I]);
   Close(F)
End;  (* ReadFile *)


(*********************************************************************)

Procedure ReadKeyBoard;
Var I,
    Sx,Sy :  Integer; (* Screen Coordinates *)
Begin (* ReadKeyBoard *)
   CLRSCR;
   Write('What is the number of X and Y pairs you wish to enter?  ');
   Readln(N);
   Writeln;
   For I:= 1 to N Do
     Begin (* For *)
        Write(     'X(',I,'):  ');
        Sx:= WhereX; Sy:= WhereY;
```

Figure 6–4 Least-squares program in Pascal.

```
            Read(X[I]);
            GoToXY(Sx+10,Sy);
            Write('       Y(',I,'):   ');
            Readln(Y[I]);
        End;   (* For *)
     CLRSCR;
  End;   (* ReadKeyBoard *)

  (*************************************************************)

  Procedure LeastSquares(X,Y:   Ary;
                            Var Ycalc:   Ary;
                            Var b,m  :   Real;
                                N   :   Integer);
  (* Fits a straight line to a set of N   X-Y pairs    *)
  (* The line has a slope equal to m and intercept b *)

  Var I:   Integer;
      SumX,SumY,SumXY,SumX2,SumY2,
      XI,YI,SXY,SXX,SYY:   Real;

  Begin (* LeastSquares *)
     SumX:= 0;
     SumY:= 0;
     SumXY:=0;
     SumX2:=0;
     SumY2:=0;
     For I:= 1 to N do
        Begin (* For *)
        XI:= X[I];
        YI:= Y[I];
        SumX:= SumX + XI;
        SumY:= SumY + YI;
        SumXY:= SumXY + XI*YI;
        SumX2:= SumX2 + XI*XI;
        SumY2:= SumY2 + YI*YI;
        End;   (* For *)
     SXX:= SumX2 - SumX*SumX/N;
     SXY:= SumXY - SumX*SumY/N;
     SYY:= SumY2 - SumY*SumY/N;
     m:= SXY/SXX;
     b:= ((SumX2*SumY - SumX*SumXY)/N)/SXX;
     CorrelCoef:= SXY/SQRT(SXX*SYY);
     Sigmam:= (SQRT((SumY2 - b*SumY -m*SumXY)/(N - 2)))/SQRT(SXX);
     Sigmab:= Sigmam*SQRT(SumX2/N);
     For I:= 1 to N do
        Ycalc[I]:= b + m*X[I];
  End;   (* LeastSquares *)

  (*************************************************************)

  Procedure WriteData;
  (* Writes out the answers on the screen *)

  Const BL = '  ';
```

Figure 6–4 (continued)

```
       sp = '        '        ;

Var I: Integer;

Begin
   CLRSCR;
   Write('  I                    X ');
   Write('                         Y');
   Writeln('                    Ycalc');
   Writeln;
   For I:= 1 to N do
       Writeln(I:3,BL,X[I],BL,Y[I],BL,Ycalc[I]);
   Writeln;
   Writeln(sp,sp,'Intercept is ',b);
   Writeln(sp,sp,'Sigma is      ',Sigmab);
   Writeln;
   Writeln(sp,sp,'Slope is ',m);
   Writeln(sp,sp,'Sigma is ',Sigmam);
   Writeln;
   Writeln(sp,sp,'Correlation Coefficient is ',CorrelCoef);
   Writeln;
   Readln;
End; (* WriteData *)

(***************************************************************)

Procedure PrintData;
(* Sends calculated values to printer *)

Const sp5 = '     ';
      sp10 = '          ';
      sp15 = '               ';

Var I:  Integer;

Begin (* PrintData *)
   Writeln(Lst,'  I',sp15,'X',sp15,sp5,'Y',sp15,sp5,'Ycalc');
   Writeln(Lst);
   For I:= 1 to N do
       Writeln(Lst,I:3,sp5,X[I],sp5,Y[I],sp5,Ycalc[I]);
   Writeln(Lst);
   Writeln(Lst,sp5,sp15,'The intercept =  ',b);
   Writeln(Lst,sp5,sp15,'sigma =          ',Sigmab);
   Writeln(Lst);
   Writeln(Lst,sp5,sp15,'The slope =      ',m);
   Writeln(Lst,sp5,sp15,'sigma =          ',Sigmam);
   Writeln(Lst);
   Writeln(Lst,sp5,sp15,'The correlation coefficient = ',CorrelCoef);
End; (* PrintData *)

(***************************************************************)

Procedure Hardcopy;
Var reply:  Char;
Begin
```

Figure 6–4 (continued)

```
      Writeln;Writeln('Do you want a printed copy?');
      Writeln('If Yes, be sure your printer is turned on!');
      Write('Please enter a Y or N:  ');
      Readln(reply);
      If (reply = 'Y') OR (reply = 'y') Then
          PrintData;
   end; (* Hardcopy *)


   (*****************************************************************)


   Begin (*  M A I N   P R O G R A M  *)
      CLRSCR;
      Writeln('Is your input from File LS.DAT or from the Keyboard?');
      Write('Enter an F or K, please.  ');
      Read(Answer);
      If (Answer = 'F') OR (Answer = 'f') Then ReadFile (X,Y,N)
                                         Else ReadKeyBoard;

      LeastSquares(X,Y,Ycalc,b,m,N);
      WriteData;
      Hardcopy;
   End.  (*  M A I N   P R O G R A M  *)
```

Figure 6–4 (continued)

from the keyboard or from a data file, which can easily be created with the Turbo Pascal Editor or with nearly any word processing software, such as WordStar or WordPerfect.

LINEAR RELATIONSHIPS IN PHYSICAL CHEMISTRY

You will constantly run into linear equations in your work in physical chemistry and other areas of science. In addition, calibration curves for many instruments are linear equations of the form $y = mx + b$. For example, the wavelength λ of diffracted radiation in a grating spectrograph depends in a linear manner on the distance S across the film:

$$\lambda = mS + b \qquad (6\text{–}16)$$

The dielectric constant ϵ of a dipole meter is a linear function of the dial reading D (capacitance):

$$\epsilon = mD + b \qquad (6\text{–}17)$$

In the first example, the film is calibrated by measuring the position S on the film of several calibration spectral lines of known wavelength λ, and fitting Equation 6–16 to the data by least squares. In the second example, the instrument is calibrated by determining the null point in dial divisions D for a few solutions of known dielectric constant ϵ. When these data are fit by least squares to Equation 6–17, the calibration curve is found.

In the experimental determination of transference numbers by the moving boundary method, the transference number t_+ depends on the value of the Faraday F, the concentration c of the solution, the current I, and the volume V swept out in time t:

$$t_+ = \frac{FcV}{It} \tag{6–18}$$

From a slightly different point of view, we can look at this equation in terms of what is observed to change: the volume of the solution swept out and the time; that is, V is a linear function of t:

$$V = \frac{t_+ It}{Fc} \tag{6–19}$$

When V is plotted against t, a straight line results, the slope m of which is equal to $t_+ I/Fc$, so that the transference number is calculated from the slope, the concentration, and the current:

$$t_+ = \frac{Fcm}{I} \tag{6–20}$$

By plotting the data as the experiment progresses, one can constantly monitor the quality of the experiment since any deviations from linearity are easily detected and are presumably caused by poor regulation of the constant current I.

Even when the independent variable is not related to the dependent variable in a simple linear manner, it is often possible to find a transformation that is linear. For example, the relationship might be

$$y = ax^b \tag{6–21}$$

The dependent variable is y, the independent variable is x, and the constants m and n are to be determined. Generally, it is better to transform such an equation than to try to fit the curves directly. Taking logs of both sides of Equation 6–21 gives

$$\ln y = b \ln x + \ln a \tag{6–22}$$

Now it is a simple matter to fit pairs of $\ln y$ and $\ln x$ data by least squares to a straight line, the slope of which equals b and the intercept of which equals $\ln a$. Equations of the form of Equation 6–21 arise frequently in physical chemistry; for example, rate laws for reactions, viscosities of high-polymer solutions, and adsorption from solution.

Even more common to physical chemistry are relationships of the type

$$y = ae^{bx} \tag{6–23}$$

Again, taking logs of both sides of the equation, we obtain a linear equation:

$$\ln y = bx + \ln a \qquad (6\text{--}24)$$

and, similarly,

$$y = ae^{b/x} \qquad (6\text{--}25)$$

Its linear transformation is

$$\ln y = \frac{b}{x} + \ln a \qquad (6\text{--}26)$$

For Equation 6–24, a plot of $\ln y$ versus x gives a straight line of slope b and intercept $\ln a$, while for Equation 6–26, a plot of $\ln y$ versus $1/x$ gives a straight line of slope b and intercept $\ln a$. The familiar plots from thermodynamics and kinetics of $\ln P$, $\ln K$, and $\ln k$ versus $1/T$ are essentially linear transformations of this type. When ΔC_p is not zero or small, then a plot of $\ln K$ versus $1/T$ is not a straight line, but the sigma plot (Experiment 5) is a linear transformation to handle this problem.

Plotting the reciprocal of a variable to achieve linearity is one of the more common tricks of the trade. As a very simple example, look at Boyle's Law:

$$V = \frac{k}{P} \qquad (6\text{--}27)$$

A plot of V versus P is clearly nonlinear, while a plot of V versus $1/P$ results in a straight line, from which it is very easy to extract the value of the constant k (the slope). The integrated rate equation for a second-order rate equation provides us with another example of plotting the reciprocal of a variable to give linearity to the plot:

$$\frac{1}{x} = \frac{1}{x_0} + kt \qquad (6\text{--}28)$$

Clearly x is not a linear function of t, but if $1/x$ is plotted against t, the relationship is linear. When pairs of $1/x$ and t data are treated by least squares, the slope of the resulting line equals k and the intercept equals $1/x_0$.

Sometimes a little rearranging is necessary before a plot of reciprocals of variables results in a straight line from which the desired parameters can be obtained. For example, the viscosity η of a liquid of density ρ measured in an Oswald viscometer is given by

$$\eta = A\rho t - B\frac{\rho}{t} \qquad (6\text{--}29)$$

where A and B are apparatus constants. The viscometer is calibrated by measuring the flow times of a few liquids of known density and viscosity.

A rearranged form of this equation permits a linear plot (the slope equals the constant B and the intercept equals the constant A):

$$\frac{\eta}{\rho t} = A + \frac{B}{t^2} \tag{6-30}$$

As a final example, let us examine the Langmuir equation for the relative amount y of a substance adsorbed as a function of concentration c:

$$y = \frac{ac}{1 + bc} \tag{6-31}$$

A little rearrangement leads to

$$\frac{c}{y} = \frac{1}{a} + \frac{b}{a}c \tag{6-32}$$

A plot of c/y against c gives a straight line, the slope of which equals b/a and the intercept of which equals $1/a$. The best method of obtaining the constants a and b is to fit the data to a straight line by the method of least squares.

Because a linear relationship between the dependent and independent variables can usually be found, the method of least squares is an

TABLE 6–3

Some Examples of Equations That Can Be Transformed[a] into the Simple Proportionality Y = AX

Function	Y =	X =	A =	Comments
$y = ax$	y	x	a	
$y = a/x$	y	$1/x$	a	$x = 0$[b]
$y = ax^r$	y	x^r	a	$x > 0$
$y = a \ln x$	y	$\ln x$	a	$x > 0$
$y = a \sin x$	y	$\sin x$	a	
$y = a \sin^r x$	y	$\sin^r x$	a	
$y = \sin ax$	$\arcsin y$	x	a	$-1 < y < +1$
$y = \sin ax^r$	$\arcsin y$	x^r	a	$-1 < y < +1$
$y = e^{ax}$	$\ln y$	x	a	$y > 0$
$y = e^{a/x}$	$\ln y$	$1/x$	a	$y > 0; x \neq 0$[b]
$y = e^{ax^r}$	$\ln y$	x^r	a	$y > 0$
$y = x^{ax}$	$\ln y$	$x \ln x$	a	$y > 0; x > 0$
$y = x^{ax^r}$	$\ln y$	$x^r \ln x$	a	$y > 0; x > 0$
$y = xe^{ae}$	$\ln (y/x)$	x	a	$(y/x) > 0$
$y = x^s e^{ax^r}$	$\ln (yx^{-s})$	x^r	a	$yx^{-s} > 0$

Source: de Levie, R., "When, Why and How to Use Weighted Least Squares," *J. Chem. Educ.* 63: 10, 1986.
[a] Transformed parameters are denoted by capitals.
[b] Use a sufficiently small number instead of $x = 0$.

TABLE 6–4

**Some Examples of Equations That Can Be Transformed[a]
into the Linear Relation $Y = AX + B$**

Function	$Y =$	$X =$	$A =$	$B =$	Comments
$y = ax + b$	y	x	a	b	
$y = a/x + b$	y	$1/x$	a	b	$x = 0$[b]
$y = ax^r + b$	y	x^r	a	b	$x > 0$
$y = a \ln x + b$	y	$\ln x$	a	b	$x > 0$
$y = ax^r + bx^s$	yx^{-s}	x^{r-s}	a	b	$x > 0$
$y = 1/(ax + b)$	$1/y$	x	a	b	$y \neq 0$[b]
$y = x/(a + bx)$	$1/y$	$1/x$	a	b	$x \neq 0$[b]; $y \neq 0$[b]
$y = ba^x$	$\ln y$	x	$\ln a$	$\ln b$	$y > 0$
$y = ba^{1/x}$	$\ln y$	$1/x$	$\ln a$	$\ln b$	$x > y; y > 0$
$y = ba^{x^r}$	$\ln y$	x^r	$\ln a$	$\ln b$	$x \neq 0$[b]
$y = be^{ax}$	$\ln y$	x	a	$\ln b$	$y > 0$
$y = be^{a/x}$	$\ln y$	$1/x$	a	$\ln b$	$y \neq 0$[b]; $y = 0$
$y = be^{ax^r}$	$\ln y$	x^r	a	$\ln b$	$y > 0$
$y = bx^{ax}$	$\ln yx$	$\ln x$	a	$\ln b$	$x > 0; y > 0$
$y = bx^{a/x}$	$\ln y$	$1/x \ln x$	a	$\ln b$	$x > 0; y > 0$
$y = bx^{ax^r}$	$\ln y$	$x^r \ln x$	a	$\ln b$	$x > 0; y > 0$
$y = bx^a$	$\ln y$	$\ln x$	a	$\ln b$	$x > 0; y > 0$

Source: de Levie, R., "When, Why and How to Use Weighted Least Squares," *J. Chem. Educ.* 63: 10, 1986.
[a] Transformed parameters are denoted by capitals.
[b] Use a sufficiently small number instead of $x = 0$.

important technique for data reduction. It eliminates bias on the part of the experimenter, since the method is statistical in nature and can be used to calculate various measures of the experimental errors associated with the slope and intercept of the line. The nature of experimental errors is discussed in the next chapter.

TABLE 6–5

**Some Examples of Equations That Can Be Transformed[a]
into the Quadratic Form $Y = AX^2 + BX + C$**

Function	$Y =$	$X =$	$A =$	$B =$	$C =$
$H = a + b/u + cu$	Hu	u	c	a	b
$y = ae^{(x-b)^2/c}$	$\ln y$	x	$1/c$	$-2b/c$	$\ln a + 2b/c$
$y = ae^{(b-\ln x)^2/c}$	$\ln y$	$\ln x$	$1/c$	$-2b/c$	$\ln a + 2b/c$
$y = 1/(a(x - b)^2 + c)$	$1/y$	x	a	$-2ab$	$c + ab^2$
$y = 1/(a(x + b)^2 + c)$	$1/y$	x	a	$2ab$	$c + ab^2$

Source: de Levie, R., "When, Why and How to Use Weighted Least Squares," *J. Chem. Educ.* 63: 10, 1986.
[a] Transformed parameters are denoted by capitals.

Some general linear transformations are summarized in Tables 6–3, 6–4, and 6–5. The use of weighted least squares for these transformations is explained by de Levie (1986).

SIMULTANEOUS LINEAR EQUATIONS

Linear equations, which occur frequently in science and engineering, have the form

$$Ax + By + Cz = D \qquad \textbf{(6–33)}$$

where A, B, C, and D are constants and x, y, and z are variables that must be to the first power. A system of n equations in n unknowns (variables) can generally be solved. Systems of linear equations of two or three unknowns are by far the most common in physical chemistry and may be solved by a number of methods.

As an example, let us consider the two equations

$$3x + 2y = 4 \qquad \textbf{(6–34)}$$

$$2x - 3y = 7 \qquad \textbf{(6–35)}$$

Since the graph of a linear equation is a straight line lying in the x–y plane, we might expect that a solution to a system of two linear equations is the value of x and y at the common point of intersection of the two lines, as shown in Figure 6–5a. However, not every pair of lines defined by two linear equations intersects. The two equations

$$2x - 3y = 6 \qquad \textbf{(6–36)}$$

$$2x - 3y = 2 \qquad \textbf{(6–37)}$$

represent two parallel lines, as shown in Figure 6–5b. The two equations

$$2x - 3y = 6 \qquad \textbf{(6–38)}$$

$$6x - 9y = 18 \qquad \textbf{(6–39)}$$

represent the same line.

We can summarize these three cases as follows:

1. The equations are *consistent* if they intercept.
2. The equations are *inconsistent* if they are parallel.
3. The equations are *equivalent* if they are the same line.

If the equations are consistent, they are said to have a simultaneous solution. Although a solution may be found graphically, it is more convenient and more exact to find a solution analytically.

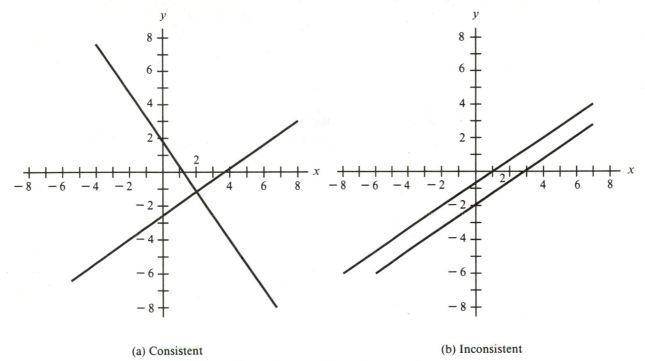

(a) Consistent   (b) Inconsistent

Figure 6–5 Consistent and inconsistent equations.

Solution by Addition or Subtraction

A constant factor can always be found that permits eliminating one of the variable terms by adding or subtracting after each term of one of the equations is multiplied by the factor. For example, consider Equations 6–34 and 6–35. If one of the equations is multiplied by 2 and the other by −3,

$$2(3x + 2y = 4) \qquad\qquad \textbf{(6–40)}$$

$$-3(2x - 3y = 7) \qquad\qquad \textbf{(6–41)}$$

the results are

$$6x + 4y = 8 \qquad\qquad \textbf{(6–42)}$$

$$-6x + 9y = -21 \qquad\qquad \textbf{(6–43)}$$

which, when added, give

$$13y = -13$$

or

$$y = -1$$

Now, if you substitute this value for y in either of the original equations, you can solve for the other variable:

$$3x + 2(-1) = 4$$

$$3x = 6$$

$$x = 2$$

Solution by Substitution

To solve by substitution, solve either of the two linear equations for one of the variables, say y in Equation 6–34:

$$y = \frac{4 - 3x}{2} \qquad\qquad \textbf{(6–44)}$$

Then substitute in the other equation,

$$2x - 3\left(\frac{4 - 3x}{2}\right) = 7 \qquad\qquad \textbf{(6–45)}$$

and simplify:

$$2x - \frac{12 + 9x}{2} = 7$$

$$4x - 12 + 9x = 14$$

$$13x = 26$$

$$x = 2$$

As before, substitution of this value for x in either of the original equations permits solving for y.

Cramer's Rule

Cramer's Rule is a more elegant way to find solutions of simultaneous linear equations by using certain properties of matrices. Before investigating Cramer's Rule let us review a few necessary definitions concerning matrices. An array is an ordered collection of elements. In experimental physical chemistry, the elements are usually real numbers, resulting from experimental measurements. A one-dimensional array is called a list or a **vector** and is enclosed in square brackets, for example,

$$\begin{bmatrix} 3.72 \\ 0.95 \\ 8.11 \\ 2.43 \\ 5.67 \end{bmatrix}$$

Such a vector is called a **column vector.** When a vector is written with its elements arranged horizontally, it is called a **row vector:**

$$[9 \quad 1 \quad 7 \quad 6 \quad -1 \quad 0 \quad 8]$$

A two-dimensional array is called a **matrix;** the elements are arranged either as a square or a rectangle and are enclosed in square brackets:

$$\begin{bmatrix} -2.37 & 0.63 \\ 0.88 & 9.13 \\ 4.43 & -3.33 \\ 7.29 & 0.3 \end{bmatrix} \qquad \begin{bmatrix} 4 & 7 & 11 \\ 45 & 31 & 8 \\ 5 & 2 & 84 \end{bmatrix}$$

The elements of a matrix are arranged into R horizontal rows and C vertical columns according to the general pattern

$$\begin{bmatrix} a_{11} & a_{12} & a_{13} & \cdots & a_{1C} \\ a_{21} & a_{22} & a_{23} & \cdots & a_{2C} \\ a_{31} & a_{32} & a_{33} & \cdots & a_{3C} \\ \vdots & \vdots & \vdots & \vdots & \vdots \\ a_{R1} & a_{R2} & a_{R3} & \cdots & a_{RC} \end{bmatrix}$$

The elements of a matrix are referenced by the row and column indices of the element.

Square matrices (but not rectangular matrices) have an associated quantity called the **determinant.** The determinant of a matrix is enclosed between two vertical lines instead of square brackets:

$$\text{If} \quad [A] = \begin{bmatrix} a_{11} & a_{12} & a_{13} \\ a_{21} & a_{22} & a_{23} \\ a_{31} & a_{32} & a_{33} \end{bmatrix} \quad \text{then} \quad |A| = \begin{vmatrix} a_{11} & a_{12} & a_{13} \\ a_{21} & a_{22} & a_{23} \\ a_{31} & a_{32} & a_{33} \end{vmatrix} \quad \textbf{(6–46)}$$

The determinant of a two-by-two matrix is given by

$$|A| = \begin{vmatrix} a_{11} & a_{12} \\ a_{21} & a_{22} \end{vmatrix} = a_{11}a_{22} - a_{12}a_{21} \qquad \textbf{(6–47)}$$

If the elements of a matrix are real numbers, then the determinant is just a single real number. The determinant of a three-by-three matrix is given by

$$|A| = \begin{vmatrix} a_{11} & a_{12} & a_{13} \\ a_{21} & a_{22} & a_{23} \\ a_{31} & a_{32} & a_{33} \end{vmatrix} = a_{11}a_{22}a_{33} - a_{11}a_{23}a_{32} - a_{12}a_{21}a_{33} \qquad \textbf{(6–48)}$$

$$+ a_{12}a_{23}a_{31} + a_{13}a_{21}a_{32} - a_{13}a_{22}a_{31}$$

The solutions to systems of linear equations according to Cramer's Rule are given in terms of the quotients of determinants. The determinants are of matrices formed from the coefficients and constants of the system of equations when written in matrix form (Barrante, 1974).

For example, we can write a system of three linear equations in three unknowns as follows:

$$a_{11}x_1 + a_{12}x_2 + a_{13}x_3 = c_1 \qquad \textbf{(6–49)}$$

$$a_{21}x_1 + a_{22}x_2 + a_{23}x_3 = c_2$$

$$a_{31}x_1 + a_{32}x_2 + a_{33}x_3 = c_3$$

The coefficients form a matrix, the determinant of which is defined as

$$|D| = \begin{vmatrix} a_{11} & a_{12} & a_{13} \\ a_{21} & a_{22} & a_{23} \\ a_{31} & a_{32} & a_{33} \end{vmatrix} \qquad (6\text{–}50)$$

In experimental applications the elements are all real numbers; the value of this determinant can be evaluated according to Equation 6–48.

The constants of the system of equations form an array, or constant vector:

$$\begin{bmatrix} c_1 \\ c_2 \\ c_3 \end{bmatrix}$$

The solutions to a system of n linear equations are given by equations of the form

$$x_n = \frac{D_n}{D} \qquad (6\text{–}51)$$

where D_n is formed by replacing the nth column of the matrix of the coefficients by the constant vector:

$$|D_1| = \begin{vmatrix} c_{11} & a_{12} & a_{13} \\ c_{21} & a_{22} & a_{23} \\ c_{31} & a_{32} & a_{33} \end{vmatrix} \qquad (6\text{–}52)$$

As an example, let us use Cramer's Rule to solve the system of equations given by

$$x_1 + x_2 + x_3 = 2 \qquad (6\text{–}53)$$

$$2x_1 - x_2 - x_3 = 1$$

$$x_1 + 2x_2 - x_3 = -3$$

$$|D| = \begin{vmatrix} 1 & 1 & 1 \\ 2 & -1 & -1 \\ 1 & 2 & -1 \end{vmatrix} \qquad |D_1| = \begin{vmatrix} 2 & 1 & 1 \\ 1 & -1 & -1 \\ -3 & 2 & -1 \end{vmatrix}$$

$$|D_2| = \begin{vmatrix} 1 & 2 & 1 \\ 2 & 1 & -1 \\ 1 & -3 & -1 \end{vmatrix} \qquad |D_3| = \begin{vmatrix} 1 & 1 & 2 \\ 2 & -1 & 1 \\ 1 & 2 & -3 \end{vmatrix}$$

Use Equation 6–48 to evaluate each determinant and Equation 6–51 to determine the three solutions:

$$x_1 = \frac{9}{9} = 1 \qquad x_2 = \frac{-9}{9} = -1 \qquad x_3 = \frac{18}{9} = 2$$

Because setting up the matrices of the equations and evaluating their determinants can be tedious when real experimental numbers are involved, it is handy to have a computer program available for solving systems of linear equations. An example of such a program, written in Pascal, that can be used to solve a system of three equations in three unknowns is show in Figure 6–6. Input to the program is interactive; after the coefficients and constants have been read in, the equations are echoed back so you can see if the data have been entered without errors. The procedures for reading data into a matrix or vector, writing data out, and copying matrices are of general utility and can be used in any Pascal program.

Gauss Elimination

Cramer's Rule is not an effective method for solving systems of more than two or three equations. One of the most effective alternative techniques for handling larger sets of linear equations is the Gauss elimination method. Essentially this method is the same method discussed earlier for solving a set of two simultaneous equations by eliminating the first variable, solving for the second, and substituting back to find the first variable. The Gauss elimination method is a systematic procedure for eliminating all the variables except one, and subsequently finding the other variables by substituting back.

It would be worthwhile to look at a specific example and solve it step by step. Consider the set of three simultaneous linear equations

$$2x_1 + 3x_2 + 8x_3 = 84 \qquad \text{(6–54)}$$

$$x_1 + 7x_2 - 3x_3 = 65 \qquad \text{(6–55)}$$

$$5x_1 - 2x_2 + x_3 = 41 \qquad \text{(6–56)}$$

An equivalent representation of this set of simultaneous linear equations is the matrix equation

$$\begin{bmatrix} 2 & 3 & 8 \\ 1 & 7 & -3 \\ 5 & -2 & 1 \end{bmatrix} \begin{bmatrix} x_1 \\ x_2 \\ x_3 \end{bmatrix} = \begin{bmatrix} 84 \\ 65 \\ 41 \end{bmatrix}$$

It is possible to perform a number of operations on the equations (or their matrix representation) without changing the solution. For example, you may do the following:

1. **Interchange rows.** This is equivalent to interchanging the equations, that is, to writing them down in a different order. Obviously, the order in which you write the equations does not affect the solution.

2. **Interchange columns.** A comparison of the simultaneous equations and their matrix representation reveals that interchanging columns is equivalent to renaming the variables; this, of course, does not affect their solution.

3. **Multiply a row by a constant.** Multiplication of a row (i.e., an equation) by a constant does not change the solution to the equation. An example should make this clear: consider the equations

(Text continues on page 95.)

```
Program Simlineq;
(**********************************)
(* A program to solve simultaneous  *)
(*linear equations by Cramer's Rule *)
(**********************************)

Const Max = 3;
Type Matrix = Array [1..Max,1..Max] of Real;
Vector = Array [1..Max] of Real;
Var Mx:  Matrix;
    Vec:  Vector;
    N:  Integer;

(************************************************************)

Function Determinant(M:  Matrix):  Real;
(* Calculates the determinant of matrix M *)
Var CrossProduct:  Real;
Begin (* Determinant *)
    CrossProduct:= M[1,1]*(M[2,2]*M[3,3]-M[3,2]*M[2,3])
                 - M[1,2]*(M[2,1]*M[3,3]-M[3,1]*M[2,3])
                 + M[1,3]*(M[2,1]*M[3,2]-M[3,1]*M[2,2]);
    Determinant:= CrossProduct;
End; (* Determinant *)

(************************************************************)

Procedure ReadEquations (Var M:  Matrix;
                         Var V:  Vector;
                             N:  Integer);
Var I, J: Integer;
Begin
  Writeln;Writeln('Please enter the coefficients and constants below:');
  Writeln;
  N:= Max;
  For I:= 1 to N do
  Begin
    Writeln('Equation', I:3);
    For J:= 1 to N do
      Begin
      Write('a',J:1,': ');
      Read(M[I,J]);
      End; (* For *)
    Write('c',I:1,': ');
    Readln(V[I]);
  End; (* For *)
End; (*ReadEquations *)

(************************************************************)
```

Figure 6–6 A Pascal program for solving three equations in three unknowns with Cramer's Rule.

```
Procedure EchoData(M:   Matrix;
                    V:   Vector;
                    N:   Integer);
(* Prints out the matrix of the equations *)

Var I,J: Integer;

Begin (* EchoData *)
  Writeln;Writeln('Your coefficients and constants are as follows');
  Writeln;
  N:= Max;
  For I:= 1 to N do
    Begin (* For I *)
    For J:= 1 to N do
      Write(M[I,J]:8:3);
    Write('*',V[I]:8:3);
    Writeln;
    End;  (* For I..*)
  Writeln;
End; (* EchoData *)

(*******************************************************)

Procedure WriteMatrix(M:   Matrix);
(* Write out matrix M *)
Var I,J,N: Integer;
Begin
  Writeln;
  N:= Max;
  For I:= 1 to N do
    Begin
    For J:= 1 to N do
      Write(M[I,J]:8:3);
    Writeln;
    End; (* For *)
End;  (* WriteMatrix *)

(*******************************************************)

Procedure WriteVector(V:   Vector);
(* Writes out a vector(list)*)
Var I:  Integer;
Begin
  Writeln;
  N:= Max;
  For I:= 1 to N do
    Write(V[I]:8:3);
  Writeln
End; (* WriteVector *)

(*******************************************************)
```

Figure 6–6 (continued)

```
Procedure CopyMatrix(Var M2:  Matrix;  M1:  Matrix);
Var I,J,N:  Integer;
Begin   (* CopyMatrix *)
   N:= Max;
   For I:= 1 to N do
      For J:= 1 to N do
         M2[I,J]:= M1[I,J];
End;   (* CopyMatrix *)


(************************************************************)

Procedure Cramer(Var B:  Matrix; A:  Matrix;
                 V:  Vector;
                 Column:  Integer);

Var Row, Col, N:  Integer;
Begin   (* Cramer *)
   N:= Max;
   CopyMatrix(B,A);
   For Col:= Column to Column do
      For Row:= 1 to N do
         B[Row,Col]:= V[Row];
End;   (* Cramer *)


(************************************************************)

Procedure SolveEquations;
Var Dx,Dx1, Dx2, Dx3:  Real;
     Mx1, Mx2, Mx3:  Matrix;
              List:  Vector;
Begin  (* SolveEquations *)
   Cramer(Mx1,Mx,Vec,1);
   Cramer(Mx2,Mx,Vec,2);
   Cramer(Mx3,Mx,Vec,3);
   Dx:= Determinant(Mx);
   List[1]:= Determinant(Mx1)/Dx;
   List[2]:= Determinant(Mx2)/Dx;
   List[3]:= Determinant(Mx3)/Dx;
   Writeln('The three solutions to the equations are:  ');
   WriteVector(List);
End;


(************************************************************)

Begin (* M A I N   P R O G R A M *)

   ReadEquations(Mx, Vec, N);
   EchoData(Mx, Vec, N);
   SolveEquations
End.  (* M A I N   P R O G R A M *)
```

Figure 6–6 (continued)

$$2x_1 + x_2 = 4 \qquad 4x_1 + 2x_2 = 8$$

$$3x_1 + x_2 = 5 \qquad 3x_1 + x_2 = 5$$

Multiplication of the first equation by 2 does not change the solution, since the solution to both sets of equations is

$$x_1 = 1$$

$$x_2 = 2$$

4. **Replace a row by the sum of that row and another.** This operation does not affect the solution to the set of equations *and is the basis for the Gauss elimination method for solving systems of simultaneous linear equations.*

Now let us apply these principles to the solution of the example system of simultaneous linear equations, dividing the problem into two parts: the *forward elimination* and the *backward substitution.* Finally, we shall examine a Pascal computer program written to carry out the same task. The key procedures in the program, *Procedure Eliminate* and *Procedure Substitute*, are the heart of the program.

Forward Elimination. Referring to Equations 6–54, 6–55, and 6–56, multiply the second equation by −2, add the product to the first equation, and replace the second equation by the sum. The result is

$$2x_1 + \quad 3x_2 + \quad 8x_3 = \quad 84$$

$$-11x_2 + 14x_3 = -46$$

$$5x_1 - \quad 2x_2 + \quad x_3 = \quad 41$$

We have eliminated x_1 from the second equation. Now let us eliminate x_1 from the third equation. We can do this by dividing it by −2.5 (= −5/2), adding the quotient to the first equation, and replacing the third equation by the sum. The result is

$$2x_1 + \quad 3x_2 + \quad 8x_3 = \quad 84$$

$$-11x_2 + 14x_3 = -46$$

$$3.8x_2 + 7.6x_3 = \quad 67.6$$

Now x_1 has been eliminated entirely. It will be clear in a moment that we can complete the forward elimination process with one more step by eliminating x_2 from the third equation. We do this by multiplying the third equation by 2.8947 (=11/3.8), adding the result to the second equation, and replacing the third equation by the sum. The result is

$$2x_1 + \quad 3x_2 + \quad 8x_3 = \quad 84$$

$$-11x_2 + 14x_3 = -46$$

$$36x_3 = 149.6782$$

At this point the forward elimination process is completed. It should be clear that the solution to the last equation is immediately obtainable, leading us to the second half of the Gauss elimination method, the backward substitution part.

Backward Substitution. Beginning with the last (third) equation, we solve for x_3:

$$x_3 = 149.6782/36 = 4.15789$$

Working backward, we substitute this value for x_3 into the second equation:

(Text continues on page 100.)

```
How many equations do you wish to solve?   3


Please enter the coefficients and constants below:

Equation  1

  a11:   2
  a12:   3
  a13:   8
  c1:   84

Equation  2

  a21:   1
  a22:   7
  a23:   -3
  c2:   65

Equation  3

  a31:   5
  a32:   -2
  a33:   1
  c3:   41


Your coefficients and constants are as follows:

   2.000     3.000     8.000    84.000
   1.000     7.000    -3.000    65.000
   5.000    -2.000     1.000    41.000

Your answers are:

  x(1)  =    1.1157894737E+01
  x(2)  =    9.4736842105E+00
  x(3)  =    4.1578947369E+00

(shell)  A:\>
```

Figure 6–7 Output of *Program Gauss.*

```
Program Gauss;
(************************************************************)
(*                                                        *)
(*  A Program to solve simultaneous linear equation by the *)
(*  Gaussian elimination method.  The algorithm was       *)
(*  suggested by Sedgewick, R., "Algorithms,"             *)
(*  Addison-Wesley, Reading  Massachusetts, 1983          *)
(*                                                        *)
(************************************************************)

uses crt,graph;

Const MaxN = 10;
      ScreenLeft =    35;
      ScreenRight =  285;
      ScreenTop =     15;
      ScreenBottom = 175;
Type Matrix = Array [1..MaxN,1..MaxN] of Real;
     Vector = Array [1..MaxN] of Real;
Var Mx:  Matrix;
    Vec:  Vector;
    i,j,k,N:  Integer;
    answer:  char;

(*********************************************************************)

Procedure GiveInstructions;
Begin
  Clrscr;
  Writeln('Your set of linear equations should be in the form:');
  Writeln;
  Writeln('       a11x1 + a12x2 + a13x3 + ... a1nxn = c1');
  Writeln('       a21x1 + a22x2 + a23x3 + ... a2nxn = c2');
  Writeln('       a31x1 + a32x2 + a33x3 + ... a3nxn = c3');
  Writeln('        .         .         .              .');
  Writeln('       an1x1 + an2x2 + an3x3 + ... annxn = cn');
  Writeln;
  Writeln('First you will be asked to enter the no. of equations.');
  Writeln;
  Writeln('Then you will be asked to enter the coefficients');
  Writeln('beginning with a11, a12.  You will be prompted for each.');
  Writeln;
  Writeln('Please enter a return to begin.');
  Readln;
  Readln;
End;

(*********************************************************************)

Procedure HeaderPage;

Var GrDriver,
    GrMode,
    ErrCode:  Integer;
```

Figure 6–8 *Program Gauss.*

```
   Procedure DrawBorder;
   Begin (* DrawBorder *)
      line(ScreenLeft,ScreenTop,ScreenRight,ScreenTop);
      line(ScreenRight,ScreenTop,ScreenRight,ScreenBottom);
      line(ScreenRight,ScreenBottom,ScreenLeft,ScreenBottom);
      line(ScreenLeft,ScreenBottom,ScreenLeft,ScreenTop);
   End;  (* DrawBorder *)

   Procedure Credits;
   Begin (* Credits *)
      SetTextJustify(CenterText,CenterText);
      OutTextXY(160,25,'This is a');
      OutTextXY(160,50,'LINEAR EQUATIONS PROGRAM');
      OutTextXY(160,80,'Written by');
      OutTextXY(160,105,'Rodney J. Sime');
      OutTextXY(160,130,'CALIFORNIA STATE UNIVERSITY');
      OutTextXY(160,155,'SACRAMENTO');
      SetTextJustify(LeftText,TopText);
   End;  (* Credits *)

Begin (* HeaderPage *)
   GrDriver:= Detect;
   InitGraph(GrDriver, GrMode,'');
   ErrCode:= GraphResult;
   SetGraphMode(3);
   If ErrCode = GrOK then
      Begin
        DrawBorder;
        Credits;
        Readln;
        CloseGraph;
      End
   Else
      Writeln('Graphics error:  ',GraphErrorMsg(ErrCode));
End;  (* HeaderPage *)


(*************************************************************)

Procedure ReadEquations (Var M:  Matrix);
(* This procedure accepts the coefficients of n equations interactively *)

Var I, J: Integer;
Begin
  CLRSCR;
  Write('How many equations do you wish to solve?  ');
  Readln(N);Writeln;
  Writeln;Writeln('Please enter the coefficients and constants below:');
  Writeln;
  For I:= 1 to N do            (* for row 1 to n *)
  Begin
    Writeln('Equation', I:3);
    Writeln;
    For J:= 1 to N+1 do          (* for col 1 to n+1 *)
      Begin
      If  (J=(N+1)) then
```

Figure 6–8 (continued)

```
            Begin
            Write(' c',I,':   ');
            Readln(M[I,J]);
            End
          Else
            Begin
            Write(' a',I,J,':   ');
            Readln(M[I,J]);
            End;
        End; (* For *)
      Writeln;
    End; (* For *)
    Clrscr
  End; (*ReadEquations *)

  (****************************************************)

  Procedure EchoData(M:   Matrix);
  (* Prints out the matrix of the equations *)

  Var I,J: Integer;

  Begin (* EchoData *)
    Writeln;Writeln(' Your coefficients and constants are as follows:');
    Writeln;
    For I:= 1 to N do
      Begin
        For J:= 1 to N+1 do
        Write(M[I,J]:8:3,' ');
        Writeln;
      End;   (* For I..*)
  End; (* EchoData *)

  (****************************************************)

  Procedure eliminate;
  (* This procedure eliminates all the unknows except the last *)
  (* according to the Gaussian elimination method for solving  *)
  (* simultaneous linear equations                             *)

    Var i,j,k, max:   Integer;
                  t:   Real;
    Begin (* eliminate *)
    For i:= 1 to N Do
      Begin  (* First For *)
        Max:= i;
        For j:= i + 1 to N Do
          if ABS(Mx[j,i])>ABS(Mx[Max,i]) then max:= j;
        For k:= i to N + 1 Do
          Begin
          t:= Mx[i,k];
          Mx[i,k]:= Mx[Max,k];
          Mx[max,k]:= t
          End;
        For j:= i + 1 to N do
```

Figure 6–8 (continued)

```
        For k:= N + 1 Downto i Do
          Mx[j,k]:= Mx[j,k]-Mx[i,k]*Mx[j,i]/Mx[i,i];
     End;  (* First For *)
   End;  (* eliminate *)

(*************************************************************************)

Procedure Substitute;
(* This procedure does the back substitution routine of the Gaussian *)
(* elimination method: it substitutes the last unknow back into the  *)
(* simplified equations to generate more answeres                    *)

  Var j,k:  Integer;
      t:    Real;
  Begin (* Substitute *)
    For j:= N downto 1 Do
      Begin  (* 1st For *)
      t:=0.0;
      For k:= j + 1 to N Do
        t:= t + Mx[j,k]*Vec[k];
      Vec[j]:= (Mx[j,N+1]-t)/Mx[j,j]
      End;  (* 1st For *)
    Writeln;
    Writeln('  Your answers are:');
    Writeln;
    For j:= 1 to N do
      Writeln('    x(',J:1,') = ',Vec[j]);
  End;  (* Substitute *)


Begin;  (* M A I N   P R O G R A M *)
   HeaderPage;
   Clrscr;
   Write('Do you want instructions?  ');
   Read(answer);
   If (answer = 'y') OR (answer = 'Y') Then GiveInstructions;
   ReadEquations(Mx);
   EchoData(Mx);
   Eliminate;
   Substitute;
End.     (* M A I N   P R O G R A M *)
```

Figure 6–8 (continued)

$$-11x_2 + 14(4.15789) = -46$$

$$x_2 = 9.47368$$

Continuing the backward substitution process, we substitute the now known values for x_2 and x_3 into the first equation and solve for x_1:

$$2x_1 + 3(9.47368) + 8(4.15789) = 84$$

$$x_1 = 11.157915$$

The solution to the set of three simultaneous linear equations is now complete.

Programming the Gauss Elimination Method. It is a programming convenience to use the matrix representation of a set of simultaneous linear equations. If the equations are given by

$$a_{11}x_1 + a_{12}x_2 + \cdots + a_{1j}x_j + \cdots + a_{1n}x_n = b_1$$

$$a_{21}x_1 + a_{22}x_2 + \cdots + a_{2j}x_j + \cdots + a_{2n}x_n = b_2$$

$$\vdots \qquad \vdots \qquad \vdots \qquad \vdots \qquad \vdots \qquad \vdots \qquad \vdots$$

$$a_{i1}x_1 + a_{i2}x_2 + \cdots + a_{ij}x_j + \cdots + a_{in}x_n = b_i$$

$$\vdots \qquad \vdots \qquad \vdots \qquad \vdots \qquad \vdots \qquad \vdots \qquad \vdots$$

$$a_{n1}x_1 + a_{n2}x_2 + \cdots + a_{nj}x_j + \cdots + a_{nn}x_n = b_n$$

then the augmented matrix is defined as

$$\begin{bmatrix} a_{11} & a_{12} & \cdots & a_{1j} & \cdots & a_{1n} & b_1 \\ a_{21} & a_{22} & \cdots & a_{2j} & \cdots & a_{2n} & b_2 \\ \vdots & \vdots & \vdots & \vdots & \vdots & \vdots & \vdots \\ a_{i1} & a_{i2} & \cdots & a_{ij} & \cdots & a_{in} & b_i \\ \vdots & \vdots & \vdots & \vdots & \vdots & \vdots & \vdots \\ a_{n1} & a_{n2} & \cdots & a_{nj} & \cdots & a_{nn} & b_n \end{bmatrix}$$

In *Procedure ReadEquations*, the coefficients a_{ij} and the constant terms b_n are read interactively into an N by $N + 1$ matrix, where N is the number of equations. *Procedure EchoData* then echoes to the screen what has just been input to check for accuracy. Next, a call to *Procedure Eliminate* carries out the forward elimination of the variables. Finally, *Procedure Substitute* does the back substitution to find the values of the N variables, which it stores in a list called *Vec*. After the statement "Your answers are:", the N values of x_i are printed as shown in Fig. 6–7. The program itself is given in Figure 6–8.

POLYNOMIAL CURVE FITTING BY LEAST SQUARES

Earlier we saw how experimental data could be fitted to a straight line represented by an equation of the form

$$y = mx + b$$

It sometimes happens that the data do not fit a straight line, but rather a polynomial expressed in general by

$$y = c_0 + c_1x + c_2x^2 + c_3x^3 + \cdots + c_px^p \qquad \text{(6–57)}$$

As already described for a linear curve fit, the residuals r_i are defined as the differences between the experimental values y_i and the calculated values:

$$r_i = y_i - c_0 - c_1x_i - c_2x_i^2 - c_3x_i^3 - \cdots - c_mx_i^m \qquad \text{(6–58)}$$

Recall that with the linear case, c_0 was called b, and c_1 was called m; and, as before, we assume that the most probable values of x_i are those that minimize the sum of the squares of the residuals. Define

$$R = \sum r_i^2 \tag{6-59}$$

The conditions for the minimum in R are that the derivatives of the function R with respect to the c_i are all equal to zero:

$$\left(\frac{\partial R}{\partial c_0}\right) = 0, \left(\frac{\partial R}{\partial c_1}\right) = 0, \ldots; \left(\frac{\partial R}{\partial c_k}\right) = 0, \ldots, \left(\frac{\partial R}{\partial c_p}\right) = 0$$

We have seen that if the polynomial is of degree one (linear), the derivatives give rise to two equations in two unknowns, c_0 and c_1 (previously called b and m). For the general case, the derivatives give rise to $p + 1$ equations in $p + 1$ unknowns, namely, $c_0, c_1, c_2, \ldots$. In order to evaluate these constants it is necessary to solve a system of $p + 1$ simultaneous linear equations. The equations are

$$c_0 n + c_1 \sum x_i + c_2 \sum x_i^2 + \cdots + c_k \sum x_i^k$$
$$+ \cdots + c_p \sum x_i^p - \sum y_i = 0$$

$$c_0 \sum x_i + c_1 \sum x_i^2 + c_2 \sum x_i^3 + \cdots + c_k \sum x_i^{k+1}$$
$$+ \cdots + c_p \sum x_i^{p+1} - \sum x_i y_i = 0$$

$$c_0 \sum x_i^k y_i^k + c_1 \sum x_i^{k+1} + c_2 \sum x_i^{k+2} + \cdots + c_k \sum x_i^{k+k}$$
$$+ \cdots + c_p \sum x_i^{p+k} - \sum x_i^k y_i = 0$$

$$c_0 \sum x_i^p + c_1 \sum x_i^{p+1} + c_2 \sum x_i^{p+2} + \cdots + c_k \sum x_i^{p+k}$$
$$+ \cdots + c_p \sum x_i^{p+p} - \sum x_i^p y_i = 0$$

The solution to this system of simultaneous linear equations is the set of values of the coefficients c_i. The augmented matrix is written

$$\begin{bmatrix} n & x_i & x_i^2 & x_i^k & \cdots & x_i^p & \cdots & y_i \\ x_i & x_i^2 & x_i^3 & x_i^{k+1} & \cdots & x_i^{p+1} & \cdots & x_i y_i \\ x_i^k & x_i^{k+1} & x_i^{k+2} & x_i^{k+k} & \cdots & x_i^{p+k} & \cdots & x_i^k y_i \\ x_i^p & x_i^{p+1} & x_i^{p+2} & x_i^{p+k} & \cdots & x_i^{p+p} & \cdots & x_i^p y_i \end{bmatrix}$$

The x_i and y_i are the experimental points we wish to fit to a polynomial of degree p. The number of x_i, y_i pairs equals n, so the summations are from $i = 1$ to n. The number of pairs must be greater than the degree of the polynomial and is often much greater.

Programming a Least-Squares Polynomial Fit of n Data Pairs

In *Program Polyfit* (Fig. 6-9), the above matrix is constructed when *Procedure BuildMatrix* is called. First, however, the groundwork for the

(Text continues on page 108.)

```pascal
Program Polyfit;
(* A program to fit n x,y pairs to a polynomial of degree m *)
(* The algoritm follows T. R. Dickson, "The Computer and   *)
(* Chemistry," Freeman, San Francisco, 1968, page 157      *)

uses crt,graph;

Const MaxN = 10;
      ScreenLeft =     25;
      ScreenRight =   295;
      ScreenTop =      15;
      ScreenBottom = 175;

Type Matrix = Array [1..MaxN,1..MaxN] of Real;
     Vector = Array [1..MaxN] of Real;

VAR  X,Y,XYvector,Vec,Temp:  Vector;
     Mat,Mx:  Matrix;
     J,K,Num,N,p:  Integer;
     Answer:  Char;

(**************************************************************)

Function Power (number: real; exponent:  integer):  real;
   Begin
      Power:= EXP(exponent*Ln(number));
   End;

(**************************************************************)

Procedure GiveInstructions;
Begin
  Clrscr;
  Writeln('Given as interactive input a set of n x,y pairs,');
  Writeln('the program will fit these data to a polynomial');
  Writeln('of any desired power p as long as n > p.');
  Writeln;
  Writeln('The polynomial is of the form:');
  Writeln;
  Writeln('                                 2      3            p');
  Writeln('            y = c  + c x + c x  + c x  + ... + c x');
  Writeln('                 o    1    2      3            p');
  Writeln;
  Writeln('The output is a list of c''s.');
  Writeln;
  Writeln('Please enter a return to begin.');
  Readln;
End;

(**************************************************************)

Procedure HeaderPage;

Var GrDriver,GrMode,ErrCode:  Integer;
```

Figure 6–9 *Program Polyfit.*

```
   Procedure DrawBorder;
   Begin (* DrawBorder *)
      Line(ScreenLeft,ScreenTop,ScreenRight,ScreenTop);
      Line(ScreenRight,ScreenTop,ScreenRight,ScreenBottom);
      Line(ScreenRight,ScreenBottom,ScreenLeft,ScreenBottom);
      Line(ScreenLeft,ScreenBottom,ScreenLeft,ScreenTop);
   End;  (* DrawBorder *)

   Procedure Credits;
   Begin; (* Credits *)
      SetTextJustify(CenterText,CenterText);
      OutTextXY(160,25,'This is a');
      OutTextXY(160,50,'NON-LINEAR LEAST SQUARES PROGRAM');
      OutTextXY(160,85,'Written by');
      OutTextXY(160,110,'Rodney J. Sime');
      OutTextXY(160,135,'CALIFORNIA STATE UNIVERSITY');
      OutTextXY(160,160,'SACRAMENTO');
      SetTextJustify(LeftText,TopText);
   End;  (* Credits *)

Begin (* HeaderPage *)
   GrDriver:= Detect;
   InitGraph(GrDriver,GrMode,'');
   ErrCode:= GraphResult;
   SetGraphMode(3);
   DrawBorder;
   Credits;
   Readln;
   TextMode(BW80)
End;  (* HeaderPage *)


(***************************************************************)

Procedure GetXYpairs;
Var I,
    Sx,Sy:  Integer;  (* Screen Coordinates *)
Begin (* GetXYpairs *)
   CLRSCR;
   Write('What is the number of X and Y pairs you wish to enter?  ');
   Readln(Num);
   Writeln;
   For I:= 1 to Num Do
     Begin (* For *)
        Write(      'X(',I,'):  ');
        Sx:= WhereX; Sy:= WhereY;
        Read(X[I]);
        GoToXY(Sx+10,Sy);
        Write('      Y(',I,'):  ');
        Readln(Y[I]);
     End;  (* For *)
   Write('What is the highest power of the polynomial?  ');
   Readln(p);
   Writeln;
End;  (* GetXYpairs *)
```

Figure 6–9 (continued)

```
(****************************************************************)

Procedure CalcXYvector;
Var I,J:  Integer;
    Sum:  Real;
  Begin (* CalcXYvector *)
    For I:= 0 To p Do
        Begin (* outer For *)
          Sum:= 0;
          For J:= 1 to Num Do
            Sum:= Sum + Y[J]*Power(X[J],I);
          XYvector[I]:= Sum;
        End;  (* outer For *)
  End;  (* CalcXYvector *)


(****************************************************************)

Procedure SumXiToPower;
(* This procedure calculates the sum of the Xi's raised to the *)
(* power 0 to p and stores the values in array temp           *)

VAR I,J,Row, Col:  Integer;
     S,T:  Real;
Begin (* SumXiToPower *)
   For I:= 0 To (p + p) DO              (* set powers of Xi  *)
      Begin (* outer For *)
        S:= 0;
        For J:= 1 to Num Do            (* calculate sums    *)
          S:= S + Power(X[J],I);
        Temp[I]:= S;                   (* store sums in array temp *)
      End;  (* outer For *)
(* Next, build the p by p matrix *)
   For Row:= 0 to p Do
      For Col:= 0 to p Do
        Mat[Row,Col]:= Temp[Row + Col];
(* Finally, make the augmented matrix *)
   Col:= p + 1;              (* this is the augmented column *)
   For Row:= 0 to p Do
      Mat[Row,Col]:= XYvector[Row];
End;  (* SumXiToPower *)


(****************************************************************)

Procedure WriteMx;
Var Row,Col,RowMax,ColMax:  Integer;
Begin
   Writeln('Mx');
   RowMax:= p + 1;
   ColMax:= p +2;
   For Row:= 1 to RowMax Do
     Begin;
        For Col:= 1 to  ColMax Do
          Begin
            Write(Mx[Row,Col]:10:0);
          End;
```

Figure 6–9 (continued)

```
      Writeln;
    End;
  Writeln;
 End;

(***************************************************************)

Procedure ConvertMatrices;
(* Needed to match indices of arrays Mat and Mx *)

Var Row,Col:  Integer;
Begin (* ConvertMatrices *)
  For Row:= 0 To p Do
    For Col:= 0 To p + 1 Do
        Mx[Row + 1,Col + 1]:= Mat[Row,Col];
End;  (* ConvertMatrices *)

(**********************************************************)

Procedure eliminate;
(* This procedure eliminates all the unknows except the last *)
(* according to the Gaussian elimination method for solving  *)
(* simultaneous linear equations                             *)

   Var i,j,k, max:  Integer;
               t:  Real;
   Begin (* eliminate *)
   For i:= 1 to N Do
     Begin  (* First For *)
       Max:= i;
       For j:= i + 1 to N Do
         if ABS(Mx[j,i])>ABS(Mx[Max,i]) then max:= j;
       For k:= i to N + 1 Do
         Begin
         t:= Mx[i,k];
         Mx[i,k]:= Mx[Max,k];
         Mx[max,k]:= t
         End;
       For j:= i + 1 to N do
         For k:= N + 1 Downto i Do
           Mx[j,k]:= Mx[j,k]-Mx[i,k]*Mx[j,i]/Mx[i,i];
     End;  (* First For *)
   End;  (* eliminate *)

(*******************************************************************)

Procedure Substitute;
(* This procedure does the back substitution routine of the Gaussian *)
(* elimination method: it substitutes the last unknown back into the  *)
(* simplified equations to generate more answers.                     *)

  Var j,k:  Integer;
        t:  Real;
  Begin (* Substitute *)
    For j:= N downto 1 Do
```

Figure 6–9 (continued)

```
         Begin   (* 1st For *)
         t:=0.0;
         For k:= j + 1 to N Do
           t:= t + Mx[j,k]*Vec[k];
         Vec[j]:= (Mx[j,N+1]-t)/Mx[j,j]
         End;   (* 1st For *)
      Writeln;
      Writeln('  Your answers are:');
      Writeln;
      For j:= 1 to N do
        Writeln('    c(',(J-1):1,') = ',Vec[j]);
   End;   (* Substitute *)

 (*************************************************************)

Begin (* M A I N   P R O G R A M *)
   HeaderPage;
   Clrscr;
   Write('Would you like instructions?  ');
   Readln(answer);
   If (answer = 'y') OR (answer = 'Y') then GiveInstructions;
   (* First built set of simultaneous equations *)
   GetXYpairs;
   CalcXYvector;
   SumXiToPower;
   ConvertMatrices;
   N:= p + 1; (* no. of equations = Polynomial power + 1 *)
   (* Next do Gaussian elimination to solve equations *)
   Eliminate;
   Substitute;
End.  (* M A I N   P R O G R A M *)P
```

Figure 6–9 (continued)

```
What is the number of X and Y pairs you wish to enter?  9

X(1):  1              Y(1):  2.07
X(2):  2              Y(2):  8.60
X(3):  3              Y(3):  14.42
X(4):  4              Y(4):  15.80
X(5):  5              Y(5):  18.92
X(6):  6              Y(6):  17.96
X(7):  7              Y(7):  12.98
X(8):  8              Y(8):  6.45
X(9):  9              Y(9):  0.27
What is the highest power of the polynomial?  2

   Your answers are:

   c(0) = -7.8266666636E+00
   c(1) =  1.0590045453E+01
   c(2) = -1.0829545453E+00
```

Figure 6–10 Sample run of *Program Polyfit.* (For a comparison with a similar program also fitting the identical data to a second-degree polynomial, see Figure 7.2 in A. R. Miller, *Pascal Programs for Scientists and Engineers,* Sybex, Berkeley, California, 1981, page 198.)

construction is provided by a call to *Procedure GetXYpairs* to input the experimental data interactively, followed by a call to *Procedure CalcXYvector*, which calculates the right-hand column of the augmented matrix.

The solution to the p simultaneous equations represented by the augmented matrix is carried out exactly as done by *Program Gauss*. In fact, the same procedures are called: *Eliminate* and *Substitute*.

A short procedure, *ConvertMatrices*, serves to adjust the indices in order to match up the matrices. This is simpler than rewriting procedures *Eliminate* and *Substitute*, as the programming of the indices in these two procedures is a bit tricky. As in *Program Gauss*, *Procedure Substitute* prints out the answers, which are now the coefficients c_i in the polynomial (Eq. 6–57). A sample run is shown in Figure 6–10.

NUMERICAL INTEGRATION

The definite integral A represents the area under the curve $y = f(x)$ between the integration limits $x = a$ and $x = b$:

$$A = \int_a^b f(x)\, dx \qquad (6\text{–}60)$$

Three cases are of interest:

1. You have had considerable experience with the first case. If $f(x)$ is a simple mathematical function that can be integrated by the usual methods of integral calculus, numerical methods are not necessary. For example, if

$$f(x) = 3x^2 + 2x - 9$$

the integration can be done directly:

$$A = \int_1^3 (3x^2 + 2x - 9) = (x^3 + x - 9x)\big|_1^3 = 10$$

For a more complex function, it might be necessary to use integration tables, integrate by parts, change variables, or use other tricks of the trade.

2. In principle, any continuous function can be integrated, but occasionally $f(x)$ is a mathematical function that is too complex to give a result consisting of a sum of simple algebraic and transcendental functions. Numerical methods are useful here.

3. Finally, $f(x)$ might not be a function at all, but rather a collection of discrete experimental points presented in tabular form:

y	1.2	4.3	9.5	17.1	26.1	37.0	85.4
x	10	15	20	25	30	35	40

An experiment often consists of a simple relationship between a dependent and an independent variable, for example, $z = xy$. To take a more specific example, consider the relationship between charge Q, current I, and time t. If the current is constant the relationship is

$$Q = It \tag{6-61}$$

If the current is not constant, but is a function of time, the charge is given by

$$Q = \int_a^b f(t)\, dt \tag{6-62}$$

Experimentally, $I = f(t)$ is a collection of I, t data pairs collected at regular intervals during the execution of the experiment, as illustrated in Table 6–6.

The determination of the total charge requires the evaluation of the integral defined by Equation 6–62. This corresponds to the area under the smooth curve drawn through the points in Table 6–6 and plotted in Figure 6–11a. You can compare this area with the area under the curve (straight line) in Figure 6–11b, which illustrates the calculation of the charge when the current is constant at 0.880 A. The area in Figure 6–11b is just the product of the height of the rectangle and the length of the base $(b - a)$:

$$Q = \int_a^b f(t)\, dt = 0.88 \int_{10}^{70} dt = 0.88(70 - 10) = 52.80 \text{ C} \tag{6-63}$$

While this is a common situation in the experimental sciences, you will occasionally encounter a smoothly varying independent variable, as illustrated in Figure 6–11a, where the current is decreasing regularly with time. These data are from Table 6–6. The area under the curve, which is the value of the integral shown in Equation 6–63, can be determined a number of ways. If the plot is carefully drawn, the rectangle enclosed by the dotted lines can be cut out with a scissors and weighed. Then, if the area under the curve is cut out and weighed, the area under the curve can be calculated.

As an example, the shaded areas in Figure 6–11 were cut out with a scissors and weighed. The resulting Q for Figure 6–11a is

TABLE 6–6

Nonconstant Current Data

I	0.98	0.97	0.95	0.92	0.87	0.82	0.73	0.60	0.45
t	0	10	20	30	40	50	60	70	80

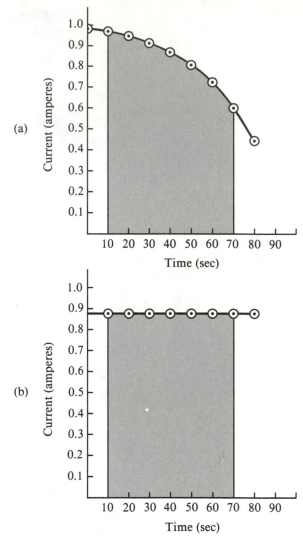

Figure 6–11 Integrals as areas.

$$Q = Q_{\text{rect}} \times \text{Wt}_2/\text{Wt}_1$$

$$= 0.880 \text{ A}(70 \text{ s} - 10 \text{ s}) \times 0.8831 \text{ g}/0.9235 \text{ g} = 50.5 \text{ C}$$

This method depends on the uniformity of the paper and the skill with which the paper is cut.

Trapezoidal Rule

For a more objective approach, let us investigate the Trapezoidal Rule and Simpson's Rule for numerical integration. If we divide the area under the curve into an infinite number of parallel segments, each segment approaches rectangularity, and the sum of their areas equals the area under the curve. For a finite number of points, however, we may choose to draw rectangular segments between the points. One way to draw the segments is shown in Figure 6–12a. The height of the single representative segment in Figure 6–12b is the average value of y at the points at $x = a$ and $x = b$, namely,

$$\text{average height} = (y_a + y_b)/2 \tag{6-64}$$

Since the width of the rectangle is $(b - a)$, and the average height is $(y_b + y_a)/2$, the area of the rectangle is

$$A = (b - a) \times \left(\frac{y_a + y_b}{2} \right) \tag{6-65}$$

Because the area of this rectangle equals the area of the trapezoid drawn through a–b–y_b–y_a (Fig. 6–12b), Equation 6–65 is known as the Trapezoidal Rule. The total area (McCracken and Dorn, 1964) of $n - 1$ trapezoidal segments defined by n points is given by

$$A = \frac{\Delta x}{2} (y_0 + 2y_1 + 2y_2 + \cdots + 2y_{n-2} + 2y_{n-1} + y_n) \tag{6-66}$$

The interval between points is Δx; $y_0 = y_a$, the value of y at $x = a$; and $y_n = y_b$, the value of y at $x = b$. For the data in Table 6–6 and Figure 6–11a, the integral from $t = 10$ to 70 is

$$A = \frac{10}{2} [0.97 + 2(0.95) + 2(0.92) + 2(0.87)$$

$$+ 2(0.82) + 2(0.73) + 0.60]$$

$$= 50.750 \ C$$

The Trapezoidal Rule works fairly well; its main virtue is its simplicity. A superior method is Simpson's Rule.

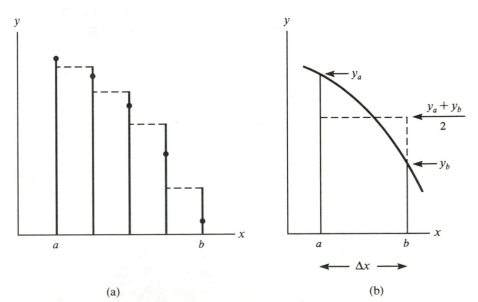

(a)   (b)

Figure 6–12 Trapezoidal segments.

Simpson's Rule

Simpson's Rule is quite similar to the Trapezoidal Rule. Instead of interpolating with a straight line between two points, however, Simpson's Rule fits a parabola to successive sets of three points as shown in Figure 6–13.

It can be shown that the area under a parabola defined by three *equally* spaced points is given by

$$A = \frac{10}{3}(y_1 + 4y_2 + y_3) \qquad (6\text{–}67)$$

Suppose that the three points in Figure 6–13b actually define the parabola $y = x^2$ so that the points are $(0, 0)$, $(1, 1)$ and $(2, 4)$. In this case we can find the area under the curve by Simpson's Rule or by direct integration between the limits $x = 0$ and $x = 2$:

Simpson's Rule: $\qquad A = \frac{1}{3}(0 + 4 + 4) = \frac{8}{3}$

Integration: $\qquad A = \int_0^2 x^2\, dx = \left.\frac{x^3}{3}\right|_0^2 = \frac{8}{3}$

In this case, Simpson's Rule is not an approximation, but exact, since the curve really is a parabola.

Notice that in this simplest approximation of Simpson's Rule the three points define two underlying segments or panels, the areas of which are added. In general n points define $n - 1$ segments. The number of points n must be an odd number so that the number of segments is even. The last restriction is that the x interval between points must be equal; that interval is called Δx here. If you anticipate using Simpson's Rule, you should design your experiment so that the data are collected at equal intervals of the independent variable x. With the Trapezoidal Rule, n can be odd or even and Δx need not be constant.

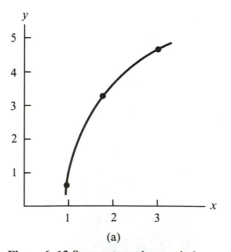 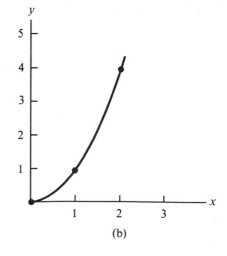

| (a) | (b) |

Figure 6–13 Segments under parabolas.

The total area under all the parabolas is given by Simpson's Rule for a set of n points (n must be odd),

$$A = \frac{\Delta x}{3}(y_0 + 4y_1 + 2y_1 + 4y_3 + \cdots + 4y_{n-1} + y_n) \quad \textbf{(6–68)}$$

where Δx is the interval between equally spaced points, $y_0 = y_a$, the value of y at $x = a$; and $y_n = y_b$, the value of y at $x = b$, where a and b are the integration limits. Let us compare the integral calculated by Simpson's Rule (Eq. 6–68) with the integral calculated by the Trapezoidal Rule (Eq. 6–66), using the same data as before:

$$A = \frac{\Delta x}{3}[0.97 + 4(0.95) + 2(0.92) + 4(0.87)$$

$$+ 2(0.82) + 4(0.73) + 0.60]$$

$$= 50.833 \text{ C}$$

This agrees quite well with the value previously calculated with the Trapezoidal Rule, 50.750 C.

Integration with a Digital Computer

When it is necessary to treat large numbers of data points, it is useful to use a computer program to carry out the summations. The Trapezoidal Rule is the algorithm for the program listed in Figure 6–14. It is especially useful when the points are not equally spaced. The listing of a Pascal program (Simpson1) for finding the area under a set of discrete points according to Simpson's Rule is given in Figure 6–15. Generally Simpson's Rule gives better results than the Trapezoidal Rule, but the points must be equally spaced and uneven in number.

Simpson's Rule, or the Trapezoidal Rule for that matter, can be used to integrate functions as well as to find the area under a set of discrete experimental points. The function can be one that cannot be integrated by ordinary means. Let us rewrite Simpson's Rule (Eq. 6–68) with $f(x_i)$ replacing the y_i to represent the value of the function at each x_i. With the series

$$A = \frac{\Delta x}{3}[f(x_0) + 4f(x_1) + 2f(x_2) + 4f(x_3) \quad \textbf{(6–69)}$$

$$+ \cdots + 4f(x_{n-1}) + f(x_n)]$$

we can choose as many points as necessary to achieve whatever accuracy we desire. Thus, Program Simpson2, shown in Figure 6–16, first calculates the area with just two points (one segment) and then proceeds to double the number of segments until the area differs from the last area calculated by an insignificant amount: 0.001%. The limits of the integration are input interactively; the function to be integrated is a declared Pascal

(Text continues on page 116.)

```
Program TrapezoidIntegration;
(* This program finds the area under a curve defined by N *)
(* experimental points input interactively.  Integration  *)
(* is by the trapezoidal rule                             *)

uses crt;

VAR X,Y:  Array [1..50] of Real;  (* lists of X-Y pairs  *)
      N:  Integer;                (* number of X-Y pairs *)

(*****************************************************************)

Procedure GetXYpairs;
Var I,
    Sx,Sy:  Integer; (* Screen Coordinates *)
Begin (* GetXYpairs *)
   CLRSCR;
   Write('What is the number of X and Y pairs you wish to enter?  ');
   Readln(N);
   Writeln;
   For I:= 1 to N Do
     Begin (* For *)
        Write(     'X(',I,'):  ');
        Sx:= WhereX;  Sy:= WhereY;
        Read(X[I]);
        GoToXY(Sx+10,Sy);
        Write('     Y(',I,'):  ');
        Readln(Y[I]);
     End;  (* For *)
End;  (* GetXYpairs *)

(******************************************************************)

Procedure CalculateArea;
Var I:  Integer;
    Area:  Real;
Begin  (* CalculateArea *)
   For I:= 1 to N - 1 Do
     Area:= Area + (X[I+1] - X[I]) * (Y[I+1] + Y[I])/2;
   Clrscr;
   Writeln('     X            Y');
   Writeln;
   For I:= 1 to N do
     Writeln(X[I]:10:3,Y[I]:20:3);
   Writeln;
   Writeln('     The Area under the curve defined');
   Writeln('     by ',N,' X-Y pairs equals ',Area:10)
End; (* CalculateArea *)

(******************************************************************)

Begin (* M A I N   P R O G R A M *)
   GetXYpairs;
   CalculateArea
End.  (* M A I N   P R O G R A M *)
```

Figure 6–14 *Program TrapezoidIntegration.*

```
Program Simpson1;
(**********************************************************)
(* Integrates, i.e., calculates the area under a curve *)
(* defined by N points.  N must be an odd number.      *)
(* Uses Simpson`s Method (equal segments).             *)
(**********************************************************)

Const Max = 99;
Type Table = Array [1..Max] of Real;

Var Y:  Table;
    N:  1..Max;
    DeltaX, Area, Xfirst,Xlast:  Real;

(**********************************************************)

Procedure ReadData;

Var I:  Integer;
Begin (* ReadData *)
  Repeat
     Writeln('Please enter the number of points (N must be odd)');
     Write('N: ');
     Readln(N);
     If N MOD 2 = 0 Then
        Writeln('N must be an odd number; try again.');
  Until N MOD 2 = 1;
  Writeln;Writeln('Please enter the initial and final values of x.');
  Write('Initial x equals: ');
  Read(Xfirst);
  Write('     Final x equals:  ');
  Readln(Xlast);
  DeltaX:= (Xlast - Xfirst)/(N - 1);
  Writeln;Writeln;
  Writeln('Now please enter the values of Y');Writeln;
  For I:= 1 to N do
     Begin
     Write('Y[',I:1,']:  ');
     Readln(Y[I]);
     End; (* For *)
END;  (* ReadData *)

(**********************************************************)
```

Figure 6–15 *Program Simpson1:* a Pascal program to find the area under a curve defined by a set of discrete experimental points.

```
Procedure CalcSum;
(* Calculate sum according to Simpson's rule *)

Var I,K:  1..Max;
    Sum:  Real;

Begin (* CalcSum *)
   Sum:= Y[1];      (* first term is fo in usual notation *)
   For I:= 2 to N - 1 do (* Begin sum loop with 2nd term *)
      Begin (* For *)
      If I MOD 2 = 1 then
         K:= 2
      Else
         K:= 4;
      Sum:= Sum + K*Y[I];
      End; (* For *)
   Sum:= Sum + Y[N];  (* Add last term separately since here K:= 1 *)
   Sum:= Sum*DeltaX/3;(* Sum is integral or area under points       *)
   Writeln;
   Writeln('The integral is:  ',Sum:8:3);
End; (* CalcSum *)

(*************************************************)

Begin (* M A I N   P R O G R A M *)
   ReadData;
   CalcSum
End.  (* M A I N   P R O G R A M *)
```

Figure 6–15 (continued)

function. In the program listing, a simple function is chosen: $f(x) = 1/x$. In this example, be careful not to choose $x = 0$ as an integration limit!

Matrix Inversion

Although division in the ordinary algebraic sense is not defined for matrices, the inverse M^{-1} of a square matrix M is defined such that multiplication of M by M^{-1} results in the identity matrix I. The identity matrix is a square matrix in which the elements along the diagonal equal unity, and everywhere else equal zero. For example, a four-by-four identity matrix is

$$I = \begin{bmatrix} 1 & 0 & 0 & 0 \\ 0 & 1 & 0 & 0 \\ 0 & 0 & 1 & 0 \\ 0 & 0 & 0 & 1 \end{bmatrix}$$

(Text continues on page 129.)

```
Program Simpson2;
(*******************************************************)
(* Integrates any desired function by Simpson's method  *)
(* Starts with 1 segment; then doubles the number of    *)
(* segments until the integrands agree within 0.001%    *)
(*******************************************************)

uses crt;

Var X, Low, High, DeltaX, Sum:  Real;
    I,K,N:  Integer;

(*******************************************************)

Function Func(X: Real): Real;
(* This is Psi for n = 1 for a particle in a box       *)
(* a has been set equal to 1                           *)
Begin
  Func:= 2*Sqr(Sin(Pi*x));
End; (* Func *)

(*******************************************************)

Procedure GetLimits(Var L,H:  Real);
(* Gets the upper and lower integration limits *)

Begin (* GetLimits *)
   ClrScr;
   Writeln;
   Writeln('Please enter the integration limits.');
   Write('The lower limit is:  ');
   Readln(L);
   Write('The upper limit is:  ');
   Readln(H);
   Writeln;
   Writeln;
End;   (* GetLimits *)

(*******************************************************)
```

Figure 6–16 *Program Simpson2:* a Pascal program to find the area under a curve defined by a function.

```pascal
Procedure Integrate(Low, High:  Real);
(* Integrate a previously defined function by Simpson's Rule *)

Const PerCentError = 1.0E-3;
      Spaces = '               ';  (* 15 spaces *)

Var I,Segments:  Integer;
        X, DeltaX, LastSum:  Real;

Begin (* Integrate *)
   Segments:= 1;
   Sum:= 0;;
   Repeat
      LastSum:= Sum;  (* For checking progress at end of loop *)
      Segments:= Segments*2;  (* double the no. of segments each pass *)
      N:= Segments + 1;
      DeltaX:= (High - Low)/(N - 1);
      X:= Low;
      Sum:= Func(X);     (* This is the first term of the sum *)
      For I:= 2 to N - 1 do
        Begin
        X:= X + DeltaX;   (* sum loop begins with 2nd term of sum *)
        If I MOD 2 = 1 then
           K:= 2
        Else
           K:= 4;
        Sum:= Sum + K*Func(X);
        End; (* For; the sum loop ends with the (N-1)th term *)
      X:= X+DeltaX;
      Sum:= Sum + Func(X);  (* This is the last or Nth term *)
      Sum:= Sum*DeltaX/3;
      Writeln(spaces,Segments:8,Sum:12:5);
   Until ABS(Sum - LastSum)*100/(Sum) <= PercentError;
   Writeln;
   Writeln(Spaces,'The integral is:  ',Sum);
End;  (* Integrate *)

(*****************************************************)

Begin (* M A I N   P R O G R A M *)
   GetLimits(Low,High);
   Integrate(Low,High);
End.  (* M A I N   P R O G R A M *)
```

Figure 6–16 (continued)

```
Program Invert;
(****************************************************)
(*    A program to invert a Matrix, adapted from        *)
(*    Radford and Haigh, "Turbo Pascal for the IBM PC *)
(****************************************************)

uses crt;

Const Max = 10;
Type Matrix = Array [1..Max,1..Max] of Real;
Var Mx,xM:  Matrix;
        N:  Integer;
    Error:  Boolean;

(***************************************************************************)

Procedure ReadEquations (Var M:  Matrix;
                             N:  Integer);
Var I, J: Integer;
Begin
  Writeln;Writeln('Please enter the coefficients and constants below:');
  Writeln;
  For I:= 1 to N do
  Begin
    Writeln('Row', I:3);
    For J:= 1 to N do
      Begin
      Write('X',I:1,',',J:1,': ');
      Readln(M[I,J]);
      End; (* For *)
    Writeln;
  End; (* For *)
End; (*ReadEquations *)

(*********************************************************)

Procedure WriteMatrix(M:  Matrix; N: Integer);
(* Write out matrix M *)
Var I,J: Integer;
Begin
  Writeln;
  For I:= 1 to N do
      Begin
      For J:= 1 to N do
         Write(M[I,J]:8:3);
      Writeln;
      End; (* For *)
End;   (* WriteMatrix *)

(*********************************************************)
```

Figure 6–17 *Program Invert,* a program for the inversion of a square matrix.

```
Procedure InvertMatrix(M: Matrix; Var Z:  Matrix; N:  Integer);
Var I, J, K:  Integer;
         D, S:  Real;
Begin (* InvertMatrix *)
   For I:= 1 to N Do
     For J:= 1 to N Do
       Z[I,J]:= M[I,J];
     For I:= 1 to N Do
     Begin (* For *)
       D:= Z[I,I];
       If (D = 0) Then
         Begin (* IF *)
         Writeln;
         Writeln('Stop program because of division by zero!');
         Error:= True;
         Halt
         End;  (* IF *)
       For J:= 1 to N Do
         If (J <> I) then Z[I,J]:= Z[I,J]/D
         else  Z[I,J]:= 1/D;
       For J:= 1 to N Do
         if (J<>I) then
           Begin (* If *)
           S:= -Z[J,I];
           For K:= 1 to N Do
             if (K<>I) then Z[J,K]:= Z[I,K]*S + Z[J,K]
             else Z[J,K]:= Z[I,K]*S
           End   (* If *)
     End  (* For *)
   End;  (* InvertMatrix *)

(*********************************************************)

Begin (* Main Program *)
   N:=3;
   ReadEquations(Mx,N);
   InvertMatrix(Mx,xM,N);
   clrscr;
   Writeln;
   Writeln('Your matrix is:');
   Writeln;
   WriteMatrix(Mx,N);
   Writeln;
   Writeln('The inverse of your matrix is:');
   Writeln;

   WriteMatrix(xM,N);
End.  (* Main Program *)
```

Figure 6–17 (continued)

```
Program LstSqr;

uses crt,graph,printer;

Const ScreenRight =  278;
      ScreenLeft  =   28;
      ScreenTop   =   15;
      ScreenBottom = 175;
      Max         = 100;
      Sp          = '          '; (* 10 blank spaces *)
      Bl          = '     ';      (*  5 blank spaces *)

Type Index = 1..Max;
     Ary = Array[Index] of Real;

Var X,Y,            (* arrays of input Xi and Yi      *)
    Ycalc: Ary;  (* values of Y calc'd at each Xi *)
    N,GrDriver,GrMode,ErrCode:  Integer;
    b,m          : Real;        (* intercept and slope of line *)
    CorrelCoef, Sigmab, Sigmam:  Real;
    Answer, Response:  Char;
    Rescale:  Boolean;
    Title:    String[40];

(*****************************************************************)

Procedure GiveInformation;
Begin
  CLRSCR;
  Writeln('This progam will read your data interactively from the keyboard');
  Writeln('or from a file, as you request.  The file must have the file ');
  Writeln('name LS.DAT. It can be created with the editor of Turbo Pascal');
  Writeln('or any wordprocessor that can create a text file.  The first');
  Writeln('line of the file must give N the number of X,Y pairs to follow.');
  Writeln('Each successive line should contain a single X,Y pair, e.g.,');
  Writeln('                        4');
  Writeln('                        0.00366   -8.75');
  Writeln('                        0.00358   -8.13');
  Writeln('                        0.00315   -1.58');
  Writeln('                        0.00301   -2.60');
  Writeln('After reading the data, the program calculates a least square line');
  Writeln('of the form Y = mX + b, and writes m, b, sigmas and data to the');
  Writeln('screen.  You will then be asked to input the min and max values');
  Writeln('of the GRAPH in the X and Y direction for scaling purposes. ');
  Writeln('The graph will be plotted and you will get a chance to rescale.');
  Writeln('To dump the graph to the printer, enter Shift-PrtSc.  Be sure');
  Writeln('that you enter GRAPHICS, a DOS command, before booting the ');
  Writeln('program.  You may also print the data and results to the');
  Writeln('printer.  The various options are given interactively.');
  Writeln;
  Writeln('Press Return');
  Readln;
  CLRSCR
End;

(*****************************************************************)
```

Figure 6–18 *Program LstSqr,* a linear-least-squares program written in Turbo Pascal. It provides for interactive input or input from a data file. The calculated-least-squares line and the data pairs are presented graphically. If the user wishes, the graph can be rescaled without recalculating the line. The interactive input is user friendly, and further directions can be requested by the user.

```
Procedure HeaderPage;

   Procedure DrawBorder;
   Begin (* DrawBorder *)
      Line(ScreenLeft,ScreenTop,ScreenRight,ScreenTop);
      Line(ScreenRight,ScreenTop,ScreenRight,ScreenBottom);
      Line(ScreenRight,ScreenBottom,ScreenLeft,ScreenBottom);
      Line(ScreenLeft,ScreenBottom,ScreenLeft,ScreenTop);
   End;   (* DrawBorder *)

   Procedure Credits;
   Begin; (* Credits *)
      SetTextJustify(CenterText,CenterText);
      OutTextXY(153,25,'This is a');
      OutTextXY(153,50,'LINEAR LEAST SQUARES PROGRAM');
      OutTextXY(153,75,'Written by');
      OutTextXY(153,100,'Rodney J. Sime');
      OutTextXY(153,125,'CALIFORNIA STATE UNIVERSITY');
      OutTextXY(153,150,'SACRAMENTO');
      SetTextJustify(LeftText,TopText);
   End;   (* Credits *)

Begin (* HeaderPage *)
   GrDriver:= Detect;
   InitGraph(GrDriver,GrMode,'');
   ErrCode:= GraphResult;
   SetGraphMode(3);
   DrawBorder;
   Credits;
   Readln;
   CloseGraph;
   TextMode(BW80)
End;   (* HeaderPage *)

(*************************************************************)

Procedure GetTitle;
Var CharNum:  Integer; (* number of characters in title *)
Begin
  CLRSCR;
  Repeat
    Writeln('Please enter a title for your graph -- ');
    Writeln('Use 35 characters or less.');
    Writeln('1--------10--------20--------30---35');
    Readln(Title);
    Writeln;
    Charnum:= Length(Title);
    If CharNum > 35 Then
      Begin
      Writeln;
      Writeln('Sorry - your title must have 35 characters or less.');
      Writeln
      End; (* If *)
  Until CharNum <=35;
End; (* GetTitle *)

(*************************************************************)
```

Figure 6–18 (continued)

```
PROCEDURE ReadFile(VAR X,Y:  Ary;
                    VAR N:  Integer);

   Var I:  Integer;
       F:  Text;

Begin (* ReadFile *)
   Assign(F,'LS.DAT');
   Reset(F);
   Readln(F,N);
   For I:= 1 to N Do
     Readln(F,X[I],Y[I]);
   Close(F)
End;  (* ReadFile *)


(*************************************************************)

Procedure ReadKeyBoard;
Var I,
    Sx,Sy:  Integer;     (* Screen Coordinates *)
Begin (* ReadKeyBoard *)
   CLRSCR;
   Write('What is the number of X and Y pairs you wish to enter?  ');
   Readln(N);
   Writeln;
   For I:= 1 to N Do
     Begin (* For *)
       Write(      'X(',I,'):  ');
       Sx:= WhereX; Sy:= WhereY;
       Read(X[I]);
       GoToXY(Sx+10,Sy);
       Write('      Y(',I,'):  ');
       Readln(Y[I]);
     End;  (* For *)
   CLRSCR;
End;  (* ReadKeyBoard *)

(*************************************************************)

Procedure LeastSquares(X,Y:  Ary;
                        Var Ycalc:  Ary;
                        Var b,m  :  Real;
                            N    :  Integer);
(* Fits a straight line to a set of N  X-Y pairs    *)
(* The line has a slope equal to m and intercept b *)

Var I:  Integer;
    SumX,SumY,SumXY,SumX2,SumY2,
    XI,YI,SXY,SXX,SYY:  Real;
```

Figure 6–18 (continued)

```
Begin (* LeastSquares *)
    SumX:= 0;
    SumY:= 0;
    SumXY:=0;
    SumX2:=0;
    SumY2:=0;
    For I:= 1 to N do
        Begin (* For *)
        XI:= X[I];
        YI:= Y[I];
        SumX:= SumX + XI;
        SumY:= SumY + YI;
        SumXY:= SumXY + XI*YI;
        SumX2:= SumX2 + XI*XI;
        SumY2:= SumY2 + YI*YI;
        End;   (* For *)
    SXX:= SumX2 - SumX*SumX/N;
    SXY:= SumXY - SumX*SumY/N;
    SYY:= SumY2 - SumY*SumY/N;
    m:= SXY/SXX;
    b:= ((SumX2*SumY - SumX*SumXY)/N)/SXX;
    CorrelCoef:= SXY/SQRT(SXX*SYY);
    Sigmam:= (SQRT(ABS((SumY2 - b*SumY -m*SumXY)/(N - 2))))/SQRT(SXX));
    Sigmab:= Sigmam*SQRT(SumX2/N);
    For I:= 1 to N do
        Ycalc[I]:= b + m*X[I];
End;   (* LeastSquares *)

(***************************************************************)

Procedure WriteData;
(* Writes out the answers on the screen *)

Var I: Integer;

Begin
    Write('  I                   X ');
    Write('                        Y');
    Writeln('                  Ycalc');
    Writeln;
    For I:= 1 to N do
        Writeln(I:3,BL,X[I],BL,Y[I],BL,Ycalc[I]);
    Writeln;
    Writeln('Intercept is ',b,sp,'Sigma is ',Sigmab);
    Writeln;
    Writeln('Slope is      ',m,sp,'Sigma is ',Sigmam);
    Writeln;
    Writeln(sp,sp,'Correlation Coefficient is ',CorrelCoef);
    Writeln;
End; (* WriteData *)

(***************************************************************)
```

Figure 6–18 (continued)

```
Procedure PrintData;
(* Sends calculated values to printer *)

Const sp5 = '     ';
      sp10 = '          ';
      sp15 = '               ';

Var I:  Integer;

Begin (* PrintData *)
   Writeln(Lst,'  I',sp15,'X',sp15,sp5,'Y',sp15,sp5,'Ycalc');
   Writeln(Lst);
   For I:= 1 to N do
      Writeln(Lst,I:3,sp5,X[I],sp5,Y[I],sp5,Ycalc[I]);
   Writeln(Lst);
   Writeln(Lst,sp5,sp15,'The intercept =   ',b);
   Writeln(Lst,sp5,sp15,'sigma =           ',Sigmab);
   Writeln(Lst);
   Writeln(Lst,sp5,sp15,'The slope =       ',m);
   Writeln(Lst,sp5,sp15,'sigma =           ',Sigmam);
   Writeln(Lst);
   Writeln(Lst,sp5,sp15,'The correlation coefficient = ',CorrelCoef);
End; (* PrintData *)

(**************************************************************)

Procedure Hardcopy;
Var Reply:  Char;
Begin
   CLRSCR;
   Write('Do you want a printed copy?  ');
   Writeln('If Yes, be sure your printer is turned on!');
   Writeln;Write('Please enter a Y or N.  ');
   Readln(Reply);
   If (Reply = 'Y') OR (Reply = 'y') Then
   Begin
      PrintData
   End; (* IF *)
end; (* Hardcopy *)

(**************************************************************)

Procedure ScaleScreen(Var Xscale,Yscale,Xleft,Ybottom:  Real);
VAR Ytop,Xright:  Real;

Begin (* ScaleScreen *)
  Write('Please enter the maximum Y value of the graph:  ');
  Read(Ytop);Writeln;
  Write('Please enter the minimum Y value of the graph:  ');
  Read(Ybottom);Writeln;
  Write('Please enter the maximum X value of the graph:  ');
  Read(Xright);Writeln;
  Write('Please enter the minimum X value of the graph:  ');
  Readln(Xleft);Writeln;
  Xscale:= (ScreenRight - ScreenLeft)/(Xright-Xleft);
  Yscale:= (ScreenBottom - ScreenTop)/(Ytop-Ybottom);
End; (* ScaleScreen *)
```

Figure 6–18 (continued)

```
(**********************************************************************)

Procedure DrawGrid;

Var Xinterval, Yinterval:  Integer;
    Xscale,Yscale,Xleft,Ybottom:   Real;

(*----------------------------------------------------------------*)

   Procedure DrawBorder;
   Begin (* DrawBorder *)
      Line(ScreenLeft,ScreenTop,ScreenRight,ScreenTop);
      Line(ScreenRight,ScreenTop,ScreenRight,ScreenBottom);
      Line(ScreenRight,ScreenBottom,ScreenLeft,ScreenBottom);
      Line(ScreenLeft,ScreenBottom,ScreenLeft,ScreenTop);
   End;

(*----------------------------------------------------------------*)

   Procedure WriteTitle(ScreenX,ScreenY:  Integer);
   Begin
      SetTextJustify(CenterText,CenterText);
      OutTextXY(ScreenX,ScreenY,Title);
      SetTextJustify(LeftText,TopText);
   End;   (* WriteTitle *)

(*----------------------------------------------------------------*)

   Procedure MarkX;
   Var I,J, ScreenX,ScreenY:  Integer;
   Begin (* MarkX *)
      ScreenY:= ScreenBottom;
      Xinterval:= (ScreenRight - ScreenLeft) DIV 10;
      Yinterval:= (ScreenBottom - ScreenTop) DIV 10;
      For I:= 1 to 9 do
         Begin  (* For *)
         ScreenX:=ScreenLeft;
         ScreenY:= ScreenY - Yinterval;
         Line(ScreenX,ScreenY,ScreenX + 2,ScreenY);
         ScreenX:= ScreenX - 2;
         For J:= 1 to 9 do
            Begin (* For J *)                   (* to bottom of screen  *)
            ScreenX:= ScreenLeft + J*Xinterval -1;
            Line(ScreenX,ScreenY,ScreenX + 2,ScreenY);
            ScreenX:= ScreenX - 1;
            End;
         ScreenX:= ScreenRight;
         Line(ScreenX,ScreenY,ScreenRight - 2,ScreenY);
         End; (* For *)
   End;   (* MarkX *)

(*----------------------------------------------------------------*)
```

Figure 6–18 (continued)

```
    Procedure MarkY;
    Var I,J,ScreenX,ScreenY:  Integer;
    Begin (* MarkY *)
       ScreenX:= ScreenLeft;
       Xinterval:= (ScreenRight - ScreenLeft) DIV 10;
       Yinterval:= (ScreenBottom - ScreenTop) DIV 10;
       For I:= 1 to 9 do
          Begin  (* For *)
          ScreenY:=ScreenTop;
          ScreenX:= ScreenX + Xinterval;
          Line(ScreenX,ScreenY,ScreenX,ScreenY + 2);
          ScreenY:= ScreenY + 2;
          For J:= 1 to 9 do
             Begin (* For *)                   (* to bottom of screen  *)
             ScreenY:= ScreenTop + J*Yinterval + 1;
             Line(ScreenX,ScreenY,ScreenX,ScreenY - 2);
             ScreenY:= ScreenY + 1;
             End;  (* For *)
          ScreenY:= ScreenBottom;
          Line(ScreenX,ScreenY,ScreenX,ScreenY - 2);
          End; (* For *)
    End;  (* MarkY *)

(*-------------------------------------------------------------*)

    Procedure DrawLine;
    VAR Xmax,Xmin,Ymax,Ymin,
        LeftY,RightY:  Real;
        I,ScreenX1,ScreenY1,ScreenX2,ScreenY2:  Integer;
    Begin (* DrawLine *)
       Xmin:= 1.0E37; Xmax:= 1.0E-37;
       For I:= 1 to N do
          Begin (* For *)
          If X[I] < Xmin Then Xmin:= X[I];
          If X[I] > Xmax Then Xmax:= X[I];
          End; (* For *)
       LeftY:= m*Xmin + b;
       ScreenX1:= ScreenLeft + Trunc(Xscale*(Xmin-Xleft));
       ScreenY1:= Screenbottom - Trunc(Yscale*(LeftY-Ybottom));
       RightY:= m*Xmax + b;
       ScreenX2:= ScreenLeft + Trunc(Xscale*(Xmax-Xleft));
       ScreenY2:= ScreenBottom - Trunc(Yscale*(RightY-Ybottom));
       Line(ScreenX1,ScreenY1,ScreenX2,ScreenY2);
    End; (* DrawLine *)

(*-------------------------------------------------------------*)
```

Figure 6–18 (continued)

```
  Procedure DrawChar(X,Y:  Integer);
  (* draws a square shaped char at Screen coordinates X,Y *)

  VAR I, X1,Y1,X2,Y2:  Integer;

  Begin (* DrawChar *)
     X1:= X - 2;
     Y1:= Y + 2;
      For I:= 1 to 5 do
        Begin (* For *)
        X2:= X1 + 5;
        Line(X1,Y1,X2,Y1);
        Y1:= Y1 + 1
        End; (* For *)
  End;  (* DrawChar *)

(*-------------------------------------------------------------*)

  Procedure Plotpoints;
  (* Plots a square char at the experimental  X[I],Y[I] *)

  VAR I,ScreenX,ScreenY:  Integer;

  Begin; (* PlotPoints *)
  For I:= 1 to N do
     Begin (* For *)
     ScreenX:= ScreenLeft + TRUNC(Xscale*(X[I] - Xleft));
     ScreenY:= Screenbottom  - TRUNC(Yscale*(Y[I] - Ybottom));

     ScreenX:= ScreenX -3;   (* This and next command centers *)
     ScreenY:= ScreenY -3;   (* The square char at X[I],Y[I]  *)
     DrawChar(ScreenX,ScreenY);
     End;  (* For *)
  End;  (* PlotPoints *)

(*-------------------------------------------------------------*)

  Begin (* DrawGrid *)
     ScaleScreen(Xscale,Yscale,Xleft,Ybottom);
     GrDriver:= Detect;
     InitGraph(GrDriver,GrMode,'');
     SetGraphMode(3);
     DrawBorder;
     WriteTitle(153,10);
     MarkX;
     MarkY;
     DrawLine;
     PlotPoints;
     Readln;
     CloseGraph;
     Textmode(BW80)
  End;  (* DrawGrid *)

(***********************************************************************)
```

Figure 6–18 (continued)

```
Begin  (* M A I N   P R O G R A M *)
   HeaderPage;
   Writeln;Write('Do you Want Information?  Enter a Y or N, please.  ');
   Readln(Answer);
   If (Answer = 'Y') OR (answer = 'y') then GiveInformation;
   Repeat
     CLRSCR;
     Writeln('Is your input from File LS.DAT or from the Keyboard?');
     Write('Enter an F or K, please.  ');
     Readln(Answer);
     If (Answer = 'F') OR (Answer = 'f') Then ReadFile (X,Y,N)
                                         Else ReadKeyBoard;
     GetTitle;
     LeastSquares(X,Y,Ycalc,b,m,N);
     Repeat
        WriteData;
        Writeln;Writeln('Press Return');
        Readln;
        Rescale:= False;
        DrawGrid;
        CLRSCR;
        Write('Do you want to Rescale?  Enter a Y or N, please.  ');
        Readln(Response); CLRSCR;
        If (Response = 'Y') OR (Response = 'y') Then Rescale:= True;
     Until Rescale = False;
     Hardcopy;
     Writeln;Write('Run again?  Enter Y or N, please.  ');
     Readln(Answer);
   Until (Answer = 'N') OR (Answer = 'n')
End.
```

Figure 6–18 (continued)

Thus,

$$M \times M^{-1} = M^{-1} \times M = I$$

For example, if

$$M = \begin{bmatrix} 1 & 4 & 3 \\ 1 & 3 & 3 \\ 1 & 3 & 4 \end{bmatrix}$$

then

$$M^{-1} = \begin{bmatrix} -3 & 7 & -3 \\ 1 & -1 & 0 \\ 0 & -1 & 1 \end{bmatrix}$$

and

$$\begin{bmatrix} 1 & 4 & 3 \\ 1 & 3 & 3 \\ 1 & 3 & 4 \end{bmatrix} \begin{bmatrix} -3 & 7 & -3 \\ 1 & -1 & 0 \\ 0 & -1 & 1 \end{bmatrix} = \begin{bmatrix} 1 & 0 & 0 \\ 0 & 1 & 0 \\ 0 & 0 & 1 \end{bmatrix}$$

Many BASIC compilers provide for matrix inversion, but Turbo Pascal does not. Matrix inversion is discussed by Mortimer (1981) and

FORTRAN subroutines are given by Wiberg (1965) and Dickson (1968). A Pascal program for inverting a matrix is given in Figure 6–17. The algorithm for the inversion is from Radford and Haigh (1986).

REFERENCES

1. Baird, D. C., *Experimentation: An Introduction to Measurement Theory and Experiment Design,* Prentice-Hall, Englewood Cliffs, New Jersey, 1962.

2. Barrante, J. R., *Applied Mathematics for Physical Chemistry,* Prentice-Hall, Englewood Cliffs, New Jersey, 1974.

3. Brinkworth, B. J., *An Introduction to Experimentation,* C. Tinling & Co. Ltd., Liverpool, 1968.

4. Cooper, D., and M. Clancy, *Oh! Pascal!,* Norton, New York, 1982.

5. De Levie, R., "When, Why and How to Use Weighted Least Squares," *J. Chem. Educ.* 63:10, 1986.

6. Dickson, T. R., *The Computer and Chemistry,* Freeman, San Francisco, 1968.

7. McCracken, D. D., and W. S. Dorn, *Numerical Methods and Fortran Programing,* Wiley, New York, 1964.

8. Miller, A. R., *Pascal Programs for Scientists and Engineers,* Sybex, Inc., Berkeley, California, 1981.

9. Mortimer, R. G., *Mathematics for Physical Chemistry,* Macmillan, New York, 1981.

10. Noggle, J. H., *Physical Chemistry on a Microcomputer,* Little, Brown, Boston, 1985.

11. Radford, L. E., and R. W. Haigh, *Turbo Pascal for the IBM-PC,* PWS Computer Science, Boston, 1986.

12. Sedgewick, R., *Algorithms,* Addison-Wesley, Reading, Massachusetts, 1983.

13. Weatherall, M., *Scientific Method,* English Universities Press, London, 1968.

14. Wiberg, K. B., *Computer Programming for Chemists,* Benjamin, New York, 1965.

15. Wilson, E. B., Jr., *An Introduction to Scientific Research,* McGraw-Hill, New York, 1952.

16. Young, H. D., *Statistical Treatment of Experimental Data,* McGraw-Hill, New York, 1962.

CHAPTER 7

Analyzing the Errors

Errors are generally classified as systematic or random errors. **Systematic errors** arise from the design and execution of experiments. Consequently, they are sometimes called determinable errors or corrigible errors, since one should be able (hopefully) to identify them through a careful analysis of the experiment and associated instruments, and to take measures to correct them. Because systematic errors occur with the same magnitude and sign every time the experiment is executed, they affect the accuracy of the results but not the precision. This insidious trap may catch the unwary experimenter, since the results may appear excellent (precise, reproducible), but in fact may differ greatly from the correct value (be inaccurate). We examined the nature of systematic errors when we discussed the design and execution of experiments in Chapters 4 and 5. If an experiment has small systematic errors, it is **accurate.**

Random errors occur with a different sign and magnitude each time an experiment is executed. Random errors arise from a large number of indeterminate causes, such as unpredictable mechanical and electrical fluctuations affecting the operation of instruments and experimental apparatus. In addition, even when prejudice is eliminated, random human errors arise from unpredictable psychological and physiological limitations inherent in human experimenters. With experience, the magnitude of random errors arising from human limitations may decrease. If an experiment has small random errors, it is **precise.**

When a physical chemist seeks a relationship between a property and what is observed, it is necessary to report not only the quantitative relationship and the value of a calculated property, but also some measure of the reliability of the measurements and subsequent calculations. Thus the errors (du) must also be reported along with the calculated quantity (u).

THE STATISTICAL TREATMENT OF ERRORS

It is a matter of experience that whenever a measurement is performed repeatedly, the observed values are not identical but tend to cluster about a mean value. Based on such experiences, we often design experiments with the expectation that we will repeat measurements to test the reproducibility of the experiment and consequently gain confidence in the accuracy of the results. It is useful to examine how errors distribute themselves and how the distribution itself depends on the sample size, that is, on the number of measurements or observations.

Small Samples

Suppose you make a single measurement of a single quantity and the result is 475. With just a single measurement you cannot make an *objective* statement about the quality of that single measurement. Based on your intuition and previous experience you might make a subjective statement about the accuracy of your measurement. It is not possible, however, to say anything about the precision, since precision is a measure of the reproducibility and is meaningless for a single measurement.

Duplicate measurements permit a statement about the reproducibility of the measurements. Suppose the results of duplicate measurements are 475 and 481. It is now possible to state the range of the measurements, which equals the highest minus the lowest, or $481 - 475 = 6$ in this example.

The range is a measure of the precision of the observations or measurements. Given a choice of two values, intuition tells us to accept the average of the two, since either value is equally likely to be correct:

$$(477 + 481)/2 = 479$$

At this point, you have more confidence in your measurements than you had when only one measurement was completed. It would be desirable if you could assess the degree of confidence you are entitled to on the basis of your measurements. Fortunately, statisticians have provided various ways of assessing the quality of multiple measurements and defining several measures of dispersion.

Large Samples

Buoyed up by the confidence given by two sample measurements, suppose our investigator makes 198 more measurements for a total of 200 measurements! This is not likely in a laboratory where physical and chemical measurements are made; more likely, 1 or 2, or maybe as many as 5 or 6, duplicate sample measurements would be made. Nevertheless, let us examine the typical outcome of a set of 200 sample measurements, listed

TABLE 7–1

Sample Frequencies

Frequency (f_i)	Sample Value (x_i)
2	486
4	485
7	484
9	483
13	482
19	481
25	480
29	479
27	478
23	477
14	476
15	475
6	474
4	473
3	472

in Table 7–1. Notice that a sample value of 486 was obtained twice, a sample value of 485 was obtained four times, and that the sample frequency increases and then drops off again at high sample values. A graphical representation of these data, shown in Figure 7–1, shows quite dramatically that the sample values cluster mostly about a middle value and that both very high and very low sample values occur with low frequency.

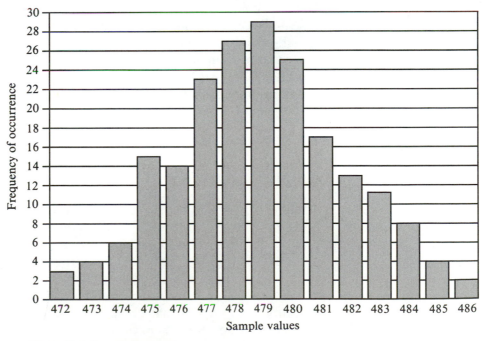

Figure 7–1 Sample frequencies.

The middle value is, of course, the average or sample mean calculated from

$$m = \frac{x_1 + x_2 + x_3 + \cdots x_n}{n} = \frac{\sum x_i}{n} \tag{7-1}$$

where m is the sample mean, x_i are the sample values, and n is the total number of samples. To calculate the sample mean of the 200 values listed in Table 7–1, it is convenient to use the frequencies f_i so that Equation 7–1 becomes

$$m = \frac{f_1 x_1 + f_2 x_2 + f_3 x_3 + \cdots + f_n x_n}{f_1 + f_2 + f_3 + \cdots + f_n} = \frac{\sum f_i x_i}{\sum f_i} \tag{7-2}$$

and $m = 479$ for the samples listed in Table 7–1.

PROBABILITIES

Examination of Figure 7–1 shows that very large and very small sample values are infrequent. Our intuition tells us that very large and very small sample values are improbable. The most probable (most frequent) sample values are clustered about the sample mean m, which equals 479.

In the absence of systematic errors, approximately equal numbers of errors occur on the high and low side. Small errors occur more frequently than large errors. In other words, small errors are more *probable* than large errors. Our intuition tells us that there is a connection between the probability of occurrence of an error and the frequency of occurrence of an error.

To review some ideas about probability, consider a jar filled with a well-mixed collection of marbles, of which 13 are white and 37 are red. A total of 50 marbles are in the jar. The probability of drawing out a white marble is $13/50 = 0.26$. This calculation is based on the following definition of probability:

$$P(A) = \frac{h}{n}$$

where h is the number of favorable outcomes of event A, n is the total number of possible outcomes, and $P(A)$ is the probability that event A will occur. It is from this concept of probability that we calculate that the probability of getting a "tail" when we flip a two-sided coin is $1/2$ (0.5). In both of these examples we calculate the probability without actually doing an experiment or making an observation.

An equivalent way of defining probability is in terms of relative frequency. To make our description more exact, let us define the probability of an event A as the relative frequency of occurrence of the event when the number of samples is very large:

TABLE 7–2

Sample Probabilities

Probability	Sample Value
0.01	486
0.02	485
0.035	484
0.045	483
0.065	482
0.095	481
0.125	480
0.145	479
0.135	478
0.115	477
0.07	476
0.075	475
0.03	474
0.02	473
0.015	472

$$P(\mathrm{A}) = \frac{f_i}{\sum f_i} \qquad (7\text{–}3)$$

With this definition, we can calculate the probabilities for the occurrence of the sample values listed in Table 7–1. For the first entry in Table 7–1, 200 measurements of 200 samples resulted in two sample values of

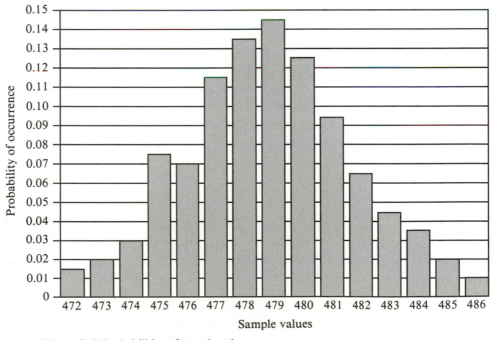

Figure 7–2 Probabilities of sample values.

486. Thus, $P = 2/200 = 0.01$. The remaining probabilities are listed in Table 7–2, and a graph is shown in Figure 7–2.

The shapes of the bar graphs in Figures 7–1 and 7–2 are identical. The bar heights in Figure 7–1 give the frequency of occurrence of sample values, and the bar heights in Figure 7–2 give the probability of occurrence of a sample value. It is readily seen that the measured values cluster about a mean value, and that values far from the mean occur with low frequency (probability).

FINITE ERROR DISTRIBUTION

An error in the measurement of a sample is the difference between the sample value and the true value, so that

$$\epsilon_i = x_i - \text{true value} \tag{7-4}$$

where ϵ_i is the error in the ith sample measurement and x_i is the ith observed sample value.

What, however, is the true value of a physical measurement? The true value is never known; consequently, the error of a measurement cannot be known. If all the errors in a set of sample measurements are random, then the sample mean is taken as the best estimate of the true value. If the errors are random and an infinite number of samples are measured, then the true value does indeed equal the mean of the infinite set of samples in the absence of systematic errors. We shall distinguish between the real finite sample mean m and the mean of an infinitely large sample set:

$$m = \frac{\sum x_i}{n} = \text{mean of a finite sample set} \tag{7-5}$$

$$\mu = \text{mean of an infinitely large sample set} \tag{7-6}$$

As the sample set becomes increasingly large, m becomes a better and better estimate of μ, the true value:

$$\lim_{n \to \infty} m = \mu \tag{7-7}$$

For the moment, let us assume that the set of 200 samples described above is sufficiently large that we can take the mean m, which equals 479, to be the true value. That being the case, we can calculate the errors and the probability of their occurrence as listed in Table 7–3 and shown in Figure 7–3.

Figure 7–3 shows that the errors range from -7 to $+7$. The large errors have low probabilities and measurements with small errors have higher probabilities. The bar graphs in Figures 7–1, 7–2, and 7–3 have identical shapes. From Figure 7–3 we see, for example, that the probability of occurrence of an error of magnitude $+3$ is 0.07 and the probability of

TABLE 7–3

Sample Error Probability Distribution

Probability	Error
0.01	−7
0.02	−6
0.035	−5
0.045	−4
0.065	−3
0.095	−2
0.125	−1
0.145	0
0.135	1
0.115	2
0.07	3
0.075	4
0.03	5
0.02	6
0.015	7

an error of −6 is 0.02. The probability that the error lies between −7 and +7 is unity (1), that is, certainty.

The area of each bar represents the probability of occurrence of an error. The probability equals the height P, times the width $d\epsilon$, which equals 1 for all bars. The sum of the areas of all the bars in Figure 7–3 should equal exactly one, that is,

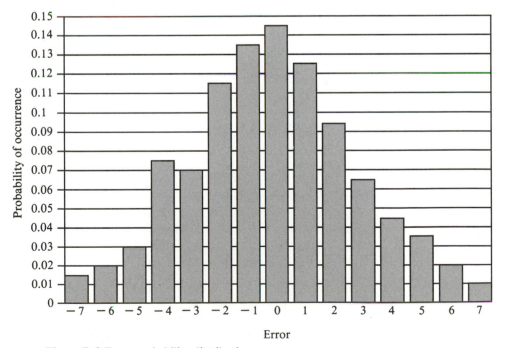

Figure 7–3 Error probability distribution.

$$\sum P(\epsilon_i) \Delta\epsilon_i = 1 \qquad (7\text{--}8)$$

This is true of the values listed in Table 7–3 since $0.01 \times 1 + 0.02 \times 1 + 0.035 \times 1 + 0.065 \times 1 + \cdots + 0.015 \times 1 = 1$. The probability of distribution is said to be **normalized** since Equation 7–8 is satisfied.

It appears that one can draw a smooth curve through the tops of the bar graphs. Such a curve could then be represented by a continuous mathematical function instead of a bar graph.

The Infinitely Large Sample Set

When a smooth curve is drawn through the tops of the bars in Figure 7–3, we assume that, as the number of observations becomes infinite, the curve can be represented by the **Gaussian** or **normal error distribution:**

$$P(\epsilon) = \frac{1}{\sigma\sqrt{2\pi}}\, e^{-\epsilon^2/2\sigma^2} \qquad (7\text{--}9)$$

where $P(\epsilon)$ is the error probability function, σ is the standard deviation, ϵ is the error, $\pi = 3.14159$, and $e = 2.71828$. Since $P(\epsilon)$ is normalized, as in Equation 7–8,

$$P = \int_{-\infty}^{+\infty} P(\epsilon)\, d\epsilon = 1$$

where P is the probability that the error ϵ lies between $-\infty$ and $+\infty$. For any x, the error is the difference between the quantity measured x and the mean of an infinite number of measurements μ:

$$P(\epsilon) = \frac{1}{\sigma\sqrt{2\pi}}\, e^{-(x-\mu)^2/2\sigma^2} \qquad (7\text{--}10)$$

For a very large (approaching infinite) number of measurements,

$$\sigma = \sqrt{\frac{\sum (x_i - \mu)^2}{n}} = \sqrt{\frac{\sum \epsilon_i^2}{n}} \qquad (7\text{--}11)$$

where x_i is the value of the ith sample measurement, μ is the mean of the infinite sample set, and n is the number of observations.

Significance of the Standard Deviation. The area under the curve between a range of errors gives the probability that the error falls within this range. The area under the curve between $-\infty$ and $+\infty$ equals 1.00; that is, it is certain that the error must lie between $-\infty$ and $+\infty$. The probability P that the error falls between a and b is shown as the shaded area in Figure 7–4:

$$P = \int_{b}^{a} P(\epsilon)\, d\epsilon = \text{shaded area} \qquad (7\text{--}12)$$

The probability P that the error lies within $\pm\sigma$ (one standard deviation) is shown in Figure 7–5:

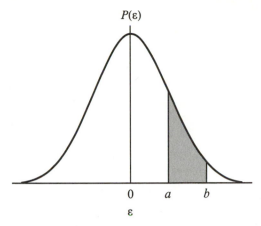

$P(\varepsilon)$

0   a   b

ε

Figure 7–4 The probability of an error as an area.

$$P = \int_{-\sigma}^{+\sigma} P(\epsilon)\, d\epsilon = 0.683 \qquad (7\text{–}13)$$

This means that there is a 68% chance that the error lies between $+\sigma$ and $-\sigma$.

What is the probability P that the error in a series of measurements lies between $+2\sigma$ and -2σ? The probability (area) is shown in Figure 7–6.

$$P = \int_{-2\sigma}^{+2\sigma} P(\epsilon)\, d\epsilon = 0.954 \qquad (7\text{–}14)$$

Similarly (but not shown), the probability P that an error lies between -3σ and $+3\sigma$ is 0.997. In other words, the probability that a measurement will deviate from the average by more than three standard deviations is only 0.003, or 0.3%.

Confidence Levels and Confidence Factors. Let us rephrase the meaning of a standard deviation. From the above discussion it follows that we can be 68.26% confident that the error lies between the true value plus one standard deviation and the true value minus one standard deviation. The **confidence level** in this example is 68.26% and the **confidence**

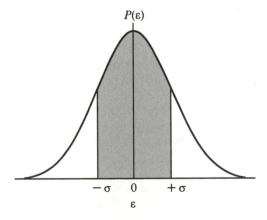

$P(\varepsilon)$

$-\sigma$ 0 $+\sigma$

ε

Figure 7–5 The probability of an error between $-\sigma$ and $+\sigma$.

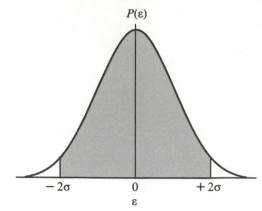

Figure 7–6 Probability of an error between -2σ and $+2\sigma$.

factor is one, that is, one standard deviation. If we choose a larger confidence limit, for example, two (standard deviations), we can be more confident that the true value lies within our reported confidence limit. The confidence levels for integral confidence factors are listed in Table 7–4. It is more usual to describe errors in terms of integral confidence levels, especially the 95% confidence level. These values are shown in Table 7–5.

These assignments of errors are valid when the number of observations n is infinite, the only errors are random, and the error distribution is Gaussian (normal). In principle, then, none of the confidence levels and confidence factors can be calculated, since σ is not known except for an infinitely large sample set. In practice, the situation is not so dismal because it is possible to calculate *approximate* values for the standard deviation σ of the finite sample set.

Limitations of the Infinite Sample Set. Obviously we can never carry out an infinite number of measurements. Consequently we can *never* know μ or σ exactly. In actual practice, we almost always carry out only a few measurements, probably less than 6 or 8. It would be unusual to carry out 200 measurements, the number used in our earlier example of a large sample. Although we would like to know μ, we can only estimate its value. The best estimate of μ is $\bar{x}$, which we can calculate for a finite sample:

$$\bar{x} = \frac{\sum x_i}{n} \tag{7–15}$$

where n is the number of samples or measurements, x_i are the individual sample measurements, and $\bar{x}$ is the sample mean.

TABLE 7–4

Integral Confidence Factors

Confidence factor	1σ	2σ	3σ	4σ
Confidence level	68.268	95.450	99.730	99.994

TABLE 7–5

Integral Confidence Levels

Confidence factor	0.674σ	1.28σ	1.96σ	2.58σ	3.29σ
Confidence level	50	80	95.00	99.00	99.90

Estimating the Standard Deviation of the Sample of Finite Size

For a real sample of finite size, a number of measures of precision seem possible. The range has already been mentioned. The average deviation has been used,

$$\bar{d} = \frac{\sum |d_i|}{n} = \frac{\sum |x_i - \bar{x}|}{n} \tag{7–16}$$

where d_i is the deviation from average, x_i is the ith sample value, n is the sample size, and the sum is of the absolute values.

A more suitable estimate of the standard deviation of the infinite sample set is the square root of the variance. The variance is s^2, and the sample standard deviation s is the square root of the variance. Thus,

$$s = \sqrt{\frac{\sum d_i^2}{n-1}} = \sqrt{\frac{\sum (x_i - \bar{x})^2}{n-1}} \tag{7–17}$$

It can be shown that Equation 7–17 may be written

$$s = \sqrt{\frac{\sum x_i^2 - n(\bar{x})^2}{n-1}} \tag{7–18}$$

a form that is sometimes more convenient for calculations.

If n is greater than 20 or 30, then s is a valid estimate of σ. Note that s is the **sample standard deviation** and σ is called simply the standard deviation, or more exactly the **standard deviation of the infinite sample set**.

Interpretation of s. For a large but finite sample, the probability that a *single* measurement will fall between $\bar{x} + s$ and $\bar{x} - s$ is 68%. This is analogous to the statement that for the infinite sample set, the probability that a single measurement will fall between $\mu + \sigma$ and $\mu - \sigma$ is 68%. The corresponding probabilities for $2s$, $3s$, and $4s$ are the same as the probabilities listed in Table 7–4 for 2σ, 3σ, and 4σ. Thus, to the extent that s is a valid approximation of σ, the confidence limits listed in Table 7–5 may be used.

The Standard Deviation of the Mean. In the previous paragraph we have interpreted s in terms of probabilities associated with a single measurement. Now suppose our sample size is finite but large, say, between 20 and 200. It is an important concept of statistics that every finite set of measurements may be considered a sample from an infinite set of similar data. Up to this point we have discussed ways to estimate σ and μ, parameters of the infinite set. Now let us see how to handle large but finite sample sets.

Suppose we measure several finite sample sets of, say, 20 samples each; a typical set is listed in Table 7–6. It turns out that the distribution of the $\bar{x}$'s of these sets follows a normal distribution with the sample standard deviation s_m, called the sample standard deviation of the mean. It can be shown (Baird, 1962, p. 64; Young, 1962, p. 95) that

$$s_m = \frac{s}{\sqrt{n}} \tag{7–19}$$

where n is the number of sample sets, s is the sample standard deviation as calculated with Equation 7–17 or 7–18, and s_m is the **standard deviation of the mean.**

It makes sense that the standard deviation of the mean is less than the sample standard deviation. This is consistent with our intuitive idea that the more measurements we make the more confidence we can have in the results. Notice that the number of samples enters as the square root. Thus, if you make 4 measurements and wish to decrease s_m by one half, you would be required to make 16 measurements. Thus, a couple of extra runs will have little effect on the experimental precision (as measured by s_m). Your time would be more advantageously spent improving the design of the experiment.

Now let us compare the meaning of s with the meaning of s_m. If we were to make *one* additional measurement to the set listed in Table 7–

TABLE 7–6

Sample of Large Data Set

i	x_i	i	x_i
1	19.9	11	20.0
2	20.1	12	19.2
3	20.2	13	19.6
4	20.3	14	19.5
5	19.8	15	19.6
6	20.3	16	20.3
7	20.7	17	20.3
8	19.5	18	19.8
9	19.9	19	19.9
10	20.0	20	19.4

$$\bar{x} = 19.92 \qquad s = 0.3750$$

$$s_m = \frac{s}{\sqrt{n}} = \frac{0.375}{\sqrt{20}} = 0.084$$

6, there is a 68% probability that the sample value would lie between $19.92 + 0.37$ and $19.92 - 0.37$ ($\bar{x} \pm s$).

Now consider the 20-sample set. There is a 68% probability that the true value x lies between $19.92 + 0.08$ and $19.92 - 0.08$ ($\bar{x} \pm s_m$). It is to be expected that the uncertainty calculated for a single measurement is greater than the uncertainty calculated for the entire set of 20 samples.

Confidence Limits

Instead of reporting s_m as a measure of precision, the value of s_m times a factor required to give a desired confidence level can be reported. The value of s_m times a factor is called the **confidence limit** and is given the symbol λ. Thus, in the previous example, we reported $\bar{x} \pm s_m$. If we wish to report $\bar{x} \pm \lambda_{99}$, we calculate λ_{99} using the confidence factors listed in Table 7–7:

$$\lambda_{99} = F_c \times s_m = 2.58 \times 0.084 = 0.22$$

and we report $\bar{x} = 19.92 \pm 0.22$, where 0.22 is the 99% confidence limit.

It is most usual to report the 95% confidence limit:

$$\lambda_{95} = F_{95} \times s_m$$

$$= 1.96 \times 0.084 = 0.16$$

We would report $\bar{x} = 19.92 \pm 0.16$ where 0.16 is the 95% confidence limit. This might also be written as $\bar{x} = 19.92(16)$, where (16) refers to the uncertainty in the last digits. One could argue that the presence of two significant figures in the confidence limit is meaningless, so what really should be reported is 19.9(2); however, it is customary to retain one extra significant figure.

In any case, it is important to *explain* what measure of precision is being reported. If it is the sample standard deviation, s_m, then that should be stated. If it is the 95% confidence limit, that fact must be explicitly stated by the writer/experimenter.

TABLE 7–7

Confidence Factors for Large (>20) Samples

F_c	Confidence Level (%)
1.00	68.268
1.28	80.000
1.96	95.000
2.00	95.450
2.58	99.000
3.00	99.730
3.29	99.900
4.00	99.994

SMALL SAMPLES

The previous discussions have centered around the infinitely large sample set ($n = \infty$) and large sample sets ($n > 20$). Of greatest interest to the experimental scientist are small sample sets ($1 < n < 20$). Most of our work probably involves n equal to three or four.

For small sample sets we can calculate the average x exactly as for large sample sets with Equation 7–1. The sample standard deviation, s, and the standard deviation of the mean, s_m, are also calculated with the same equations that were used to calculate these quantities for large sample sets: Equations 7–17 and 7–19.

For small samples, however, the confidence factors discussed in the previous discussion depend on the sample size itself. The confidence factors listed in Tables 7–4 and 7–5 are valid and essentially constant for any sample size larger than about 20. Below $n = 20$, however, the confidence factors depend not only on the confidence interval, but also on n, the sample size.

Another way of looking at the problem is in terms of the distribution itself. The distribution for small samples is low and broad compared to a normal or Gaussian distribution, as illustrated in Figure 7–7.

Because the t distribution is broader, the t factor is somewhat greater than the corresponding confidence factor for a given confidence limit. The solution to this problem was first solved by Gosset writing under the pseudonym "Student," hence the name "Student's t distribution." In Table 7–8 are listed t factors for various sample sizes and confidence limits. Let us illustrate the calculation of confidence limits with Student's t factors for a typical small sample set ($n = 5$), as shown in Table 7–9.

From Table 7–8, for $n - 1 = 4$ and 95% confidence level, $t_{95} = 2.776$. The confidence limit $\lambda_{95} = t_{95}s_m = 2.776 \times 0.038 = 0.106$. For

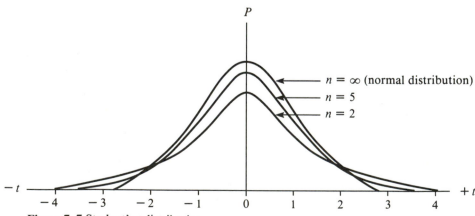

Figure 7–7 Student's t distribution.

TABLE 7–8

Critical Values of t

n − 1	Confidence Limits (%)						
	50	60	70	80	90	95	99
1	1.000	1.376	1.963	3.078	6.314	12.706	63.657
2	0.816	1.061	1.386	1.886	2.920	4.303	9.925
3	0.765	0.978	1.250	1.638	2.353	3.182	5.841
4	0.741	0.941	1.190	1.533	2.132	2.776	4.604
5	0.727	0.920	1.156	1.476	2.015	2.571	4.032
6	0.718	0.906	1.134	1.440	1.943	2.447	3.707
7	0.711	0.896	1.119	1.415	1.895	2.365	3.499
8	0.706	0.889	1.108	1.397	1.860	2.306	3.355
9	0.703	0.883	1.100	1.383	1.833	2.262	3.250
10	0.700	0.879	1.093	1.372	1.812	2.228	3.169
11	0.697	0.876	1.088	1.363	1.796	2.201	3.106
12	0.695	0.873	1.083	1.356	1.782	2.179	3.055
13	0.694	0.870	1.079	1.350	1.771	2.160	3.012
14	0.692	0.868	1.076	1.345	1.761	2.145	2.977
15	0.691	0.866	1.074	1.341	1.753	2.131	2.947
16	0.690	0.865	1.071	1.337	1.746	2.120	2.921
17	0.689	0.863	1.069	1.333	1.740	2.110	2.898
18	0.688	0.862	1.067	1.330	1.734	2.101	2.878
19	0.688	0.861	1.066	1.328	1.729	2.093	2.861
20	0.687	0.860	1.064	1.325	1.725	2.086	2.845
21	0.686	0.859	1.063	1.323	1.721	2.080	2.831
22	0.686	0.858	1.061	1.321	1.717	2.074	2.819
23	0.685	0.858	1.060	1.319	1.714	2.069	2.807
24	0.685	0.857	1.059	1.318	1.711	2.064	2.797
25	0.684	0.856	1.058	1.316	1.708	2.060	2.787
26	0.684	0.856	1.058	1.315	1.706	2.056	2.779
27	0.684	0.855	1.057	1.314	1.703	2.052	2.771
28	0.683	0.855	1.056	1.313	1.701	2.048	2.763
29	0.683	0.854	1.055	1.311	1.699	2.045	2.756
30	0.683	0.854	1.055	1.310	1.697	2.042	2.750
40	0.681	0.851	1.050	1.303	1.684	2.021	2.704
50	0.680	0.849	1.048	1.299	1.676	2.008	2.678
60	0.679	0.848	1.046	1.296	1.671	2.000	2.660
120	0.677	0.845	1.041	1.289	1.658	1.980	2.617
∞	0.674	0.842	1.036	1.282	1.645	1.960	2.576

Source: Fisher, R. A., and F. Yates, *Statistical Tables for Biological, Agricultural and Medical Research,* Hafner, New York, 1963. Used with permission.

our sample of 5 then we could report that $\bar{x} = 3.37 \pm 0.11$ with 95% confidence. Notice that the bottom line of Table 7–8 corresponds to Table 7–5, valid for a large sample. Thus Student's t factors equal confidence factors for $n = \infty$. Before Gosset, only the bottom line of Table 7–8 was available.

TABLE 7-9

Small Sample Data Set

i	x_i	d_i
1	3.47	0.10
2	3.29	0.08
3	3.46	0.11
4	3.34	0.03
5	3.31	0.06

$$\bar{x} = \frac{\sum x_i}{n} = \frac{3.47 + 3.29 + 3.46 + 3.54 + 3.31}{5} = 3.374$$

$$s = 0.085 \qquad s_m = 0.038$$

MAXIMUM ERROR

In general an experimental physical chemist searches for a simple mathematical dependence of a numerical property of a system on one or more experimental parameters. If the numerical property is symbolized by u and the experimental parameters by $x, y, z, \ldots$, then the dependence of u on $x, y,$ and z may be written as

$$u = f(x, y, z, \ldots) \tag{7-20}$$

Usually a physical chemist has or is seeking a mathematical equation relating the dependent variable (u) to the independent variables ($x, y,$ and z). The independent variables are often the experimental parameters or experimentally observed and measured quantities. The way in which small changes in the independent variables cause changes in the dependent variable is analogous to the way in which small errors in experimental parameters cause errors in a calculated result. For this reason, the total differential of the function may be used to calculate the propagated error—that is, the error in a calculated value caused by errors in experimental parameters. The total differential is given by

$$du = \left(\frac{\partial u}{\partial x}\right)_{y,z} dx + \left(\frac{\partial u}{\partial y}\right)_{x,z} dy + \left(\frac{\partial u}{\partial z}\right)_{x,y} dz \tag{7-21}$$

Mathematically, the total differential gives the small change du in u caused by small changes dx in x, dy in y, and dz in z. The first term on the right gives the partial change in u from the product of the rate of change of u with respect to x $(\partial u/\partial x)_{y,z}$ and a small change dx in x. The sum of the three terms on the right gives the total change, that is, the total differential in the function.

The total differential defined by Equation 7-21 is useful in determining the **maximum error** in the quantity u (the dependent variable) caused by errors in the experimental parameters (independent variables) dx, dy, and dz. In summary, x, y, and z are measured experimentally, and from them u is calculated. Then if dx, dy, and dz are the maximum errors in experimentally measured quantities, the maximum error in u, du, can be calculated.

An Example Error Calculation

To illustrate some of these ideas, consider the calculation of the volume of a small right cylinder from the experimental measurement of the height and diameter. The design of this simple experiment consists of selecting a measuring instrument, say, a ruler with a millimeter scale. Suppose that, upon a single execution of the experiment, the length of the cylinder is 2.9 mm and the radius 1.6 mm with an uncertainty of 0.1 mm. The volume of the cylinder may be calculated:

$$V = \pi r^2 h = 3.14 \times 1.6^2 \times 2.9 = 23.3 \text{ mm}^3 \qquad \text{(7–22)}$$

The error in the volume V may be calculated by determining the total differential of Equation 7–22. Errors in the measurement of the radius (dr) and height (dh) cause an error (dV) in V, the calculated value. The total differential of the function (Eq. 7–22) may be written to calculate the propagated error. According to Equation 7–21, the total differential of Equation 7–22 is

$$dV = \left(\frac{\partial V}{\partial r}\right)_h dr + \left(\frac{\partial V}{\partial h}\right)_r dh = 2\pi rh \, dr + \pi r^2 \, dh \qquad \text{(7–23)}$$

$$= 2 \times 3.14 \times 1.6 \times 2.9 \times 0.1 + 3.14 \times 1.6^2 \times 0.1$$

$$= 2.9 + 0.8 = 3.7 \text{ mm}^3$$

Reporting Errors. The value reported for the volume is 23.3 ± 3.7 mm^3. More frequently you will see this written 23.3(3.7), which is much easier to type. In this example the value is written with one more significant figure than the error calculation warrants. It might be more logical to round 23.3 down to 23 and 3.7 up to 4 and write the calculated volume as 23(4). It is usual, however, to carry one more significant figure than the error calculation logically justifies. Open a copy of one of the international scientific journals at random to a table of calculated data, and you will probably see one extra significant figure.

In tables of data, the errors for the tabulated data should be shown. Often, the errors for all the entries in a given column are the same. In this case, indicate the error in the first entry only; it will be understood that the same error follows for the entries below. Be sure that the calculated values and their errors are rounded off consistently. It would be senseless to report $V = 23.3(3.724)$ or even 23(3.7). Whenever you present any numerical result, you should carry the proper number of significant figures. To do this, you may just have to carry out a calculation of the propagated error as in the above example.

In any case, it is the responsibility of the researcher and writer to calculate the uncertainty in important results and to communicate this information to the reader. Furthermore, the nature of the error should be stated; that is, whether the error is the maximum error (as it is in the example we have just calculated) or is the probable error, standard deviation, or confidence interval. The magnitude of the errors in the experimental parameters (dr and dh in this example) should also be reported, probably in the experimental section. In journal articles the details of the calculations are not ordinarily presented nor are the details of the error calculations shown; however, in a physical chemistry report the sample calculations of both the results and their associated errors should be presented in detail.

Errors and the Design of Experiments. The calculation of errors can play an important role in the design and refinement of experiments. In the example with the cylinder just shown, the relative error expressed as a percentage is

$$\frac{3.7 \times 100}{23.3} = 16\%$$

Chances are that an error of such magnitude is not satisfactory. The error calculation tells us that the experiment was poorly designed. In this example, redesigning the experiment is simple. The choice of a micrometer caliper instead of a millimeter ruler would certainly improve the accuracy.

To take an example more closely related to physical chemistry, suppose that the pressure of a sample of nitrogen gas contained in a glass bulb of a known volume is to be calculated by weighing the gas sample at a given temperature. Some typical data are listed below:

$g = 0.0524(3)$ g $\qquad\qquad\qquad M = 28.0134$ g/mole

$T = 301.0(5)$ K $\qquad\qquad\qquad V = 0.027(2)$

$R = 0.08205$ liter atm mole^{-1} K^{-1}

From these data the pressure may be calculated with the Ideal Gas Law:

$$P = \frac{gRT}{MV} = \frac{0.0524 \times 0.08205 \times 301.0}{28.0134 \times 0.027} = 1.711 \text{ atm} \qquad (7\text{-}24)$$

To calculate the maximum error in P, the total differential of Equation 7–24 is written according to Equation 7–21, giving

$$dP = \left(\frac{\partial P}{\partial g}\right)_{T,V} dg + \left(\frac{\partial P}{\partial T}\right)_{g,V} dT + \left(\frac{\partial P}{\partial V}\right)_{g,T} dV$$

Notice that the contributions to the error in P caused by errors in M and R are considered to be negligible. The maximum error can now be calculated:

$$dP = \left(\frac{RT}{MV}\right) dg + \left(\frac{Rg}{MV}\right) dT + \left(\frac{gRT}{MV^2}\right) dV \tag{7-25}$$

$$= \frac{0.082}{28} \left(\frac{301 \times 0.0003}{0.027} + \frac{0.052 \times 0.5}{0.027} + \frac{0.052 \times 301 \times 0.002}{0.027^2}\right)$$

$$= 0.003(3 + 1 + 43) = 0.14 \text{ atm}$$

The pressure should be reported as $P = 1.71(14)$. It is immediately clear that the contribution to the total error from the error in the measurement of volume (dV) is far greater than the contribution from the errors in the measurement of temperature (dT) or in the measurement of mass (dg). The errors in the measurement of mass and temperature are conquerable, but the error in volume is comparatively too large. This kind of calculation can assist the experimentalist in improving the design of the experiment. In this case, an effort should be made to improve the measurement of volume. In a well-designed experiment, the contributions to the total error from the errors in each independent variable should be roughly equal.

Relative Errors

The errors calculated so far have been labeled the maximum error; they are also called the absolute maximum error to distinguish them from relative errors and percent errors. If Equation 7–25 is divided by Equation 7–24, the result is the relative or fractional error in P:

$$\frac{dP}{P} = \frac{dg}{g} + \frac{dT}{T} + \frac{dV}{V} \tag{7-26}$$

The fractional error in P is just the sum of the fractional errors in g, T, and V. If both sides of Equation 7–26 are multiplied by 100, then we see that the percent error in P is equal to the sum of the percent errors in g, T, and V. Thus, the fractional errors in the dependent variable are very simply related to the fractional errors in the dependent (experimental) variables.

Similarly, if Equation 7–23 is divided by Equation 7–22 to obtain the fractional error in the volume, the result is

$$\frac{dV}{V} = 2\frac{dr}{r} \frac{dh}{h} \tag{7-27}$$

Notice the factor of 2 preceding the fractional error in the radius. This arises from the r^2 in Equation 7–22. In general, for each independent variable of the form x^n, the contribution to the error in the dependent variable is n times the fractional error in x, that is, $n(dx/x)$. Suppose that we calculate a quantity u by measuring x, y, and z, and that the quantitative relationship between our independent variable u and the experimental parameters has the form

$$u = \frac{x^4 y}{z^2} \qquad (7\text{-}28)$$

Application of Equation 7–21 to u gives du, the total differential of u. The quantity du is the maximum absolute error in u:

$$du = \frac{4x^3 y}{z^2}\, dx + \frac{x^4}{z^2}\, dy + \frac{2x^4 y}{z^3}\, dz \qquad (7\text{-}29)$$

Division of Equation 7–28 by Equation 7–29 gives the maximum relative error in u, du/u:

$$\frac{du}{u} = 4\frac{dx}{x} + \frac{dy}{y} + 2\frac{dz}{z} \qquad (7\text{-}30)$$

For the function Equation 7–28, Equation 7–29 gives the maximum absolute error and Equation 7–30 gives the maximum relative error. Note the relationship between the integral factors in Equation 7–30 and the exponents in Equation 7–28.

It is common practice to express the relative error du/u as parts per hundred (percent), parts per thousand (ppt), or parts per million (ppm):

$$
\begin{aligned}
\text{fractional} &= \quad du/u \\
\text{percent} &= 10^2\, du/u \\
\text{ppt} &= 10^3\, du/u \\
\text{ppm} &= 10^6\, du/u
\end{aligned}
$$

Probable Propagated Error

In all of the above examples we have calculated the *maximum* propagated error. In effect, it has been assumed that the errors dx, dy, and dz have all been of the same sign, so that the total effect is cumulative. In other words, we have assumed that the values measured for x, y, and z have all been either too high or too low. Common sense suggests that some of the measurements would turn out high, while some would be low. If the errors dx, dy, and dz in a set of measurements are truly random errors, then it certainly is true that there is an equal probability that an error is plus or minus—too high or too low. Consequently, we expect intuitively that the maximum error is likely to be too large. A more realistic measure of the error is the probable propagated error. The **probable propagated error** is also called the **standard deviation of the calculated value.** Before examining the probable error in more detail, let us examine a simple but important example.

One of the more common laboratory measurements consists of obtaining a value from the difference between two measured quantities:

$$u = x - y \qquad (7\text{-}31)$$

The maximum error is given very simply as

$$du = dx + dy \qquad (7\text{-}32)$$

TABLE 7–10

Error Distributions

Both too high	+	+
Both too low	−	−
One high, one low	+	−
One low, one high	−	+

You have used Equation 7–31 many times, for example,

$$\text{buret volume} = \text{final volume} - \text{initial volume}$$

$$\text{weight} = \text{weight full} - \text{weight empty}$$

$$\text{film distance} = \text{scale left} - \text{scale right}$$

Suppose you wish to calculate a scale value V from the difference between an observation of the right end of a scale R and the left end of a scale L:

$$V = R - L$$

In a set of a large number of scale observations, it is to be expected that the values of R would sometimes be higher, sometimes lower, than a mean value. The same is true with observations of the value of L, which, in the absence of systematic errors, would cluster about a mean value. Table 7–10 shows the possible ways in which errors in the observations might occur. Sometimes both measurements might have the same sign, so the maximum error would occur. At other times the errors would have opposite signs and would tend to cancel each other. If a large number of measurements were made, half of the errors would cancel, since it can be assumed that if the errors are random, the *a priori* probability of occurrence of each of the above pairs will be identical. Such considerations reinforce our intuitive belief that the maximum propagated error is too large. To calculate the probable propagated error, we must focus on the dx and dy, the maximum errors of the independent variables.

THE STANDARD DEVIATION OF A COMPUTED VALUE

In the total differential (Eq. 7–21), dx and dy represent the maximum errors in the measurement of the independent variables x and y. The establishment of their values is a judgment call by the experimenter, based on an honest assessment of the errors associated with the experimental components used in the experiment, such as thermometers, barometers, and other scaled instruments.

Suppose, however, that the experimenter were to measure x and y each n times before calculating u, the value of the dependent variable.

In that case it would be possible to calculate the individual sample standard deviations s_x and s_y, where

$$s_x^2 = \frac{\sum\limits_{i=1}^{n} dx_i^2}{n} \qquad s_y^2 = \frac{\sum\limits_{i=1}^{n} dy_i^2}{n} \tag{7-33}$$

This is an application of Equation 7–17, where the approximation that n is approximately equal to $n - 1$ has been made.

Having made n measurements, the experimenter can use Equation 7–17 to calculate the sample standard deviation s_u for the dependent variable u

$$s_u = \frac{\sum\limits_{i=1}^{n} du^2}{n} \tag{7-34}$$

where du is given by Equation 7–21 written for two independent variables x and y:

$$du = \left(\frac{\partial u}{\partial x}\right)_y dx + \left(\frac{\partial u}{\partial y}\right)_x dy \tag{7-35}$$

Substitution of the differential expression for du given by Equation 7–35 into Equation 7–34 results in an expression for the variance s_u^2 of u, the dependent variable:

$$s_u^2 = \frac{1}{n} \sum\limits_{i=1}^{n} \left[\left(\frac{\partial u}{\partial x}\right)_y dx_i + \left(\frac{\partial u}{\partial y}\right)_x dy_i\right]^2 \tag{7-36}$$

When the expression in the square brackets is expanded, the result is

$$s_u^2 = \sum\limits_{i=1}^{n} \left[\left(\frac{\partial u}{\partial x}\right)_y^2 \frac{dx_i^2}{n} + \left(\frac{\partial u}{\partial y}\right)_x^2 \frac{dy_i^2}{n} + \frac{2}{n}\left(\frac{\partial u}{\partial x}\right)_y\left(\frac{\partial u}{\partial y}\right)_x dx_i\, dy_i\right] \tag{7-37}$$

If x and y are independent, then the dx_i and dy_i in the last term of Equation 7–37 are equally likely to be positive or negative, so that the sum over their products is probably very small or nearly zero. Consequently we shall assume that

$$\sum\limits_{i=1}^{n} dx_i\, dy_i = 0$$

In addition, we recognize in Equation 7–37 the expression for the variances given by Equation 7–33. It follows that Equation 7–37 may be rewritten as follows:

$$s_u = \sqrt{\left(\frac{\partial u}{\partial x}\right)_{y,z}^2 s_x^2 + \left(\frac{\partial u}{\partial y}\right)_{x,z}^2 s_y^2 + \left(\frac{\partial u}{\partial z}\right)_{x,y}^2 s_z^2 + \cdots} \tag{7-38}$$

Suppose that for the simple case of Equation 7–31, $dx = dy = 2$, then the probable propagated error is given by

$$s_u = \sqrt{s_x^2 + s_y^2} = \sqrt{2^2 + 2^2} = 2.8$$

which should be compared to the maximum propagated error given by

$$du = dx + dy = 2 + 2 = 4$$

In the same way, compare the maximum propagated error in the volume of a cylinder given by Equation 7–23 with the probable propagated error. The probable propagated error is given by

$$s_u = \sqrt{(2\pi rh)^2 s_r^2 + (\pi r^2)^2 s_h^2}$$

$$= \sqrt{29^2 \times 0.1^2 + 8^2 \times 0.1^2}$$

$$= \sqrt{8.4 + 0.6} = 3.0$$

As expected, the probable propagated error s_u (3.0) is smaller than the maximum propagated error dV (3.7). Thus we would report our calculated results as $V = 23(3)$ rather than $V = 23(4)$, and it should be noted in the body of the report what kind of propagated error has been calculated. The difference between the two is not especially great in this example, which involves just two independent variables; however, when three or four independent variables contribute to the propagated error, the maximum error often turns out to be excessively large. This is because the assumption is being made that all the errors are off in the same direction, an unlikely situation, and one that becomes more and more unlikely as the number of independent variables increases.

At first glance, Equation 7–38 appears cumbersome, especially when compared with Equation 7–21. Remember, however, that the error s_u need only be calculated to one or at most two significant figures. With a hand calculator, the use of Equation 7–38 involves merely pressing a few more keys for squaring the terms before summing them and finally taking a square root. In addition, the data that are entered can be quite severely

TABLE 7–11

Maximum and Probable Propagated Errors for Some Common Functions

Function $u =$	Maximum $du =$	Probable $s_u =$
$x + y$	$dx + dy$	$\sqrt{s_x^2 + s_y^2}$
$x - y$	$dx + dy$	$\sqrt{s_x^2 + s_y^2}$
xy	$y\,dx + x\,dy$	$\sqrt{y^2 s_x^2 + x^2 s_y^2}$
$1/x$	dx/x^2	s_x/x^2
$\ln x$	dx/x	s_x/x
x^n	$nx^{n-1}\,dx$	$nx^{n-1}s_x$
e^x	$e^x\,dx$	$e^x s_x$

rounded off, which also saves time. It is strongly recommended that the probable propagated error be calculated, rather than the maximum propagated error. A comparison of maximum and probable propagated errors is shown in Table 7–11. When the experiment cannot be repeated many times, the propagated error is sometimes the only measure of the quality of the experimental value that can be reported. When the experiment can be repeated many times, statistical methods can often be used to advantage.

REJECTION OF DATA FOR SMALL SAMPLES

As an experimental scientist you will frequently encounter in your data one or two measurements that quite obviously deviate markedly from the average. Not liking the idea that one bad apple can spoil the whole barrel, you may be tempted to throw out the "bad" measurement. Is it good science to discard such suspect measurements? When is it permissible and when is it not? Objectivity is an extremely important part of the scientific method; scientists must constantly guard against such subjective temptations.

Let us examine a set of data (Table 7–12) consisting of several measurements of a specific rate constant for a reaction. The deviations from average are also tabulated, although they would not normally be tabulated in a report. If these were your data, would you be suspicious of any one of them? An examination of the deviations from average in Table 7–12 immediately brings to our attention the result for run 4, for which the deviation is 0.14, a substantially larger value than the others. Is it good science to discard it? Is it bad science to keep it?

TABLE 7–12

Sample Data with a Doubtful Value

Run	k_i	$k_i - k_{ave}$
1	4.51	0.01
2	4.54	0.02
3	4.52	0.00
4	4.66	0.14
5	4.51	0.01
6	4.50	0.02
7	4.48	0.04
8	4.49	0.03
9	4.52	0.00
10	4.51	0.01
Average	4.524	0.028

$$s = 0.051 \qquad s_m = 0.016$$

Using the Laboratory Notebook to Reject Data

If these were your data, and if, upon examining your notebook, you found that at the time you did run 4 you had entered a notation such as, "This run is suspect because the temperature controller seems erratic . . . ," then you may be justified in disregarding the run 4 data. The notation must, however, be made at the time of the run and should be based on some actual observation at that time, not at the time the data are being worked up and compared. In the absence of such an *a priori* notation on your data sheet or in your notebook, you should not reject any datum, unless you use some of the criteria developed by statisticians for the objective rejection of data.

Using the Standard Error to Reject Data

Earlier it was shown that the probability that a measurement deviates by more than three standard deviations (3σ) from the average is only 0.003 or 0.3%. In some circumstances, some investigators routinely reject data that deviate by more than 3σ from the average. With this criterion, should run 4 be rejected? The standard deviation is 0.051, so $3\sigma = 0.15$, but run 4 deviates from the average by 0.14 and so is not quite deviant enough to reject, if you want to be 99.7% confident that you are doing the right thing. If you use two standard deviations as the criterion for rejecting data, then $0.15 > 0.10$, so you could reject run 4.

Using the Q Test to Reject Data

The range (R) of a set of measurements is defined simply as the largest value measured ($X_{largest}$) minus the smallest value measured ($X_{smallest}$):

$$R = X_{largest} - X_{smallest} \qquad (7\text{--}39)$$

If a set of measurements includes a doubtful value, then the ratio Q_{expt} of the difference between a doubtful value and its nearest neighbor to the range provides an objective criterion for rejecting the doubtful value. When Q_{expt} is given by

$$Q_{expt} = \frac{X_{doubtful} - X_{nearest\ neighbor}}{X_{largest} - X_{smallest}} \qquad (7\text{--}40)$$

then $X_{doubtful}$ may be rejected if Q_{expt} is greater than Q_{crit}. Critical values of Q are listed in Table 7–13.

The measurements listed in Table 7–12 provide data for an example of the Q test. By substituting the appropriate values in Equation 7–40, we may calculate Q_{expt}:

$$Q_{expt} = \frac{4.66 - 4.54}{4.66 - 4.48} = 0.67$$

Since 0.67 is greater than 0.41, the critical value of Q listed in Table 7–13, the doubtful value 4.66 may be rejected from the set of data.

TABLE 7–13

Rejection Quotients Based on the Range of a Set of Measurements (90% confidence)

n	Q_{crit}
2	—
3	0.94
4	0.76
5	0.64
6	0.56
7	0.51
8	0.47
9	0.44
10	0.41

Source: Dean, R. B., and W. J. Dixon, "Simplified Statistics for Small Numbers of Observations," *Anal. Chem.* 23:636, 1951.

After rejecting the datum 4.66, a new average and a new standard error should be calculated. For small numbers of data, the Q test is highly recommended for the rejection of data (Skoog and West, 1986).

The analysis and reporting of errors is of paramount importance in scientific writing. The details of the error calculations must always be included in a physical chemistry report.

REFERENCES

1. Baird, D. C., *Experimentation: An Introduction to Measurement Theory and Experiment Design,* Prentice-Hall, Englewood Cliffs, New Jersey, 1962.

2. Barford, N. C., *Experimental Measurements: Precision, Error and Truth,* 2nd ed., Wiley, New York, 1985.

3. Barrante, J. R., *Applied Mathematics for Physical Chemistry,* Prentice-Hall, New York, 1974.

4. Beers, Y., *Introduction to the Theory of Errors,* Addison-Wesley, Reading, Massachusetts, 1957.

5. Brinkworth, B. J., *An Introduction to Experimentation,* C. Tinling & Co. Ltd., Liverpool, 1968.

6. Freedman, P., *The Principles of Scientific Research,* Pergamon Press, New York, 1960.

7. Hall, C. W., *Errors in Experimentation,* Matrix, Champaign, Illinois, 1977.

8. Lacey, O. L., *Statistical Methods in Experimentation,* Macmillan, New York, 1953.

9. Montgomery, D. C., *Design and Analysis of Experiments,* Wiley, New York, 1984.

10. Mortimer, R. G., *Mathematics for Physical Chemistry,* Macmillan, New York, 1981.

11. Rabinowicz, E., *An Introduction to Experimentation,* Addison-Wesley, Reading, Massachusetts, 1970.

12. Skoog, D. A., and D. M. West, *Analytical Chemistry,* Saunders, Philadelphia, 1986.

13. Topping, J., *Errors of Observation and Their Treatment,* Chapman and Hall, London, 1962.

14. National Bureau of Standards, *Precision Measurement and Calibration,* Special Publication 300, U.S. Government Printing Office, Washington, D.C., 1969.

15. Weatherall, M., *Scientific Method,* English Universities Press, London, 1968.

16. Wilson, E. B., Jr., *An Introduction to Scientific Research,* McGraw-Hill, New York, 1952.

17. Youden, W. J., *Statistical Methods for Chemists,* Wiley, New York, 1951.

18. Young, H. D., *Statistical Treatment of Experimental Data,* McGraw-Hill, New York, 1962.

CHAPTER 8

Writing the Report

Why write reports? If our work is of any value or interest, we must be able to communicate what we have accomplished to other scientists. To communicate, we must be clear, concise, and accurate. In addition, scientists have another responsibility: we must critically evaluate our own work and tell others, quantitatively if possible, how good or bad, how accurate or inaccurate our data are. It is necessary to evaluate the errors of our measurements and how those errors are reflected in the results we calculate. It is the responsibility of the writer, not the reader, to evaluate the quality of the experimental work, and for this reason, we learn how to evaluate experimental errors and how to describe errors to our readers.

ORGANIZATION

Clearly, some strategic planning is necessary if one is to minimize such odious tasks as rearranging, revising, and rewriting. Contemporary computer programmers use the term **top-down programming** to describe a method of planning that in many ways is useful in planning the structure of a report. In this method, the design develops from a simple overall list of tasks to be performed, to a series of modules of increasing detail. Physical chemistry reports tend to follow the structure of a scientific journal article, so the first level of writing might be something like this:

> Abstract
> Introduction
> Experimental
> Results
> Conclusions and Errors
> Appendices

References
Data Sheets

The second level of writing consists of building from the available components of a report the more detailed modules, namely the text, tables, and figures.

Some guidelines are available to take some of the mystery out of constructing good tables of data, drawing readable graphs and figures, and even writing reasonably clear and accurate text. The idea to keep constantly in mind is that we are trying to communicate what we have done to someone who knows less about our work than we do. When we construct a table, we should keep asking ourselves, "Is it clear, is it readable, does information flow smoothly and effortlessly from the table to the reader?" Before examining some of the guidelines for text, tables, and figures, let us return to the top-down design of a report by filling in the modules with brief general descriptions of the usual contents.

Title

by Mary P. Student

Abstract

In a very few lines, give a statement identifying the experiment or investigation and perhaps the method or apparatus. Always give a numerical result of the experiment. Be *brief*.

Introduction

With a short summary of relevant theory and important chemical and mathematical equations, put the investigation into perspective for the reader. Number each equation near the right margin, for example, (1). Use the present tense. Clearly define terms and symbols to follow. Be *original* and be *brief*.

Experimental

Describe the apparatus, the purity and purification of materials, the method of chemical analysis, how the system was defined, how independent variables (e.g., T and P) were measured and controlled, and errors in the measurement of raw data. Include your observations of what you saw happen or change. Tables of raw data and calibration curves might be appropriate here. The presentation is almost always chronological and is written in the past tense.

Results

Here the actual results are presented and discussed, and the experimental results are compared with theoretical considerations. Use at least a sentence or two to prepare the reader in a general way for what is to follow. Never begin a section with a table. Tables, if they are not too long, should be embedded in the text in a logical manner. Ideally, text and tables should alternate smoothly, so that information flows clearly and logically to the reader. The reader should not be forced to page forward or backward to find the data under discussion. Coordinate the numerical relationship between graphs and the tables from which they are prepared.

Errors and Sample Calculations

Two calculations should be given here: first, a representative calculation of the physical chemical quantity being sought, and second, a calculation of the propagated probable error. All independent variables and their associated errors should be clearly defined. The output of any computer program used should be verified for a typical data pair with a sample calculation.

Appendices

Here is the place to put hard copies of computer output. These should be cut and pasted onto $8\frac{1}{2}$-by-11-inch paper. Each appendix should have a number and a title and be referred to in the body of the report. Informative labels should be inked in where appropriate to make the data more readable. For example, the output of a least squares is often labeled X and Y. You might ink in log P and $1000/T$ if those were the actual refined parameters.

Recorder chart paper can be included as an appendix if it is cut and pasted to fit on $8\frac{1}{2}$-by-11-inch paper. The report should not contain folded-up chart paper and computer hard copy.

References

Present a numbered list of references to texts, monographs, standard computer programs, journal articles, and collections like Landolt-Börnstein.

Data Sheets

For the physical chemistry laboratory, observed data are recorded in bound notebooks that have duplicate numbered pages, with a sheet of carbon paper placed between them. The carbon copies are handed in to the instructor at the end of each laboratory period. The original data sheets are attached to the end of the report.

TABLES

Physical chemical investigations tend to generate groups of numerical data that are usually best organized into tables. A table consists of a matrix of rows and columns of data. The table itself must be identified with a title and a number and be referred to in the text. In addition, all the rows and columns of data must be labeled and correct physical units must be indicated. The flow of information in a table is from left to right and from top to bottom. This means that primitive data fall into the left-most columns, while numerical data calculated from primitive data go into the columns to the right. The column on the far right, then, contains the final calculated results.

The title should be informative and tell something specific about the contents of the table. Titles such as "Experimental Data" or "Pressures and Temperatures" are not satisfactory. If possible, the system under investigation as well as some of the experimental conditions should be identified; for example, "The Vapor Pressure of Acetone from 30 to 50°C." The labels for columns are particularly important, since these are the actual observed and calculated quantities. These quantities should be defined in the report before the table is encountered. This is often done in the introduction, where some mathematical equation is given and the associated parameters are named, symbols are explained, and units are given. With this background, it is often sufficient to head each column with a symbol (P, T, M), with the unit separated by parentheses (), an underline, or a slash /.

$$P\,(atm) \qquad \frac{P}{atm} \qquad P/atm \qquad T/K \qquad M\,(moles/liter)$$

The same method for indicating units should be used for all the columns in a single table. When numbers are presented in a column, the decimal points should be aligned. A zero should precede the decimal point if the number is less than one (e.g., 0.123). If the numbers are very large or very small, all entries in a column may be multiplied by an appropriate power of ten, indicated in the column heading, for example,

$$10^5\,P\,(atm)$$

$10^5\,P\,(atm)$
3.58
5.77
etc.

For the first entry, this means that $P = 3.58 \times 10^{-5}$ atm, not that $P = 3.58 \times 10^{+5}$ atm.

Various scientific journals sometimes have their own particular method of labeling tabular column heads. You should look at a few by studying the "Note to Authors" section, which usually appears once or twice a year. The table of contents of a current journal indicates when the most recent "Note to Authors" appeared.

The labels for the rows in a table are usually not so important, since these are often just run numbers, which sometimes are omitted, or the names of substances or conditions.

Although the aforementioned conventions are widely used, you should page through a few journals such as the *Journal of the American Chemical Society,* the *Journal of Physical Chemistry,* or the *Journal of Physical and Chemical Reference Data* to get a feel for the variety and ingenuity authors use to communicate their data in a clear, unambiguous way to their readers. Table 8–1 from the *Journal of the American Chemical Society* illustrates some of the points just made.

Note that the table must have both a number and a brief but descriptive name. The two columns on the left present the raw experimental observations: the rates of effusion and the various experimental temperatures. The flow of information is from left to right. From the effusion rates the pressure of the system was calculated, and then the molecular weight of the effusing species.

The average molecular weight from the experiment is 62. The nine in parentheses (9) is the uncertainty in the average; and, if not otherwise noted, it is almost always the standard deviation. This method of presenting the error in a measured value is preferred over the alternative 62 ± 9, which is more difficult to type. Although the errors in the other quantities are not specifically stated, the author has paid careful attention to the correct number of significant figures, which give a rough indication of the uncertainty in these quantities. By inspection it appears that the relative uncertainty in the effusion rate is somewhat less than 0.1%.

While the uncertainty in the temperature measurements is not given in the table, the author must (and does) discuss the method of attaining, controlling, and measuring the temperature, as well as the uncertainty in the temperature. The uncertainty in the device used to measure the temperature may be, say, one degree, but the uncertainty in the temperature of the system reported in Table 8–1 may be five degrees. Can you see why? The distinction is an important one.

TABLE 8–1

Title of Table

T	Effusion Rate	P	M
K	10^{-6} g cm^{-2} sec^{-1}	dynes cm^{-2}	g mol^{-1}
1406	14.0	1.78	46
1462	53.3	5.84	71
1501	99.2	12.72	48
1545	264.2	29.2	66
1564	363.6	41.2	64
1587	476.5	61.8	49
1626	1182	119.8	82
			Average = 62(9)

Keep in mind what this table is being used for. The author is communicating to the reader what he did experimentally, what he calculated from those measurements, and an honest assessment of the quality of all of these data.

GRAPHS

Even more than a table, a graph serves to inform the reader very quickly of the relationship between an experimental parameter (the independent variable) and the calculated physical chemical quantity (the dependent variable). The independent variable should be plotted along the horizontal axis (x axis or abscissa) and the dependent variable should be plotted along the vertical axis (y axis or ordinate). The graph should be prepared on a full-size ($8\frac{1}{2}$-by-11-inch), regular-weight, commercial graph paper.

Like tables, graphs must have a number—e.g., Figure 3–1 (even if there is only one)—and a descriptive title. Titles such as "P vs T" are not satisfactory. Every graph should be preceded in the body of the report

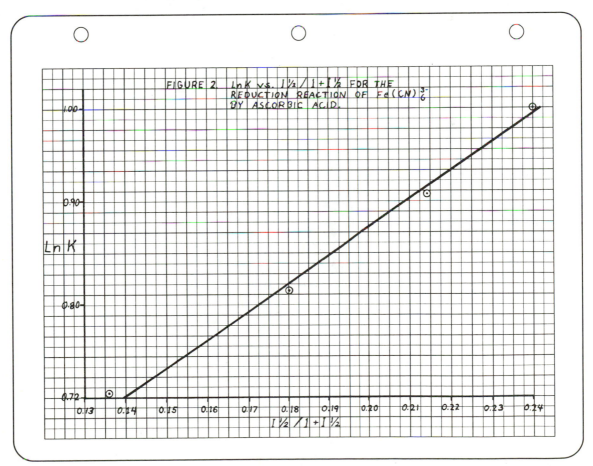

Figure 8–1 A good example of a linear plot from Experiment 28. (Gabriel M. Ruiz, May 1987)

by the table of data from which it is prepared, and it is helpful to the reader if their titles are either identical or very similar. The numbers need not correspond, since the data in some tables are not graphed and because not all figures are graphs. Some are illustrations or drawings of apparatus. In numbering, the word "Figure" is used, not the word "Graph."

Choosing the scales for the dependent and independent variables causes the most difficulty. Remember, our goal with this graph is to present to the reader as much information as possible and to do it in a simple, clear manner. The graph should nearly fill the page, running from corner to corner. The smallest divisions of both axes should be divisible

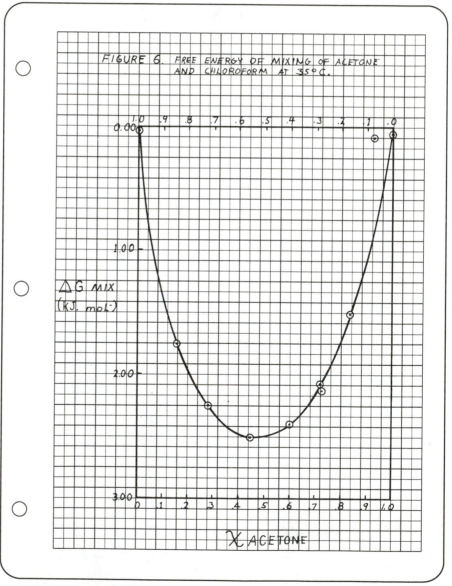

Figure 8–2 A good example of a curved plot from Experiment 10. (Vincella Zaccone, May 1988)

by 2, 5, or 10, if possible, for ease of interpolation. The same is true for the numbers printed periodically along the axes. The axes can be labeled either with a symbol or with a word followed by the appropriate units: for example, "P (atm)" or "Pressure (atm)."

The graph itself consists of a line and a set of points. The line may be straight or curved, but in either case it should be smooth, not drawn from point to point. Each point should consist of a small, precise dot surrounded by a simple geometric figure, usually a circle. The circle is often drawn in a size that corresponds approximately to the uncertainties in the variables. When a smooth curve is drawn through the points, the line should not go through the circle, but should be drawn up to the circle on each side.

If more than one set of points is plotted on the same graph, squares and triangles, filled and unfilled, may be used in addition to circles to distinguish between the curves. A legend labeling the curves should be provided on the page. Normally, no calculations or information appears within the graph other than the line(s), points, and legend. If it seems desirable to state the slope of a line, or a quantity calculated from the slope of a line, then such information should be made part of the title or subtitle. Extended discussion of the graph belongs in the text of the report itself. Some journals and books, however, use extended captions (separate from the figure title) to describe the meaning of a figure or graph.

Some of these points are illustrated in the samples shown in Figures 8–1 and 8–2.

TEXT

The principal parts of your report are the Abstract, Introduction, Experimental, Results, and Discussion sections. The components of these sections are the tables, figures, and text.

Continuity

A section should never begin abruptly with a table or a figure. Write at least a sentence or two to introduce the reader to the material. If your report is well organized, it will practically write itself. Your task is to provide a smooth flow of information to the reader. The reader should not be forced to look back or look ahead several pages to relate to the point you are making. Use a transitional sentence to provide continuity between ideas or sections of your report.

If a table is about to appear, prepare the reader by stating clearly the source or equation used to calculate the data in the table. The symbols for the table column headings and the symbols used in corresponding equations should be identical. Explain or summarize why the table is presented. The text should refer by number at least once to every table and figure in the report.

Similarly, introduce the reader to your figures. If the figure is a graph, state by table number the source of the graphical data. Do not assume that the reader can guess what you mean because it is so obvious to you, the writer. As the writer, you can *remember* how things fit together, even if you do not state such details in your text. Your reader, however, has no such advantage and you must provide the organizational connections that will lead the reader smoothly from beginning to end.

Overall, the flow of information in a report should be chronological. For example, tables of experimental data should precede, not follow, tables of calculated results. Similarly, a graph should follow the table of data from which it is constructed. Figures and tables should not simply be bunched together at the end of the report. Instead, text, tables, and figures should flow chronologically from the beginning to the end of the report. Small tables should be embedded in the text where possible, slightly indented or centered. If a table is too long to be embedded conveniently, it should appear on the very next page, and it should be referred to by number and page in the text, so that the reader is not forced to search back and forth through the text to find it.

Conjunctive adverbs and phrases are helpful in giving continuity and a feeling of flow to your writing. Conjunctions help us provide a smooth flow of information to the reader, connecting sentences, paragraphs, and ideas. Try a few from the following list and see if you don't find them helpful (Potera, 1984):

above all	in particular
accordingly	instead
and so	in summary
again	likewise
also	moreover
besides	more specifically
but	nevertheless
consequently	nonetheless
finally	on the other hand
first	rather
for example	second
furthermore	similarly
hence	so
however	still
in addition	then
in conclusion	therefore
indeed	though
in fact	thus

If your scientific work is good and original, you can be assured that your reader will be interested. Even so, you would do well to avoid monotony in your writing by occasionally varying the length of your sentences, and by once in a while beginning a sentence with a subordinate clause instead of the main clause. Use conjunctive adverbs to provide smooth connections and transitions.

Style

In all writing, the intended audience determines the writing style. The audiences for scientific reports and journal articles are essentially the same, and so is the style: formal. Because formal scientific writing is not literature, it should not be written with flair, humor, or emotion. Instead, you should strive for brevity and clarity. The following examples from student physical chemistry laboratory reports illustrate the kind of informal, colloquial writing that is not appropriate for reports or journal articles:

> Plugging these values into equation 19 give. . . .

> This compares pretty well with the literature. . . .

> The data on water was thrown out. . . .

You may notice more errors in these short fragments than just the use of colloquial expressions.

Students are frequently tempted to use pompous, complex words when writing about complex subjects like thermodynamics, spectroscopy, and kinetics. It would be better to reverse this tendency and to use the simplest words for the most complex subjects. Complex and pompous writing is largely an attempt to cover up a lack of confidence. With simplicity comes clarity.

Eventually, many of you will be writing scientific articles for publication in major scientific journals. You should bear in mind that journals such as the *Journal of the American Chemical Society,* the *Journal of Physical Chemistry,* and even the *Journal of Chemical Education* are international journals. English is not the first language of many readers. Colloquial expressions and humorous interjections do not clarify the material for these readers. In fact, it's not a bad idea to keep in mind that your audience knows about as much physical chemistry as you do, but has mastered English at about the level of an American high-school student. Keep your language simple. Avoid unusual words. Almost any nonscientific word that you would avoid in conversational English probably would be best avoided in your written work. A few examples, and their more easily understood alternatives, are shown below:

ascertain	find out
assistance	aid
commence	begin
encounter	meet
endeavor	try
expedite	hasten
finalize	finish

initiate	begin
interface	connect
modification	change
subsequent	next
terminate	end
utilize	use

A few phrases tend to be popular with beginning science writers trying to be sophisticated. Here are a few of those fuzzy phrases and their more concrete equivalents:

in order to	to
for the purpose of	for
by means of	with, by
in accordance with	with, by
under the circumstances	because
due to the fact that	because
in view of the fact that	because
as a matter of fact	in fact
prior to the start of	before
on a few occasions	occasionally
in the majority of cases	usually
it is obvious that	obviously
in the event that	if
in order that	so
on account of	because
with the end in view of heating	to heat
for the purpose of heating	to heat
in relation to	about (or omit)
with regard to	about (or omit)
in connection with	about (or omit)
on the subject of	about (or omit)
with reference to	about (or omit)

These phrases are not forbidden. Their occasional use legitimately adds interest or breaks up the monotony. Just be careful not to overuse them.

You can also avoid monotony by varying the sentence length and structure. Every once in a while begin a sentence with a subordinate clause instead of the main clause. None of your sentences should be very long, say 20 words or so. From time to time, use a rather short sentence for clarity, for effect, or for interest. The repetitive use of sentences of the same length and structure has an anesthetic effect on the reader. In one student report, five sentences in a row began

A discharge was used. . . . A lens was used. . . . The spectrograph was capable. . . . The film was developed. . . . A comparator was used. . . .

All are passive, all begin with the main clause, all are of about the same length, and altogether they are deadly dull. Another student became infatuated with the word "for":

> For the triatomic molecule. . . . For any molecule. . . . For a linear molecule. . . . For the asymmetrical stretching. . . . For the bending modes. . . . For sulfur dioxide. . . . For both of these modes. . . .

If either of these reports had been read out loud, it might never have been submitted.

It is the responsibility of the writer to make the report easy for the reader to understand. Because the report may cover material that is difficult or new to the reader, it is important to write in a simple, clear style. Generally, you should avoid esoteric words and long, complicated sentences. As in any expository writing, however, it is a good idea to vary sentence length to avoid monotony. If a statement can be written in few words, then that is the better way of writing it. Try to find ways of shortening what you have already written. For example (Baker, 1969),

> *Before:* Many scientists find it helpful to accumulate a list, in the form of a card index, of promising research problems from which selections may be made. It is advantageous to make a tentative analysis of each subject and to indicate briefly the object, scope, general plan of investigation, and probable nature of the results that might be obtained.

> *After:* Many scientists jot down their promising ideas on file cards: object, scope, technique, probable results.

This may be a bit too terse, but it is certainly an improvement over the original. Let's look at one more example.

> *Before:* Cardiovascular functions and general bodily efficiency relationships have formed the subject of a great deal of research in order to gauge the general health of individuals.

> *After:* The relation between cardiovascular and bodily efficiency provides a testable index of health.

Unfortunately, it is easier for most of us to use many unnecessary words, resulting in bloated, stilted, and fuzzy writing. Even some rather gifted writers have found good writing difficult. Charles Darwin complained, "I have as much difficulty as ever in expressing myself clearly and concisely; . . . but it has had the compensating advantage of forcing me to think long and intently about every sentence, and thus I have been led to see errors in my reasoning and in my observations of those of others." That was written 100 years ago. Have things become any easier? Consider the words of Barbara Tuchman (1981), the distinguished American historian (twice winner of the Pulitzer Prize): "Research is endlessly seductive; writing is hard work. One has to sit down on that chair and think and transform thought into readable, conservative, interesting sentences that both make sense and make the reader turn the page. It is laborious, slow, often painful, sometimes agony. It means rearrangement, revision, adding, cutting, rewriting. But it brings a sense of excitement, almost of rapture; a moment on Olympus. In short, it is an act of creation."

Tense

The proper choice of tense invariably causes great difficulty to inexperienced writers of scientific reports and articles. Don't worry, however, for it's really quite simple. About 95% of the time, just two tenses satisfy the requirements of good style: the simple past and the present. The rules are quite simple. You must report your work in the *past* tense and the work of other scientists in the *present* tense. Further generalizations follow from these rules. Since the introduction is essentially a summary of other scientists' work, it is written in the present tense. On the other hand, the experimental section, a presentation of your work, is written in the past tense and is usually chronologically ordered. Take care to avoid the future tense; it is almost never appropriate in scientific report writing.

The generally accepted properties of material substances are always reported in the present tense:

> Sulfur *is* a yellow crystalline solid, which *melts* at 112°C to form a brown, viscous liquid.

> Because titanium(III) ion in aqueous solution *absorbs* radiation at 500 nm, it *appears* light blue in solution.

On the other hand, if you are reporting your own experimental observations, you might write:

> After the residue *was* washed with ether, a dark brown precipitate *was* collected. When dry, it *melted* at 225°C. The product *showed* evidence of decomposition at the melting point, but *was* stable toward moisture at room temperature.

The following excerpt, taken from *Science,* illustrates the use of the past tense in reporting the work of the writer:

> The HCl *was* decanted and the sample rinsed in slowly running water for 10 hours (Fig. 1d). The sample *was* next placed in an acetone bath under vacuum at 60 torr for 1 hour (Fig. 1e) and then allowed to dry at room temperature. To ensure thorough removal of moisture, the sample *was* dried in a microwave oven at 500 watts for 30 minutes or in a conventional oven at 130° for 12 hours (Fig. 1f) (7).[1]

The (7) here refers to a reference at the end of the paper; this method is required by the editors of *Science* and is convenient for typed reports since it allows the avoidance of superscripts.

---

[1] Zapasnik, H. T., and P. A. Johnston, "Replication in Plastic of Three-Dimensional Fossils Preserved in Indurated Clastic Sedimentary Rocks," *Science* 224:1425, 1984.

The following two examples, taken from student physical chemistry laboratory reports, illustrate the improper use of the future tense:

A plot of B_v against $v + 1/2$ *will yield* a_e as the slope and B_e as the y-intercept.

Although the bromide *will crystallize,* the chloride forms a gel.

The present tense should have been used in the first example ("yields"). In the second example, if the sentence is from the Introduction and the properties described are well established, then the sentence should be in the present tense ("crystallizes," "forms"). However, if the sentence describes the experimental observations of the writer, it should be in the past tense ("crystallized," "formed"). Also notice that in this example future and present tenses are combined. This is undesirable; tense should be consistent within a statement.

To demonstrate how tense is properly used in scientific writing, let us examine a few paragraphs from an article chosen at random from a recent issue of the *Journal of Physical Chemistry.* The article treats the formation of the triplet excited state in some organic compounds generated by laser photoexcitation. Let us begin with the first paragraph of the introduction:

Porphycene, a structural isomer of porphin,[3] formally derives from the hypothetical [16]-51-annulene dianion, whereas porphin is related to the [16]-85-annulene dianion.[4] Hence, the difference in structure and symmetry should be manifested by modified photophysical and photochemical properties of these compounds. Preliminary spectroscopic studies of porphycenes indeed show noticeable changes in the ground-state absorption and fluorescence spectra as compared to their porphyrin analogues.[5,6] In addition, triplet detection by EPR at low temperatures[5] results in spectra that are sensitive to perturbations of the porphycene basic structure, such as substitution at selected peripheral sites. For structures of these compounds see ref 3 and 5.[2]

Notice that every verb in this paragraph from the introduction is in the present tense. The introduction deals with what *is* known about the system of interest, so it is consistent to use the present tense to describe the investigation in terms of what *is* believed to be true. Now let us skip to the second paragraph of the experimental section of the same paper and notice how the tense has changed from the present to the simple past tense:

Sensitization-Pulse Radiolysis. These experiments utilized α-acetonaphthone (Aldrich) recrystallized from ethanol. The optical-pulse-radiolysis setup and data gathering is described elsewhere. . . . All pulse radiolysis experiments employed benzene solutions consisting of the same concentration of sensitizer (20 mM), and the

[2] Levanon, H., et al., "Triplet-State Formation of Porphycenes: Intersystem Crossing Versus Sensitization Mechanism," *J. Phys. Chem.* 92:2430, 1988.

porphycene concentrations were on the order of 3–5 M. . . . These experiments were carried out by using a flow system, with the solutions flushing a cell of 1.0-cm path length. Prior to and during each experiment, the sample was thoroughly deaerated by continuously bubbling Ar gas which had been presaturated with benzene. This procedure was used to avoid any triplet quenching by traces of oxygen.[3]

Except for the second sentence, the authors use the simple past tense to describe the experimental work that they *did do* some time in the past before writing this article. Notice that one verb after another is in the simple past tense: "utilized," "employed," "were," "were carried out," "was deaerated," and "was used."

Finally let us skip again, this time to the results-and-discussion section of the paper. Once again the primary tense is the present tense because the authors are presenting data considered to reflect the true properties of the system. These new data *are* the properties of the system.

The success in observing triplet–triplet absorption at low temperatures together with the apparent common room temperature behavior, i.e., negligible triplet absorption and fluorescence intensity, prompted us to search for a temperature dependence of the luminescence of H_2PC3. The fluorescence properties of H_2PC1 and H_2PC2 have been reported elsewhere.[6] It is noteworthy, within the present context, to indicate that both compounds exhibit strong fluorescence intensities at room temperature. This does not occur for H_2PC3, where its fluorescence intensity strongly depends on the temperature; i.e., its fluorescence is extremely weak at room temperature, and it is smaller by approximately an order of magnitude, at 186 K, that of H_2PC1 recorded at 300 K. The steady-state fluorescence spectrum of H_2PC3 at 186 K exhibits two bands peaked at about 673 and 735 nm, originating from the S_1 state, with an intensity ratio of about 4:1.

The predominate tense in this paragraph of the results-and-discussion section is the present tense. Occasionally, when the authors refer to some particular experimental procedure they carried out, they use the past tense just as they did in the experimental section. However, what is true *is* true, and what was done experimentally *was* done experimentally.

Voice

Scientists writing articles for scientific journals are generally required to avoid the first person, especially "I" (and also "we" when the article is coauthored). Unfortunately, this requirement generates the use of a great number of passive constructions. If "I measured. . . ." is forbidden, the usual substitution is the passive verb: ". . . was measured (with, by. . . .)." Since the passive voice is often unavoidable, use the active voice whenever possible. Usually you can avoid

---

[3] Ibid., p. 2431.

> It was found that. . . .
> Tests were made. . . .
> Spectra were recorded. . . .

if you think about alternative constructions. Use transitive verbs like "show," "indicate," "prove," instead of "It was found. . . ." Try "Tests indicate that. . . ." or "The spectra showed a strong peak at. . . ." With a little imagination you can cut your passive constructions by half. Your writing will be more lively, concrete, and convincing. If you cannot quite find the right word, do not be afraid to use *Roget's Thesaurus,* a dictionary of synonyms, available in inexpensive paperback editions.

Another problem that arises with the passive voice is the dangling participle, and the participle that dangles most frequently is the word "using." The problem is most easily corrected by changing the dangling participle to a subordinate clause or prepositional phrase. For example, one student wrote:

> Using a Gouy balance, five transition metal complexes were weighed.

"Using" dangles because it has no noun to modify. (It may look as though "using" modifies the noun "complexes," but this doesn't work since the complexes are not doing the using.) If the first person were permitted, the solution would be to rewrite the sentence in the active voice so that the participial phrase modified the noun "I":

> Using a Gouy balance, I weighed solutions of five transition metal complexes.

(It is also an improvement to weigh solutions rather than complexes!) However, the first person is not recommended, so another approach must be used:

> Solutions of five transition elements were weighed with a Gouy balance.

Sometimes an entirely different approach solves the dangling participle problem and also improves the writing:

> A Gouy balance was used to determine the weight change with and without a magnetic field for solutions of five transition element complexes.

Grammar

This is not the place to review English grammar, since you have been studying grammar to some degree for the past 15 years or so. Rereading your draft is usually sufficient to reveal grammatical mistakes. Nevertheless, the following may serve as a reminder of some common problems:

The Ten Commandments of Good Writing

1. Each pronoun should agree with their antecedent.
2. Just between you and I, case is important.

3. A preposition is a poor word to end a sentence with.
4. Verbs has to agree with their subject.
5. Don't use no double negatives.
6. A writer mustn't shift your point of view.
7. When dangling, don't use participles.
8. Join clauses good, like a conjunction should.
9. Don't write a run-on sentence it is difficult when you got to punctuate it so it make sense when the reader reads what you wrote.
10. About sentence fragments.

Numbers

It is considered poor style to begin a sentence with a symbol, letter, or number. Be careful not to change the symbol you have chosen to represent some variable or parameter. The same symbol should be used in the text, equations, tables, and figures. If you use D_1 for the dielectric constant of the solvent, then be sure the D_1 does not inadvertently appear elsewhere meaning something else. If a number in the text begins a sentence, it is written out: "Nine spectra were photographed. . . ." If the number is so large or so small that it must be expressed in exponential notation, for example, 6.626×10^{-34}, then you must rearrange the sentence. Since computer notation for exponential numbers should not be used in a report or manuscript, it is improper to write 6.626E-34.

When you use integers for counting, use figures for numbers above ten and spell out numbers below 11. Use figures for numbers with units, whether below or above 10:

A 5-ml aliquot. . . .
The first and second run. . . .
The 15th attempt. . . .
. . . discussed on pp. 9–12

However, if small and large numbers are mixed, use figures for both:

From the 1st to the 92nd element. . . .

Spell out the first number of a compound number adjective:

Three 500-ml volumetric flasks. . . .

Proofreading

One of the most important ways of improving your writing is to reread and rewrite your own material. If you do, you will avoid submitting material like these samples taken from student reports, which were obviously not reread and rewritten:

The visible spectrum of nitrogen consists of a series of band heads under low resolution corresponding to, within experimental error of this approximation, the vibrational shifts between adjacent electronic states.

Then, the data contained in Table 2 was tabulated and next, determination of N and a were made. This data is tabulated in Table 3. The data is self explanatory except of N and a.

Random errors in readings were minimal by the potentiometer and clock being digital and the galvanometer mirrors, eliminating parallax.

These embarrassingly bad passages should have been spotted by rereading and corrected by rewriting. If rereading silently doesn't work, try reading your writing out loud to yourself. If it sounds good, it probably is good; if it sounds bad, it surely is bad. Be your own critic.

Reading for Writing

In the long run, reading large doses of good writing will improve your own writing. Get in the habit of reading at least one or two journal articles every week. *Science,* published by the American Association for the Advancement of Science, *Discover,* published by Time, Inc., and *Scientific American* are excellent models to follow. A number of contemporary scientists are writing books about science for a general audience. Try to find time to read some of these to pick up elements of style:

1. Asimov, Isaac, *The Noble Gases,* Basic Books, New York, 1966.
2. Bronowski, J., *The Ascent of Man,* Little, Brown, Boston, 1973.
3. Bowen, M. E., and J. A. Mazzeo, eds., *Writing About Science,* Oxford University Press, New York, 1979.
4. Carson, Rachel, *The Sea Around Us,* Oxford University Press, New York, 1951.
5. Gould, Stephen Jay, *Hen's Teeth and Horse's Toes,* Norton, New York, 1983.
6. Hawking, Stephen, *A Brief History of Time,* Bantam Books, New York, 1988.
7. Morris, Desmond, *The Naked Ape,* Dell, New York, 1967.
8. Sagan, Carl, *Broca's Brain,* Ballantine Books, New York, 1974.

Reading good science will be a source of pleasure to you for a lifetime; the immediate challenge is to write reports and write them well.

REFERENCES

1. Baker, S., "Clarity," *Science* 166:365, 1969.
2. Chambers, J. M., W. S. Cleveland, B. Kleiner, and P. A. Tukey, *Graphical Methods for Data Analysis,* Wadsworth, Belmont, California, 1983.
3. *The Chicago Manual of Style,* 13th ed., University of Chicago Press, Chicago, 1982.
4. Cleveland, W. S., *The Elements of Graphing Data,* Wadsworth, Belmont, California, 1984.
5. Day, R. A., *How to Write and Publish a Scientific Paper,* ISI Press, Philadelphia, 1979.
6. Dodd, J. S., *The ACS Style Guide,* American Chemical Society, Washington, D.C., 1967.

7. Embergen, M. R., and M. R. Hall, *Scientific Writing,* Harcourt, Brace, New York, 1955.

8. Fieser, L. F., and M. Fieser, *Style Guide for Chemists,* Reinhold, New York, 1960.

9. Gensler, W. J., and K. D. Gensler, *Writing Guide for Chemists,* McGraw-Hill, New York, 1961.

10. Marder, D., *The Craft of Technical Writing,* Macmillan, New York, 1960.

11. Potera, C., "The Basic Elements of Writing a Scientific Paper: The Art of Scientific Style," *J. Chem. Ed.* 61:246, 1984.

12. Rhodes, F. H., *Technical Report Writing,* McGraw-Hill, New York, 1961.

13. Sandman, P. M., C. S. Klompus, and B. G. Yarrison, *Scientific and Technical Writing,* CBS College Publishing, New York, 1985.

14. Tuchman, B., *Practicing History,* Knopf, New York, 1981.

15. Tufte, E. R., *The Visual Display of Quantitative Information,* Graphics Press, Cheshir, Connecticut, 1983.

16. Ward, R. R., *Practical Technical Writing,* Knopf, New York, 1968.

TECHNIQUES

CHAPTER 9

Microcomputers in the Physical Chemistry Laboratory

All computers store information on disks and have an associated disk operating system (DOS). IBM-PCs use PC-DOS[1] and most PC compatibles use MS-DOS, a version written by Microsoft. The two versions are similar and many commands are identical. In this chapter we will look at some of the more important features of these disk operating systems. The most common use of microcomputers being for word processing, we will examine some of the functions of WordStar, a widely used word-processing program, the command set of which is the basis of the Turbo Pascal Editor. (Turbo Pascal is described in the next chapter.) Finally, we will see how the spreadsheet LOTUS 1-2-3 can be used to do a linear regression and graph the results.

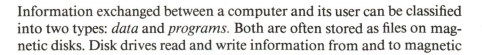

USING THE DISK OPERATING SYSTEM

Information exchanged between a computer and its user can be classified into two types: *data* and *programs*. Both are often stored as files on magnetic disks. Disk drives read and write information from and to magnetic

---

[1] IBM's newest computers have a new operating system called OS/2. You can, however, run any program you currently use on PC-DOS or MS-DOS on OS/2, because OS/2 has a "built-in DOS application environment."

disks. A disk and a disk drive are quite analogous to a tape and a tape deck, except that a disk–disk-drive combination is a more sophisticated unit because it can "play back" any file the user desires; tape decks play back the musical pieces in the sequence in which they were written on the tape and the user has no control over that sequence.

In order for the user to select particular files on a disk, two additional components are required. The first is obvious: a *name*. In order to select and distinguish one file from another, each file must have a unique name. The second component is the *disk operating system*, often referred to as *DOS*, pronounced "dee oh ess" or "doss." The disk operating system is a set of program files that carry out a number of uninteresting but important file housekeeping chores, such as copying files from one disk to another and permitting the computer to *read, write, erase,* and *rename* files.

Naming Disk Files and Disk Drives

Most microcomputer systems have two disk drives that are distinguished from each other by naming the first A: and the second B:.

A file name consists of three parts: a drive name, a file identifier, and a file extension:

[*drive name*][*file identifier*] . [*file extension*]

The **drive name** is optional; if it is omitted, it is assumed by the disk operating system to be the logged drive, which is usually drive A:; that is, the **default drive** is the A: drive. The file extension and the period separating it from the file identifier are optional.

The **file identifier** consists of one to eight contiguous characters (no spaces), followed by an optional period, followed by an optional **file extension.** The extension consists of any three characters, some of which are used for special purposes and should be avoided for ordinary text or data files, that is, *.COM, .BAS,* and *.PAS.* Finally, it is strongly suggested that file names be informative. The file name should suggest something about its content or function. Some examples are:

Z	(legal, but poor because it is not informative)
PCREPORT.1	(OK)
LETTER.FRD	(OK)
HEATCAPACITY	(illegal; more than eight characters)
REPORT1.PCHEM	(illegal; extension has more than three characters)
BOMBCAL.	(legal, but poor because the period is easy to miss or forget)
GAUSS.PAS	(legal; used by Pascal to indicate a Pascal source program file)
GRAPH3.BAS	(legal; used to indicate a BASIC source program)

GAUSS.COM (legal; used to indicate a compiled program)

LS.EXE (legal; used to indicate a compiled program)

Common PC-DOS and MS-DOS Commands

The files that constitute the disk operating system are usually stored on a back-up disk, not the original disk. The operation of transferring the disk operating system files from the disk to the computer memory is called "booting" the system. When the computer is off, booting is accomplished by inserting the DOS disk into drive A:, closing the drive door, and turning on the power to the computer.

The computer responds by turning on a red light on disk drive A:, and finally showing the following display on the screen:

```
Current date is Mon  1-01-1980

Enter new date:  7-2-1989

Current time is  0:00:31.91

Enter new time:  17:10:22


The IBM Personal Computer DOS
Version 2.10 (C) Copyright IBM Corp 1981, 1982, 1983


A>
```

There are many slight variations on this, depending on whether you are using an IBM-PC and PC-DOS or a PC clone and MS-DOS. They will, however, all ask for the time and date.

If you don't care about the time or date, you can press the return key twice. When the *A>* (read "A prompt") is on the screen, the computer is ready to accept commands entered from the keyboard. When the DOS disk is booted, some, but not all, of the disk operating system files are loaded into the computer's memory. Those that are loaded into memory are called **internal** and are available for use even if the DOS disk is removed. Those DOS files that are not loaded into memory are called **external** and can be executed only when the DOS disk is in a disk drive.

Note that when the *A>* is on the screen, the A: drive is the logged disk drive. If you should enter from the keyboard *B:* followed with a return, a *B>* will appear on the screen and the B: drive will become the logged disk drive.

DISKCOPY. This may well be the first DOS command you use. It is used to copy all the files on one disk to another, which is what you do when you prepare a back-up disk of the disk you have been using. The usual form of the command is

DISKCOPY A: B:

With this command, the system copies all the files from the disk in drive A: to the disk in drive B:. The disk in drive A: is called the source disk and the disk in drive B: is called the target disk. Because the *DISKCOPY* program formats the target disk as it copies, it is not necessary to format the target disk prior to copying. The *DISKCOPY* program is an external file, so the DOS disk must be in the logged drive in order to use *DISKCOPY.*

DISKCOMP. If you have just made a duplicate copy of an entire disk using *DISKCOPY,* you will probably want to know if the copying process was successful. The DOS command *DISKCOMP* serves to compare one disk with another. Since *DISKCOMP* is an external program, the DOS disk must be in a drive to call it. Insert the DOS disk in drive A: and enter *DISKCOMP A: B:.* The screen will respond with

```
Insert FIRST diskette in drive A:

Insert SECOND diskette in drive B:

Strike any key when ready. . . .
```

Remove the DOS disk, insert the disks to be compared in drives A: and B:, and strike any key. The two drives will run for a few moments, comparing the information on the two disks track by track. If the *DISKCOPY* process is successful, *DISKCOMP* will respond:

```
Diskettes compare ok

Compare more diskettes (Y/N)?
```

If something goes wrong, however—perhaps the target diskette is dirty or has a manufacturing flaw in one or more tracks—then *DISKCOMP* will respond quite differently; for example:

```
Compare error(s) on track 23

Compare error(s) on track 24

Compare error(s) on track 25
```

The usual remedy is simply to try again: recopy the original source disk onto a new target disk.

CHKDSK. This command allows you to learn more about the files on a disk, including their sizes and dates of generation, the total space taken up by the files, and the total space left. Usually you check a disk with *CHKDSK* to learn whether the disk contains sufficient space to store one or more files you wish to transfer from another source. Since *CHKDSK.COM* is an external file, the DOS disk must be in a drive, usually the A: drive, when you execute the *CHKDSK* command to the disk operating system. If DOS is in drive A:, then the disk to be checked should be in drive B:, and the command is

CHKDSK B:

An example of a response is

```
A>CHKDSK B:

      362496   bytes total disk space

      340992   bytes in 11 user files

       21504   bytes available on disk

      655360   bytes total memory

      630560   bytes free
```

If you intend to transfer several files to a disk that already has some files written on it, it is a good idea to run *CHKDSK* to determine whether or not space is available.

FORMAT. To **format** a disk is to prepare it so that it is capable of having files copied on to it. With the exception of the *DISKCOPY* command, all programs that transfer a file to a disk can only transfer files to a formatted disk. Each disk operating system has its own particular format; for example, a disk formatted for Apple-DOS cannot store files generated by a program operating under, say, MS-DOS. Because the *FORMAT* command is external, the DOS disk must be in a drive, usually drive A:. If the DOS disk is in drive A: and the disk to be formatted is in drive B:, then the command to format is

FORMAT B:

After you press the return key, drive A: runs for a fraction of a second and the following request appears on the screen:

```
Insert new diskette for drive B:

and strike ENTER when ready
```

After you strike any key, the red light on drive B: lights up, and the drive runs for about 30 seconds. The light goes out and the screen delivers the following message:

```
Formatting . . . Format complete

Format another (Y/N)?
```

If a disk is intended only for storing files, the method described above suffices for formatting disks. However, if program files are stored on a disk, then it is convenient if the disk is bootable, making it unnecessary to boot the system from the DOS disk. To get a bootable, formatted disk, use the command

FORMAT B:/S

The screen message now reads

```
Formatting . . . Format complete

System transferred

Format another (Y/N)?
```

The phrase "system transferred" means that just enough of the disk operating system has been transferred to the newly formatted disk so that the new disk is bootable. This means that if you write a program and store it on this disk, you can boot your system with this disk without using the DOS disk and then immediately execute your program, if you wish.

So why not always use *FORMAT/S*? Simply because the system files take up space on the disk that could otherwise be used for file storage. The difference is shown by running *CHKDSK* after a *FORMAT* command and after a *FORMAT/S* command:

FORMAT then *CHKDSK*:   *FORMAT/S* then *CHKDSK*:

```
362496 bytes total disk        362496 bytes total disk
       space                          space
362496 bytes available on       22528 bytes in 2 hidden
       disk                            files
                                18432 bytes in 1 user file
                               321536 bytes available on
                                      disk
```

Those two hidden files and one user file generated by the *FORMAT/S* command not only make the disk bootable, but also contain the internal DOS files. Consequently, these valuable utility DOS files are available for use without reinserting the DOS disk. The price you pay is about 40,960 bytes or about 11% of the total disk space.

DIR. Entering the command *DIR* returns to the screen a directory of the files in the logged disk drive, which is usually drive A:. If the logged drive is drive A:, then entering

DIR

or

DIR A:

would return the same information. The directory message consists of five vertical columns, listing for each file (1) the file name, (2) the file extension, (3) the file size (bytes), (4) the date the file was stored on the disk, and (5) the time at which the file was stored on the disk.

If the directory is very long, the names of the top files may scroll off the screen before you have a chance to see them. If you enter the command

DIR A:/P

the listing of file names will pause when the screen is full. Pressing any key will scroll in another screenful of file names.

If you are not interested in all that information, but want only the list of file names, you have that option by entering, for example,

DIR B:/W

With this command, the screen lists from drive B: only the file names and their extensions from left to right across the width of the screen. The *W* stands for a wide directory.

Suppose you have a disk full of files that are letters back home to Mom, letters to your boyfriend, and some Pascal programs. If you had chosen the file extensions *.MOM, .LUV,* and *.PAS,* then entering

DIR ∗.LUV

would return to the screen only the file names of letters to your boyfriend. If you have two boyfriends, you could use the extensions *.LV1* and *.LV2.* If you wanted to see a list of files containing both extensions, you would enter

DIR ∗.LV?

and a list of files with the extension *.LV* and any other character would be returned to the screen. The "?" is appropriately called a **wild card.** The *DIR* command is an internal file and is available anytime the screen shows a system prompt, for example, the *A>*.

COPY. Like the *DIR* command, the *COPY* command is an internal file, so it is not necessary to have the DOS disk in a drive to use it. The *COPY* command is used to copy files or groups of files from one disk to another, already-formatted disk. Copying a single file from the logged drive, which is usually A:, to a disk in drive B: is the most common use of the *COPY* command. The general form of the *COPY* command is

COPY [*source file name*] [*target file name*]

Suppose you have a file on a disk in drive A:, which happens to be the logged drive, and you wish to make a copy of it on a disk in drive B:. If the name of the file is *XMAS86.LV1* and you want the copy to have the same name, the *COPY* command has the form

COPY XMAS86.LV1 B:

Because in this case the logged drive is drive A:, it is not necessary to include it in the name of the source file. It is exactly as though you had entered

COPY A:XMAS86.LV1 B:

It's a good idea, however, to use explicit drive names rather than relying on the system to default to the logged drive. If you use a microcomputer with a hard disk drive in addition to the two floppy disk drives, you need to keep track of a third drive name (C:). Including the drive name as part of the file name decreases the chance of an error in the copy operation.

You can also copy a file using a new name for the copied file. Suppose you would like to copy the file *XMAS86.LV1* from disk drive A: to disk drive B: and save it with the name *BOB.LV1*. In this case the *COPY* command is

COPY A:XMAS86.LV1 B:BOB.LV1

Remember that the source file name comes first and the target file name comes second.

You can transfer groups of files from one disk to another with the *COPY* command using the * or wild cards, exactly as shown above with the *DIR* command. Suppose you wish to transfer all the files with the extension *.LV1* from disk A: to disk B:. You would enter

COPY A:.LV1 B:*

The files copied to B: would have the same names as they had on the disk in drive A:.

Another common use of the *COPY* command is the transfer of all the files from one disk to another without changing the names. To do this, enter

COPY A:.* B:*

If you feel a little insecure about using the * and/or wild cards in copying files from one disk to another, practice using the *DIR* command with the combination of * and wild cards you want to try. When the directory that appears on the screen correctly lists the files you wish to copy, use the same combination of * and wild cards with the *COPY* command.

Remember that the target disk must be a formatted disk. Notice the difference between *DISKCOPY* and *COPY A:*.* B:*. Because the *DISK-COPY* command formats the target disk as it copies all the files to it, any files that are on the disk will be erased and written over. If files are already present on the target disk, the *COPY* command just adds the copied files to the files already present.

ERASE. This is another internal file that is available for use anytime the system prompt (*A>* or *B>*) shows on the screen. The *ERASE* command (you guessed it!) is used to erase a file or group of files. Since erasing is irreversible, it is especially important that you name the file exactly correctly. An example of erasing a single file is

ERASE A:XMAS86.LV1

As usual, the * and wild cards can be used with the *ERASE* command. If you want to erase all the files on a disk so you can start over with an empty but formatted disk, you can enter

ERASE A:.**

but if you do, the screen will respond with

```
               Are you sure (Y/N)?
```

because this is a rather drastic (and irreversible) action.

RENAME. This internal file is self-explanatory; it is used to change the name of a file without taking any other action. The old name comes first and the new name follows:

RENAME [old file name] [new file name]

The *RENAME* command is usually used with the logged drive; for example,

RENAME VIBROT.RP1 IRSPEC.RP1

If the file *A:VIBROT.RP1* exists on the disk in the default drive (drive A:), it will be renamed *IRSPEC.RP1*. Note that the correct file name must include the file extension, if there is one, and the new name must not be the name of an already-existing file. For either of those transgressions, you would be chastised with the following message on the screen:

```
      Duplicate file name or File not found
```

Usually when you name a file and you get this message or a similar one indicating that the file cannot be found, it is because you omitted the extension from the file name. So don't panic when it appears that a file it took you hours to create has disappeared. Instead, just try again and be sure that you have given the file a complete and accurate name including drive name, file identifier, and file extension.

WORD PROCESSING WITH WORDSTAR

Dozens of word-processing programs are available. WordStar was one of the first full-featured word-processing programs to be made available for use on a microcomputer operating under the MS-DOS, PC-DOS, and CP/M operating systems. AppleWriter was an early entry for the Apple series of computers. A glance at a computer trade magazine today shows offerings of Multimate, WordPerfect, Microsoft Word 4, WordStar, Volkswriter, PFS, and many others. The words in this book were first composed with WordPerfect. So why choose WordStar?

"To kill two birds with one stone" is as good an answer as any.

WordStar is a fine word-processing program. It has withstood the tests of time and of many critical users. It is powerful and consequently can be complex; but employed with only a small subset of its commands, it is simple, easy to learn, and entirely adequate for writing physical chemistry reports and other general writing. Furthermore, this subset of commands is essentially identical to the full-screen editor commands of Turbo Pascal. Turbo Pascal is a set of utility programs that permit writing programs in Pascal on a microcomputer; it will be discussed in Chapter 10. Thus, by using WordStar for word processing and writing physical chemistry lab reports, we do not need to learn a new word-processing program for Turbo Pascal.

Getting Started with WordStar

We will assume that your instructor has furnished you with a WordStar disk that was formatted with the system on it and a mostly blank, formatted disk for storing the files you write. It is helpful to have one or two practice files already on the mostly blank, formatted disk.

Place the WordStar disk in drive A: and the practice file disk in drive B: and turn on the microcomputer. Drive A: will operate, and after a few messages the Opening Menu appears on the screen as shown in Figure 9–1.

The Opening Menu of WordStar consists of two columns of eight letters corresponding to the commands that can be given to WordStar when this menu is on the screen. Of the 16 commands, only about 5 require a brief explanation before trying out the program:

```
J help

L change logged drive/directory

D open a document

P print a document

X exit
```

```
            WordStar Professional Release 4
          ═══════  O P E N I N G   M E N U  ═══════
┌─────────────────────────────────┬──────────────────────────────────┐
│  D open a document              │  L change logged drive/directory │
│  N open a nondocument           │  C protect a file                │
│  P print a file                 │  E rename a file                 │
│  M merge print a file           │  O copy a file                   │
│  I index a document             │  Y delete a file                 │
│  T table of contents            │  F turn directory off            │
│  X exit WordStar                │  Esc shorthand                   │
│  J help                         │  R run a DOS command             │
└─────────────────────────────────┴──────────────────────────────────┘

DIRECTORY    Drive B   332k free
```

Figure 9–1 Opening Menu.

J Help. If you forget the meaning of any of these commands, you can always enter a *J*, then one of the command letters on the Opening Menu screen for which you seek help. WordStar will write a brief explanation on the screen for you. After reading the information, press the *Esc* key to get back to the Opening Menu.

L Change Logged Drive/Directory. Since your WordStar disk was in drive A: when you entered *WS* to run it, WordStar assumes that all your files are also on a disk in drive A:. Because the WordStar program disk is so full of its own files, little room remains for you to store your files; they should be stored on a formatted disk in drive B:. To inform WordStar that you will retrieve and save your files on a disk in drive B:, enter *L* (with the Opening Menu on screen) to change the logged drive; in response to the query on the screen "What would you like the new drive to be?" answer by entering the letter *B* and pressing the return key. Drive B: will run for a moment and a directory of the files (if any) present on the disk in drive B: will appear on the screen.

D Open a Document. With this command you can retrieve one of the files you have previously saved on the disk in drive B:, or you can create a new file. After you enter *D* to execute this command, WordStar responds with a blinking cursor after the on-screen query "Document to open?"

If you respond by entering the name of a file on the logged disk drive (presumably drive B:), the drive will run momentarily and load the file into the computer's memory. The top portion (if it is a long file) will be displayed on the screen below the ruler line, and the blinking cursor will be located at line 1, column 1. The Opening Menu will be replaced by the Edit Menu (Figure 9–2).

If, in response to the query "Document to open?" you enter the name of a file that is *not* on the logged drive, WordStar will continue with the statements "Can't find that file. Create a new one (Y/N)?" If you enter a *Y*, the Opening Menu is replaced by the Edit Menu, the blinking cursor is at line 1, column 1, and you can begin writing.

P Print a Document. After you have typed a few lines—don't worry about typographical errors—try out the printer. First you must save your document by entering $^\wedge KD$, which will save your document, with the name you gave it, to the disk in the logged drive and return you to the Opening Menu. You enter $^\wedge KD$ by depressing the *Ctrl* key ($^\wedge$ is a shorthand symbol for *Ctrl*) and, while it is depressed, entering a *K* and then a *D*. The logged drive will run momentarily, and your file will be saved.

To print any file stored on a disk in the logged drive, just enter *P* when the Opening Menu is on screen. WordStar will respond with the query "Document to print?" Enter the full name of the file you wish to print. WordStar will then respond with "Number of copies?" You will probably answer *1;* if you do, about six more queries will follow. Normally, you will use the default settings to these queries to save time, which you

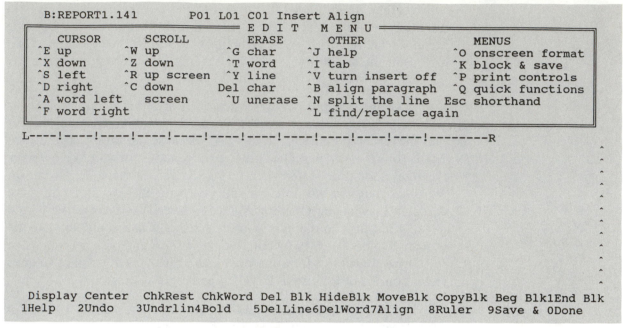

```
  B:REPORT1.141          P01 L01 C01 Insert Align
                            = E D I T   M E N U =
      CURSOR        SCROLL          ERASE        OTHER              MENUS
 ^E up         ^W up          ^G char      ^J help        ^O onscreen format
 ^X down       ^Z down        ^T word      ^I tab         ^K block & save
 ^S left       ^R up screen   ^Y line      ^V turn insert off  ^P print controls
 ^D right      ^C down        Del char     ^B align paragraph  ^Q quick functions
 ^A word left     screen      ^U unerase   ^N split the line  Esc shorthand
 ^F word right                             ^L find/replace again

 L----!----!----!----!----!----!----!----!----!----!----!-------R
```

Figure 9–2 Edit Menu.

can do by just pressing the *Esc* key as an answer. At this point the drive will run a bit and your file will be printed on the printer. (Be sure it is turned on!)

X Exit WordStar. With the Opening Menu on the screen, pressing the *X* key results in an *A:* showing at the bottom of the screen, indicating that you are out of WordStar and back in DOS. If you want to do more with WordStar, just enter *WS* and the Opening Menu will appear on the screen once again.

Enter a *D* from the Opening Menu and respond with the name of a new file. If you can't think of a name, use *MYFILE1*. Notice that the Opening Menu has disappeared and a new menu has appeared: the Edit Menu. Dividing the Edit Menu from the text file appearing on the screen is a horizontal line of dashes and exclamation points:

```
 L----!----!----!----!----!----!----!----!----!----!----!-------R
```

This is the **ruler line,** which indicates with an *L* and an *R* where the left and right margins are set. The exclamation points indicate the default tab settings. Since you probably will have no need to change these default settings for a while, let's not worry about them for now and get on with writing something.

When your new file is opened and the Edit Menu is on the screen, the cursor is positioned on line 1, column 1, which corresponds to the upper left corner of a printed page (allowing for margins). To enter text, simply begin typing. Do not press the return key when you reach the end of a line, because WordStar automatically "wraps around" to the next line. To see how this works, type in a series of words long enough to reach past the end of the line. Press the return key only at the end of each paragraph. For now, don't worry about correcting mistakes.

After you have typed a few lines and are finished with this new file for the present, you have three choices. First, you can save the file to the logged disk and come back to it later. This is accomplished by entering the command $^\wedge KD$.[2] This will save your file and return you to the Opening Menu. The name you gave to the file will be added to the disk directory.

Second, you can abandon the file by entering the command $^\wedge KQ$. If you have never previously saved the file, it will be abandoned and will not appear on the disk directory. If you have previously saved the file with a $^\wedge KD$, then you will have abandoned only the editing you just did. This command will also return you to the Opening Menu.

Third, you can save what you have done, and continue writing and correcting your errors by editing the file. To do this, enter $^\wedge KS$. The disk drive will operate, saving what you have done so far, but the file will remain on screen and you will be free to continue your writing. When you are doing a lot of writing, you should get in the habit of entering $^\wedge KS$ every 10 or 15 minutes. This is a safety measure to minimize the possibility of losing your work due to a malfunction such as an electrical failure. If such a failure occurs, you lose everything that is in the computer's memory, but nothing that is on a disk.

Save what you have done with $^\wedge KD$, which will return you to the Opening Menu. Get a practice file (about two pages long) on the screen by entering D followed by the name of the practice file. With a practice file you can try out some commands discussed below.

Moving the Cursor

When a file is opened, the initial position of the cursor is at line 1, column 1, and its current position is given by the status line at the very top of the screen. The status line also reminds us of the name of the file we are currently editing and the page number. The line (and page) number indicates the vertical position, and the column number indicates the horizontal position, of the cursor. The L on the ruler line is located at column 1 and the R is located at column 65. When the file is printed, the text on column 1 is indented 8 spaces from the left. A page consists of 55 lines, allowing for margins of about 3 lines at the top of the printed page and 5 lines at the bottom of the page.

The four arrow keys on the right side of your keyboard provide the simplest way of moving the cursor one line up, one line down, one char-

---

[2] The $^\wedge$ character will symbolize the control (*Ctrl*) key on the PC keyboard. To enter this command string, hold the *Ctrl* key down continuously, while pushing first the *K* key momentarily and then the *D* key momentarily.

acter to the right, or one character to the left. Try them out with the sample file. With long files, two more keys in this area of the keyboard are useful and self-explanatory: *PgUp* and *PgDn* move the cursor a page at a time.

In addition, WordStar provides a number of main keyboard commands that are used with the *Ctrl* located at the middle left of the keyboard. For example, holding the *Ctrl* key down and pressing the *F* key results in the cursor moving one word to the right. Figure 9–3 shows a number of WordStar's cursor movement commands. Reminders for some of these cursor movements are listed on the Edit Menu. To aid in remembering the cursor movement keys, their positions are somewhat analogous to the direction of the movements they cause. Compare the keys shown in Figure 9–3 with their positions on the keyboard shown below:

```
          E  R
       A  S  D  F
          X  C
```

Go back to your practice file now and try out these commands. They may seem a little awkward at first, but they have become second nature to thousands of WordStar users. You are about to join them.

Deleting

In order for you to correct mistakes, improve style, change the order of words, and update files, WordStar provides a variety of deletion commands. Let's begin with the smallest deletable unit, a single character, and work up to the largest unit, an entire file.

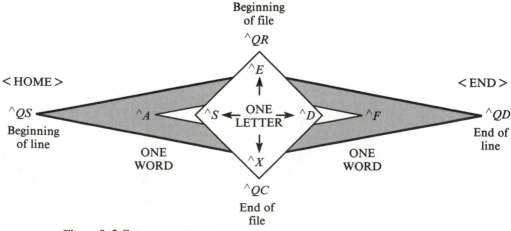

Figure 9–3 Cursor movements.

Deleting One Character at a Time. WordStar provides two methods for deleting one character at a time, and these are probably the most used commands in all of WordStar. Pressing the *Del* key, located at the bottom right of the keyboard, results in the deletion of the character that the cursor is currently on. Alternatively, enter a $^\wedge G$. More than one character can be deleted by simple holding down one of these keys for an extended period of time; the characters to the right of the cursor are deleted one after another.

Deleting Words and Lines. To delete the word to the right of the cursor, enter $^\wedge T$ with the cursor on the first character of the word. If a space lies to the immediate right of the cursor, then it will be deleted.

To delete the entire line on which the cursor is located, enter the command $^\wedge Y$. The cursor may be anywhere on the line. A variation on this command is the $^\wedge QY$ command, which deletes all the words from the cursor's position to the end of the line.

The Paragraph-Align Command. If you have practiced some of the deletion commands just discussed, you may have noticed that the deletion of a word or sentence left the remaining paragraph badly aligned, with blank spaces in the middle or at the ends of lines. The paragraph-align command is used to align paragraphs that have been "bent out of shape" by editing. To align a paragraph, place the cursor anywhere on the first line of the paragraph, then enter the paragraph-align command: $^\wedge B$. That's all there is to it. The cursor will race through the paragraph—that is, race to the next place where the return or enter key was pressed. As the cursor moves through the paragraph, the lengths of the lines will be readjusted and superfluous spaces removed.

Occasionally the cursor will halt at the end of a line and flash on and off. It is asking you if you want to hyphenate that word. To answer yes, press the - (hyphen) key; to answer no, enter $^\wedge B$ again. Normally, words at the ends of lines in a manuscript are not hyphenated unless they have permanent hyphens.

Manuscripts are also normally not right-justified. If your WordStar is configured so that it comes up with right justification turned on, turn it off by placing the cursor at line 1, column 1, and entering the command $^\wedge OJ$. This command toggles right justification on and off. If some of your text has already been entered and is right-justified, and you want to turn off the right justification with $^\wedge OJ,$ place the cursor on the first line of that paragraph and enter $^\wedge B$. The right edge will become ragged as the cursor sweeps through. For further information about the paragraph-align commands and how they are used to change the right or left margin or line spacing, enter $^\wedge JB$ to get the paragraph-align help menu.

Most often you will use the paragraph-align command as you insert or delete text in an unfinished text file, and its use will become automatic. This is one of the few commands that is very important in WordStar but is not used at all in the Turbo Pascal Editor.

Deleting Bigger Chunks. Suppose you would like to delete a sentence, a paragraph, or even several paragraphs. WordStar defines such a group of characters as a **block.** Naturally before you can ask WordStar to delete a block, you must tell WordStar where the block begins and where the block ends. To do this move the cursor to the beginning of the block and enter the command $^\wedge KB$. Then move the cursor to the end of the block and enter the command $^\wedge KK,$ at which time the marked block may visibly dim on the screen. Even if the marked block does not dim on the screen, it has been marked. To delete the block, enter the command $^\wedge KY,$ which results in the disappearance of the block. The cursor can be anywhere in the file when the command $^\wedge KY$ is entered.

Occasionally, you will want to get rid of entire files that are no longer needed but are taking up space on the disk. Normally, these are deleted when the Opening Menu is on screen, with its accompanying disk directory (Fig. 9–1). In the File Commands portion of the Opening Menu is the *Y DELETE A FILE* command. If you enter a *Y,* you will be prompted for a file name. Enter the file name exactly as it appears in the disk directory, including the extension. The disk drive will operate a few seconds and the file will be deleted. Forever. So be careful.

Moving and Copying Blocks

You have just seen how a block is defined by marking its beginning and end with the commands $^\wedge KB$ and $^\wedge KK.$ A marked block can be moved or copied, which permits rather drastic revisions of your manuscript with ease. Sometime in the course of your writing you will realize that the paragraph you have just completed really should appear two paragraphs earlier. With WordStar the remedy is simple. First, mark the paragraph as a block. Then move the cursor to the point in your manuscript where you would like the marked block to be moved. Finally, enter the command $^\wedge KV.$ The marked block will be inserted at the point where you positioned the cursor and will disappear from its previous position.

Notice the *V* in $^\wedge KV;$ it is to remind you of the word mo*V*e and that all the commands dealing with bloc*K*s use $^\wedge K$–. You can probably guess what the command is for *C*opying a block: $^\wedge KC.$ This command also moves the block to the current cursor position, but unlike the move command, this command leaves the original block unchanged. This is an especially useful command for replicating blocks. Perhaps you have several similar tables in a report. You can copy the first few lines containing the title and column heading, and with a small amount of editing you can create a new table beginning.

All of the commands, moves, and copies that have been described so far take place in the memory of the computer. You can also create a new file on the disk containing the marked block. Simply mark the block and enter $^\wedge KW,$ the command to *W*rite the file currently in memory to a disk for permanent storage. You will naturally be prompted to give a name to the file. As usual, it is a good idea to include the name of the disk drive with the name of the file; for example, *B:PARTII* instead of just *PARTII*. If you omit the name of the drive, the file will be read to the currently logged disk drive.

In a similar fashion you can retrieve into your document an entire file that is stored on your disk. Position the cursor at the place in your document where you want the new file insertion to begin, and enter $^\wedge KR$. You will be prompted for the name of the file, for example, *B:MYNEW-SEC*. The B: drive will run momentarily and the file *MYNEWSEC* will be inserted at the cursor into your current file.

Print Control

Although none of the print control commands is used by the Turbo Pascal Editor, several are of particular interest to chemists using WordStar, especially *subscript* and *superscript*. WordStar (and other word-processing programs) sends special commands to the printer to carry out certain tasks that only the printer can do and that serve to make the printed text clearer, more elegant, and more accurate.

Print control commands are toggles that switch printer functions on and off. Consequently, print control commands are always used twice: to toggle on and to toggle off. Thus, to underline a word, the print control command $^\wedge PS$ is entered, followed immediately by the characters making up the word, which is terminated by a second $^\wedge PS$ to turn off the underlining function of the printer. If you enter the following string of characters,

Einstein was a $^\wedge PSgreat^\wedge PS$ scientist.

the screen will show

```
         Einstein was a ^Sgreat^S scientist.
```

and the printed output will be

```
             Einstein was a great scientist.
```

Notice that the *P* part of the print command does not appear on the screen and none of the command is printed by the printer. The following commands are used in the same manner:

$^\wedge PB$	boldface
$^\wedge PD$	double strike
$^\wedge PS$	underline
$^\wedge PT$	superscript
$^\wedge PV$	subscript

Writing chemical formulas is no problem. When the character sequence

$H^\wedge PV2^\wedge PVSO^\wedge PV4^\wedge PV$

is entered from the keyboard,

$$H^\wedge V2^\wedge VSO^\wedge V4^\wedge V$$

shows on the screen and the printer prints

$$H_2SO_4$$

The most common error in using print commands is the omission of the second member of the pair. If you enter $^\wedge PS$ once and forget to enter a second $^\wedge PS$ where you want the underlining to stop, then the text will be underlined from the occurrence of the $^\wedge PS$ to the end of the printed file.

If you try the preceding example and it prints out *H2SO4,* then your WordStar program has not been configured to your printer. WordStar, like other word-processing programs, is designed to function with many different commercially available printers. Unfortunately, manufacturers have done little to standardize such features as superscripts and subscripts. Consequently, it is necessary to configure each user's WordStar to a particular printer with a program furnished by WordStar for that purpose. If you have a problem, call your instructor.

WYSIWYG

What You See Is What You Get. Almost. WordStar was one of the earliest word-processing programs to provide a printed copy virtually identical to what appears on the screen. When the text is right-justified, not only is it printed right-justified, but the text on the screen appears right-justified and every printed character is in the same relative position on the screen as on the printed page.

What about those print control characters that appear on the screen but not on the printed page? WordStar provides a number of commands to format your text so that the *on-screen* text coincides with the *printed* text. These commands all have an *O* in them. A beginner's list is given below:

$^\wedge OC$	*C*enters the line on which cursor is positioned
$^\wedge OD$	causes the print control commands to *D*isappear (an on/off toggle)
$^\wedge OJ$	right *J*ustification (an on/off toggle)
$^\wedge OL$	sets *L*eft margin (interactive)
$^\wedge OR$	sets *R*ight margin (interactive)
$^\wedge OS$	sets the line *S*pacing

To center a line, just place the cursor anywhere on the line, enter $^\wedge OC,$ and the line will center itself between the current margins.

With the screen showing some print commands, enter $^\wedge OD$ a few times. This action toggles the appearance and disappearance of the print commands. Note that even if you toggle the print commands off the screen, they are still in the file and will send a message to the printer.

When you boot up your WordStar disk, the margins are preset—for example, with the left margin at 6 and the right margin at 69. Suppose you would like to change the margins to 15 and 50. To reset the left margin, enter $^\wedge OL$. WordStar responds by writing

> New left margin?

just above the ruler line. Enter *15* to change the left margin to 15. Similarly, change the right margin by entering $^\wedge OR$ *50* (return). Alternatively, either margin can be reset to the current cursor position by entering $^\wedge OR$ followed by *Esc*. If existing text is present, it will be necessary, as usual, to align the paragraph with $^\wedge B$.

To change the line spacing, enter $^\wedge OS$. Just above the ruler line appears the message

> Enter new spacing?

Enter *1* for single spacing, *2* for double spacing, up to *9*. If, for example, you have an existing file that you wish to change from single to double spacing, enter *2* when you see this message. Then align each paragraph with $^\wedge B$. The actual spacing will appear on screen and the file will be printed as it appears. If the file is a new file, it is most convenient to set the spacing before beginning to enter text.

WordStar offers over 20 $^\wedge O$ commands, but the 6 listed on page 196 are the most frequently used.

FIND

Have you ever written a report only to discover that you have misspelled one word over and over again from beginning to end? Word processing programs like WordStar have a simple solution to this kind of problem. To find a word, place the cursor somewhere before the first occurrence of the word or phrase you wish to find and enter $^\wedge QF$. WordStar responds with

> Find what?

Enter your target word, press return, and the screen reads

> Option(s)?

Ignore the extended message for now and enter another return. The cursor will move through the text and come to rest, blinking, to the right of the last character in the target word. The target word, incidentally, may be *any* string of characters, including punctuation marks or spaces at the beginning, middle, or end of the string.

Want to do it again? Just enter $^\wedge L$, a frequently used function that means "repeat the *L*ast command." If the last command was the above $^\wedge QF$ sequence, WordStar will repeat the operation, sending the cursor to the next occurrence of your target word. The $^\wedge L$ command may be entered as many times as desired. Note that $^\wedge L$ repeats any previous command, not just the *FIND* command demonstrated here.

A number of options are available for finding a target word. The desired option is entered as a single letter, upper or lower case.

Option B. This option commands WordStar to search for the target string *B*ackward through the file from the current cursor position to the beginning of the file.

Option W. This option limits the search to *W*hole *W*ords. Suppose you wish to find all occurrences of the word "table." Without the *W* option, WordStar would pause at such strings as "comfortable," "stable," and "tables."

Option U. Without the *U* option entered, WordStar distinguishes between *U*pper and lower case. With the *U* option in effect, WordStar would find both "table" and "Table."

FIND AND REPLACE

When you need it, this is a powerful command. Suppose you write a 20-page report and then discover that about half the time you spelled "necessary" as "neccessary." After you enter $^\wedge QA$, the exchange between you and WordStar is

```
Find what?      [Enter your target word or string here]

Replace with?   [Enter your replacement word or string here]

Option(s)?      [Enter return, B, G, W and/or U]
```

The *B, U,* and *W* options used with the $^\wedge QF$ (*FIND*) function are the same with $^\wedge QA$. The $^\wedge L$ ("repeat last command") function may also be used with the find-and-replace sequence. An additional option is available for use with $^\wedge QA$. **Option G** is the *G*lobal replace, which has the effect of automatically entering $^\wedge L$ after each target word is found. The message *Replace (Y/N)* still appears in the upper right corner of the screen. The use of this option speeds up the operation in place of entering $^\wedge L$ after each target word is located.

Note that, with both $^\wedge QF$ and $^\wedge QA$, more than one option may be invoked. If you should opt to use *BU* with $^\wedge QF$, then WordStar will search *B*ackward through the file without distinguishing between *U*pper- and lower-case letters until it finds your target string.

In addition, both $^\wedge QF$ and $^\wedge QA$ and the options function identically when used with Turbo Pascal. The ability to *FIND* or *FIND AND RE-PLACE* is just as important in writing and editing programs as it is in writing up reports.

Except for the print control commands and $^\wedge B$ (paragraph align), all of the above WordStar commands are used in the Turbo Pascal Editor to enter programs. Even though you may eventually choose a word-processing program other than WordStar, familiarity with the WordStar set of commands permits immediate entry into the use of Turbo Pascal. Other word-processing programs all have more or less the same kind of capabilities, and once you have used one word-processing program it is quite a simple task to shift to another.

By now you've noticed that a number of the double-letter WordStar commands occur in groups with a common letter. When the Edit Menu is on screen, a menu of these commands appears on the right side of the Edit Menu:

```
^O onscreen format

^K block & save

^P print controls

^Q quick functions
```

If you invoke one of these partial commands, for example, $^\wedge O$, the Edit Menu disappears and is replaced by a new menu listing the various $^\wedge O$ commands. Entering the second letter of the command at this time causes the command to be invoked at the point where the cursor is currently blinking. When you know the command well enough to enter it completely and quickly, for example, $^\wedge OC$, no menu appears and the command is invoked immediately at the cursor. If you hesitate a few moments after entering the $^\wedge$ and the letter, WordStar very cleverly decides that you need a little help and writes the appropriate menu to the screen.

COMMERCIAL SPREADSHEETS

A spreadsheet allows the input of text, data, and equations expressed in BASIC. Input is to cells arranged in a two-dimensional matrix shown in Figure 9–4, the columns of which are referenced by letters and the rows of which are referenced by numbers. Thus, the cell in the upper left corner is located at A1, the cell to its right is at B1, and the cell immediately below it is at A2. This arrangement is similar to an accountant's balance sheet, and people in the business world have found spreadsheets invaluable. Because second-generation spreadsheets such as LOTUS 1-2-3 also include powerful data reduction and graphics capabilities, they have gained popularity in the scientific community (Coe, 1987; Levkov, 1987; Hilgeman, 1988). As an example, let us see how to use LOTUS 1-2-3 to carry out a linear least squares and generate a graphic representation of the results. For a comparison, let us use the data from Table 6–1 in Chapter 6. The text output is shown in Figure 9–4 and the graphic output in Figure 9–5. These may be compared with Figure 6–3.

```
C5: +$G$10*A5+$H$4                                              MENU
Fill  Table  Sort  Query  Distribution  Matrix  Regression  Parse
Fill a range with numbers
          A         B         C       D      E       F       G       H
 1                Linear Least Square
 2
 3        X         Y       Ycalc                    Regression Output:
 4                                       Constant                 4.306631
 5      -3.8       8.2    8.805524       Std Err of Y Est         0.739202
 6      -1.2       6      5.727334       R Squared                0.983947
 7      0.75       3.3    3.418692       No. of Observations             6
 8       3.1       1.6    0.636483       Degrees of Freedom              4
 9       5.6      -2     -2.32331
10         8      -6     -5.16472        X Coefficient(s)  -1.18391
11                                       Std Err of Coef.   0.075610
12
13
14
15
16
17
18
19
20
27-Mar-89  09:38 AM
```

Figure 9–4 Least-squares data on a LOTUS 1-2-3 spreadsheet.

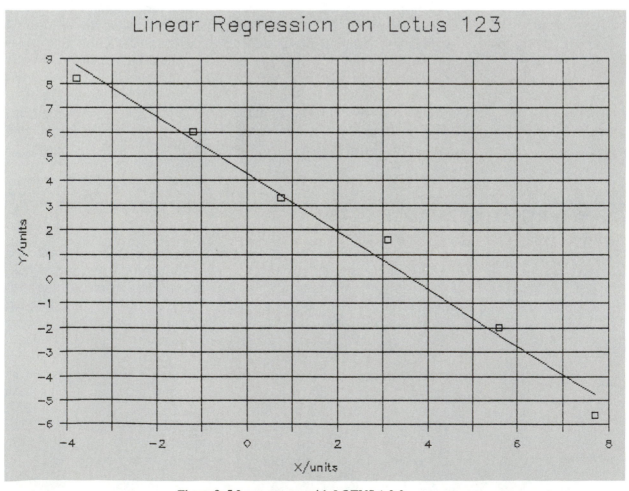

Figure 9–5 Least squares with LOTUS 1-2-3.

200

Linear Regression with LOTUS 1-2-3

Boot up LOTUS 1-2-3 by entering *LOTUS* after the prompt and then enter *1* to enter the spreadsheet facility. The data in Table 6–1 can be recognized in Figure 9–4. To duplicate this spreadsheet, place the cursor at B1 and enter a suitable title: *Linear Least Squares.* Place the cursor at A3 and enter $^\wedge X$. The caret[3] before the *X* causes the *X* to be centered in the cell. With the cursor at B3, enter $^\wedge Y$. Place the cursor at A5 and enter *−3.8*. Enter the remaining *X* values in cells A6, A7, A8, A9, and A10. Beginning with 8.2, place the *Y* values in cells B5 through B10 as shown in Figure 9–4. One more title can be entered by placing the cursor at C3 and entering *Ycalc.* At this point the spreadsheet looks like Figure 9–4 (columns A through D) but with no numbers under *Ycalc.*

All the remaining text and numbers on the spreadsheet are generated by LOTUS 1-2-3 after running the LOTUS 1-2-3 command *Regression.* Get the command line on the sceeen by entering a /, which results in the main *Worksheet* menu line

```
Worksheet Range Copy Move File Print Graph Data System Quit
   Fill, Table, Sort, Query, Distribution, Matrix, Regression, Parse
```

On all the command menus of LOTUS 1-2-3, the command is most simply invoked by entering the first letter of the command. Alternatively, you can use the arrow keys to move the cursor to the command word and invoke it by pressing the return key. So enter *D* for *Data reduction* and then *R* for *Regression* to give the following submenu line:

```
   X-Range  Y-Range  Output-Range  Intercept  Reset  Go  Quit
   Set independent variable(s), or X, range
```

Incidentally, you can go backward through the submenus by repeatedly pressing the *Esc* key or the *Q* (for *Quit*) key. Next we must tell the regression routine where the data are. With the above menu in place, enter an *X* for *X-Range,* enter *A5..A10,* and press the enter key. With the same menu in place, enter *Y* for the *Y-Range* and then enter *B5..B10* to establish the corresponding *Y* values that will be used in the linear regression.

In Figure 9–4 the output appears in a block of cells bounded by E3 in its upper left corner and by H11 in its lower right corner. To duplicate this output, enter an *O* for *Output range* and enter *E3.* This locates the upper left corner of the output at E3 as shown in Figure 9–4. If you don't like it there, you could, for example, enter *A12,* which would result in the output being located under the input data. To run the linear regression, enter *G* for *Go.* The screen should now look like Figure 9–4, except that the *Ycalc* values are still missing. The *X, Y* data pairs have been fit to a

---

[3] The caret in LOTUS 1-2-3 must not be confused with the caret in WordStar. Recall that the WordStar caret symbolizes the *Ctrl* key.

straight line of the form $Y = mX + b$. In the block labeled *Regression Output*, b is the *constant*, equal to 4.306631, and the slope m is the X *coefficient*, equal to -1.18391. These are, of course, the same values we calculated in Chapter 6 with the data in Table 6–1.

The *Ycalc* values are calculated from m, b, and the corresponding values of X that were input (cells A5 through A10). Put the cursor at C5 and enter the expression *$G$10*A5+$H$4*. When you press "enter," the value of this expression (8.805524) appears in cell C5. The value 8.805524 is the result of the value of the cell G10 (m) multiplied by the value of cell A5 (X) plus the value of the cell H4 (b). To get the remaining values of *Ycalc*, we need similar expressions in cells C6 through C10. To accomplish this, we copy the contents of cell C5 into the cells C6..C10 with the *COPY* command. Place the cursor on C5 (if it isn't already there), and enter /C. The range you wish to copy from is just the single cell C5, the range of which is denoted by *C5..C5*. Just enter return. Now enter *C6..C10* to establish the range to which you want the expression in cell C5 copied. When the expression is copied, the correct *relative* cell locations will be copied. However, *$* on each side of the *G* and *H* keeps the contents of cells G10 and H4 *absolute*. Run the cursor over each cell from C5 through C10 to verify that the correct expression is in place, as shown in the upper left corner of your monitor. As soon as the enter key is pressed, the remaining values of *Ycalc* should appear on the screen. The screen will now look exactly like Figure 9–4.

Graphing the Linear Regression with LOTUS 1-2-3

Enter / to get the main command line on the screen. Then enter *G* for *Graphics*. The screen now displays the Graphics Menu line above the column-heading letters:

```
Type  X  A  B  C  D  E  F  Reset View Save Options Name Quit
Set graph type
```

Enter a *T* for *Type*, and a new menu appears as follows:

```
        Line  Bar  XY  Stacked-Bar  Pie
        Line graph
```

Enter an *X* for *XY*, the type of graph.

Back on the previous menu, Enter *X* to select the range of *X* values to be graphed. For the *X* range, use the *X* values we used for the linear regression by entering *A5..A10*. Press the enter key, then enter *A* for the *A* range. Enter the *Y* values by entering *B5..B10*. Finally, enter a *B* for the *B range*, and supply the *Ycalc* values by entering *C5..C10*.

Now we want LOTUS 1-2-3 Graphics to draw the linear regression line that passes through all the *X*, *Ycalc* values. But we also want some little circles drawn at each of the experimental *X*, *Y* values. These are our options, and we select them by entering *O* for *Options* when the Graphics Menu (see above) is displayed. This causes the Options Menu to be displayed as shown below:

```
Legend Format Titles Grid Scale Color B&W Data-Labels Quit
Specify data-range legends
```

With the Options Menu on screen, enter *F* for *Format.* This brings up the Format Menu as shown below:

```
Graph  A  B  C  D  E  F  Quit
Set format for all ranges
```

Enter *A* and then select *S* for *Symbols (only).* The possibilities are shown in the following submenu:

```
Lines   Symbols   Both   Neither
Draw lines between data points
```

Then, for the *B* data range, enter *L* for *Lines (only).* This selection results in a symbol drawn on the graph for each *X*-*Y* pair input and a straight line for the *X*-*Ycalc* pairs.

The other options, such as *Title (1st, 2nd, X axis,* and *Y axis*) and *Grid,* are self-explanatory. Entering *Q* for *Quit* a few times brings you back to the Main Graphics Menu. Enter *V* for *View* to see what the graph of the linear regression looks like. When it is satisfactory, enter *S* for *Save* and give the picture a file name, which will have the extension *.PIC.*

Enter *Q* again and *Y* to end the LOTUS 1-2-3 session, which will bring you back to the LOTUS 1-2-3 Main Menu; then you can select *P* for *PrintGraph.* The *Image-Select* command gives a directory of the files with the extension *.PIC* for selection and printing.

If you have carried out a linear regression with *Program LstSqr* (Fig. 6–18) *and* with LOTUS 1-2-3, you have seen that the use of LOTUS 1-2-3 for linear regression is much trickier and less user-friendly. In general, as software becomes more versatile and more powerful, it invariably becomes more difficult to use. The user must necessarily compromise in seeking low cost, high power, and ease of use. Sometimes the best solution to a small problem is to write your own software or modify existing software, such as *Program LstSqr.* In Chapter 10 we will introduce the Pascal programming language. Pascal is simple, readable, and easy to maintain; consequently, Pascal programs are easy to modify and customize.

REFERENCES

1. Chertok, B. L., D. Rosenfeld, and J. H. Stone, *IBM PC and XT Owner's Manual,* Robert J. Brady Co., Bowie, Maryland, 1984.

2. Coe, D. A., "Using a Spreadsheet To Calculate the Fugacity of a Van der Waals Gas," *J. Chem. Educ.* 64:137, 1987.

3. Hilgeman, F. R., and R. H. Richter, "A Curve-Fitting Method for Experimental Data," *J. Chem. Educ.* 65:A96, 1988.

4. LeBlond, G. T., and T. Carlton, *Using 1-2-3,* Que Corporation, Indianapolis, 1985.

5. Levkov, J. S., "The Use of Commercial Spreadsheet Programs in the Science Laboratory," *J. Chem. Educ.* 64:31, 1987.

6. Puotinen, C. J., *Using the IBM Personal Computer: WordStar,* Holt, Rinehart and Winston, New York, 1983.

Pascal for Physical Chemistry

A computer program is a list of instructions that can be understood by both people and computers. A computer can understand only one language, called machine language. People have invented a number of languages for giving instructions to computers, among them FORTRAN, ALGOL, C, BASIC, APL, and Pascal. Pascal is not written in capital letters because it is not an acronym. The Pascal programming language was named in honor of Blaise Pascal, the French mathematician, by its developer, Professor Niklaus Wirt of the Eidgenossche Technische Hochschule (ETH), Zurich, Switzerland.

In order for a people's programming language to be translated into a computer's machine language, it must be processed by a special program called a compiler. In this chapter, we will examine a version of Pascal called Turbo Pascal, a sophisticated and powerful programming language developed by Borland International, Scotts Valley, California. Although Turbo Pascal Version 5.0 is a complete programming environment, in a short time a beginner can easily master enough of the Turbo Pascal environment and a subset of the Pascal language to write programs for solving quite complex scientific problems.

Writing a program in Turbo Pascal is very much like writing a letter with WordStar. In fact, Turbo Pascal contains within it a *text editor* that is really just a simple word processor for entering the program, and the Turbo Pascal Editor commands are nearly identical to WordStar commands.

THE COMPONENTS OF TURBO PASCAL

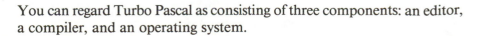

You can regard Turbo Pascal as consisting of three components: an editor, a compiler, and an operating system.

Pascal Operating System

The operating system carries out the same sort of tasks as the PC operating system does. It saves and retrieves files. It gets you back and forth among the editor, the compiler, and the operating system. It provides for and maintains a directory of your program files. Finally, it gets you out of Turbo Pascal and back into the PC operating system.

Pascal Editor

The editor is where you spend most of your time when you are writing a program. You can spend hours here, whereas you will spend only minutes with the Turbo Pascal operating system, and only seconds with the compiler. It really is fast! The Turbo Pascal Editor is fashioned after WordStar. Most of the editor commands are identical to those of WordStar, so if you are familiar with WordStar, you are already familiar with the Turbo Pascal Editor.

Occasionally you will take a break from writing the source program in the Pascal language with the Turbo Pascal Editor. You will then want to save what you have written on a disk so that you can retrieve it and continue later. When you do so, you may give the source program an eight-character name to which the operating system will add the extension *.PAS* (e.g., *ProgName.PAS*). The rules for naming Pascal program files are essentially the same as the rule for naming DOS files, described in the previous chapter.

Pascal Compiler

Before a Pascal program is run, it must be compiled, which Turbo Pascal does for you upon request. The compiled program may then be stored in compiled form on a floppy disk, in which case its name is the same as that of the source program file—except that its extension, automatically added, is *.EXE* (e.g., *ProgName.EXE*). Once compiled, it is run from the computer operating system (DOS) at the sign of the usual prompt (*A>*) by simply entering the file name (*ProgName* in this example), and pressing enter.

THE NATURE OF THE PASCAL LANGUAGE

Like any language, whether it be French or FORTRAN, Pascal has its own syntax and rules for grammar and punctuation. One of the useful features of Pascal is the identifiers allowed for variable names. You could use *A* and *B* or *X* and *Y*, but Pascal allows meaningful names like *HeatCapacity, Mole, Density,* and *Time.* Pascal also allows the use of both upper and lower cases, which is to be encouraged, as it aids in the readability of identifiers. (Compare, for example, *HeatCapacity* and *HEATCAPACITY*.) This helps enormously in making your source program more understandable to you and to others who use it. Pascal does not distinguish between upper and lower cases in identifiers.

In some ways, Pascal is just like any other programming language. For example, the arithmetic operators are the same:

Operation	Symbol
Addition	+
Subtraction	−
Multiplication	*
Division	/

On the other hand, Pascal classifies its variable into more types than do other languages. In addition to integer numbers, real numbers, and arrays, Pascal recognizes other types: characters, Booleans, records, sets, files, and even variable types defined by you, the user! For this reason Pascal is often described as a highly "typed" language. Most of the more modern programming languages are also highly typed; this is sometimes inconvenient to those who have already programmed in BASIC or FORTRAN, but it leads to clear thinking and to ease in debugging and maintaining programs.

Pascal is also described as a highly structured language because of the way it can be divided up into blocks of code, that is, blocks of statements written to carry out a particular small task. Pascal by its nature encourages solving problems by breaking them down into smaller and smaller parts until the solution to each part becomes simple. The grammar, syntax, and punctuation of Pascal make it easy to recognize the component parts of the program. Together, these features make Pascal programs readable and understandable.

A SIMPLE EXAMPLE OF A PASCAL PROGRAM

Even before seeing how the Turbo Pascal operating system works, let us look at a very short Pascal program listed in Figure 10–1. *Program Birthday* calculates the year of your birth if you tell it how old you are. Although short and simple, it does illustrate the major sections of a Pascal program:

1. The *header,* which is only one line long.
2. The *definitions,* where you define the *constants* and *data types* that will be used later in the program.
3. The *declarations,* where all *variables* and *procedures* to be used in the *main program* are declared and their types given.
4. The *statements,* where the program's statements are given sequentially.

```
Program Birthday;

Const ThisYear = 1989;

Var Birthyear,Age:  Integer;

Begin  (* Main Program *)
   Write('How old are you?  ');
   Readln(Age);
   BirthYear:= ThisYear - Age;
   Writeln('You were born in ',Birthyear)
End.    (* Main Program *)
```

Figure 10–1 An example of a Pascal program.

Only section 4 is absolutely necessary, but our first example program (Fig. 10–1) has all four.

The block structure of Pascal is already evident in *Program Birthday.* The four main sections of a Pascal program are clearly confined to their respective blocks.

The Semicolon as a Block Separator

Pascal uses the semicolon as a block separator. How many block separators are required to separate four blocks? Just three. One follows the word *Birthday,* a second follows *1989,* and a third follows *Integer.*

Notice that the statement section can be further divided into smaller blocks, each consisting of a single statement. How many semicolons are required to separate the statements (blocks) that make up the statement section? Again we happen to have four statements, which require three semicolons to separate them. For this reason, no semicolon follows the) in the next-to-last line of the program. Actually, a semicolon could be placed there, but the compiler would ignore it.

The compiler also ignores the statements enclosed by the symbols (∗ and ∗). These statements are simply *comments* added to help understand the program. Comments are an important component of more complicated programs and help increase their readability. Comments can be placed anywhere in the program: between lines, at the ends of lines, even in the middle of a line, although that would be unusual. You should use comments freely to make the program more readable and to clarify what is going on. Often, code that you write is clear at the moment, but when you come back to it after several months, you may not quite understand what you yourself have written. Comments help keep code fresh in your mind and understandable in the minds of others. Even as you begin to write simple programs, use comments freely, so that their use becomes an unshakable habit.

Input and Output (I/O)

It is safe to say that all programs have information of some kind that is output, and that most programs accept some kind of information as input, which they then process. In small programs information is usually put into the program by means of a keyboard[1] and output to a cathode ray tube (CRT) screen. The I/O instructions used for this purpose by Pascal are the statements

> *Read* and *readln:*   these accept data from the keyboard.
> *Write* and *writeln:*   these write data to the CRT screen.

When *Program Birthday* is run, the words *How old are you?* are output to the screen. Because the next statement encountered is the *readln* statement, the program then stops and waits for data to be input from the keyboard. The fourth statement carries out a simple calculation, a subtraction to obtain a value for the variable *BirthYear.* The last statement of the program, a *writeln* statement, again outputs information to the screen. This time there are two parameters in the list of parameters to be printed. The first parameter is the string of characters between the quotation marks: *you were born in.* The second parameter in the list is the value of the integer variable *BirthYear.*

The combination of characters with a colon followed immediately by an equal sign at its end is called the *assignment statement.*[2] In the third line of the statement block, the value of the variable *Age* is subtracted from the value of the constant *ThisYear:*

$$ThisYear - Age$$

The difference resulting from this subtraction (an arithmetic operation) is then *assigned* as a value to the variable *BirthYear:*

$$BirthYear := ThisYear - Age$$

The value of the constant *ThisYear* is defined in the definition section of the program, while the value for the variable *Age* is input from the keyboard with the *readln* statement.

Finally, after the last **end** statement of the entire program, a period signals the compiler that the end of the program has been reached. Every Pascal program contains one period, and it is located at the very end of the program.

Identifiers

Even in this very small program, several identifiers appear. The word *Birthday* was chosen as an identifier for the program. In the definition

---

[1] When the program must handle large amounts of data, the data may be read from or written to external files stored permanently on floppy disks. The use of files is discussed in more detail at the end of this chapter.

[2] In FORTRAN and BASIC, the assignment statement ends with only an equal sign.

section of the program, a constant was given the identifier *This Year.* The program's two variables have the identifiers *Birth Year* and *Age.* Turbo Pascal is quite flexible when it comes to naming identifiers. An identifier consists of any combination of letters, digits, and underscores. An identifier must begin with a letter, however, and it is limited to 127 characters. Upper- and lower-case letters can be mixed and are not distinguished, that is, *Age, age,* and *AGE* are not different identifiers.

Some legal identifiers are

> *Vector*
> *Read_Equations*
> *Slope*
> *X*
> *SumX3*

Some illegal identifiers are

> *This Year*   (contains a space)
> *2ndEquation*   (begins with a digit)
> *Years/Months*   (contains the division operator)

You are encouraged to use meaningful identifiers to make your programs easy to read. As you will see, if you choose an illegal identifier, the Pascal Compiler will give you an appropriate error message and even locate the point in the program where the error occurs.

Reserved Words

By now you have probably already guessed that certain words must be reserved by the Pascal language for special purposes. These reserved words, of course, may not be used as identifiers. In our sample program (Fig. 10–1), the reserved words are **program, const, var, begin,** and **end.** A complete list of reserved words is given in Table 10–1. Reserved words may be used in any combination of upper and lower cases. Most of them

TABLE 10–1

Reserved Words in Turbo Pascal

absolute	end	inline	procedure	type
and	external	interface	program	unit
array	file	interrupt	record	until
begin	for	label	repeat	uses
case	forward	mod	set	var
const	function	nil	shl	while
div	goto	not	shr	with
do	if	of	string	xor
downto	implementation	or	then	
else	in	packed	to	

will become very familiar to you, so you will have little trouble in finding proper identifiers.

Before firing up Turbo Pascal, let us look once more at the statements beginning with *read, readln, write,* and *writeln.* When a program runs, the instructions are carried out the way you read and write: sequentially, from left to right and top to bottom. On a more detailed scale, you can think of an invisible pointer or cursor scanning sequentially through the program. When that invisible pointer encounters a *readln,* it reads all the data on that line and then it executes a carriage return and a form feed. In other words, the pointer advances to the beginning of the next line. On the other hand, when the invisible pointer encounters a *read* statement, it reads through the end of the data and then stops dead in its tracks. It won't proceed to the beginning of the next line unless it encounters a statement that carries out that so-called carriage return and line feed, namely a subsequent *readln* or *writeln.*

The differences between *read* and *readln* and *write* and *writeln* may seem unnecessarily complicated at this time, but in practice the differences provide you with great power to control the way data show up on the screen. This in turn will help you to process and provide information more clearly, as well as more quickly. And that is what computers are designed to do.

GETTING STARTED WITH TURBO PASCAL

Turbo Pascal Version 5.0 consists of two diskettes containing several files: the editor, the compiler, the operating system, error messages, some utility programs, and some sample Pascal programs. To get started, boot the IBM-PC operating system diskette as usual. When the prompt *A>* appears, insert the Turbo Pascal diskette in drive A: and a formatted disk in drive B:. Type in *Turbo* and hit the enter key. After a few seconds you should see the display shown in Figure 10–2.

Tap any key and the trademark window disappears, leaving a mostly blank screen divided by a horizontal line labeled *Output.* The Main Menu is the single line at the top of the screen:

```
File  Edit  Run  Compile  Options  Debug  Break/watch
```

As a beginner, you will need only the first four commands. Let us assume that you have no programs on your disk and that you wish to write one. The Main Menu commands you use are **F**ile to create a new file for your program, **E**dit to enter, change, and correct your program, and **R**un to run it. The **R**un command actually compiles *and* runs the program. Only the first letter of each command need be entered to initiate the Main Menu commands.

To begin, enter an **F** for File and immediately the File Menu window appears as shown in Figure 10–3. The command you will use to begin a

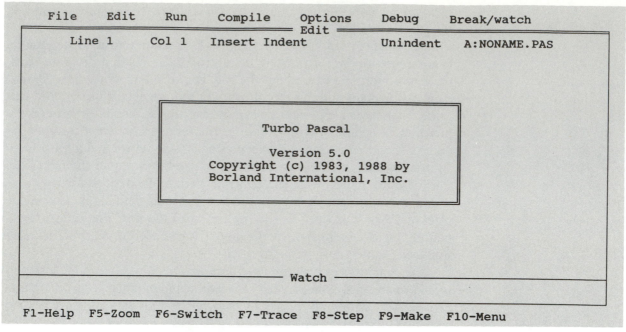

Figure 10–2 Turbo Pascal logo.

new file is New. The other commands in the File Menu that you will use frequently, even as a beginner, are Load, Save, and Quit.

Quit is the last command you use at each programming session. Entering a **Q** when the File Menu is up returns you to DOS, the com-

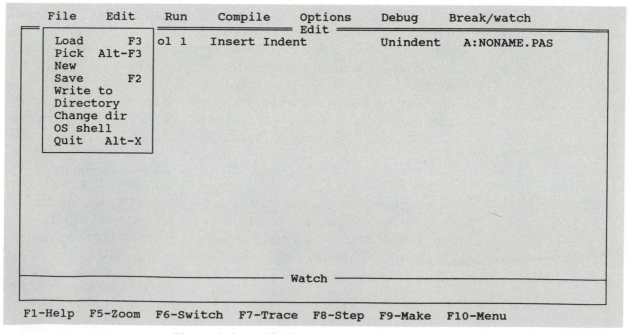

Figure 10–3 The File Menu.

puter's disk operating system. (Another way to exit Turbo Pascal is to enter *Alt-X,* which is equivalent to entering **F**, **Q**.)

At this time, however, enter **N** for New. As a result, the File Menu disappears, and a blinking cursor appears at the upper left corner of the blank part of the screen. The Pascal operating system has activated the editor so that you can enter a program. *Whenever the blinking cursor is on the screen somewhere, the editor is active.* The Main Menu still shows at the top of the screen, but it is now inactive. The cursor is located at column 1, line 1, as indicated on the editor line, which reads

Line 1	Col 1	Insert	Indent	A:NONAME

The visible part of the screen is 78 columns wide and 21 lines deep. Touching the *F5* key will toggle the screen so that either the full 21-line region is visible or a lower region for output is visible. If you have used WordStar, you are at the same point you would be if you had entered the WordStar *D* command. In other words, you have a blank page on which to write a Pascal program.

Entering and Editing the Program

Figure 10–1 lists a very short program that gives you a chance to try out the Pascal Editor. It is possible to write an even shorter program than the one shown in Figure 10–1, but *Program Birthday* is long enough to illustrate a few important characteristics of the Pascal language. So, with your cursor at line 1, column 1 (the upper left corner of the page), type in the program in Figure 10–1 exactly as written. Use colons where colons are used, use semicolons where semicolons are used, and use spaces where spaces are used. When you are finished copying the program in Figure 10–1, your screen will appear as shown in Figure 10–4 and the cursor will be at line 12, column 26.

Spaces are used to separate reserved words from identifiers; at least one is required, but extra spaces are generally ignored. The two spaces after the question mark are for readability only. If you need to correct any typographical errors, just use the same commands for moving the cursor and for editing that you have used with WordStar, described in Chapter 9.

When you are satisfied that the program looks like Figure 10–4, you can save your file by pressing the *F2* key. The *F2* saves your work, but you remain in the editor. To leave the editor and go back to the Main Menu, press the *F10* key. Pressing the *F10* key causes the blinking cursor to disappear, which is your cue that you are no longer in the editor and that the Main Menu is waiting for your command.

Saving a Pascal Program

The program you have entered from the keyboard that now appears on the screen resides in the computer's memory, where it would disappear

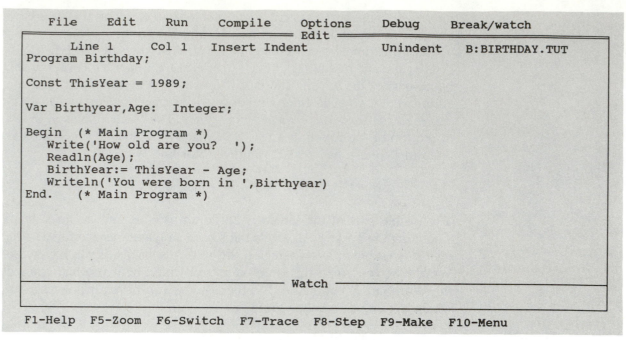

```
    File      Edit      Run      Compile      Options      Debug      Break/watch
══════════════════════════════════════════ Edit ══════════════════════════════
        Line 1       Col 1      Insert Indent              Unindent     B:BIRTHDAY.TUT
Program Birthday;

Const ThisYear = 1989;

Var Birthyear,Age:  Integer;

Begin  (* Main Program *)
   Write('How old are you?  ');
   Readln(Age);
   BirthYear:= ThisYear - Age;
   Writeln('You were born in ',Birthyear)
End.    (* Main Program *)




══════════════════════════════════════════ Watch ═════════════════════════════
F1-Help   F5-Zoom   F6-Switch   F7-Trace   F8-Step   F9-Make   F10-Menu
```

Figure 10–4 *Program Birthday* on screen.

if you turned off the computer. To save the program you have written
as a file on a diskette, enter an **F** for File now that the Main Menu is
active (no blinking cursor). The File Menu now appears just as in Figure
10–3 except that part of your program appears beneath it. To save the
program on a diskette, enter **S** for **S**ave, and a small new window will
appear over the File Menu suggesting that you rename the file, replacing
the name temporarily given to your program. Move the blinking cursor
to the first *N* of *NONAME*, press the delete key six times, and then enter
BIRTHDAY as a file name. The *Rename NONAME* window should now
contain

A:\BIRTHDAY.PAS

When you press the return key, the A: disk drive runs a few seconds as
your file is sent from the memory of the computer to the disk in
drive A:.

To demonstrate that your file was indeed saved, enter **D** for **D**irectory
(the File Menu should still be on screen). A new window appears (labeled
Enter Mask) containing ∗.∗. You can ignore this by pressing the enter
key. If you want to see a directory listing only those files with the extension
.PAS, edit the window entry to read ∗*.PAS,* then press the enter key. A
new window labeled *A:\∗.PAS* will appear, and in its list of file names
BIRTHDAY.PAS will appear. While the screen lists the directory of
A:∗.PAS, the bottom of the screen displays a one-line menu of possible
commands. Let us return to the Main Menu by selecting the *Esc-Abort*

command, which aborts the window on the screen. Pressing *Esc* twice causes the two windows to disappear, leaving the program on the screen and the Main Menu waiting for a command.

Compiling and Running a Pascal Program

Normally at this point we would simply enter **R** for **R**un to see if our programming is successful. Entering **R** does two things: it causes the program to be compiled and then, if the compiling is successful, it causes the program to run. In order to get a better idea of how Turbo Pascal functions, it is instructive at this point to compile and run separately. To do this, press *F10* to get the Main Menu active, then just enter **C** for **C**ompile. When **C** is entered, the Compile Menu appears:

```
Compile  Alt-F9
Make         F9
Build
Destination       Memory
Find error
Primary file:
Get info
```

The first menu item is **C**ompile, so enter **C** again. Now the compiler takes off, and if your program was entered exactly as shown in Figure 10–1, the screen will appear as in Figure 10–5. A window titled *Compiling to Memory* and displaying the compiling information covers part of the Compile Menu. Press any key and the *Compiling to Memory* window disappears, leaving your source code on screen, the editor active, and the cursor blinking at column 1, line 1. Press *F10* to return to the Main Menu. The source code stored in memory was compiled, and the matching code resulting from compiling is now in memory. To run the program, enter **R** for **R**un, and a little run menu appears on the screen. Enter **R** again (or the *F9* key), and the line

```
How old are you?
```

appears on screen. If you respond by entering *20,* the screen displays the following output:

```
How old are you?  20
You were born in 1969
```

Before you can read these lines of output, however, Turbo Pascal instantly switches back to the source code screen in editor mode. To get your output back on the screen so you can read it, you must press *Alt-F5,* which enables you to toggle back and forth between the output screen and the editor.

If you originally entered **R** for **R**un instead of **C** for **C**ompile, the result would be the same if no typographical errors were present. Com-

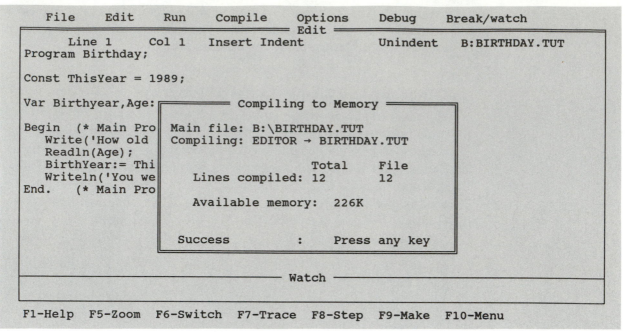

Figure 10–5 Compiling to memory.

piling would take place and the program would run immediately. Thus, entering **R** is equivalent to entering **C** then **R**. In either case the screen appears as shown in Figure 10–4, which displays both the program and its output.

Dealing with Errors

There is a good chance that you have already encountered an error. Perhaps you made a typographical error in copying the program in Figure 10–1 with the Turbo Pascal Editor, or in running the program you accidentally entered a letter or decimal number instead of an integer for your age. When an error occurs, Turbo Pascal is very helpful not only in identifying the kind of error, but also in locating the point in the program where the error occurred. Let us classify three kinds of errors and examine how they arise: compile-time errors, run-time errors, and logical errors.

Compile-Time Errors. Let us assume that you have correctly copied *Program Birthday* onto the screen and that it compiles and runs correctly. Introduce an error in the program by changing the spelling of the identifier *BirthYear* in the main block of the program to *BirtYear*. Press *F10* to leave the editor, then **R** or **C** (in this case the result will be the same). The disk drive will run to access the compiler, and if you watch carefully you will see the compiler message window appear momentarily on the screen just before the system goes back to the editor. Notice that the blinking cursor is located on the *B* of *BirtYear,* locating the point in the program where the error occurred. Moreover, the edit line near the screen top has been replaced with the message

> Error 3:   Unknown identifier

The message is consistent with the error, since the variable *BirthYear* is declared in the declaration section. When the compiler meets the identifier *BirtYear,* which has never been declared, the compiler does not recognize it and delivers the appropriate message. The ability of Turbo Pascal to identify the type and location of an error is exceedingly helpful in debugging a program.

Run-Time Errors. Once again let us assume that your *Program Birthday* compiles and runs without any errors. Run it, but enter your exact age as a *real* (decimal) number instead of an *integer,* for example, *20.4* instead of *20.* The screen will now appear as follows (after toggling with *Alt-F10*):

> How old are you?   20.4
>
> Run time error 106 at 1D9A:0060

Upon toggling again with *Alt-F10,* you are returned to the editor and again the blinking cursor is located at the position of the error in your program, this time at *Readln(Age).* Again a message replaces the edit line near the top of the screen:

> Error 106:   Invalid numeric format

The *Readln(Age)* statement expects an *integer* to be input since the variable *Age* is declared to be of type *integer.* Inputting a *real* number like 20.4 causes a run-time error, even though the program compiled without error. Try entering *XX* for *20* and the same error results, since the *X*'s are characters, not integers.

Logical Errors. Suppose that a student entered *Program Birthday* (Fig. 10–1) into her computer and it compiled without error, so she ran the program and entered *20* for her age. The program ran with no run-time error message, but the screen read

> You were born in 2009

What happened? Upon typing in the statement

$$BirthYear := ThisYear - Age;$$

which is the algorithm for the program, she inadvertently entered a plus sign instead of a minus sign, so the statement read

BirthYear:= ThisYear + Age;

Such an error is called a *logical error*. It generates neither a compile-time nor a run-time error message; nevertheless, the output is wrong because the algorithm is illogical.

Because the Pascal compiler understands neither the chronology of birthdays nor the theories of physical chemistry, its operating system cannot detect logical errors. Logical errors are more insidious and more difficult to detect than other errors, which the compiler often detects and locates. In the above example, the error in the output was so large that it was easy for the programmer to detect it and subsequently correct her input. Often, however, a logical error can lead to relatively small errors in the output that are not so obvious. In this example the logical error arose from a typographical error. A logical error might also arise from a misunderstanding of a more complex algorithm. The only way to ensure that a program is free from logical errors is to run it with data for which correct output is known. Correct data might come from the literature or from a hand-calculator calculation. A computer calculation and a hand-calculator calculation involving real numbers should agree exactly to as many digits as the devices carry—about nine digits.

Logical errors and poor input data are the origins of the well-known acronym GIGO: Garbage In, Garbage Out.

A list of the error numbers and messages, along with more detailed explanations, is given in *The Turbo Pascal Reference Guide,* pages 451–472.

Compiling to Memory Versus Compiling to Disk. Eventually you will write a program and want to save the machine code on a disk so that the program can be run directly without the need for the Turbo Pascal environment. This is simple with Turbo Pascal. Just get your debugged program, which you have demonstrated to compile successfully, on the screen. With the Main Menu active, enter **C** for **C**ompile just as before. The Compile Menu comes up and the third line down reads

```
                    Destination Memory
```

With the Compile Menu on screen, press **D** for **D**estination and notice that this line changes to

```
                    Destination Disk
```

Press **D** several times and observe how the line toggles between *Memory* and *Disk.* Leave it so that it reads *Destination Disk* and then press **C**. Immediately the disk drive runs and a familiar window appears over your program. This time it is titled *Compiling to Disk.* If all goes well, the bottom line of the window reads

```
        Success              :         Press any key
```

What happened? This time your source program was compiled, and the machine code was saved as a new file on your disk with the same name as the source program file, but with the extension *.EXE*. To prove it, just enter **F** for File to the File Menu, then enter **D** for **D**irectory. Your directory now lists a file with the name *BIRTHDAY.EXE*. After you quit Turbo Pascal and get back to DOS, enter the name *BIRTHDAY* followed by a return. The program will now run immediately.

A SCIENTIFIC PROGRAM AND AN INTRODUCTION TO ITERATION

Program Birthday is a nice conversation piece for beginning, but it doesn't get us into number crunching, which is the most common use of computers in the scientific community. The most important capacity of a computer running uner the instruction of a program is its ability to carry out similar calculations repetitively and quickly. The execution of *Program Birthday* proceeds absolutely sequentially from top to bottom, from beginning to end. One statement is executed after the other. Nothing is repeated. Everything is sequential.

Pascal (and other programming languages) includes some powerful tools for taking control of the order of execution. Normally each statement is executed only once. It is often desirable to execute a statement or block of statements repetitively, usually with a change of a variable within the block. This is called *iteration* or *looping*. Looping is so important that Pascal has three control statements for looping:

for . . . to . . . do

while . . . do

repeat . . . until

Because the last two statements require the use of conditional statements or Boolean variables, we shall postpone their use and instead examine the **for** structure at this time.

Program RotationalEnergies

Program RotationalEnergies (Fig. 10–6) illustrates a **for** loop. Except for the **for** loop, this program consists of a sequence of simple statements. Execution proceeds sequentially through the program until the **for** statement is reached. The loop statement consists of a loop variable (*J* in this example), an initial and final value of the loop variable (*1* and *10*), and a loop block. In this example the loop block consists of the indented **begin . . . end** block containing three statements. For more practice, enter the Turbo Pascal Editor and type in *Program RotationalEnergies*. When you are finished, tell Turbo Pascal by pressing the *F10* key. This

```
Program RotationalEnergies;
(* A Program to calculate rotational energies (Joules)  *)
(* of HCl.  This is a starting program for learning     *)

uses crt;

Const h = 6.626186E-34;    (* Planck const, joule sec   *)
      c = 2.9979250E8;     (* speed of light, meter/sec *)

Var EnergySI,             (* Rot Energy in joules        *)
    EnergyCGS,            (* Rot Energy in wavenumbers   *)
    B,                    (* Rot const in wavenumbers    *)
    BSI:  Real;           (* Rot const in 1/meter (SI units *)
    J:  Integer;          (* Rot quantum number          *)

Begin (* Main Program *)
    Write('What is the rotational constant B in Wavenumbers?  ');

    Readln(B);
    BSI:= 100*B;             (* convert cgs to SI units  *)
    CLRSCR;
    Writeln('       Rotational Energies');
    Writeln;
    Writeln(' J     E x Cm.             E/Joules');
    Writeln;
    For J:= 1 to 10 do
      Begin (* For *)
         EnergyCGS:= J*(J + 1)*B;
         EnergySI:= J*(J + 1) * h * c * BSI;
         Writeln(J:3,EnergyCGS:10:2,'            ',EnergySI:9);
      End;  (* For  *)
    Writeln;
    Writeln(' The rotational constant is ',B:5:2,' wavenumbers');

End.  (*  Main Program *)
```

Figure 10–6 *Program RotationalEnergies.* Notice the generous use of comments to identify the physical units of the calculated quantities.

brings the Main Menu up on the screen. Enter a **C**ompile command to find any typographical errors. Go back and forth between the editor and the compiler until the program compiles without error. Then enter **R**un and compare your output with that shown in Figure 10–7.

More on the Block Structure of Pascal

The loop block may consist of a simple statement or a compound statement, that is, a sequence of simple statements. The compound statement always starts with the reserved word **begin** and terminates with the reserved word **end.** Compound statements are used frequently in Pascal. If you look again at the overall structure of *Program RotationalEnergies,* you

```
              Rotational Energies

      J      E x Cm.              E/Joules

      1        21.10              4.19E-22
      2        63.30              1.26E-21
      3       126.60              2.51E-21
      4       211.00              4.19E-21
      5       316.50              6.29E-21
      6       443.10              8.80E-21
      7       590.80              1.17E-20
      8       759.60              1.51E-20
      9       949.50              1.89E-20
     10      1160.50              2.31E-20

 The rotational constant is 10.55 wavenumbers
```

Figure 10–7 Output of *Program RotationalEnergies.*

will see that the main program itself is a compound statement. When you reach the **for** statement, the **begin** and **end** reveal that the **for** loop is a compound statement. The indentations are not required by Pascal, but are required by good programming style to improve the readability of the program. The block structure for *Program Rotational Energies* is outlined in Figure 10–8.

Pascal's block structure permits a great deal of freedom in the use of spaces, which are ignored by the compiler. The Pascal programmer uses spaces to make Pascal programs clear, readable, and understandable. *Program Birthday* as written in Figure 10–9 is syntactically correct and will compile and run error free. The output appears the same, and the program user is not aware that the program differs *stylistically* from *Program Birthday* in Figure 10–1. However, one program is much easier to read than the other. In larger programs style makes an even greater difference.

Walking Through the Loop

When the **for** loop is first entered, *J* is assigned the value *1*. The program then proceeds sequentially through the loop block. Notice how the (∗ **for** ∗) comments delineate the beginning and end of the loop block. Since *J* has not yet reached its final value, Pascal increments *J* from *1* to *2* and the loop block is executed again. The loop counter is incremented until it is satisfied, that is, until it reaches its final value. When the loop control variable (or loop counter) *J* is satisfied, execution again proceeds sequentially out of the loop to the next series of statements.

It is the loop that gives computers the power to carry out tedious calculations relentlessly and repetitively, without complaining, without boredom, and without errors.

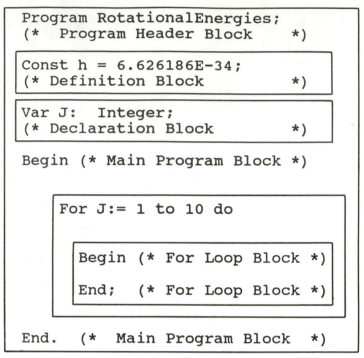

Figure 10–8 Block structure and the compound statement. The **for** statement operates on the next *block,* which in this example is a *compound statement* delineated by a **begin** and an **end.**

More on Style

Notice the use of comments in *Program RotationalEnergies.* Immediately after the program header you should insert a few lines explaining what the program does, what is input, and what sort of output to expect.

In Pascal, the main program constants are defined and variables are declared at the beginning of the program. Comments should be used freely to explain or clarify their meaning or use. It is often helpful to specify the units that will be used for constants and variables. The units should also be given whenever the user is prompted for input.

The program looks better and is easier to read if the comment delimiters (∗ and ∗) are lined up over each other. Notice that comments

```
Program Birthday;Const ThisYear = 1989;Var
Birthyear,Age:Integer;Begin Write('How old are you?   ');
Readln(Age);BirthYear:= ThisYear-Age;Writeln
('You were born in ',Birthyear)End.
```

Figure 10–9 *Program Birthday* without style. This program compiles and runs exactly like *Program Birthday* with style.

can follow a short statement on the same line or follow a long statement on the next line.

Indent blocks of statements that logically belong together in the same block. Pascal's syntax does not require indentation but permits it for program readability. At first, just imitate the examples of indentation that you see in sample programs. After a while, you'll get a feel for writing your programs in a clear and readable manner.

FORMATTING THE OUTPUT OF NUMBERS: INTEGERS AND REALS

The output of *Program RotationalEnergies* is typical of a scientific program, consisting of a mixture of integers and real numbers in decimal form and in scientific or exponential notation. Pascal has its own way of formatting integers and real numbers, but it also provides the means for programmers to customize the way numbers appear on the screen and in print.

Before seeing how the output of numbers in *Program Rotational-Energies* is formatted, let us examine *Program FormattingNumbers* (Fig. 10–10) and its output (Fig. 10–11). This program was written to introduce two features of Pascal: formatting numbers and the use of the **procedure.**

The output in Figure 10–11 shows the default formatted output of five sample numbers, two integers (365 and -9876) and three real numbers (365, $-45,689,768$, and 6.626186×10^{-34}). The default format is the format Pascal gives you if you do not take action to customize the output format. Notice how 365 can be either an *integer* or a *real* number depending on what we declare integers and reals to be (in the declaration block).

The Default Output Format

The five sample numbers are shown output in their default format in the top half of Figure 10–11. The field in which the numbers appear begins immediately after the colon. Pascal prints integers (or their signs) beginning in the first available space. Consequently all integers appear to be pushed to the left as far as possible. In the jargon of computer science this is called *left justification.*

On the other hand, Turbo Pascal *right-justifies* real numbers in a 17-position field when the output is in default format. The first position is blank or holds a negative sign if there is one. The number is always written in exponential notation with the last four positions holding the letter *E* followed by the sign of the exponent and then its value. The decimal point is always in the third position. Thus, the number is written

```
Program FormattingNumbers;

uses crt;

VAR Int1,
    Int2:  Integer;
    Real1,
    Real2,
    Real3: Real;

Procedure AssignValues;
   Begin (* AssignValues *)
   Int1:= 365;
   Int2:= -9876;
   Real1:= 365;
   Real2:= -456.89768;
   Real3:= 6.626186E-34
   End; (* AssignValues *)

Procedure WriteDefaultValues;
Begin
   Writeln('These are default format values.');
   Writeln;
   Writeln('Integer 1:',Int1);
   Writeln('Integer 2:',Int2);
   Writeln('   Real 1:',Real1);
   Writeln('   Real 2:',Real2);
   Writeln('   Real 3:',Real3);
   Writeln('            |   |    |     |     |');
   Writeln('            1---5----10---15---20   (Use this to count field width)');
End;

Procedure WriteFormattedValues;
Begin
   Writeln;
   Writeln('These are formatted values with:   Field Width   Decimal Places');
   Writeln;
   Writeln('Integer 1:',Int1:5,'                       5          not defined');
   Writeln('Integer 2:',Int2:10,'                      10          not defined');
   Writeln('   Real 1:',Real1:10:0,'                   10             0');
   Writeln('   Real 2:',Real2:10:2,'                   10             2');
   Writeln('   Real 3:',Real3:10,'                     10          omitted');
   Writeln('            |   |    |     |     |');
   Writeln('            1---5----10---15---20   (Use this to count field width)');
   Writeln;
   Writeln('The format for the writeln procedure is:');
   Writeln;
   Writeln('            Writeln(VariableValue:FieldWidth:DecimalPlaces)');
End;

Begin (* M A I N   P R O G R A M  *)
  AssignValues;
  Clrscr;              (* Clear the Screen *)
  WriteDefaultValues;
  WriteFormattedValues;
End.
```

Figure 10–10 *Program FormattingNumbers.*

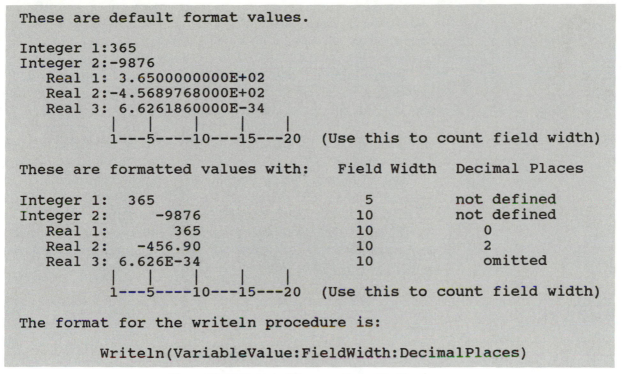

```
These are default format values.

Integer 1:365
Integer 2:-9876
    Real 1: 3.6500000000E+02
    Real 2:-4.5689768000E+02
    Real 3: 6.6261860000E-34
            |   |    |    |    |
            1---5----10---15---20   (Use this to count field width)

These are formatted values with:    Field Width   Decimal Places

Integer 1:    365                        5          not defined
Integer 2:         -9876                10          not defined
    Real 1:         365                 10          0
    Real 2:      -456.90                10          2
    Real 3: 6.626E-34                   10          omitted
            |   |    |    |    |
            1---5----10---15---20   (Use this to count field width)

The format for the writeln procedure is:

        Writeln(VariableValue:FieldWidth:DecimalPlaces)
```

Figure 10–11 Output of *Program FormattingNumbers.*

with 11 significant figures.[3] Look at the output of *Real 1, 2,* and *3* in Figure 10–11 to verify the default output format of real numbers.

Custom-Formatted Numerical Output

You have already seen how the statement *writeln(Int1)* results in the output of a left-justified integer in a field just long enough to hold the number and its sign if there is one. The format for customizing the output of integers is *writeln(Int1:FieldWidth)*. The integer *Int1* is now written in a field of width specified by the value of *FieldWidth* and is now *right-justified.* In *Program FormattingNumbers* (Fig. 10–10), the field width for the output of 356 is 5 and the field width for −9876 is 10. Check the output in Figure 10–11. The integer 365 is right-justified in a field of width 5, and the integer −9876 is right-justified in a field of length 10.

As with integers, you can specify the field width for outputting real numbers. In addition, you can specify the position of the decimal place. The format for customizing the output of real numbers is *writeln(Real1: FieldWidth:DecimalPlaces)*. In *Program FormattingNumbers* (Fig. 10–

[3] The format may vary slightly with different implementations of Pascal. The number of significant figures also varies. Turbo Pascal on an Apple has 6, while standard Pascal on a Control Data Corporation Cyber 173 has 18 significant figures. With an 8087 coprocessor, Turbo Pascal supports 19 significant figures.

10), the *FieldWidth* of the two sample integers is 5 and 10. The *FieldWidth* of all three real numbers is 10, and the numbers of decimal places specified are 0 and 2 for the real numbers 365 and −456.89768, respectively. The output is shown in the lower half of Figure 10–11.

For the third real number, the *FieldWidth* is 10, but no value for *DecimalPlaces* is given. In this way the number is kept in exponential form, but the number of decimal places can also be controlled. If the *FieldWidth* for *Real3* is changed from 10 to 20, the number will be printed exactly as it is in Figure 10–11 but will be right-justified in a field of width equal to 20.

Type in *Program FormattingNumbers* and run it with different integers and different real numbers. Change the values of the *FieldWidth* and *DecimalPlaces;* experiment and see how the output changes.

It is easy to get a computer to crank out large volumes of numbers. However, it takes care to organize the display of numbers and to format numbers so they can be interpreted clearly.

ARRAYS

In experimental physical sciences and engineering, an experiment often results in a series of measurements that lead to one or more lists of data of the same type. The **array** is the natural Pascal variable for handling lists of data. The array is a structured data type; that is, it consists of a collection of values of the same simple type, which are called *elements*.

Compare the following two declarations:

var *Temp: real;*

This statement declares that *Temp* is the identifier of a variable of type *real*. The statement

var *X: **array**[1..15] **of** real;*

declares that X is the identifier of an array containing 15 elements that are real numbers. Each element in the array has its own identifier:

X[1],
X[2],
X[3],

and so on until $X[15]$. All the elements of an array must be of the same type, for example, all integers or all real numbers. The names of the array elements are similar to the algebraic names we might choose: $X_1, X_2, X_3,$ .... The declaration of the simple variable *Temp* sets aside a place in the computer memory for the storage of a single real number. The declaration of the array variable X sets aside a collection of 15 places in the computer memory for the storage of a list of real numbers that are referenced element by element.

Declaring Arrays

Arrays can be declared directly, similarly to the way we declare simple variables:

var   *R,t:*   **array[1..15] of** *real;*

The maximum number of elements must be specified and the **type** must be specified.

Alternatively, arrays can be declared indirectly by first defining a **type** of **array** and subsequently a variable of that **type,** for example,

const *Max = 15;*
type *List =* **array[1..Max] of** *real;*
var *R,t: List;*

The results of the two declarations are identical. In either case, *R* and *t* are the names of one-dimensional arrays, each consisting of 15 elements capable of being assigned the value of a real number.

Input to and Output from an Array

Although it is not usual, you can assign a value directly to the element of an array in the same way that you sometimes assign a value to a simple variable. For example, you can assign the value *5* to a variable *N* with the assignment statement

N:= 5;

In the same manner you can assign values to the elements of an array identified by *X :*

X[1]:= 7.31;
X[2]:= 9.11;
X[3]:= −2.28;

and so on.

The initial input to a program, however, usually gets there by means of someone entering the data on a keyboard; the data are read into the program with a *read* or *readln* statement. If *N* is a simple variable, then

Readln(N);

suffices.

If the variable is an array, however, a **for** loop is the most common control structure for assigning values to the elements of an array, since we usually know in advance how many data are to be assigned. This number must be less than or equal to the declared dimension of the array.

Input to an Array. If *X* is the name of the array, then the following loop would assign values from the keyboard to the elements of the array:

for *J:= 1* **to** *7* **do**
    Readln (X[J]);

In practice, the program should prompt the user to input each value. Without some kind of prompting, the user is likely to forget how many values have already been assigned. Let us put a statement just before the *Readln* to help out the user, so that the loop looks like this:

> **for** *J:= 1* **to** *7* **do**
> > **begin**
> > > *Write('Please enter X[',J,']:   ');*
> > > *Readln X[J]*
> > **end;**

Notice that now the **for** loop operates on a compound statement consisting of the *write* and *readln* statements. Consequently these statements must be delimited by a **begin** and **end**.

If the value input on the first iteration of the loop is, say, *25.7*, the screen appears as follows after *27.5* is entered on the keyboard and the return key is pressed:

```
                    Please enter X[1]:   25.7
                    Please enter X[2]:
```

```
PROGRAM ArrayDemo;
(* A demonstration program to show how to read data into     *)
(* an array, do something with it, store the results in       *)
(* another array, and print out the contents of the two arrays *)

uses crt;

CONST Max = 15;

TYPE MyList = ARRAY [1..Max] of Real;

VAR J,N:  INTEGER;
    Xin,Yout:  MyList;

BEGIN
   WRITELN('Please input your list of pressures in torr.');
   WRITELN('How many pressures do you wish to input?');
   READLN(N);
   For J:= 1 TO N DO
      BEGIN
         WRITE('Pressure ',J,' = ');
         READLN(Xin[J]);      (* from keyboard to an array element *)
         Yout[J]:= Xin[J]/760;(* from calcn. to another array element  *)
      End;
   WRITELN;WRITELN('Press enter key to get printed output');
   READLN;
   CLRSCR;
   WRITELN('    N      P/torr      P/atm');
   WRITELN;
   FOR J:= 1 TO N DO
      WRITELN(J:5,Xin[J]:10:1,Yout[J]:12:4)
END.
```

Figure 10–12 *Program ArrayDemo.*

N	P/torr	P/atm
1	23.8	0.0313
2	760.0	1.0000
3	1500.0	1.9737

Figure 10–13 Output of *Program ArrayDemo.*

This code is user-friendly because it helps the user keep track of the progress of entering the data. The user should not have to know what is going on inside the program; the material on the screen should be self-explanatory.

Output from an Array. The value of a single element of an array can be output to the screen with a *writeln* statement:

$$Writeln(X[1]:4:1);$$

This statement would write *25.7* on the screen if *27.5* had been input to the first element of the array X as shown above. If several values have been assigned to the elements of the array X, then a **for** loop is a good method of writing the contents of the array to the screen:

$$Writeln(' Run   X');$$
$$\textbf{for } Run:= 1 \textbf{ to } 7 \textbf{ do}$$
$$Writeln(Run:6,X:4:1);$$

Program ArrayDemo, Figure 10–12, demonstrates the use of an array. The output of the program is shown in Figure 10–13.

PROCEDURES

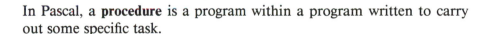

In Pascal, a **procedure** is a program within a program written to carry out some specific task.

Predeclared Procedures

Pascal has within it some predeclared procedures. These are already written and available for your use if you call them. Some procedures that you have already encountered are *read, readln, write,* and *writeln.* These predeclared procedures perform specific input and output tasks that have already been discussed. Another predeclared procedure, used in some of the sample programs, is *Clrscr.* A call to this procedure clears the screen and puts the pointer in the upper left corner of the screen.

You call a procedure by writing its name identifier followed by any parameters used by the standard procedure: for example,

$$Read(X);$$

Readln(A,B);

Write('X equals ',X);

Clrscr;

The predeclared procedure *Clrscr* has no parameters; it just clears the screen. Both Turbo Pascal and standard Pascal contain large numbers of predeclared procedures that help make the language powerful.

User-Declared Procedures

If you can write a Pascal program, you can write a Pascal procedure. The syntax for a procedure is the same as for a program, except that the reserved word **procedure** is used instead of the reserved word **program** in the header block. As in a program, the definition block and declaration block of a procedure are optional.

Look at *Program FormattingNumbers* (Fig. 10–10) again. The program consists of three procedures. The main program (bottom of Figure 10–10) consists of just four procedure calls, one of them to the predeclared procedure *Clrscr*. Notice that the names of the procedures are descriptive of the tasks they perform. Other than **begin** and **end,** the main program consists of just four words—the procedure calls. This is typical of a Pascal program. The main work of the program is accomplished by the procedures. The main program is like the head office: it calls the shots, manages, and directs the overall operation.

The block structure of *Program FormattingNumbers* is shown in Figure 10–14.[4] Since the program has no constants to define, it contains no definition block. Notice, however, that the procedures are declared, along with the variables, in the declaration block. Because the declaration block contains all the program's procedures, it is normally the largest block in a typical Pascal program. The main program should be much smaller and consequently immediately understandable.

Scope of Variables and Procedures

The block of a program in which an identifier is accessible is called its *scope. Identifier* refers to the name of a variable, constant, procedure, or function.

Scope of Variables and Constants. Identifiers for constants and variables declared or defined in the main program are *global* identifiers. They are known to and can be accessed by any block in the program. All other identifiers are *local* and are known only within the block in which they are declared.

---

[4] The program shown in Figure 10–14 can actually be compiled and run without any error messages; however, there is no output in this abbreviated form.

```
Program FormattingNumbers;
(* Program Header Block      *)

(* Definition Block          *)
(* not used here             *)

VAR Int1: Integer;
(* Declaration Block         *)

┌─────────────────────────────────────────┐
│ Procedure AssignValues;                  │
│    Begin                                 │
│    End;                                  │
└─────────────────────────────────────────┘
┌─────────────────────────────────────────┐
│ Procedure WriteDefaultValues;            │
│    Begin                                 │
│    End;                                  │
└─────────────────────────────────────────┘
┌─────────────────────────────────────────┐
│ Procedure WriteFormattedValues;          │
│    Begin                                 │
│    End;                                  │
└─────────────────────────────────────────┘

Begin (* M A I N   P R O G R A M   B L O C K  *)
   AssignValues;
   WriteDefaultValues;
   WriteFormattedValues;
End.  (* M A I N   P R O G R A M   B L O C K  *)
```

Figure 10–14 The block structure of *Program FormattingNumbers.*

If a variable is declared both globally (in the main program) and locally (in a procedure), the local declaration takes precedence. This means that changes in the value of the locally declared variable do not affect the value of the globally declared variable of the same name. Nevertheless, this can be a source of confusion, so it is best to give names to globally declared variables and constants that differ from the names of those declared locally.

Scope of Procedures. The scope of a procedure is the block in which it is declared. Procedures can call only the procedures previously declared *in that same block.*

Consider the block diagram of a program outlined in Figure 10–15 and the possible procedure calls:

1. The main program can call A, B, and E.
2. *Procedure A* can call nothing.
3. *Procedure B* can call A, C, and D.
4. *Procedure C* can call nothing.
5. *Procedure D* can call C.
6. *Procedure E* can call B and A (but not C or D).

Now apply the scope rule to the variables declared in each unit:

1. Variables declared in the main program are global and known everywhere.

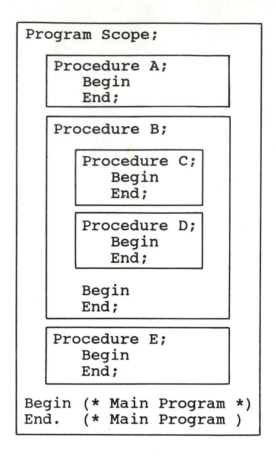

```
Program Scope;

    Procedure A;
        Begin
        End;

    Procedure B;

        Procedure C;
            Begin
            End;

        Procedure D;
            Begin
            End;

        Begin
        End;

    Procedure E;
        Begin
        End;

Begin (* Main Program *)
End.  (* Main Program )
```

Figure 10–15 Blocks to illustrate scope.

2. Variables declared in *A* are accessible only in *A*.
3. Variables declared in *B* are accessible in *B*, *C*, and *D*.
4. Variables declared in *C* are accessible only in *C*.
5. Variables declared in *D* are accessible only in *D*.
6. Variables declared in *E* are accessible only in *E*.

In addition, it is good programming practice to declare variables and constants locally whenever possible. This tends to make the procedure more modular and easier to use in other programs with little change.

Passing Parameters into a Procedure

We have already seen how a procedure is written to carry out a specific task. Each of the three procedures in *Program FormattingNumbers* does a certain job when simply called. No values or parameters are sent to them, nor are any returned by the procedures to the calling unit, the main program in this example.

It is useful to vary the task done by a procedure by passing one or more parameters to the procedure when it is called. The procedure does a slightly different task each time a different set of parameters is sent to it.

Program GrafDemo (Fig. 10–16) illustrates how parameters are passed from the main program block into a procedure. In addition, some elementary features of Turbo Pascal graphics are demonstrated. In this program the procedure *DrawBox* is written to perform a single task: it draws a rectangular box at a specified location on the screen. To draw several boxes at several locations on the screen, it is desirable to use the same procedure. This can be accomplished by sending certain parameters to the procedure, namely the location on the screen and the height and width of the box. The output of *Program GrafDemo* is shown in Figure 10–17.

Declaring Procedures When Parameters Are Passed. When parameters are passed into a procedure, the procedure heading has the form

Procedure identifier (formal parameter list);

The formal parameters consist of a sequence of identifiers separated by commas and followed by their type, for example,

procedure *DrawBox(Xstart, Ystart, Height, Width: integer);*

The formal parameters may be of any number and type, such as

procedure *Example2(P,R: Real;J: Integer;Point: char);*

Calling Procedures When Parameters Are Passed. Pascal syntax requires that the number and type of actual parameters in the procedure call be identical to those in the parameter heading. Thus, a legal call to *DrawBox* is

DrawBox(100, 175, 30, 40);

or

DrawBox(X, Y, H, W);

This call is legal if X, Y, H, and W have previously been delcared to be variables of type *integer*. Similarly, the call

Example2(Left, Right, Number, Star);

is legal if it has previously been declared that *Left* and *Right* are variables of type *real,* that *Number* is of type *integer,* and that *Star* is of type *char*.

A skeleton of *Program GrafDemo* (Fig. 10–18) shows how the parameters are passed from the main block into the procedure *DrawBox*. Notice that variables H and L are declared in the main program and it is in the main program that they are assigned values. The variables *Height* and *Width* are declared in the procedure *DrawBox,* and it is to these variables that parameter values are passed from the main program.

Enter and run *Program GrafDemo*. Edit the parameters in the parameter list and run it again. Experiment a little. Change the assignments to the variables X, Y, H, and W in the main program, run it, and see if you get the expected results.

```pascal
Program GrafDemo;
(* This program illustrates some elementary Turbo Pascal   *)
(* graphics commands.  It also demonstrate the use of value *)
(* parameters in calling a procedure:  this is the way     *)
(* Pascal passes a value to a procedure                    *)

uses graph,crt;

Var H,                 (* Height of box        *)
    W,                 (* Width  of box        *)
    X,                 (* X screen coordinate *)
    Y,                 (* Y screen coordinate *)
    I, GraphDriver,
    GraphMode, ErrorCode:  Integer;

(***********************************************************)

Procedure DrawBox(Xstart,Ystart,Height,Width:  Integer);
(* Draw a four sided figure starting        *)
(* at screen coordinates Xstart and Ystart *)

VAR X1,X2,Y1,Y2:  Integer; (* coordinates of a point  *)
                           (* on the screen           *)
Begin (* DrawBox *)
   X1:= Xstart;
   Y1:= Ystart;
   X2:= Xstart + Width;
   Y2:= Ystart + Height;
   Line(X1,Y1,X2,Y1);
   Line(X2,Y1,X2,Y2);
   Line(X2,Y2,X1,Y2);
   Line(X1,Y2,X1,Y1);
End;  (* DrawBox *)

(***********************************************************)

Begin (* M A I N   P R O G R A M *)
      (* Prepare to do graphics *)
      GraphDriver:= Detect;
      InitGraph(GraphDriver,GraphMode,'');
      ErrorCode:= GraphResult;
      SetGraphMode(3);
      If ErrorCode <> GrOK then
         Begin
            Writeln('Graphics error:  ',GraphErrorMsg(ErrorCode));
            Writeln('Program Aborted...');
            Halt(1);
         End;
      (* Now do Graphics *)
      DrawBox(0,0,199,319);   (* This is the largest possible box *)
      DrawBox(50,100,25,75);
      DrawBox(10,60,125,15);
      (* Next, a different way to assign the values *)
      X:= 150;
      Y:= 5;
      H:= 165;
      W:= 150;
      DrawBox(X,Y,H,W);
      Readln;  (* Stops execution to see boxes drawn so far *)
      CloseGraph(* leaves graph mode & goes back to text mode *)
End.  (* M A I N   P R O G R A M *)
```

Figure 10–16 *Program GrafDemo.*

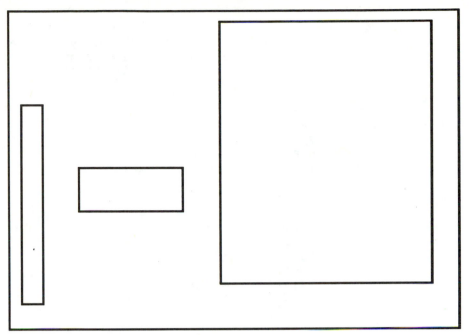

Figure 10–17 Output of *Program GrafDemo.*

Passing Parameters from a Main Program Loop
to Procedures

The calculation of thermodynamic properties of a diatomic molecule from the equations of statistical thermodynamics is a problem that arises frequently in physical chemistry. The calculation is straightforward and involves adding up the translational, rotational, and vibrational contri-

```
Program GrafDemo;
(* This program illustrates some elementary Turbo Pascal    *)
(* graphics commands.  It also demonstrate the use of value *)
(* parameters in calling a procedure:  this is the way      *)
(* Pascal passes a value to a procedure                     *)


Var H,                 (* Height of box *)
    L: Integer;        (* Length of box *)

Procedure DrawBox(Height,Length:  Integer);
Begin (* DrawBox *)
End;  (* DrawBox *)

Begin (* M A I N  P R O G R A M *)
      DrawBox(H,L);
End.  (* M A I N  P R O G R A M *)^Z
```

Figure 10–18 Passing parameters to a procedure.

```
Program EnergyCv;
(* a program to calculate the interal energy and the *)
(* heat capacity at constant volume of a diatomic     *)
(* molecule between a low and a high temperature       *)

uses crt;

Const h = 6.6262E-34;   (* Planck constant            *)
      k = 1.3807E-23;   (* Boltzmann constant         *)
      c = 2.997925E8;   (* Speed of light             *)
      R = 8.3144;       (* Gas constant               *)

Var T,Thigh, DelTemp,
    V,x,               (* wavenumbers and x = 100hcV/kT *)
    Etot,Cvtot:  Real;
    N, Nint:  Integer;
    Molecule:   String[20];

(**************************************************)

Procedure GetData;
Begin
   Write('Enter the lower temperature in deg K, please:  ');
   Readln(T);
   Write('Enter the upper temperature in deg K, please:  ');
   Readln(Thigh);
   Write('Enter the number of temperature intervals, please:  ');
   Readln(Nint);
   DelTemp:= (Thigh - T)/Nint;   (* calculate temp differences *)
   Write('Enter the fundamental vibration (wavenumbers), please:  ');
   Readln(V);
   Write('What is the name of the diatomic molecule:  ');
   Readln(Molecule);
End;

(**************************************************)

Procedure Energy(T,x:  Real);
Var Etran,Erot,Evib:  Real;
Begin
   Etran:= 1.5*R*T;
   Erot:= R*T;
   Evib:= R*T*x/(Exp(x) - 1);
   Etot:= Etran + Erot + Evib;
End;

(**************************************************)
```

Figure 10–19 *Program EnergyCv.*

butions to the thermodynamic property as illustrated in *Program EnergyCv* (Fig. 10–19), the output of which is shown in Figure 10–20.

The desired output is the calculated values of the internal energy $E - E_0$, calculated in *Procedure Energy,* and the heat capacity at constant

```
Procedure Cv(T,x: Real);
Var Cvtran,Cvrot,Cvvib:  Real;
Begin
   Cvtran:= 1.5*R;
   Cvrot:= R;
   Cvvib:= R*(Sqr(x))*Exp(x)/(Sqr(Exp(x) - 1));
   Cvtot:= Cvtran + Cvrot + Cvvib;
End;

(*************************************************)

Begin (* M A I N   P R O G R A M *)
GetData;
Clrscr;
Writeln('    T/K         (E - Eo)/J     Cv/(J/K)');
Writeln;
For N:= 1 to Nint + 1 do
   Begin
      x:= 100*h*c*V/(k*T); (* 100 converts 1/cm to 1\meter *)
      Energy(T,x);
      Cv(T,x);
      Writeln(T:8:1,Etot:14:1,Cvtot:12:2);
      T:= T + Deltemp;
   End; (* For *)
Writeln;
Writeln('  For ',Molecule,' V = ',V:1:2,' wavenumbers.');
End.  (* M A I N   P R O G R A M *)
```

Figure 10–19 (continued)

volume C_v, calculated in *Procedure Cv*. The input is the upper and lower temperature limits, the number of temperature intervals, and the fundamental vibration frequency, which are all input with a call to *Procedure GetData*.

The structure of the Pascal program flows naturally from the structure of the physical chemistry problem. Just as the solution to the physical chemistry problem dictates that the problem be broken down into simpler parts, so does Pascal cry out for procedures for the physical chemistry parts of the problem.

The calculation of the energy and heat capacity of a diatomic molecule demonstrates the striking similarity of problem solving in physical chemistry and in Pascal (Table 10–2). Because the use of procedures is so natural in Pascal, programming in Pascal gets one thinking in small, understandable blocks, which makes problem solving easier in both Pascal and physical chemistry.

The comparison also demonstrates what programmers call *top-down programming*. Top-down programming involves looking at the overall problem to be solved, breaking it down into simple parts (procedures), and finally calling the procedures as needed from the main program.

```
     T/K           (E - Eo)/J        Cv/(J/K)

     300.0           6245.5           21.03
     500.0          10595.4           22.65
     700.0          15314.2           24.47
     900.0          20350.1           25.80
    1100.0          25603.9           26.68
    1300.0          31002.6           27.27
    1500.0          36499.5           27.68
    1700.0          42065.1           27.96
    1900.0          47680.1           28.18
    2100.0          53331.7           28.33
    2300.0          59011.1           28.46

  For oxygen V = 1580.36 wavenumbers.
```

Figure 10–20 Output of *Program EnergyCv.*

Returning Parameters from a Procedure

If programs have both input and output, then why not procedures, which, after all, are programs within a program? We have just seen how parameters are passed *into* a procedure. Let us now see how parameters can be

TABLE 10–2

Comparing Problem Solving in Physical Chemistry and Pascal

Physical Chemistry Problem	Pascal
State the problem so that it is clear in your mind what data are given and what needs to be calculated.	Write the program header and a comment stating the purpose of the program. Determine exactly what is input and what is output.
Collect the necessary constants to be used in the calculations.	Define the constants to be used in the program.
Decide on the temperature range of the calculations, the fundamental vibration of the molecule, and the number of calculations.	Declare *Procedure GetData* in order to enter the upper and lower temperature, the number of temperature intervals, and the fundamental vibration of the molecule.
Learn the algorithm for calculating the total internal energy from physical chemistry.	Declare *Procedure Energy* with algorithm from physical chemistry.
Learn the algorithm for calculating the total heat capacity from physical chemistry.	Declare *Procedure Cv* with algorithm from physical chemistry.
At each temperature, calculate $x = hc\tilde{v}/kT$, then the total energies and heat capacities.	In the main program, call *Procedure GetData*. Then write a loop to call Procedure *Energy* and *Procedure Cv* the desired number of times.

passed back *from* a procedure *to* the calling unit. The unit can be the main program block or another procedure block.

When a procedure both receives parameters and returns them to the calling unit, the procedure head looks like the following example, taken from *Program ConvertCartesianToPolar* (Fig. 10–21):

```
Program ConvertCartesianToPolar;
(* This program  illustrates how parameters are     *)
(* passed into a procedure and how parameters are    *)
(* returned from a procedure                         *)


VAR    Xcartesian,Ycartesian,              (* Cartesian coordinates *)
       RadiusVector,VectorAngle:  real;    (* Polar coordinates     *)
       answer:                    char;
       finished:                  boolean;


(***************************************************************)

Procedure Convert(X,Y:  Real;  VAR r,theta:  Real);
(* This procedure accepts X and Y coordinates as input        *)
(* and returns polar coordinate r and theta as output         *)

CONST  Pi = 3.14159265;

VAR radians:  Real;  (* plays a strictly local role           *)

Begin (* Convert *)
   r:= ABS(SQRT(X*X + Y*Y));
   radians:= ARCTAN(Y/X);
   theta:= 360*radians/2/Pi  (* convert radians to degrees     *)
End;  (* Convert *)


(***************************************************************)

Begin (* M A I N   P R O G R A M   *)
   Finished:= false;
   Repeat
      Write('Please input your X coordinate:  ');
      Readln(Xcartesian);
      Write('Please input your Y coordinate:  ');
      Readln(Ycartesian);
      Convert(Xcartesian,Ycartesian,RadiusVector,VectorAngle);
      Writeln;
      Writeln('The Radius Vector equals:  ',RadiusVector:6:2);
      Writeln('The Vector Angle equals:   ',VectorAngle:6:2,' degrees');
      Writeln;  Writeln;
      Write('Run again?  Please enter a Y or N:  ');
      Readln(answer);
      If (answer = 'N') or (answer = 'n') then finished:= true;
   Until finished;
End.  (* M A I N   P R O G R A M   *)
```

Figure 10–21 *Program ConvertCartesianToPolar.*

procedure *Convert(X, Y:* **real;**   **var** *r, theta:*   *real);*

            Input part of   Output part of formal
         formal parameter   parameter list
               list

The input part of the formal parameter list is the same as in a procedure that only receives parameters without returning parameters. The formal parameters that receive values (*X* and *Y* in this example) are called *value parameters*. The formal parameters that return values to the calling unit are called *variable parameters* (*r* and *theta* in this example). The variable parameters are distinguished from value parameters by the reserved word **var,** which precedes them. The procedure carries out its action and then returns the values of the **var**iable parameters *r* and *theta* to the calling unit. The formal parameter list may contain only variable parameters, only value parameters, or both. Both may be of any type, but the type must be declared in the formal parameter list.

     The procedure call is the same as before; that is, the parameters in the argument of the procedure call must agree in number and type with the parameters in the formal parameter list. For example,

Convert(DeltaX, DeltaY, radius, angle);

```
Please input your X coordinate:   50
Please input your Y coordinate:   100

The Radius Vector equals:   111.80
The Vector Angle equals:     63.43 degrees


Run again?  Please enter a Y or N:  y
Please input your X coordinate:   -50
Please input your Y coordinate:   100

The Radius Vector equals:   111.80
The Vector Angle equals:    -63.43 degrees


Run again?  Please enter a Y or N:   y
Please input your X coordinate:   111.8
Please input your Y coordinate:   -63.43

The Radius Vector equals:   128.54
The Vector Angle equals:    -29.57 degrees


Run again?  Please enter a Y or N:
```

Figure 10-22 Output of *Program ConvertCartesianToPolar.*

```
Program ConvertCartesianToPolar;
(* This programs illustrates how parameters are      *)
(* passed into a procedure and how parameters are    *)
(* returned from a procedure                         *)


VAR    Xcartesian,Ycartesian,                 (* Cartesian coordinates *)
       RadiusVector,VectorAngle:  Real;       (* Polar coordinates     *)

(******************************************************************)

Procedure Convert(X,Y:  Real;  VAR r,theta:  Real);
Begin (* Convert *)
End;  (* Convert *)

(******************************************************************)

Begin (* M A I N     P R O G R A M    *)
      Convert(Xcartesian,Ycartesian,RadiusVector,VectorAngle);
End.  (* M A I N    P R O G R A M    *)
```

Figure 10–23 Passing parameters to and from a procedure.

If *DeltaX, DeltaY, radius,* and *angle* are previously declared variables of **type** *real,* then this is a legal call to *Procedure Convert.* The screen display of typical input and output for *Program ConvertCartesianToPolar* is shown in Figure 10–22. For another legal call of *Procedure Convert,* see Figure 10–23, which indicates the directions in which the parameters are passed.

Notice that the program would run fine if the procedure call were

$$Convert(DeltaX, DeltaY, angle, radius);$$

However, the wrong values would be received because *radius* and *angle* are not in the intended positions. This is an example of a *logical* error in programming. The error is not in Pascal syntax. Just because a program compiles and runs does not mean that it will output correct answers. The output must always be checked against independently calculated values.

FUNCTIONS

Functions are very similar to procedures and, like procedures, are either predeclared or user declared. Turbo Pascal has over 50 predeclared functions. The word **function** is a reserved word in Pascal.

Predeclared Functions

The most common functions that arise in numerical calculations are

Abs(x)	Returns the absolute value of x
ArcTan(x)	Returns the arctan of x

Cos(x)	Returns the cosine of *x* (*x* in radians)
Exp(x)	Returns the exponential of *x*
Ln(x)	Returns the natural log of *x*
Round(x)	Returns *x* rounded up or down to nearest integer
Sin(x)	Returns the sine of *x* (*x* in radians)
Sqr(x)	Returns the square of *x*
Sqrt(x)	Returns the square root of *x*

```
Program TryOutFunctions;
(* A program to demonstrate some predefined functions *)

uses crt;

Var Y,X:   Real;
Begin  (* TryOutFunctions *)
   clrscr;
   Write('Enter a pos or neg real number between -88 and +88:  ');
   Readln(X);
   If (X < - 88) or (X > 88) then
     Begin (* If *)
         Writeln('Sorry, but your number is out of range.');
         Writeln('Please try again.')
     End    (* If *)
   Else
     Begin (* Else *)
       Writeln;
       Y:= ABS(X);
       Writeln('The absolute value of X is:  ',Y:7:3);
       Y:= EXP(X);
       Writeln('                      EXP(X) is:  ',Y:9);
       Y:= ROUND(X);
       Writeln('              X rounded off is:  ',Y:7:0);
       Y:= SQR(X);
       Writeln('                 x squared is:  ',Y:7:3);
       If X > 0 then
         Begin (* If *)
           Y:= SQRT(X);
           Writeln('   The square root of X is:  ',Y:7:3);
           Y:= LN(X);
           Writeln('   The natural log of X is:  ',Y:7:3);
         End;  (* If *)
       Writeln;
       Writeln('X equals:  ',X);
     End  (* Else *)

End.  (* TryOutFunctions *)
```

Figure 10–24 *Program TryOutFunctions.*

Whether predeclared or user declared, a **function** is a program within a program that returns a single value in the name of the function.

A call to a predeclared function is made by using it in an expression in a simple Pascal statement. Suppose we want the square root of x, and the current value of x is 27.3. A call to the square-root function is

$$y := Sqrt(x);$$

The current value of y now equals 5.224940191. The quantity within the parentheses is called the *argument of the function*. A few of these predeclared functions are illustrated in *Program TryOutFunctions,* Figure 10–24, and some outputs of this program are shown in Figure 10–25.

User-Declared Functions

Like a **procedure,** a **function** must have an identifier with which it is called. Unlike a procedure, a function always returns a value. Both functions and procedures have the same block structure as the main program itself: header section, definition block, declaration block, and statement block.

Declaring and Calling a Simple Function. The arguments of the predeclared Pascal functions for calculating sine and cosine must be in radians. Since Pascal does not have a predeclared function to convert degrees to radians, let's write a user-declared function to do just that, shown in Figure 10–26. Notice that even this simple function is block structured like a program or a procedure. It has a header block, a definition block (for one constant), and a statement block (for one statement). It happens that this function has no variable (other than the function parameter), so it has no declaration block. Only the header block differs from what we've experienced so far. The function heading begins with the reserved word **function,** which is separated by a space from an identifier. The identifier is followed by a parameter list enclosed in parentheses:

function *identifier(param1, param2: type;param3: type): type*

```
Enter a pos or neg real number between -88 and +88:   23.81634

The absolute value of X is:     23.816
             EXP(X) is:     2.20E+10
        X rounded off is:         24
          x squared is:    567.218
  The square root of X is:      4.880
  The natural log of X is:      3.170

X equals:    2.3816340000E+01
```

Figure 10–25 Output of *Program TryOutFunctions.*

```
Program FunctionDemonstration;
(* A program demonstrating a user defined function to *)
(* convert degrees to radians and how that function   *)
(* is called from the main program                    *)

uses crt;

Var R,              (* angle in radians *)
    Theta,          (* angle in degrees *)
    Value:  Real;

(*******************************************************************)

Function Radian(Degrees: Real): Real;
(* Converts degree to radians       *)

Const Pi = 3.141592654;

Begin (* Function Radian *)
   Radian:= Degrees * Pi/180;
End;   (* Function Radian *)

(*******************************************************************)

Begin (* M A I N   P R O G R A M *)
   clrscr;
   Write('What is the angle theta in degrees?  ');
   Readln(Theta);
   Writeln;
   R:= Radian(Theta); (* This is a call to a user declared function *)
   Writeln('The angle theta is: ',theta:6:2,' degrees.');
   Writeln('The angle theta is: ',R:6:4,' radians.');
   Value:= Sin(R);    (* This is a call to a pre-declared function  *)
   Writeln('The Sine of theta is: ',Value:6:4,'.');
   Value:= Cos(R);    (* This is a call to a pre-declared function  *)
   Writeln('The Cosine of theta is: ',Value:6:4,'.');
End.  (* M A I N   P R O G R A M *)
```

Figure 10–26 A program to demonstrate a user-declared function.

The parameters are the values that are input to the function when it is called. A function may have any number of parameters. The type of each parameter is declared in the parameter list. Finally, the type of the value returned by the function is declared at the very end of the function header block. Some examples are

```
What is the angle theta in degrees?  38.37

The angle theta is:  38.37 degrees.
The angle theta is: 0.6697 radians.
The Sine of theta is: 0.6207.
The Cosine of theta is: 0.7840.
```

Figure 10–27 Output of *Program FunctionDemonstration.*

function *Radian(Degrees: real):   real;*

function *VibEnergy(nubar: real, V: integer):   real;*

function *dSpacing(h, k, l: integer; a: real):   real;*

The statement block of a function is conventional, but there is one restriction. The assignment to the function must be made in the last line of the statement block and the identifier for the function must appear there exactly as written in the header. No statements are allowed to follow the assignment to the function.

The output of *Program FunctionDemonstration* is shown in Figure 10–27.

BOOLEAN EXPRESSIONS

A *Boolean* expression can have either value *true* or *false*. In this way it is identical to a Boolean variable. In fact, a Boolean variable is the simplest form of a Boolean expression. Expressions represent values; Boolean expressions represent Boolean values (*true* or *false*).

Relational Operators

Simple Boolean expressions are represented by a relation between two values. For example,

$$J <> K$$

$$Answer \ = \ 'N'$$

$$Pressure \ < \ Po$$

These expressions represent values, and the only possible values are *true* and *false*.

All the usual *relational operators* are available in Pascal:

Pascal Relational Operator	Meaning
=	Is equal to
<>	Is not equal to
<	Is less than
>	Is greater than
<=	Is less than or equal to
>=	Is greater than or equal to

The relational operator (also called the *conditional operator*) compares values. Notice that only values of the same type can be compared:

$$5 > 3 \qquad \text{has the value } true$$

$$'z' < 'd' \qquad \text{has the value } false$$

$$5 < 'd' \qquad \text{is nonsense and not allowed by Pascal}$$

Since arithmetic expressions represent values, they can also be compared:

$$(X*SQR(Y)) < 125$$

$$(150*X) <= 1.5$$

$$3 > EXP(X)$$

are all legal Boolean expressions.

Logical Operators

More complex Boolean expressions can be constructed with the logical operators **and, or,** and **not,** which are reserved words in Pascal. Some legal Boolean expressions that include logical operators are

$$(T > Tice) \textbf{ and } (T < Tboil)$$

$$(theta <= 90) \textbf{ or } (theta >= 270)$$

$$((B < A) \textbf{ or } (C > B)) \textbf{ and } (D >= 2*A)$$

$$X > 100 \textbf{ and not } finished$$

In the last example, *finished* must have been declared a Boolean variable. Whether the last example has the value *true* or the value *false* depends on the value of *finished* and the value of *X*. If *X* has been assigned the value *200*, and *finished* has been assigned the value *true,* is the expression true or false?

A Sample Application of Boolean Expressions

In *Program ConvertCartesianToPolar* (Fig. 10–21), a Boolean variable given the identifier *finished* is declared in the declaration section of the main program. In the first statement of the main program the variable *finished* is assigned the value *false,* certainly a logically correct assignment, since the first time the program is run, the use of the program is not finished!

After the calculations have been carried out for the first time through the program, the user is interactively asked, "*Run again? Please enter Y or N:.*" The variable *answer,* declared earlier to be of type *char,* is assigned a character value by the statement *Readln(answer).* If the value of *answer* becomes either *N* or *n*, then the Boolean variable *finished* is assigned the new Boolean value *true;* otherwise, the value of *finished* remains what it was assigned on the first pass through the program.

As long as the value of *finished* remains *false,* the program loops back to the first statement of the main program. Once the value of *finished* becomes *true,* the loop becomes satisfied, then the program drops out of the **repeat . . . until** block to the next statement, which in this case is the **end** statement signifying the end of the program.

Repetition, Control, and Loops

Boolean expressions, functions, and variables are used in Pascal to control the flow of programs by making decisions. The control statement **repeat . . . until** and the control statement **while . . . do** are always used with Boolean expressions to control a loop. The **if . . . then . . . else** statement also uses Boolean expressions.

At this point it is useful to review the overall nature of a computer program. A computer program consists of *action, repetition,* and *control.*

Action. In the simplest computer program, statements are translated into action sequentially from top to bottom. Statements in a computer program are commands to carry out actions, which may be simple or complex.

A simple assignment statement:	$X := 2*Y + 3.5/Z$
A compound assignment statement:	**begin**
	.
	end;
An assignment of a **function** call:	$R := Sin(ang)$;
A **procedure** call:	*GetData;*

Repetition. We have already seen the **for** statement: it carries out an unconditional number of repetitions on the next statement, which can be any combination of the above actions. The Pascal syntax for its three-loop structures is as follows:

```
for (variable):= (value1) to (value2) do
    begin
        (action);
    end;

while (Boolean expression) do
    begin
        (action);
    end;

repeat
    (action);
until (Boolean expression);
```

When the number of iterations is known in advance, the **for . . . to . . . do** control statement is the proper choice. Pascal requires that the loop counter variable must be declared in the procedure in which it is used. The program in Figure 10–28 demonstrates a Pascal **for** statement. If a 5 is entered the output consists of the following six lines:

```
            1
            2
            3
            4
            5
Loop finished.
```

```
Program ForDemo1;
(* A program to demostrate a FOR loop *)

Var I,N:  Integer;

Begin
   Write('Please enter a small integer:  ');
   Readln(N);
   FOR I:= 1 to N do
      Writeln(I);
   Writeln('Loop finished.');
End.
```

Figure 10–28 The **for** statement.

It is common in physical chemistry to read a known number of x, y data pairs into an array for further data processing; because the user is the experimenter, the user knows how many runs (N) were made. Thus, the **for . . . to** N **do** statement is the natural choice, as illustrated in *Procedure ReadKeyBoard* (Fig. 10–29). The program prompts the user for the number N of x, y pairs to be input, establishing the number of iterations the **for** statement will execute. The program then prompts the user for each data pair, assigned to the elements of arrays X and Y. (The *WhereX, WhereY,* and *GoToXY* are Turbo Pascal predefined graphics procedures used here to make the program user-friendly.) The arrays X and Y are globally declared in the main program.

When the number of iterations is not known in advance, the **repeat**

```
Procedure ReadKeyBoard;
Var I,
    Sx,Sy:  Integer;      (* Screen Coordinates *)
Begin (* ReadKeyBoard *)
   CLRSCR;
   Write('What is the number of X and Y pairs you wish to enter?  ');
   Readln(N);
   Writeln;
   For I:= 1 to N Do
     Begin (* For *)
       Write(      'X(',I,'):  ');
       Sx:= WhereX; Sy:= WhereY;
       Read(X[I]);
       GoToXY(Sx+10,Sy);
       Write('     Y(',I,'):  ');
       Readln(Y[I]);
     End;  (* For *)
   CLRSCR;
End;  (* ReadKeyBoard *)
```

Figure 10–29 Data input with a **for** statement.

or **while** statements are useful for iterating a section of a program. The main difference between the two is that the **while** statement checks the Boolean expression at the beginning of the loop, but the **repeat** statement checks the Boolean statement at the end of the loop (Fig. 10–30). Consequently, the **repeat** statement executes the loop at least once, but the **while** statement might not execute the loop at all. Compare the outputs of *Program WhileDemo* (Fig. 10–31) and *Program RepeatDemo* (Fig. 10–32).

If an integer from 1 to 10 is input, the output is the same for either program. However, if *11* or a larger integer is input to each, the loop is not executed with *Program WhileDemo* at all, but is executed once with *Program RepeatDemo,* as shown in Table 10–3.

For another example, consider two programs that will determine the sum of all odd numbers up through some desired odd number input by the user (Figs. 10–33 and 10–34). With either program the output is *9* when *5* is input or *2500* when *99* is input. What is the output if *1* is input? How many times is the loop executed when *1* is input?

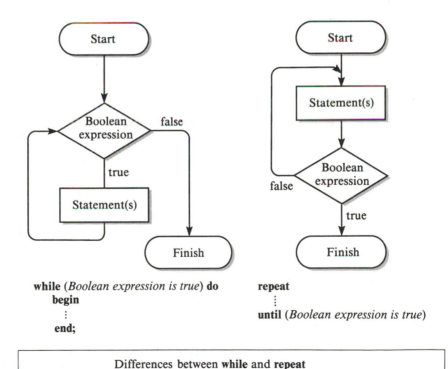

while (*Boolean expression is true*) **do**
    begin
       ⋮
    end;

repeat
   ⋮
until (*Boolean expression is true*)

Differences between **while** and **repeat**	
1. Control test is at top	1. Control test is at bottom
2. Executes on true	2. Executes on false
3. Terminates on false	3. Terminates on true
4. Executes zero times as minimum	4. Executes one time as minimum

Figure 10–30 Flow chart for **while** and **repeat** statements.

```
Program WhileDemo;
(* A program to demonstate a while loop *)

Var I:  Integer;

Begin
   Write('Please enter a small integer:  ');
   Readln(I);
   While I <= 10 Do
      Begin
         Writeln(I);
         I:= I + 1
      End;
   Writeln('Loop finshed.')
End.
```

Figure 10–31 *Program WhileDemo.*

Control. Under certain conditions it may be necessary to change the sequence in which actions are carried out. Pascal provides two control statements: the **if . . . then . . . else** and the **case** statements.

The If . . . Then . . . Else Control Statement. The syntax of **if . . . then . . . else** is as follows:

> **if** (*Boolean expression*) **then**
> > **begin**
> > > (*action1*)
> > **end**
> **else**
> > **begin**
> > > (*action2*)
> > **end;**

```
Program RepeatDem1;
(* A program to demonstrate a repeat loop *)

Var I:  Integer;

Begin
   Write('Please enter a small integer:  ');
   Readln(I);
   Repeat
      Writeln(I);
      I:= I + 1;
   Until I > 10;
   Writeln('Loop finshed.')
End.
```

Figure 10–32 *Program RepeatDem1.*

TABLE 10–3

"While" Output Versus "Repeat" Output

	Output	
Input	Program WhileDemo	Program RepeatDemo
5	5	5
	6	6
	7	7
	8	8
	9	9
	10	10
	Loop finished	*Loop finished*
11	*Loop finished*	11
		Loop finished

The **if . . . then** statement is just a special case of the **if . . . then
. . . else** statement, since the **else** part of the latter is optional. The **if
. . . then** is written as follows:

> **if** (*Boolean expression*) **then**
> > **begin**
> > > (*action*)
> > **end;**

It is frequently necessary to check whether a number is negative,
and the **if . . . then . . . else** provides a simple solution. Suppose you
want to avoid taking the square root of a negative number. You might
include a statement such as

```
Program SumOddNumbers;
(* written with a WHILE loop *)

Var  OddNum, HiOdd, Sum:  Integer;

Begin
   Write('Please input an odd number:  ');
   Readln(HiOdd);
   Sum:= 0;
   OddNum:= 1;
   While OddNum <= HiOdd Do
      Begin  (* While loop *)
         Sum:= Sum + OddNum;
         OddNum:= OddNum + 2
      End;    (* While loop *)
   Writeln;
   Writeln('The sum of all odd numbers from 1 through ',HiOdd,' is ',sum,'.');
End.
```

Figure 10–33 Summing odd numbers with **while**.

```
Program SumOddNumbers;
(* written with a REPEAT loop *)

Var  OddNum, HiOdd, Sum:  Integer;

Begin
   Write('Please input an odd number:  ');
   Readln(HiOdd);
   Sum:= 0;
   OddNum:= 1;
   Repeat
      Sum:= Sum + OddNum;
      OddNum:= OddNum + 2
   Until OddNum > HiOdd;
   Writeln;
   Writeln('The sum of all odd numbers from 1 through ',HiOdd,' is ',sum,'.');
End.
```

Figure 10–34 Summing odd numbers with **repeat.**

Write('Please enter a number: ');
Readln(Number);
if *Number* <= *0* **then**
    Writeln('Sorry but your number must be greater than zero.');
else
    begin
        Root:= SQRT(Number);
        Writeln('The square root of your number is: ',Root:10:5);
    end;

Look again at *Program TryOutFunctions* (Fig. 10–24). In the ninth line of the program, a Boolean expression is used to limit the value of the input to a number between -88 and $+88$. Why 88? Because *EXP(88)* exceeds the range of Turbo Pascal's real data type: 2.9×10^{-39} and 1.7×10^{38}. Near the end of the program, the square root of the number input is taken, but only if the number is greater than zero.

In Figures 10–33 and 10–34 (*Program SumOddNumbers*) it is assumed that an odd number is input by the user, which might not be the case. To ensure that an odd number is input, we can check on the input integer by using an **if** statement with a **mod** operator, as illustrated in Figure 10–35.

Pascal handles arithmetic operations differently for real numbers and integers, as shown in Tables 10–4 and 10–5. Thus, in the case of integer arithmetic, **div** gives the quotient without the remainder, while **mod** gives the remainder without the quotient. Note also that **div** and **mod** cannot be used with real variables and "/" cannot be used with integers (unless the result is assigned to a real). The **mod** operator provides a simple check on whether an integer is even or odd, since *EvenNumber* **mod** *2* has the value zero.

```
Program SumOddNumbers;
(* demonstrates the IF with the MOD operator for control *)
(* and the REPEAT statement for repetition               *)

Var  OddNum, HiOdd, Sum:  Integer;
                    Odd:  Boolean;
Begin
   Odd:= False;
   Repeat
      Write('Please input an odd number:  ');
      Readln(HiOdd);
      If HiOdd MOD 2 = 0 Then
         Begin
            Writeln('Sorry, but ',HiOdd:1,' is even.  Try again.');
            Writeln
         End
      Else
         Begin (* Else *)
            Odd:= True;
            Sum:= 0;
            OddNum:= 1;
            Repeat
               Sum:= Sum + OddNum;
               OddNum:= OddNum + 2
            Until OddNum > HiOdd;
            Writeln;
            Write('The sum of all odd numbers from 1 through ',HiOdd);
            Writeln(' is ',sum,'.')
         End (* Else *)
   Until Odd
End.
```

Figure 10–35 Summing only odd numbers.

Thus in the last version of *Program SummOddNumbers* (Fig. 10–35), the output is *2500* when *99* is input. However, if *98* is input the output is *Sorry, but 98 is even. Try again.*

The **if . . . then** statement provides one choice of action before continuing; the **if . . . then . . . else** statement provides two choices of action before continuing, as illustrated in the flow diagrams of Figure

TABLE 10–4

Real Operators, $X:= 5.2$; $Y:= 2.1$;

Operator	Meaning	Example	Current Value of Z
+	Addition	$Z:= X+Y$;	7.3
−	Subtraction	$Z:= X-Y$;	3.1
*	Multiplication	$Z:= X*Y$;	10.92
/	Division	$Z:= X/Y$;	2.476190476

TABLE 10–5

Integer Operators,   A:= 7;   B:= 3;

Operator	Meaning	Example	Current Value of C
+	Addition	$C := A+B$;	10
−	Subtraction	$C := A-B$;	4
*	Multiplication	$C := A*B$;	21
mod	Remainder division	$C := A$ **mod** B;	1
div	Quotient division	$C := A$ **div** B;	2

10–36. To avoid one common error, note that no semicolon follows the **end** that precedes the **else**.

The Case Statement. The **case** statement provides for multiple choices. This powerful option is especially useful in constructing a menu from which the user can choose the desired action. The syntax for the **case** statement is as follows:

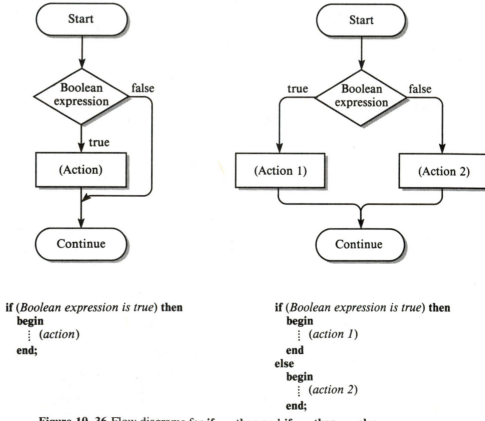

```
if (Boolean expression is true) then
   begin
      ⋮  (action)
   end;
```

```
if (Boolean expression is true) then
      begin
         ⋮  (action 1)
      end
   else
      begin
         ⋮  (action 2)
      end;
```

Figure 10–36 Flow diagrams for **if** . . . **then** and **if** . . . **then** . . . **else**.

```
Program CaseDem1;

uses crt;

Var Choice:  Integer;

Procedure GetMenu;
Begin
  Writeln('Enter the number below to get examples');
  Writeln('of different kinds of MATTER.');
  Writeln('1.  Elements');
  Writeln('2.  Compounds');
  Writeln('3.  Mixtures');
  Writeln('4.  Solutions');
  Writeln;
End;

Begin (* Main Program *)
  Clrscr;
  GetMenu;
  Write('Enter the number of your choice:  ');
  Readln(Choice);
  Case Choice of
    1:   Begin
            Writeln('oxygen');
            Writeln('iron');
            Writeln('aluminum');
            Writeln('tin');
         End;
    2:   Begin
            Writeln('salt');
            Writeln('water');
            Writeln('baking soda');
            Writeln('monosodium glutamate');
         End;
    3:   Begin
            Writeln('sand');
            Writeln('wood');
            Writeln('blood');
            Writeln('brass');
         End;
    4:   Begin
            Writeln('salt water');
            Writeln('rubbing alcohol');
            Writeln('honey');
            Writeln('Cocoa Cola');
         End;
    Else
         Begin
            Writeln('Sorry, your choice is 1,2,3 or 4.');
            Writeln('Run again to try again.');
         End;
  End; (* of Case *)
End.   (* Main Program *)
```

Figure 10–37 Demonstration of the **case** statement.

```
case (expression) of
    value1:    action1;
    value2:    action2;
    value3:    action3;
              ⋮
    else       action;
end;
```

The expression, called the *selector,* is usually an identifier for an *integer* or *char* variable. Each action is preceded by a *case label,* which must be of the same **type** as the selector. The **else** at the end is optional, just as it is in the **if . . . then . . . else** statement. The **else** statement is useful when the selector is assigned a value outside the range of the *case label.* The **case** statement is demonstrated in Figure 10–37.

FILES

Pascal considers a **file** to be a structured data type in the same way that an **array** is a structured data type. Each consists of a sequence of components of identical type. Thus Pascal recognizes a file of *integer,* a file of *real,* a file of **array,** a file of *char,* or any other data type. Because all components of one file must be of the same type, it is not possible to have a file of both *integer* and, say, *real,* or a file of both *char* and *integer.* Files differ from all other data types in one unique manner: files are permanently stored on diskettes. All other data types exist only in the memory of the computer and vanish when the computer is electrically disconnected.

Creating a Disk File of Data

Of particular importance is the file of *char,* also known as a *text* file. Computer programs (source programs) are just text files created by an editor. The files created by word-processing programs are text files. The Turbo Pascal Editor creates text files. Text files can also be created by programs. A nontext file (that is, a file of integer, real, or array) *must* be created by a Pascal program. Text files are so important that Turbo Pascal predefines this type, which means that it is not necessary to define

type *Text* = **file of** *char;*

Writing to and Reading from a Disk File

Whether data are input to a program from the keyboard or from a file, the output can be directed to either the screen or a file (or both). The familiar call to the predefined procedure *writeln* directs output to the screen. Before looking at an example program that writes to a disk file, let us compare the standard set of file-handling procedures for reading from and writing to disk files. Like any variable, the file must be declared. Let us assume that we have previously written the declaration as "**var** *f: Text*;" so that *f* is the identifier for the file variable of **type** *text*. Let us also assume that the disk file name shall be *LS.DAT,* a legal file name.

Reading from a Disk File	Writing to a Disk File
Assign(f, 'ls.dat');	*Assign(f, 'ls.dat');*
Reset(f);	*Rewrite(f);*
. . .	. . .
Readln(f, . . .);	*Writeln(f, . . .);*
. . .	. . .
Close(f);	*Close(f);*

Figures 10–38 and 10–39 list two short programs that illustrate writing to a file and reading from a file.

Writing to a Disk. *Program WriteToAFile* declares a file variable with the identifier *f* to be of type *text*. The *Assign* procedure assigns the file name *MyFile.dat* to an external file to be written on your disk drive. Any legal DOS file name can be used here. As usual, the extension (a period plus zero to three letters) is optional. If you want the file to show up on your B: drive, then just use the file name *B:MyFile.dat*. The *rewrite* procedure creates and opens a new file on your disk with the name you gave it with the *Assign* procedure. The *rewrite* statement is always used following an *Assign* procedure.

Notice the difference between *writeln('string')* and *writeln(f, 'string')*: The first *writeln* sends the string to the screen, where it is immediately visible. The second *writeln* sends the string to the file *f* that has the name *MyFile.dat* and is stored on an external device, the disk drive; the only indication that something is happening is the indicator light on the disk drive lighting up momentarily.

Finally, after assigning, rewriting, and writing, you close the file with the *close* procedure. The *close* procedure closes a file that was opened

```
Program WriteToAFile1;

Var f: text;

Begin
   Writeln('Watch the light of your disk drive turn on.');
   Assign(f, 'MyFile1.dat');
   Rewrite(f);
   Writeln(f,'This sentence will be on disk file ''MyFile1.dat''.');
   Close(f);
   Writeln('Now check your directory for the new file: MyFile1.dat');
End.
```

Figure 10–38 Writing to a disk file.

```
Program ReadFromAFile;

Var f:   text;
    ch:  char;

Begin
   Assign(f,'Myfile1.dat');
   Reset(f);
   While not EOF(f) do
     Begin
       Read(f,ch);
       Write(ch);
     End;
   Close(f);
End.
```

Figure 10–39 Reading from a disk file.

with the *rewrite* procedure. The *close* procedure is just part of the tidy file operations that Pascal and DOS team up to manage. If you worked in a small office with a large filing cabinet, you would open the file drawer, do something with the file, and then close the file drawer. If you left the drawer open, someone (your boss?) might come along, bump into the drawer, and bad things could happen. The same is true with Pascal files. Don't forget to close them or bad things could happen.

After checking your disk directory, run *Program WriteToAFile*. The light on your disk drive should light up and two statements should appear on your screen:

```
Watch the light of your disk drive turn on.
Now check your directory for the new file:  MyFile1.dat.
```

If you do, you should see among all the other files a new file with the name *MyFile1.dat:*

```
MYFILE1  DAT        52    5-26-88      9:57a
```

Reading from a Disk. *Program ReadFromAFile* illustrates the basic Pascal procedures for reading data from a disk and writing those data to the screen. Along with the procedures *Assign* and *Close,* a new procedure, *Reset,* is used to open a file for reading. *Reset* is always preceded by *Assign* so that the *Reset* will know which file is to be prepared for reading by going to its beginning.

Program ReadFromAFile is very similar to *Program WriteToAFile:* both begin with an *Assign* statement and end with a *Close* statement. The *Reset* statement following the *Assign* statement prepares the file for reading. Since a text file contains a sequence of text characters, a loop is the logical way to read the characters. In principle, either a **while,** a **repeat,** or a **for . . . to . . . do** construction would work. Since a text file (a file of characters) could contain any number of characters, the **while** loop is a choice since it will read the characters one at a time from the file f (*Read(f, ch)*) and write the characters one at a time to the screen (*Write(ch)*). Remember that the **while** construction requires a Boolean value of *false* to stop the loop. The last component of a text file is *always* the predefined Pascal Boolean **function** *EOF* (*End of File*). This **function** returns the value *false* until the last character of the file (preceding the *EOF* itself) is read. Then *EOF* returns a value of *true* so that execution drops out of the loop to the next sequential statement. If *Program ReadFromAFile* (Fig. 10–39) is run after *Program WriteToAFile* (Fig. 10–38), the following statement will appear on the screen:

> This sentence will be on disk file 'MyFile1.dat'.

The equivalence of the **for . . . to . . . do** and **while** statements is shown by a comparison of the programs in Figures 10–39 and 10–40. Run the programs in Figures 10–39 and 10–40 and the same output appears on the screen. Change the final value of the **for** statement to *20, 50,* and *150* in the program in Figure 10–40, run it, and examine the output.

Using a Pascal Text File to Handle Numbers

Let us see why files are so useful in physical chemistry and data reduction. In the laboratory we frequently have lists of parameters, often dependent and independent variables, that we often call X and Y, as illustrated in Table 10–6. It is a simple matter to use the Turbo Pascal Editor to create a text file containing the data listed in Table 10–6. Boot up Turbo Pascal

```
Program ReadFromAFile;

Var f:   text;
    ch:  char;
     i:  integer;

Begin
   Assign(f,'Myfile1.dat');
   Reset(f);
   For i:= 1 to 51 do
      Begin
         Read(f,ch);
         Write(ch);
      End;
   Close(f);
End.
```

Figure 10–40 Reading text with a **for** statement.

TABLE 10–6

Sample Data for Data Reduction

Run	X	Y
1	1.20	2.30
2	2.04	4.33
3	3.52	7.21
4	5.55	11.01

and go into the Turbo Pascal Editor. Then enter the tabulated numbers followed by returns, or spaces and returns (R stands for return).

```
4  R
1.20    2.30  R
2.04    4.33  R
3.52    7.21  R
5.55    11.01  R
^KD or F2 or F10
```

The numbers can be separated by any number of spaces, but at least one space is necessary. The first number in each pair can be preceded by any number of spaces or by no space. Just make the file look clear by lining by the decimal points and inserting as many spaces as you find pleasing. The final entry, $^\wedge KD$, indicates the end of file (*EOF*), which is necessary for any Pascal program that eventually will read the data on this file. To save this file on the disk, just enter **S** for **S**ave; when prompted by Turbo Pascal for a file name, you can enter any legitimate file name. In this case enter *ls.dat*. The extension *.dat* is not necessary but will be helpful to identify this file as containing some experimental data. A listing of the file just created is shown in Figure 10–41.

Notice that the file created is a text file, even though the characters stored on the file are the *char* representations of *numbers*. In fact, both *numbers* and *text* are stored in a text file; it may seem as though this violates Pascal syntax, but different types are not being mixed in a text file since all the characters on the keyboard are stored as their *char* representations.

Entering raw numerical data as components of a text file is a convenience for two reasons. First, it is simple to read such data from a file

```
4
1.2    2.3
2.04   4.33
3.52   7.21
5.55  11.01
```

Figure 10–41 Listing of a text file containing four pairs of numbers that can be read by *Program FileTry1*.

into your program, as shown in the next section. Second, it is easy to edit a text file with the Turbo Pascal Editor. In fact, a text file whose components are the character representations of numbers can be created not just with the Turbo Pascal Editor but with most word-processing programs, such as WordStar and WordPerfect. They all do the same thing: they create a text file.

Reading Numerical Data from a Text File Stored on a Disk. Now let us examine a simple but typical Pascal program (Fig. 10–42) that can be called to read the numerical data in the text file created as described above.

In any program that reads a file, three predeclared Turbo Pascal procedures are always called: *Assign, Reset,* and *Close.* The *Assign* procedure has two value parameters. The first is the identifier for the file variable, in this case f. Any nonreserved word could be used. If you don't like the identifier f, you could change it everywhere to, say, *MyFile, DataList, XYdata,* or anything you like. The second parameter is the string, which is the name of the file stored on a diskette. The file name is the one already chosen when we created the file, *ls.dat.* Note that this string is enclosed between apostrophes. The statement *Assign (f, 'ls.dat')* assigns the file name *ls.dat* to the file variable f. This is always the first of the three file-handling procedures used in reading files.

The second file-handling statement, which should come immediately after the first, is the statement *Reset(f),* which prepares the file f for reading by putting the "file pointer" to the beginning of the file. Files are always read from the beginning. We know that, but a computer must be told.

Next come the remaining statements of the program, including those that read the file. In the sample program (Fig. 10–42), the statement *Readln(f,N)* reads the first line of the file f (Fig. 10–41) and assigns the value it finds to the integer variable N. In the example file, N is assigned the value *4.* Next comes a **for** loop, which will be executed N times. Each time the loop is executed, a line is read. Since two parameters follow the file variable named f, two parameters are expected on each line—which is the reason we created the file with two numbers separated by a return. The return determines the end of the line in the file. The first time *Readln(f,X,Y)* is executed, the value *1.20* is assigned to X and the value *2.3* is assigned to Y. The product of X and Y equals *2.58,* which is assigned to Z. The *Writeln* then displays the values of X, Y, and Z on the screen. The second time the loop is executed, *2.04* is assigned to X and *4.33* is assigned to Y. Execution of the loop continues until the loop counter is satisfied. In this example, this occurs when the loop counter I equals *4.*

As the program reads the data from the file on the disk, the indicator light on the disk drive lights up just as it does when you load a program or carry out some file management task with the DOS.

Finally, sometime after the last datum is read from the file, the file is closed with the third of the three required file-handling statements:

```
Program Filetry1;

uses crt;

Var      f:  Text;
      X,Y,Z:  Real;
        N,J:  Integer;

Begin
  CLRSCR;
  Writeln('         X            Y                Z');
  Writeln;
  Assign(f,'LS.DAT');
  Reset(f);
  Readln(f,N);
  For J:= 1 to N Do
     Begin (* For *)
        Readln(f,X,Y);
        Z:= X*Y;
        Writeln(X:10:2,Y:10:5,Z:10:5);
     End; (* For *)
  Close(f)
End.
```

Figure 10–42 A program for reading numerical data from a text file stored on a disk.

Close(f). The predeclared procedure *Close* causes the operating system to update the file and the disk directory. After *Program FileTry* (Fig. 10–42) is run with the input shown in Figure 10–41, the output is as shown in Figure 10–43.

Reading Numerical Data from a Text File into Arrays. It is often convenient to read numerical data from a text file into program arrays for further data reduction. *Procedure ReadFile,* shown in Figure 10–44, would be part of a program in which the array variables *Xry* and *Yry* are globally declared. The counter integer *I* and the text file *f* are declared locally. This procedure is very similar to the program listed in Figure 10–42.

Reading from a File to an Array and Writing from an Array to a File. Suppose you have a file containing some raw data, say, pressures and temperatures (°C), and you want to convert them to a file of ln *P* and $1/T$ (K) so that you can run a least-squares fit of these data to a line. *Program FileConversion,* shown in Figure 10–45, reads the data from an input file (*lsraw.dat*) into an array, converts the data as desired, and writes the converted data into an output file (*ls.dat*).

At the beginning of the program the file *f* is assigned the name *lsraw.dat* and opened for reading with the *Reset(f)* statement. After read-

X	Y	Z
1.10	2.30000	2.53000
2.04	4.33000	8.83320
3.52	7.21000	25.37920
5.55	11.01000	61.10550

Figure 10–43 Output of a file-reading program.

ing n (the number of data pairs to follow), a **for** loop reads the data pairs from the text file into the arrays x and y. The elements of the arrays are converted, in this example from $t\,°C$ to $1/T$ and from P to $\ln P$.

After the conversion, the file f is assigned the name $ls.dat$ and opened for writing with the $Rewrite(f)$ statement. After the value of n is written to the file, a **for** loop writes the new converted pairs of data to the file $ls.dat$. Subsequently file $ls.dat$ can be read by a least-squares program such as $LSPChem.pas$ (Fig. 6–4). By editing the two existing conversion lines, you can use this utility program to linearize data according to the many algorithms described in Chapter 6.

An example of $lsraw.dat$ is shown in Figure 10–46. After $Program$ $FileConversion$ is run with $lsraw.dat$ as the input file, $ls.dat$ appears as shown in Figure 10–47. Finally, after $Program$ $LSPChem$ is run with $ls.dat$ as input, the output of $LSPChem$ is as shown in Figure 10–48.

Review: File Reading and Passing Arrays

$Program$ $LSPChem$ (Fig. 6–4) provides an example of a Pascal procedure used to read numerical data from a text file into an array. The call to the procedure illustrates another example of passing the value of a variable from a procedure to the call. When $Procedure$ $ReadFile$ (Fig. 6–4) is called, the values of two arrays (Xry and Yry) are passed to arrays declared

```
Procedure ReadFile;

   Var I:   Integer;
       f:   Text;

Begin (* ReadFile *)
   Assign(f,'LS.DAT');
   Reset(f);
   Readln(f,N);
   For I:= 1 to N Do
      Readln(f,Xry[I],Yry[I]);
   Close(f)
End;   (* ReadFile *)
```

Figure 10–44 Reading from a text file to arrays.

in the main block (*X*, *Y*); the value of the integer *Num* is passed to *N*. A summary of this call is shown in Figure 10–49. Notice how the number and type of parameters in the procedure call match the number and type of parameters in the procedure parameter list, although the identifiers are different. Compare this call with the call shown earlier in Figure 10–23.

For the purpose of review, a second call is included in Figure 10–49, a call to the least-squares procedure. For the sake of brevity, the complete least-squares procedure is omitted from Figure 10–49, but it is included in Figure 6–4. Notice that there are six parameters in the call and six in the procedure parameter list. In this example the identifiers used are the same, but they need not be. The parameter identifiers must agree only in their number and type. In *Procedure LeastSquares, X* and *Y* are value parameters. The values of arrays *X* and *Y* and the integer *N* are passed from the call in the main block to the procedure. Compare

```
Program FileConversion;
(* A program to read raw data from file 'lsraw.dat' and convert (x,y) *)
(* values to some other functions f(x) and g(y).  For example, if     *)
(* you want to plot ln(y) vs 1/x, then after the next comment line     *)
(* just add the two statements x[i]:=1/x[i]; and y[i]:=ln(y[i]);       *)
(* The converted data are written on a file named ls.dat              *)
(* The file lsraw.dat must be on a disk in the B: drive               *)

Type list = array[1..100] of real;

Var n,i: integer;
    name,f: text;
    x,y: list;

Begin
    Assign(f,'B:lsraw.dat');
    Reset(f);
    Readln(f,n);
    For i:=1 to n do
        Begin
            Readln(f,x[i],y[i]);
(* Add your conversion statements below this comment line. *)
(* If someone else left their statements here, delete them first.*)
            x[i]:=1/(x[i] + 273.15);
            y[i]:=Ln(y[i]);
(* End of conversion statements *)
        End;
    Assign(f,'B:ls.dat');
    Rewrite(f);
    Writeln(f,n);
    For i:=1 to n do
        writeln(f,x[i],'  ',y[i]);
    Close(f);
    Writeln('Conversion completed.  Press Enter key');
    Readln;
End.
```

Figure 10–45 *Program FileConversion.*

```
4
   750   9.1E-5
   830   3.8E-4
  1065   1.18E-2
  1132   2.6E-2
```

Figure 10–46 Input (*lsraw.dat*) to *Program FileConversion.*

```
4
    9.7737379661E-04    -9.3046510514E+00
    9.0649503694E-04    -7.8753393052E+00
    7.4730037739E-04    -4.4396557475E+00
    7.1166779347E-04    -3.6496587410E+00
```

Figure 10–47 Output (*ls.dat*) from *Program FileConversion* with *lsraw.dat* as input.

with Figure 10–23. The remaining parameters in *Procedure LeastSquares* are declared to be variable parameters so that their values may be passed from the procedure to the main block. These include an array (*Ycalc*) and two integers (*m* and *b*, the slope and intercept, respectively).

The abbreviated program in Figure 10–49 is syntactically correct and will run without error if a correctly formatted text file *ls.dat* is present for reading. The only output is the text at the end of the main block.

Summary of Pascal Input/Output. Files are the second of two methods described in this chapter for exchanging numerical data between a computer and the outside world. In the first method the user inputs data from the keyboard using the *read* and *readln* statements. Compare procedures *ReadFile* and *ReadKeyBoard,* which are adjacent in Figure 6–4. If the user has only a small amount of data to input, then the keyboard is the appropriate tool, but if a large amount of data are to be

I	X	Y	Ycalc
1	9.7737379661E-04	-9.3046510514E+00	-9.3405999054E+00
2	9.0649503694E-04	-7.8753393052E+00	-7.8279709779E+00
3	7.4730037739E-04	-4.4396557475E+00	-4.4305858053E+00
4	7.1166779347E-04	-3.6496587410E+00	-3.6701481587E+00

```
        The intercept =    1.1517607612E+01
        sigma =            1.7236934484E-01

        The slope =       -2.1341075022E+04
        sigma =            2.0449560238E+02

    The correlation coefficient = -9.9990819276E-01
```

Figure 10–48 Output of *Program LSPChem* with *ls.dat* as input.

```
Program LSPChem;
(********* An abbreviated version for demonstration ********)

Const Max = 50;

Type Ary = Array [1..Max] of Real;

VAR
  X,Y,Ycalc        :  Ary;
  N                :  Integer;
  A,b,m            :  Real;
  CorrelCoef, Sigmam, Sigmab:  Real;

(******************************************************************)

PROCEDURE ReadFile(VAR Xry,Yry:  Ary;
                       VAR Num:  Integer);

   Var I:  Integer;
       F:  Text;

Begin (* ReadFile *)
   Assign(F,'LS.DAT');
   Reset(F);
   Readln(F,N);
   For I:= 1 to Num Do
      Readln(F,Xry[I],Yry[I]);
   Close(F)
End;  (* ReadFile *)

(******************************************************************)

Procedure LeastSquares(X,Y:  Ary;
                        Var Ycalc:  Ary;
                        Var b,m :  Real;
                            N  :  Integer);

Begin (* LeastSquares *


End;  (* LeastSquares *

(******************************************************************)

Begin (*  M A I N  P R O G R A M  *)
   ReadFile (X,Y,N);
   LeastSquares(X,Y,Ycalc,b,m,N);
   (* See comment below *)
   Writeln('If you read this line, it means that this demo ');
   Writeln('program compiled and ran.  The disk light lit, the');
   Writeln('file LS.dat was read, and the parameters were passed.');
   Writeln;
End.  (*  M A I N  P R O G R A M  *)
```

Figure 10–49 File reading and parameter passing in a least-squares program.

processed, then a disk drive is a better tool for transferring numerical information to the computer. Note that use of the disk drive requires two steps: first, creating the file with the Turbo Pascal Editor and storing it on a diskette; and second, reading the file from the diskette with the program statements just described. Since the data are stored on a diskette, the file may be retrieved at any time with the Turbo Pascal Editor to edit or correct erroneous data in the file.

Part Three (Experiments) includes several Pascal programs for the reduction of experimental data.

REFERENCES

1. Cooper, D., and M. Clancy, *Oh! Pascal!,* Norton, New York, 1982.

2. Peters, J. F., *Problem Solving with Pascal,* Holt, Rinehart and Winston, New York, 1986.

3. Radford, L. E., and R. W. Haigh, *Turbo Pascal for the IBM PC,* Prindle, Weber & Schmidt, Boston, 1986.

4. *Turbo Pascal Version 5.0 Reference Guide,* Borland International, Inc., Scotts Valley, California, 1988.

5. *Turbo Pascal Version 5.0 User's Guide,* Borland International, Inc., Scotts Valley, California, 1988.

Working with Glass

Other than repairing an occasional cracked flask or constructing an electrochemical cell, nearly all of the glass working a physical chemist encounters involves a high-vacuum rack. The construction and maintenance of a high-vacuum rack requires only a few simple skills for working with glass: cutting, bending, and forming straight seals and T-seals. After you feel comfortable with these elementary operations, you may want to try some slightly more advanced techniques: ring seals, coils from tubing, and glass-to-metal seals. Working with glass on a high-vacuum rack differs considerably from the techniques used by professional glass blowers. Professionals use a torch mounted in a fixed position on a bench, moving or rotating the glass between their fingers to ensure uniform heating. Rotating two pieces of semifluid glass at the same speed with the proper configuration requires a great deal of skill and practice. In contrast, the tubing of which a vacuum rack is constructed, and consequently the glass, is clamped in position, while the torch is held by hand and moved to ensure uniform heating of the glass. Fortunately, this method, described in this chapter, is easy to learn, although the results may not be as aesthetic as those attained by professionals.

PROPERTIES OF GLASS

With the exception of fused silica (SiO_2), glasses are mixtures of silica and light-element oxides. The compositions of a few typical glasses are given in Table 11–1.

Glasses are noncrystalline substances that fracture conchoidally rather than along crystal planes. Glasses have about the same degree of order (on a molecular scale) as liquids and may be considered to be

TABLE 11-1

Approximate Compositions of Some Glasses (%)

	Soft	Pyrex	Vycor	Fused Silica
SiO_2	72	80.1	96.3	100
B_2O_3	1	12.0	2.9	—
Al_2O_3	1	3.1	0.4	—
CaO	6	0.2	—	—
MgO	2	—	—	—
Na_2O	18	4.2	0.04	—
As_2O_3	—	0.4	0.01	—

supercooled liquids of very high viscosity. As a result, glasses have no definite melting point, but instead soften over a relatively wide range of temperature as their viscosity decreases. Like all substances, they expand when heated. Because glasses are very poor thermal conductors, large temperature differences can easily develop between different points in a piece of glass. This, in turn, causes different amounts of thermal expansion, accompanied by large shearing forces, which result in a thermal fracture. The thermal expansion a glass undergoes upon heating depends on the linear expansion coefficient, which varies considerably from glass to glass (Table 11–2). The linear expansion coefficient is defined as $(1/\ell_0)(\Delta\ell/\Delta t)$, where ℓ_0 is the length at 0°C. Glasses with greatly different expansion coefficients cannot be sealed to each other. Thus, Nonex and Pyrex can be sealed to each other, but not to soft glass or Vycor. Some metals can be sealed to glass if their linear expansion coefficients (Table 11–3) are not too different from the glass and if the soft glass wets the metal. Platinum wire less than 1 mm in diameter can be sealed into the end of a soft glass tube about 6 to 8 mm in diameter to make an electrode. Insert the wire, heat until the softened glass collapses down around the wire, and then anneal carefully. In the same manner tungsten wire can be sealed into Nonex, which can be sealed directly to Pyrex.

The annealing temperature is the temperature at which strains within

TABLE 11-2

Physical Properties of Some Glasses

Glass	Linear Expansion Coefficient $\times 10^9$	Annealing Temperature (°C)	Softening Temperature (°C)	Refractive Index
Soft	92	510	696	1.512
Nonex	36	526	756	1.484
Pyrex 7740	32	553	819	1.474
Vycor	8	890	1510	1.458
Fused silica	5	1140	1650	1.544

TABLE 11–3

Physical Properties of Some Metals

Metal	Linear Expansion Coefficient $\times 10^9$	Melting Temperature (°C)
Copper	166	1083
Platinum	89	1773
Molybdenum	54	2620
Tungsten	45	3370

a piece of glass are relieved. The softening temperature is the minimum temperature at which the glass can be worked. Soft glass can be worked in a Bunsen burner flame, small sizes of Nonex in a Meeker burner, but a gas–oxygen flame is necessary to work Pyrex tubing of about 8- to 10-mm diameter or larger. Working with Vycor or fused silica is more like welding than glass blowing, and an oxy-hydrogen or oxy-acetylene flame is needed, as well as dark welder's goggles to protect the eyes from the intense incandescence of glass heated above 1500°C.

Pyrex is the brand name given to a number of borosilicate glasses manufactured by the Corning Glass Works, Corning, New York. The glass used in Pyrex-brand laboratory glassware and tubing is Pyrex 7740. Because the properties of various glasses are very different, it is important to be able to distinguish between them. A quick and effective method of identifying Pyrex 7740 is based on its refractive index of 1.474. Pyrex 7740 becomes virtually invisible when immersed in a liquid with a matching refractive index. Prepare a mixture of 16 parts methanol and 84 parts benzene. Immerse the ends of several pieces of glass tubing and/or rod into the solution. The submerged sections of Pyrex 7740 become nearly invisible, while the other pieces of tubing are clearly visible. Pure trichloroethylene or pure dimethysulfoxide have been suggested as alternatives to the methanol and benzene mixture (Drake, 1969). With practice you can also identify Pyrex by simply holding the glass in a flame and noting the flame color and the comparative rate of softening. Soft glass softens almost instantly. Vycor becomes white hot before it softens.

EQUIPMENT

The equipment you need for working glass on a vacuum rack is considerably simpler than the array of apparatus required by a professional glass blower. Instead of a bench burner you will need a hand torch with tips of various sizes, normally for burning oxygen and natural gas from the house supply. The tips can give flames from about one-half inch in diameter down to the size of a pencil point. You can use any soft rubber tubing to connect the burner to the source of gas and oxygen, but twin tubing is available and makes a neat installation less likely to get tangled. Be sure that the tubing is long enough to reach over the entire area of the vacuum rack, with enough play for reaching around the work.

The first thing to do with your hand torch is to silver solder a brass hook to the straight shank of the burner (Fig. 11–1). Don't use heavy copper wire, it's too soft. Then, when you need two hands, you can hang the burner from one of the crosspieces of the vacuum rack without extinguishing the flame. Just be sure that the flame is not accidently directed at the rack or its components. Besides a hand torch, you should always keep a small Bunsen burner burning within easy reach of your work in case your flame accidently blows out. The Bunsen burner serves as a pilot light.

The oxygen source is normally a No. 1 cylinder equipped with a two-stage regulator valve (Fig. 11–2). The gauge closest to the cylinder measures the total gas pressure in the cylinder, which is about 2200 psi at room temperature when the tank is full. This gauge is a measure of the amount of gas remaining in the tank. If the pressure measures about 500 psi, then a little more than three fourths of the gas has been used. The second gauge measures the controlled delivery pressure, that is, the pressure between the regulator valve and the needle valve. The delivery pressure can be changed by turning the bar-shaped handle on the regulator. Note that **turning the regulator valve clockwise increases the delivery pressure.** This is the opposite effect of most valves, including the main valve and the needle valve on the gas cylinder, or an ordinary water faucet.

When the system is not in use, both gauges should indicate zero pressure, the main valve should be closed (clockwise), the regulator valve

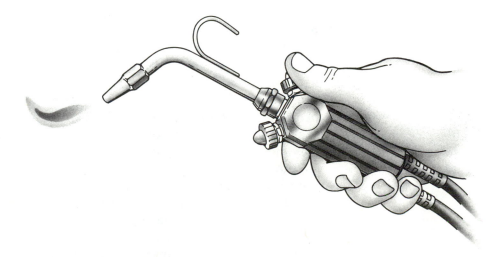

Figure 11–1 The hand torch.

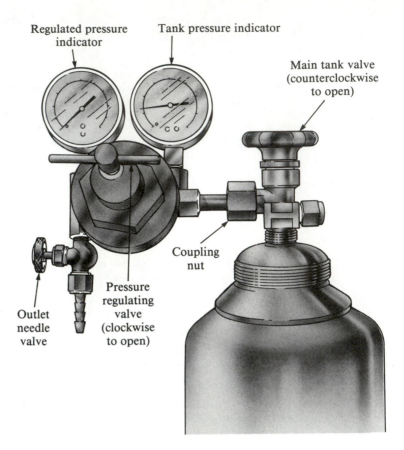

Figure 11–2 The two-stage gas regulator valve.

should be set to zero delivery pressure (counterclockwise), and the needle valve should be closed (clockwise).

To use the system, check that the needle valve is closed and that the regulator valve is closed (set to zero pressure). Check the regulator valve by turning it counterclockwise; if it is very loose it may unscrew and come out in your hand. Don't worry about this; just screw it back in a turn or two. Open the main cylinder valve by turning it about a half-turn counterclockwise. The main cylinder pressure gauge will jump immediately to 2200 psi if the tank is full. Next, turn the regulator valve clockwise; the pressure reading on the delivery pressure gauge will gradually increase. Set the pressure to about 10 psi. Then open the needle valve to admit gas to the torch lines. Open the oxygen valve on the torch momentarily to see if you can hear oxygen issuing smoothly.

To light the hand torch, be sure that its oxygen valve is off; then open the torch gas valve slightly and light with a striker. If the flame does not maintain itself, reduce the gas flow and try again. When the gas alone is burning stably, slowly open the oxygen valve. At first, the pure yellow flame becomes tinged with blue; this is a reducing flame. With an increase in oxygen, the flame becomes pale blue and burns quietly. With a further increase in oxygen, the flame becomes smaller, hotter, noisier, and less stable. If the oxygen is increased beyond this point, the flame usually extinguishes itself with a loud "pop." To relight, reduce or turn off the oxygen valve on the torch.

To turn off the torch, first turn off the oxygen, then the gas. The reverse order guarantees a loud "pop." Then, turn off the main valve of the oxygen cylinder (clockwise). Next, open the oxygen valve of the torch to bleed out the oxygen from the two-stage regulator valve. The regulator gauge and main cylinder gauges will slowly drop to zero. Finally, close the needle valves on the torch and the regulator (clockwise). Close the regulator by turning it counterclockwise until it is quite loose.

The gas cylinder must be chained or otherwise securely fastened to one end of the vacuum rack. If a cylinder is tipped over, the valve or outlet connection can be broken, resulting in the sudden release of gas at high pressure. This can suddenly turn the cylinder into a rocket capable of penetrating a brick wall if it has a chance to accelerate the length of a laboratory. Don't take chances with unchained gas cylinders.

In addition to a hand torch, a supply of natural gas, and a cylinder of oxygen with a two-stage regulator, you should assemble and have within easy reach of your work a number of small items:

1. Pyrex cane, 2- and 4-mm diameter. As we shall see, the large size is used as a temporary handle for short pieces of work, and the small cane is used to seal tiny holes in imperfect seals.
2. Triangular file, 6-inch. This is used for cutting glass.
3. Two or three sets of assorted solid cork stoppers, 2 to 24 mm. These are used to plug the ends of pieces of tubing so that you can blow into the piece being worked.
4. Assorted cork stoppers, 6 to 24 mm, bored and fitted with 2-inch lengths of 4-mm Pyrex tubing. These should be mounted on a rack consisting of a 4-by-12-by-1/2-inch board into which a row of 2-inch small-headed nails have been hammered. One of these stoppers, with a rubber tube attached, is used to blow a slight positive pressure into the piece being worked, when necessary.
5. Rubber tubing, to fit 4-mm glass tubing, about 6 to 8 feet long, with a pipe stem in one end to serve as a mouthpiece. This is attached to one of the cork stoppers fitted with a short glass tube.
6. Biological specimen forceps, about 12 inches long. These are used frequently for pulling off excess soft glass.
7. Glass blower's goggles, Didymium glass (Corning No. 5120). Didymium glass, containing neodymium and praseodymium oxides, nearly completely absorbs the intense yellow radiation arising from the sodium D line of hot glass. It is virtually impossible to work with hot glass without Didymium goggles. They are also available as clip-on glasses for those who normally wear corrective glasses.

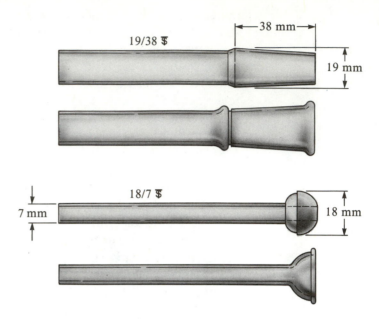

Figure 11–3 Standard taper joints.

8. Pyrex glass tubing. A large stock is not necessary for building and maintaining a vacuum rack. You might consider standardizing on 24-mm tubing for the manifold and 8-mm tubing for connections. Many high-vacuum stopcocks have 8-mm connecting tubes, and the 8/15 standard taper ⅋ joint is a convenient size for connecting to the manifold. Also, the 24/40 standard taper joint fits the 24-mm tubing perfectly. The tubing should be stored horizontally in dust-free bins.

9. Standard taper ⅋ joints and stopcocks. The joints are of two types: conical, and ball and socket (Fig. 11–3). The first number on a conical standard taper joint, for example, ⅋ 24/40, shows that the width of the top of the joint is 24 mm. The second number indicates that the length of the ground surface is 40 mm. The first number of a ball-and-socket joint, such as ⅋ 28/15, indicates that the diameter of the ball is 28 mm. The second number shows that the bore of tubing is 15 mm.

10. Vacuum stopcocks, which are distinguished by hollow rather than solid plugs. Solid-plug stopcocks should not be used in vacuum racks because the plug and the barrel expand differently with changes in temperature, resulting in leaks.

CUTTING GLASS

Nearly always, your first operation in glass working consists of cutting glass tubing to approximate size. You can cut tubing less than 12 to 15

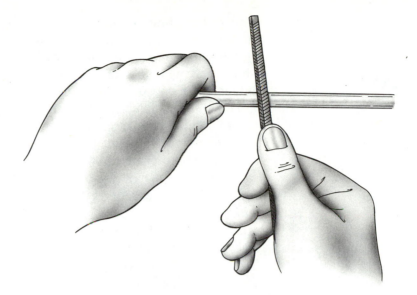

Figure 11–4 Scratching glass.

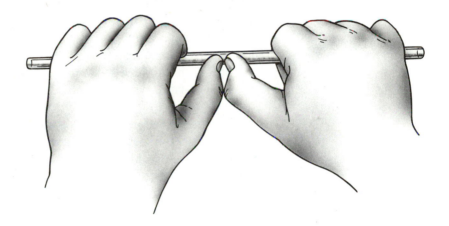

Figure 11–5 Preparing to break glass.

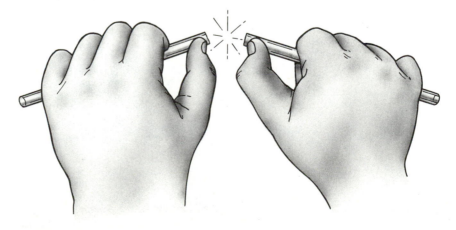

Figure 11–6 Breaking glass.

mm by scratching it with a short triangular file and breaking the piece in two by hand. Hold the tubing with your left hand, and place your right thumb on top of one surface of the file to apply downward pressure while applying two or three short forward strokes (Fig. 11–4). Files cut only in the forward direction. Grasp the tubing as shown in Figure 11–5, with thumbs on each side of the scratch. Pull the tubing with only a slight bending force (Fig. 11–6). If the tubing does not break, scratch it once more and try again. You may want to wrap large tubing, 10 to 15 mm, in a rag to protect your hands in case something goes wrong.

For larger pieces of glass tubing, alternative methods must be used. Scratch the tube exactly as before. Then apply a tiny but intense flame tangent to the tube, parallel to the circumference and about 1/2 to 1 mm from the end of the scratch (Fig. 11–7). Usually this induces a fine crack, which runs nearly around the tube. A second application of the flame 1/2 to 1 mm beyond the crack usually leads the crack the rest of the way around. At this time the piece separates in two voluntarily or does so with the slightest touch. This method works well with small tubing as well. In fact, for modifying or removing sections of tubing from an existing vacuum rack, where the tubing is clamped in place, it is about the only practical method. A variation on this method is to scratch the tube all the way around and then to apply the tip of a small piece of cane, which has been heated very hot.

Another method used with large glass tubing is the hot-wire method (Fig. 11–8). Again, make a scratch around the entire circumference of the tube. Wrap a loop of 28-gauge nichrome wire around the scratch. Then heat the wire to red heat electrically and apply water to the scratch; if all goes well, the tube will separate cleanly into two pieces. A step-down transformer furnishes the required low-voltage supply of electrical power. Be sure not to short-circuit the loop.

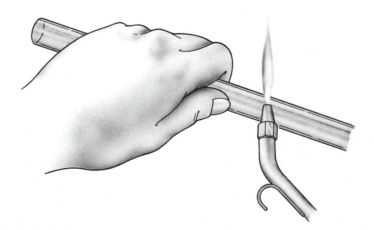

Figure 11–7 Crack cutting with a flame.

None of the above methods works with fused silica or Vycor because they have low coefficients of expansion. For Vycor or even large-diameter Pyrex, a cutoff wheel is effective. The thin wheel, coated with an abrasive and cooled with flowing water, is turned with a motor. The tubing is moved into the wheel with a jig as shown in Figure 11–9.

Another method for "cutting" tubing, which is effective for both glass and small-diameter Vycor (less than 24-mm diameter), is to heat the tubing and pull slowly while it is in the flame, causing the hot glass to become constricted. With further heating and pulling, the tube will separate into two pieces. With gentle blowing and heating, the end can be formed into a test-tube end.

Straight Seal Between Equal Tubes

After cutting glass tubing, the most common operation is forming a straight seal between two tubes of the same diameter. This is a fundamental operation and you should practice it until you have confidence in your ability. Learning to work with glass is something like learning to swim or ride a bicycle: you never forget it and you never know when you will need it.

Since most of your work will probably be on a vacuum rack, clamp two pieces of 8- to 10-mm tubing, each about 6 inches long, as shown in

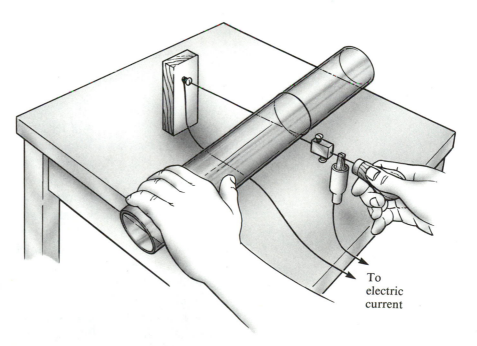

To
electric
current

Figure 11–8 The hot-wire method.

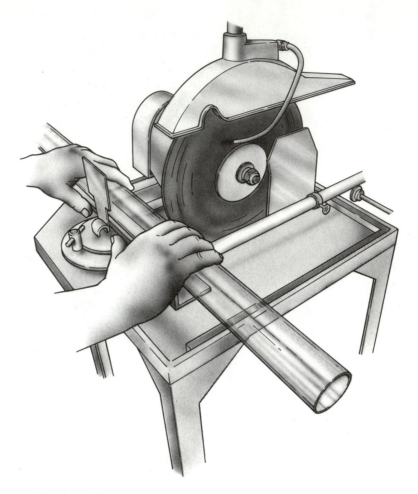

Figure 11–9 The cutoff wheel.

Figure 11–10. Slip a 3-inch length of 3/8-inch rubber tubing over the end of one tube and stopper the other tube with a small cork stopper. Use lighter rubber tubing as a blow tube and connect it to the short 3/8-inch rubber tube with a small cork fitted with a short piece of 4-mm glass tubing. Clamp the tube with the blow tube tightly; clamp the other tube, but not so tightly that it cannot be pushed slightly to the left or right in the clamp. Line up the two tubes so that their axes are collinear. Have a lighted Bunsen burner nearby as a pilot light and light your hand torch. Test the tubes to verify that one is clamped tightly and that the other is movable. The tubes should be about 1 mm apart.

Heat the ends of the two tubes with a blue flame. Move the torch constantly from above the tubes to 180° below the tubes. Hold the torch so that the flame heats the back side as well as the front, top, and bottom. When the tubes have visibly softened, hang up the torch momentarily, and grasp the tubes as shown in Figure 11–10 and Figure 11–11a. Slide

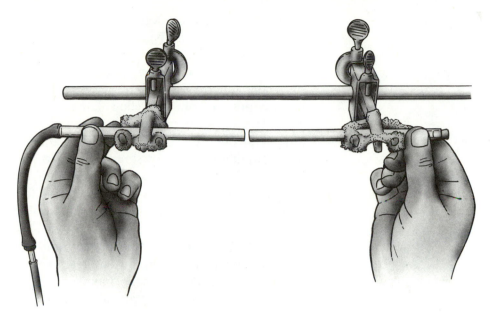

Figure 11-10 The straight seal on a vacuum rack.

the tube on the right toward the tube on the left until contact between the soft glass is made and a uniform ridge is pushed up around the seal as shown in Figure 11-11b.

The glass will have cooled a bit by now, so reheat the seal until it is soft. Again, grasp the tubes, but now pull them slightly apart in order to thin the thick ridge formed upon initial contact (Fig. 11-11c). Next, reheat the seal and blow gently until the diameter of the seal is close to the diameter of the tubing (Fig. 11-11d). While the seal is still soft,

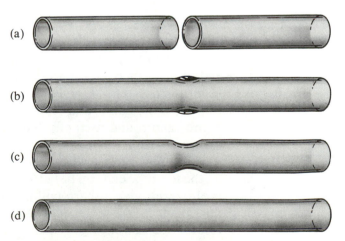

(a)

(b)

(c)

(d)

Figure 11-11 Forming a straight seal.

tighten the clamp that was not completely tight. If the work was set up carefully, there should be no movement of the tubing on each side of the still-soft seal. Next reduce the oxygen until the flame is somewhat, but not entirely, yellow. Anneal the seal by heating in this flame a minute or so. Then shut off the oxygen and continue annealing until the seal is black with soot from the yellow flame. After the seal has cooled, wipe it clean with a piece of paper towel. The soot itself has no annealing action; however, it does not condense on tubing above the annealing point. Thus, the accumulation of soot simply serves to indicate that you have been patient and taken sufficient time to properly anneal the seal.

The following points serve as a guide for controlling the diameter of a tube and the thickness of the tubing walls:

1. Pulling decreases the diameter and the wall thickness.
2. Blowing increases the diameter and decreases the wall thickness.
3. Heating decreases the diameter and increases the wall thickness.
4. Pushing an enlarged tube increases the diameter and the wall thickness.
5. Pushing a constricted tube decreases the diameter and increases the wall thickness.

Straight Seal Between Unequal Tubes

In order to make a straight seal between two tubes of different diameters, you will need to master two additional operations: making a test-tube end (Figs. 11–12a and 11–12b) and blowing a kidney, that is, a kidney-shaped bubble of exceedingly thin glass that can easily be chipped away to leave a tube with a slightly jagged flare. Further heating of the end results in a nicely rounded and smooth flare, as shown in Figures 11–12c and 11–12d.

Heat one end of the larger tube about an inch or so from the end; when the glass has softened, grasp the solid end with a forceps or tongs and pull out a point as shown in Figure 11–12a. Further heating permits the point to be pulled off. If a little too much glass still adheres to the end, bring a piece of 4-mm rod close to the end, touch it to the excess glass, and pull it off. Finally, heat the end uniformly around, remove the tube from the flame, and blow a gentle puff; the end will tend to expand slightly and round itself off. Repeat until the end is nicely rounded as in Figure 11–12b.

To blow out a kidney, adjust the flame so that a very small region on the test-tube end can be heated. When an area about 5 mm in diameter is quite soft, give a sharp puff of air. A kidney-shaped bubble will blow out. When this is chipped away and the area is heated until soft, the result is a smooth, slightly flared opening, smaller than the diameter of the tube. With a little practice, you can adjust the area heated on the

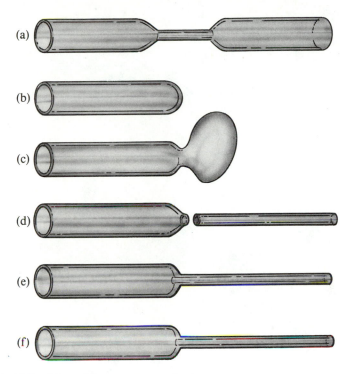

Figure 11–12 Straight seal between unequal tubes.

large test tube end so that the resulting small opening just fits the desired diameter of the smaller tube to be jointed to it.

Complete the operation just as you would do for a straight seal. The glass is a bit too thick where the two pieces are pushed together, as shown in Figure 11–12e. Soften this area and pull and blow slightly; the final result is shown in Figure 11–12f.

The T-Seal

No new operations are involved in the T-seal. Whether the tubes are equal or unequal in diameter, a T-seal requires blowing a kidney, chipping it away, fire polishing the edges, and sealing the perpendicular tube in place. If we assume that the T-seal is made on a rack, then it is convenient to clamp the work in place as shown in Figure 11–13a. The clamp near the air inlet tube should be tight, but the clamp at the closed end should be loose. In fact, it is often best just to rest the end of the glass tube on the bottom two prongs of a three-prong clamp as shown in Figure 11–13a. The tube to be joined may be clamped above or below the horizontal tube. This tube should be clamped tight enough so that it does not slip through the clamp from its own weight, but not so tight that it cannot be pushed up against the hole formed after the kidney is chipped away

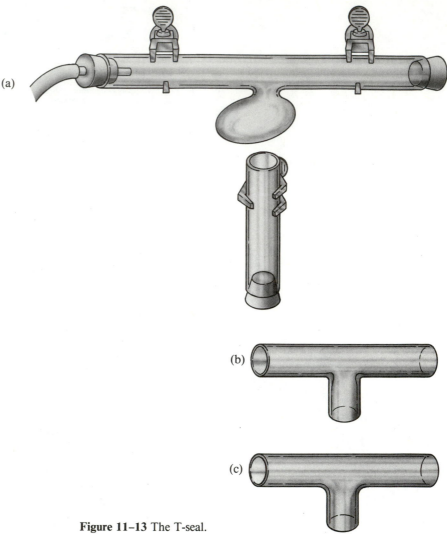

(a)

(b)

(c)

Figure 11–13 The T-seal.

and fire polished. Be sure that it is initially placed far enough away from the horizontal tube so that there is room for the kidney to form.

The important part is to get everything carefully in place before you begin. First, blow the kidney and be sure that the heated area is not too large. Beginners tend to heat too large an area, so that the hole formed is larger than the tube to be connected. It is better that the hole be a little too small than a little too large. Subsequent pulling and blowing permits a small hole to be enlarged.

In any case, after the hole in the horizontal tube has been formed, slide the lower tube to within about 1 mm of the hole; then heat the rim and the end of the tube simultaneously. When the glass softens, push the lower tube against the hole. The result looks something like Figure 11–13b. The glass is a little too thick. Soften the joint, pull the lower tube down slightly, and blow just enough to round up the corners between the two tubes and to thin the glass a bit at the joint (Fig. 11–13c). If the second clamp on the horizontal tube is to be tightened, soften the whole seal area and tighten the clamp while the seal is soft; then anneal the seal with a yellow reducing flame as already described.

A T-seal between a small and a large tube is made exactly the same way—by blowing out a kidney from the top or bottom side of the large tube to form a hole the size of the tube to be joined.

Using Cane

Occasionally, things just don't work out the way we plan them. You will usually discover this when you are finishing up a seal by blowing into it in order to round off the shoulders and you notice that a tiny draft of air is disturbing the torch flame when you blow into the apparatus. If you look carefully and you are lucky enough that the leak is on the front side of the seal, you may be able to see it with the naked eye. If the leak is less than about 1-mm diameter, you can generally repair it with 2-mm glass cane. It is a good idea to have a piece of 2-mm cane prepared and within reach when you are about to make glass-to-glass seals of any kind.

Prepare several pieces of cane, each about 6 to 10 inches long, by pulling the end off to form a fairly short point, tapering over 5 to 10 mm. To seal a visible pinhole, warm the entire seal until it is at the softening point; then heat the pinhole very sharply with a small hot flame. Touch the cane point directly on the pinhole, but very lightly. Then pull it off, heat strongly again, and blow to even out the thickness of the glass. Finally, anneal the entire seal (Fig. 11–14).

If the seal is part of a vacuum system, it should be tested after annealing by evacuating the system. Most leaks in new glass seals can be found with a Tesla coil. A Tesla coil is a hand-held, high-voltage, step-up transformer, capable of forming a spark that can jump across about 1 cm of air to a suitable conductor—including your finger, so be careful! When the tip of a Tesla coil is held near an evacuated glass tube in a vacuum system, a spark jumps to the tube, causing the residual air to

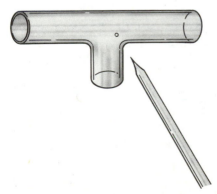

Figure 11–14 Using cane to seal a pin hole.

glow with a faint but visible light blue color. The gas discharge occurs when the gas pressure is less than about 1 torr, increasing in intensity as the pressure becomes lower, but finally disappearing when the pressure is below about 10^{-4} torr.

If a pinhole leak is present, even when it is invisible to the eye, the application of a spark from a Tesla coil causes a very bright, nearly white, discharge to pass through the point of the pinhole. If the leak is so small that it is invisible to the naked eye, but shows up when tested with a Tesla coil, it can often be repaired by simply heating the point where the leak is located with a very small, very hot flame, and blowing just enough to reform the tube wall as it shrinks under its own weight. If intense heating at the point of the leak does not correct it, then it may be necessary to apply the point of a piece of cane to the heated area as described above.

You should also realize that the spark from a Tesla coil is capable of punching a hole in glass that has inadvertently been made too thin in a poor seal, especially if the spark intensity is set at maximum. However, if the glass in a seal is that thin, you are advised to remove it and start over. One other possible source of trouble is the tip of the Tesla coil. This cylindrical piece of brass, conical at one end, slip fits into a socket at one end of the coil. Because it does not screw in and has no set screw, it can come loose, and if it does it can drop from the top of a vacuum rack to the bottom, where it can land with devastating results if it should strike a mercury diffusion pump, a trap, a McLeod gauge, etc., especially when the system is evacuated. This unforgettable experience happened to me many years ago and I wouldn't want it to happen to you. Be sure that the tip is firmly seated in the coil before using it!

Bending Glass Tubing

As usual on a vacuum rack, the glass tube to be bent is clamped in place and the hand torch must be moved evenly around the work. Typically, as shown in Figure 11–15, a horizontal glass tube is clamped firmly in a three-prong clamp. To begin the bending operation, direct the flame back and forth along the top of the tube, then back and forth along the bottom. Heat over a length of about 70 to 80 mm for tubing of 8- to 10-mm diameter. If the length of tubing beyond the bend is more than 25 to 30 cm, the unsupported end will begin to bend from its own weight and will, with continued heating between the areas indicated by the arrows in Figure 11–15, drop nearly 90° to form a perfect bend; if the piece of tubing is short it may be necessary to give the tube a helping hand, as indicated in Figure 11–15, to complete the final few degrees of bending. If the heating is not uniform, the bend may not lie in the plane of the paper; small corrections can be made by hand, so be sure to watch not only for the right angle of the bend but also for the planarity of the bend: it should lie in a plane parallel to the paper of Figure 11–15—or the wall of the laboratory. These kinds of bends are really quite easy to make and, unlike seals, they never leak!

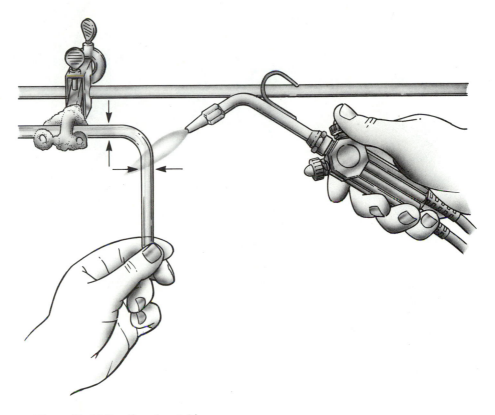

Figure 11–15 Bending glass tubing.

Bending works quite well for tubing of less than 18- to 20-mm diameter. You may wish to avoid bending tubing larger than this by forming a modified T-seal, that is, an L-seal, as shown in Figure 11–16. First form a test-tube end on the tubing at the point where you wish to have a bend. Then, as near as possible to the end, blow out a kidney so that the resulting hole matches the tubing to be connected. Chip off the kidney, fire polish the edges, bring the tube to be joined up close so that it and the edge of the hole can be simultaneously heated until soft. After pushing the lower piece into place, you will find the junction is usually a bit too thick, as indicated in Figure 11–16c. You can remedy this by pulling and blowing, so that the final product looks something like Figure 11–16d. The appearance is not as attractive as a bend formed by a professional, but it probably looks better than a bend done with large tubing by you or me. Unfortunately, unlike a bend, it might leak, so check it with a Tesla coil. It is especially important with large tubing to heat the end of the tube and edge of the hole evenly all the way around.

Ring Seals

If you have made a few T-seals and joined some unequal pieces of tubing, you are probably ready to try to make a ring seal. As usual, success

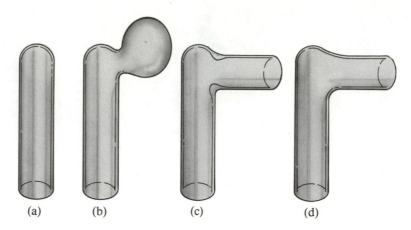

(a)   (b)   (c)   (d)

Figure 11–16 Sealing an L-bend.

depends upon setting up the components properly in advance. One way of setting up the work is shown in Figure 11–17a. Form an opening at the end of the larger outer tube, as usual, by forming a test-tube end, blowing a kidney, chipping it away, and fire polishing the edge. Form a ridge or bulge on the inner tube by heating uniformly around the tube and pushing slightly, taking care to keep the tube straight. These two

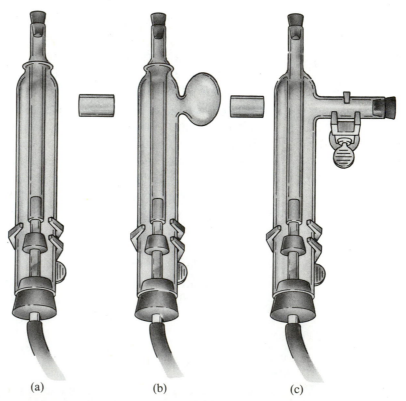

(a)   (b)   (c)

Figure 11–17 The ring seal with large-bore tubing.

operations can be done leisurely before assembling the components shown in Figure 11–17a.

The outer tube is clamped firmly with a three-prong clamp and fitted with a blow-tube inlet. The inner tube is clamped lightly at its upper end and inserted into the outer tube until its bulge nearly touches the slightly flared opening at the upper end of the outer tube. The blow-tube inlet is long enough and small enough so that it can be inserted into the inner tube. It is also fitted with a small cork stopper, positioned so that it almost touches the inner tube and thus prevents the inner tube from sagging when the upper ring seal is soft. The presence of the blow tube inside the inner tube prevents the inner tube from drifting sideways, away from collinearity with the outer tube. Just the end of the side-arm tube is shown; it is clamped lightly in place and positioned sufficiently distant so that there is room for a kidney to be formed. If you set up the components properly, you won't have any trouble making a ring seal. The secret lies more in the preparation than in actually manipulating the soft glass.

Make the ring seal by first uniformly heating around the inner tube bulge and the outer tube opening. When the glass is soft, push the inner tube down to make contact and then pull back slightly if necessary to thin the initial seal, which is probably a bit too thick. Heat evenly again and blow lightly to lift the shoulder of the outer tube. Ideally, contact between the walls of the outer and inner tubes should be perpendicular. While the glass is still above the annealing point, blow out a kidney just below the shoulder of the ring seal (Fig. 11–17b). Chip it away, fire polish the opening, and keep the ring seal above the annealing point. Slide the side arm up to the opening left by the kidney and heat it and the slightly flared opening until the glass is soft. Push the side arm into contact and then pull back very slightly. Blow slightly to round off the corners (Fig. 11–17c). Do not forget to keep the back side of the ring seal above the annealing point while you are sealing on the side arm.

Anneal the work carefully, with a yellow flame, until a good layer of soot is deposited. Because of its circular symmetry, the strains in a ring seal are evenly distributed. Nevertheless, a great deal of care is needed when a ring seal is reheated. For this reason, it is advisable to attach a side arm at the time that the ring seal is first fabricated, rather than later. The test-tube end is made as already described and can be done any time. The side arm and the extension of the inner tube should be long enough so they can be bent to the desired shape. If the bends are close to the ring seal, they should also be made when the ring seal is made. Making a ring seal is easier than it looks. Try it.

Glass Coils

Glass coils, such as those in condensers, can be made by wrapping the softened glass around a mandrel as shown in Figure 11–18. The mandrel is made of metal or glass tubing that has been wrapped with wet asbestos

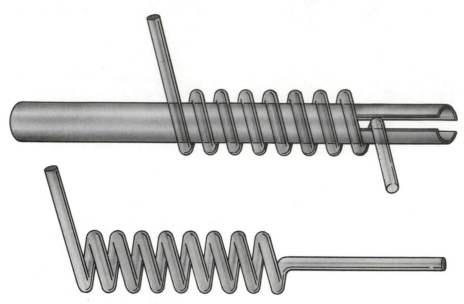

Figure 11–18 Making a glass coil.

paper. The seams should be smoothed out carefully and then the mandrel should be allowed to dry overnight. A metal mandrel should have a slot cut into it to hold the first turn of the coil in place. If a glass mandrel is used, the end piece, bent perpendicular to the tubing axis, can be temporarily sealed to the glass mandrel.

Place the mandrel on a horizontal metal rod and clamp it securely in place. Hold the burner in your right hand and the length of tubing in your left. You will need another person to turn the mandrel as you heat the glass tubing next to the mandrel while bending it and maintaining an even spacing between turns. To remove the coil, soak the coil, mandrel, and asbestos in water. The asbestos will soften and can be shredded and pulled out, permitting the coil to slip off the mandrel. Unlike making a ring seal, this process is not as easy as it looks. But try it anyway.

In fact, try any of the glass working techniques described in this chapter, beginning with cutting a tube and then rejoining the two halves. Be patient, practice, and after a while you will indeed get the feel for working with glass.

REFERENCES

1. Drake, R. F., *J. Chem. Educ.* 46:48, 1969.

2. Heldman, J. D., *Techniques of Glass Manipulation in Scientific Research,* Prentice-Hall, Englewood Cliffs, New Jersey, 1946.

3. Housekeeper, W. G., *Elect. Engineering* 42:954, 1923.

4. *Laboratory Glass Blowing with Corning's Glasses,* Corning Glass Works, Corning, New York, 1969. This booklet is sent at no charge upon request by phoning 1-607-737-1667.

5. Robertson, A. J. B., D. J. Fabian, A. J. Crocker, and J. Dewing, *Laboratory Glass-Working for Scientists,* Butterworths, London, 1957.

Vacuum Technique

High-vacuum systems are useful in many areas of chemistry. A high vacuum provides an inert environment for the protection of nonvolatile substances. Small amounts of volatile substances or gases can be transferred quantitatively from one container to another. The investigation of the physical chemistry of gases—how they conduct, effuse, diffuse, and flow—requires the use of a high vacuum. Air is a ubiquitous substance, a mixture of many substances capable of reacting chemically or physically with a system of interest (Table 12–1). In addition to the substances listed in Table 12–1, air contains variable amounts of water. With a vapor pressure of 20 to 25 torr at room temperature, water is an important and reactive constituent of air. An ordinary glass laboratory high-vacuum system can easily achieve a vacuum of 10^{-6} torr, at which pressure only 10^{-9} of the air remains.

The heart of a high-vacuum system is the diffusion pump. It can routinely achieve a pressure of 10^{-6} to 10^{-7} torr. However, a diffusion pump functions only when the pressure is below about 10^{-2} torr, so it requires a fore pump or backing pump. The final component of a high-vacuum system is the line or manifold, which allows the experimenter access to the high-vacuum system. In practice, many other accessories are useful, including traps, gauges, and stopcocks.

THEORY

In order to achieve a low pressure in a vacuum line, some air must be removed by pumping; as it is removed it must flow from one end of the tube to the other. The rate of flow of a gas, called the **throughput** Q, is defined as

TABLE 12–1

The Composition of Dry Air

Component	Volume Percent
Nitrogen	78.084 (4)
Oxygen	20.946 (2)
Argon	0.934 (1)
Carbon dioxide	0.033 (1)
Neon	0.001818 (4)
Helium	0.000524 (4)
Krypton	0.000114 (1)

The numbers in parentheses represent the uncertainties in the last digits.
Source: Hand Book of Chemistry and Physics, 63rd ed., CRC Press, Boca Raton, Florida, 1982.

$$Q = P \frac{dV}{dt} \tag{12-1}$$

where P is the pressure at which it is measured, and dV/dt is the volume flow rate across a plane (Fig. 12–1). Notice that throughput does not have the same units as ordinary gas flow rate (unit volume/unit time). The units of throughput are energy/time, that is, L atm min^{-1}, or in SI units, Pa m^3 s^{-1}, or J s^{-1}, or watts.

The throughput depends on the resistance to flow and the pressure drop between the entrance and exit to a tube or channel:

$$Q = \frac{P_2 - P_1}{Z} = F(P_2 - P_1) \tag{12-2}$$

where P_1 is the downstream pressure (measured at the exit), P_2 is the upstream pressure (measured at the entrance), Z is the resistance, and F is the conductance. (The nomenclature in this chapter follows that of Dushman, 1962, quite closely. Other sources may use W or Z for resistance and C, U, or F for conductance. The use of Q for throughput seems to be more universal.) The conductance is the throughput per unit pressure difference between the tube entrance and exit. The SI unit of conductance is m^3 s^{-1}; the unit of resistance is m^{-3} s.

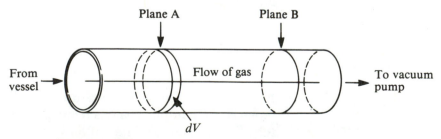

Figure 12–1 Throughput in a tube.

The quantities in Equation 12–2 are quite analogous to Ohm's Law, which relates the flow of current I through a resistance R under the influence of a potential difference E:

$$I = \frac{E}{R} \qquad (12\text{--}3)$$

Equation 12–2 is the Ohm's Law of gas flow through a tube: it relates the gas flow Q (throughput) through a tube of resistance Z under the influence of a pressure difference $P_2 - P_1$. Just as the resistance of electrical resistors in series is given by

$$R_T = R_1 + R_2 + R_3 + \cdots \qquad (12\text{--}4)$$

and in parallel by

$$\frac{1}{R_T} = \frac{1}{R_1} + \frac{1}{R_2} + \frac{1}{R_3} + \cdots \qquad (12\text{--}5)$$

so the resistance to gas flow is given by

$$Z_T = Z_1 + Z_2 + Z_3 + \cdots \qquad (12\text{--}6)$$

where Z_T is the total resistance and Z_i are the series resistances of the various components in a vacuum line, that is, the traps, baffles, stopcocks, and tubes of different diameters. It is somewhat more common in discussing gas flow to speak of conductance rather than resistance, so Equation 12–6 is frequently written

$$\frac{1}{F_T} = \frac{1}{F_1} + \frac{1}{F_2} + \frac{1}{F_3} + \cdots \qquad (12\text{--}7)$$

Viscous Flow Versus Molecular Flow

The nature of gas flow through a tube is quite different at low pressures than at high pressures. In addition, the flow characteristics depend on the flow rate and the geometry of the tube, pipe, or channel through which the gas flows. Three kinds of flow are recognized: turbulent, viscous (laminar), and molecular. The Reynolds number R is useful in expressing the boundary between turbulent and viscous flow; similarly, the Knudsen number K_n helps define the boundary between viscous and molecular flow. The Reynolds number, a dimensionless number, is defined as

$$R = \frac{a\rho U}{\eta} \qquad (12\text{--}8)$$

where a is the tube radius, ρ is the gas density, η is the viscosity, and U is the flow velocity across a plane in the tube, defined as

$$U = \frac{Q}{\pi a^2 P} \qquad (12\text{--}9)$$

The Knudsen number K_n is purely empirical and is defined as the ratio of the mean free path L to "a characteristic dimension" of the system, say, the radius a of the tube:

$$K_n = \frac{L}{a} \qquad (12\text{--}10)$$

The Knudsen number is also dimensionless. The mean free path is given by

$$L = \frac{1}{\sqrt{2}\,\pi d^2 N^*} \qquad (12\text{--}11)$$

where d is the molecular diameter and N^* is the number of molecules per unit volume. Since N^* is related to the pressure and temperature, it is convenient to express Equation 12–11 for air at 25°C as

$$L \approx \frac{0.005}{P} \qquad (12\text{--}12)$$

where L is the mean free path in centimeters and P is the pressure in torr. This is useful for getting a rough value for the Knudsen number for a tube or bulb in a vacuum line, the dimensions of which are usually measured in centimeters. The rough ranges of flow are summarized in Table 12–2. As an example, let us calculate the maximum pressure (torr) at which molecular flow is observed in a long glass tube 25 mm in diameter:

$$L = K_n a = 1.0 \times 1.25 \text{ cm} = 1.25 \text{ cm}$$

$$P = 0.005/L = 0.005/1.25 = 4 \times 10^{-3} \text{ torr}$$

Viscous Flow

Above about 10^{-3} torr, gas properties depend upon collisions between molecules, which occur much more frequently than between molecules and their container. At pressures below 10^{-3} torr, viscosity is not a property of a gas, since collisions between molecules are infrequent. In the region of viscous flow, the Poiseuille equation gives the throughput through a straight tube of circular cross section:

TABLE 12–2

Types of Gas Flow

Flow Type	Pressure	Reynolds No.	Knudsen No.
Turbulent	High	>2200	—
Viscous	Medium	<1200	<0.01
Molecular	Low	—	>1.00

$$Q = \frac{\pi a^4}{8\eta\ell} P_{\text{ave}}(P_2 - P_1) \qquad (12\text{–}13)$$

where a and ℓ are the tube radius and length, η is the gas viscosity, and P_{ave} is the average of P_2 and P_1.

If we combine equations 12–13 and 12–2, we obtain an equation for the viscous flow conductance in a tube of circular cross section:

$$F = \frac{\pi a^4}{8\eta\ell} P_{\text{ave}} \qquad (12\text{–}14)$$

Note that the most widely tabulated unit of viscosity is the cgs unit, the poise: 1 poise = 1 g cm^{-1} sec^{-1}. The SI unit is the Pa s: 1 Pa s = 1 kg m^{-1} s^{-1}. Thus 1 poise = 0.1 Pa s. The viscosity of air at 25°C is 1.845 $\times$ 10^{-4} poise = 1.845 $\times$ 10^{-5} Pa s.

If a and ℓ are given in centimeters and P_{ave} in torr, then Equation 12–14 may be written to give F in L/sec:

$$F = \frac{2840 a^4 P_{\text{ave}}}{\ell} \qquad (12\text{–}15)$$

In Figure 12–2a, because the mean free path L is small compared to the radius of the tube a, collisions are more frequent between molecules than they are between molecules and the walls of the container. Consequently, the properties of the gas are quite constant over several mean free paths and the gas acts like a continuous viscous fluid. The Knudsen number is defined as L/a, and when $L/a < 0.01$, the gas flow is considered viscous.

When the mean free path is large compared to the diameter of the tube (high Knudsen number or $L/a > 1.0$), the gas molecules collide with the walls of the container more frequently than with each other.

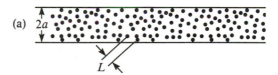

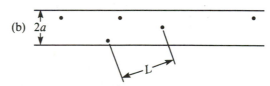

Figure 12–2 Viscous and molecular flow.

(a) Viscous flow ($L/a < 0.01$). The molecules collide much more frequently with each other than with the wall of the containing tube.

(b) Molecular flow ($L/a > 1.0$). The molecules collide much more frequently with the walls of the containing tube than with each other.

Viscosity is then undefined, since there can be no shearing forces between layers of molecules nor is any momentum transferred between molecular layers. These are the conditions for molecular flow, shown in Figure 12–2b. To compare these two figures, visualize the mean free path in comparison to the dimensions of the tube and to the average distance between molecules.

Pumping Speed

The pumping speed S at any point in the vacuum system is defined as the ratio of the throughput Q to the pressure at that point:

$$S = \frac{Q}{P} \tag{12–16}$$

The units of S are unit volume/unit time, the same as the units of conductance. To design a vacuum line properly, it is useful to know how the pumping speed S_L in the line differs from the pumping speed of the pump S_P. The effect of line resistance in reducing the pumping speed is given by

$$\frac{1}{S_L} = \frac{1}{S_P} + \frac{1}{F} \tag{12–17}$$

Notice that if the conductance equals the speed of the pump, the pumping speed in the line is just one half that of the pump, since

$$\frac{1}{S_L} = \frac{1}{S_P} + \frac{1}{F} = \frac{2}{S_P}$$

$$S_L = \frac{S_P}{2}$$

Consequently, it is important to make the conductance of the line as large as possible in comparison to the speed of the pump. Only when F is infinite does $S_L = S_P$.

For a real vacuum line, the resistance to pumping is the sum of the resistances of all the components (traps, baffles, stopcocks, etc.) that make up the line. This relationship is usually written in terms of conductances ($F = 1/Z$):

$$\frac{1}{S_L} = \frac{1}{S_P} + \sum \frac{1}{F_i}$$

where the F_i are the conductances of the components. Thus, it is important to minimize the number of components and not to clutter up the vacuum line with unnecessary or unused traps and stopcocks. According to Equation 12–15, the viscous conductance depends on the fourth power of the tube radius. Consequently, tubing between the fore pump and the diffusion pump should be of as large a diameter as possible. Reducing the radius by one half reduces the conductance by one sixteenth.

Equation 12–17 may be rearranged to give

$$S_L = \frac{S_P F}{S_P + F} \qquad \textbf{(12–18)}$$

The graphical dependence of the pumping speed S_L in the line on the conductance F of the line for various pumping speeds S_P is shown in Figure 12–3.

Molecular Flow

At pressures below approximately 10^{-3} torr, collisions between molecules and the container walls are much more frequent than between molecules. As with viscous flow, conductance is strongly dependent upon the tube radius, but is independent of pressure:

$$F = \frac{2\pi a^3 u_{\text{ave}}}{3\ell} \qquad \textbf{(12–19)}$$

where u_{ave}, the average molecular speed, is given by

$$u_{\text{ave}} = \left(\frac{8RT}{\pi M}\right)^{1/2} \qquad \textbf{(12–20)}$$

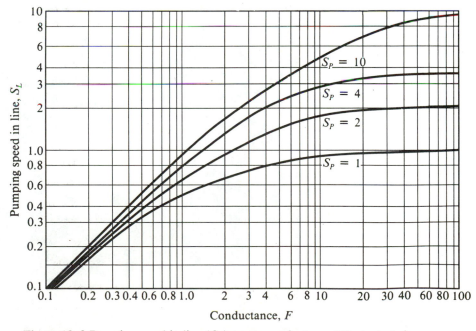

Figure 12–3 Pumping speed in line (S_L) versus conductance (F) at selected pump speeds (S_P) (Eq. 12–8). The pumping speed falls off as the conductance decreases.

At 25°C the conductance of air in a cylindrical tube is given approximately by

$$F = 100\ \frac{a^3}{\ell} \qquad\qquad (12\text{--}21)$$

where a and ℓ are in centimeters and F is in L/sec.

VACUUM PUMPS

Mechanical Fore Pumps

In 1905, Dr. Wolfgang Gaede, German scientist and inventor, built the first rotary vacuum pump. Prior to 1905, scientists used a variety of piston pumps that could attain a pressure of about 0.25 torr.

Typical Mechanical Pumps. The Boekel Cenco Hyvac pump (Fig. 12–4), based on Gaede's original design, consists of a cylindrical rotor that revolves inside a larger stationary cylinder (Fig. 12–5). Mounted on the rotor 180° apart are two sliding vanes, held in close contact with the

Figure 12–4 Boekel Cenco Hyvac 7 vacuum pump. (Courtesy of Boekel Industries, Inc., Philadelphia.)

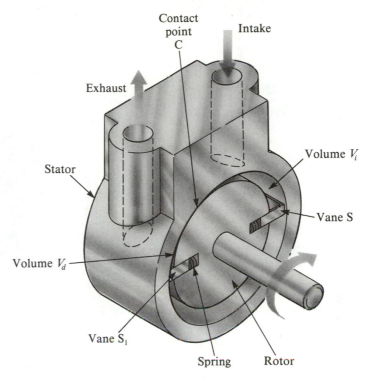

Figure 12-5 Operating principle of an internal-vane mechanical vacuum pump. (Courtesy of Boekel Industries, Inc., Philadelphia.)

stationary cylinder by springs and centripetal force. The vanes, free to move back and forth along their long axis, divide the volume between the rotor and stationary cylinder into two parts. As the rotor rotates, the intake volume V_i increases, causing gas from the system to enter through the intake port. Simultaneously, volume V_d decreases, compressing the gas and expelling it through the exhaust port into the atmosphere. As soon as vane S_1 passes the exhaust port, volume V_d becomes the intake volume and is compressed and expelled. After every 180° of rotation the intake volume becomes the exhaust volume and vice versa. The Sargent-Welch DuoSeal pump (Fig. 12-6) operates on the same principle.

From atmospheric pressure down to about 10^{-3} torr, the pumping speed of the typical mechanical vacuum pump is quite constant, but at lower pressures the pumping speed drops off rapidly as seen in Figures 12-7 and 12-8. In a real laboratory vacuum line, the ultimate vacuum with a mechanical pump is not much below 0.1 torr. For a better vacuum it is necessary to use a second pump in conjunction with a mechanical fore pump. In the laboratory the second pump is usually a diffusion pump, described later. All these pumps are manufactured to very close tolerances between the rotor, vanes, and stationary cylinder, and are sealed in oil.

Because condensable vapors, such as water vapor, cannot be removed by a mechanical pump that compresses a gas before expelling it,

Figure 12–6 Sargent-Welch DuoSeal 1400 vacuum pump. (Courtesy of Sargent-Welch
Scientific Company, Skokie, Illinois.)

some pumps are equipped with a gas ballast valve (called a vented exhaust
by Sargent-Welch Scientific Company). A gas ballast valve allows a measured quantity of fresh air (the gas ballast) to be admitted to the compression chamber. With gas ballast added to the compression chamber, the

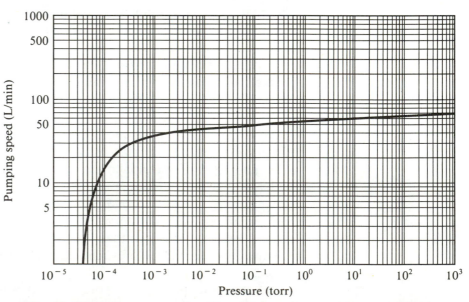

Figure 12–7 Performance curve for a Hyvac 7 vacuum pump. (Courtesy of Boekel
Industries, Inc., Philadelphia.)

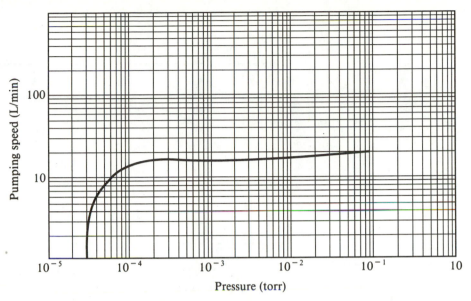

Figure 12–8 Performance curve for a Sargent-Welch DuoSeal 1400 vacuum pump. (Courtesy of Sargent-Welch Scientific Company, Skokie, Illinois.)

partial pressure of the unwanted condensable vapors becomes only a small fraction of the total pressure of gas, vapor, and gas ballast. The vapor is thus prevented from reaching its saturation pressure and is finally expelled from the pump.

In order to achieve a lower ultimate pressure, some pumps are constructed with two stages. The two stages are connected internally so that the effect is that of two pumps in series, as indicated in Figure 12–9. Although the pumping speed is not increased, the ultimate pressure is lowered by two orders of magnitude, as shown in Figure 12–10.

The ultimate vacuum of rotary vane pumps is about 10^{-2} torr for single-stage pumps and about 10^{-4} torr for two-stage pumps. The ultimate vacuum also depends on the age and mechanical condition of the pump and the nature of the pump oil, its age, and whether it has been contam-

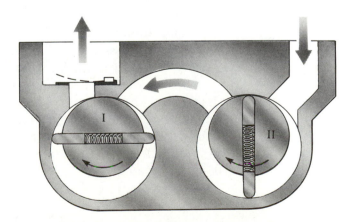

Figure 12–9 Operation of a two-stage mechanical vacuum pump.

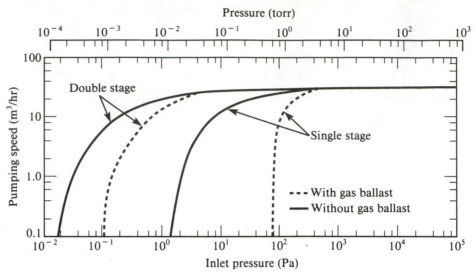

Figure 12–10 Effect of doubling pump stages and ballast.

inated or otherwise abused. Pumping speed, which depends primarily on the size of the pump, ranges from about 20 to 400 L/min.

Pumpdown Time. By combining Equations 12–1 and 12–16, the pumping speed may be written as

$$S = \frac{Q}{P} = \frac{-V}{P}\frac{dP}{dt} \qquad (12\text{–}22)$$

$$dt = \frac{-V}{S}\frac{dP}{P} \qquad (12\text{–}23)$$

The time required to reduce the pressure from initial P_i to final pressure P_f can be calculated by integrating Equation 12–23:

$$t = \frac{V}{S}\ln\frac{P_i}{P_f} \qquad (12\text{–}24)$$

In the integration it is assumed not only that the volume is constant, but that the pumping speed S is constant as well. Consequently, Equation 12–24 is valid only in the horizontal portion of Figures 12–6 and 12–8.

Suppose we have a vacuum system with a volume estimated to be 8 L and we wish to know how long it takes the smallest Welch two-stage pump to evacuate the system from atmospheric pressure to 10^{-3} torr. Examination of the performance curve (Fig. 12–8) shows that the pumping speed at 1 atm is 20 L/min and drops to 10 L/min at a pressure of 10^{-3} torr, for an average speed in this pressure range of 15 L/min:

$$t = \frac{8}{15}\ln\frac{760}{0.001} = 7.2 \text{ min}$$

Diffusion Pumps

The first diffusion pumps were made by Wolfgang Gaede in Germany in 1915 and by Irving Langmuir in 1916 in the United States. Contemporary pumps are closely related to the Langmuir pump but extend and improve upon the original design, shown in Figure 12–11.

In this prototype pump, liquid mercury is boiled at a pressure of about 10^{-2} torr provided by a mechanical fore pump. This provides a vapor stream directed by the nozzle downward toward the condenser walls, on which the mercury vapor condenses, returns to the boiler, and is recycled. Gas molecules diffuse from the system into the pump, where they encounter the stream of mercury vapor molecules. The mercury vapor molecules collide with the diffusing gas molecules and sweep the gas toward the fore pump port, where they are removed from the system by the mechanical fore pump. Because the room-temperature vapor pressure of mercury is relatively high ($\sim 10^{-3}$ torr), it is necessary to position a trap, cooled with dry ice or liquid nitrogen, between the pump and the system to prevent mercury vapor from entering the system (Fig. 12–11).

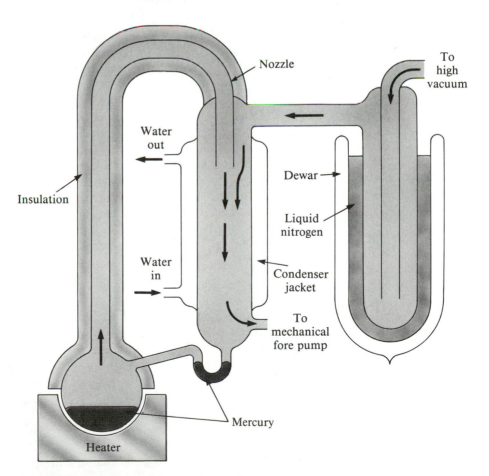

Figure 12–11 Single–stage Langmuir diffusion pump.

In the same way that the ultimate vacuum can be lowered by increasing the number of stages of a mechanical pump, so can the ultimate vacuum of a diffusion pump be lowered by increasing the stages to two or three. The three-stage glass pump shown in Figure 12–12 uses oil instead of mercury as a working fluid. The design appears to be similar to that of a three-stage oil diffusion pump formerly made by Consolidated Vacuum Corporation, a company still in the vacuum business that no longer manufactures glass pumps.

In 1928, C. R. Burch in England and K. C. D. Hickman in the United States substituted low-vapor-pressure organic fluids for mercury. When the pump-fluid vapor pressure is below 10^{-6} torr, a trap between the pump and the system is unnecessary. It is a great convenience not to have to maintain liquid nitrogen around a trap. A typical small, laboratory-size metal oil diffusion pump is shown in Figure 12–13. Its principles of operation (Fig. 12–14) and performance (Fig. 12–15) are similar to those of mercury diffusion pumps.

Oil diffusion pumps are constructed of both metal and glass, while mercury diffusion pumps are constructed of glass because of mercury's tendency to amalgamate metals. For large-scale industrial applications, metal oil diffusion pumps are preferred because they are rugged and need no cold traps. For small-scale laboratory vacuum lines, however, glass mercury diffusion pumps have the edge on metal oil diffusion pumps. Glass is easily cleaned, is corrosion resistant, and seals simply to a glass vacuum line. Because glass is transparent, you can constantly monitor the operation of the pump and the condition of the mercury.

Mercury itself has certain useful properties. It is not subject to thermal decomposition, it is relatively inert, and gases are insoluble in it. If

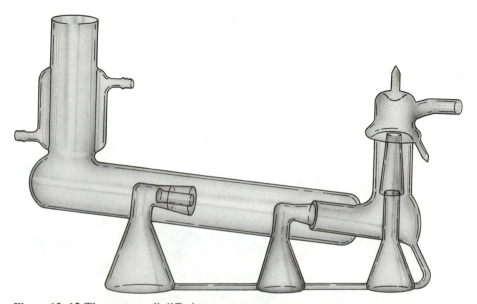

Figure 12–12 Three-stage oil diffusion pump.

Figure 12–13 Metal oil diffusion pump (CVC model MCF-60). (Courtesy of CVC Products, Inc., Rochester, New York.)

contaminated, mercury is easily cleaned. It is, however, toxic and should be handled with care. Mercury has a simple mass spectrum and is easy to recognize. Oils, on the other hand, are subject to thermal decomposition and form ill-defined products. They react with air, possibly explosively if suddenly exposed to oxygen at high temperatures. Oils decompose in electrical discharges, sometimes causing electrical problems.

Oils, however, have very low vapor pressures and require no cold trap. Oils are not as toxic as mercury, but should not be regarded as completely benign. Unlike mercury, oils are not harmful to mechanical pumps and other metal components. Some data on commercially available pump oils are summarized in Table 12–3.

The cost of the diffusion pump oils listed in Table 12–3 ranges from about \$50 to \$500/L. Phenyl ethers such as Neovac-SY and Santovac-5 are quite resistant to oxidation and to static charge buildup. Consequently they are especially useful around electronic devices. The silicones are chemically inert and relatively inexpensive. The esters combine low vapor pressure and low cost. Design considerations suggest that considerable care be exercised in selecting diffusion pump oils.

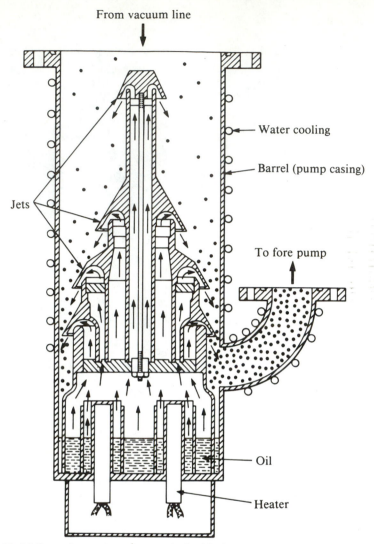

Figure 12–14 Operating action of a metal oil diffusion pump. (Courtesy of CVC Products, Inc., Rochester, New York.)

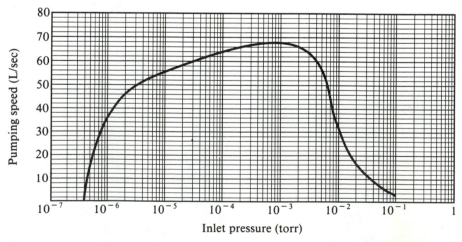

Figure 12–15 Performance curve for the MCF-60. (Courtesy of CVC Products, Inc., Rochester, New York.)

TABLE 12–3

Diffusion Pump Fluids

Commercial Name	Vapor Pressure (torr) at 25°C	Chemical Name	Supplier
Apiezon Oil A	5×10^{-7}	Hydrocarbon	a
Apiezon Oil B	1×10^{-7}	Hydrocarbon	a
Convalex-10	1×10^{-10}	Phenyl ether	b
Convoil-20	5×10^{-8}	Hydrocarbon	b
DC-704	1×10^{-8}	Tetramethyl tetraphenyl trisiloxane	c
DC-705	3×10^{-10}	Tetramethyl pentaphenyl trisiloxane	c
Fomblin (Krytox)	1×10^{-8}	Fluorinated polyether	d
Neovac-SY	1×10^{-7}	Phenyl ether	e
Octoil-S	5×10^{-8}	Hydrocarbon	b
Santovac-5	1×10^{-7}	Phenyl ether	f

a. James G. Biddle Co., Blue Bell, Pennsylvania 19422.
b. CVC Products, Inc., Rochester, New York 14603.
c. Dow Corning Corp., Midland, Michigan 48640.
d. Montedison USA, Inc., New York, New York 10036.
e. Varian Associates, Inc., Lexington, Massachusetts 02173.
f. Mansanto Co., St. Louis, Missouri 63166.

Matching the Fore Pump and Diffusion Pump

The fundamental equation for a dynamic pumping system in equilibrium is

$$Q = PS \qquad (12\text{-}25)$$

where Q is the throughput, P is the pressure, and S is the pumping speed (dV/dt). When the fore pump is matched to the diffusion pump, the throughput of the fore pump equals the throughput of the diffusion pump:

$$P_f S_f = P_d S_d \qquad (12\text{–}26)$$

Where P_f is the fore pump pressure, S_f is the fore pump speed, P_d is the diffusion pump operating pressure, and S_d is the diffusion pump speed. Suppose we wish to establish whether a Welch No. 1400 is a suitable fore pump for a CVC Type MCF-60 metal oil diffusion pump. We solve Equation 12–26 for S_f and substitute the manufacturer's data:

$$S_f = \frac{P_d S_d}{P_f} \qquad (12\text{–}27)$$

For P_f we use the value for tolerable pressure for the diffusion pump of 0.40 torr. From the performance curve for the MCF-60, we use S_d equals 36 L/sec when P_d equals 10^{-6} torr:

$$S_f = \frac{10^{-6} \text{ torr} \times 36 \text{ L/sec} \times 60 \text{ sec/min}}{0.40 \text{ torr}}$$

$$= 0.005 \text{ L/min}$$

At P_f equals 0.40 torr, the performance curve for the Welch 1400 shows that S_f equals about 18 L/min, which is more than adequate.

Miscellaneous Pumps

Certain specialized applications require particular pump characteristics, giving rise to a wide variety of vacuum pumps. Examples of such applications are analytical instruments, computer chip manufacture, electron tube manufacture, evaporation deposition, laser manufacture, mass spectrometers, optics, outer space simulation, particle accelerators, surface analysis, and x-ray spectroscopy. The list is far from complete. Let us take a brief look at some of the more important developments in vacuum pumps.

Turbo Molecular Pumps. Turbo molecular pumps (Fig. 12–16) operate just about the same as a diffusion pump. In a diffusion pump, momentum is transferred from streaming vapor molecules to gas molecules that diffuse into the vapor stream. In a turbo molecular pump, momentum is transferred from the blades of a metal turbine to the gas

Figure 12–16 Turbo molecular pump (Edwards ETP 80). (Courtesy of Edwards High Vacuum Inc., Grand Island, New York.)

molecules that diffuse between the blades. So that the tip velocity of the turbine blades reaches the thermal velocity of the air molecules, the turbine rotates at exceedingly high speeds: 20,000 to 60,000 rpm.

Turbo molecular pumps achieve pumping speeds comparable to large oil diffusion pumps: 80 to 9000 L/sec is typical (Fig. 12–17). Their ultimate pressure of 10^{-11} torr tops that of the best diffusion pumps. Turbo molecular pumps are lubricated with low-pressure hydrocarbon diffusion pump oils and require no cold traps. As with diffusion pumps, turbo molecular pumps operate most efficiently in the molecular flow region and consequently require a backing pump to lower the system pressure to about 10^{-3} torr before operating. In addition, a controller to control the startup, running speed, and shutdown, as illustrated for the ETP 80 in Figure 12–18, is required. The ETP 80, the smallest in the Edwards series, pumps air at about 80 L/sec when running at 60,000 rpm. Starting time to operating speed is about 20 seconds. For very large turbo molecular pumps, 15 to 25 minutes may be required when the pumping speed is 9000 L/sec. The ultimate vacuum for the ETP 80 is about 5×10^{-9} mbar (3.7×10^{-9} torr).

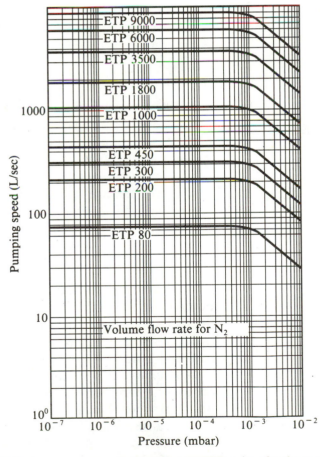

Figure 12–17 Performance curves for the Edwards ETP series of turbo molecular pumps. (Courtesy of Edwards High Vacuum Inc., Grand Island, New York.)

Figure 12–18 Turbo molecular pump with controller and fore pump. (Courtesy of Edwards High Vacuum Inc., Grand Island, New York.)

Entrainment Pumps. Entrainment pumps operate by chemically or physically adsorbing gas molecules on a surface. One of the earliest applications of entrainment was in the use of getters in the manufacture of vacuum tubes. When vacuum tubes were first manufactured, it was difficult to evacuate the tube to less than 10^{-4} torr. To remove the last traces of gas from the tube, a small amount of a reactive metal such as barium or titanium was sealed into the tube. Upon baking, the vaporized metal reacted with traces of remaining air to "get" it out. Thus, the reactive metal became known as a "getter." Nowadays you will come across the verb "to getter" and the gerund "gettering."

Titanium Sublimation Pumps (TSP). A titanium sublimation pump (Fig. 12–19) is simply a titanium getter. The Varian TSP model 916-0009 consists of a hollow sphere of titanium about 3 cm in diameter, inside of which is located a small resistance heater. The entire apparatus is placed inside the vacuum system. Access to the outside is provided with a suitable flange mounting. After the system is pumped down to about 10^{-5} torr with another pump, say, a turbo molecular pump, the titanium sphere is heated electrically so that it vaporizes at the required rate. Because of the spherical shape, the titanium vapor deposits uniformly in all directions on the walls of the vacuum system, where the titanium getters residual gases by chemisorbtion, for example,

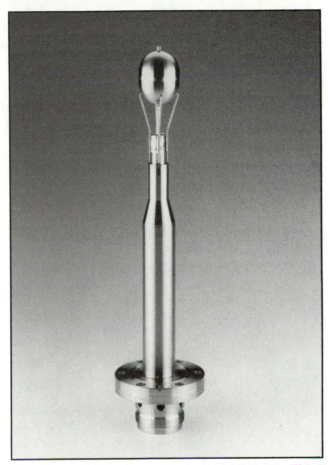

Figure 12–19 Titanium sublimation pump (Varian 916-0009). The Ti-Ball source—consisting of 35 g titanium that can be sublimed at rates from 0.01 to 0.5 g/h—can pump up to 2500 L/sec at 10^{-5} torr. (Courtesy of Varian, Vacuum Products Division, Lexington, Massachusetts.)

$$4 \text{ Ti} + 3 \text{ O}_2 = 2 \text{ Ti}_2\text{O}_3$$

$$2 \text{ Ti} + \text{N}_2 = 2 \text{ TiN}$$

The sphere in the aforementioned TSP contains 35 g titanium; this is sufficient to react with 11.7 g oxygen, which would occupy 65 million L at 10^{-4} torr. At a sublimation rate of 0.01 g/h, the lifetime of the titanium ball is 3500 hours, after which it is replaceable. Depending on the tubulation size, the pumping speed can be very high, as seen in Figure 12–20. Although effective in pumping hydrogen, oxygen, nitrogen, carbon monoxide, carbon dioxide, and water vapor, TSPs do not pump methane, argon, or helium. If this drawback is not objectionable, the TSP is a relatively economical tool for achieving an ultrahigh vacuum at high pumping speeds.

Ion Sputter Pump. The ion sputter pump, also known as the ion getter pump or simply ion pump, is related in its operation to the TSP,

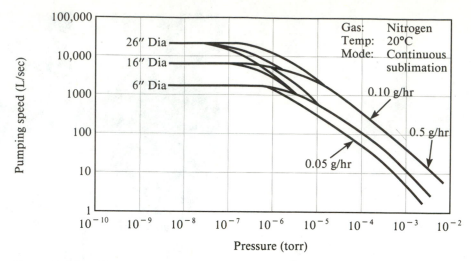

Figure 12–20 Performance curve for a titanium sublimation pump. The pumping speed depends on the pressure and sublimation rate for a system of a given size. (Courtesy of Varian, Vacuum Products Division, Lexington, Massachusetts.)

but is more sophisticated and has the distinct advantage that it pumps inert gases. The simplest unit of a diode ion pump is shown in Figures 12–21 and 12–22.

Subject to a potential difference of 5000 to 7000 V between the anode and cathode, gas molecules diffusing into the anode cylinder form positive ions and electrons. Because the electrons are formed in a strong magnetic field, they travel on a long spiral path to the anode, increasing their chance of an encounter with other gas molecules. Such encounters cause further ionization of gas molecules.

The positive ions thus formed accelerate toward the titanium cathode; they collide with the cathode sufficiently violently to dislodge titanium atoms, which deposit (sputter) on the anode or elsewhere on the

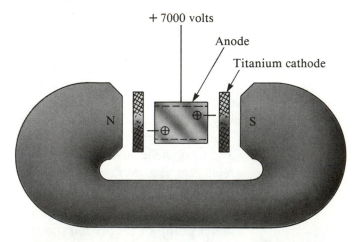

Figure 12–21 Operation of an ion pump.

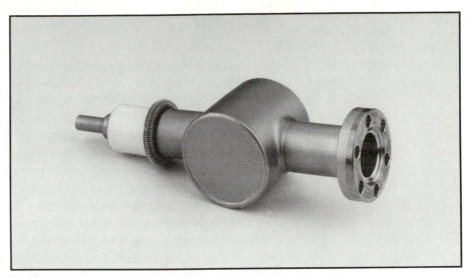

Figure 12–22 Ion pump (Varian Vacion 913, 2 L/sec), shown here without its permanent magnet in place. (Courtesy of Varian, Vacuum Products Division, Lexington, Massachusetts.)

cathode in a very finely divided and reactive state. The sputtered titanium then getters reactive gas molecules such as O_2, O_2^+, etc. Nonreactive molecules such as He and Ar are also entrained, not by reaction, but by "burial" under sputtered titanium atoms on both the anode and cathode.

You are probably wondering how such a pump was developed in the first place, since the pumping mechanism is a bit unusual. In the late 1930s, it was observed that ionization gauges designed to measure pressure actually exerted a pumping action of their own, resulting in a noticeable pressure reduction in a small vacuum system. During the 1950s, investigations of this ion gauge pumping action led to the development of commercial ion sputter pumps.

Because the ion pump can routinely achieve an ultimate pressure of 10^{-11} torr (Fig. 12–23), it is widely used in research and industry,

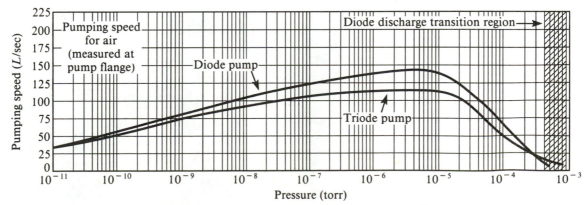

Figure 12–23 Ion pump performance curve. (Courtesy of Varian, Vacuum Products Division, Lexington, Massachusetts.)

especially where a bakeable, oil- and vibration-free environment is required.

Sorption Pump. The sorption pump is essentially a very efficient cold trap used in parallel rather than in series with the system (Fig. 12–24). Whereas physical chemists usually use a cold trap with a diffusion pump to prevent oil or mercury from diffusing into the system, the sorption pump is a roughing pump used to evacuate a system from atmospheric pressure down to the millitorr region, as shown in Figure 12–25.

The Varian VarSorb pump, shown in Figure 12–24, has no moving parts. It is simply a canister that holds 3 lb (1.4 kg) of Type 5A molecular sieve, which has a surface area of nearly 300 acres (1.2 km^2). The 5A refers to the sieve pore size of 5 Å, which is just large enough to soak up nitrogen, oxygen, argon, and water, the main constituents of air. The pump shown in Figure 12–24 has overall dimensions of 38.3 by 11.4 cm. In use it is surrounded by liquid nitrogen held in a stainless-steel Dewar flask and is regenerated by electrically heating the canister.

Cryopumps. The cryopump illustrated in Figure 12–26 is shown without the refrigeration unit, which operates with liquid nitrogen (50

Figure 12–24 Sorption pump (Varian). (Courtesy of Varian, Vacuum Products Division, Lexington, Massachusetts.)

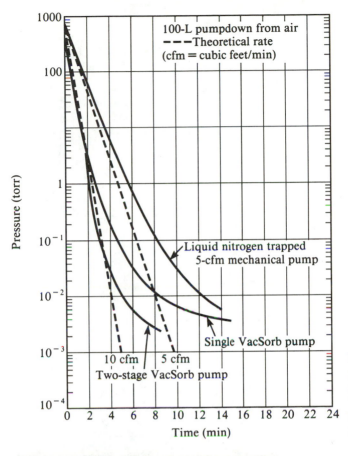

Figure 12–25 VacSorb pumpdown curves. (Courtesy of Varian, Vacuum Products Division, Lexington, Massachusetts.)

Figure 12–26 Cryopump (Air Products AP-8S). (Courtesy of Air Products and Chemicals, Inc., Allentown, Pennsylvania.)

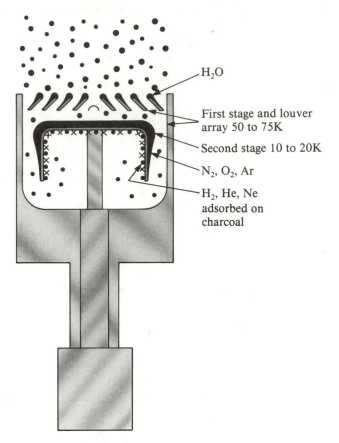

H$_2$O

First stage and louver
array 50 to 75K

Second stage 10 to 20K

N$_2$, O$_2$, Ar

H$_2$, He, Ne
adsorbed on
charcoal

Figure 12–27 Operation of a basic cryopump. (Courtesy of Air Products and Chemicals, Inc., Allentown, Pennsylvania.)

to 70 K) to cool the outer cryopanel (concentric rings in Fig. 12–26) and with liquid helium (10 to 20 K) to cool the inner cryopanel. The outer panel (Fig. 12–27) freezes out water vapor and the inner panel freezes out nitrogen, oxygen, and argon. A coating of high-surface-area charcoal on the inside of the inner panel cryosorbs hydrogen, helium, and neon. A cryopump operates in the pressure range from 10^{-3} to 10^{-12} torr, and although it requires a fore pump operating at about 5×10^{-3} torr, the fore pump is not required continuously after the initial pumpdown has been achieved. Because it has no moving parts, maintenance costs are low and maintenance intervals long—up to 10,000 hours. Initial costs are higher than for a diffusion pump, but lower than for a turbo molecular pump.

MEASUREMENT OF PRESSURE: BAROMETERS AND MANOMETERS

A **barometer** is an instrument for measuring the pressure of the atmosphere, whereas a **manometer** is a device for measuring the pressure of a contained gas. In either instrument, mechanical equilibrium is achieved by balancing the force F arising from an unknown pressure against the force exerted by the barometer or manometer.

The balancing force in either case may arise from the pressure P exerted by a column of liquid ($F = PA = \rho ghA$) or from the elastic force of a diaphragm or spring ($F = -kx$), where A is the cross-sectional area of the liquid column, ρ is the density of the liquid, h is its height, g is the acceleration of gravity, k is the Hook's Law constant, and x is the linear displacement.

Fortin Barometer

The Fortin barometer (Fig. 12–28) consists of a glass tube about 90 cm long, 0.6 to 0.8 cm in diameter, closed on one end and open on the other end. After it is cleaned, degassed, and filled with distilled mercury, the tube is inverted in a cistern of mercury, creating a Torricellian vacuum at the closed end. The cistern consists of a leather bag filled with mercury and arranged so that its level can be changed by manual adjustment of a screw that pushes against the leather bag.

The tube of mercury is mounted vertically against a brass millimeter scale, permitting measurement of the height of the column of mercury above the mercury in the cistern. Because the height of the mercury column changes with barometric pressure, so does the mercury level in the cistern. Consequently, an ivory pointer is provided so that the level of mercury in the cistern can be adjusted to the same reference level each time the barometer is used.

Using the Fortin Barometer. The following steps are followed in measuring with the Fortin barometer.

1. The level of the mercury in the cistern is adjusted until it is just touching the ivory reference peg. The barometer should be tapped lightly.
2. The vernier scale is manually adjusted so the mercury meniscus appears in line with the front and back of the sighting ring.
3. The pressure in millimeters of mercury is read from the brass vernier scale to 0.1 mm.
4. The temperature of the scale is read from a thermometer attached to the barometer.
5. A correction from Table 12–4 is subtracted from the scale reading to convert the scale reading from pressures to torr. At 0°C, 1 mm Hg = 1 torr.

Errors in Fortin Barometry. The largest source of error arises from the linear thermal expansion of the mercury column and the brass scale. Corrections for this error are listed in Table 12–4. This correction must always be carried out (unless the ambient temperature of your laboratory is 0°C!).

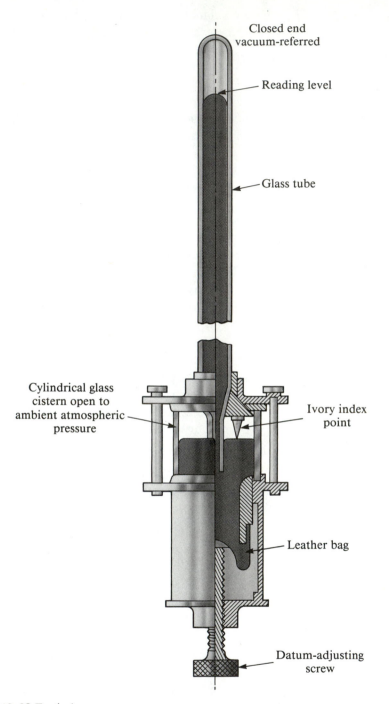

Figure 12–28 Fortin barometer.

The correction terms in Table 12–4 were calculated for the formula

$$C_t = \frac{(s - m)tR_s}{1 + mt} \qquad \textbf{(12–28)}$$

For Table 12–4 the value for the coefficient of linear expansion s of yellow brass is taken to be $18.4 \times 10^{-6}/°C$ and the cubical coefficient

TABLE 12–4

**Temperature Corrections[a] for Fortin Barometer
with Brass Scales Accurate at 0°C**

t (°C)	Barometer Reading (mm Hg)				
	740	750	760	770	780
16	1.93	1.96	1.98	2.01	2.03
17	2.05	2.08	2.10	2.13	2.16
18	2.17	2.20	2.23	2.26	2.29
19	2.29	2.32	2.35	2.38	2.41
20	2.41	2.44	2.47	2.51	2.54
21	2.53	2.56	2.60	2.63	2.67
22	2.65	2.69	2.72	2.76	2.79
23	2.77	2.81	2.84	2.88	2.92
24	2.89	2.93	2.97	3.01	3.05
25	3.01	3.05	3.09	3.13	3.17
26	3.13	3.17	3.21	3.26	3.30
27	3.25	3.29	3.34	3.38	3.42
28	3.37	3.41	3.46	3.51	3.55
29	3.49	3.54	3.58	3.63	3.68
30	3.61	3.66	3.71	3.75	3.83
31	3.73	3.78	3.83	3.88	3.93
32	3.85	3.90	3.95	4.00	4.05
33	3.97	4.02	4.07	4.13	4.18
34	4.09	4.14	4.20	4.25	4.31
35	4.21	4.26	4.32	4.38	4.43
36	4.32	4.38	4.44	4.50	4.56
37	4.44	4.50	4.56	4.62	4.68
38	4.56	4.62	4.69	4.75	4.81
39	4.68	4.75	4.81	4.87	4.94
40	4.80	4.87	4.93	5.00	5.06

Source: W. G. Brombucher, D. P. Johnson, and J. L. Cross, *Mercury
Barometers and Manometers,* National Bureau of Standards Monograph
8, Washington, D.C., 1960.

[a] These corrections should be *subtracted* from the observed brass scale
reading.

of expansion m of mercury is taken to be $181.8 \times 10^{-6}/°C$. The temperature t in Equation 12–28 is in °C and R_s is the observed scale reading. The same equation may be used to calculate scale temperature corrections for materials other than brass.

Small errors also arise from the change in the density of mercury with temperature, the change in gravity acceleration with location, and the capillary depression of mercury. These are treated in detail by Brombacher et al. (1960).

Because mercury does not wet glass, the mercury level is depressed slightly—the amount varying with the surface tension of the mercury, the age of the barometer, the nature of the glass surface, and the diameter of the tube. If the tube diameter is greater than approximately 2 cm, capillary depression is negligible. Corrections for the capillary depression of mercury in a round tube are listed in Table 12–5. The use of large-

TABLE 12–5

Capillary Depression in a Mercury Column at 20°C

Bore of Tube (mm)	Meniscus Height (mm)			
	0.2	0.6	1.0	1.4
Surface Tension 400 dynes/cm				
8	0.11	0.32	0.49	0.63
10	0.06	0.17	0.27	0.35
16	0.01	0.03	0.05	0.07
Surface Tension 450 dynes/cm				
8	0.13	0.37	0.58	0.74
10	0.07	0.20	0.32	0.42
16	0.01	0.04	0.07	0.09

Source: W. G. Brombucher, D. P. Johnson, and J. L. Cross, Mercury Barometers and Manometers, National Bureau of Standards Monograph 8, Washington, D.C., 1960.

bore tubing is the only way to avoid capillary error; the corrections are not particularly reliable, since they depend so much on surface effects.

McLeod Gauge

The McLeod gauge, a common fixture in many vacuum racks, is useful as an absolute manometer for measuring pressures from about 1 to 10^{-5} torr. It is a simple gauge based on Boyle's Law ($P_1 V_1 = P_2 V_2$). A sample of a gas of known volume (V_1) but unknown pressure (P_1) is compressed hydraulically (with mercury) to a known final volume (V_2) and a measured final pressure (P_2).

Construction of a McLeod Gauge. The McLeod gauge is illustrated in Figure 12–29. The lower flask serves as a reservoir for mercury. Its volume is about 300 cm^3.

A capillary about 10 to 15 cm long and with 0.5- to 1.0-mm-diameter bore is sealed to the top of a bulb about 250 cm^3 in volume. An identical, but slightly longer capillary is sealed parallel to the first capillary. The stopcock connecting the gauge to the vacuum line should be more than 76 cm above the bottom of the reservoir.

When the gauge is open to an evacuated vacuum system and the reservoir is also evacuated, all the mercury is contained in the lower reservoir bulb, as shown in Figure 12–29a. A roughing pump is convenient for evacuating the reservoir volume. To measure the pressure of the system, air is let into the reservoir through the lower three-way stopcock, forcing mercury to rise through the tube extending into the reservoir.

When the mercury reaches point A, a definite volume of the gas whose pressure is to be measured is trapped in the bulb and capillary. The mercury is allowed to rise until its level in the open-ended capillary

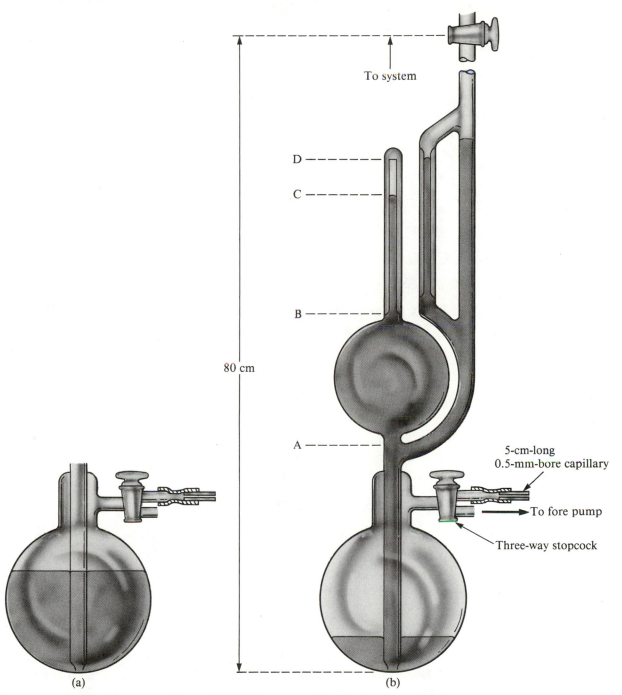

To system

D

C

B

80 cm

A

5-cm-long
0.5-mm-bore capillary

To fore pump

Three-way stopcock

(a)

(b)

Figure 12–29 McLeod gauge.

is at the same level as in the closed end of the capillary, point D in Figure 12–29b.

At this point all the gas originally in the volume between points A and D has been compressed into the tiny volume in the capillary between points C and D. Applying Boyle's Law,

$$P_1V_1 = P_2V_2 \qquad (12\text{--}29)$$

where P_1 is the pressure to be measured,

> $P_2 = P_1 + h$, where h is the hydrostatic head of mercury between points C and D (mm)
>
> $V_2 = bh$, where b is the volume per unit length of capillary (cm^3/mm)
>
> V_1 = volume from point A to point D, i.e., the volume of the bulb plus the total capillary volume.

When these quantities are substituted into Boyle's Law (Eq. 12–29),

$$P_1V_1 = (P_1 + h)bh$$

Since $P_1 \ll h$,

$$P_1 = (b/V_1)h^2 \qquad (12\text{--}30)$$

$$P = Kh^2 \qquad (12\text{--}31)$$

where the apparatus constant K equals b/V_1.

Calibration of a McLeod Gauge. The volumes of the measuring bulb (A to B, Fig. 12–29b) and capillary (B to D, Fig. 12–29b) should be measured before assembling and sealing at points A and B.

Measuring the Capillary Tube. The parameter b (cm^3/mm) of the capillary should be determined on a piece of the same length of stock used for the gauge capillary. The radius of a capillary can be determined quite accurately by filling a 10-cm length with mercury and weighing on an analytical balance to 0.1 mg before and after filling. From the density of mercury (13.54 g/cm^3) the radius can be calculated.

In one experiment, the length of mercury in a capillary was 7.56 cm and its weight w was 0.9906 g:

$$w = \pi r^2 h\rho \qquad (12\text{--}32)$$

$$r = (w/\pi h\rho)^{1/2} = (0.9906/\pi \times 13.54 \times 7.56)^{1/2} = 0.0555 \text{ cm}$$

The parameter b, the volume per unit length of the capillary (cm^3/mm), is given by

$$b = \pi r^2 h = \pi(0.0555)^2(0.1) = 9.68 \times 10^{-4} \text{ cm}^3/\text{mm}$$

The Apparatus Constant. The total volume of the measuring bulb plus the capillary (A to D, Fig. 12–29b) is found by filling this volume with water and weighing before and after filling. The volume in cubic centimeters equals the mass of water contained since the density of water equals 1.00 g/cm^3 with sufficient accuracy.

If the volume V_1 in our sample equals 284 cm^3, then the apparatus constant is given by

$$K = b/V_1 = 9.68 \times 10^{-4}/284 = 3.41 \times 10^{-6}$$

TABLE 12–6

McLeod Gauge Scale Data

Pressure P (mm)	Scale h (mm)
1×10^{-6}	0.54
1×10^{-5}	1.71
1×10^{-4}	5.42
5×10^{-4}	12.11
1×10^{-3}	17.12
5×10^{-3}	38.29
1×10^{-2}	54.14

From the apparatus constant a scale can be prepared. This consists of calculating the values of h in millimeters, which correspond to the values P chosen to appear on the scale. The values of h are given by

$$h = (P/K)^{1/2} \tag{12–33}$$

and are listed in Table 12–6. The resulting scale is shown drawn on millimeter graph paper in Figure 12–30. The scale is glued to a piece of stiff paper and attached to the back side of the capillary tube so that the zero of the scale lines up with the top of the closed capillary.

Diaphragm Gauges

In these devices the force arising from the gas pressure causes a mechanical deflection of a thin area of metal or glass, and this deflection is proportional to the pressure.

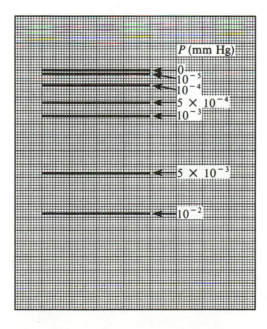

Figure 12–30 McLeod gauge scale.

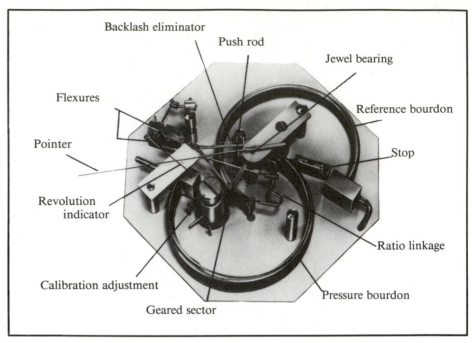

Figure 12–31 Metal Bourdon gauge. (Courtesy of the Wallace and Tierman Division, Pennwalt Corp., Newark, New Jersey.)

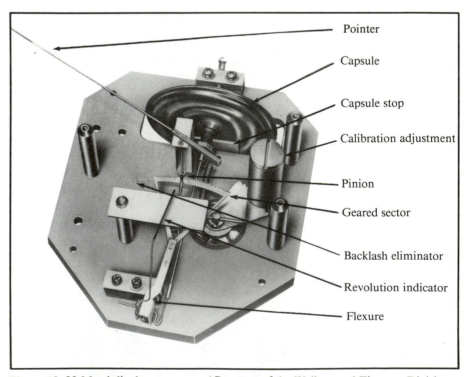

Figure 12–32 Metal diaphragm gauge. (Courtesy of the Wallace and Tierman Division, Pennwalt Corp., Newark, New Jersey.)

Metal Gauges (Bourdon and Diaphragm). The common industrial pressure gauges on tanks and boilers are of this type, illustrated in Figures 12–31 through 12–34. Aneroid (without liquid) barometers for measuring atmospheric pressure are also Bourdon gauges. The curled tube in a Bourdon gauge is of elliptical cross section, open at one end to the pressure to be measured, and attached to a movable needle at the other end. Pressure changes cause the needle to move. If the gauge scale range is from 0 to 760 mm, the accuracy is about 1 to 2 mm. These gauges are rugged and easily read from a distance.

Glass Diaphragm Gauges. A diaphragm gauge constructed of glass is frequently used to isolate and protect mercury or oil manometers from corrosive gases whose pressure must be measured. For this purpose, the diaphragm gauge is a null instrument placed between the system being

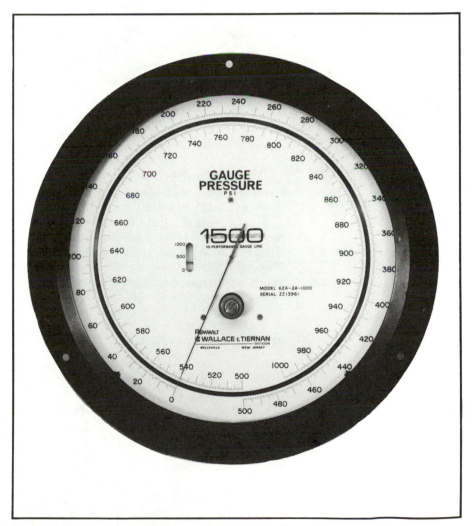

Figure 12–33 Metal pressure gauge scale. (Courtesy of the Wallace and Tierman Division, Pennwalt Corp., Newark, New Jersey.)

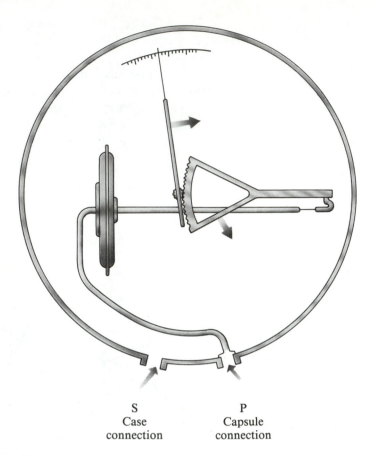

S
Case
connection

P
Capsule
connection

Figure 12–34 Operation of a metal diaphragm gauge. (Courtesy of the Wallace and Tierman Division, Pennwalt Corp., Newark, New Jersey.) **Gauge pressure:** pressure is applied to capsule (P); case (S) is open to atmosphere. **Differential pressure:** high pressure to capsule (P); low pressure to case (S). **Absolute pressure:** pressure to capsule (P); case (S) held at full vacuum with a pump. **Vacuum** (clockwise pointer): capsule (P) open to atmosphere; vacuum to case (S). **Vacuum** (counterclockwise pointer movement): case (S) open to atmosphere; vacuum to capsule (P). **Compound** (pointer moves both ways from center zero): one pressure to capsule (P), the other to case (S). When pressure to capsule (P) is higher than pressure to case (S), pointer gives positive reading; when pressure to capsule (P) is lower than pressure to case (S), pointer gives negative reading.

measured and a mercury manometer to which air at various pressures can be admitted (Fig. 12–35).

The diaphragm gauge is constructed by blowing a thin bubble on the end of an 8- to 10-mm-diameter glass tube (Fig. 12–36). By carefully touching a flame to the bubble, you can make it collapse at an angle to the tube. Seal a glass fiber about 0.2 to 0.3 mm in diameter, drawn from a 2-mm glass rod, to the flattened bubble. This is done by allowing the fiber to rest against the flattened bubble and touching a tiny but hot hand-torch flame to the diaphragm at the point of contact.

The tube should be tilted so that the full weight of the fiber does not rest on the diaphragm. By inserting a small glass tube and clamping

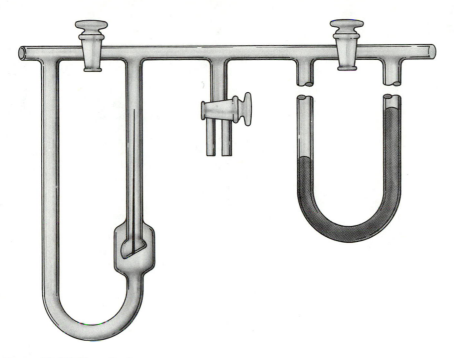

Figure 12–35 Glass diaphragm gauge.

it carefully in position, you can position the top end of the needle in the center of the tube before sealing it to the diaphragm.

If a puff of air results in sufficient distortion to cause the needle to touch the side of the tube, its sensitivity is about right. When used, the null position of the needle should be observed under a low-power microscope. A sensitivity of about 0.5 to 1 mm is not difficult to achieve— once you have the knack of blowing a bubble of just the correct thickness. It takes some practice. Once the tube, bubble, and needle are assembled, the ring seal and joints are done as usual (see Chapter 11).

It is also possible to make the diaphragm gauge from fused silica or Vycor. To work quartz it is necessary to use an acetylene–oxygen torch; however, the usual gas–oxygen torch should be used to seal the fiber to the diaphragm.

Such a quartz diaphragm gauge can be used to measure vapor pressures or decomposition pressures at high temperatures—up to about 1000 K in practice. The sample is sublimed under vacuum into the outer bulb, which is then sealed off. The diaphragm gauge is then heated in a tube furnace so that the end of the fiber is just visible outside the furnace when the sample region is centered in the furnace.

Indirect-Reading Gauges

Most indirect-reading gauges are electronic gauges. For their operation they rely on some property of a gas that depends upon pressure and can be translated into a measurable electric current or voltage.

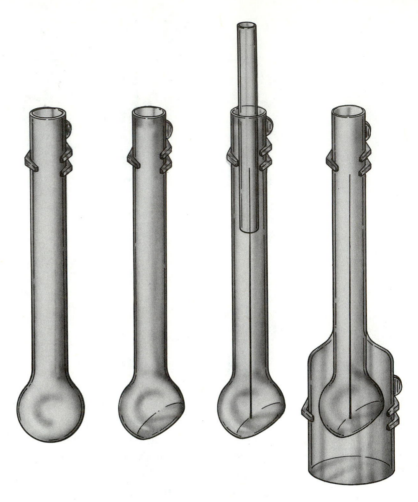

Figure 12–36 Construction of a glass diaphragm gauge.

Thermal Conductivity Gauges. If an electric current is passed through a metal wire, the temperature of the wire rises due to the electrical energy (I^2RT) dissipated in the wire. The resistance of the wire depends on the temperature, which in turn depends on the heat transfer from the wire to the surroundings. The heat transfer or conductance depends on the pressure. Both the measurement of the resistance of the wire and the temperature of the wire form the basis for the operation of practical vacuum gauges.

Pirani Gauge. In the Pirani gauge, a wire (R_1) is exposed to the gas pressure to be measured (Fig. 12–37). It is heated with a constant voltage and is electrically placed in one arm of a Wheatstone bridge. An identical compensating wire (R_2) is placed in a second arm. They are physically adjacent and sealed in the Pirani tube jacket. In a typical Pirani gauge the resistance of R_3 equals that of R_4 equals 15.0 ohm, and R_5 is a 2-ohm zero adjust. With a constant voltage applied, the resistance of R_1 depends upon the gas pressure. The microammeter measures the extent

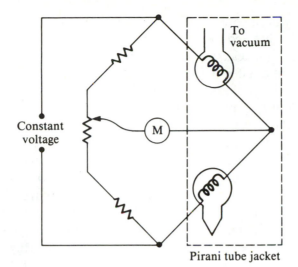

Figure 12–37 Operation of the Pirani gauge.

Pirani tube jacket

of unbalance in the bridge and is calibrated in torr. The calibration depends on the nature of the gas, since gases have different conductivities, as shown in Figure 12–38.

This apparent weakness turns out to be an advantage since it permits the Pirani gauge to function as a leak detector. A fine stream of a gas

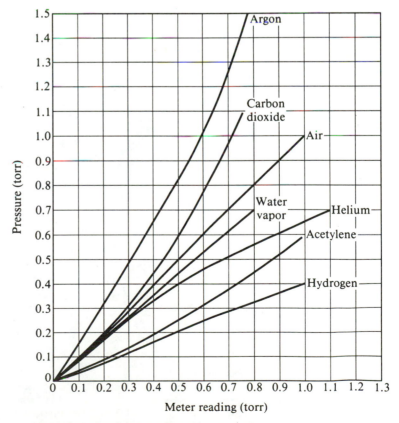

Figure 12–38 Typical Pirani gauge calibration.

whose conductivity is greatly different from air (e.g., helium or argon) is directed over the area suspected of leaking; if the gas stream impinges on the leak, the meter shows a marked fluctuation. Acetone is also sometimes effective.

In the ACCUVAC-500, shown in Figure 12–39, modern digital circuitry replaces and extends the simple diagram of Figure 12–38. Both the sensor tube bridge and the ambient temperature compensation windings voltage are converted to digital values via a 15-bit analog-to-digital converter. Built-in calibration curves may be selected for N_2 and Ar. Resolution is 1 torr above 100 torr and 0.1 millitorr below 100 millitorr with a digital display of measured pressure. The Pirani gauge is simple and rugged, and it provides a continuous indication of pressure.

Thermocouple Gauge. As in a Pirani gauge, a wire is heated with a regulated power supply. Instead of measuring the resistance of the wire, a tiny thermocouple welded to the center of the heated wire measures its temperature (Fig. 12–40). The electromotive force of the thermocouple depends on the wire temperature, which in turn depends on the gas pressure and its conductivity. The range of the Varian 531 thermocouple gauge (Fig. 12–41), which is typical, is 2.5 to 0.001 torr (Fig. 12–42). The instrument is calibrated for air and must be recalibrated for other gases for precise work. As with a Pirani gauge, the thermocouple gauge also functions as a leak detector.

Ionization Gauges. Ionization gauges all have one feature in common: they form positive ions from the gas whose pressure is to be measured. The positive ions are collected on a negatively charged cathode

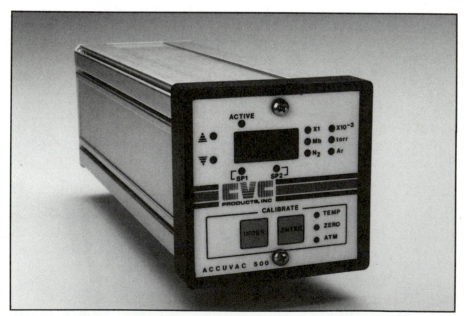

Figure 12–39 CVC ACCUVAC 500 (Pirani gauge). (Courtesy of CVC Products, Inc., Rochester, New York.)

To vacuum

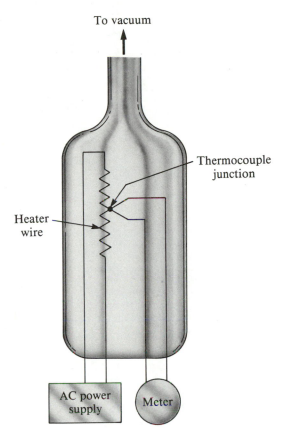

Thermocouple
junction

Heater
wire

AC power
supply

Meter

Figure 12–40 Operation of a
thermocouple gauge.

and the resulting positive ion current is measured. The positive ion current
is linearly dependent on the pressure. Ionization gauges differ in the manner in which they generate positive ions from gas molecules.

   Thermionic Ionization Gauges. The thermionic ionization gauge is
a slightly modified triode vacuum tube. The VG-1A, an early form, consists of a central filament at ground potential and heated to incandescence

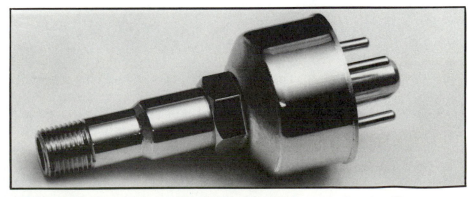

Figure 12–41 Thermocouple gauge (Varian 524-2). (Courtesy of Varian, Vacuum
Products Division, Lexington, Massachusetts.)

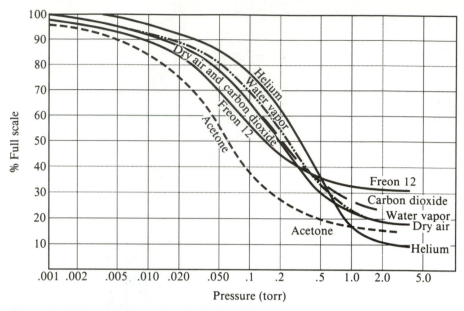

Figure 12–42 Response curve for the Varian 531 thermocouple gauge. (Courtesy of Varian, Vacuum Products Division, Lexington, Massachusetts.)

with a low-voltage alternating current (Fig. 12–43). Thermionic electrons emitted from the hot filament accelerate toward the positively charged (approximately +150 volts) grid, where they are collected, causing a grid current i_g to flow. As the electrons stream from the filament to the grid, they collide at pressure P with gas molecules, which form positive ions. The plate, or positive ion collector, is a coating of platinum deposited on the inside of the glass envelope. Positive ions are attracted to the negatively charged (approximately −50 volts) plate, leading to a small positive ion current i_p that depends linearly on pressure P and the number of electrons/sec (i_g) emitted by the hot filament. The proportionality constant S is defined as the sensitivity of the gauge:

$$i_p = Si_gP \qquad\qquad (12\text{--}34)$$

$$P = \frac{1}{S}\frac{i_p}{i_g}$$

The proportionality constant S (the sensitivity) is usually reported with units of torr^{-1}. For the VG-1A, S equals 20 torr^{-1} for nitrogen.

The useful range of the VG-1A is from about 10^{-3} to 10^{-7} torr. Above 10^{-3} torr the filament may burn out or, at least, oxidize badly, and below 10^{-7} torr the onset of x-ray production makes the tube unreliable.

An improved design (Baynard and Alpert, 1950) reduces the effect of x rays, extending the useful range to nearly 10^{-10} torr. The Baynard-Alpert tube (Fig. 12–44) differs from the VG-1A tube (Fig. 12–43) in

To vacuum

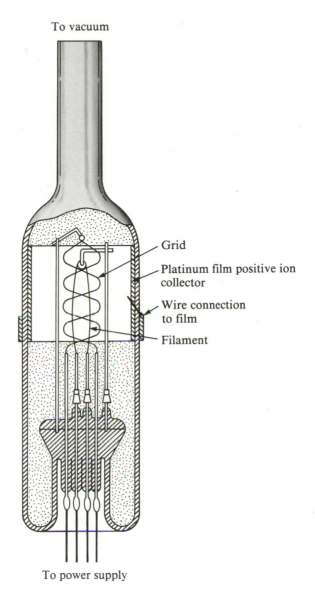

Grid

Platinum film positive ion
collector

Wire connection
to film

Filament

To power supply

Figure 12–43
Thermionic ionization
gauge (VG-1A).

the placement of the filament and plate (positive ion collector), which
are reversed. The bias voltages remain approximately the same in sign
and magnitude, however, and the principle of operation is identical. The
small surface area of the central positive collector wire in the Baynard-
Alpert tube, however, presents a much smaller target for x rays generated
by electrons striking the grid. Because the filament is outside the grid,
there is room for a spare filament on the opposite side in case the first
filament should burn out. In naked gauge tubes, constructed without the
envelope, the filaments are replaceable. The electrodes are attached to a
metal base, which in turn seals to the metal system with an O-ring or
similar flange.

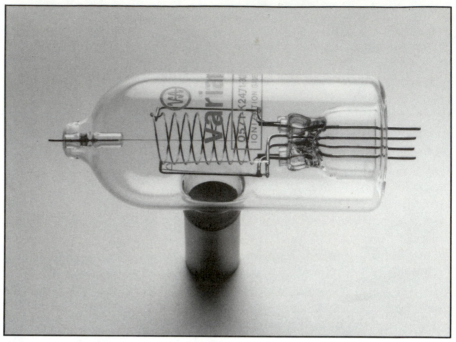

Figure 12–44 Baynard-Alpert ion gauge (Varian 571). (Courtesy of Varian, Vacuum Products Division, Lexington, Massachusetts.)

The ionization gauge requires a controller to furnish the needed voltages and to measure the current. The Varian Baynard-Alpert 571 ionization gauge operates with the collector at 0 volts to ground, the grid at +150 volts, and the filament at +30 volts. Its sensitivity is 10 torr^{-1} (twice that of the VG-1A) and its pressure range is from 1×10^{-3} to 2×10^{-10} torr.

Cold Cathode Gauge. This gauge, also known as the Penning or Philips gauge, has no filament. The anode consists of a wire loop or short open-ended cylinder. It is placed between two grounded cathodes. Ions are generated by applying a high voltage, 2000 to 5000 volts, between the anode and cathode (Fig. 12–45). The electrodes are sealed in a glass envelope, which is fixed in position between the poles of a permanent magnet (1000 to 2000 gauss).

Electrons leaving the cathode migrate toward the positively charged anode in a spiral path due to the parallel applied magnetic field. On their way the electrons collide with residual gas molecules to form positive ions, which migrate toward the cathode. The long spiral path increases the probability of a collision between electrons and molecules, thus increasing the sensitivity of the gauge.

There is no filament to burn out and the construction of the cold cathode gauge is simple and rugged. The long spiral path of the electrons generates a much higher positive ion current than occurs in hot filament

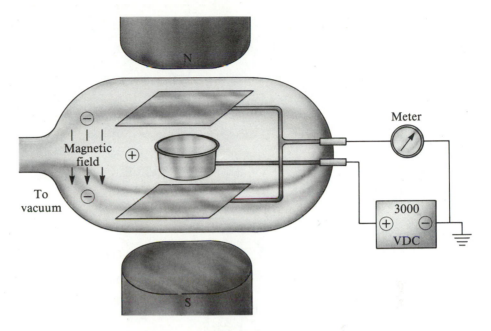

Figure 12–45 Operation of a cold cathode gauge.

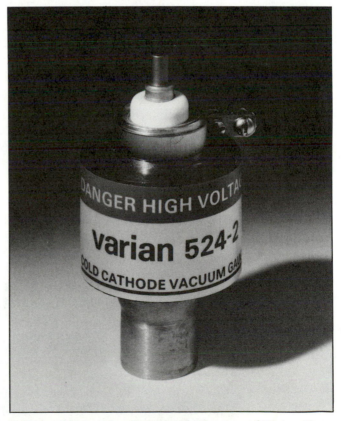

Figure 12–46 Varian 524-2 cold cathode gauge. (Courtesy of Varian, Vacuum Products Division, Lexington, Massachusetts.)

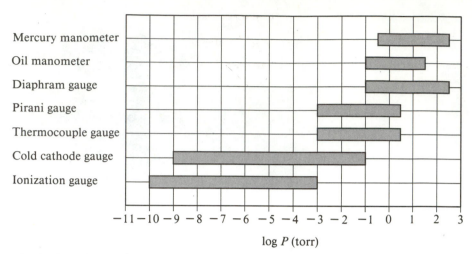

Figure 12–47 Comparison of vacuum gauge ranges.

ionization gauges (milliamps versus microamps). This permits simpler auxiliary electronics.

On the other hand, the accuracy of a cold cathode gauge is less than that of ionization gauges. The magnet must be kept away from iron and other ferromagnetic materials. The voltage is dangerously high and may be stored in power supply condensers some time after the gauge is turned off, presenting a hazard to the unwary.

Because ionized residual gas molecules may be trapped permanently when they eventually collide with the cathode, the gauge acts as a getter for these molecules; that is, it effectively removes residual gas from the system and so acts as a pump. Consequently, the gauge must be connected to the system by very large tubing (2- to 2.5-cm diameter). If the gauge is connected by small tubing, the flow rate from the system to the gauge is slow, getting action by the gauge lowers the pressure in the gauge to values lower than the system, and errors result. In other words, $P(\text{measured}) < P(\text{true})$. The getting action of this gauge is the basis of the ion getter pump (ion sputter pump).

The Varian 524-2 cold cathode gauge, shown in Figure 12–46, has a range of 1×10^{-2} to 1×10^{-7} torr when used with an 860A series cold cathode gauge controller.

As shown in Figure 12–47, it is often necessary to use two or more gauges to cover the measurement of a wide range of pressures.

REFERENCES

1. Baynard, R. T., and D. A. Alpert, *Rev. Sci. Instr.* 21:571, 1950.
2. Benedict, R. P., *Fundamentals of Temperature, Pressure, and Flow Measurements,* 3rd ed., Wiley, New York, 1984.
3. Brombacher, W. G., D. P. Johnson, and J. L. Cross, *Mercury Barometers and Manometers,* National Bureau of Standards Monograph 8, Washington, D.C., 1960.

4. Carlson, K. D., "The Knudsen Effusion Method," in *The Characterization of High Temperature Vapors,* J. L. Margrave, ed., Wiley, New York, 1967.

5. Diels, K., and R. Jaeckel, *Leybold Vacuum Handbook,* Pergamon Press, Oxford, 1966.

6. Dushman, S., *Scientific Foundations of Vacuum Technique,* 2nd ed., Wiley, New York, 1962.

7. Guthrie, A., *Vacuum Technology,* Wiley, New York, 1963.

8. Kennard, E. H., *Kinetic Theory of Gases,* McGraw-Hill, New York, 1938.

9. Kubachewski, O., and E. L. Evans, *Metallurgical Thermochemistry,* Wiley, New York, 1956.

10. O'Hanlon, J. F., *A User's Guide to Vacuum Technology,* Wiley, New York, 1980.

11. Rondeau, R. E., "Design and Construction of Glass Vacuum Systems— Part One," *J. Chem. Educ.* 42:A445, 1965.

12. Rondeau, R. E., "Design and Construction of Glass Vacuum Systems— Part Two," *J. Chem. Educ.* 42:A511, 1965.

13. Thomson, G. W., and D. R. Douslin, "Determination of Pressure and Vacuum," in Weissberyer, A., ed., *Physical Method of Chemistry,* Wiley-Interscience, New York, 1971.

14. Yarwood, J., *High Vacuum Technique,* 4th ed., Chapman and Hall, London, 1967.

Temperature

Of all the variables upon which the properties of a system depend, temperature is without a doubt the most important one to measure and control. Temperature is an everyday concept familiar to everyone. It was probably the physiological response to hot and cold that formed the basis for the earliest concept of temperature. Below the temperature of the human body one could recognize and describe various degrees of cold: cool as a cucumber, cool, chilly, biting, frigid, frozen, fresh, keen, bleak, glacial, frosty, wintry, boreal, arctic, Siberian, or blue with cold. In other seasons and other places the environment might be hot, roasting, burning, scalding, incandescent, flaming, fiery, boiling, glowing, broiling, torrid, feverish, steaming, incinerating, sweltering, stewing, red-hot, white-hot, simmering, seething, smoking, blazing, volcanic, or hotter than Hades.

HISTORY

Early scientists next observed that many physical properties depend on temperature: the volume of a gas, liquid, or solid, or the length of a wire or column of liquid. Credit for the earliest thermometer, based on the expansion of a material substance with an increase in temperature, should be given to Galileo Galilei, who in 1592 invented a glass thermometer containing both air and water. Since his thermometer was open to the atmosphere, it was sensitive to changes in barometric pressure, and consequently was not very effective. Nevertheless, it was an improvement over "cool as a cucumber" or "hotter than Hades" thermometry.

In 1654, Ferdinand II, Grand Duke of Tuscany, made the first hermetically sealed thermometer, which he filled with alcohol rather than water. His thermometer was crude, but independent of pressure.

Early Temperature Scales

Sir Isaac Newton was one of the first scientists to define a temperature scale based upon two reproducible fixed points. For his lower point he chose the melting point of ice and assigned it a value of zero. As a second fixed point, he chose the "armpit temperature of a healthy Englishman," and assigned this temperature the value 12, or a dozen degrees between the melting point of ice and normal body temperature. Newton used the term "equal parts of heat" to define what we would now call a degree. The distinction between heat and temperature was not yet clear.

In 1714, Gabriel Daniel Fahrenheit, a German instrument maker from Danzig and a resident of Holland, invented the first mercury-in-glass thermometer. Following the example of Newton, Fahrenheit defined a temperature scale based upon two points. He chose as the lower point the lowest temperature obtainable with a mixture of ice and common salt and assigned this the value of zero. As an upper temperature point he chose the temperature "found in the blood of a man." Like Newton, Fahrenheit divided his scale into a dozen parts, but, probably upon noticing the large size of his degree, later divided his scale into eight dozen, or 96 degrees. On his thermometer, he found that the melting and boiling points of water had approximately the values 32 and 212, respectively. Later, these two points were found to be very reproducible, and so the melting point of ice was defined as 32°, the boiling point of water was defined as 212°, and a degree Fahrenheit was defined as 1/180 of the temperature difference between the two points.

In 1742, Anders Celsius, professor of astronomy at the University of Uppsala, Sweden, proposed a temperature scale consisting of exactly 100 degrees between two fixed points: the melting point of ice, defined as zero, and the boiling point of water, defined as 100 degrees. The same scale was independently suggested by Christin of Lyons, France, in 1743, who called it the centigrade scale.

Careful experimentation with liquid-in-glass thermometers over the next hundred years revealed that the expansive properties of a liquid in a glass tube were not satisfactory for defining an acceptable temperature scale. Even if two thermometers could be constructed to agree exactly at the ice point and the boiling point, it was impossible to get them to agree exactly at intermediate temperatures. The problem was caused not only by the nonuniformity of the glass bore, but also by the nonlinearity of the thermal coefficients of expansion of different liquids. In other words, temperature scales defined in terms of the expansive properties of liquids were not independent of the materials of construction.

Gas Thermometry

Parallel to the development of liquid-in-glass thermometers and the concept of a temperature scale based upon two fixed points were the evolution of thermodynamics and the discovery of the gas laws by Boyle (1662), Charles (1787), Gay-Lussac (1802), and Regnault (1845). These investigators realized that, unlike liquids and solids, all gases had the same volume coefficient of expansion. Working with different gases, Gay-Lussac found a value of 1/267 per degree Celsius. Regnault's careful investigations led to a value of $1/273$; he also found that even gases, because of nonideal behavior, have slightly different coefficients of expansion. Gas thermometry, pioneered by Gay-Lussac and Regnault, is still important today in establishing temperature scales since the expansivities of (ideal) gases are identical. The basic equation for gas thermometry is the familiar relationship

$$\frac{P_1 V_1}{T_1} = \frac{P_2 V_2}{T_2} \tag{13-1}$$

From this equation it is possible to define a temperature scale and establish the size of a unit of temperature interval.

You can see that the situation is going to get out of hand if everyone who invents a better thermometer decides on his own fixed points and establishes a temperature scale that agrees well with his instrument. This is the stuff of which international agreements are made. Indeed, in 1887, about 40 years after Regnault's work, the first conference of the International Bureau of Weights and Measures (BIPM) announced its endorsement of a temperature scale. At that time the scale was defined with two fixed points: the normal freezing point of water was *defined* to be 0°C and the normal boiling point of water was *defined* to be 100°C. The size of the degree was then set equal to $1/100$ the difference between the temperatures of water at its freezing and boiling points. The temperatures of a number of fixed points were then determined by gas thermometry measurements:

$$T = 100 \, \frac{\lim\limits_{P \to 0} (PV)_T}{\lim\limits_{P \to 0} (PV)_{100} - \lim\limits_{P \to 0} (PV)_0} \tag{13-2}$$

To eliminate nonideal behavior, PV products are measured at successively low pressures and the data are extrapolated to zero pressure. Alternatively, an equation of state may be used to correct for the slight nonideal behavior of the gas. In either case, hydrogen or helium are used because they are less nonideal than other gases and may be used at very low temperatures.

Thermodynamic Temperature Scales

Occasionally you will find a temperature scale defined in terms of the behavior of "ideal" gases, known as a **thermodynamic** temperature scale. Strictly speaking, this is not quite correct, since no real gases behave ideally. A truly thermodynamic definition of temperature is absolutely independent of the properties of any substance. For example, temperature can be defined thermodynamically as

$$T = \left(\frac{\partial E}{\partial S} \right)_V \tag{13-3}$$

Not very practical, though, is it? The Carnot cycle also can be considered a thermodynamic definition of temperature:

$$\frac{Q_1}{T_1} = \frac{Q_2}{T_2} \tag{13-4}$$

$$\frac{T_1}{T_2} = \frac{Q_1}{Q_2}$$

At best, this equation provides a thermodynamic definition of the ratio of two temperatures. Thermodynamics provides no definition of the magnitude of a degree or the magnitude of a temperature difference. In other words, there is no such thing as a truly thermodynamic temperature *scale*. Temperature scales and the magnitude of a degree are and must be inventions of human beings—and international committees.

INTERNATIONAL PRACTICAL TEMPERATURE SCALE

Gas thermometry still plays an important role in the precise measurement of temperatures as low as about 10 K and as high as around 1500 K. Precision gas thermometry in practice, however, is very complex, and only a dozen or so laboratories in the world are equipped to carry out gas thermometric measurements with high accuracy. Consequently, working scientists need a more practical method for measuring temperature in their laboratories. For this reason the International Practical Temperature Scale (IPTS) was adopted in 1927 by the Comité International des Poids et Measures (CIPM).

The International Practical Temperature Scale was revised in 1948 and 1968, and amended in 1975 (Comité, 1969, 1976) and is expected to be revised again by 1990. As shown in Figure 13–1, the current scale, IPTS-68 as amended, differs only slightly from previous scales. Acting upon a suggestion first made by Clausius or Lord Kelvin (1854), the International Committee now defines the size of a degree with one experimental point: The triple point of water is *defined* to be *exactly* 273.16 K. The IPTS-68 equivalent of Equation 13–2 is

$$T = 273.16 \frac{\lim_{P \to 0} (PV)_T}{\lim_{P \to 0} (PV)_{tp}} \tag{13-5}$$

The second fixed point is absolute zero, but it is not necessary to measure that point. The triple point of water is a more reproducible point than the normal freezing point of water, since problems with pressure variations and dissolved air are avoided.

In addition to the magnitude of the unit of temperature interval, a *practical* temperature scale consists of three components, all determined by international agreement:

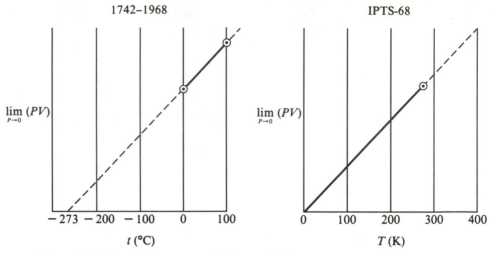

Figure 13-1 The temperature scale before and after 1968.

1. Some fixed temperature points with which to calibrate the instruments used to measure the temperature. The primary fixed points (IPTS-68) are the oxygen point, the triple point of water, the steam point, the zinc point, the silver point, and the gold point. These and some secondary points are listed in Table 13-1.

2. Some convenient instruments with which to measure temperature, that is, to interpolate between fixed points. The IPTS-68

TABLE 13-1

Defined Primary Fixed Points of the IPTS-68

Equilibrium State[a] of Fixed Point	Assigned Value of IPT	
	°C	K
tp hydrogen	−259.34	13.81
lv hydrogen at 25/76 atm	−256.108	17.042
bp hydrogen	−252.87	20.28
bp neon	−246.048	27.102
tp oxygen	−218.789	54.361
tp argon	−189.352	83.798
bp oxygen	−182.962	90.188
tp water	0.01	273.16
bp water	100	373.15
fp tin	231.9681	505.1181
fp zinc	419.58	692.73
fp silver	961.93	1235.08
fp gold	1064.43	1337.58

Source: Comité International des Poids et Mesures, "The International Practical Temperature Scale of 1968, Amended Edition of 1975," *Metrologia* 5:7–17, 1976.

[a] tp = triple point, lv = liquid vapor equilibrium, bp = boiling point, fp = freezing point.

instruments are the platinum resistance thermometer, the platinum and platinum–10% rhodium thermocouple, and the optical pyrometer.

3. Some equations with which to interpolate between fixed points. Unlike the gas thermometer, which is linear with respect to temperature ($PV = RT$), all other instruments are nonlinear. Since a variety of polynomials can be found, it is necessary to select suitable equations by international agreement.

In the same way that the triple point of water is defined to be 273.16 K in order to determine the magnitude of a kelvin, the temperatures of a number of equilibrium states are assigned defined temperatures to establish a working practical temperature scale that agrees within current experimental error with the thermodynamic temperature scale. In this sense, the thermodynamic temperature scale is taken to mean the temperature scale determined by the best possible gas thermometric methods.

Interpolation Between Fixed Points

The temperature scale determined by the 13 fixed points listed in Table 13–1 is so wide that it is necessary to use three different instruments to interpolate between fixed points.

Resistance Thermometer Interpolation. From 13.81 K (triple point of hydrogen) to 630.74°C (melting point of antimony), the CIPM agreed upon the platinum resistance thermometer as the most suitable instrument. Certain construction details are also spelled out. For example, the thermometer resistor must be strain-free, annealed pure platinum. The resistance ratio W, defined by

$$W = \frac{R_T}{R_{273.15}} \tag{13–6}$$

where R is the resistance, must not be less than 1.39250 at the normal boiling point of water ($T = 373.15$°C). The upper boundary for using the platinum resistance thermometer, 630.74°C, corresponds to the melting point of antimony, which is neither a primary nor secondary fixed point substance.

Thermocouple Interpolation. From 630.74 to 1064.43°C (the gold point), a thermocouple of platinum and platinum–rhodium alloy is the chosen instrument for interpolating between fixed points. The alloy must contain 90% platinum and 10% rhodium by weight.

Radiation Pyrometry Interpolation. Above 1064.43°C, temperatures are measured with an optical pyrometer by matching the brightness (J) of a blackbody at temperature T and a blackbody at the gold point

(1064.43°C) according to Planck's Law at wavelength λ. The Planck radiation equation is given by

$$J = \frac{c_1 \lambda^{-5}}{\exp(c_2/\lambda T) - 1}$$

Interpolation Equations

Not only the instruments, but also the specific mathematical equations to be used for interpolating between fixed points on the international temperature scale are determined by international agreement.

Above the Gold Point. The interpolation equation for temperatures above the gold point is established by comparing the Planck radiation equation at two temperatures: the gold point ($T_{Au} = 1064.43$°C) and the temperature to be measured. The constant c_2 in the Planck equation is defined to be 0.014388 m K:

$$\frac{J_T}{J_{Au}} = \frac{\exp(0.014388/\lambda T_{Au}) - 1}{\exp(0.014388/\lambda T) - 1} \tag{13–7}$$

This is the interpolating equation above the gold point. The brightness J has units of energy per unit wavelength interval at wavelength λ, emitted per unit time, per unit solid angle, per unit area, of a blackbody at temperature T.

Between 630 and 1064°C. In the range from 630.74 to 1064.43°C (the antimony point to the gold point) temperatures are defined and interpolated by

$$E = a + bt + ct^2 \tag{13–8}$$

where E is the electromotive force of a standard thermocouple of platinum and platinum/10% rhodium alloy, when one junction is at 0°C and the other is at temperature t. The three constants a, b, and c are calculated from the values of E at the freezing points of gold and silver and at 630.74 ± 0.2°C. The temperature 630.74°C is to be determined with a platinum resistance thermometer (at the temperature of but not physically at the equilibrium melting point of antimony).

Below 630°C. The temperature range from 13.81 to 630°C is broken into five subranges. Within each subrange, temperatures are also defined by quadratic or cubic equations in T, the constants of which are determined by calibration at two (if quadratic) or three (if cubic) primary fixed temperatures listed in Table 13–1. The actual method of interpolating the temperatures and making them conform with the thermodynamic temperature scale is quite complex and involves a 20-term series of 16-digit constants. In addition, some complex functions are required to correct the deviations from the thermodynamic temperature scales. It

TABLE 13–2

Comparison of Primary Fixed Point Values

Equilibrium State of Fixed Points[a]	IPTS-27 (°C)	IPTS-48 (°C)	IPTS-68	
			°C	K
tp hydrogen			−259.34	13.81
bp hydrogen			−256.108	17.042
bp neon			−246.048	20.28
tp oxygen			−218.789	54.361
bp oxygen	−182.97	−182.97	−182.962	90.188
fp water	0.000			
tp water		0.01	0.01	273.16
bp water	100.000	100	100	373.15
fp zinc			419.58	692.73
bp sulfur	444.60	444.6		
fp silver	960.5	960.8	961.93	1235.08
fp gold	1063.0	1063	1064.43	1337.58

Sources: Comité International des Poids et Mesures, "The International Practical Temperature Scale of 1968," *Metrologia* 5:35–44, 1969, and Benedict, R. P., *Fundamentals of Temperature, Pressure, and Flow Measurements,* 3rd ed., Wiley, New York, 1984.

[a] tp = triple point, bp = boiling point, fp = freezing point.

TABLE 13–3

Secondary Fixed Points

Equilibrium State[a] of Fixed Points	T (°C)	T (K)
fp mercury	234.314	−38.836
tp diphenyl ether	300.02	26.87
tp benzoic acid	395.52	122.37
mp indium	429.784	156.634
fp tin	505.1181	231.9681
fp bismuth	544.592	271.442
fp cadmium	594.258	321.108
fp lead	600.652	327.502
fp sulfur	717.824	444.674
fp antimony	903.905	630.755
fp aluminum	933.61	660.46
fp copper	1358.03	1084.88
fp nickel	1728	1455
fp cobalt	1768	1495
fp palladium	1827	1554
fp platinum	2042	1769
fp rhodium	2236	1963
fp iridium	2720	2447
fp tungsten	3695	3422

Sources: Adapted from Comité International des Poids et Mesures, "The International Practical Temperature Scale of 1968, Amended Edition of 1975," *Metrologia* 12:7–17, 1976, and Benedict, 1984.

[a] fp = freezing point, tp = triple point, mp = melting point.

would not be surprising if a simpler interpolation method were adopted at the next CIPM meeting, probably in the early 1990s.

Although the IPTS primary fixed points are said to be defined, they are, after all, just experimentally determined temperatures (based upon Eq. 13–2). The only point that is really defined is the triple point of water. Work continues to improve the fit of experimental equilibrium temperatures to the IPTS-68 scale. Fortunately, the changes made periodically by international agreement are small, so for most purposes the temperatures reported 40 years ago are very close to those measured today, as indicated in Table 13–2.

In addition to the 13 IPTS primary fixed points, a number of secondary points are available for purposes of experimental convenience (Table 13–3).

EXPANSION THERMOMETERS

Virtually all solids, liquids, and gases expand with an increase in temperature.[1] Gas thermometry is based on the expansion of gases; liquid-in-glass thermometers depend upon the difference in the expansive properties of glass and liquids such as ethanol and mercury. The coefficient of volume expansion β is defined by

$$\beta = \frac{1}{V_1} \frac{V_2 - V_1}{T_2 - T_1} = \frac{1}{V} \frac{dV}{dT} \qquad (13\text{–}9)$$

where V_1 and V_2 are initial and final volumes and T_1 and T_2 are initial and final temperatures.

The coefficient of expansion of mercury is about 50 times that of glass. Although organic liquids may be used at lower temperatures than mercury and give greater sensitivity (Table 13–4), they require the addition of a dye, which can lead to chemical reactions that can cause the liquid volume to change gradually with time. Consequently, mercury thermometers are preferred because of their great stability. Scientists use mercury-in-glass thermometers more than any other device for the measurement of temperature (Fig. 13–2). They are inexpensive, reliable, compact, and portable.

Total- and Partial-Immersion Thermometers

The thermometer shown in Figure 13–2 is a typical partial-immersion thermometer, calibrated for immersion exactly to a line etched between the bulb and the scale. Immersion to a different point leads to erroneous temperature readings. A thermometer without a line is calibrated to be

---

[1] Since water is most dense at about 4°C, it expands with a decrease in temperature between 0 and 4°C.

TABLE 13–4

Coefficients of Volume Expansion

Material	$10^4 \beta$ (K^{-1})	Freezing Point (°C)
Ethanol	11.2	−117.3
Toluene	9.6	−95.0
Pentane	16.08	−131.5
Water	2.07	0
Mercury	1.818	−38.836
Pyrex glass	0.036	—

correct when the thermometric fluid—usually mercury—is totally immersed in the system whose temperature is being measured, as shown in Figure 13–3. If a total immersion thermometer is used under conditions of partial immersion, an emergent stem correction must be calculated

$$t_c = t_o + kn(t_o - t_s) \qquad (13\text{--}10)$$

where t_c is the correct temperature, t_o is the observed temperature read from the scale, t_s is the average temperature of the emergent stem, and n is the number of degrees on the emergent stem, that is, the actual number of degrees on that part of the thermometric fluid (usually mercury) not in the bath or system being measured. The constant k equals 0.00016 for mercury in Pyrex or Jena glass. Since k depends on the difference in volume coefficients of expansion of glass and the thermometric liquid, it is larger for organic liquid; a value of 0.001 is recommended (Busse, 1941).

As an example of the emergent stem correction, suppose that a total-immersion thermometer is immersed up to the 16°C mark on the scale and the observed temperature is 78.69°C, so that $n = 79 - 16 = 63$. It is assumed that the emergent stem is at room temperature, 23°C. According to Equation 13–10, the correct temperature is given by

$$t_c = 78.69 + 0.00016 \times 63(79 - 23) = 78.69 + 0.56 = 79.25°C$$

Notice that the average stem temperature may be quite different from room temperature. Also, the emergent stem correction can be negative if the average temperature of the stem is higher than that of the bulb.

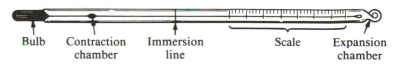

| Bulb | Contraction chamber | Immersion line | Scale | Expansion chamber |

Figure 13–2 Typical laboratory mercury-in-glass thermometer. (Adapted from Wise, 1976.)

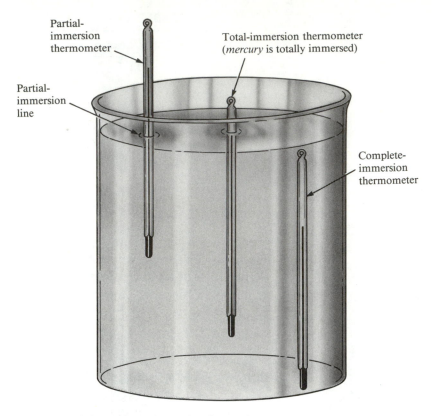

Partial-immersion thermometer

Partial-immersion line

Total-immersion thermometer (*mercury* is totally immersed)

Complete-immersion thermometer

Figure 13–3 Partial-, total-, and complete-immersion thermometers.

Differential Thermometers

The Beckman and Philadelphia thermometers are representative of thermometers that permit the accurate measurement of small temperature differences: the range is 5 to 6°C and the smallest divisions are 0.01°C, within which the temperature can be estimated to about 0.002°C. The scales, which read from 0 to 6°, are arbitrary. The setting refers to the bulb temperature when the mercury is at 0° on the scale. The setting is changed by adding mercury or removing it from the bulb to the reservoir at the top of the thermometer. The reservoirs and filling arrangements are slightly different in the Beckman and Philadelphia thermometers, as indicated in Figure 13–4.

The bulb contains less mercury when set to a high temperature than it does when set to a low temperature. Thus, to raise the setting, you must remove mercury from the bulb by warming the bulb until droplets of mercury flow out of the capillary at the top of the thermometer and fall into the reservoir. Then place the thermometer in a thermostat at the desired temperature to determine if the mercury is on scale or not. You may have to repeat the procedure a few times to get the mercury level close to the desired setting.

To lower the setting, bring mercury from the reservoir into the bulb.

(a) Beckman   (b) Philadelphia

Figure 13–4 Beckman and Philadelphia thermometers.

Warm the mercury until droplets begin to fall into the reservoir. Then invert the thermometer so that the reservoir mercury is in continuous contact with the thermometer capillary mercury. Now if you let the thermometer bulb cool slowly, mercury will be drawn from the reservoir as the bulb contracts upon cooling. When the desired amount of mercury has been drawn from the reservoir into the bulb, tap the reservoir end of the thermometer sharply (but not too sharply!) with a snap of your forefinger. This should break the mercury thread. Now bring the thermometer into an upright position. In practice, it is a little easier to reach the desired setting by heating and removing mercury than by cooling. Once adjusted, the thermometer should not be inverted or laid on its side or the setting may be changed. Setting the Philadelphia thermometer is similar, except that it is not necessary to break the thread by tapping. Inverting is sufficient.

Because the Beckman and Philadelphia thermometers are total-immersion thermometers, it is necessary to apply an emergent stem correction as described above. In addition, it is necessary to apply a setting factor, because the volume coefficient of expansion of mercury is not constant but varies with temperature. The thermometer is constructed so that the setting factor equals unity when set to 20°C (Table 13–5).

TABLE 13–5

Setting Factors for Beckman Thermometers

Setting (°C)	Factor	Setting (°C)	Factor
0	0.9931	55	1.0094
5	0.9950	60	1.0105
10	0.9968	65	1.0115
15	0.9985	70	1.0125
20	**1.0000**	75	1.0134
25	1.0015	80	1.0143
30	1.0029	85	1.0152
35	1.0043	90	1.0161
40	1.0056	95	1.0169
45	1.0069	100	1.0177
50	1.0082		

Source: Busse, J., "Liquid-In-Glass Thermometers," in *Temperature, Its Measurement and Control in Science and Industry,* vol. 1, American Institute of Physics, Reinhold, New York, 1941, p. 246.

Finally, each thermometer comes with a certificate giving small scale corrections.

As an example, consider the following observations made with a Beckman thermometer:

$$\text{setting} = 25°C \qquad \text{upper reading} = 5.127°C$$

$$\text{stem temperature} = 24°C \qquad \text{lower reading} = 2.058°C$$

First, the correction from the calibration certificate:

	Lower Reading	Upper Reading
Observed	2.058°C	5.127°C
Certificate correction	+0.005°C	−0.008°C
Corrected readings	2.063°C	5.119°C

Thus,

$$\text{corrected difference} = 5.119 - 2.063 = 3.056$$

$$\text{setting factor correction} = 3.056 \times 1.0015 = +0.005$$

$$\text{emergent stem correction} = +0.004$$

$$\text{final corrected difference} = 3.065$$

RESISTANCE THERMOMETRY

The resistance of metals increases slowly with temperature, and the resistance of semiconductors (thermistors) decreases rapidly with temperature. Resistance thermometry is based on these generalizations. Platinum is nearly always the metal of choice for resistance thermometers, but nickel and copper are also used. Thermistors are nonstoichiometric mixtures of transition element oxides.

Experimental Establishment of Fixed-Point Temperatures

All the IPTS fixed points are equilibrium temperatures between two or three phases: solid, liquid, and gas. Most of the fixed points are established by the equilibrium temperature between the two condensed phases, solid and liquid. A few are determined by the equilibrium temperature between the liquid and gas phases, and three are fixed by the equilibrium temperature between three phases at the triple point. Since most physical chemistry measurements are made between, say, 0 and 1000°C, let us examine this region in detail.

The Triple Point and Normal Freezing Point. Of particular importance are the triple point and normal freezing point of water, shown in Figure 13–5. The triple point of water is the point at which solid, liquid, and gaseous water are in thermodynamic equilibrium. The vapor pressure of water at this point is 4.58 torr. The temperature at the triple point is defined (IPTS-68) to be exactly 273.16 K. The Celsius temperature at the triple point is 0.01°C, since the relationship between the Kelvin tem-

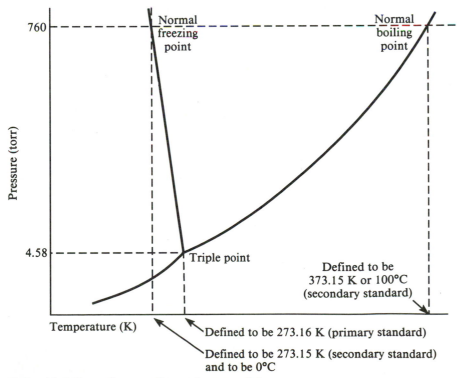

Figure 13–5 Phase diagram of water (not to scale).

perature and Celsius temperature is also established by international agreement:

$$t(°C) = T(K) - 273.15$$

The IPTS-68 normal (at $P = 760$ torr) freezing point of water is defined by international agreement to be 0.01 K lower than the triple point. Thus the IPTS-68 normal freezing point of water is exactly 0°C, or 273.15 K.

You may find it strange that an international body can define temperatures such as the fixed points in Tables 13–1 and 13–2 or the normal freezing point of water. After all, the melting point of a substance must have some natural, physically correct temperature. The word "define," as used in IPTS, really means that a critically evaluated numerical value, representative of the careful, repeated experimental measurements of many workers, has been accepted by international agreement.

The physical construction of cells for establishing the normal freezing point and the triple point of water is shown in Figure 13–6. It is quite simple to establish a temperature of 273.15 K. Just get a 500-ml Dewar flask, fill it with cracked ice from the ice machine down the hall, add water (previously cooled to 0°C), and let the system come to thermal equilibrium for a few minutes. The error caused by the impurities in tap

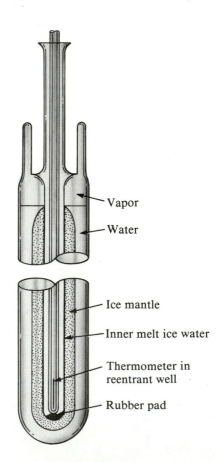

Figure 13–6 Triple point cell. (Adapted from Benedict, 1984.)

water is less than 0.01°C, which is negligible for most laboratory work. It is very important, however, to maintain solid ice to the bottom of the Dewar flask. As ice melts, water at a temperature of 4°C, the temperature at which water is most dense, sinks to the bottom of the flask. Because the resistance thermometer being calibrated is probably located at the bottom of the flask, this can cause an intolerably large error. A glass siphon with a short piece of rubber tubing to which a pinch clamp is attached provides a simple remedy for this problem. Periodically open the pinch clamp to siphon out a little water from the bottom of the flask. When the water level at the top drops just a little below the level of the ice, enough has been removed. Occasionally add a little ice to keep the Dewar flask full to the top.

A triple point cell is more difficult to construct, but nearly as easy to use. The cell shown in Figure 13–6 contains highly purified, gas-free water that is vacuum distilled into the cell. The cell is then sealed under vacuum, so that it contains nothing but liquid and gaseous water. Before it is filled, the glass is cleaned with concentrated sulfuric acid, distilled water, and finally steam.

To use the cell, pour powdered dry ice into the cell well, which causes a mantle of ice 3 to 10 mm thick to freeze on the outside of the well. The inside of the cell now contains the triple point phases: solid, liquid, and gaseous water. Then place the entire cell in a Dewar flask containing a mixture of flaked ice and water; allow the mixture to fill the well to provide good thermal contact between the cell wall and the sensor being calibrated.

Metal Melting Points as Fixed Points. The melting points of zinc (Preston-Thomas, 1955) and tin (Furukawa et al., 1972) are particularly convenient for the calibration of resistance thermometers and thermocouples, not only in national standards laboratories but also in university laboratories. In addition to the sensor being calibrated, the equipment required consists of a furnace, a cylindrical crucible for containing the metal, and a protection tube. A metal block surrounds the crucible to ensure a uniform temperature. The top of the metal is covered with graphite powder to inhibit oxidation of the melt. This arrangement permits a reproducibility of about 0.01°C.

Calibration of Resistance Thermometers

Because the dependence of resistance upon temperature of a platinum resistance thermometer is very reproducible and stable, platinum is the metal of choice for the most exacting work. We have seen in this chapter that a thermometer is just a device for interpolating between fixed points on a temperature scale. Until IPTS-68, the Callendar equation was used to interpolate temperatures between the ice point and the antimony point with a platinum resistance thermometer (Weber, 1950):

$$t = \frac{R_t - R_0}{R_{100} - R_0} \, 100 + \delta\left(\frac{t}{100} - 1\right)\frac{t}{100} \qquad \text{(13-11)}$$

Since this equation is a quadratic in t(°C), three constants, R_t, R_0, and δ must be determined, so three temperatures must be measured: the triple point of water, the boiling point of water, and one other, usually the boiling point of sulfur or the melting point of zinc. The resistance is measured and the temperature of interest is extracted from Equation 13–11 either by solving the quadratic for t or by iteration. In either case, a programmed calculator or a computer is convenient.

However, if you can tolerate a deviation between your calibration and IPTS-68 of 0.045° or less (Benedict, 1984), then you can use for calibration three condensed-phase equilibria: the normal freezing points of water, tin, and zinc. Boiling sulfur is messy; in addition, its boiling point and that of water are very pressure dependent. This is not true of condensed-phase equilibria; their temperatures are essentially independent of pressure.

Furthermore, an equivalent form of the Callendar equation is

$$R_t = R_0 + bt + ct^2 \qquad \text{(13-12)}$$

where R_0 is the resistance at 0°C and the constants c and b are given by (Benedict, 1984)

$$c = \frac{R_0(t_{Zn} - t_{Sn}) - R_{Sn}t_{Zn} + R_{Zn}t_{Sn}}{t_{Sn}t_{Zn}(t_{Zn} - t_{Sn})} \qquad \text{(13-13)}$$

$$b = \frac{R_{Zn} - R_0 - ct_{Zn}^2}{t_{Zn}} \qquad \text{(13-14)}$$

where t_{Sn} is the temperature at the tin point, t_{Zn} is the temperature at the zinc point, R_{Sn} is the measured resistance at the tin point, R_{Zn} is the measured resistance at the zinc point, and R_0 is the measured resistance at the ice point. As an example, consider the observations in Table 13–6; substituting −5.539 in Equations 13–13 and 13–14 results in the following three constants for Equation 13–11:

$$R_0 = 25.56600$$

$$b = 0.1019045$$

$$c = -1.502488 \times 10^{-5}$$

TABLE 13–6

Resistance Thermometer Calibration

Measured Resistance R (ohm)	Normal Freezing Point t (°C)	Condensed-Phase Equilibria
25.56600	0.00	$H_2O(s) = H_2O(\ell)$
48.39142	231.9681	$Sn(s) = Sn(\ell)$
65.6700	419.58	$Zn(s) = Zn(\ell)$

For careful work, each of the R, t pairs listed in Table 13–6 is determined independently several times and an average value is calculated.

Measurement of Resistance of Resistance Thermometers

Although a potentiometer and standard resistor can be used to calibrate a platinum resistance thermometer (Weber, 1950), it is necessary to use a special Wheatstone bridge for accurate work. The main problem is the indeterminant and variable resistance of the leads between the platinum resistance thermometer and the bridge.

Figure 13–7a shows an ideal resistance thermometer with two leads. The leads, however, have their own resistance, labeled C and T in Figure 13–7b, the electrical equivalent of Figure 13–7a. With a four-lead (Fig. 13–7c) platinum thermometer (Smith, 1912) and a Mueller bridge (Mueller, 1941), as shown in Figure 13–8a and b, the effect of the lead resistances can be eliminated by making two observations at each temperature. It is conventional to label the four lead wires, C, c, T, and t. When the bridge is balanced, $R_1 = R_2$ in Figure 13–8a and b. However, a single switch in a Mueller bridge permits connecting the four lead wires in two configurations, shown in the two parts of the figure. When no current flows through the galvanometer G, the bridge is balanced and the usual Wheatstone bridge conditions for balance hold:

$$\text{In Figure 13–8a:} \qquad R_a + C = R_{Pt} + T \qquad (13\text{–}15)$$

$$\text{In Figure 13–8b:} \qquad R_b + T = R_{Pt} + C \qquad (13\text{–}16)$$

$$\text{Addition gives:} \qquad R_{Pt} = \frac{R_a + R_b}{2} \qquad (13\text{–}17)$$

(a) R_{Pt}

(b) R_{Pt}

(c) R_{Pt}

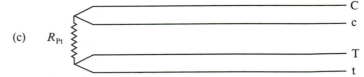

Figure 13–7 Lead resistance in resistance thermometers.
   (a) Two wires connected to a resistor R_{Pt}.
   (b) The electrical equivalent of (a) showing the lead resistances C and T.
   (c) The four-wire connection to a Mueller bridge. The resistance to be measured is R_{Pt}. Two leads connect to the same side of R_{Pt} having lead resistances C and c. On the other side of R_{Pt} two more leads connect having lead resistances T and t.

(a)   (b)

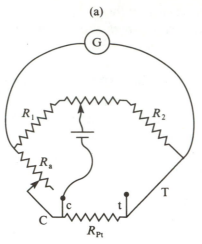
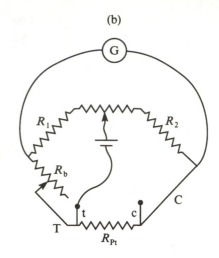

Figure 13–8 Mueller bridge circuit.

In Equations 13–15 and 13–16, C and T represent the resistances of the leads labeled C and T. Notice that the resistances of the leads t and c play no roll in the calculation, since they are either disconnected or not in a balancing arm of the bridge. Simple, but ingenious.

A suitable Mueller bridge, matching galvanometer, and 25-ohm platinum resistance thermometer are available commercially[2] that permit the measurement of temperature with an uncertainty of slightly less than 0.01°C.

THERMOCOUPLES

In 1821, Thomas Johann Seebeck discovered that an electric current flows continuously in an electric circuit consisting of two different metals when the two junctions between the metals are maintained at different temperatures (Fig. 13–9). If a potentiometer is inserted into this circuit, an electrical potential arises, the magnitude of which is proportional to the temperature difference of the two junctions. The proportionality is somewhat nonlinear and depends on the metals, but it is highly reproducible. Consequently, a thermocouple is a useful temperature-measuring device. In practice, the temperature of one junction is held constant, normally at 0°C. This is called the **reference junction.**

The choice of metals depends upon a number of factors such as cost, ductility, ease of welding (forming a junction), magnitude of the potential developed, resistance to physical abuse, and chemical resistance under oxidizing conditions (in the presence of oxygen, air, etc.) or reducing conditions (in the presence of hydrogen, hydrocarbons, etc.) Ap-

[2] Leeds Northrup, North Wales, Pennsylvania.

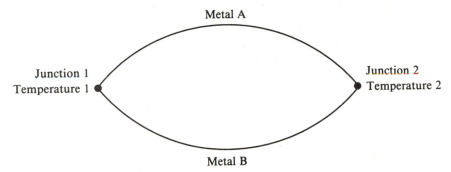

Figure 13–9 Simplified thermocouple.

proximate electromotive force and temperature relationships for the four most commonly used thermocouples are given in Table 13–7.

In practice, the usual setup for measuring temperatures includes more than one thermocouple, an equivalent number of reference junctions, copper extension wires, and a selector switch between these components and a suitable potentiometer. A typical arrangement is shown in Figure 13–10, where two Chromel–Alumel thermocouples are to be

TABLE 13–7

Relationship Between Temperature and Electromotive Force in Common Thermocouples

Temperature (°C)	Electromotive Force (mV)			
	Platinum to Platinum/ 10% Rhodium	Chromel to Alumel[a]	Iron to Constantan	Copper to Constantan
−200		−5.75	−8.27	−5.539
−100		−3.49	−4.82	−3.349
0	0.000	0.00	0.00	0.00
100	0.643	4.10	5.40	4.276
200	1.436	8.13	10.99	9.285
300	2.315	12.21	16.56	14.859
400	3.250	16.39	22.07	20.865
500	4.219	20.64	27.58	
600	5.222	24.90	33.27	
700	6.260	29.14	39.30	
800	7.330	33.31	45.72	
900	8.434	37.36	52.29	
1000	9.569	41.31	58.22	
1100	10.736	45.14		
1200	11.924	48.85		
1300	13.120	52.41		
1400	14.312	55.81		
1500	15.498			
1600	16.674			
1700	17.841			

Source: Roeser, W. F., "Thermometric Thermometry," in *Temperature, Its Measurement and Control in Science and Industry,* Reinhold, New York, vol. 1, 1941.
[a] The words "Chromel" and "Alumel" are trade names of the Hoskins Manufacturing Company.

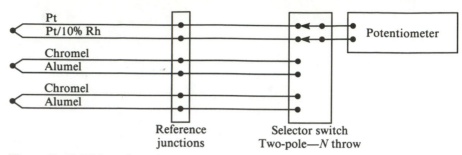

Figure 13–10 Wiring of multiple thermocouples.

calibrated against a standard platinum–platinum/10% rhodium thermocouple.

The construction of a typical thermocouple is shown in Figure 13–11. The wire is 0.3 to 0.5 mm in diameter. If the metals are platinum and platinum/10% rhodium, then the ends should not be twisted, but made into simple butt welded together without flux by means of a carbon arc, resulting in a small, smooth bead. Other metals and alloys (Chromel, Alumel, copper, iron, and constantan) should be twisted together in a few turns. All but one and one-half turns are snipped off; the end is dipped in paste flux and welded together by heating with an ordinary glass-blowing torch until the end of the twisted pair melts, forming a smooth, round bead.

Calibration of Thermocouples

Depending upon the accuracy required, two methods of calibrating thermocouples are in general use. For the most accurate work, calibration at fixed points, already described for platinum resistance thermometers, is required. For most laboratory work, however, it is adequate to calibrate a thermocouple by comparison with another, previously calibrated temperature sensor, such as a platinum resistance thermometer, a platinum–platinum/10% rhodium thermocouple, or a base metal thermocouple. In addition, tables and equations are available for most common ther-

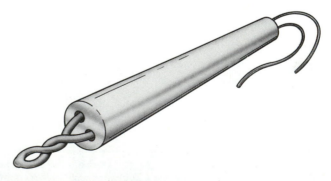

Figure 13–11 Thermocouple construction.

TABLE 13–8

Thermocouple Table Errors

Thermocouple Type	Temperature Range (°C)	Error Limits
Iron–constantan	0–277	±2 K
	277–760	±0.75%
Chromel–Alumel	0–277	±2 K
	277–1260	±0.75%
Pt–Pt/10% Rh	0–538	±1.4 K
	538–1482	±0.25%
Copper–constantan	−101–−59	±2%
	−59–93	±0.8 K
	93–371	±0.75%

mocouples that permit their use without any calibration at all, if the error is tolerable; this is possible because the electromotive force of a homogeneous thermocouple of a definite composition is a particular physical property of the junction, independent of its size and shape. It is an intensive property of the system. Nevertheless, junction strain, metal fatigue, and minor impurities do cause small errors, so the use of tables without calibration is a less accurate method; agreement with the tables depends on how well the wire manufacturer has exercised quality control and how carefully you construct your thermocouple. These errors are summarized in Table 13–8.

The electromotive force developed by a thermocouple at the freezing point of a pure substance is constant and reproducible with the exercise of some precautionary measures. The thermocouple must be protected from contamination by a thermocouple protection tube. Closed-end Pyrex tubes are satisfactory at temperatures below 750 K. At higher temperatures, thermocouple protection tubes of porcelain[3] or Alundum[4] are necessary. Before thermocouple protection tubes are used for the first time, they should be heated to the temperature at which they will be used in order to clean them of any contamination. Most thermocouples function most satisfactorily in an oxidizing atmosphere, but can be damaged in a reducing atmosphere.

The thermocouple must be immersed sufficiently deeply into the molten metal so that the junction is not cooled by heat flow along the thermocouple wire. This can be determined experimentally by raising and lowering the thermocouple about 1 cm from its normal position in the melt. If the electromotive force does not change, the depth of immersion is satisfactory. The National Bureau of Standards used 10-cm immersion, but 5 cm is probably satisfactory.

[3] Coors Porcelain Company, 600 9th Street, Golden, Colorado 80401.
[4] Norton Company, Worcester, Massachusetts 01606.

Figure 13–12 Thermocouple calibration furnace.

    A. End piece of Fiberfrax board, 1/2 inch thick, 10 inches in diameter, with a 2-inch-diameter hole.

    B. Core holder of Fiberfrax board, 2 inches thick, 5 inches in diameter, with a hole large enough to receive the core coated with Alundum cement.

    C. Stainless-steel tube 11 inches in diameter, 24 inches long, with a 1/2-inch-wide flange to mount end pieces. Spot welded.

    D. Alundum core with a 2-inch bore, 24 inches in length, 6 turns per inch.

    E. Core with wire wound on grooves and covered with Alundum cement.

It is advisable to protect the molten metals from oxidation by floating a layer of powdered graphite on their surface. Copper oxide forms a eutectic mixture with copper metal, which melts about 20° lower than pure copper. Silver, aluminum, and antimony should also be protected from air with powdered graphite, since small amounts of impurities can cause intolerable temperature calibration errors. The National Bureau of Standards recommends graphite crucibles because they are inert, refractory, and easily machined to the proper shape. They also have used porcelain, silica, Alundum (aluminum oxide), and Pyrex glass for crucibles (Roeser and Lonberger, 1958). However, aluminum attacks any material containing silica, so it should be contained in a graphite crucible. In all cases you should check for the possibility of a reaction between the melt and the crucible.

Figure 13–12 shows a furnace useful to about 1000°C for calibrating thermometric sensors. A wide variety of Alundum cores with grooves for the wire winding are available from Norton Company,[5] which also man-

[5] Norton Company, Worcester, Massachusetts 01606.

ufactures Alundum cement. Nichrome or Kanthal[6] is a suitable wire. After the wire is wound on the grooves, Alundum cement is mixed with enough water to form a thick paste, which is troweled over the winding and smoothed out. The space between the core and the outer metal container is filled with powdered diatomaceous earth, which is available from swimming pool suppliers as a filtering medium. Fiberfrax,[7] a mixture of alumina and silica, is available in a variety of forms. As a bulk fiber it is useful to stuff in around the end of the core to prevent the diatomaceous earth from leaking out. Diatomaceous earth has the consistency of flour. As a high-density pressed board, Fiberfrax can be used as a replacement for Transite, a mixture of concrete and asbestos. In board form Fiberfrax is available in thicknesses from 1/4 inch to 2 inches and may be used to 1538°C.

To prevent the wire leads from shorting, they should be insulated with porcelain beads.[8] The wire leads are then attached to ordinary screw-anchored electrical feed-through connectors. The furnace will operate continuously at 1000°C or less with an ordinary Variac variable transformer and temperature controller. Commercially available furnaces operate comparably but cost 10 to 50 times as much.

Calibration by Comparison Methods

For most purposes, it is adequate to calibrate a thermocouple by comparison, usually with a platinum–platinum/10% rhodium thermocouple that has been calibrated at fixed points. It may be convenient to have the National Bureau of Standards do the fixed-point calibration for you; this service is provided for a reasonable fee. The major source of error in the comparison calibration lies not in the reference thermocouple, but in bringing the two thermocouples to thermal equilibrium with each other. At low temperatures this is not a severe problem, because immersion of the two thermocouples (in their protection tube) in a well-stirred liquid bath ensures that the two thermocouples are at the same temperature. Water is a suitable bath fluid between 0 and 100°C and oil can be used up to about 300°C.

At very low temperatures, mixtures or organic liquids serve well as bath fluids. A mixture of carbon tetrachloride and chloroform freezes below −75°C. A five-component mixture containing 14.5% chloroform, 25.3% methylene chloride, 33.4% ethyl bromide, 10.4% *trans*-dichloroethylene, and 16.4% trichloroethylene freezes below −140°C.

Interpolation Methods for Thermocouples

After a thermocouple has been calibrated at a number of temperatures—either at fixed melting points or by comparison with a standard ther-

---

[6] The Kanthal Corporation, Bethel, Connecticut 06801.

[7] The Carborundum Company, Insulation Division, P.O. Box 808, Niagara Falls, New York 14302.

[8] Coors Porcelain Company, 600 9th Street, Golden, Colorado 80401.

mocouple—it is still necessary to interpolate between these points. The choice of interpolation method depends upon the relative importance of convenience and accuracy. The electromotive force of thermocouples tends to follow an equation of the form

$$E = a + bt + ct^2 + dt^3 \tag{13–18}$$

As already mentioned, the IPTS-68 temperatures are defined between 630 and 1064°C by this equation ($d = 0$) when the calibration is carried out at the melting points of antimony, silver, and gold. The addition of one more point, such as zinc, allows Equation 13–18 to be used from 400 to 100°C with an agreement to IPTS-68 of less than 0.5°C. Substitution of aluminum for antimony and copper for gold introduces an additional uncertainty of less than 0.1°C. If a standard calibrated reference thermocouple is available, a comparison calibration should be carried out at three or four widely separated temperatures.

Copper–constantan thermocouples are often used to measure temperatures below room temperature. Between −190° and 0°C, the electromotive force of a copper–constantan thermocouple follows Equation 13–18. The calibration is conveniently carried out in a stirred bath. From 0 to 100°C, b and d equal zero and c equals 0.04, so that calibration at a single temperature (100°C) is adequate. From 100°C to the highest useful operating temperature of about 300°C, Equation 13–18 is also followed ($d = 0$), and three widely separated temperatures should be used.

RADIATION PYROMETRY

The measurement of temperature by radiation pyrometry is based on the fact that the color of radiation emitted by an incandescent material depends on its temperature. The actual spectral radiance from such an object is given by Planck's radiation equation—which, one could argue, Japanese sword makers anticipated by 1200 years. Unlike copper and bronze, steel must be tempered by being heated to just the correct temperature and annealed by being plunged into water. "The temperature for this final moment has to be judged precisely, and . . . 'it is the practice to watch the sword being heated until it glows to the colour of the morning sun.' "[9]

Theory of Radiation Pyrometry

Spectral radiance J is defined as the energy radiated by a blackbody in a particular direction per unit time, per unit wavelength, per unit projected area of the body, and per unit solid angle. Its units are J m^{-3} sec^{-1}, and

[9] Bronowski, J., *The Ascent of Man*, Little, Brown and Company, Boston, 1973, p. 132.

the wavelength distribution of its values is given by the Planck radiation equation

$$J = \frac{c_1 \lambda^{-5}}{\exp(c_2/\lambda T) - 1} \tag{13-19}$$

where (Cohen and Taylor, 1986) the first radiation constant

$$c_1 = 2\pi c^2 h = 3.7417749(22) \times 10^{-16} \text{ J m}^2 \text{ sec}^{-1}$$

and the second radiation constant

$$c_2 = hc\backslash k = 0.01438769(12) \text{ m K}$$

Experimental Radiation Pyrometry

If you have ever sat by a campfire after night has fallen and poked at the glowing coals, you will readily understand the principle of the disappearing-filament pyrometer. You can tell which coals are the hottest by comparing their color and brightness with other glowing coals.

The disappearing-filament optical pyrometer measures temperature by comparing the color and brightness of an object at an unknown temperature with the color and brightness of a tiny internal tungsten filament, which is heated electrically to incandescence. The instrument is calibrated so that the variable filament current corresponds to a temperature scale engraved on the instrument. The heated filament is the central component of a disappearing-filament optical pyrometer. The supporting components are shown in Figure 13-13.

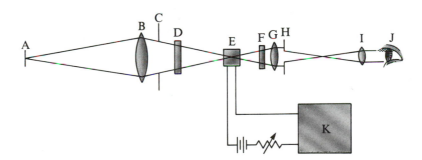

Figure 13–13 Schematic diagram of an optical pyrometer. (Adapted from Kostkowski and Lee, 1962.)

    A. Source
    B. Objective lens
    C. Objective aperture
    D. Absorption filter (used for temperatures above 1300°C)
    E. Pyrometer lamp
    F. Red filter
    G. Microscope objective lens
    H. Microscope aperture stop
    I. Microscope ocular
    J. Eye
    K. Current measuring instrument

The optical system is essentially a low-power telescope that allows the user to see an image of the heated filament superimposed over an image of a small area of the source whose temperature is being measured (Fig. 13–14). When the filament is at a higher temperature than the source, it appears brighter than the background (source), and when it is at a lower temperature than the source, it appears darker than the background. When the filament current has been adjusted so that its temperature equals the source temperature, its image fades to the source background color and brightness so that it disappears. The temperature of the source is then read off a calibrated scale.

When the filament disappears, the spectral radiance from the source and filament are equal, so that the following equality holds:

$$\frac{J}{J_x} = \frac{\exp(c_2/\lambda T_x)}{\exp(c_2/\lambda T)} \qquad (13\text{--}20)$$

Blackbodies. A blackbody is a material object that completely absorbs incident radiation. Lampblack is a substance that behaves very much like a blackbody since it absorbs about 99% of the radiation falling upon it. The emissivity of a real object is the ratio of the real radiance to that of a blackbody at the same temperature:

$$\epsilon = \frac{J}{J_b}$$

A good absorber is a also a good radiator, which can be demonstrated by heating to red heat a white porcelain crucible with dark lettering on it. When incandescent, the letters appear brighter than the background.

Emissivity. The emissivity ϵ of a blackbody is unity, while the emissivity of real material ranges between zero and one (Table 13–9). When the emissivity of a material is known, the correct temperature of an object can be calculated from the apparent temperature T_a (Benedict, 1984):

$$\frac{1}{T} - \frac{1}{T_a} = \frac{\lambda}{c_2} \ln \epsilon \qquad (13\text{--}21)$$

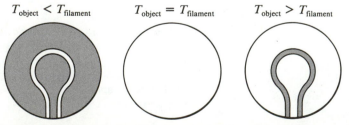

$T_{\text{object}} < T_{\text{filament}}$ $T_{\text{object}} = T_{\text{filament}}$ $T_{\text{object}} > T_{\text{filament}}$

Figure 13–14 Telescopic image of optical pyrometer filament and source.

TABLE 13–9

Emissivities of Some Metals and Oxides

Material	Emissivity ϵ	Material	Emissivity ϵ
Aluminum oxide	0.30	Molybdenum	0.37
Carbon	0.95	Nickel	0.36
Chromium	0.34	Nickel oxide	0.70
Chromium oxide	0.70	Tantalum	0.50
Cobalt	0.36	Tantalum oxide	0.50
Cobalt oxide	0.75	Tungsten	0.43
Copper	0.10	Uranium	0.54
Copper oxide	0.70	Uranium oxide	0.30
Gold	0.55	Tungsten	0.43
Iron	0.35	Zirconium	0.32
Iron oxide	0.70	Zirconium oxide	0.40

Source: Metal data from Weber, R. L., *Heat and Temperature Measurement,* Prentice-Hall, New York, 1950.

Suppose that the apparent temperature T_a of an Alundum crucible is 1300 K. The emissivity of Alundum is that of aluminum oxide, 0.35, and the effective wavelength is 666 nm:

$$\frac{1}{T} - \frac{1}{1300} = \frac{6.66 \times 10^{-7}}{0.014388} \ln 0.35 \qquad (13\text{–}22)$$

$$T = 1388 \text{ K}$$

The emissivities of a number of substances are listed in Table 13–9. Emissivities are used to correct the measured temperature when the radiation is emitted from the *surface* of the object whose temperature is being measured. If the emissivity of an object is unknown, it is possible

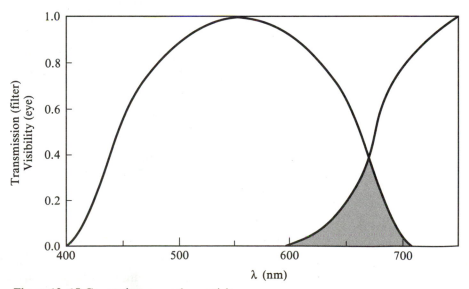

Figure 13–15 Generating monochromaticity.

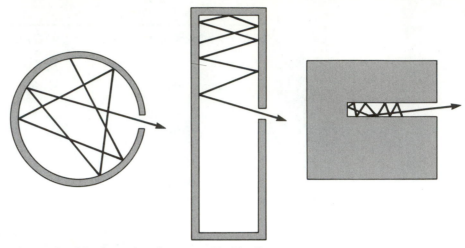

Figure 13–16 Isothermal enclosures as blackbodies.

to approximate blackbody radiation by drilling a hole in the object and sighting the optical pyrometer into the hole. A hole five times as deep as its diameter approximates a blackbody.

 The presence of a red filter (Corning Red Glass) in the optical path achieves the effect of nearly monochromatic radiation, as shown in Figure 13–15. The range visible to the human eye drops off rapidly beyond about 600 nm; at the same time the red glass transmission increases beyond about 600 nm so that the net visibility to the eye peaks at about 666 nm, where the two curves intersect.

 It can be argued that there is no such thing as a blackbody, since a blackbody is really a cavity in which radiation is emitted and absorbed at equal rates. Thus, a "blackbody" is not a material object at all, but is a region of space—an isothermal enclosure or cavity. The English word "blackbody" translates into the German word *Hohlraum,* which means "cavity." (There is no literal translation to, say, something like *Schwarzkorper.*) Fortunately, it is possible to approximate blackbody conditions experimentally. Figure 13–16 shows some experimentally attainable configurations in which radiation is approximately at equilibrium.

THERMISTORS

The resistance of metals is very low and increases gradually with an increase in temperature. The resistance of nonmetals, in contrast, is relatively high and decreases gradually with an increase in temperature (a negative coefficient of resistivity). At room temperature the specific resistance of a typical metal is around 10^{-6} ohm cm, while the specific resistance of nonmetals is around 10^{14} ohm cm. Semiconductors have an intermediate specific resistance ranging from about 10^{-2} to 10^{9} ohm cm. Semiconductors include such elements as silicon, germanium, and tellurium and compounds between Group III and Group V elements (e.g., GaAs). Certain binary and ternary transition metal oxides fall into the semiconductor class (e.g., Cu_2O and $NiMn_2O_4$).

Composition and Fabrication

A **thermistor** (*therm*ally sensitive re*sistor*) is a semiconductor characterized by a high negative coefficient of resistivity. The composition is often nonstoichiometric—for example, $Mi_{2-x}Mn_{2+x}O_4$, where x ranges from 0 to about 0.7 (Zurburchen and Case, 1982). Small amounts of other transition element oxides are sometimes added to achieve the desired characteristics.

Thermistors are fabricated by thorough mixing of finely divided oxides and additives. These ingredients are then sintered at high pressure and temperature into the desired form, usually a tiny bead (1- to 2-mm diameter) or a thin disk. Copper wire leads are attached either in the sintering process or subsequently, as shown in Figure 13–17.

Obviously, the electrical characteristics of a thermistor are sensitive to the composition, particle size, and sintering pressure and temperature used in the manufacturing process. In recent years, however, carefully controlled manufacturing methods have led to the development of thermistors of exceptionally reproducible electrical and thermal properties. For example, the interchangeability tolerance of a Yellow Springs Instrument (YSI) 44000 series thermistor is less than $\pm 0.2°C$, and the stability, as measured by the drift per 10 months service at $100°C$, is also less than $\pm 0.2°C$ (LaMers et al., 1982).

Relationship Between Resistance and Temperature

The resistance of a thermistor appears to decrease exponentially with an increase in temperature (Fig. 13–18), suggesting that a suitable interpolation equation for a thermistor is of the form

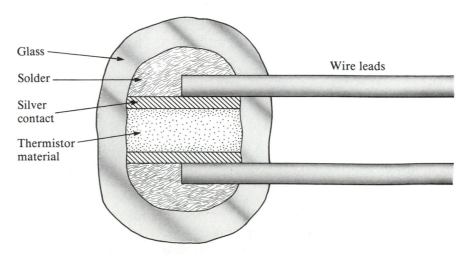

Figure 13–17 Cross section of a bead thermistor.

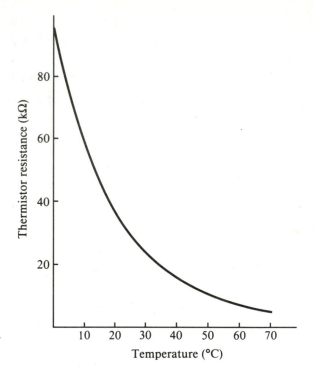

Figure 13–18 Resistance-versus-temperature curve for a thermistor.

$$R = Ae^{B/T}$$

or

$$\ln R = \frac{B}{T} + \ln A \qquad \text{(13–23)}$$

Typical values of A range from -3 to -6; values of B range from 3000 to 5000 K (Benedict, 1984, p. 57). Between -20 and $+120°C$, such a two-parameter equation produces a deviation of about 1 K. For a better approximation, the exponential equation may be expanded empirically:

$$\ln R = \frac{B}{T} + \frac{C}{T^2} + \frac{D}{T^3} + \cdots \qquad \text{(13–24)}$$

The Steinhart and Hart equation (YSI, 1986), a three-parameter interpolating equation, is usually written explicitly in $1/t$ rather than R:

$$1/t = a + b(\ln R) + c(\ln R)^3 \qquad \text{(13–25)}$$

Calibration. The three parameters a, b, and c of this interpolating equation are determined by calibrating the thermistor at three reference temperatures at which R is measured. For thermistors, the three usual temperatures are the ice point ($0°C$), the melting point of gallium ($29.7714°C$), and the boiling point of water ($100°C$ at 1 atm). Temperatures interpolated with the Steinhart and Hart equation are accurate to less than $0.01°C$ if widely separated temperatures are used in the calibration.

Calculations. The coefficients in the Steinhart and Hart equation are found by solving simultaneously the three equations defined by the three R, t calibration pairs:

$$1/t_1 = a + b(\ln R_1) + c(\ln R_1)^3$$

$$1/t_2 = a + b(\ln R_2) + c(\ln R_2)^3$$

$$1/t_3 = a + b(\ln R_3) + c(\ln R_3)^3$$

The Steinhart and Hart equation may be transformed so that it is expressed explicitly in R (YSI, 1986):

$$R = \exp\left\{\left[-\frac{\alpha}{2} + \left(\frac{\alpha^2}{4} + \frac{\beta^3}{27}\right)^{1/2}\right]^{1/3} + \left[-\frac{\alpha}{2} - \left(\frac{\alpha^2}{4} + \frac{\beta^3}{27}\right)^{1/2}\right]^{1/3}\right\}$$

(13–26)

where $\alpha = (a - 1/t)/c$ and $\beta = b/c$.

Because thermistors are often used between 0 and 100°C, it is useful to have a reliable third secondary fixed point between these two easily reproduced primary standard temperatures. Two that have recently been suggested are the triple points of gallium, 29.77398 ± 0.00014°C (Magnum, 1982) and succinonitrile, 58.0805 ± 0.0004°C (Glicksman and Voorhees, 1982).

Figure 13–19 is a Pascal program for calibrating a thermistor with the Steinhart and Hart equation. The interactive input consists of three R,t pairs (ohm and °C), and the output lists the three Steinhart and Hart parameters a, b, and c. A table of R,t pairs is also output after interactive input of the table range and increment size (Fig. 13–20).

Advantages and Disadvantage of Thermistors

Compared to thermocouples and resistance thermometers, thermistors are useful over a short temperature range, approximately −80 to +250°C.

(Text continues on page 371.)

TABLE 13–10

Typical Stability for YSI 44005 Thermistors

Operating Temperature (°C)	Thermometric Drift (°C)	
	10 months	**100 months**
0	<0.01	<0.01
25	<0.01	<0.02
100	0.20	<0.32
150	1.50	Not recommended

Source: Yellow Springs Instrument Company, Yellow Springs, Ohio.

```
Program ThermistorRT;
(* This is a program to calibrate thermistors according to      *)
(* to the Steinhart-Hart equation                               *)

uses crt;

CONST K = 273.15;

TYPE Ary3  = ARRAY [1..3] OF REAL;
     Ary20 = ARRAY [1..20] OF REAL;

VAR R,T,TK:  Ary3;
    a,b,c:  Real;
    Tout,Rout:  Ary20;
(************************************************************)

Function Power (X,n:  Real):  Real;
   Begin
      Power:= EXP(n*Ln(X));
   End;  (* Power *)


(************************************************************)

Procedure GetTempAndResistance;
Var I,
    X,Y:  Integer;    (* Screen Coordinates *)
Begin (* GetTempAndResistance *)
   CLRSCR;
   Writeln('Please enter three calibration temperatures (deg C)');
   Writeln(' and three thermistor resistances (ohm)');
   Writeln;
   For I:= 1 to 3 Do
     Begin (* For *)
        Write(      'R(',I,'):  ');
        X:= WhereX; Y:= WhereY;
        Read(R[I]);
        GoToXY(X+10,Y);
        Write('      T(',I,'):  ');
        Readln(T[I]);
     End;  (* For *)
End;  (* GetTimeAndResistance *)

(************************************************************)
```

Figure 13–19 A Pascal program for thermistor calibration and interpolation.

```
Procedure CalculateABC(VAR A,B,C:  Real);
(* This procedure calculates the calibration constants          *)
(* in the Steinhart-Hart Equation                               *)

Var L:  Ary3;
    J:         Integer;
    S,U,V,W,X1,X2,X3:  Real;

Begin (* CalculateABC *)
   FOR J:= 1 TO 3 DO
       Begin (* FOR *)
          TK[J]:= T[J] + K;
          L[J]:= Ln(R[J]);
       End;  (* FOR *)
   S:= L[1] - L[2];
   U:= L[1] - L[3];
   V:= 1/TK[1] - 1/TK[2];
   W:= 1/TK[1] - 1/TK[3];
   X1:= Power(L[1],3);
   X2:= Power(L[2],3);
   X3:= Power (L[3],3);
   c:= (V - S*W/U)/((X1-X2)-S*(X1-X3)/U);
   b:= (V-c*(X1-X2))/S;
   a:= 1/TK[1] - c*X1 - b*L[1];
End;  (* CalculateABC *)

(************************************************************)

Procedure CalcTempTable;

Var I:  Integer;
    T,J,L,N,P,X,Trange,DeltaT:  Real;

Begin (* CalcTempTable *)
   Trange:= TK[3] - TK[1];
   DeltaT:= Trange/7;
   T:= TK[1] -DeltaT;
   For I:= 1 to  10 DO
     Begin  (* For *)
       T:= T + DeltaT;
       J:= a/c - 1/((T)*c);
       L:= SQRT((Power((b/(c*3)),3)) + SQR(J)/4);
       N:= Power((-J/2+L),0.3333333333);
       P:= -Power((J/2+L),0.3333333333);
       X:= N + P;
       Rout[I]:= EXP(X);
       Tout[I]:= T;
     End;  (* For *)
End; (* CalcTempTable *)


(************************************************************)
```

Figure 13-19 (continued)

```
Procedure WriteData;

VAR J:   INTEGER;

BEGIN   (* WriteData *)
   CLRSCR;
   Writeln('The Steinhart & Hart parameters are:');
   Writeln;
   Writeln('a = ',a);
   Writeln('b = ',b);
   Writeln('c = ',c);
   Writeln;
   Writeln('Resistance/ohm    Temperature/deg C');
   Writeln;
   For J:= 1 to 8 Do
       Writeln(Rout[J]:7:2,Tout[J] - k:18:2)
END;   (* WriteData *)

(*****************************************************************)

BEGIN (* M A I N   P R O G R AM *)
   GetTempAndResistance;
   Calculateabc(a,b,c);
   CalcTempTable;
   WriteData;
END.   (* M A I N   P R O G R AM *) 18
```

Figure 13-19 (continued)

```
Please enter three calibration temperatures (deg C)
 and three thermistor resistances (ohm)

R(1):   7355            T(1):   0
R(2):   1200            T(2):   40
R(3):   394.5           T(3):   70


The Steinhart & Hart parameters are:

a =   1.4740799675E-03
b =   2.3704159446E-04
c =   1.0839894582E-07

Resistance/ohm    Temperature/deg C

7355.00                 0.00
4485.18                10.00
2815.79                20.00
1815.67                30.00
1200.00                40.00
 811.36                50.00
 560.27                60.00
 394.50                70.00
```

Figure 13-20 Output of thermistor program.

For applications near room temperature, thermistors are almost as interchangeable as thermocouples or platinum resistance thermometers. Above 150°C, however, thermistors show considerable thermal drift and should be recalibrated frequently, as indicated in Table 13–10. Because they have extraordinarily high negative coefficients of resistance, thermistors are unsurpassed for measuring very small changes in temperature. At 0°C, the negative coefficient of resistance is about 500 ohm/°C.

REFERENCES

1. Benedict, R. P., *Fundamentals of Temperature, Pressure, and Flow Measurements,* 3rd ed., Wiley, New York, 1984.

2. Busse, J., "Liquid-in-Glass Thermometers," in *Temperature, Its Measurement and Control in Science and Industry,* vol. 1, American Institute of Physics, Reinhold, New York, 1941, p. 228.

3. Cohen, E. R., and B. N. Taylor, *J. Res. Natl. Bur. Stand.* 92:85, 1986.

4. Comité International des Poids et Mesures, *Metrologia* 5:35, 1969.

5. Comité International des Poids et Mesures, *Metrologia* 12:7, 1976.

6. Coxon, W. F., *Temperature Measurement and Control,* Macmillan, New York, 1960.

7. Dike, P., *Thermoelectric Thermometry,* Leeds Northrup Company, Philadelphia, 1954.

8. Eggers, D. F., Jr., N. W. Gregory, G. D. Halsey, Jr., and B. S. Rabinovitch, *Physical Chemistry,* Wiley, New York, 1964.

9. Furukawa, G. T., and W. R. Bigge, "Reproducibility of Some Triple Points of Water Cells," in *Temperature, Its Measurement and Control in Science and Industry,* vol. 5, pt. 1., American Institute of Physics, Reinhold, New York, 1982, p. 291.

10. Furukawa, G. T., Riddle, J. L., and Bigge, W. R., "Investigation of Freezing Temperatures of National Bureau of Standards Tin Standards," in *Temperature, Its Measurement and Control in Science and Industry,* vol. 4, pt. 1, American Institute of Physics, Reinhold, New York, 1972, p. 247.

11. Glicksman, M., and P. Voorhees, "The Triple-Point Equilibria of Succinonitrile: Its Assessment as a Temperature Standard," in *Temperature, Its Measurement and Control in Science and Industry,* vol. 5, American Institute of Physics, Reinhold, New York, 1982, p. 321.

12. Hust, J. G., "A Compilation and Historical Review of Temperature Scale Differences," in *Evolution of the International Practical Temperature Scale of 1968,* ASTM STP 565, American Society for Testing and Materials, Philadelphia, 1974, pp. 20–49.

13. Klotz, I. M., *Chemical Thermodynamics,* Prentice-Hall, Englewood Cliffs, New Jersey, 1950.

14. Kostkowski, H. J., and R. D. Lee, "Theory and Methods of Optical Pyrometry," National Bureau of Standards Monograph No. 41, U.S. Government Printing Office, Washington, D.C., 1962.

15. Kubachewski, O., and E. L. Evans, *Metallurgical Thermochemistry,* Wiley, New York, 1956.

16. LaMers, T. H., J. M. Zurburchen, and H. Trolander, "Enhanced Stability in Precision Interchangeable Thermistors," in *Temperature, Its Measurement and Control in Science and Industry,* vol. 5, American Institute of Physics, Reinhold, New York, 1982, pp. 865–873.

17. Lewis, G. N., and M. Randall, revised by K. S. Pitzer and L. Brewer, *Thermodynamics,* McGraw-Hill, New York, 1961.

18. Magnum, B. W., "Triple-Point of Gallium as a Temperature Fixed Point," in *Temperature, Its Measurement and Control in Science and Industry,* vol. 5, American Institute of Physics, Reinhold, New York, 1982, p. 299.

19. *Manual on the Use of Thermocouples in Temperature Measurement,* ASTM STP 470B, American Society for Testing and Materials, Philadelphia, 1981.

20. Middleton, W. E. K., *A History of the Thermometer,* Hopkins Press, Baltimore, 1966.

21. Mueller, E. F., "Precision Resistance Thermometry," in *Temperature, Its Measurement and Control in Science and Industry,* vol. 1, American Institute of Physics, Reinhold, New York, 1941.

22. Preston-Thomas, H., "The Zinc Point as a Thermometric Fixed Point," in *Temperature, Its Measurement and Control in Science and Industry,* vol. 2, American Institute of Physics, Reinhold, New York, 1955, p. 153.

23. Roeser, W. F., and S. T. Lonberger, "Methods of Testing Thermocouples and Thermocouple Materials," National Bureau of Standards Monograph No. 590, U.S. Government Printing Office, Washington, D.C., 1958.

24. Quinn, T. J., *Temperature,* Academic Press, New York, 1983.

25. Riddle, J. L., G. T. Furukawa, and H. H. Plumb, "Platinum Resistance Thermometry," National Bureau of Standards Monograph No. 126, U.S. Government Printing Office, Washington, D.C., 1972.

26. Smith, F. E., *Phil. Mag.,* 24:541, 1912.

27. Sparks, L. L., "Reference Tables for Low Temperature Thermocouples," National Bureau of Standards Monograph No. 124, U.S. Government Printing Office, Washington, D.C., 1974.

28. Stimson, H. H., "Precision Resistance Thermometry and Fixed Points," in *Temperature, Its Measurement and Control in Science and Industry,* vol. 2, American Institute of Physics, Reinhold, New York, 1955, p. 170.

29. Sturtevant, J. M., "Temperature Measurement," in A. Weissberger and B. W. Rossiter, eds., *Physical Methods of Chemistry,* vol. I, pt. V, Wiley-Interscience, New York, 1971, pp. 1–21.

30. Weber, R. L., *Heat and Temperature Measurement,* Prentice-Hall, New York, 1950.

31. Wise, J., "Liquid in Glass Thermometry," National Bureau of Standards Monograph No. 150, U.S. Government Printing Office, Washington, D.C., 1976.

32. "Basic Concepts of Thermistors for Thermometry," Yellow Springs Instrument Company, Yellow Springs, Ohio, 1986, p. 4.

33. Zurburchen, J. M., and D. A. Case, "Aging Phenomena in Nickel-Manganese Oxide Thermistors," in *Temperature, Its Measurement and Control in Science and Industry,* vol. 5, American Institute of Physics, Reinhold, New York, 1982, pp. 889–896.

Instruments

This chapter treats a few of the common instruments found in the physical chemistry laboratory for measuring absorbance, spectra, refractive index, and density. Measurement of some of these physical properties is necessary in several experiments described in Part Three. Specific manufacturer's operating instructions or local guides to equipment should also be consulted *before* using the instrument.

ABSORPTION OF LIGHT

Light—that portion of electromagnetic radiation to which the human eye is sensitive—is a narrow band of wavelengths between about 400 and 800 nm. Figure 14–1 shows the major divisions of the electromagnetic spectrum from 10^2 to 10^{-12} m. When electromagnetic radiation of initial intensity I_0 passes through a non-opaque sample, absorption reduces the intensity to I. At a given wavelength, the reduction in intensity depends on the concentration of the absorbing species and the path length through the sample. Beer's Law (Swinehart, 1962) gives the relationship between these parameters as follows:

$$A = \log \frac{I_0}{I} = \epsilon bc \qquad (14\text{--}1)$$

When the path length b is in centimeters and the concentration c is in moles per liter, the proportionality constant ϵ is the **molar absorptivity,** which then has units of liters per mole per centimeter. The dimensionless quantity A is called the **absorbance,** which is measured by comparing the electrical signals generated by I_0 and I falling on a suitable detector.

Ultraviolet/Visible Spectrophotometry

The conventional spectrophotometer consists of a source, a dispersive element, an exit slit, a sample, and a detector, differing slightly in the

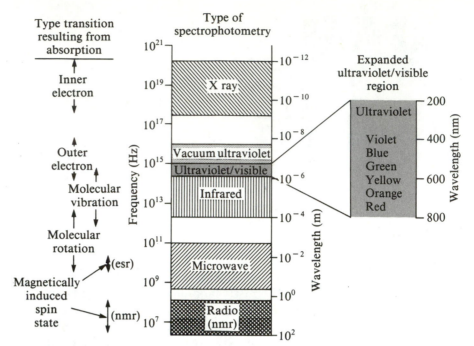

Figure 14–1 The electromagnetic spectrum.

materials of construction depending on whether operation is in the ultraviolet/visible (UV/VIS) range or in the infrared range. In the visible region, the source is an incandescent lamp and the detector is a photomultiplier or similar device for converting visible electromagnetic radiation to an electric current. The electrical output is displayed on a meter, recorder, or cathode ray tube by means of a computer. Dispersion of the polychromatic source radiation is accomplished with either a prism or grating (Fig. 14–2).

The Spectronic 20 shown in Figure 14–3 is one of the world's most widely used spectrophotometers. The operational layout is shown in Figure 14–4. The Spectronic 20 is a double-beam instrument that has a range from 340 to 625 nm and a spectral slitwidth of 20 nm. A reference phototube measures I_0, a measuring phototube measures I, and the sample absorbance is displayed on a meter. The instrument has only three controls:

1. Wavelength to set wavelength (top right).
2. Zero control to set meter to 0 at 0% transmission (bottom left).
3. Light control to set meter to 100 at 100% transmission (bottom right).

To use the instrument, turn it on with the zero control, which also functions as an on/off switch. Permit the instrument to warm up for 20 to 30 minutes. Select the desired wavelength. With the cuvette holder (top left) empty, zero the meter with the zero control.

Insert into the cuvette holder a cuvette three-fourths full of the solvent blank. Be sure to wipe the outside clean and dry with lens tissue or

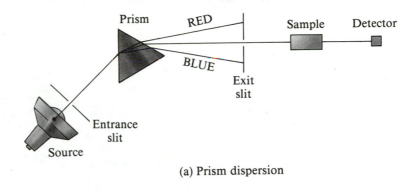

(a) Prism dispersion

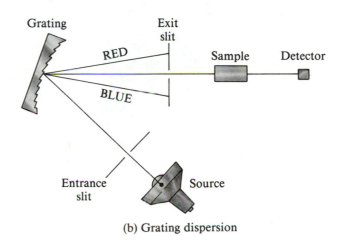

(b) Grating dispersion

Figure 14–2 Conventional spectrophotometers.

Kimwipes. Set the meter to 100% transmission with the light control. Replace the blank with the sample to be measured and read the absorbance from the meter.

The monochrometer of a conventional spectrophotometer consists of the dispersive element (prism or grating) plus the exit slit. The narrower the exit slit, the more monochromatic the radiation. However, as resolution is gained, intensity is lost. In the UV/VIS region, the diode array spectrophotometer overcomes the slit problem by replacing the slit with a diode array (Fig. 14–5). A linear array of photodiodes, which can be made very small, permits simultaneous sampling of the entire spectrum arising after white light impinges on the sample. With no slit, the intensity and corresponding signal-to-noise ratio are very high.

The Perkin-Elmer 3840 UV/VIS lambda array spectrophotometer shown in Figure 14–6 is a computer-controlled, single-beam instrument. The background, run with a blank cell, is stored in the computer memory and subtracted from the spectrum run on a sample. The sample spectrum is digitized and may be transferred from the computer memory to a floppy disk for storage and later manipulation (editing, labeling, ex-

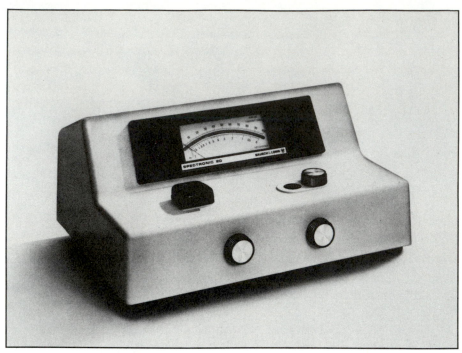

Figure 14–3 The Spectronic 20 spectrophotometer. (Courtesy of Milton Roy Company, Rochester, New York.)

panding, and printing). Figure 14–7 shows the optical layout of the Perkin-Elmer 3840. This instrument has a range from 190 to 900 nm. In survey mode its resolution is 1.5 nm, but in high-performance mode its resolution is 0.25 nm.

The ability of a photodiode array spectrophotometer to produce a usable spectrum at very low intensities is demonstrated in Figure 14–8. The spectrum was taken of gaseous iodine ($P < 1$ torr) in equilibrium

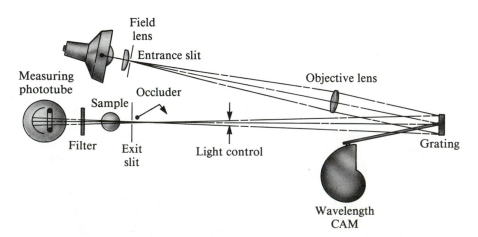

Figure 14–4 Operational layout of the Spectronic 20. (Courtesy of Milton Roy Company, Rochester, New York.)

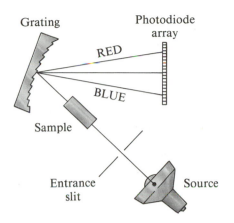

Grating

Photodiode array

RED

BLUE

Sample

Entrance slit

Source

Figure 14–5 Photodiode array spectrophotometer.

with the solid at room temperature in a 1-cm cell. The maximum absorbance is 0.0377, yet the spectrum is quite clear.

Glass cells for UV/VIS spectrophotometers are not usable in the UV region. For UV radiation, cells are made of quartz (silica). Whether glass or quartz, cells should be regarded as optical instruments and handled carefully, touched only on the frosted sides. The cells should not be allowed to knock against each other, but should be cradled in soft, non-scratching tissue when not in use.

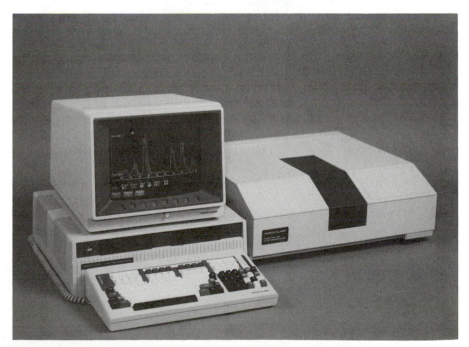

Figure 14–6 Perkin-Elmer 3840 UV/VIS lambda array spectrophotometer with the Perkin-Elmer 7300 computer. (Courtesy of Perkin-Elmer Corporation, Norwalk, Connecticut.)

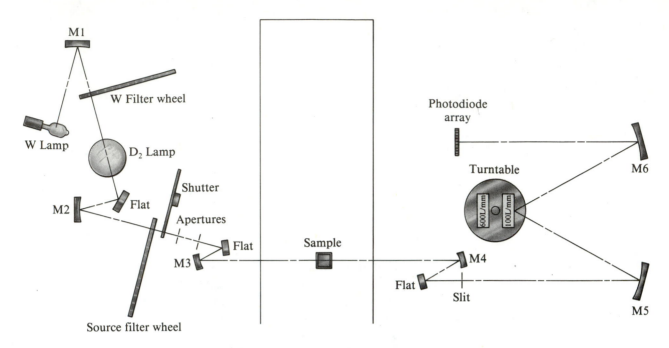

Figure 14–7 Optical layout of the Perkin-Elmer 3840 UV/VIS lambda array spectrophotometer. (Courtesy of Perkin-Elmer Corporation, Norwalk, Connecticut.)

The cleaning of cells is important in their use and critical in their maintenance. Never use alkali, abrasives, or hot concentrated acids. Mild liquid detergents that are free of particulate matter may be used. For stubborn deposits or films, use a solution of 50% 3 M HCl and 50%

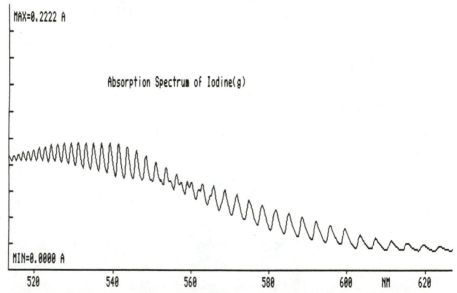

Figure 14–8 Spectrum of gaseous iodine taken with the Perkin-Elmer 3840 spectrophotometer. (Vapor in equilibrium with the solid at room temperature.)

ethanol. Rinse with distilled water. An organic solvent may be used to rinse the cell, but if it is not of spectrograde purity, it may leave a film. The cell should be rinsed with two or three portions of the sample solution just before it is filled and the spectrum or absorbance is measured.

Infrared Spectrophotometry

The absorption of electromagnetic radiation in the infrared (IR) region of the spectrum arises from transitions between vibration–rotation levels in molecules. Measurement of the energy of these transitions is of interest to physical chemists because they are related to the structure of molecules on the one hand and the values of molecular vibrational and rotational energy levels on the other. The region of IR radiation on the electromagnetic spectrum is shown in Figure 14–1.

Dispersive Spectrophotometry in the Infrared. Figure 14–2a and b also show the basic structure of an IR spectrophotometer, but the source is an electrically heated resistor made of sintered oxides of zirconium, yttrium, and thorium (Nernst glower) or of silicon carbide (Globar). The simple resistance heating of a coil of nichrome wire is also used. Optical components, fabricated from single crystals of alkali or alkali earth metal halides, transmit IR radiation satisfactorily. Otherwise, surface-reflecting mirrors and gratings are used. Some of the more commonly used materials for cell windows are listed in Table 14–1. It is of utmost importance not to touch cell windows, which are easily fogged and fingerprinted by exposure to moisture in perspiration. The detector consists of a material that can convert heat (IR radiation) into an electrical signal: for example, deuterated triglycine sulfate (DTGS) or mercury cadmium telluride (MCT).

Conventional dispersive IR spectrophotometers suffer from some of the same disadvantages as those operating in the UV/VIS region. The narrow slits that are necessary to define the wavelength of the monochromator severely reduce the amount of radiation falling on the detector. To sample a range of wavelengths, the monochromator must scan over

TABLE 14–1

Infrared-Transmitting Materials

Material	Useful Range (cm^{-1})
AgCl	25,000–435
BaF$_2$	50,000–870
CaF$_2$	6670–1110
CsBr	10,000–270
CsI	10,000–200
Ge	20,000–870
KBr	40,000–400
NaCl	45,000–625

the range, a time-consuming mechanical process leading to slow and insensitive operation.

Because the Fourier transform infrared spectrophotometer (FT-IR) requires no slits, its ability to gather radiant energy is much greater than that of dispersive IR spectrophotometers of comparable resolution. In addition, an FT-IR is much faster because all the radiation from one end of the spectrum to the other is incident on the detector simultaneously.

Fourier Transform Infrared Spectrophotometry. An FT-IR spectrophotometer bears little resemblance to a conventional dispersive instrument. At one end, both use a glowbar as a source of IR radiation, and at the other end, both produce an IR spectrogram. In between, there is nothing in common except for the sample itself, as shown in Figure 14–9. The key new elements are the interferometer, the analog-to-digital (A/D) converter, and the computer to compute the Fourier transform.

Figure 14–10 illustrates the optical path in an FT-IR. The stationary mirror, the beam splitter, and the movable mirror together form a Michelson interferometer. The interferogram formed is the Fourier transform of the spectrum. The interferogram is digitized and sent to a computer, where the Fourier transform (the spectrum) is calculated and the spectrum plotted. First, let us examine the interferometer in more detail.

The beam splitter consists of a KBr mirror coated with a layer of germanium so thin that about half the beam intensity is reflected upward toward the stationary mirror, while the other half passes through toward the movable mirror. In this manner, the beam is split into two legs: one reflected back from the stationary mirror (labeled "Leg S"), the other reflected back from the movable mirror (labeled "Leg M"). Since one mirror is movable, the path lengths along leg M are variable. For the moment, let us suppose that the beam is monochromatic. When the two legs rejoin back at the beam splitter, the beams from leg S and leg M differ in phase, depending on the position of the movable mirror. If the distance between the beam splitter and the movable mirror is x, then the length of leg M is $2x$. Consequently, the relationship between the phase shift δ and the mirror movement is $\delta = 2\Delta x$.

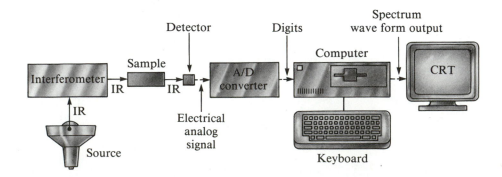

Figure 14–9 Block diagram of an FT-IR spectrophotometer.

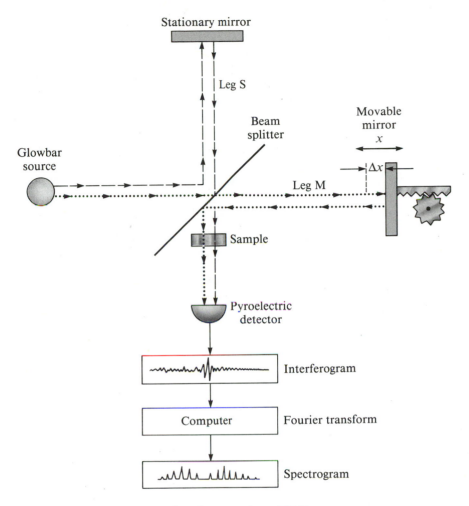

Figure 14–10 Optical path and interferometer in an FT-IR.

Suppose that the mirror is positioned so that the two beams are exactly in phase. If the mirror is moved by $\Delta x = \lambda/4$, then the path length along leg M changes by $\lambda/2$, so that the phase difference changes by 180° and the resultant beam is wiped out by destructive interference. In Figure 14–11a, the mirror is positioned so that the path lengths from the stationary mirror and the movable mirror are identical. In this position, consider the mirror to be situated at $x = 0$ and free to move along an x axis. At $x = 0$, the beams from the stationary and movable mirrors are exactly in phase, so that constructive interference occurs. In Figure 14–11b, the mirror has moved a quarter wavelength, that is, $\Delta x = \lambda/4$, so that $\delta = \lambda/2$, destructive interference occurs, and the intensity drops to zero. Figure 14–12 shows how the intensity of a monochromatic beam varies with the phase shift caused by a change in position of the movable mirror by Δx.

Next, let us examine how the intensity varies as the mirror is moved over several wavelengths. In a typical FT-IR spectrophotometer the mirror

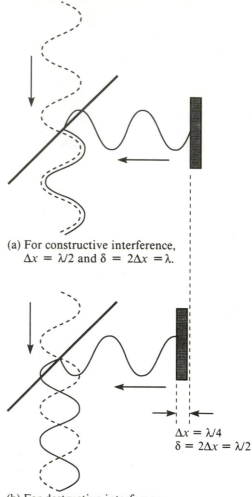

(a) For constructive interference,
$\Delta x = \lambda/2$ and $\delta = 2\Delta x = \lambda$.

$\Delta x = \lambda/4$
$\delta = 2\Delta x = \lambda/2$

Figure 14–11 Constructive and
destructive interference.

(b) For destructive interference,
$\Delta x = \lambda/4$ and $\delta = 2\Delta x = \lambda/2$.

moves about 1 cm, corresponding to many wavelengths of IR radiation.
In Figure 14–13, the origin corresponds to the mirror at $x = 0$, where
leg S = leg M, $\Delta x = 0$, constructive interference occurs, and the intensity
is at a maximum. As the mirror moves (in either direction), the intensity
decreases, reaching the first minimum at $\Delta x = \lambda/4$ and $\delta = \lambda/2$ as
described above and in Figure 14–12. Further movement of the mirror
generates a beam of radiation, the intensity of which follows a cosine
curve. Because the intensity varies in the same manner whether the mirror
moves toward plus x or minus x, the intensity $I(\delta)$ follows a cosine curve
and is centrosymmetric about the origin. This is the interferogram generated by radiation from a source of a single frequency, as shown in
Figure 14–13a. If the frequency of a monochromatic source is lower,
then the interferogram appears as shown in Figure 14–13b.

    If the source is not monochromatic, but consists of two frequencies
related by a factor of two, the resulting interferogram appears as shown

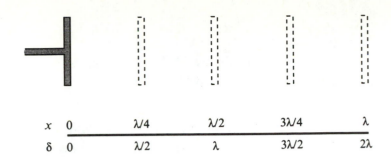

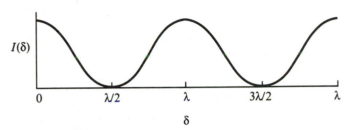

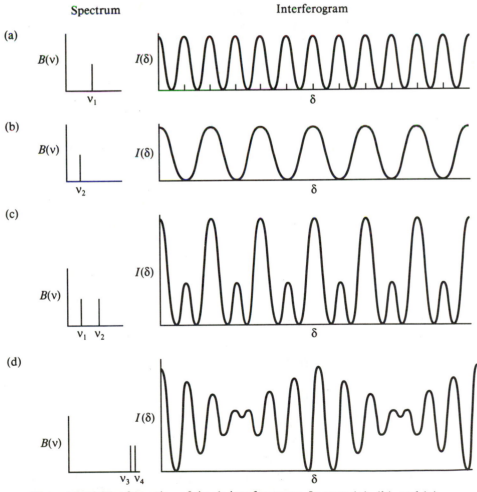

Figure 14–12 A simple interferogram caused by a moving mirror.

Figure 14–13 The formation of simple interferograms. In parts (a), (b), and (c), $\nu_1 = 2\nu_2$. In part (d), $\nu_4 = 7\nu_3/6$.

in Figure 14–13c. This can be regarded as the algebraic sum of the first two interferograms, or

$$I = \cos(2\pi x) + \cos(4\pi x) \qquad \text{(14–2)}$$

plotted in Figure 14–13c from $x = 0$ to $x = 6$. Since the radiation from the source of an FT-IR is not dispersed by a prism or grating, the radiation is polychromatic and can be regarded as a large number of closely spaced frequencies. First, let us consider the interferogram of just two closely spaced frequencies, as shown in Figure 14–13d. This is a plot of

$$I = \cos(6\pi x) + \cos(7\pi x) \qquad \text{(14–3)}$$

from $x = 0$ to $x = 4$. Notice that the plot is still centrosymmetric, since it is the sum of cosine terms. The frequency distribution of a glowbar ranging from about 200 to 4000 cm^{-1} is polychromatic, as shown in Figure 14–14a. The resulting interferogram, shown in Figure 14–14b, arises from one pass of the mirror over a distance of about 1 cm^{-1}. The buildup of intensity in the middle is called the center burst. When a sample that absorbs IR radiation is placed in the optical path of the interferometer, the intensity of the interferogram is further modulated by the varying intensity of the absorption of the sample. The resulting distribution of frequencies and their intensities is shown in Figure 14–14c, and the corresponding interferogram is shown in Figure 14–14d. When the background is subtracted and the Fourier transform of the interferogram is calculated, the final FT-IR spectrum appears as shown in Figure 14–14e.

When the source consists of a single frequency, the general equation for the interferogram is

$$I(\delta) = B(\nu)\cos(2\pi\delta\nu) \qquad \text{(14–4)}$$

where $B(\nu)$ is the intensity of the source as a function of frequency ν. In Figure 14–13a and b, $B(\nu)$ is a constant—that is, there is only one frequency. In Figure 14–13c and d, the intensity distribution of frequencies is just a little more complex—there are two frequencies. If the source consists of two or more frequencies, the equation for the interferogram is

$$I(\delta) = \sum B(\nu)\cos(2\pi\delta\nu_i)\, d\nu \qquad \text{(14–5)}$$

If the source is polychromatic, consisting of a large number of infinitesimally spaced frequencies ν_i, the summation may be replaced by an integral:

$$I(\delta) = \int_{-\infty}^{+\infty} B(\nu)\cos(2\pi\delta\nu_i)\, d\nu \qquad \text{(14–6)}$$

In Figure 14–14a and c the intensity distribution of frequencies is much more complex and is described by Equation 14–6. *Notice that the intensity distribution of frequencies, $B(\nu)$, is the spectrum.* This is what we want.

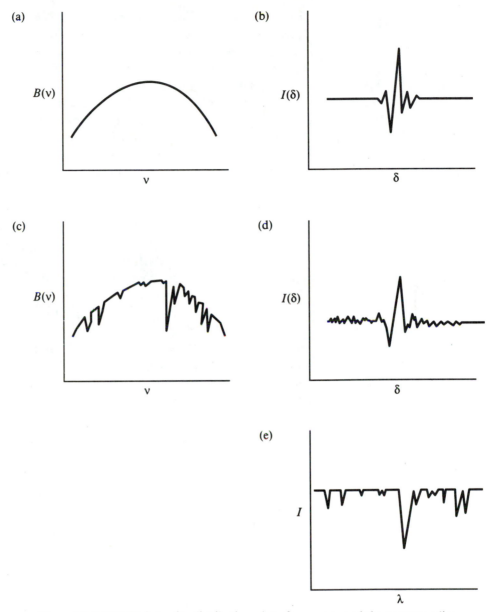

Figure 14–14 FT-IR intensity distributions, interferograms, and the corresponding spectrum.

However, an FT-IR spectrophotometer measures an interferogram $I(\nu)$ (Fig. 14–14d), produced by a frequency intensity distribution $B(\nu)$ (Fig. 14–14c) containing the spectrum (Fig. 14–14e). How can we get $B(\nu)$, the spectrum, from $I(\nu)$, the interferogram?

Mathematicians would recognize Equation 14–6 as one half of a Fourier transform pair. The other half is

$$B(\nu) = \int_{-\infty}^{+\infty} I(\delta) \cos(2\pi\delta\nu_i) \, d\delta \qquad (14\text{–}7)$$

Thus, $B(\nu)$ is the Fourier transform of $I(\delta)$, and $I(\delta)$ is the Fourier transform of $B(\nu)$. Since $I(\delta)$, the interferogram, is measured, $B(\nu)$, the spectrum, can be calculated. The right-hand side of Equation 14–7 is integrated by separating into a large number of terms over the range of x, the distance the mirror moves.

The resolution of the FT-IR spectrophotometer is inversely proportional to the distance the moving mirror travels. To produce a spectrum with 2-cm^{-1} resolution, the mirror travels 0.38 cm in the Perkin-Elmer 1800 (Fig. 14–15), and for 0.50-cm^{-1} resolution the mirror travels four times as far, or 1.52 cm. The system is constantly calibrated automatically with a helium–neon laser operating at 632.8 nm. The optics of the Perkin-Elmer 1800 are shown in Figure 14–16.

The interferogram is digitized and the thousands of individual terms are summed with a microcomputer at each of the sampling points. For accurate digital representation of the analog data contained in the IR interferogram, the interferogram is sampled 10,535 times in the Perkin-Elmer 1800 for each centimeter of optical path difference, where the optical path difference equals two times the mirror travel distance. It is only in recent years that microcomputers have become so small, fast, and inexpensive that they are a practical component of a spectrophotometer. The development of special computer programs to calculate Fourier transforms exceedingly quickly has also facilitated the practical development of FT-IR spectrophotometers. The fast Fourier transform or Cooley-Tukey (1965) algorithm is used to calculate the FT-IR spectrum.

Figure 14–15 Perkin-Elmer Model 1800 FT-IR spectrophotometer. (Courtesy of Perkin-Elmer Corporation, Norwalk, Connecticut.)

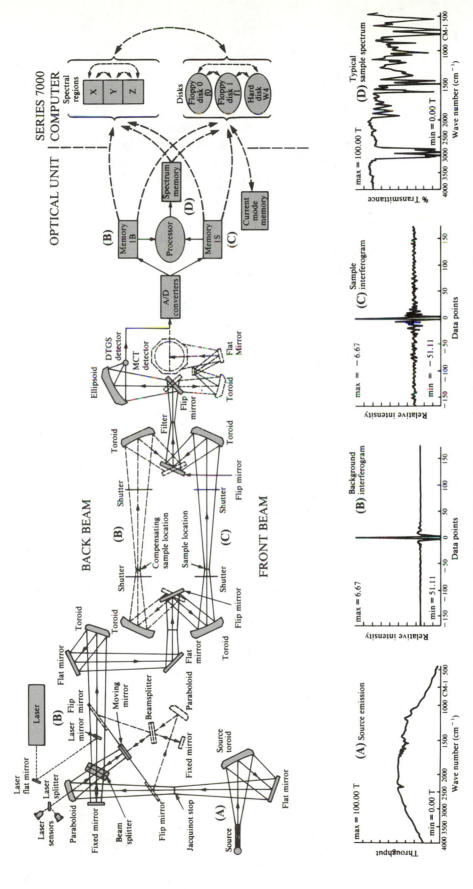

Figure 14–16 Functional diagram of the Perkin-Elmer Model 1800 FT-IR. (Courtesy of Perkin-Elmer Corporation, Norwalk, Connecticut.)

This algorithm is also widely used in the analysis of crystal structures by x-ray diffraction methods.

In fact, calculating the spectrum from the interferogram is very similar to calculating the structure of a crystal from its x-ray diffraction pattern. After all, an x-ray diffraction pattern is also an interferogram, and the crystal structure is the Fourier transform of the diffraction intensities (actually, their square roots, i.e., the structure factors, $F_{hk\ell}$). This is especially evident if you compare the equation for the structure of a one-dimensional crystal

$$\rho(x) = (1/L)(L_0 + 2 \sum F_h \cos 2\pi hx) \qquad \textbf{(14–8)}$$

with Equation 14–6, an equation for the "structure" of the spectrum, and the equation for the crystal structure factors (the crystal interferogram)

$$F_h = 2 \sum f_j \cos 2\pi hx \qquad \textbf{(14–9)}$$

with Equation 14–7, an equation for the FT-IR interferogram. Equations 14–6 and 14–7 form a Fourier pair, as do Equations 14–8 and 14–9.

The infrared spectrum of DCl recorded with a Perkin Elmer 1800 FT-IR is shown in Figure 14–17. The digitized spectrum, stored on a disk, was expanded with the instrument's computer to show the resolution of the $P(1)$, $P(2)$, $R(0)$, and $R(1)$ vibration–rotation transitions (Fig. 14–18). The signal-to-noise ratio is further improved by averaging multiple spectra obtained by moving the mirror several times. Since each pass of the mirror only requires a second or two, little time is required to average several spectra.

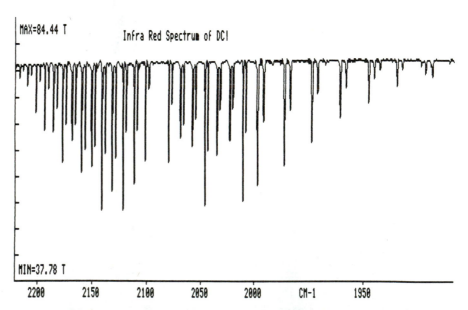

Figure 14–17 Infrared spectrum of DCl taken with the Perkin-Elmer Model 1800 FT-IR.

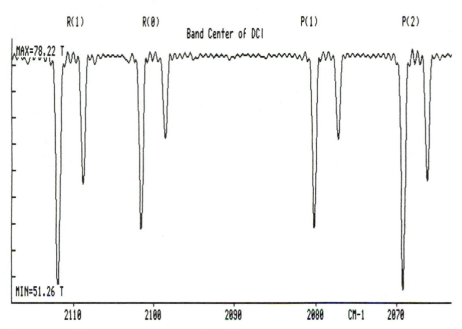

Figure 14–18 Computer expansion of the DCl spectrum showing separation of the $P(1)$, $P(2)$, $R(0)$, and $R(1)$ vibration–rotation bands in DCl. The further splitting arises from the Cl-35 and Cl-37 isotopes.

MEASUREMENT OF REFRACTIVE INDEX

A light beam incident on a liquid surface is both reflected and bent as it enters the liquid. The bending of such a light beam is called **refraction.** In Figure 14–19 the angle of incidence θ_1 equals the angle of reflection θ_2. The index of refraction is given by

$$n = \frac{\sin \theta_1}{\sin \theta_3}$$

As the angle of incidence increases from 0 to 90°, the angle of refraction θ_3 increases from zero to θ_C, the critical angle.

Abbé Refractometer

From a different point of view, a source of light *in the liquid* lying between θ_C and 90° is totally reflected within the liquid and no refraction occurs. The operation of the Abbé refractometer depends on observing the total reflection boundary.

Because the observed index of refraction depends on both the wavelength of the incident light and the temperature, it is necessary to thermostat the apparatus and to use monochromatic radiation, conventionally the sodium D line at 589.0 nm. The symbol n_D^{25} stands for the index of refraction for the D line of sodium at 25°C. The Abbé refractometer

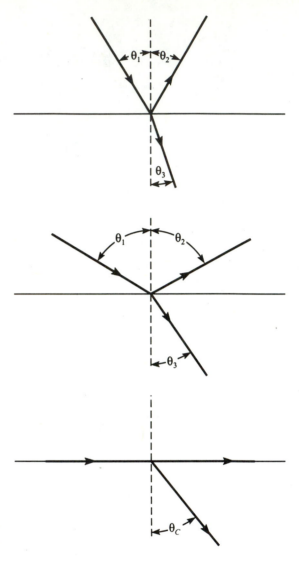

Figure 14–19 Relationship between angle of incidence and angle of refraction. From top to bottom, as the angle of incidence θ_1 increases, the angle of refraction θ_3 increases from 0 to the critical angle θ_C.

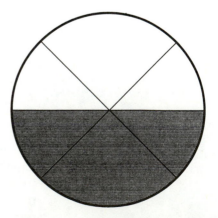

Figure 14–20 View of the reflection borderline.

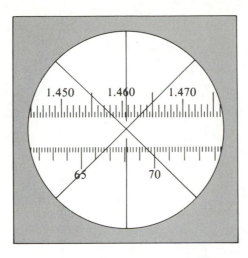

Figure 14–21 View of internal scales. (Courtesy of Milton Roy Company, Rochester, New York.)

measures the index of refraction by determining the position of the critical ray observed as a boundary between a light and a dark field. When the critical ray is lined up in the cross hairs (Fig. 14–20) the index of refraction is read directly from a scale (Fig. 14–21). The Abbé refractometer uses white light as a source, but internal Amici compensating prisms make it appear as if the source were the sodium D line. A widely used refractom-

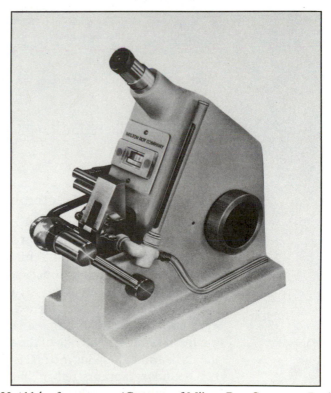

Figure 14–22 Abbé refractometer. (Courtesy of Milton Roy Company, Rochester, New York.)

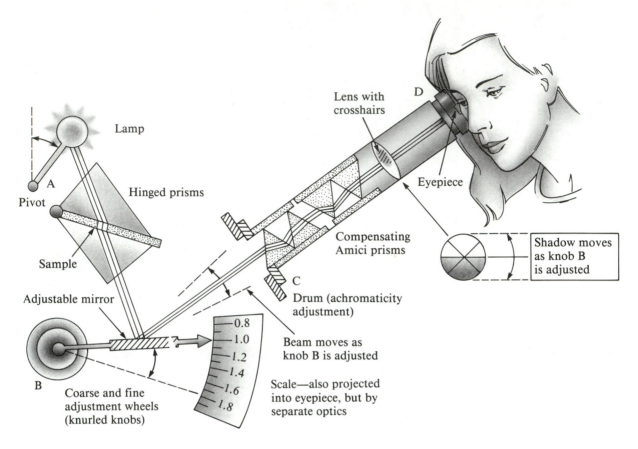

Figure 14–23 An optical diagram of the Abbé refractometer.

eter is shown in Figure 14–22, and its operation is shown in Figure 14–23.

Using the Abbé Refractometer

The thermostat should have been running long enough so that the refractometer sample prisms are at thermal equilibrium. Allow a few drops of the sample to drip into the small opening at the junction of the hinged prisms. Be careful not to touch the prisms with a glass eyedropper since the optical surfaces are easily scratched.

With the light source turned on, look in the eyepiece and adjust the light position for maximum brightness. Adjust the coarse- and fine-adjustment knobs on the side of the instrument until the boundary between the dark and light fields intersects the cross hairs exactly (Fig. 14–20). If the boundary appears red or blue, adjust the compensator. When the compensator is properly adjusted, the boundary seen through the eyepiece is achromatic in the center (at the cross hairs), with a faint red color at one end and a faint blue color at the other end.

To see the index of refraction scale, depress the contact switch on the left side of the instrument. The scale is visible through the same eyepiece. Some older Abbé refractometers have separate eyepieces for viewing the scale and boundary, but otherwise operate similarly.

If the initial adjustment of the refractometer is difficult, place on the prism a few drops of a substance of known index of refraction. With the scale set to the known value, adjust the light for maximum brilliance and contrast of the boundary between the light and dark fields.

MEASUREMENT OF DENSITY

Density is an important property of all substances and its value is frequently required in physical chemical calculations. The measurement of the density of a gas is described in Experiment 16. The measurement of the density of liquids and solids is described below.

Westphal Balance

The Westphal balance is useful for determining the density of nonviscous liquids (Fig. 14–24). The buoyancy of a glass plummet immersed in a liquid is linearly dependent on the density of the liquid. Consequently, the beam of the Westphal balance is calibrated linearly in density units (g/cm^3) relative to water, which is used to calibrate the apparatus. The density of water at its measured calibration temperature must be used (Table 14–2).

The water level and/or height of the cylindrical water container is adjusted until the plummet is completely submerged. A small thermometer sealed inside the plummet permits recording the water temperature,

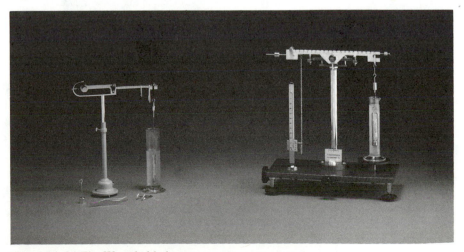

Figure 14–24 The Westphal balance.

TABLE 14–2

The Density of Water (g/cm³) at 0.1 °C Temperature Intervals

t (°C)	0.0	0.1	0.2	0.3	0.4	0.5	0.6	0.7	0.8	0.9
0	0.999840	847	853	860	866	872	878	883	889	894
1	899	904	909	913	917	922	926	930	933	937
2	940	943	946	949	952	954	956	959	960	962
3	964	966	967	968	969	970	971	971	972	972
4	972	972	972	971	971	970	969	968	967	966
5	964	962	960	958	956	954	951	949	946	943
6	940	937	933	930	926	922	918	914	910	906
7	901	896	892	887	881	876	871	865	860	854
8	848	842	836	829	823	816	809	803	796	788
9	781	774	766	758	750	742	734	726	717	708
10	700	691	682	673	664	654	645	635	625	615
11	605	595	584	574	563	553	542	531	520	508
12	497	485	474	462	450	438	426	414	401	389
13	376	363	350	337	324	311	298	284	271	257
14	243	229	215	201	186	172	157	143	128	113
15	098	083	068	052	037	021	006	990[a]	974[a]	958[a]
16	0.998942	926	909	893	876	860	843	826	808	791
17	774	757	739	721	704	686	668	650	632	613
18	595	576	558	539	520	501	482	463	444	425
19	405	385	366	346	326	306	286	265	245	225
20	204	183	163	142	121	100	079	057	036	015
21	0.997993	971	949	928	906	883	861	839	816	794
22	771	748	725	702	679	656	633	609	586	563
23	539	515	492	468	444	420	396	371	347	323
24	298	273	248	224	199	173	148	123	097	072
25	046	020	995[a]	969[a]	943[a]	917[a]	891[a]	864[a]	838[a]	812[a]
26	0.996785	758	732	705	678	651	624	596	569	542
27	514	486	459	431	403	375	347	319	291	262
28	234	205	177	148	119	091	062	033	004	974[a]
29	0.995945	916	886	857	827	797	768	738	708	678
30	648	62	59	56	53	50	47	43	40	37

Source: Landolt-Börnstein, 6th ed., *Numerical Data and Functional Relationships in Physics, Chemistry, Geophysics and Technology,* H. Borchers, K.-H. Heilwege, and K. Schäfer, eds., vol. IV, pt. 1, Springer-Verlag, New York, 1980, pp. 101–102.

[a] The first three digits are the same as in the *next* entry of the 0.0 column.

so that the calibration density of the water may be determined. Care must be taken that the plummet does not touch the sides of the liquid container.

The balance is calibrated by setting the rider at 0.9 and adjusting the chain until the reading coincides with the density of water taken from Table 14–1. Agreement between the scale and the density is achieved by fine adjustment of the screws on the balance arm. If the scale reading is quite close to the true density, it is sometimes convenient to record the difference and add or subtract this correction term from the scale reading to get the true density.

After the balance is calibrated with water of known density, the density of an unknown liquid is found by rinsing the plummet and wire with the liquid of unknown density and filling the cylinder. If the liquid and water are immiscible, it may be necessary to rinse with a mutually miscible organic solvent such as acetone first. Readjustment of the rider and chain gives the new density at the recorded temperature.

It is recommended that the platinum wire be immersed to the same depth for calibration and measurement. Contact between the plummet and the liquid container is the most common source of error. Duplicate measurements should be made to avoid such errors.

Top-Loading Analytical Balance

A density determination kit is available for some top-loading analytical balances such as Mettler top loaders. The accessory, shown in Figure 14–25, consists of a sheet-metal bridge that fits over the balance pan. A plummet hangs from a bracket clamped to the balance pan, and a 150-ml beaker holding a small thermometer sits on the bridge.

With this configuration the balance is tared. The liquid of unknown density is poured into the beaker until its level is about 10 mm above

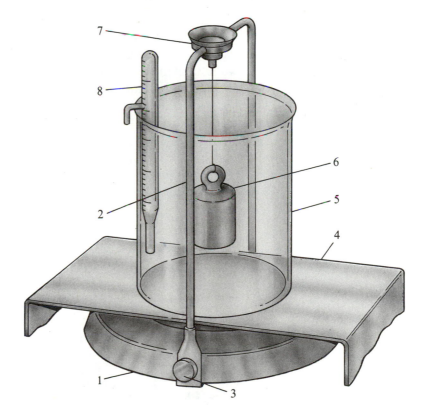

1. Weighing pan of balance	5. 150-ml beaker on bridge
2. Bracket attached to weighing pan	6. Plummet
3. Clamping screw holding bracket in place	7. Suspension point for plummet
4. Bridge placed over weighing pan	8. Thermometer

Figure 14–25 Measuring the density of a liquid with a toploader.

the plummet's eyelet. The balance now shows the weight loss due to the buoyancy of the plummet, P_s (grams). The volume of the plummet is 10.0 cm^3. The density is given by

$$d = \frac{P_s}{10} + 0.001$$

Thus, for the determination of the density of acetone at 25°C,

$$d = \frac{7.898}{10} + 0.001 = 0.791 \text{ g/cm}^3$$

The accuracy with this method is about three significant figures. For higher accuracy, pycnometry should be used, as described in Experiment 11.

The use of a gem holder instead of the plummet permits the determination of the density of solids.

Density of Solids

Even when only a few tiny crystals are available it is possible to measure the density of a solid by the flotation method. In this method, a few crystals are added to a test tube or graduated cylinder containing a mixture of two liquids in which the solid is insoluble. Heptane ($d = 0.684$ g/cm^3) and carbon tetrachloride ($d = 1.589$ g/cm^3) are frequently used for salts and polar organic compounds. For nonpolar organic compounds, a concentrated solution of KI ($d = 1.636$ g/cm^3 for a solution containing 818 g/L) is used. The composition of the solution is adjusted until the crystals remain suspended, neither sinking nor floating. The density of the solid and solution are then equal and the density of the liquid is measured as usual. The method is quite sensitive, and an accuracy of better than 1% can be achieved.

REFERENCES

1. Cooley, J. W., and J. W. Tukey, "An Algorithm for the Machine Calculation of Complex Fourier Series," *Math. Comput.* 19:297, 1965.

2. Glasser, L., "Fourier Transforms for Chemists: Part II. Fourier Transforms in Chemistry and Spectroscopy," *J. Chem. Educ.* 64:A260, 1987.

3. Griffiths, P. R., *Chemical Infrared Fourier Transform Spectroscopy*, Wiley, New York, 1975.

4. Horlick, G., "Introduction to Fourier Transform Spectroscopy," *Applied Spectroscopy* 22:617–626, 1968.

5. Hurley, W. J., "Interferometric Spectroscopy in the Far Infrared," *J. Chem. Educ.* 43:236, 1966.

6. Low, M. J. D., "Multiple-Scan Infrared Interference Spectroscopy," *J. Chem. Educ.* 43:637, 1966.

7. Perkins, W. D., "Fourier Transform Infrared Spectroscopy: Part II. Advantages of FT-IR," *J. Chem. Educ.* 64:A269, 1987.

8. Swinehart, D. F., "The Beer-Lambert Law," *J. Chem. Educ.* 39:333, 1962.

EXPERIMENTS

Gases

Calorimetry

Phase Equilibria

Properties of Solutions

Surface Effects

Equilibrium Constants

Electrochemistry

Kinetics

Spectroscopy

Electric and Magnetic Properties

X-Ray Diffraction

Gases

Experiment I

Knudsen Effusion

The Knudsen effusion method is widely used to determine equilibrium vapor pressures, especially at high temperatures. The vapor pressure is calculated by measuring the rate at which molecules in a system at equilibrium effuse through an opening to a vacuum. Gas pressures must be in the molecular flow region (below 10^{-3} torr). The lower pressure limit depends only on the sensitivity of the method used to measure the number of molecules effusing per unit time. For gravimetric methods, the practical lower limit is approximately 10^{-4} or 10^{-5} torr. At lower vapor pressures it simply takes too long to observe measurable amounts of effused gas. At a constant temperature, the number of molecules striking a surface per unit area per unit time is constant.

Kinetic Molecular Theory of Gases

According to the kinetic molecular theory of gases, molecules confined to a container undergo random and frequent collisions with each other and with the walls of the container. If the container is separated from a vacuum by a thin wall with a small hole in it, gas molecules that "collide" with the hole leave the container and are transported into the vacuum or region of lower pressure. This transport process is called **effusion**. At a pressure sufficiently low so that the mean free path is large compared to the dimensions of the container, it is possible to calculate the relationship between the effusion rate Z (molecules per unit area per unit time) and the pressure.

A schematic representation of a Knudsen effusion cell is shown in Figure 1–1. We shall assume that the molecules in the cell are in equilibrium with a condensed phase, so that the pressure in the cell represents the equilibrium vapor pressure, which is constant at constant temperature.

In the cell, the distribution of molecular speeds is given by the Maxwell-Boltzmann distribution equation

$$F(u) = 4\pi(m/2\pi kT)^{3/2} u^2 \exp(-mu^2/2kT) \, du \qquad \text{(1–1)}$$

where $F(u)$ is the probability that a molecule has a speed (without regard to direction) between u and $u + du$ and

m = mass of molecule (kg)
π = 3.1459
k = 1.3807×10^{-23} J/K
T = temperature (K)

The average speed of a molecule equals

$$\bar{u} = \int_{-\infty}^{+\infty} F(u)u \, du = (8kT/\pi m)^{1/2} \qquad \text{(1–2)}$$

In addition to the average speed (without regard to direction) of the molecules in the Knudsen cell, we are especially interested in the average velocity in the $+z$ direction as it is this component of the *velocity* that transports molecules up through the hole in the wall (Fig. 1–1). The one-dimensional distribution of velocities along the z axis is given by

$$F(u_z) = (m/2\pi kT)^{1/2} \exp(-mu_z^2/2kT) \, du_z \qquad \text{(1–3)}$$

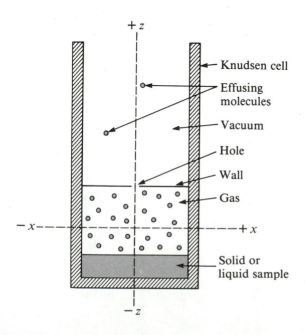

Figure 1–1 Knudsen effusion cell schematic.

The average speed in the $+z$ direction is

$$\bar{u}_{+z} = \int_0^\infty F(u_z) u_z \, du_z = (kT/2\pi m)^{1/2} \qquad (1\text{--}4)$$

If the hole is small and no molecules collide with each other on the way to the hole, then Z_w, the number of collisions with the wall per unit time per unit area, is given by

$$Z_w = \bar{u}_{+z} \text{ m/sec} \times N^* \text{ molecules/m}^{-3} \qquad (1\text{--}5)$$

where N^* is the molecular concentration.

Comparison of $\bar{u}_{+z}$ (Eq. 1–4) and u (Eq. 1–2) shows that

$$\frac{\bar{u}}{\bar{u}_{+z}} = \left(\frac{8kt}{\pi m}\right)^{1/2} \times \left(\frac{2\pi m}{kT}\right)^{1/2} = 4 \qquad (1\text{--}6)$$

Consequently, the effusion rate can be expressed simply in terms of the average molecular speed and the concentration of molecules by combining Equations 1–5 and 1–6:

$$Z_w = \frac{N^* \bar{u}}{4} \qquad (1\text{--}7)$$

The Knudsen Effusion Equation

Next, we must connect the theoretical equations derived so far to experimentally determinable parameters: the pressure, temperature, and molecular weight of the gas, the area of the hole, and the weight loss during the effusion time.

From the Ideal Gas Law we can write

$$n/V = P/RT \quad (\text{moles/unit volume})$$

or, on a molecular level (N_A = the Avogadro constant),

$$N^* = N_A P/RT \quad (\text{molecules/unit volume})$$

In terms of weight loss, the effusion rate Z' is

$$Z' = mN^* \bar{u}/4 \quad (\text{weight loss/unit volume/unit time})$$

where m is the mass per effusing molecule, or

$$Z' = M\bar{u}P/4RT \qquad (1\text{--}8)$$

where M is the molecular weight of the effusing gas. Experimentally, Z' is found by measuring the weight loss g in time t of molecules effusing through a hole of area A:

$$Z' = g/At \qquad (1\text{--}9)$$

By combining the last two equations, substituting for $\bar{u}$ (Eq. 1–2), and solving for the pressure, we arrive at the Knudsen effusion equation:

$$P = (g/At)(2\pi RT/M)^{1/2} \tag{1–10}$$

where

g = weight loss (kg)
A = area of hole (m^2)
t = time of effusion (sec)
R = gas constant (J K^{-1} mol^{-1})
T = temperature (K)
M = molecular weight (kg/mol)
P = pressure (N/m^2 = Pa)

The pressure in torr or atmospheres can be obtained by using the equivalencies

$$1 \text{ atm} = 760 \text{ torr} = 101{,}325 \text{ N/m}^2$$

or, alternatively (for example), you can easily show that

$$P(\text{torr}) = (17.55 \, g/At)(T/M)^{1/2} \tag{1–11}$$

where

g = weight loss (g)
t = effusion time (sec)
T = temperature (K)
M = molecular weight (g/mol)
A = area of hole (cm^2)

The Clausing Factor

It has been assumed in this derivation that the hole through which effusion occurs has infinitely thin edges. Since it is impossible to fabricate such thin materials, collision of the molecules with the edges is a source of error, because some of the molecules colliding with an edge recoil back into the cell and do not escape. In the derivation of Equation 1–10, we assume that every molecule that "strikes" the hole leaves the cell, which is true only if the cell wall is infinitely thin.

Consider the situation depicted in Figure 1–2. Most molecules, like B and C, effuse cleanly through the hole; some, like A, hit the edge but still make it out; but others, like D, hit the edge and return back into the cell. Thus, the true effusion rate is higher than calculated by Equation 1–9 and the true pressure is higher than calculated by Equation 1–10.

Clausing (1932) has calculated factors to correct the measured pressure to account for back reflection due to channel-shaped holes of radius r and length ℓ (Table 1–1).

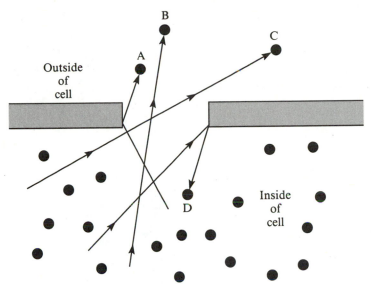

Figure 1-2 Origin of the Clausing factor.

Apparatus

The apparatus consists of a Knudsen effusion cell and a cell holder that can be attached to a vacuum line at one end and immersed in a thermostat bath at the other end (Fig. 1–3). The cell, machined from brass, is a hollow cylinder, permanently closed at one end and threaded to accept a screw cap at the other end. A thin (0.0008-mm) brass disk with a small hole in the center is seated between the top edge of the cell and the inner surface of the screw cap.

TABLE I-I

Clausing Factor (K) as a Function of the Length/Radius Ratio of an Effusion Hole: $P_{true} = P_{meas}/K$

ℓ/r	K	ℓ/r	K
0	1	1.1	0.6514
0.1	0.9524	1.2	0.6320
0.2	0.9092	1.3	0.6319
0.3	0.8699	1.4	0.5970
0.4	0.8341	1.5	0.5810
0.5	0.8013	1.6	0.5659
0.6	0.7711	1.7	0.5518
0.7	0.7434	1.8	0.5384
0.8	0.7177	1.9	0.5256
0.9	0.6940	2.0	0.5136
1.0	0.6720	2.2	0.4914

Source: Dushman, S., *Scientific Foundations of Vacuum Technique,* 2nd ed., Wiley, New York, 1962.

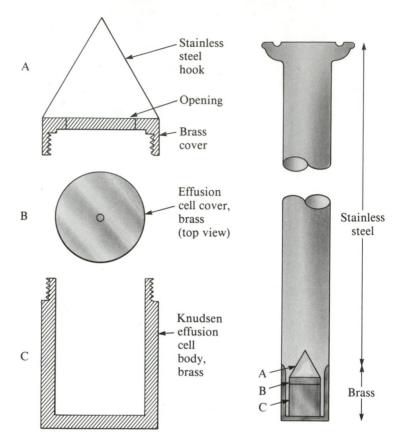

Figure 1–3 Practical effusion cell.

The cell is shown in exploded view in Fig. 1–4. It is 32.8 mm high and has a 25.4-mm outer diameter (OD) and a 19-mm inner diameter (ID). The cell holder is constructed from 40-mm-diameter, thin-wall stainless steel tubing terminating at the top with a stainless steel O-ring fitting.[1] The bottom 20 mm is fabricated from brass with an OD of 42 mm and an ID of 27 mm. The brass bottom of the cell holder transmits heat efficiently to the brass effusion cell, but the stainless steel section insulates the cell from the surroundings since the conductivity of stainless steel is about the same as that of glass.

The cell is placed into the cell holder by being lowered with a nylon string terminating with a small stainless steel wire hook. The cell is shown readied for loading in Figure 1–5.

Naphthalene is suitable for studies between 0 and 25°C. In addition,

---

[1] Type 303, available from Ace Glass Company, 1430 Northwest Boulevard, Vineland, New Jersey 08360.

Figure 1–4 Exploded view of an effusion cell. (Karen R. Sime)

p-chloroaniline, *p*-chloronitrobenzene, and *p*-bromonitrobenzene have also been investigated by Knudsen effusion in this temperature range (Swan and Mack, 1925).

SAFETY CONSIDERATIONS

Safety goggles should always be worn when working with a high-vacuum apparatus. If a cold trap filled with liquid nitrogen or dry ice–acetone is used, it should be wrapped with electrical tape. Keep flames away from the dry ice–acetone mixture. If a mercury diffusion pump is used, be sure that the cooling water is running before the diffusion pump heater is turned on.

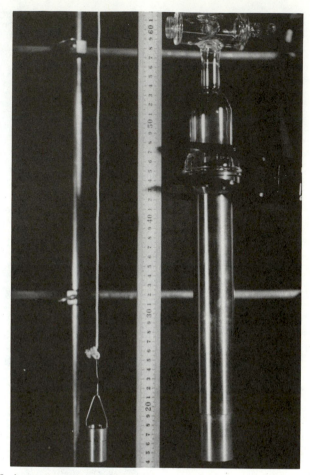

Figure 1–5 Effusion cell readied for loading into cell holder. (Karen R. Sime)

EXPERIMENTAL PROCEDURE

It is convenient to make three runs simultaneously, using three different sizes of holes in the Knudsen cell and three different temperatures: 1.5 mm at 0°C, 1.0 mm at 20°C, and 0.5 mm at 25°C. For the 25°C run use a thermostat bath, but for the 20°C run, several liters of water equilibrated to room temperature are adequate. A water–ice slush bath provides the 0°C temperature. For the room temperature run (~20°C), allow several liters of water in an appropriate container to come to thermal equilibrium by standing overnight.

The Knudsen cells should be clean, dry, and free of grease and fingerprints. Fill the cell with powdered naphthalene or a substance of similar vapor pressure to a depth of 1 cm and seal the hole cover in place with the screw cap. Do not touch the cell with bare fingers.

After weighing the cell plus the sample, lower it into the cell holder, which in turn is clamped to the vacuum system. Then put the constant-temperature baths in place around the bottom half of the cell holder. Record the temperature periodically to ±0.1°C.

As long as the cells are full of air, effusion of naphthalene through the cell hole is negligible, since collisions between air molecules and naphthalene molecules block the entrance to the hole. Start a run by opening the stopcock to the cell holder and recording the beginning of the effusion time.

After about 2 hours, let air into the cells and record the end of the effusion time. Carefully remove the cells without touching them, and weigh them to ±0.1 mg.

Results and Calculations

From the measured weight loss, the known area of the hole, and the effusion time, calculate the effusion rate and the vapor pressure. Tabulate Z', P_{torr}, P_{atm}, and T. If we assume that ΔC_P for the process

$$C_8H_{10}(s, 1 \text{ atm}) = C_8H_{10}(g, 1 \text{ atm})$$

equals zero, then

$$\ln P_{atm} = \frac{-\Delta H^0}{RT} + \frac{\Delta S^0}{R}$$

At 25°C, calculate ΔG^0, ΔH^0, and ΔS^0 by plotting $\ln P_{atm}$ versus $1/T$.

Analyze the propagated error in P calculated with Equation 1–10 or 1–11. Compare your results with the corresponding values reported in the literature.

REFERENCES

1. Balson, E. W., "Some Aspects of Molecular Effusion," *J. Phys. Chem.* 65: 1151, 1961.

2. Carlson, K. D., "The Knudsen Effusion Method," in J. L. Margrave, ed., *The Characterization of High Temperature Vapors,* Wiley, New York, 1967.

3. Clausing, P., "The Flowing of Very Dilute Gases Through Tubes of Any Length," *Ann. Physik.,* 12:961, 1932.

4. Hollahan, J. H., "Molecular Effusion: Its Newer Features and Applications," *J. Chem. Educ.* 39:23, 1962.

5. Kennard, E. H., *Kinetic Theory of Gases,* McGraw-Hill, New York, 1938.

6. Knudsen, M., "Effusion and the Molecular Flow of Gases Through Openings," *Ann. Physik.,* 29:179, 1909.

7. Kubachewski, O., and E. L. Evans, *Metallurgical Thermochemistry,* Wiley, New York, 1956.

8. Smith, N. K., G. Gorin, W. D. Wood, and J. P. McCullough, "The Heats of Combustion, Sublimation, and Formation of Four Dihalobiphenyls," *J. Phys. Chem.* 68:940, 1964.

9. Swan, T. H., and E. Mack, Jr., "Vapor Pressures of Organic Chrystals by an Effusion Method," *J. Am. Chem. Soc.* 47:2112, 1925.

Experiment 2

Measurement and Calculation of the Heat Capacity of Gases

It may be deduced from the First Law of Thermodynamics that the difference between heat capacities measured at constant volume and at constant pressure equals the gas constant: $C_P - C_V = R$. The ratio $C_P/C_V = \gamma$ and is directly measurable. Consequently, we can calculate a measured value of C_V by eliminating C_P from these two equations:

$$C_V = \frac{R}{\gamma - 1} \tag{2-1}$$

In addition, we can calculate C_V for any gas with the theoretical equations of statistical thermodynamics at any temperature if we know just two molecular parameters: (1) the shape of the molecule, that is, whether it is linear or nonlinear; and (2) its set of vibrational energies. The number of vibrational modes in a molecule containing n atoms equals $3n - 5$ if the molecule is linear and $3n - 6$ if it is nonlinear. Spectroscopic measurements, usually in the infrared region, permit the experimental determination of molecular vibrational energies.

A comparison of the heat capacity of a gas measured experimentally with the heat capacity of the gas calculated by statistical thermodynamics therefore provides us with a test of the First Law of Thermodynamics and of the postulates of statistical thermodynamics—which for our purposes might be simply given as a variation of the equipartition of energy:

$$C_V(\text{total}) = C_V(\text{trans}) + C_V(\text{rot}) + C_V(\text{vib})$$

Thus, the total heat capacity is calculated as the sum of the translational, rotational, and vibrational contributions.

Classical Thermodynamics of the Heat Capacities of Gases

For the First Law of Thermodynamics we write

$$\Delta U = q + w$$

or

$$dU = \delta q + \delta w \tag{2-2}$$

in differential form. In words, the increase in the internal energy U of a system equals the heat absorbed (q) by the system plus the work (w) done on the system.

Work is defined as force acting through a distance,

$$\delta w = f \, ds$$

We are concerned only with pressure–volume work, so that

$$\delta w = -P \, dV \qquad (2\text{--}3)$$

The negative sign is required by our choice of signs in the First Law. Had we chosen to write $\Delta U = q - w$, as is sometimes done, then we would define PV work as $\delta w = +P \, dV$.

If a process takes place at constant volume, $dV = 0$; no work is done, so $\Delta U = q_V$, where q_V is the heat absorbed at constant volume. If a process takes place at constant pressure, $w = -P \, \Delta V$ and $\Delta U = q - P \, \Delta V$. By defining $H = U + PV$, it follows that at constant pressure $\Delta H = \Delta U - P \, \Delta V$, so that $q_P = \Delta H$. In differential form we can write $\delta q_V = dU$ and $\delta q_P = dH$.

The heat capacity of a substance is defined as $C = q/\Delta T$, but since the value of q depends on the path (constant volume or pressure), we can see the need for defining two heat capacities:

$$C_V = \frac{\partial q_V}{\partial T} = \left(\frac{\partial U_V}{\partial T} \right)_V \qquad (2\text{--}4)$$

$$C_P = \frac{\partial q_P}{\partial T} = \left(\frac{\partial H_P}{\partial T} \right)_P \qquad (2\text{--}5)$$

The Relation Between C_P and C_V. According to the phase rule, two properties must be specified in order to determine the state of a one-component system. It is conventional and convenient to specify T and V so that

$$U = f(T, V) \qquad (2\text{--}6)$$

Because U is a state function, the differential dU is given by

$$dU = \left(\frac{\partial U}{\partial T} \right)_V dT + \left(\frac{\partial U}{\partial V} \right)_T dV \qquad (2\text{--}7)$$

The first term gives the partial change in U due to a small change dT in the temperature while the volume V remains constant. The second term gives the partial change in U due to a small change dV in the volume at constant pressure. The sum of these two changes equals the total differential change dU in U.

When Equation 2–7 is combined with the first law (Equation 2–2) and Equation 2–3, the incremental heat δdq exchanged between the system and its surroundings is given by

$$\delta q = \left(\frac{\partial U}{\partial T}\right)_V dT + \left[P + \left(\frac{\partial U}{\partial V}\right)_T\right]dV \qquad (2\text{–}8)$$

Recognizing that $(\partial U/\partial T)_V = C_V$, we can write

$$\delta q = C_V dT + \left[P + \left(\frac{\partial U}{\partial V}\right)_T\right]dV \qquad (2\text{–}9)$$

Dividing Equation 2–9 by δT at constant pressure and recognizing that $\delta q_P/\delta T = C_P$, we can rewrite Equation 2–9 as

$$C_P - C_V = \left[P + \left(\frac{\partial U}{\partial V}\right)_T\right]\left(\frac{\partial U}{\partial T}\right)_P \qquad (2\text{–}10)$$

This equation for the difference $C_P - C_V$ is valid for real and ideal gases. The term $(\partial U/\partial V)_T$ gives the dependence of the internal energy U on the volume. The internal energy of an ideal gas is independent of its volume, so $(\partial U/\partial V)_T = 0$ for an ideal gas; for this special case Equation 2–10 reduces to

$$C_P - C_V = P\left(\frac{\partial V}{\partial T}\right)_P \qquad (2\text{–}11)$$

For an ideal gas,

$$V = \frac{RT}{P} \qquad (2\text{–}12)$$

and

$$\left(\frac{\partial V}{\partial T}\right)_P = \frac{R}{P} \qquad (2\text{–}13)$$

so that for an *ideal* gas,

$$C_P - C_V = R \qquad (2\text{–}14)$$

Adiabatic Expansions, C_P and C_V. Let us begin with the First Law of Thermodynamics:

$$dU = \delta q + \delta w$$

An adiabatic process is one in which q equals zero, that is, no heat is exchanged between the system and its surroundings, so that the First Law reduces to

$$dU = \delta w$$

In an adiabatic expansion, since no heat is absorbed to compensate for the work done by the system, the temperature drops by an amount proportional to the work done and the First Law becomes

$$C_V \, dT = -P_{op} \, dV \tag{2-15}$$

The proportionality constant is C_V, and P_{op} is the opposing pressure against which the expansion takes place. If the expansion takes place reversibly, the opposing pressure and the internal pressure of the gas never differ by more than an infinitesimal amount, that is, the opposing pressure essentially equals the internal pressure of the gas. If the gas is ideal, then the internal pressure equals RT/V, and we may write

$$P_{op} = P_{int} = RT/V$$

Substitution in Equation 2–15 and integration after separating variables results in

$$C_V \ln \frac{T_2}{T_1} = -R \ln \frac{V_2}{V_1} \tag{2-16}$$

It would appear that if we know enough about the initial and final states, we could calculate C_V from this equation. This is essentially what we do, although somewhat indirectly and under irreversible conditions (Bertrand and McDonald, 1986). The method we use (Désormes and Clément, 1819) is an adiabatic expansion carried out in such a way as to eliminate the need to actually measure the associated change in temperature. Because the gas, slightly pressurized in a large bottle, is allowed to expand adiabatically against the constant opposing pressure of the atmosphere, the process occurs irreversibly. With $P_{op} = P_2$, the atmospheric pressure, integration of Equation 2–15 leads to

$$nC_V(T_2 - T_1) = -P_2(V_2 - V_1) \tag{2-17}$$

If V_2 and V_1 are given by the Ideal Gas Law ($V = RT/P$), then Equation 2–17 becomes

$$C_V(T_2 - T_1) = -P_2(RT_2/P_2 - RT_1/P_1) \tag{2-18}$$

Since the gases are assumed to be ideal, $C_P - C_V$ may be substituted for R, resulting in

$$C_V(T_2 - T_1) = -(C_P - C_V)(T_2 - T_1 P_2/P_1) \tag{2-19}$$

Upon rearrangement, Equation 2–18 may be written

$$C_V(P_2 - P_1)/P_1 = C_P(P_2/P_1 - T_2/T_1) \tag{2-20}$$

The temperature T_2 to which the gas cools upon expansion is not known. However, if the gas is allowed to warm isochorically back to the initial temperature T_1, and the final pressure P_3 is measured, then according to Charles's Law,

$$P_2/T_2 = P_3/T_1 \qquad (2\text{--}21)$$

Combination of Equations 2–20 and 2–21 results in

$$\gamma = \frac{C_P}{C_V} = \frac{(P_1/P_2) - 1}{(P_1/P_3) - 1} \qquad (2\text{--}22)$$

This will be our working equation for the experimental determination of γ; as already pointed out, combining γ with $C_V - C_P = R$ permits the calculation of C_V, since

$$C_V = \frac{R}{\gamma - 1} \qquad (2\text{--}23)$$

Statistical Thermodynamic Calculation of C_V

The heat capacity C_V of a molecule equals the sum of the external (translational) and internal (rotational, vibrational, and electronic) contributions. Since the electronic contribution is zero for most ordinary molecules having no unpaired electrons, we can write that sum as

$$C_V(\text{tot}) = C_V(\text{trans}) + C_V(\text{rot}) + C_V(\text{vib}) \qquad (2\text{--}24)$$

Linear molecules have $3n - 5$ vibrational modes (1, if diatomic), and nonlinear molecules have $3n - 6$ vibrational modes. Consequently, we must sum $3n - 5$ or $3n - 6$ terms to calculate the total vibrational contribution $C_V(\text{vib})$ in Equation 2–24. The translational contribution $C_V(\text{trans})$ equals $3R/2$ for all molecules, but the rotational contribution $C_V(\text{rot})$ equals R if the molecule is linear and equals $3R/2$ if the molecule is nonlinear. For the purpose of calculating $C_V(\text{tot})$, it is convenient to consider linear and nonlinear molecules separately.

$$\text{If linear:} \qquad C_V = \frac{5R}{2} + R \sum_{i=1}^{3n-5} \frac{x_i^2\, e^{x_i}}{(e^{x_i} - 1)^2} \qquad (2\text{--}25)$$

$$\text{If nonlinear:} \qquad C_V = 3R + R \sum_{i=1}^{3n-6} \frac{x_i^2\, e^{x_i}}{(e^{x_i} - 1)^2} \qquad (2\text{--}26)$$

The summed functions are called Einstein functions, and the quantity $x_i = hc\tilde{\nu}_i/kT$ where h is Planck's constant, c is the speed of light, $\tilde{\nu}_i$ is the ith vibration energy expressed in units of reciprocal length, k is the Boltzmann constant, and T is the temperature. The constants have their usual values and are given in the declarative section of the Pascal computer program listed in Figure 2–1 that can be used for calculating C_V from these equations.

(Text continues on page 415.)

```pascal
Program Cv2;(* a program to calculate heat capacities at constant volume  *)
            (* for diatomic, polyatomic, linear or nonlinear molecules by *)
            (* calculating the translational, rotational and vibrational  *)
            (* contributions from the usual equations of statistical      *)
            (* thermodynamics, e.g. Lewis and Randall, rev Pitzer and     *)
            (* Brewer, "Thermodynamics," 2nd ed. McGraw-Hill, New York,    *)
            (* 1961, page 59-60)                                          *)

Const h = 6.66186E-34;      (* joule sec              *)
      c = 2.99790250E8;     (* meters/sec             *)
      k = 1.380621E-23;     (* joules/deg/molecule *)
      R = 8.31434;          (* joules/deg/mole        *)

Var  DegreesOfFreedom, (* No. of vib. degrees of freedom in molecule      *)
     NumberOfTemperatureIntervals: Integer;
     Linear:  Boolean; (* A molecule is either linear or nonlinear        *)
     TempLow,          (* Calculate Cv's from here up                     *)
     TempHigh,         (* up to this maximum temperature                  *)
     DeltaT:  Real;    (* Temperature increment between temp calculations *)
     T,                            (* a list of up to 50 temperatures     *)
     Cv:      Array[1..50] of Real; (* a list of up to 50 heat capacities *)
     Vib:     Array[1..10] of Real; (* a list of vibrational energies     *)
                                    (* in wavenumbers.  The number in the *)
                                    (* list = the number of vibrational   *)
                                    (* degrees of freedom                 *)

   (**********************************************************************)

Function Einstein (TempKelvin, WaveNumber:  Real):  Real;
Var    x,                    (* x = hv/kT = hcw/kT  *)
       T,                    (* kelvins             *)
       w:  Real;             (* 1/meter             *)
Begin
   w:= 100*WaveNumber;               (* convert 1/cm to 1/m  *)
   T:= TempKelvin;
   x:= h*c*w/(k*T);
   Einstein:= Sqr(x)*Exp(x)/(Sqr(Exp(x) - 1));
End;

   (**********************************************************************)

Procedure GetMolecularData;
Var answer:  char;
    Mode:       integer;
Begin  (* GetMolecularData *)
   Writeln; Write('Is the molecule linear? (Y or N)   ');
   Readln(answer);
   If (answer = 'Y') or (answer = 'y') Then
      linear:= true
   Else linear:= false;
   Writeln;
   Write('Please enter the number of vibrational degrees of freedom:  ');
   Readln(DegreesOfFreedom);
   Writeln;
   Writeln('Please enter the vibrational energy in wavenumbers.');
   For Mode:= 1 to DegreesOfFreedom Do
   Begin  (* For *)
      Writeln;
      Write('Nubar[',Mode:1,']:   ');
      Readln(Vib[Mode]);
   End;  (* For *)
   Writeln;
End; (* GetMolecularData *)

   (**********************************************************************)
```

Figure 2–1 *Program Cv2.*

```
Procedure GetTempRange;
Begin
   Write('Please enter the lowest temperature(K)   ');
   Readln(TempLow);Writeln;
   Writeln('Please enter the temperature interval and ');
   Writeln('the number of intervals (20 will fill a screen)');
   Write('The temperature interval is:  ');
   Readln(DeltaT);Writeln;
   Write('The number of intervals is:  ');
   Readln(NumberOfTemperatureIntervals);Writeln;
End;

(**********************************************************************)

Procedure CalculateHeatCapacities;
Var Interval,         (* counter for temperature intervals *)
    Mode,             (* counter for vibrational modes      *)
    N, F:  Integer;
    CvVib: Real;
Begin (* CalculateHeatCapacities *)
   N:= NumberOfTemperatureIntervals;
   F:= DegreesOfFreedom;
   For Interval:= 1 to N Do  (* same in number as 1 to N *)
   Begin (* For Interval *)
      T[Interval]:= TempLow + (Interval - 1)*DeltaT;
      CvVib:= 0;   (* Initialize sum of CvVib to zero *)
      For Mode:= 1 to F Do
      Begin
         CvVib:= CvVib + R*Einstein(T[Interval],Vib[Mode]);
         (* Writeln('Cvib = ',CvVib:3:1);*)
      End;
      If linear
         Then Cv[Interval]:= 2.5*R + CvVib
         Else Cv[Interval]:= 3.0*R + CvVib
   End; (* For Interval *)
TempHigh:= T[Interval];  (* Gets last, i.e, highest temperature *)
End;  (* CalculateHeatCapacities *)

(**********************************************************************)

Procedure WriteOutput;
Const Sp5  = '     ';                     (* 5  blank spaces *)
      Sp20 = '                    '; (* 20 blank spaces *)
Var Interval:  Integer;                   (* counter          *)
Begin
Write(Sp5,Sp5,'Heat Capacities (Cv) from ');
Writeln(TempLow:7:2,' to ',TempHigh:7:2,'K');
Writeln;
Writeln(Sp20,'Temperature',Sp5,'Heat Capacity');
Writeln;
For Interval:= 1 to NumberOfTemperatureIntervals Do
   Writeln(Sp20,'   ',T[Interval]:7:2,Sp5,Sp5,Cv[Interval]:5:2)
End;

(**********************************************************************)

Begin (* M A I N   P R O G R A M *)
   GetMolecularData;
   GetTempRange;
   CalculateHeatCapacities;
   WriteOutput;
End.  (* M A I N   P R O G R A M *)
```

Figure 2–1 (continued)

Apparatus

The apparatus, which is quite simple, is illustrated in Figure 2–2. It consists of a large (40- to 50-liter) glass carboy or bottle B, stoppered with a four-hole rubber stopper and connected to a source of a dry gas through stopcock S1 and to an open-end manometer M through another stopcock S3. Stopcock S2 permits sweeping residual air to the atmosphere. An optional gas flow meter between the tank and the apparatus is convenient for adjusting the flow rate. The manometer is filled with an organic liquid ("oil") of low vapor pressure. Some suitable oils are listed in Table 2–1.

The connection to a few centimeters of mercury in the Erlenmeyer flask E acts as a safety valve to prevent the liquid in the manometer from being blown out by the accidental application of too much gas pressure

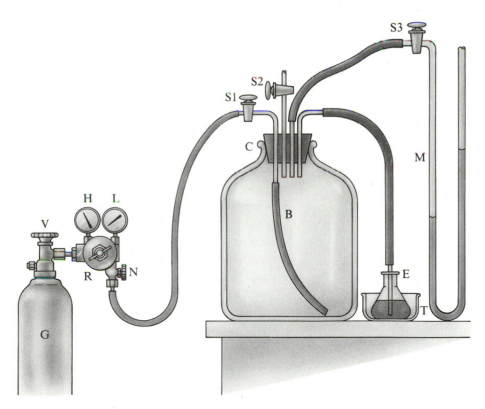

Figure 2–2 Apparatus for measuring heat capacity. S1, S2, and S3 are 4-mm-bore stopcocks, C is a rubber stopper bored with four holes, and B is a 40- to 50-L glass carboy. M is an open-end manometer, about 1 m overall, half filled with a dense organic liquid (Table 2–1). The apparatus is attached to a tank of compressed gas with a two-stage pressure regulator valve. G is the gas tank, H is the high-pressure indicator, L is the low-pressure indicator, R is the regulator valve, N is a nonregulating output needle valve, E is an Erlenmeyer flask filled with mercury to a depth of about 2 cm, and T is a tray.

TABLE 2–1

Low-Vapor-Pressure Organic Oils

Name	Density (g/ml)
Benzoyl benzoate	1.12
α-Bromonaphthalene	1.48
Dibutyl phthalate	1.05
Methyl salicylate	1.18

---

from the connection to the gas tank G. The Erlenmeyer flask E is contained in a tray T to collect any spilled mercury. The depth of the mercury in E should be less than the mercury equivalent of the length of the manometer M, calculated from the ratio of the densities of the organic liquid and the density of mercury (13.6 g/ml).

Various parts of the apparatus are connected by flexible rubber tubing, since the stopper C must be removed and reseated in the course of an experiment.

SAFETY CONSIDERATIONS

Because mercury is toxic and volatile, the Erlenmeyer flask should be stoppered when the apparatus is not in use. The gas tanks should be chained securely to the table adjacent to the apparatus and moved with a dolly designed for the safe movement of high-pressure gas tanks. Since the pressure in the carboy never reaches more than about 1.1 atm, it is not particularly dangerous. Nevertheless, safety glasses should be worn.

EXPERIMENTAL PROCEDURE

The experiment is designed to carry out the changes of state shown in Figure 2–3. The system is brought to state I, then expanded adiabatically and irreversibly to state II, and finally allowed to warm isochorically to state III.

Preliminary Preparations. To prepare to sweep the air from the system, open stopcocks S1 and S2. Check that the tank needle

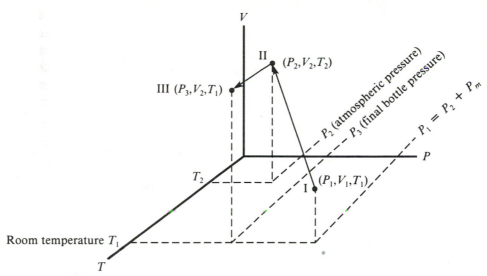

Figure 2–3 Changing the state from I to II to III. Initially the system is at state I (P_1, V_1, and T_1), where P_1 is greater than atmospheric pressure P_2 and T_1 equals room temperature. The gas is expanded adiabatically from state I to state II, where the pressure equals atmospheric pressure P_2 and the gas has cooled to T_2. In going from state II to state III, the gas warms back to room temperature and the pressure drops correspondingly to the final pressure in the bottle P_3. The manometric pressure is P_m.

valve N is closed (clockwise) and that the regulator valve R is closed (counterclockwise). When the main tank valve V is opened, the full-tank high pressure registers on gauge H. When the regulator valve R is screwed in (clockwise), the low regulated pressure registers on gauge L. Set this pressure to about 1 or 2 divisions in gauge L (2 to 5 psi). Double-check that S1 and S2 are open.

To sweep air from the system, open the needle valve N and allow 5 or 6 carboy volumes of gas to pass through the system at about 5 to 6 L/min.

Getting to the Initial State. To get the system to the initial state (I in Fig. 2–3), decrease the flow rate by nearly closing needle valve N, close stopcock S2, and allow the pressure registered by the oil manometer to reach about 500 to 600 mm of oil. Close the needle valve N and stopcock S1 (in that order). Allow the system to stand about 15 minutes so the gas can come to thermal equilibrium with its surroundings, as indicated by a constant reading of the oil manometer. Notice that the pressure of the gas in the system equals the atmospheric pressure plus the manometric pressure, all expressed in the same units, for example, torr:

$$P_{\text{gas}} = P_{\text{atmosphere}} + P_{\text{manometer}} \qquad (2\text{–}27)$$

Record the atmospheric pressure with a barometer, since the pressure in the system equals the atmospheric pressure plus the manometric pressure. Convert the manometric pressures to torr by multiplying by the ratio of the manometric liquid density to the density of mercury.

The Adiabatic Expansion. To carry out the adiabatic expansion from state I to state II (Fig. 2–3), remove the stopper from the carboy and immediately replace it tightly. The carboy should be open to the atmosphere for about 1 second. If this operation is done too quickly, the bottle pressure does not reach atmospheric pressure. If it is done too slowly, the gas remaining in the bottle begins to warm back to room temperature so that the expansion is not completely adiabatic. You are more likely to be too slow than too quick. As soon as the rubber stopper is replaced, the system is at state II.

At state II, the gas has expanded adiabatically and the temperature has dropped from T_1 to T_2. For an instant the oil levels in the manometer are equal. However, as the gas now comes to thermal equilibrium with its surroundings, the temperature rises from T_2 back to the original T_1 and the pressure increases to the slightly higher pressure P_3, the final bottle pressure at the end of the run. This system is now at state III and the run is finished.

Results and Calculations

From the measured values of P_1, P_2, and P_3, calculate γ with Equation 2–22 and the heat capacity with Equation 2–23. The initial total pressure of the gas (torr) is P_1, the atmospheric pressure (torr) is P_2, and the final pressure of the gas after warming back to the initial temperature is P_3.

Measure the heat capacities of a monatomic gas (helium or argon), a diatomic molecule (nitrogen or oxygen), and a polyatomic molecule (carbon dioxide or nitrous oxide). Make each measurement at least in duplicate. Calculate the heat capacities of each gas with Equation 2–25 and the computer program listed in Figure 2–1. Use vibrational energies from the literature.

Tabulate the experimental pressures, γ's, heat capacities, and literature values for heat capacities. Compare C_V(measured) with C_V(calculated), which can be done with the Pascal program listed in Figure 2–1. Discuss the magnitude of the heat capacities in terms of the degree of contribution of the various vibrational modes, the rotational degrees of freedom, and the translational contribution.

Consider the errors associated with the experimental procedure. How would the observed pressures be affected by restoppering the flask too quickly? Too slowly? What would be the effect of reading P_3 too quickly or too slowly? How would these affect the calculated value of γ and of C_V?

REFERENCES

1. Barrow, G., *Physical Chemistry,* 4th ed., McGraw-Hill, New York, 1979.

2. Bertrand, G. L., and H. O. McDonald, "Heat Capacity Ratio of a Gas by Adiabatic Expansion," *J. Chem. Educ.* 63:252, 1986.

3. Désormes and Clément, "Détermination Expérimental du Zéro absolu de la chaleur et du calorique spécifique des Gaz," *Journ. de Phys.,* 1819, pp. 428–455.

4. Gill, S. J., "Maintained Oscillations for the Bouncing-Ball Determination of C_P/C_V," *J. Chem. Educ.* 37:586, 1960.

5. Herzberg, G., *Spectra of Diatomic Molecules,* 2nd ed., Van Nostrand, Princeton, New Jersey, 1950.

6. Lewis, G. N., and M. Randall, revised by K. Pitzer and L. Brewer, *Thermodynamics,* 2nd ed., McGraw-Hill, New York, 1961.

7. Orchard, S. W., and L. Glasser, "Rüchardt's Method for Measuring the Ratio of Heat Capacities of Gases," *J. Chem. Educ.* 65:824–826, 1988.

Calorimetry

Experiment 3

Bomb Calorimetry

Calorimetry is the science of measuring the heat q evolved or absorbed for a chemical or physical change of state. Experimentally, this is generally accomplished under conditions of constant volume or constant pressure. When a change of state occurs in an open vessel, flask or Dewar, the heat measured directly is q_P, since the pressure exerted on a system exposed to the atmosphere is considered constant. If a change in state is allowed to occur under conditions of constant volume, the heat measured is q_V. Because the volume of a bomb is considered constant during a change of state, changes of state occurring in a bomb are considered constant-volume processes.

Many substances, particularly hydrocarbons, react readily with oxygen, that is, they burn. Taking advantage of this, Berthelot in 1881 devised a closed vessel filled with oxygen in which to burn organic compounds and to measure the heat evolved. Present-day calorimeters burn compounds in a vessel called a **bomb** because it is filled with oxygen at high pressure, and is constructed with thick walls of stainless steel to maintain constant volume. Although the combustion reaction probably takes place with explosive violence, there is no external evidence of such a reaction.

In a calorimeter bomb, the hydrocarbon and oxygen are initially at the same temperature, and the products (CO_2 and H_2O) are cooled to within a few degrees of the initial temperature; also, the water vapor formed by the combustion is condensed to the liquid state. Although the initial and final temperatures are not quite the same—differing by the temperature rise of the calorimeter—this difference can be neglected in most work because the heat of reaction is virtually independent of temperature for small temperature changes.

In a bomb calorimeter the heat measured for the sample burned is q_V, which equals ΔU, the difference in internal energy between the initial and final states. We will assume that $\Delta U = \Delta U^0$; that is, we will assume that ΔU is independent of pressure. The enthalpy change for the combustion process, ΔH^0, can be calculated from ΔU^0 with the following equations:

$$\Delta H^0 = \Delta U^0 + \Delta(PV) \tag{3-1}$$

$$\Delta(PV) = \Delta n\, RT \tag{3-2}$$

Here, Δn is the difference in number of moles of *gaseous* products and reactants. For example, the balanced equation for the combustion of naphthalene is

$$C_{10}H_8(s) + 12\,O_2(g) = 10\,CO_2(g) + 4\,H_2O(\ell) \tag{3-3}$$

$$\Delta n = n(CO_2) - n(O_2) = 10 - 12 = -2 \tag{3-4}$$

Thus, for the example of naphthalene, the relationship between the enthalpy of combustion ΔH^0 and the measured energy of combustion ΔE^0 is given by

$$\Delta H^0 = \Delta U^0 - 2\,RT \tag{3-5}$$

Calorimetry

An oxygen bomb calorimeter consists of three essential parts:

1. A bomb, containing the sample and oxygen, in which the combustion takes place.
2. A bucket, holding a measured quantity of water, in which the bomb, thermometer, and stirring device are immersed.
3. A jacket, used to shield the bucket from the surroundings.

Two methods are used to jacket the calorimeter, which give rise to the methods called **isothermal calorimetry** and **adiabatic calorimetry.** These differ not only in the construction of the jacket, but also in the way the temperature readings are recorded.

In the isothermal system, the jacket temperature remains constant while the bucket temperature rises. It requires temperature readings before, between, and after the initial and final temperature readings in order to calculate a radiation correction.

In the adiabatic system, the temperature of the jacket is adjusted, usually automatically, so that it never differs from the temperature of the bucket, which rises during the course of a run. Since the bucket temperature and the surrounding temperature are the same, no heat is transferred between system and surroundings, that is, the process is adiabatic.

In Figure 3–1a the bucket temperature is below the jacket temperature before ignition and above the jacket temperature after ignition, with corresponding rising and falling temperatures before and after ig-

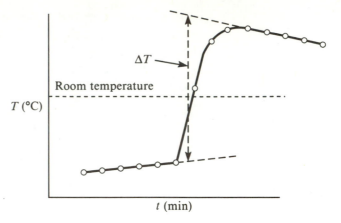

(a) Isothermal calorimeter

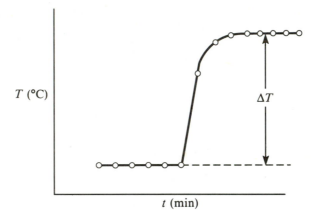

(b) Adiabatic calorimeter

Figure 3–1 Temperature-versus-time curves.

nition. In Figure 3–1b the bucket temperature remains constant until ignition; after the temperature rise due to ignition, the temperature again remains constant, since no heat is transferred between the bucket and the jacket because the jacket and bucket temperatures are equal.

Both calorimetric systems are available commercially. For routine work, the isothermal system is simpler and usually less expensive. The adiabatic system is more complex because controls are necessary to maintain automatically the same jacket and bucket temperature. Since no corrections are needed, it is somewhat more rapid and more convenient to use. Nevertheless, for work of highest precision, the isothermal method is generally more adaptable (Jessup, 1960).

For either system, the apparatus must be calibrated. The heat capacity of the apparatus (bucket, water, and bomb) is found by measuring the temperature rise caused by the combustion of a substance of known heat of combustion, namely, benzoic acid. The heat of combustion of benzoic acid equals -6318.3 cal/g (26436 J/g) (Jessup, 1960).

Corrections. The sample in the bomb is ignited by being placed in contact with a few centimeters of iron fuse wire, which is briefly heated to incandescence electrically. Naturally, the iron wire also burns in the combustion process, giving off a small quantity of heat energy that must be included in the calculations. Because the combustion of the iron is often not complete, it is necessary to weigh the iron wire before and after the combustion. Take care not to confuse iron oxide (nonmagnetic) with unburned iron (magnetic).

Another reaction that contributes to the total heat of combustion is the formation of nitric acid from oxygen and the nitrogen present in the air within the bomb when it is sealed. At the high temperature momentarily reached during combustion, high-pressure oxygen oxidizes nitrogen in the presence of water (formed by the combustion of the hydrocarbon) to aqueous nitric acid.

Calibration Calculation. The energy equivalent of the calorimeter, E (J/K), is calculated from the temperature rise when a weighed sample of pure benzoic acid is burned in a calorimeter:

$$E = \frac{Q_b \times m_b + c_{Fe} + c_N}{\Delta t} \tag{3-6}$$

where

Q_b = heat of combustion of benzoic acid (26435.8 J/g)
m_b = weight in grams of the benzoic acid burned (g)
c_{Fe} = correction (J) for the combustion of iron fuse wire, which amounts to 5858 J/g
c_N = correction (J) for the formation of nitric acid, which amounts to 57.7 kJ/mol
Δt = *corrected* temperature rise in the calorimeter (K)

The quantity of nitric acid formed is determined by washing out the bomb with distilled water after a run and titrating the washings with a standard solution of sodium hydroxide. Thus,

$$c_N = 57.7MV \tag{3-7}$$

where M is the molarity of the sodium hydroxide solution, V is the volume of the sodium hydroxide in milliliters, and c_N is the nitrogen correction in joules. When the molarity is about 0.1 M, the volume V ordinarily amounts to only a few milliliters.

The thermometer should have been previously calibrated and the temperature rise corrected accordingly. Check to determine that the serial number of the calorimeter thermometer and the serial number on the thermometer calibration certificate are the same.

Calculation of Heat of Combustion. After the energy equivalent of the calorimeter has been determined, the heat of combustion of an unknown compound is found by duplicating the calibration experiment with the unknown sample. The experiment should be duplicated as exactly as possible: same pressure of oxygen, same length of fuse wire, same weight of sample, same thermometers, and same weight of calorimeter water. The heat of combustion is given by the equation

$$Q_s = \frac{E_b \times \Delta t - c_{Fe} - c_N}{m_s}$$ (3-8)

where

Q_s = heat of combustion of the sample (J/g) determined under conditions duplicating the calibration

E = energy equivalent of the calorimeter (J/K)

Δt = *corrected* temperature rise in the calorimeter (K)

c_{Fe} = correction for the combustion of iron fuse wire (J)

c_N = correction for the formation of nitric acid (J)

m_s = mass of the sample (g)

Resonance Energy from Heats of Combustion. Cyclohexane and 1,5,9-*trans-,trans-,cis*-cyclododecatriene (TTCC) are nonplanar. Neither is stabilized by resonance energy. On the other hand, aromatic molecules such as benzene are stabilized by resonance energy. The relationship between benzene and the pair cyclohexane and TTCC is shown in Figure 3-2. Since the number of bonds (C—C, C=C and C—H) is identical on both sides of the change of state displayed in Figure 3-2, this "reaction" is exothermic by an amount exactly equal to the resonance energy of benzene. Although it is not possible to measure directly the enthalpy change of the reaction in Figure 3-2, the following reactions suggest a solution:

$$C_6H_6(g) + C_6H_{12}(g) + \tfrac{33}{2} O_2(g) \rightarrow$$ (3-9)

$$12 CO_2(g) + 9 H_2O(g) \qquad \Delta H_I$$

$$C_{12}H_{18}(g) + \tfrac{33}{2} O_2(g) \rightarrow 12 CO_2(g) + 9 H_2O(g) \qquad \Delta H_{II}$$ (3-10)

Equations 3-9 and 3-10 represent the heat of combustion of the hydrocarbons *in the gas phase.* The difference in enthalpies of Equations 3-9 and 3-10 equals the enthalpy change for the reaction in Figure 3-2, which *almost* equals the resonance energy of benzene, differing from it by RT (the difference between ΔE and ΔH for a reaction of the stoichiometry shown in Figure 3-2.

The enthalpy changes for Equations 3-9 and 3-10 differ from experimentally determined heats of combustion by the heats of vaporization of the organic molecules on the left and the heat of vaporization of water on the right. The thermodynamic cycle in Figure 3-3 illustrates the dif-

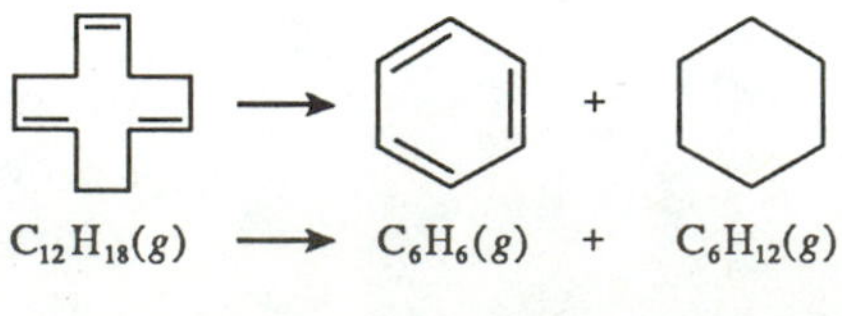

Figure 3-2 The relationship between 1,5,9-*trans-,trans-,cis*-cyclododecatriene and the pair cyclohexane and benzene.

$$C_{12}H_{18}(g) \longrightarrow C_6H_6(g) + C_6H_{12}(g)$$

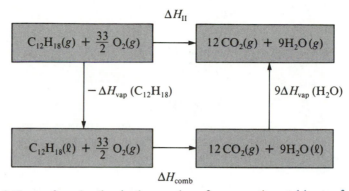

Figure 3–3 Heats of combustion in the gas phase from experimental heats of combustion.

ference between experimental heats of combustion ΔH_{comb} and gas-phase heats of combustion ΔH_{II}. A similar thermodynamic cycle can be drawn for ΔH_I.

It follows from the thermodynamic cycle (Fig. 3–3) that

$$\Delta H_I = \Delta H_{comb}(C_6H_{12}, \ell) \tag{3-11}$$

$$+ \Delta H_{comb}(C_6H_6, \ell) - \Delta H_{vap}(C_6H_{12}, \ell)$$

$$- \Delta H_{vap}(C_6H_6, \ell) + 9\Delta H_{vap}(H_2O, \ell)$$

$$\Delta H_{II} = \Delta H_{comb}(C_{12}H_{18}, \ell) - \Delta H_{vap}(C_{12}H_{18}, \ell) \tag{3-12}$$

$$+ 9\Delta H_{vap}(H_2O, \ell)$$

The resonance energy of benzene equals $\Delta H_{II} - \Delta H_I$:

$$\Delta H_{II} - \Delta H_I = \Delta H_{comb}(C_{12}H_{18}, \ell) - \Delta H_{comb}(C_6H_{12}, \ell) \tag{3-13}$$

$$- \Delta H_{comb}(C_6H_6, \ell) - \Delta H_{vap}(C_{12}H_{18}, \ell)$$

$$+ \Delta H_{vap}(C_6H_{12}, \ell) + \Delta H_{vap}(C_6H_6, \ell)$$

TABLE 3–1

Auxiliary Thermodynamic Data (kJ/mol)

Substance	ΔH_{vap}	ΔH_{comb}
$C_6H_6(\ell)$	33.03[a]	−3267.62[a]
$C_6H_{12}(\ell)$	33.85[a]	−3919.91[a]
$C_{12}H_{18}(\ell)$	67.2[b]	To be measured

Sources:
[a] Pickering, M., "A Novel Bomb Calorimetric Determination of the Resonance Energy of Benzene," *J. Chem. Educ.* 59:318, 1982.
[b] Rauh, H.-J., W. Geyer, H. Schmidt, and G. Geiseler, "Bildungsenthalpien und Mesomerieenergien von π-Bindungssystemen," *Z. phys. Chemie* 253:43–48, 1973.

For this experimental determination of the resonance energy of benzene, the only quantity measured in Equation 3–13 is the heat of combustion of liquid $C_{12}H_{18}$. The remaining required thermodynamic data are listed in Table 3–1.

Apparatus

The apparatus consists of a bomb calorimeter (e.g., Parr model 1200, Fig. 3–4); calibrated thermometers, graduated to 0.01°C; ignition unit;

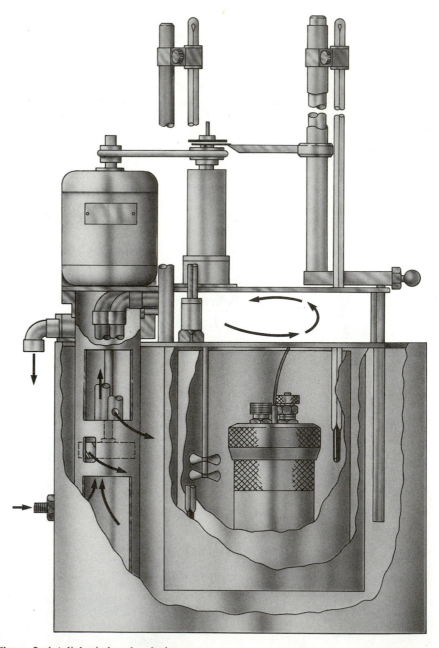

Figure 3–4 Adiabatic bomb calorimeter.

stand for bomb head; ignition cups; iron fuse wire, calorimetric grade; 1,5,9-*trans-,trans-,cis*-cyclododecatriene; benzoic acid; compressed gas tank of oxygen; analytical balance; pellet press; and a balance with a 4000-g capacity and a precision of less than 1 g.

SAFETY CONSIDERATIONS

Do not use too large a sample; the total weight of combustible material *should not exceed 1.10 g* according to the Parr manual. Do not charge the bomb with more than 30 atm of oxygen. If the bomb is overcharged, do not fire it; dismantle it and begin again. Do not fire the bomb if gas bubbles leak from it when it is submerged; dismantle it and begin again. The oxygen tank should be chained into position. Used properly, a bomb calorimeter is no more dangerous than other physical chemistry apparatus; nevertheless, it is prudent to stand back for at least 15 seconds after firing the bomb.

EXPERIMENTAL PROCEDURE

The apparatus is calibrated with duplicate samples of benzoic acid. Weigh out roughly 0.8 to 0.9 g benzoic acid and press into a pellet with a pellet press. Weigh a clean, dry ignition cup to ±0.1 mg; place the pellet in it and weigh again to ±0.1 mg. Place the bomb head on its stand (Fig. 3–5) and put the ignition cup in place. Before every run, polish the two electrodes on the bomb head with fine emery paper before attaching the fuse wire. Cut about 8 to 10 cm of fuse wire and weigh to ±0.1 mg. Loop it from electrode to electrode in such a way that in between it rests gently upon the pellet, but does not come in contact with the combustion cup, as shown in Figure 3–6.

With the pellet in place and the fuse wire attached, place the bomb head on the bomb and screw the cover ring in place tightly. Do not handle roughly; the pellet and fuse wire are in fragile contact. Close the outlet needle valve on the bomb head with the tool provided. Be careful not to use too much force as this is a needle valve.

Check that the valves on the oxygen tank are closed. Attach the inlet tube from the tank to the bomb. Open the main valve on the oxygen tank. Slowly let 25 atm oxygen into the bomb (Fig. 3–7). Do not exceed 30 atm! If you do, dismantle the bomb and begin again. Bleed the oxygen from the connecting line so that it may be detached from the bomb.

Gently place the bomb in the bucket and weigh to ±1 g. Add 2 L water, the temperature of which has previously been adjusted to between 20 and 25°C. The water should just cover the bomb completely. Weigh again and record the water weight. Place the

Figure 3–5 Bomb, cover, and sample cup. (Karen R. Sime)

bucket inside the calorimeter jacket, and attach the hot lead to the bomb. Swing the cover into place. Carefully lower the thermometers down into position in the bucket and in the jacket. Turn the water supply and the water heater on. Turn the bridge resistor all the way counterclockwise; turn the controller on. Manually inject water until

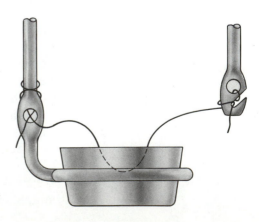

Figure 3–6 Attaching the fuse wire.

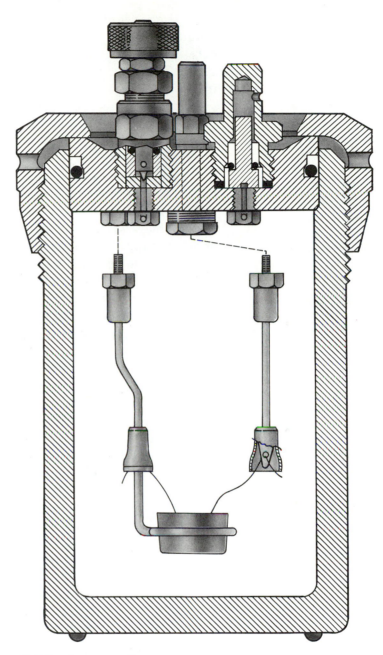

Figure 3–7 Bomb, cross section.

the jacket temperature is equal to or 0.05°C lower than the bucket temperature. Balance the bridge galvanometer and let the system run for 5 minutes to come to thermal equilibrium. If necessary, adjust the bridge resistor occasionally to maintain equality between the bridge and jacket temperatures. After 5 minutes, adjust the jacket temperature to within ±0.005°C of the bucket temperature. The galvanometer should oscillate smoothly, with hot and cold water alternately injected into the jacket.

After the calorimeter has been cycling automatically for about 5 minutes, and the jacket temperature is within $\pm0.005°C$ of the bucket temperature, read and record the bucket temperature to $\pm0.002°C$. Fire the charge by pressing the button on the ignition unit. Keep the bottom depressed until the pilot light goes out, but release it within 5 seconds regardless of the light. It should glow for not more than 2 seconds and then go out when the charge starts to burn.

The calorimeter temperature will begin to rise about 20 seconds after firing. This rise will be rapid during the first few minutes, and then become slower until a stable maximum is reached. As the bucket temperature rises, the controller automatically injects hot water into the jacket in order to maintain adiabatic conditions. When the bucket temperature levels off, the jacket temperature levels off at the same temperature. To be sure that the bucket temperature has indeed leveled off, continue to read and record the bucket temperature at 1-minute intervals until three successive readings agree to within $\pm0.005°C$. Use a thermometer reading lens.

After recording the final bucket temperature, turn off the controller, stop the stirring motor, lift the cover, and swing it out of the way. Lift the bucket out of the jacket and remove the bomb. Open the oxygen outlet valve slowly, so that at least 1 minute is required to relieve the oxygen pressure. Open the bomb and inspect for incomplete combustion indicated by traces of soot. The inside should be clean, but covered with a dewlike layer of water. Discard the run if there is any evidence of incomplete combustion. Often pieces of unburned iron wire are found in the combustion cup or the bottom of the bomb. These must be removed and weighed. Subtract their weights from the original weight of the fuse wire to determine the actual mass of iron burned. For every gram of iron burned, 5858 J should be subtracted from the heat of combustion of the sample. Be careful not to confuse small, round balls of iron oxide with unburned iron.

Calibrate the apparatus twice and determine the heat of combustion of a sample of TTCC twice. Use the same amount of water (±1 g) in the bucket in all four runs, approximately 2000 g.

Results and Calculations

Tabulate the appropriate masses, temperatures, and heats. You may want to plot the temperature data to determine if your temperature-versus-time data appear like those in Figure 3–1. Be sure to correct the thermometer reading with the thermometer calibration certificate. Use Equation 3–6 to determine the energy equivalent of the calorimeter (J/K). The heat of combustion may then be calculated with Equation 3–8. Then calculate the enthalpy change for the combustion reaction.

Calculate the resonance energy of benzene with your measured heat of combustion and the thermodynamic data furnished in Table 3–1.

Analyze the errors in ΔH^0 and compare its value with a literature value. Combine your experimental ΔH^0 with standard heats of formation from the literature for $CO_2(g)$ and $H_2O(\ell)$ to determine the heat of formation of 1,5,9-*trans-,trans-,cis*-cyclododecatriene. Analyze the errors in this calculation, and explain why the relative error in the calculation of the heat of formation may be quite large.

REFERENCES

1. Hemminger, W., and G. Hohne, *Calorimetry,* Verlag Chemie, Weinheim, West Germany, 1984.

2. Jessup, R. S., *Precise Measurement of Heat of Combustion with a Bomb Calorimeter,* National Bureau of Standards Monograph 7, U.S. Government Printing Office, Washington, D.C., 1960.

3. *Oxygen Bomb Calorimetry and Combustion Methods,* Tech. Manual 130, Parr Instrument Co., Moline, Illinois, 1960.

4. Pickering, M., "A Novel Bomb Calorimetric Determination of the Resonance Energy of Benzene," *J. Chem. Educ.* 59:318, 1982.

5. Rauh, H.-J., W. Geyer, H. Schmidt, and G. Geiseler, "Bildungsenthalpien und Mesomerieenergien von π-Bindungssystemen," *Z. phys. Chemie* 253:43–48, 1973.

6. Rossini, F. D., ed., *Experimental Thermochemistry,* Interscience, New York, 1956.

7. Stevenson, G. R., "The Determination of the Resonance Energy of Benzene," *J. Chem. Educ.* 49:781, 1972.

8. Sturtevant, J. M., in A. Weissberger, ed., *Technique of Organic Chemistry,* vol. 1, pt. 1, Interscience, New York, 1959.

9. Sturtevant, J. M., in A. Weissberger, ed., *Techniques of Chemistry,* vol. 1, pt. 5, Wiley-Interscience, New York, 1971.

10. Wheland, G. W., *Resonance in Organic Chemistry,* 2nd ed., Wiley, New York, 1961.

11. Wilhoit, R. C., "Recent Developments in Calorimetry," *J. Chem. Educ.* 44: A571, 1967.

Experiment 4

Heat of Solution

Mixing two substances to form a solution generally results in the evolution or absorption of heat. If the mixing process is carried out at constant pressure, the heat change is equal to the change in enthalpy:

$$q_P = \Delta H \tag{4–1}$$

Most commonly, heats of solution are measured with water as the solvent. Aqueous heats of solution depend not only on the nature of the solute, but also on the relative amounts of solute and solvent. The integral heat of solution is defined as the enthalpy change when one mole of solute is dissolved in a definite number of moles of water, for example,

$$H_2SO_4(\ell) + \quad 2\ H_2O(\ell) = H_2SO_4(2\ H_2O) \quad \Delta H^0 = -41{,}920\ J \quad \textbf{(4-2)}$$

$$H_2SO_4(\ell) + \quad 10\ H_2O(\ell) = H_2SO_4(10\ H_2O) \quad \Delta H^0 = -67{,}030\ J \quad \textbf{(4-3)}$$

$$H_2SO_4(\ell) + \quad 50\ H_2O(\ell) = H_2SO_4(50\ H_2O) \quad \Delta H^0 = -73{,}340\ J \quad \textbf{(4-4)}$$

$$H_2SO_4(\ell) + 200\ H_2O(\ell) = H_2SO_4(200\ H_2O) \quad \Delta H^0 = -74{,}940\ J \quad \textbf{(4-5)}$$

$$\vdots \qquad\qquad\qquad \vdots \qquad\qquad\qquad \vdots$$

$$H_2SO_4(\ell) + aq \qquad\qquad = H_2SO_4(aq) \qquad \Delta H^0 = -96{,}190\ J \quad \textbf{(4-6)}$$

In Figure 4–1, these data are plotted for sulfuric acid and some other substances. It is evident that the enthalpy change per mole solute eventually becomes constant. The integral heat of solution at infinite dilution is the value obtained when further dilution causes no further effect. For sodium chloride this occurs at $nH_2O < 200$; for sulfuric acid, at $nH_2O > 800$.

Experimentally, the integral heat of solution at infinite dilution is found by measuring the integral heats of solution at progressively higher dilutions until the enthalpy change per mole of solute no longer changes. Integral heat of solution data permit calculation of the integral heat of dilution. In this case, the initial state is a solution of some definite concentration. For example, if Equation 4–3 is subtracted from Equation 4–4, the result is

$$H_2SO_4(10\ H_2O) + 40\ H_2O = H_2SO_4(50\ H_2O) \qquad \textbf{(4-7)}$$

$$\Delta H^0 = -6310\ J$$

This is the heat evolved when 1 mole of sulfuric acid dissolved in 10 moles of water is diluted with 40 moles of water to form a new solution containing 1 mole of sulfuric acid dissolved in 50 moles of water.

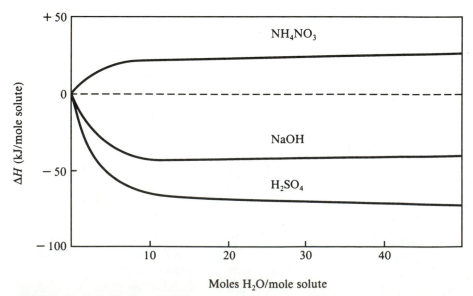

Figure 4–1 Integral heats of solution.

Integral Heats of Formation of Individual Ions

Ionic reactions involving heats of reaction and heats of formation of individual aqueous ions are important in solution calorimetry. Consider the net ionic reaction for the heat of neutralization of any strong acid and strong base:

$$H^+(aq) + OH^-(aq) = H_2O(\ell) \qquad \Delta H^0 = -55.89 \text{ kJ} \qquad (4\text{-}8)$$

In terms of heats of formation:

$$\Delta H^0 = \Delta H_f^0(H_2O) - [\Delta H_f^0(OH^-) + \Delta H_f^0(H^+)] \qquad (4\text{-}9)$$

For this or any reaction involving ions, an equal number of oppositely charged ions must be present. Consequently, there is no way to separate the contributions of the individual ions represented in the square-bracketed expression. If the heat of formation of just one ion—any ion—were known, the contribution of each member of the pair could be separated. Consequently, it is necessary to adopt the convention that the heat of formation of the hydrogen ion is zero at 25°C:

$$\tfrac{1}{2} H_2(g) + aq = H^+(aq) \qquad \Delta H_f^0 = 0 \qquad (4\text{-}10)$$

This equation, combined with the heat of formation of water and Equation 4–8, gives the heat of formation of the hydroxyl ion:

$$H_2O(\ell) \qquad\qquad = H^+(aq) + OH^-(aq) \quad \Delta H^0 = \quad +55.89 \text{ kJ} \quad (4\text{-}11)$$

$$H_2(g) + \tfrac{1}{2} O_2(g) \quad = H_2O(\ell) \qquad\qquad\qquad \Delta H^0 = -285.84 \text{ kJ} \quad (4\text{-}12)$$

$$H^+(aq) \qquad\qquad = \tfrac{1}{2} H_2(g) \qquad\qquad\quad \Delta H^0 = \qquad\quad 0 \text{ kJ} \quad (4\text{-}13)$$

$$\tfrac{1}{2} H_2(g) + \tfrac{1}{2} O_2(g) = OH^-(aq) \qquad\qquad \Delta H_f^0 = -229.95 \text{ kJ} \quad (4\text{-}14)$$

Now that the heat of formation of the hydroxyl ion is known, it can be combined with thermal data for many cations to determine their heats of formation. Similarly, the heat of formation of $H^+(aq)$, zero by convention, can be used with a variety of metals to find the heats of formation of their cations by measuring heats of solutions of metals in dilute acid, for example,

$$Zn(s) + 2 H^+(aq) = Zn^{2+}(aq) + H_2 \qquad \Delta H^0 = -152.42 \text{ kJ} \qquad (4\text{-}15)$$

The heat of this reaction in terms of heats of formation is

TABLE 4–I

Heats of Formation at 25°C (kJ/mol)

Compounds	ΔH_f	Cations	ΔH_f	Anions	ΔH_f
AgCl(s)	−127.03	$Ag^+(aq)$	105.90	$Br^-(aq)$	−120.92
$H_2O(\ell)$	−285.84	$H^+(aq)$	0.0	$Cl^-(aq)$	−167.44
KOH(s)	−425.85	$K^+(aq)$	−251.21	$OH^-(aq)$	−229.95
ZnS(s)	−189.5	$Zn^{2+}(aq)$	−152.42	$S^{2-}(aq)$	41.8

$$-152.42 \text{ kJ} = \Delta H_f(\text{Zn}^{2+}) + 0 - 0 - 0 \tag{4-16}$$

$$\Delta H_f(\text{Zn}^{2+}) = -152.42 \text{ kJ} \tag{4-17}$$

Combining heats of formation of known cations with suitable heat of solution data and heats of formation of pure substances results in further heat of formation data for anions. A few representative data are listed in Table 4–1.

Heats of Solution and Reaction from Heats of Formation of Ions

From such data, we can calculate heats of solution. For example, we can calculate the heat of solution of $\text{KOH}(s)$:

$$\text{KOH}(s) + \text{aq} = \text{K}^+(aq) + \text{OH}^-(aq) \tag{4-18}$$

$$\Delta H_{\text{soln}} = \Delta H_f^0(\text{K}^+) + \Delta H_f^0(\text{OH}^-) - \Delta H_f^0(\text{KOH}) \tag{4-19}$$

$$\Delta H_{\text{soln}} = -251.21 - 229.95 + 425.85 = -55.31 \text{ kJ}$$

From a slightly different point of view, we can calculate the heat of precipitation of a very slightly soluble substance, $\text{AgCl}(s)$:

$$\text{Ag}^+(aq) + \text{Cl}^-(aq) = \text{AgCl}(s) \tag{4-20}$$

$$\Delta H_{\text{pptn}} = \Delta H_f^0(\text{AgCl}) - \Delta H_f^0(\text{Ag}^+) - \Delta H_f^0(\text{Cl}^-) \tag{4-21}$$

$$= -127.03 - 105.90 + 167.44 = -65.49 \text{ kJ}$$

Measuring Heats of Solution

Experimental heats of solution are measured in ordinary Dewar flasks open to the atmosphere, equipped with a stirrer, a temperature-sensing device (thermometer or thermistor), and a heater. The heater is used to calibrate the calorimeter by measuring the temperature rise caused by the absorption of a definite amount of electrical energy:

$$C = \frac{I^2 R t}{\Delta T_{\text{calib}}} \tag{4-22}$$

When the current I is in amperes, the resistance R in ohms, and the time t is in seconds, the heat capacity C has units of joules per kelvin. Measurement of the temperature change caused by a sample of g grams and M molecular weight gives the heat of solution in joules per mole:

$$\Delta H = \frac{CM \, \Delta T_{\text{soln}}}{g} \tag{4-23}$$

Apparatus

A heat of solution calorimeter consists of a Dewar flask, stirrer, and heater. A control unit for passing an accurately measured constant electric current through the heater and standard resistor is used for measuring the potential developed and the time. A thermistor and Wheatstone bridge are needed, and about 20 g potassium nitrate.[1]

SAFETY CONSIDERATIONS

The calorimeter Dewar flask should be well taped and safety goggles should be worn.

EXPERIMENTAL PROCEDURE

Preliminary Preparations. Weigh 2 to 2.5 g (± 0.001 g) potassium nitrate (KNO_3) on a tared, glossy weighing paper. With a graduated cylinder, add 400 ± 0.5 ml water to the calorimeter Dewar flask. The temperature of the water should be within a few degrees of room temperature. Assemble the Dewar flask with the stirrer, heater, and cover. Be sure the stirrer moves freely, without touching the walls of the Dewar flask.

Calibration of the Calorimeter. As already pointed out (Eq. 4–22), you calibrate a calorimeter by finding its energy equivalent, C (J/K). This is nearly always accomplished by directing a known quantity of electrical energy into the system plus calorimeter. This energy is dissipated in a heater resistor R_h immersed in the calorimeter. To calculate the calibration energy, it is necessary to know the current flowing through the heater resistor. The heater current is usually not measured directly with an ammeter, because the current can be more accurately measured indirectly potentiometrically. The essential circuitry of a calorimeter is shown in Figure 4–2.

Because the two resistors, R_h and R_s, are in series, the same current I passes through both. (The subscripts h and s mean *heater* and *standard,* respectively.) The heater resistor R_h is a coil of nichrome wire of unknown resistance; the R_s is a standard resistor of accurately known resistance ($\pm 0.05\%$). The electrical energy q_h dissipated in the heater resistor is

$$q_h = I^2 R_h t \qquad (4\text{--}24)$$

[1] A commercial unit is available: Parr 1451 Solution Calorimeter, Parr Instrument Company, 211 53rd Street, Moline, Illinois 61265.

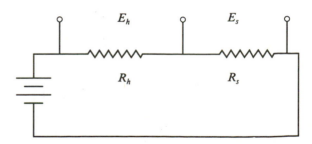

Figure 4–2 Calorimeter calibration circuit.

The current flowing through the two resistors is found by measuring the potential E_s over the standard resistor:

$$I = \frac{E_s}{R_s} \qquad (4\text{--}25)$$

Now the resistance of the heater R_h may be found if the potential E_h over the heater resistor is measured:

$$R_h = \frac{E_h}{I} \qquad (4\text{--}26)$$

Substitution of Equations 4–25 and 4–26 into Equation 4–24 gives an equation for the calibration energy in terms of the time and potentiometric measurements:

$$q_h = \frac{E_h E_s t}{R_s} \qquad (4\text{--}27)$$

The energy equivalent of the calorimeter C is

$$C = \frac{q_h}{\Delta T} \qquad (4\text{--}28)$$

It is not actually necessary to measure the temperature if a thermistor is used as a temperature-measuring device. Instead, the resistance of the thermistor may be balanced against the dial divisions of a Wheatstone bridge, and the energy equivalent of the calorimeter may then be expressed in joules per dial divisions:

$$C = \frac{q_h}{\Delta D_h} = \frac{E_n E_s t}{R_s \Delta D_h} \qquad (4\text{--}29)$$

where ΔD_h is the change in Wheatstone bridge dial divisions recorded upon balancing the bridge before and after heating the heater resistor for t seconds. The data for the calibration consist of the points lying on the line segments ABCDEF in Figure 4–3. The length of the segment ED corresponds to ΔD_h.

Measuring the Heat of Solution. In a calorimeter consisting of a 1-L Dewar flask containing about 800 ml water, a 4- to 5-g sample of potassium nitrate is about right. Determining the heat of solution consists of measuring the change in bridge dial divisions when the sample of nitrate is added to the Dewar flask. This corresponds to determining the length of line segment BC in Figure 4–3.

A Practical Calorimeter. Solution calorimeters are available commercially, but the one you will use is probably not of commercial origin. It is likely that it was built locally. Calorimeter designs differ widely in detail but have a number of features in common, as already indicated. The design of a simple, inexpensive, and convenient calorimeter is shown in Figure 4–4. It consists of three physically separable components, enclosed within the dotted lines of Figure 4–4.

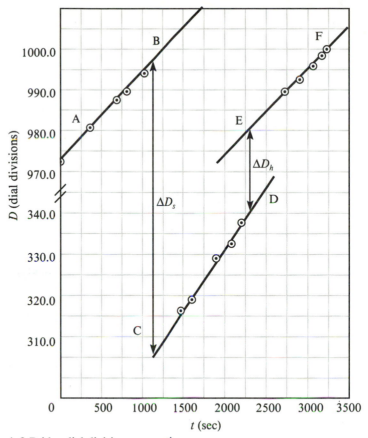

Figure 4–3 Bridge dial division versus time.

The first component, the calorimeter, is an ordinary 1-L Dewar flask fitted with a 2-cm-thick plastic cover through which four holes are drilled. Through the holes pass a stirrer, a heater resistor R_h, and a YSI[2] thermistor. The heater is constructed of 20 turns of No. 26 nichrome wire, glued to a plastic rod and contained in a piece of glass tubing extending out of the calorimeter. The tube is filled with mineral oil to improve transfer of heat from the heater to the solution. The fourth opening allows for a sample to be introduced.

The second component is a commercially available Wheatstone bridge (e.g., Sargent[3]), of which the thermistor forms one leg. The variable resistors in the bridge are helical potentiometers with 1000 divisions. The bridge balance point is detected with an electronic galvanometer (e.g., Honeywell[4]). The energy equivalent of the calorimeter C is determined in joules per division.

The third component is a control unit, the three main parts of which are a digital readout timer, a digital potentiometer, and a small regulated power supply, which supplies power to the heater

[2] Yellow Springs Instrument Company, Inc., Box 465, Yellow Springs, Ohio 45387.
[3] Sargent-Welch Scientific Company, 9520 Midwest Ave., Garfield Heights, Cleveland, Ohio 44125.
[4] Honeywell, Inc., 440 Bernardo Ave., Mountain View, California 94043.

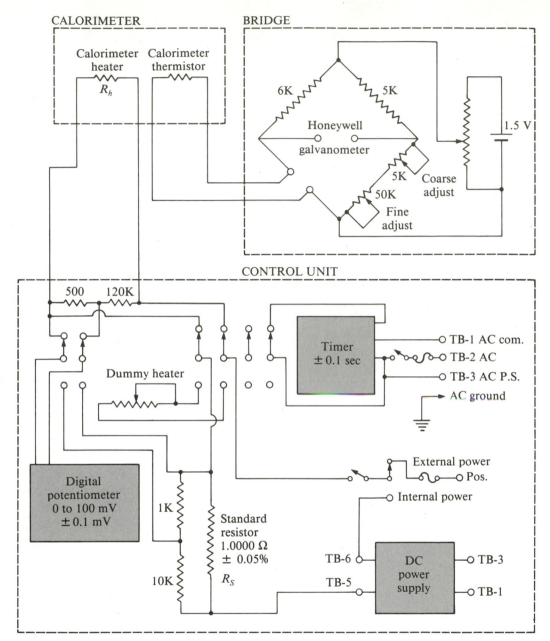

Figure 4–4 Calorimeter control circuit.

resistor. The power to the heater resistor and the timer is switched on and off simultaneously with a four-pole, two-throw switch. A second timer (not shown) allows measurement of the total run time. When the power supply is not heating the heater resistor, it is given a dummy load so that drift is minimized when it is heating the calorimeter heater resistor. A standard resistor R_s is wired in series with the heater resistor R_h. The voltage E_h over the heater resistor R_h is reduced to V_h with a voltage divider, and the voltage E_s over the standard resistor R_s is similarly reduced to V_s. The voltages V_h and V_s are measured with a digital potentiometer, which is switched back and forth between the heater and standard resistors. For the voltage divider resistor shown in Figure 4–2, the values of E_s and E_h are calculated from

$$E_s = \frac{11000V_s}{1000} \qquad (4\text{--}30)$$

$$E_h = \frac{120500V_h}{500} \qquad (4\text{--}31)$$

Making a Run. Check the control unit; the heater switch should be on "dummy," not on "heater." Turn the control unit and bridge galvanometer on and let them warm up several minutes. Zero the run timer and the heater timer. The potentiometer should be set to monitor the potential (V_s) across the standard 1-ohm ($\pm0.05\%$) resistor R_s. The value should be around 0.06 volts.

When the Dewar flask and water have come to thermal equilibrium after being stirred several minutes, set the initial dial settings on the Wheatstone bridge. Set the sensitivity dial on the bridge to maximum and lock it. Set the sensitivity dial on the Honeywell null galvanometer about a half-turn less than maximum. Do not disturb these settings for the duration of the experiment.

Set the fine-adjust dial of the Wheatstone bridge near the high end of the scale, around 950. With this initial setting on the fine-adjust dial, null the Honeywell galvanometer by adjusting the coarse-adjust dial of the bridge. Because of the extreme sensitivity of the galvanometer, it may be necessary to finish the actual nulling with the fine adjust of the bridge; when nulled, the fine adjust should have a reading over 900.

Figure 4–3 shows that making a run consists of five parts, indicated by the five line segments, plotted over the data in the figure.

1. Start the run timer and null the galvanometer every 50 seconds until enough data are gathered to determine the slope of line AB.

2. Open the calorimeter cover, quickly dump in the sample of KNO_3, and record the run time at the moment of dumping the sample. Line BC is a nearly vertical line; its ordinate value is the run time of dumping the sample.

3. Null the galvanometer again with the bridge fine adjust. Continue to null every 50 seconds until enough data are gathered to establish the slope of line CD.

4. Switch the heater switch from "dummy" to "heater" and record the run time at the moment of switching. The heater timer automatically begins to run as a constant potential is applied to the calorimeter heater. Heat for 100 to 200 seconds. While heating, switch the potentiometer back and forth between the potential over the standard resistor and the calorimeter heater. They should be constant within about ±0.05%. Record the reading of V_h and V_s. The digital readout potentiometer gives values around 60 and 70 mV, respectively. Line DE is a vertical line; it is drawn exactly at the halfway time of heating (run time + heating time/2). Note that heating more than about 200 seconds may result in exceeding the number of dial divisions on the fine-adjust dial.

5. Null the galvanometer with the fine adjust. Continue to null the galvanometer every 50 seconds until enough data are gathered to establish the slope of line EF.

Results and Calculations

The electrical heating energy q_{elect} in joules is given by Equation 4–27. When Equations 4–30 and 4–31 are combined with Equation 4–29, the energy equivalent of the calorimeter C (joules/bridge dial division) is

$$C = 2651 V_s V_h t / \Delta D_h \tag{4–32}$$

where ΔD_h is the change in bridge fine-adjust dial divisions occurring when the system is heated for t seconds with R_s set to 1.000 ohm (line segment DE in Fig. 4–3), and V_h and V_s are the potentials in volts across the voltage dividers in Figure 4–4.

The heat of solution is given by

$$\Delta H = MC \, \Delta D_{soln} / g \tag{4–33}$$

where M is the formula weight, g is the mass of the sample in grams, and ΔD_{soln} is the change in bridge fine-adjust dial divisions occurring during the solution process (line segment BC in Fig. 4–4).

Calculate the heat of solution (kJ/mol) for four samples of potassium nitrate. Compare the average with literature values. Determine the propagated errors.

Other Systems Suitable for Calorimetric Investigation

A wide variety of processes lend themselves to calorimetric study. In addition to the solution process, enthalpies of formation, mixing, neutralization, ionization, reaction, displacement, and complex ion formation have been studied (Table 4–2).

The reaction between silver ion and zinc (Hill et al., 1965) is interesting because the heat of reaction can also be measured potentiometrically as described in Experiment 22. The potentiometric heat of reaction can then be compared with the direct calorimetric heat of reaction.

To measure the heat of the reaction between silver ion and zinc with a calorimeter with the sensitivity of the apparatus described in this experiment, use 400.00 ml (four 100.00-ml aliquots) of a standard solution of silver sulfate about 0.00625 M. Because silver sulfate is the limiting reagent, the volume of its solution must be accurately known, but the calorimeter Dewar flask need not be perfectly dry. The optimum concentration and amount of zinc to be added depend on the apparatus

TABLE 4–2

Some Suitable Calorimetric Systems

1. Enthalpy of formation: MgO(s) (Mahan, 1960), LiOH (Matthews, 1963)
2. Enthalpy of mixing: chloroform–acetone, carbon tetrachloride–acetone, chloroform–carbon tetrachloride (Zaslow, 1960)
3. Enthalpy of neutralization/ionization: sulfamic, acetic, monochloroacetic, oxalic, and tartaric acids (Miller et al., 1947)
4. Enthalpy of precipitation: $Mg(OH)_2(s)$, $Al(OH)_3(s)$ (Pattison et al., 1943); $AgCl(s)$, $AgBr(s)$, $PbI_2(s)$ (Clever, 1961)
5. Enthalpy of reaction (Pattison et al., 1943):

   $H_2O_2 + MnO_2 =$

   $CaC_2 + HCl =$

   $KBrO_3 + HCl =$

   $(CH_3CO)_2O + NaOH =$
6. Enthalpy of complex ion formation: $HgCl_4^{2-}$, $Cu(NH_3)_4^{2+}$ (Pattison et al., 1943)
7. Enthalpy of displacement:

   $Cu^{2+} + Zn(s) = Zn^{2+} + Cu(s)$ (Charlesworth and Patch, 1932)

   $Ag^+ + Zn(s) = Ag(s) + Zn^{2+}$ (Hill et al., 1965)

used and are chosen to give an adequate temperature change. Allow the solution to come to thermal equilibrium and add about 10 g zinc dust. Ten grams of zinc is a large excess and does not need to be accurately measured. With a large excess of zinc the reaction takes place rapidly and to completion. Silver nitrate is not a suitable substitute for silver sulfate, because zinc reacts too slowly with the silver nitrate, resulting in an inconveniently slow temperature rise (Hill et al., 1965). However, silver sulfate dissolves in water very slowly, so the solution should be prepared in advance.

Since the reaction is exothermic, it may be necessary to cool the solution to approximately the original temperature before calibrating the calorimeter electrically. This is conveniently accomplished by adding three or four pea-sized pieces of solid dry ice to the reaction mixture. When the original temperature is reached, let the calorimeter stand several minutes to come to thermal equilibrium before the electrical calibration.

Calibrate the calorimeter; calculate the heat of reaction as described above and compare it with the heat of reaction measured potentiometrically in Experiment 22.

REFERENCES

1. Charlesworth, M. E., and E. M. Patch, "Heat of Reactions," *School Science Review* 13:256, 1932.

2. Clever, H. L., "Heat of Precipitation," *J. Chem. Educ.* 38:470, 1961.

3. Hill, D. L., S. J. H. Moss, and R. L. Strong, "Heat of Reaction in Aqueous Solution by Potentiometry and Calorimetry," *J. Chem. Educ.* 42:541, 1965.

4. Lewis, G. N., and M. Randall, revised by K. Pitzer and L. Brewer, *Thermodynamics,* 2nd ed., McGraw-Hill, New York, 1961.

5. Mahan, B. H., "A Simple Ice Calorimeter," *J. Chem. Educ.* 37:634, 1960.

6. Matthews, P. G., "A Simple Experimental Test of Hess's Law," *School Science Review* 45:194, 1963.

7. Miller, J. G., A. I. Lowell, and W. W. Lucasse, "Calorimetric Studies of Neutralization Reactions," *J. Chem. Educ.* 24:121, 1947.

8. Neidig, H. A., H. Schneider, and T. G. Teates, "Thermochemical Investigations for a First-Year College Chemistry Course," *J. Chem. Educ.* 42:26–31, 1965.

9. Pattison, D. B., J. G. Miller, and W. W. Lucasse, "Simplified Calorimetric Studies of Various Types," *J. Chem. Educ.* 20:319, 1943.

10. Rossini, F. D., ed., *Experimental Thermochemistry,* Interscience, New York, 1956.

11. Sturtevant, J. M., in A. Weissberger, ed., *Technique of Organic Chemistry,* vol. 1, pt. 1, Interscience, New York, 1959.

12. Sturtevant, J. M., in A. Weissberger, ed., *Techniques of Chemistry,* vol. 1, pt. 5, Wiley-Interscience, New York, 1971.

13. Zaslow, B., "The Heat of Mixing of Organic Liquids," *J. Chem. Educ.* 37: 578, 1960.

Phase Equilibria

Experiment 5

Liquid–Vapor Equilibrium for a Liquid

The vapor pressure of a pure liquid is a simple example of heterogeneous physical equilibrium: heterogeneous, because the equilibrium is between two phases; physical, because no chemical change occurs. The vapor pressure of a substance increases rapidly as the temperature increases, and vapor pressure measurements invariably involve equilibrium measurements over a range of temperatures; therefore, one must be aware of the dependence of the enthalpy, heat capacity, free energy, and equilibrium constants on temperature. In this experiment, we shall measure the vapor pressure of a pure liquid over a range of temperature and use the data obtained to calculate the enthalpy and entropy of vaporization.

The equilibrium between a pure liquid, substance A, and its vapor at pressure P is written

$$A(\ell, P) = A(g, P) \tag{5-1}$$

The equilibrium constant is given by

$$K = \frac{a_A(g)}{a_A(\ell)} \tag{5-2}$$

The activity of the pure liquid $a_A(\ell)$ equals unity, and the activity of the gas phase $a_A(g)$ equals P_A/P_A^0. The usual choice of standard state P_A^0 is the ideal gas at 1 atm or 1 bar. The equilibrium constant expression is then written

$$K = P_A \tag{5-3}$$

443

where P_A is the vapor pressure in atmospheres at temperature T. The equilibrium constant (Eq. 5–2) is dimensionless because it is the ratio of two activities.

When the vapor pressure of a substance is known over a temperature interval, the enthalpy, free energy, and entropy of vaporization can be calculated. How the calculation is carried out, however, depends on the value of ΔC_P for the vaporization process, that is, on whether ΔC_P is appreciable or whether ΔC_P is assumed to be zero or negligible.

Case I: ΔC_P is Zero or Negligible

Whether $\Delta C_P = 0$ or not, the free energy of vaporization is related to the vapor pressure by

$$\Delta G^0 = -RT \ln K = -RT \ln P_A \qquad (5\text{–}4)$$

If $\Delta C_P = 0$, then ΔH^0 and ΔS^0 are independent of temperature, so that the free energy at any temperature is given by

$$\Delta G^0 = \Delta H^0 - T \, \Delta S^0 \qquad (5\text{–}5)$$

Consequently, Equations 5–4 and 5–5 can be combined to give an equation for the vapor pressure as a function of temperature:

$$\ln P = \frac{-\Delta H^0}{RT} + \frac{\Delta S^0}{R} \qquad (5\text{–}6)$$

If vapor pressures P are measured over a range of temperatures, a plot of $\ln P$ against $1/T$ gives a straight line of slope equal to $-\Delta H^0/R$ and an intercept equal to $\Delta S^0/R$, permitting the evaluation of ΔH^0 and ΔS^0 for

$$A(\ell, P^0 = 1 \text{ atm}) = A(g, P^0 = 1 \text{ atm}) \qquad (5\text{–}7)$$

For vaporization, both ΔH^0 and ΔS^0 are always positive quantities.

Case II: ΔC_P Is Not Equal to Zero

If $\Delta C_P \neq 0$ for the vaporization process, then it is convenient to express ΔC_P as a power series in T, usually of the form

$$\Delta C_P = \Delta a + \Delta b T + \Delta c T^{-2} \qquad (5\text{–}8)$$

All three terms on the right are rarely known for most substances; sometimes the first two terms are known. In practice, it is often necessary to estimate the values of a for the liquid and gas phases and settle for one term.

The starting point for considering the variation of K or ΔG with temperature is

$$\frac{d(\Delta G^0/T)}{dT} = \frac{-\Delta H^0}{T^2} \qquad (5\text{–}9)$$

where ΔH^0 is the heat of reaction (or vaporization) as a function of temperature, given by

$$\Delta H^0 = \Delta H^* + \Delta aT + \Delta bT^2/2 - \Delta cT^{-1} \qquad (5\text{--}10)$$

Equation 5–10 is obtained by substituting Equation 5–8 in

$$d(\Delta H^0) = \Delta C_P^0\, dT \qquad (5\text{--}11)$$

and integrating. The integration constant is given the symbol ΔH^*. If Equation 5–10 is substituted in Equation 5–9 and the result is integrated, one obtains a general expression for the free energy as a function of temperature:

$$\Delta G^0 = \Delta H^* - aT \ln T - \Delta bT^2/2 - \Delta cT^{-1}/2 + IT \qquad (5\text{--}12)$$

The integration constant for this integration is given the symbol I. This is an equation for the free energy as a function of temperature for the general case $\Delta C_P \neq 0$.

Since Equation 5–4 is always valid, it may be combined with Equation 5–12 to give a general equation for the equilibrium constant (or vapor pressure) as a function of temperature for the general case $\Delta C_P \neq 0$:

$$-R \ln K + a \ln T + bT/2 + cT^{-2}/2 = \Delta H^*/T + I \qquad (5\text{--}13)$$

At each temperature at which K (which equals P for vapor pressure equilibria) is measured, the left side of Equation 5–13 may be evaluated. The left side of Equation 5–13 is usually given the symbol Σ:

$$\Sigma = \Delta H^*/T + I \qquad (5\text{--}14)$$

If Σ is plotted against $1/T$, the result is a straight line of slope equal to ΔH^* and intercept equal to I. Once the integration constants ΔH^* and I have been determined in this manner, ΔH^0 can be evaluated at any

TABLE 5-1

Selected Heat Capacities C_P (J K^{-1} mol^{-1})

Substance	Formula	Liquid	Gas
Acetone	C_3H_6	125.0	74.9
Benzene	C_6H_6	136.1	81.67
Carbon tetrachloride	CCl_4	131.7	83.4
Chloroform	$CHCl_4$	116.3	65.7
Cyclohexane	C_6H_{12}	156.5	106.3
Hexane	C_6H_{14}	195.0	146.7
Methanol	CH_3OH	81.6	43.9

Source: Landolt-Börnstein, *Numerical Data and Functional Relationships in Physics, Chemistry, Astronomy, Geophysics, and Technology,* 6th ed., H. Borchers, H. Hausen, K.-H. Heilwege, and K. Schäfer, eds., vol. II, pt. 4, sec. 22 132, Springer-Verlag, Berlin, 1966.

temperature with Equation 5–10, and K (or P, the vapor pressure) can be evaluated at any temperature with Equation 1–13.

The sigma plot is necessary when $\Delta C_P \neq 0$; only when $\Delta C_P = 0$ is Equation 5–6 valid, permitting a linear plot of ln K against $1/T$. The sigma plot is frequently used in studies of equilibria at very high temperatures where the temperature range is great and ΔC_P is not negligible.

Apparatus

The vapor pressure apparatus (Fig. 5–1) consists of a special reflux flask, thermometer (0.2°C divisions), cold finger, manometer, ballast bottle (20 liter), three-way stopcock, mechanical vacuum pump, Variac, and heating mantle. An organic liquid boiling between 70 and 90°C (Table 5–1) is needed.

SAFETY CONSIDERATIONS

Most organic liquids are toxic; many are inflammable; some are known or suspected carcinogens. Always wear safety goggles when working with a vacuum system. The ballast bottle is a heavy-walled 5-gal vessel; it should be shielded by a large plastic or sheet-metal screen.

EXPERIMENTAL PROCEDURE

The apparatus illustrated in Figure 5–1 allows the measurement of the boiling of a liquid at various external pressures established by the experimenter. When the three-way stopcock is positioned so that the apparatus is open to the atmosphere, the external pressure equals the current barometric pressure and the boiling temperature is approximately the normal ($P_{ext} = 1$ atm) boiling point. With the pump on, the stopcock can be turned to partially evacuate the system; this lowers the applied external pressure slightly, so that the organic liquid boils at a slightly lower temperature.

To carry out a run, fill the flask about one-third full with a suitable organic liquid. Turn on the water to the cold finger. Position the stopcock so that the system is open to the pump. With the heating mantle turned off, turn on the pump and evacuate until the system pressure P is about 80 to 100 torr. Then position the stopcock so the system is closed to the pump and to the atmosphere. In general, the pressure of the system is given by

$$P = P_b - (h_u - h_l) \qquad \text{(5–15)}$$

where P_b is the measured barometric pressure and h_u and h_l are the upper and lower levels of the mercury manometer. The barometer reading must be corrected according to Table 12–4 on page 317. With an open-end manometer, $P = P_b$ when the system is open to the atmosphere and the upper and lower levels of mercury are equal.

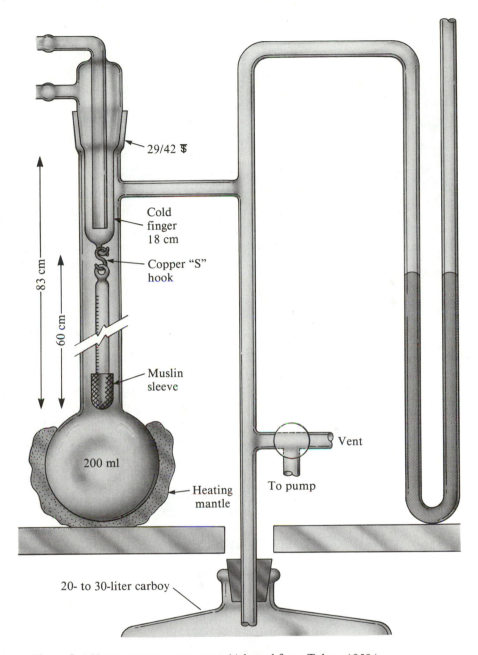

Figure 5–1 Vapor-pressure apparatus. (Adapted from Tobey, 1958.)

If the pressure of the closed system varies with time, the system has a leak and the stopcocks may need regreasing or the hose connections may need to be tightened. If no leaks are present, turn off the heating mantle and set the Variac to about 75% of capacity. After the liquid begins to boil, control the heating so that about 5 to 15 drops per minute drip fron the wick covering the thermometer bulb. Read the two manometer mercury levels. Within this drip rate, the temperature should remain constant.

After recording the mercury levels and the boiling temperature, open the stopcock and increase the pressure by about 50 torr. Record the new mercury levels and the boiling temperature. Adjust the heating and drip rate if necessary.

Continue to measure boiling points as the applied pressure is increased until finally the stopcock is open to the atmosphere. Collect 10 to 12 approximately evenly spaced P and T pairs with increasing pressure. The heating rate may have to be increased periodically.

After the highest P and T pair has been measured, open the stopcock to the pump until the pressure has dropped 50 to 100 torr. If the drip rate is too fast, decrease the heating until it is proper. Again, measure the boiling point and the mercury levels. Continue the measurement of 10 to 12 P and T pairs as the pressure is decreased to 80 to 100 torr. After collecting the data, open the system to the atmosphere and turn off the pump, Variac, and condenser water.

Results and Calculations

Tabulate P_{torr}, P_{atm}, T, $\ln P$, Σ, and $1/T$. Prepare three graphs. Plot P versus T, $\ln P$ versus $1/T$, and Σ versus $1/T$. Determine the slopes and intercepts of the graphs with a least-squares analysis of the data. Report standard deviations in calculated parameters.

From the second two plots, calculate ΔH^0, ΔG^0, and ΔS^0 for the vaporization process at 25°C. As a standard state, choose $P^0 = 1$ atm (Eq. 5–7).

Compare your measured values of the heats of vaporization from the $\ln P$ versus $1/T$ and the Σ plot with each other and with the values reported in the literature. Comment on the uncertainties.

REFERENCES

1. Lewis, G. N., and M. Randall, revised by K. S. Pitzer and L. Brewer, *Thermodynamics,* McGraw-Hill, New York, 1961.

2. Ramsey, W., and S. Young, "On a New Method for Determining the Vapour Pressures of Solids and Liquids and on the Vapour Pressure of Acetic Acid," *J. Chem. Soc.* 47:42, 1885.

3. Thomson, G. W., in A. Weissberger, ed., *Technique of Organic Chemistry,* 3rd ed., pt. 1, Interscience, New York, 1959.

4. Thomson, G. W., and D. R. Douslin, in A. Weissberger, ed., *Techniques of Chemistry,* vol. 1, pt. 5, Wiley-Interscience, New York, 1971.

5. Tobey, S. W., "Vapor Pressure Apparatus," *J. Chem. Educ.* 35:352, 1958.

Experiment 6

Liquid–Vapor Equilibrium in an Azeotropic Mixture

In a homogeneous mixture of two liquids, the vapor pressures are quite different from the vapor pressures of the pure components. Three different cases may be recognized: those that obey Raoult's Law, and those that deviate positively or negatively from Raoult's Law. For an ideal solution, Raoult's Law is written

$$P_i = X_i P_i^0 \qquad \qquad (6\text{-}1)$$

where P_i is the partial pressure of the ith component, P_i^0 is the vapor pressure of the *pure* ith component, and X_i is the mole fraction of the ith component in the liquid phase. The three cases are illustrated in Figure 6–1. The corresponding curves for the boiling point as a function of the gas and liquid phase composition are more or less the inverse of these graphs, since a mixture with a low total vapor pressure has a high boiling point, and conversely, a mixture with a high vapor pressure has a low boiling point (Fig. 6–2).

Since it is much easier to measure boiling points than it is to measure vapor pressures, we will investigate the boiling point as a function of the gas and liquid phase composition. A suitable system is the binary mixture ethyl acetate–cyclohexane, which exhibits behavior similar to that shown in Figure 6–2b. In addition to measuring the boiling point curves, we will carry out two calculations. First, we shall calculate what the T versus X_2 curve would look like if the solution were ideal, that is, if it obeyed Raoult's Law. Second, we shall calculate the T versus X_2 curve according to the Van Laar equation and compare the Van Laar curve of X_2 versus T with the experimentally determined curve of X_2 versus T.

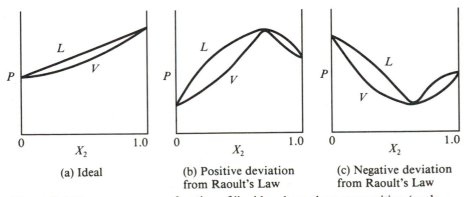

(a) Ideal

(b) Positive deviation from Raoult's Law

(c) Negative deviation from Raoult's Law

Figure 6–1 Vapor pressure as a function of liquid and gas phase composition (mole fraction).

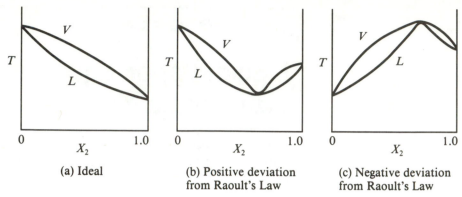

(a) Ideal

(b) Positive deviation
from Raoult's Law

(c) Negative deviation
from Raoult's Law

Figure 6–2 Boiling point as a function of liquid and gas phase composition (mole fraction).

Ideal Solution Behavior

According to Dalton's Law, the total pressure P is just the sum of the partial pressures of the two components:

$$P = P_1 + P_2 \qquad (6\text{–}2)$$

This is true whether or not the mixture obeys Raoult's Law; however, if the mixture does obey Raoult's Law, then Equation 6–2 may be written

$$P = X_1 P_1^0 + X_2 P_2^0 \qquad (6\text{–}3)$$

where X_1 and X_2 are the mole fractions in the liquid phase and P_1^0 and P_2^0 are the vapor pressures of the pure components.

Composition of the Liquid Phase. We can solve Equation 6–3 for the composition of the liquid phase, X_1 or X_2. For example, since $X_1 = 1 - X_2$, Equation 6–3 may be rewritten as

$$X_2 = \frac{P - P_1^0}{P_2^0 - P_1^0} \qquad (6\text{–}4)$$

Thus, the composition of the liquid phase is a simple function of the total pressure, which is usually atmospheric, and the vapor pressures of each pure component at the equilibrium temperature.

Composition of the Gas Phase. The composition of the gas phase is even more straightforward, as it follows directly from Dalton's Law. For example, for component 2 we may write

$$P_2 = Y_2 P \qquad (6\text{–}5)$$

where Y_2 is the mole fraction of component 2 in the gas phase:

$$Y_2 = P_2/P = X_2 P_2^0/P \qquad (6\text{–}6)$$

Once X_2 for the liquid phase is calculated with Equation 6–4, then Y_2 for the gas phase can be calculated with Equation 6–6, again with the total pressure P and the vapor pressure of pure component 2, P_2^0. Usually the total pressure $P = 1$ atm $= 760$ torr.

To calculate the temperature-versus-composition curves of both the liquid and gas phase compositions, assuming ideal solution behavior as illustrated in Figure 6–2a, one needs vapor pressure data for each component. Vapor pressure equations are available for most common organic compounds (Boublík et al., 1973), often as the Antoine equation:

$$\log P^0 = A - \frac{B}{t + C} \tag{6–7}$$

Some representative values of the Antoine constants are shown in Table 6–1 (P in torr, t in °C). The points on the ends of the curves in Figure 6–2 are just the boiling points of pure component 1 and pure component 2. These can be calculated with Equation 6–7 by letting P equal the measured atmospheric pressure and solving for T. If $P = 760$ torr, then these are the normal boiling points.

To calculate the boiling temperature-versus-composition curve between the normal boiling points, the temperature range between these two values is then divided into, say, 10 intervals and X_2 and Y_2 are calculated repeatedly at each interval.

In outline form, then, the calculation of the ideal solution curve is carried out as follows.

1. Choose a temperature between the boiling point of pure component 1 and pure component 2.
2. At this temperature, calculate P_1^0 and P_2^0 with vapor pressure equations (Eq. 6–7) for each component.

TABLE 6–1

Antoine Constants for $\log P^0 = A - \dfrac{B}{t + C}$

BP (°C)	Formula	Name	A	B	C
76.72	CCl_4	Carbon tetrachloride	6.92180	1235.172	228.937
61.204	$CHCl_3$	Chloroform	6.95465	1170.966	226.252
64.55	CH_3OH	Methanol	8.08097	1582.27	239.726
117.897	CH_3COOH	Acetic acid	7.38782	1533.313	222.309
78.298	C_2H_5OH	Ethanol	8.11220	1592.864	226.184
56.707	CH_3COCHH_3	Acetone	7.11714	1210.595	229.664
79.589	$CH_3COC_2H_5$	Methyl ethyl ketone	7.06356	1261.339	221.969
77.063	$C_4H_8O_2$	Ethyl acetate	7.10179	1244.951	217.881
65.965	C_4H_8O	Tetrahydrofuran	6.99515	1202.296	226.254
80.102	C_6H_6	Benzene	6.89272	1203.531	219.888
80.737	C_6H_{12}	Cyclohexane	6.84941	1206.001	223.148
68.740	C_6H_{14}	Hexane	6.88555	1175.817	224.867
110.622	C_7H_8	Toluene	6.95805	1346.773	219.693
97.153	C_3H_8O	1-Propanol	7.74416	1437.686	198.463

Source: Boublík, T., V. Fried, and E. Hála, *The Vapor Pressures of Pure Substances,* Elsevier, Amsterdam, 1973.

3. Calculate X_2 and Y_2 with Equations 6–4 and 6–6.
4. Plot these two points at the temperature chosen above.
5. Repeat at several more equally spaced temperatures.

Program Ideal at the end of the experiment is a Pascal computer program for doing this calculation.

Nonideal Solution Behavior

When the solution exhibits nonideal behavior, as illustrated in Figure 6–2b and c, the calculations become somewhat more complex, since activity coefficients γ_1 and γ_2 must be calculated. Let us review and compare some of the relevant definitions and relationships for ideal and nonideal behavior. For all solutions,

$$a_i = \frac{f_i}{f_i^0} \tag{6–8}$$

where a_i are the activities, f_i are the fugacities of the vapor phase over the solution, and f_i^0 are the fugacities of the pure liquids. Although the liquids may be nonideal, the gases behave sufficiently ideally at ordinary condition that the fugacities may be replaced with pressures:

$$a_i = \frac{P_i}{P_i^0} \tag{6–9}$$

or

$$P_i = a_i P_i^0 \tag{6–10}$$

If the solution behaves ideally, then according to Raoult's Law,

$$P_i = X_i P_i^0 \tag{6–11}$$

but if the solution behaves nonideally,

$$P_i = \gamma_i X_i P_i^0 \tag{6–12}$$

where γ_i is the activity coefficient. Whether the solution is ideal or nonideal, the gas phase obeys Dalton's Law:

$$P_i = Y_i P \tag{6–13}$$

By combining Equations 6–12 and 6–13,

$$P_i = Y_i P_i = \gamma_i X_i P_i^0 \tag{6–14}$$

At the azeotropic point, the compositions of the liquid and gas phases are equal, that is, $X_1 = Y_1$ and $X_2 = Y_2$. It follows immediately from Equation 6–14 that the activity coefficients *at the azeotropic point* are

$$\gamma_{i,az} = \frac{P}{P_i^0} \tag{6–15}$$

where P is the total pressure (usually the atmospheric pressure) and P_i^0 are the vapor pressures of the components of the solution at the azeotropic temperature, calculated with Equation 6–7.

At compositions other than the azeotropic composition, the activity coefficients may be calculated from the Van Laar equations (Hála et al., 1967):

$$\log \gamma_1 = \frac{A}{\left(1 + \dfrac{X_1}{X_2}\dfrac{A}{B}\right)^2} \tag{6–16}$$

$$\log \gamma_2 = \frac{B}{\left(1 + \dfrac{X_2}{X_1}\dfrac{B}{A}\right)^2} \tag{6–17}$$

The Van Laar constants A and B are given by

$$A = \log \gamma_1 \left(1 + \frac{X_2 \log \gamma_2}{X_1 \log \gamma_1}\right)^2 \tag{6–18}$$

$$B = \log \gamma_2 \left(1 + \frac{X_1 \log \gamma_1}{X_2 \log \gamma_2}\right)^2 \tag{6–19}$$

The constants A and B are calculated with Equations 6–18 and 6–19 and the activity coefficients with Equation 6–15. Equation 6–15 is valid only at the azeotropic point, but once A and B have been calculated, the activity coefficients can be calculated over the entire composition range with Equations 6–16 and 6–17. The Van Laar constants are quite insensitive to pressure and are reasonably constant over temperature changes as great as 20 or 30°C.

The calculation of the azeotrope T versus X_2 curve is quite different from the calculation of the ideal solution curve. Two experimental quantities are needed: the total (atmospheric) pressure P, and the azeotropic composition (mole fraction). The calculation should be done by dividing the composition into several parts, say, X_2 from 0.1 to 0.9 in steps of 0.1. The total pressure P is known and is related to the partial pressures and the composition of the liquid by Dalton's Law:

$$P = P_1 + P_2 = \gamma_1 X_1 P_1^0 + \gamma_2 X_2 P_2^0 \tag{6–20}$$

At each composition at which we wish to calculate the corresponding boiling temperature, we first calculate γ_1 and γ_2 with Equations 6–16 and 6–17. Then, with activity coefficients and mole fractions inserted, Equation 6–20 might typically look like

$$762.4 = 0.5711 P_1^0 + 0.8467 P_2^0 \tag{6–21}$$

The problem is to find the temperature T at which this equality holds. The vapor pressures of the two *pure* components P_i^0 are calculated with vapor pressure equations (e.g., Eq. 6–7). However, Equation 6–21 is one equation with two unknowns, so it is necessary to use successive approximations until it is satisfied. The temperature is repeatedly varied slightly and the two vapor pressures are repeatedly calculated until an equality is reached. A computer program for carrying out this repetitious task is included at the end of the experiment.

Apparatus

The distilling apparatus shown in Figure 6–3 is constructed from a 50-ml distilling flask from which the side arm is removed. A small water-cooled condenser is sealed to the neck of the flask near the junction of the neck and the flask. A small volume at the bottom of the condenser allows removal of a sample that has the composition of the vapor refluxing in the condenser. A small glass stopper sealed to the flask itself allows

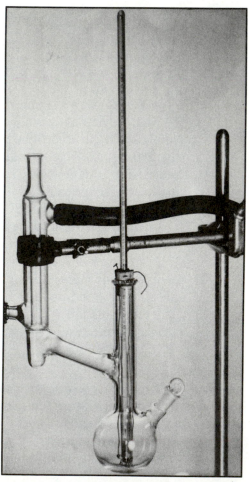

Figure 6–3 Apparatus for measuring boiling points as a function of composition. The liquid refluxing into the small glass bulb below the condenser has the composition of the gas phase. The liquid in the main boiler has the composition of the liquid in equilibrium with that gas phase. (Karen R. Sime)

removal of a sample of the liquid phase to determine its composition. The heating coil consists of about 15 cm of No. 26 nichrome wire coiled around a pencil as a form and then silver soldered to sturdy copper wires. The input should be variable from 0 to about 12 volts at 25 watts. In addition, the following items are useful: thermostated Abbé refractometer; thermometer (0.1° divisions); 25-ml volumetric pipet; 10-ml measuring pipet; barometer; 10-ml weighing bottle; eye dropper; 10 sample vials or small glass-stoppered bottles. Absolute ethanol–benzene, chloroform–acetone, and ethyl acetate–cyclohexane are suitable pairs for analysis by measuring refractive index. Toluene may be substituted for benzene and is much less toxic, but the vapor-pressure curve is quite asymmetric. Ethyl acetate and cyclohexane are the least toxic pair and give a boiling curve similar to the more toxic benzene plus ethanol. A rather complete list of binary systems is given by Hála et al. (1967).

SAFETY CONSIDERATIONS

Check the toxicity and flammability of the specific compounds you choose to investigate. Be sure the heating wire is covered with liquid before turning on the power. Turn water on to the condenser before using, and turn it off when finished.

EXPERIMENTAL PROCEDURE

Calibration. Make up a series of solutions by weight of known mole fractions approximately equal to 0.25, 0.50, and 0.75 ethyl acetate and cyclohexane. The total volume need only be 5 or 10 ml; weigh to ±0.5 mg. Keep the solutions tightly stoppered when storing or the composition may change due to evaporation. Measure the refractive indices of these solutions as well as of pure ethyl acetate and pure cyclohexane, and prepare a calibration curve of refractive index versus mole fraction. The curve is *not* generally linear. The measurements of refractive index for the calibration curve and the azeotrope should be done at exactly the same temperature, preferably with the same refractometer and thermostat.

The Boiling-Point Curve. With a graduated cylinder, place about 25 ml ethyl acetate in the reflux apparatus and measure the boiling point. With a measuring pipet, add about 2 ml cyclohexane and let the system reflux until the small sample tube below the reflux condenser has been thoroughly rinsed out and contains liquid with the composition of the gas phase (several minutes). Let the apparatus cool somewhat; then remove a sample from the sample tube (gas phase composition) and from the reflux flask (liquid phase composition) and analyze with the refractometer.

Repeat this procedure with subsequent additions of 4, 6, 8, and 12 ml cyclohexane. Then discard the mixture, rinse the flask

```
Program IdealBP;

(********************************************************************)
(*                                                                  *)
(* This program calculates the composition of the liquid and vapor  *)
(* phases (mole fractions) as functions of the boiling point for    *)
(* ideal solutions.                                                 *)
(*                                                                  *)
(********************************************************************)

uses crt;

Const A1 = 7.10179;        (* These are the constants  *)
      B1 =1244.951;        (* in the Antoine Equation *)
      C1 = 217.881;        (* for components 1 and 2    *)
      A2 = 6.84941;        (* 1 = ethyl acetate        *)
      B2 = 1206.001;       (* 2 = cyclohexane          *)
      C2 = 223.148;
      N  = 20;             (* This is the number of boiling pts calculated *)
      c = 2.302581;        (* lnX = c*logX *)

Type Ary = Array[1..N] of Real;

Var X2Vap, X2Liq,T:  Ary;  (* These are arrays of mole fractions        *)
    BP1,BP2,               (* boiling points of pure components 1 and 2 *)
    Ptotal,                (* total atmospheric pressure in torr        *)
    DeltaT:  Real;         (* Temperature interval                      *)
    Temp:  Real;           (* mixture, component 1 & 2, respec.         *)
    I:  Integer;

(****************************************************************)

Function Press1(T:  Real):  Real;
(* Press2 is the vapor pressure (torr) of pure component 1 at T *)
Begin
   Press1:= EXp(c*(A1-B1/(C1 + t)))
End;

(****************************************************************)

Function Press2(T:  Real):  Real;
(* Press2 is the vapor pressure(torr) of pure component 2 at T *)
Begin
   Press2:= EXP(c*(A2-B2/(C2 + t)))
End;

(****************************************************************)
```

Figure 6–4 *Program IdealBP.*

```
Procedure CalcTempInterval;

Begin (* CalcTempInterval *)
   Write('Please input the total (atmospheric) pressure in torr:  ');
   Readln(Ptotal);
   BP1:= B1/(A1-Ln(Ptotal)/c)-C1;   (* BP1 is Boil pt of component 1 *)
   BP2:= B2/(A2-Ln(Ptotal)/c)-C2;   (* BP2 is Boil pt of component 2 *)
   DeltaT:= ABS(BP2 - BP1)/(N-1);   (* Temp interval is div into N-1 parts*)
End;   (* CalcTempInterval *)

(****************************************************************)

Procedure CalcBoilingPoints;

Begin (* CalcBoinlingPoints *)
   CLRSCR;
   Writeln('  Temperature      X2(liq)         X2(Vap)');
   Writeln;
   Temp:= BP1 - DeltaT;
   For I:= 1 to N DO
      Begin (* For *)
      Temp:= Temp + DeltaT;
      T[I]:= Temp;
      X2Liq[I]:= abs((Ptotal - Press1(T[I]))/(Press2(T[I])-Press1(T[I])));
      X2Vap[I]:= X2Liq[I]*Press2(T[I])/Ptotal;
      Writeln(T[I]:10:2,'    ',X2Liq[I]:10:4,'      ',X2Vap[I]:10:4);
   End; (* For *)
End;   (* CalcBoilingPoints *)
(****************************************************************)

Begin (*Main Program *)
   CalcTempInterval;
   CalcBoilingPoints;
End. (*Main Program*)
```

Figure 6–4 (continued)

with pure cyclohexane, and fill it with about 25 ml cyclohexane. As above, determine the boiling point of pure cyclohexane, and mixtures of cyclohexane with 2, 4, 6, 8, and 12 ml ethyl acetate added. The analyses may be carried out as the system comes to equilibrium or the samples may be stored in tightly stoppered vials and analyzed later. Storing, however, is risky since the composition may change due to evaporation.

Results and Calculations

Tabulate and plot a calibration curve of refractive index versus mole fraction for your system. Draw a smooth curve through the points. Use this curve to determine the composition of the liquid and gas phases at the recorded boiling temperatures. Tabulate and plot boiling point versus mole fraction. Determine the azeotrope composition and compare it with the literature value, if available.

(Text continues on page 460.)

```pascal
Program VanLaar;

(*****************************************************************)
(*                                                               *)
(*    A Program to calculate the azeotropic boiling point vs.    *)
(*    composition curve for a binary pair of miscible liquids.  The *)
(*    input is the barometric pressure, the composition and      *)
(*    temperature of the azeotrope                               *)
(*                                                               *)
(*                                                               *)
(*****************************************************************)

uses crt;

Const A1 = 7.10179;     (* 1 refers to ethyl acetate *)
      B1 = 1244.951;
      C1 = 217.881;
      A2 = 6.84941;     (* 2 refers to cyclohexane *)
      B2 = 1206.001;
      C2 = 223.148;
      c  = 2.302581;    (* c*LogX = LnX *)

VAR X1azeo,X2azeo,Tazeo,Pbarometric,Gamma1,Gamma2,A,B:   Real;

(*****************************************************************)

Function P1(T: Real): Real;
   Begin
   P1:= Exp(c*(A1-B1/(C1+T)))
   End;

(*****************************************************************)

Function P2(T: Real): Real;
   Begin
   P2:= Exp(c*(A2-B2/(C2+T)))
   End;

(*****************************************************************)

Procedure GetData;
   Begin
      Writeln;
      Write('Plese enter the barometric pressure in torr:  ');
      Readln(Pbarometric);
      Write('Please enter the mole fraction ethyl acetate ');
      Write('at the azeotropic point:  ');
      Readln(X1azeo);
      X2azeo:= 1.0 - X1azeo;
      Write('Please enter the azeotropic temperature in deg C:  ');
      Readln(Tazeo);
   End;

(*****************************************************************)

Procedure CalcActivityCoefficients; (* at the azeotropic temperature*)
   Begin
      Gamma1:= Pbarometric/P1(Tazeo);
      Gamma2:= Pbarometric/P2(Tazeo);
   End;

(*****************************************************************)
```

Figure 6–5 *Program VanLaar.*

```
Procedure CalcVanLaarConstants;
VAR Term:   Real;
Begin
   Term:= X2azeo*Ln(Gamma2)/(X1azeo*Ln(Gamma1));
   A:= Sqr(1 + Term)*Ln(Gamma1)/c;
   B:= Sqr(1 + 1/Term)*Ln(Gamma2)/c;
   CLRSCR;
   Writeln('          A and B are ',A:10:4,' and ',B:7:4);
End;


(************************************************************)


Procedure VanLaarCalculation;
VAR X1liq,X2liq,X2vap,Denom1,Denom2,Deriv1,Deriv2:  Real;
    ParPres1,ParPres2,T,DeltaT,Over,Under: Real;
    I:   Integer;
Begin (* VanLaarCalculation *)
   X2liq:= 0.95;Writeln;
   Write('       Temp     X2(liq)   X2(vap)   Gamma1   ');
   Writeln('Gamma2');Writeln;
   For I:= 1 to 19 do
      Begin (* For *)
      X1liq:= 1 - X2liq;
      (* calculate activity coefficients of ethyl acetate *)
      Denom1:= (1+(A*X1liq)/(B*X2liq));
      Denom1:= Denom1*Denom1;
      Gamma1:= Exp(c*A/Denom1);
      (* calculate activity coefficients of cyclohexane *)
      Denom2:= (1+(B*X2liq)/(A*X1liq));
      Denom2:= Denom2*Denom2;
      Gamma2:= Exp(c*B/Denom2);
      (* Do successive approximations to get T and X2ap *)
      Repeat
         ParPres1:= Gamma1*X1liq*P1(T);
         ParPres2:= Gamma2*X2liq*P2(T);
         Deriv1:= c*B1/((C1+T)*(C1+T));
         Deriv2:= c*B2/((C2+T)*(C2+T));
         Over:= (Pbarometric-ParPres1-Parpres2);
         Under:= (ParPres1*Deriv1 + ParPres2*Deriv2);
         DeltaT:= Over/Under;
         T:= T + DeltaT;
      Until (Abs(DeltaT) - 0.01) < 0;
      X2vap:= ParPres2/Pbarometric;
      Writeln(    T:10:2,X2liq:10:5,X2vap:10:5,Gamma1:10:5, Gamma2:10:5);
      X2liq:= X2liq - 0.05;
   End; (* For *)
End; (* VanLaar Calculation *)


(************************************************************)


Begin (* M A I N   P R O G R A M *)
   GetData;
   CalcActivityCoefficients;
   CalcVanLaarConstants;
   VanLaarCalculation;
End.
```

Figure 6–5 (continued)

Compare your experimental values of boiling temperatures and composition with those calculated for ideal solution behavior using Equations 6–4 and 6–6. If the boiling points of the two pure liquids are very close together, you should plot the ideal solution curve on a separate paper with the temperature axis expanded. Calculate the Van Laar constants A and B for your system with Equations 6–18 and 6–19.

At equally spaced intervals from $X_2 = 0.0$ to $X_2 = 1.0$, calculate activity coefficients with the Van Laar Equations 6–16 and 6–17. By successive approximations, calculate the temperatures at these mole fractions at which an equation of the form of Equation 6–21 is satisfied for your system. Computer programs for ideal and nonideal Van Laar solutions are furnished in Figures 6–4 and 6–5. If you use these programs, you should calculate a pair of points for both the ideal and Van Laar cases by hand, and verify that these results are *identical* with those on a hard copy of the computer output.

REFERENCES

1. Boublík, T., V. Fried, and E. Hála, *The Vapor Pressures of Pure Substances,* Elsevier, Amsterdam, 1973.

2. Carlson, H. C., and A. P. Coburn, "Vapor–Liquid Equilibria of Nonideal Solutions," *Ind. Eng. Chem.* 34:581, 1942.

3. Grow, J. M., "Display of Vapor Pressure Data with a Theoretical Fit," *J. Chem. Educ.* 60:1062, 1983.

4. Hála, E., J. Pick, V. Fried, and O. Vilim, *Vapor–Liquid Equilibrium,* 2nd ed., Pergamon Press, New York, 1967.

5. Lewis, G. N., and M. Randall, revised by K. Pitzer and L. Brewer, *Thermodynamics,* 2nd ed., McGraw-Hill, New York, 1961.

6. Prausnitz, J. M., C. A. Eckert, R. V. Orye, and J. P. O'Connell, *Computer Calculations for Multicomponent Vapor–Liquid Equilibria,* Prentice-Hall, Englewood Cliffs, New Jersey, 1967.

7. Rossini, R. D., et al., *Selected Values of Physical and Thermodynamic Properties of Hydrocarbons and Related Compounds,* API Project 44, Carnegie Press, Pittsburgh, Pennsylvania, 1953.

8. Timmermans, J., *Physico-Chemical Constants of Binary Systems,* Interscience, New York, 1959.

Experiment 7

Solid–Liquid Equilibrium in Two-Component Systems

Two-component systems are a common part of our everyday lives, ranging from water–ethylene glycol mixtures in automobile radiators to ordinary brass or bronze. Bronze, one of the earliest known alloys, is a two-component mixture of copper and tin; brass is an alloy of copper and zinc. The freezing point of a mixture is lower than that of either component, a property utilized in antifreeze mixtures. The freezing point lowering of a solution can be used to determine molecular weights because it depends on the number of moles of solute present. The solubility of most substances increases as the temperature increases. All of these phenomena are intimately related and governed by the same thermodynamic treatment.

The Phase Rule

The simplest possible two-component phase diagram is one in which no solid solutions or compounds occur (Fig. 7–1). According to the phase rule, a single phase in a two-component system may possess three degrees of freedom:

$$F = C - P + 2 = 2 - 1 + 2 = 3 \qquad (7\text{--}1)$$

These are temperature, pressure, and the composition of one of the two components. Since studies of solid–liquid equilibrium are nearly always carried out at constant pressure, one degree of freedom is lost and the phase rule becomes

$$F = C - P + 1 = 2 - P + 1 = 3 - P \qquad (7\text{--}2)$$

Thus, temperature and the composition of one component are the variables, and these may be plotted on ordinary graph paper.

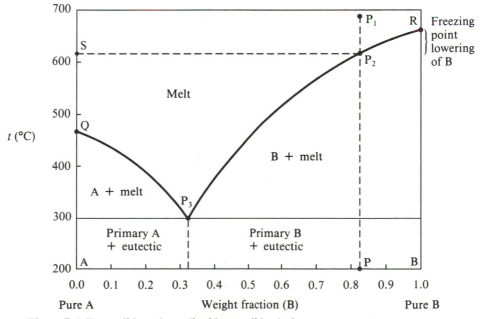

Figure 7–1 Two solids and one liquid, no solid solutions.

Graphical Representation of a Two-Component System

In a graphical representation, an area corresponds to $F = 3$, a line to $F = 1$, and a point to $F = 0$ (invariant). In a two-component system, the composition is represented by the position along the line AB in Figure 7–1. The properties of the line are such that no matter where point P lies on the line, AP + PB = AB so that

$$W_A/W_B = BP/PA \qquad (7-3)$$

where W_B and W_A are weight fractions, and BP and PA are the lengths of the line segments. Point P_1 lies in an area where a single phase is present (the homogeneous melt), and $F = 3 - 1 = 2$ degrees of freedom; both temperature and composition must be specified to define the system. Point P_2 lies on a line where two phases, the melt and solid B are in equilibrium, so that $F = 3 - 2 = 1$ degree of freedom; either the temperature or composition must be specified to define the system. At point P_3, three phases, solid A, solid B, and the melt, are in equilibrium, so that $F = 3 - 3 = 0$. There are no degrees of freedom. The system is invariant; the temperature and composition are uniquely defined when three phases are in equilibrium.

Solid–Liquid Equilibrium as Freezing Point Lowering. The curves QP_3 and P_3R are usually referred to as freezing-point curves from the experimental method used to determine them. A sample of known composition is heated until the system is homogeneously melted; it is then cooled slowly until freezing occurs: a solid phase forms. At point P_1, you can consider the system as a solution of solute A in solvent B. The presence of the solute A lowers the freezing point of the solvent B. As the temperature is lowered along P_1P_2, freezing does not occur until P_2 is reached.

Solid–Liquid Equilibrium as Solubility Equilibrium. Alternatively, we can look at the system from a different point of view. Suppose we go to point P_2 from point S. At point S, we have pure A in the liquid state. If B is added to A, we would now think of B as the solute and A as the solvent. Along line SP_2 more and more B is added until at point P_2 the solution of B in A is saturated with B, which proceeds to precipitate out of solution.

Thermodynamic Analysis

Let us continue with this point of view, and think of the system at point P_2 as a saturated solution of solid B in the solvent A. At point P_2, then, the following equilibrium is established between solid B and a solution of mole fraction X_B:

$$B(s) = B(soln) \qquad (7-4)$$

Since the system is at equilibrium, $\Delta G = 0$, so that we can write

$$\Delta G = 0 = G_B(soln) - G_B(s) \qquad (7-5)$$

If we choose as the standard state a solution of unit mole fraction ($X_B = 1$), then the relationship between the free energy per mole at any X_B and the free energy per mole in the standard state is given by

$$G_B(\text{soln}) = G_B^0(\text{soln}) + RT \ln X_B \qquad (7\text{--}6)$$

When Equation 7–6 is substituted in Equation 7–5, the result is

$$G_B(s) - G_B^0(\text{soln}) = RT \ln X_B \qquad (7\text{--}7)$$

This may be rearranged slightly to give

$$\ln X_B = \frac{-\Delta G_B(\text{soln})}{RT} \qquad (7\text{--}8)$$

Recalling that $(\partial(G/T)/\partial T)_P = -H/T^2$ at constant pressure, we differentiate Equation 7–8 to obtain

$$\left(\frac{\partial \ln X_B}{\partial T}\right)_P = \frac{\Delta H_B(\text{soln})}{RT^2} \qquad (7\text{--}9)$$

Equation 7–9 may also be written in a slightly different form:

$$\left(\frac{\partial \ln X_B}{\partial 1/T}\right)_P = \frac{-\Delta H_B(\text{soln})}{R} \qquad (7\text{--}10)$$

At this point let us write Equation 7–4 in two steps to indicate first the solution process involving the fusion of the solute, and then the *ideal* mixing of the liquid solute with the already liquid solvent, for which $\Delta H(\text{mix}) = 0$:

$$\text{B}(s) \quad = \quad \text{B}(\ell) \quad = \quad \text{B}(\text{soln}) \qquad (7\text{--}11)$$

$$\leftarrow\!\Delta H(\text{fus}) \quad \rightarrow\!\leftarrow \quad \Delta H(\text{mix})\!\rightarrow$$

$$\leftarrow \qquad \Delta H(\text{soln}) \qquad \rightarrow$$

Heat of Solution. When the solution process is ideal, $\Delta H(\text{mix}) = 0$, so that $\Delta H(\text{soln}) = \Delta H(\text{fus})$, and Equation 7–10 can be written:

$$\left(\frac{\partial \ln X_B}{\partial 1/T}\right)_P = \frac{-\Delta H_B(\text{fus})}{R} \qquad (7\text{--}12)$$

Consequently, data from phase diagrams such as Figure 7–1 can be used to determine heats of fusion of both components, since a plot of $\ln X_A$ against $1/T$ gives a straight line of slope equal to the heat of fusion of A, and a plot of $\ln X_B$ against $1/T$ gives a straight line of slope equal to the heat of fusion of B (both divided by $-R$). In this manner heats of fusion for many nonelectrolytes and metals can be determined, if the two components form an ideal solution—which is not uncommon.

Ideal Solubility. Integration of Equation 7–10 leads to another interesting interpretation of ideal solubilities:

$$\frac{\ln X_B}{\dfrac{1}{T_B^0} - \dfrac{1}{T}} = \frac{\Delta H_B(\text{fus})}{R} \tag{7-13}$$

This may be written explicitly in terms of $\ln X_B$, where X_B is the solubility of B in A at temperature T, $\Delta H_B(\text{fus})$ is the heat of fusion of B, and T_B^0 is the melting point of pure B:

$$\ln X_B = \frac{\Delta H_B(\text{fus})}{R}\left(\frac{1}{T_B^0} - \frac{1}{T}\right) \tag{7-14}$$

What is remarkable about this equation is that it states that the solubility of B expressed in mole fraction is the same in *all* solvents, and that the solubility depends only on two properties of the solute B: the heat of fusion of B and the melting point of pure B. Whether A or B is considered the solvent or solute is immaterial; it is conventional to look upon the substance present in the larger amount as the solvent. Consequently, we can rewrite Equation 7-14 for substance A:

$$\ln X_A = \frac{\Delta H_A(\text{fus})}{R}\left(\frac{1}{T_A^0} - \frac{1}{T}\right) \tag{7-15}$$

Equations 7-14 and 7-15 are valid for ideal solutions. To check the validity of these equations, let us examine the solubility of *p*-dibromobenzene in a number of solvents in which it forms nearly ideal solutions, as shown in Table 7-1. The solubility in mole fraction agrees quite well with the mole fraction calculated with Equation 7-14. Not surprisingly, the agreement is best with bromobenzene as solvent.

Calculation of the Liquidus Curve. Equations 7-14 and 7-15 as written might be described as giving the solubility (X_A or X_B) of a solute at temperature T. Alternatively, we can look at T as the temperature at which the solid and liquid phases are in equilibrium (Eq. 7-11). The phase diagram is a plot of this temperature against composition (mole fraction). Let us rearrange Equations 7-14 and 7-15 so that the solid-liquid equilibrium temperature T is expressed explicitly in terms of com-

TABLE 7-1

Solubility of *p*-Dibromobenzene at 25°C
[$\Delta H(\text{fus}) = 20.29$ kJ/mol, $T^0 = 359.1$ K]

Solvent	Measured[a]		X_A Calculated with Eq. 7-14
	g/100 g	X_A	
CS_2	47.4	0.225	0.2497
C_6H_6	45.6	0.217	0.2497
C_6C_5Br	32.5	0.243	0.2497

[a] Data from Stephen and Stephen, 1963.

position and two properties of the solvent—heat of fusion and melting point:

$$T = \left(\frac{1}{T_A^0} - \frac{R \ln X_A}{\Delta H_A(\text{fus})} \right)^{-1} \tag{7-16}$$

$$T = \left(\frac{1}{T_B^0} - \frac{R \ln (1 - X_A)}{\Delta H_B(\text{fus})} \right)^{-1} \tag{7-17}$$

Equations 7–16 and 7–17 intersect at the eutectic composition and temperature. This point is found graphically by plotting T versus X_A with both equations.

The calculation is facilitated with the computer program listed at the end of the experiment (Fig. 7–5). The input data (the heats of fusion and melting points of A and B) are input interactively. The output is a listing of T (solid–liquid equilibrium temperatures) and X_A (the mole fractions of A).

Apparatus

A number of choices for investigation are satisfactory. Some readily available substances that form simple systems are listed in Table 7–2.

For organic pairs, the simple apparatus illustrated in Figure 7–2a is satisfactory. It consists of a 400-ml beaker, a 25-by-200-mm test tube, a 0 to 110°C thermometer, and a sturdy copper wire stirrer bent into a circle at one end. The thermometer is inserted into a split one-hole rubber

TABLE 7–2

Two-Component Systems

Component A	Component B
Organic two-component systems	
Naphthalene	Diphenylamine
Naphthalene	Nitrophenol
Naphthalene	Dibromobenzene
Naphthalene	Diphenyl
p-Dichlorobenzene	Diphenyl
Beta-naphthol	p-Toluidene
Beta-naphthol	Acetamide
Salicylic acid	Acetamide
Benzoic acid	Cinnamic acid
Metal two-component systems	
Cadmium	Bismuth
Cadmium	Lead
Cadmium	Tin
Lead	Bismuth
Lead	Tin
Bismuth	Tin

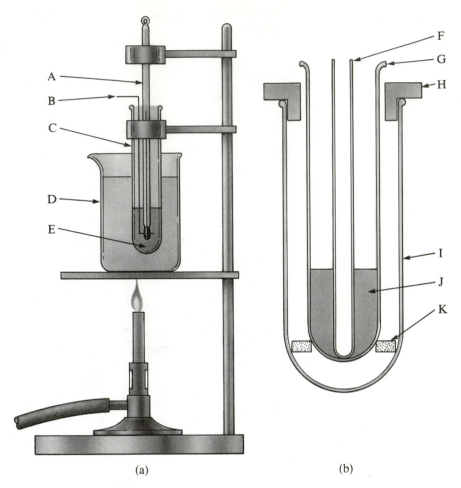

(a)   (b)

Figure 7–2 Freezing-point apparatus. A, thermometer; B, stirrer; C, 25-by-200-mm test tube; D, 400-mm beaker; E, organic sample; F, thermocouple well of 8-mm tubing; G, 25-by-200-mm test tube; H, brass 45-mm diameter to pass 25-by-200-mm test tube and to fit in I; I, 40-by-180-mm test tube; J, alloy sample; K, brass ring.

stopper supported with a clamp and ring stand. The test tube is also supported with a clamp. The thermometer may be replaced with a thermocouple inserted into a 6- or 7-mm-diameter thermocouple well.

For binary alloy systems, the apparatus shown in Figure 7–2b allows the cooling to take place more slowly. The inner 8-mm glass tube is a thermocouple well. The middle tube is a 25-by-200-mm test tube containing the mixture of metals. It is supported by a Fiberfrax washer about 5 mm thick at the bottom of a larger test tube. The middle tube is kept vertical with a brass ring, which fits into the largest tube and is drilled to pass the middle tube. If breakage of the middle tube is a problem, the alloy can be contained in a short stainless steel test tube contained in the middle tube.

The thermocouple reference junction is maintained at 0°C with the arrangement shown in Figure 7–3. The leads to the potentiometer are

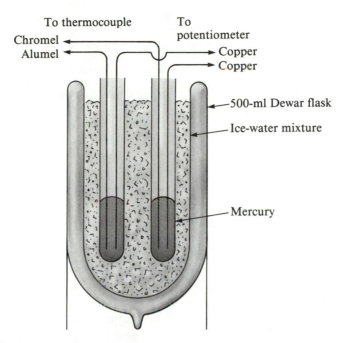

To thermocouple

To potentiometer

Chromel

Alumel

Copper

Copper

500-ml Dewar flask

Ice-water mixture

Mercury

Figure 7–3 Thermocouple connections.

insulated copper and make electrical contact with the thermocouple leads by immersion in 6-mm tubes filled with mercury to a depth of about 20 mm. The mercury-filled tubes are maintained at 0°C by immersion in an ice bath. Note that the ice must go all the way to the bottom; otherwise water at about 4°C (where it is most dense) collects at the bottom of the Dewar flask.

SAFETY CONSIDERATIONS

Some of the above compounds are known carcinogens and should not be allowed to contact skin. Others smell bad and it may be advantageous to carry out the measurements in a hood. Take care not to touch a hot thermocouple or test tube. The ice-water Dewar flask should rest in a tray to prevent mercury from spilling on the desk top.

EXPERIMENTAL PROCEDURE

Select a pair of components for investigation. Make two series of runs: an A-rich series, beginning with pure A, followed by systems made by successive additions of B to the previous run beginning with pure A; and a B-rich series similarly prepared. Weigh to ±0.01 g. Some approximate weights are given in Table 7–3. For metals, scale up the weights by a factor equal to the density of the component.

TABLE 7–3

Approximate Range of Compositions

	Start: 10 g Pure A			Start: 10 g Pure B	
Run	Add B (g)	Wt.% A	Run	Add A (g)	Wt.% B
1A	0.0	100	1B	0.0	100
2A	1.5	87	2B	1.5	87
3A	2.0	74	3B	2.0	74
4A	2.5	63	4B	2.5	63
5A	3.5	51	5B	3.5	51

Heat the mixture in a water bath until it is melted. Then remove the water bath and measure the temperature periodically (e.g., every minute) until the system is essentially solid. The relationship between the cooling curves and the phase diagram is shown in Figure 7–4.

When a pure substance cools, the temperature falls smoothly until the freezing point is reached; then the temperature halts, as shown for curve 1, Figure 7–4b, when the solid A begins to solidify. For a one-component system at constant pressure the phase rule is $F = C - P + 1 = 2 - P$. When two phases are in equilibrium, the system is invariant and the temperature is fixed, that is, at the melting point. Above and below the melting point, there is only one phase, so $F = 1$, and the temperature is no longer fixed; it varies and must be specified to fix the state of the system. Runs 1A and 1B in Table 7–3 are of this type.

For a two-component system, the phase rule is $F = C - P + 2 = 3 - P$ at constant pressure. When a homogeneous melt cools

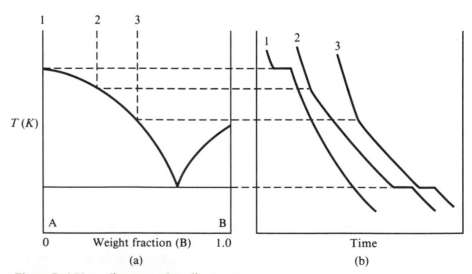

Figure 7–4 Phase diagram and cooling curves.

(curves 2 and 3 in Fig. 7–4b), a temperature is reached at which the solution is saturated with respect to one of the components, A in this case. At this temperature, A begins to precipitate out of the solution and the rate of cooling decreases. It decreases because heat is given up to the system by the exothermic process:

$$A(soln) \rightarrow A(s) \qquad \Delta H \text{ is negative} \qquad (7\text{–}18)$$

Above the two-phase line, there is one phase, so that $F = 3 - 1 = 2$; below the two-phase line, there are two phases in equilibrium, so that $F = 3 - 2 = 1$. However, at the eutectic temperature, the solution is not only saturated with respect to A, but it is also saturated with respect to B, which now begins to precipitate out of solution simultaneously with A. Since three phases are in equilibrium, $F = 3 - 3 = 0$ and the system is invariant. The temperature is fixed and the cooling halts at the eutectic temperature.

The cooling curves for all the systems in Table 7–3 are of this type except those for runs 1A and 1B, which are one-component systems. Measuring the cooling curve for a two-component system of known composition reveals the eutectic temperature as well as the temperature on the two-phase curve. The cooling curves of all the two-component systems show a halt at the same (eutectic) temperature. They also show an inflection point at a higher temperature, which depends upon the composition.

Results and Calculations

Graph the cooling curves. Tabulate the compositions, observed eutectic temperatures, and two-phase temperatures for the various runs. Construct a phase diagram from your observations. Tabulate and plot $\ln X$ versus $1/T$ for both A and B. Calculate the heat of fusion.

Compare your heats of fusion with the literature values. With the literature values for the heat of fusion and melting points of pure A and B, construct a phase diagram based upon ideal solution behavior (Eqs. 7–16 and 7–17). If it is convenient, use the computer program listed at the end of the experiment (Fig. 7–5). Compare your observed phase diagram, eutectic temperature, and composition with the calculated values and calculated phase diagram.

The algorithm for the computer programs in Figures 7–5 and 7–6 is the pair of equations 7–16 and 7–17. Input to both programs consists of the heats of fusion and the melting points of the pure components. The first program simply prepares a table of freezing points that lie above the eutectic temperature from mole fraction 0.0 to 1.0. The second program does exactly the same calculations, but also takes advantage of Turbo Pascal's graphics features to prepare a complete phase diagram, shown in Figure 7–7.

(Text continues on page 479.)

```
Program Phase2IBM;

uses crt;

Const R = 0.00831434;

VAR T,XA,XB,DelHA,DelHB,ToA,ToB:  Real;

Procedure GetData;
Begin
   Write('Please enter the freezing point of pure A. (deg K):  ');
   Readln(ToA);
   Write('Please enter the heat of fusion of A (kJ/mole):  ');
   Readln(DelHA);
   Write('Please enter the freezing point of pure B. (deg K):  ');
   Readln(ToB);
   Write('Please enter the heat of fusion of B (kJ/mole):  ');
   Readln(DelHB);
End; (* GetData *)


Function LeftTemp(XA:  Real): Real;
Begin
   LeftTemp:= 1/((1/ToA)-(R*ln(XA)/DelHA));
End;

Function RightTemp(XA: Real):  Real;
Begin
   RightTemp:= 1/((1/ToB)-(R*ln(1-XA)/DelHB));
End;

Procedure CalcData;
VAR I:  Integer;
    TL,TR:  Real;

Begin (* CalcData *)
   clrscr;
   Writeln('Mole Fraction  ','Temperature(k)');
   Writeln;
   Writeln(0:10,ToB:10:1);
   XA:= 0;
   For I:= 1 to 19 do
      Begin (* For *)
      XA:= XA + 0.05;
      TL:= LeftTemp(XA);
      TR:= RightTemp(XA);
      If TL > TR then
         T:= TL
      Else
         T:= TR;
      Writeln(XA:10:2,T:10:1);
      End; (* For *)
   Writeln(1.0:10:2,ToA:10:1)
End; (* CalcData *)

Begin (* M A I N   P R O G R A M *)
   GetData;
   CalcData
End.  (* M A I N   P R O G R A M *)
```

Figure 7–5 *Program Phase2IBM.*

```
Program TwoPhase;

uses crt,graph;

Const Screenright  = 278;
      ScreenLeft   =  28;
      ScreenTop    =  15;
      Screenbottom = 175;
      R = 0.00831434;
      Blank = '    ';

Type Index = 1..250;
     Ary   = Array[Index] of Real;
     Word  = String[40];

Var XA,T:  Ary;
    Answer:  Char;
    Title:  Word;
    ToA,ToB,DelHA,DelHB:  Real;
    Xleft,Xright,Xscale,Yscale:  Real;
    Ytop,Ybottom,
    HighTemp,LowTemp,ShoveTitle,
    GrDriver,GrMode,ErrCode: Integer;

(*************************************************************)

Function LeftTemp(XA:  Real):  Real;
   Begin
      LeftTemp:= 1/((1/ToA)-(R*ln(XA)/DelHA));
   End;

Function RightTemp(XA:  Real):  Real;
   Begin
      RightTemp:= 1/((1/ToB)-(R*ln(1-XA)/DelHB));
   End;

(*************************************************************)

Procedure HeaderPage;

   Procedure DrawBorder;
   Begin (* DrawBorder *)
      Line(ScreenLeft,ScreenTop,ScreenRight,ScreenTop);
      Line(ScreenRight,ScreenTop,ScreenRight,ScreenBottom);
      Line(ScreenRight,ScreenBottom,ScreenLeft,ScreenBottom);
      Line(ScreenLeft,ScreenBottom,ScreenLeft,ScreenTop);
   End;   (* DrawBorder *)

   Procedure Credits;
   Begin; (* Credits *)
      SetTextJustify(CenterText,CenterText);
      OutTextXY(153,25,'This is a');
      OutTextXY(153,50,'PHASE DIAGRAM PLOT PROGRAM');
      OutTextXY(153,75,'Written by');
      OutTextXY(153,100,'Rodney J. Sime');
```

Figure 7–6 *Program TwoPhase*, a graphical version of *Program Phase2IBM*.

```
            OutTextXY(153,125,'CALIFORNIA STATE UNIVERSITY');
            OutTextXY(153,150,'SACRAMENTO');
            SetTextJustify(LeftText,TopText);
      End;   (* Credits *)

Begin (* HeaderPage *)
      GrDriver:= Detect;
      InitGraph(GrDriver,GrMode,'');
      ErrCode:=GraphResult;
      SetGraphMode(3);
      DrawBorder;
      Credits;
      Readln;
      TextMode(BW80)
End;   (* HeaderPage *)

(***************************************************************)

Procedure GetData;

Procedure GetTitle;
Var CharNum:  Integer; (* number of characters in title *)
Begin
  Repeat
    Writeln('Please enter a title for your graph -- ');
    Writeln('Use 35 characters or less.');
    Writeln('1--------10--------20--------30---35');
    Readln(Title);
    Writeln;
    Charnum:= Length(Title);
    If CharNum > 35 Then
      Begin
      Writeln;
      Writeln('Sorry - your title must have 35 characters or less.');
      Writeln
      End; (* If *)
  Until CharNum <=35;
End; (* GetTitle *)

(***************************************************************)

Procedure GetThermoData;
   Begin   (* GetThermodata *)
     Write('Please enter the freezing point of pure A(deg K):  ');
     Readln(ToA);
     Write('Please enter the heat of fusion of A (kJ/mol):  ');
     Readln(DelHA);
     Write('Please enter the freezing point of pure B(deg K):  ');
     Readln(ToB);
     Write('Please enter the heat of fusion of B (kJ/mol):  ');
     Readln(DelHB);
   End;   (* GetThermoData *)

Begin (* GetData *)
    GetTitle;
```

Figure 7–6 (continued)

```pascal
      GetThermoData;
End;   (* GetData *)

(****************************************************************)

Procedure FindEutectic(Var XAeutectic,Teutectic:   Real);
Var I:   Integer;
     TempT, TempXA:   Real;
Begin (* FindEutectic *)
   Teutectic:= 9999.0;
   For I:= 1 to 249 do
      Begin (* For *)
         TempT:= T[I];
         TempXA:= XA[I];
         If TempT < Teutectic Then
            Begin (* If *)
               Teutectic:= tempT;
               XAeutectic:= TempXA;
            End (* If *)
      End; (* For *)

End;   (*  FindEutectic *)

(****************************************************************)

Procedure WriteEutectic;
VAR   XA,T:   Real;
Begin
   FindEutectic(Xa,T);
   Writeln;
   Writeln('At the eutectic, the temperature equals ',T:5:1);
   Writeln('and the mole fraction A equals ',XA:5:3);
   Writeln;
End;

(****************************************************************)

Procedure CalcArrays;
Var I:   Integer;
    TL,TR, DeltaXA,X,First,Last:   Real;
    Teutectic, Xeutectic: Real;
Begin
  X:= 0.0;
  DeltaXA:= 0.004;
  First:= 0.0;
  Last:= 1.0;
  Writeln;
  Writeln('Mole Fraction A  Temperature(K)');Writeln;
  Writeln(First:10:2,ToB:15:1);
  For I:= 1 to 249 do
    Begin
       X:= X + DeltaXA;
       XA[I]:= X;
       TL:= LeftTemp(X);
       TR:= RightTemp(X);
```

Figure 7–6 (continued)

```
              If TL > TR Then
                  T[I]:= TL
              Else
                  T[I]:= TR;
              If I MOD 25 = 0 then
                  Writeln(XA[I]:10:2,T[I]:15:1);
          End; (* For *)
      Writeln(Last:10:2,ToA:15:1);
      Writeln;
      FindEutectic(Xeutectic,Teutectic);
   End; (* CalcArrays *)

(*****************************************************************)

Procedure GetTempExtremes;
   Begin
      Writeln('Enter temperatures at graph top and bottom as integers.');
      Writeln;
      Write('Please enter the maximum temperature (K) of the graph:  ');
      Readln(Hightemp);
      Writeln;
      Write('Please enter the minimum temperature (K) of the graph:  ');
      Readln(LowTemp)
   End;

(*****************************************************************)

Procedure DrawGrid;

Var Xinterval, Yinterval:  Integer;
    Xscale,Yscale,Xleft,Ybottom:   Real;

(*-----------------------------------------------------------*)

   Procedure DrawBorder;
   Begin (* DrawBorder *)
      Line(ScreenLeft,ScreenTop,ScreenRight,ScreenTop);
      Line(ScreenRight,ScreenTop,ScreenRight,ScreenBottom);
      Line(ScreenRight,ScreenBottom,ScreenLeft,ScreenBottom);
      Line(ScreenLeft,ScreenBottom,ScreenLeft,ScreenTop);
   End;

(*-----------------------------------------------------------*)

   Procedure WriteTitle(ScreenX,ScreenY:  Integer);
   Begin
      SetTextJustify(CenterText,CenterText);
      OutTextXY(ScreenX,ScreenY,Title);
      SetTextJustify(LeftText,TopText);
   End;  (* WriteTitle *)

(*-----------------------------------------------------------*)

   Procedure MarkX;
   Var I,J, ScreenX,ScreenY:  Integer;
```

Figure 7–6 (continued)

```
   Begin (* MarkX *)
      ScreenY:= ScreenBottom;
      Xinterval:= (ScreenRight - ScreenLeft) DIV 10;
      Yinterval:= (ScreenBottom - ScreenTop) DIV 10;
      For I:= 1 to 9 do
          Begin  (* For *)
          ScreenX:=ScreenLeft;
          ScreenY:= ScreenY - Yinterval;
          Line(ScreenX,ScreenY,ScreenX + 2,ScreenY);
          ScreenX:= ScreenX - 2;
          For J:= 1 to 9 do
              Begin (* For J *)                 (* to bottom of screen  *)
              ScreenX:= ScreenLeft + J*Xinterval -1;
              Line(ScreenX,ScreenY,ScreenX + 2,ScreenY);
              ScreenX:= ScreenX - 1;
              End;
          ScreenX:= ScreenRight;
          Line(ScreenX,ScreenY,ScreenRight - 2,ScreenY);
          End; (* For *)
   End;   (* MarkX *)

(*-------------------------------------------------------------*)

   Procedure MarkY;
   Var I,J,ScreenX,ScreenY:  Integer;
   Begin (* MarkY *)
      ScreenX:= ScreenLeft;
      Xinterval:= (ScreenRight - ScreenLeft) DIV 10;
      Yinterval:= (ScreenBottom - ScreenTop) DIV 10;
      For I:= 1 to 9 do
          Begin  (* For *)
          ScreenY:=ScreenTop;
          ScreenX:= ScreenX + Xinterval;
          Line(ScreenX,ScreenY,ScreenX,ScreenY + 2);
          ScreenY:= ScreenY + 2;
          For J:= 1 to 9 do
              Begin (* While *)                 (* to bottom of screen  *)
              ScreenY:= ScreenTop + J*Yinterval + 1;
              Line(ScreenX,ScreenY,ScreenX,ScreenY - 2);
              ScreenY:= ScreenY + 1;
              End;  (* While *)
          ScreenY:= ScreenBottom;
          Line(ScreenX,ScreenY,ScreenX,ScreenY - 2);
          End; (* For *)
   End;   (* MarkY *)

(*-------------------------------------------------------------*)

   Procedure WriteXnumbers(ScreenX,ScreenY:  Integer);
      Begin
          OutTextXY(ScreenX,ScreenY,
          '   0.0    0.2    0.4    0.6    0.8    1.0');
          SetTextJustify(CenterText,CenterText);
          SetTextStyle(SmallFont,HorizDir,4);
          OutTextXY(153,ScreenY + 12,'Mole Fraction');
```

Figure 7–6 (continued)

```
              SetTextJustify(LeftText,TopText);
              SetTextStyle(DefaultFont,HorizDir,1);
          End;

  (*---------------------------------------------------------------------*)

     Procedure WriteTemps(ScreenYhi,ScreenYlo:  Integer);
     Var ScreenX:  Integer;
          CharTemp:  String[10];
     Begin
        ScreenX:= 1;
        Str(HighTemp,CharTemp);   (* Convert real number to string *)
        OutTextXY(ScreenX,ScreenYHi,CharTemp);
        Str(LowTemp,CharTemp);    (* Convert real number to string *)
        OutTextXY(ScreenX,ScreenYLo,CharTemp);
        SetTextStyle(SmallFont,VertDir,4);
        SetTextJustify(CenterText,CenterText);
        OutTextXY(15,102,'T/deg K');
     End;

  (*****************************************************************)

     Begin (* DrawGrid *)
        InitGraph(GrDriver,GrMode,'');
        SetGraphMode(3);
        DrawBorder;
        WriteTitle(153,10);
        MarkX;
        MarkY;
        WriteXnumbers(1,180);
        WriteTemps(12,172);
     End;   (* DrawGrid *)

  (*****************************************************************)

Procedure DrawEutectic(Xscale,Yscale:  Real);
Var ScreenX,ScreenY:  Integer;    (* Pixel Coordinates *)
     Xe,Te:  Real;                 (* Xe and Te are eutectic mole *)
                                   (* fraction and temperature;   *)
Begin  (* Draw Eutectic *)
   FindEutectic(Xe,Te);
   ScreenY:= ScreenBottom - Trunc(Yscale*(Te - LowTemp));
   Line(Screenleft,ScreenY,ScreenRight,ScreenY);
   ScreenX:= ScreenLeft + Trunc(Xscale*Xe);
   Line(ScreenX,ScreenY,ScreenX,ScreenBottom);
End;  (* DrawEutectic *)

  (*****************************************************************)

PROCEDURE WriteData;
Var I:  Integer;
    Xe,Te,first,second: Real;
Begin (* WriteData *)
   First:= 0.0;
   Second:= 1.0;
```

Figure 7-6 (continued)

```
   Writeln;
   Writeln('Mole Fraction Xa    Temperature(K)');Writeln;
   Writeln(first:10:2,ToA:15:1);
   For I:= 1 to 249 do
      If I MOD 25 = 0 then
         Writeln(XA[I]:10:2,T[I]:15:1);
   Writeln(second:10:2, ToB:15:1);
   Writeln;Writeln;
   FindEutectic(Xe,Te);
End;  (* WriteData *)

(*************************************************************)

Procedure ScaleScreen(Var Xscale,Yscale,Xleft:  Real;
                      Var Ybottom:  Integer);

Var Xright:  Real;
    Ytop:  Integer;

Begin
  Xright:= 1.0;
  Xleft:= 0.0;
  Xscale:=(ScreenRight-Screenleft)/(Xright-Xleft);
  Yscale:=(ScreenBottom-ScreenTop)/(HighTemp - LowTemp);
End; (* ScaleScreen *)

(*************************************************************)

Procedure PlotCurves;

Var I,ScreenX,ScreenY:  Integer;

Begin (* PlotCurves *)
  InitGraph(GrDriver,GrMode,'');
  SetGraphMode(3);
  DrawGrid;
  Ybottom:= LowTemp;
  ScreenX:= ScreenLeft;  (* Plot from left to right *)
  ScreenY:= ScreenBottom - Trunc(Yscale*(ToB-Ybottom));  (* start at left *)
  For I:= 1 to 249 do
     Begin (* For *)
        ScreenX:= ScreenX + 1;
        ScreenY:= ScreenBottom - Trunc(Yscale*(T[I]-Ybottom));
        PutPixel(ScreenX, ScreenY,3);
     End; (* For *)
  DrawEutectic(Xscale,Yscale);
End;  (* PlotCurves *)

(*************************************************************)

Begin (* M A I N   P R O G R A M *)
  HeaderPage;
  Repeat
    GetData;
    CalcArrays;
    WriteEutectic;
    GetTempExtremes;
    Scalescreen(Xscale,Yscale,Xleft,Ybottom);
    PlotCurves;
    Readln;
    Textmode(BW80);
    Write('Run again?  Enter Y or N, please.  ');
    Readln(Answer);
  Until (Answer = 'N') OR (Answer = 'n');
End.  (* M A I N   P R O G R A M *)
```

Figure 7–6 (continued)

```
Please enter a title for your graph --
Use 35 characters or less.
1--------10--------20--------30---35
Benzene-Naphthalene

Please enter the freezing point of pure A(deg K):   353.4
Please enter the heat of fusion of A (kJ/mol):   19.1
Please enter the freezing point of pure B(deg K):   278.6
Please enter the heat of fusion of B (kJ/mol):   9.90

Mole Fraction A   Temperature(K)

       0.00            278.6
       0.10            271.9
       0.20            283.3
       0.30            298.2
       0.40            309.7
       0.50            319.3
       0.60            327.7
       0.70            335.0
       0.80            341.7
       0.90            347.8
       1.00            353.4


At the eutectic, the temperature equals 269.7
and the mole fraction A equals 0.132

Enter temperatures at graph top and bottom as integers.

Please enter the maximum temperature (K) of the graph:   360

Please enter the minimum temperature (K) of the graph:   260
```

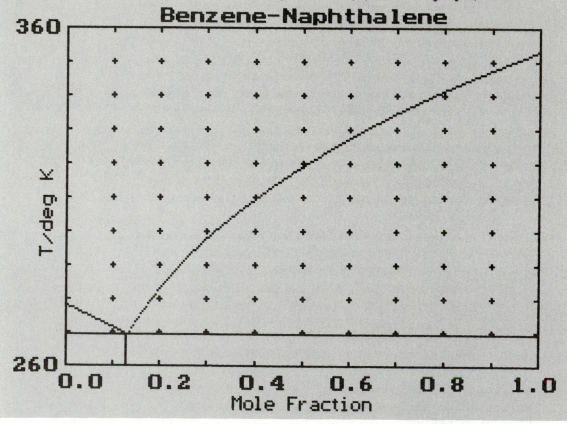

Figure 7–7 Output of *Program TwoPhase* for benzene and naphthalene. The input consisted of the two melting points (353.4 and 278.6 K) and the two heats of fusion (19.1 and 9.90 kJ/mol) for benzene and naphthalene, respectively. The upper temperature limit for graphing was chosen to be 360 K; the lower limit, 260 K.

REFERENCES

1. Findlay, A. N., A. N. Campbell, and N. O. Smith, *The Phase Rule and Its Applications,* Dover, New York, 1951.

2. Hansen, M. *Constitution of Binary Alloys,* McGraw-Hill, New York, 1958.

3. Kubachewski, O., and J. A. Catterall, *Thermochemical Data of Alloys,* Pergamon Press, London, 1956.

4. Mason, C. M., B. W. Rosen, and R. M. Swift, "Phase Rule Experiments with Organic Compounds," *J. Chem. Educ.* 18:473, 1941.

5. Skau, E. L., and J. C. Arthur, Jr., in A. Weissberger and B. W. Rossiter, eds., *Techniques of Chemistry,* 4th ed., vol. 1, pt. 5, chap. 3, Wiley-Interscience, New York, 1970.

6. Stephen, H., and T. Stephen, eds., *Solubilities of Inorganic and Organic Compounds,* Pergamon Press, Oxford, 1963, p. 38.

Experiment 8

The Three-Component System: Acetic Acid, Vinyl Acetate, and Water (three liquids)

Liquid systems of three components consist of three liquid components, two liquids and one solid, or one liquid and two solids. In this experiment, we investigate the behavior of a system of three liquids: acetic acid, vinyl acetate, and water. Acetic acid is miscible with both vinyl acetate and water, but vinyl acetate and water are quite insoluble in each other.

According to the phase rule, a single phase in a three-component system may possess four degrees of freedom:

$$F = C - P + 2 = 3 - 1 + 2 = 4 \tag{8-1}$$

These are temperature, pressure, and the compositions of two of the three components. Because of the difficulty in graphically representing so many variables, temperature and pressure are generally held constant, so the phase rule reduces to

$$F = C - P = 3 - P \tag{8-2}$$

This is the same special form of the phase rule that applies to two-component systems at constant pressure only. In graphical representations of both of these systems, an area corresponds to $F = 3$, a line to $F = 1$, and a point to $F = 0$ (invariant).

Review of Two-Component Systems

In a two-component system, the composition is represented by the position along the line AB in Figure 8–1. The properties of the line are such that no matter where point P lies on the line, AP + PB = AB, so that

$$W_A/W_B = BP/PA \qquad (8\text{--}3)$$

where W_A and W_B are weight fractions and AP and PB are the lengths of the line segments. Point P_1 lies in an area where a single phase is present, the homogeneous melt, and $F = 3 - 1 = 2$ degrees of freedom: both temperature and composition must be specified to define the system. Point P_2 lies on a line where two phases, the melt and solid B, are in equilibrium, so that $F = 3 - 2 = 1$ degree of freedom: either the temperature or composition must be specified to define the system. At point P_3, three phases, solid A, solid B, and the melt, are in equilibrium, so that $F = 3 - 3 = 0$ and there are no degrees of freedom. The system is invariant; the temperature and composition are uniquely defined when three phases are in equilibrium.

Graphical Representation of a Three-Component System

For a three-component system, the properties of an equilateral triangle provide a convenient means of representing the composition of a three-component system at constant temperature and pressure. The interpre-

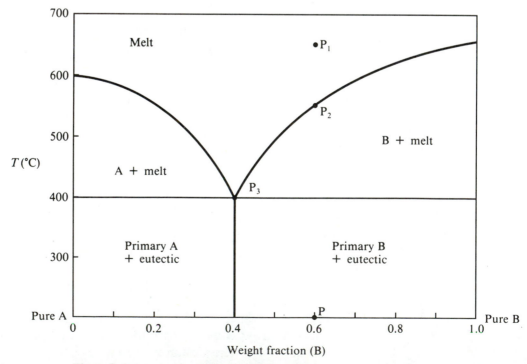

Figure 8–1 A two-component system at constant pressure.

tation of point P is the same in both Figure 8–1 and Figure 8–2. Line AB represents the compositions of all systems that can be prepared from pure A and pure B; at point P in both figures the system is 60% B and 40% A.

In Figure 8–2 the edges of the triangle represent the compositions of all the two-component systems that can be prepared from A and B, B and C, and A and C. At point Q, the system is 30% A, 70% C, and 0% B. Systems of three components lie inside the triangle. At point R the system is 30% A, 20% B, and 50% C. At point S the system is 50% A, 40% B, and 10% C.

A line parallel to an edge, for example, GH, represents all systems with a constant percent of one component (25% C in this case) and a variable ratio of the other two components. All systems that are 25% C must lie on this line.

A line from a vertex to the opposite side, for example, CP, represents all systems with a constant ratio of two components (A:B = 40:60 in this case) and a variable percent of the third component. It is useful to think of a line as the locus of all compositions that can be prepared by mixing the constituents whose compositions are indicated by the ends of the line. Thus, if we begin with a two-component system of composition P and add C, the composition changes from P to C as a large amount of C is added. Conversely, if we begin with a three-component system that lies somewhere on the line CP and gradually remove C, say, by evaporation, then the composition changes toward P and reaches it when all the C has been removed. It is also true that for point W on line PC the

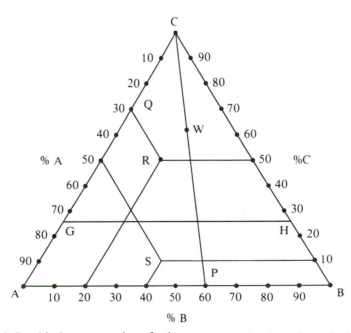

Figure 8–2 Graphical representation of a three-component system at constant pressure and temperature.

relative amounts of C and P are related by C:P = WP:WC. Either weight fractions or mole fractions may be used. When compounds are formed, it is usually more useful to use mole fractions in order to assess the stoichiometry of the system.

Typical Phase Diagrams of Three-Component Systems

In Figure 8–3a, components A and B are miscible and components A and C are miscible, but components B and C show a region of immiscibility. Three-component systems consisting of two liquid phases lie under the curved line. Above the curved line, only a single phase is present and the system is homogeneous. If sufficient A is added to a two-phase system under the curved line, the composition shifts toward A and eventually crosses the curved line toward A, and the system becomes homogeneous. The tie lines under the curved line connect the two liquid phases in equilibrium, giving their compositions.

In Figure 8–3b, only A and C are miscible, and two regions of immiscibility exist. If none of the three components is miscible, then three regions of two liquid phases may exist, as shown in Figure 8–3c. If the three components are very insoluble in each other, the situation in Figure 8–3d may occur. Near the corner, the system is homogeneous. In the tie-line areas, two liquid phases are in equilibrium, and in the center triangular area, three liquid phases are in equilibrium. In the latter case, $F = 3 - P = 3 - 3 = 0$ and the system is invariant. This means that anywhere in the center triangle the compositions of the three liquid phases are fixed and are given by the vertices of the center triangle. These can be described as an A-rich phase, a B-rich phase, and a C-rich phase, depending upon which is the closest vertex.

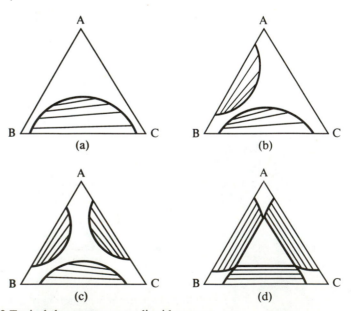

Figure 8–3 Typical three-component liquid systems.

Determination of the Two-Phase Boundary

The curved boundary between the two-phase region (below the curved line) and the one-phase region (above the curved line) is determined by a series of titrations with vinyl acetate and water as shown in Figure 8–4. Consider the system at point w_1, a homogeneous mixture of water and acetic acid. All three component systems that can be prepared from this starting material and pure vinyl acetate lie along the line w_1v. If a weighed sample of composition w_1 is titrated with vinyl acetate, the composition changes along the line w_1v until point w_1' is reached. At this point, a tiny amount of a second liquid layer just begins to separate out, which is indicated by the onset of cloudiness in the titrated liquid mixture. From the volume of the vinyl acetate used and its density, along with the weight of w_1 and its known composition, the percent of all three components at point w_1' can be calculated. Similar titrations of starting mixtures w_2, w_3, and w_4 give enough information to outline the curve from w_1' to w_4'.

In principle, one should be able to continue the titration until the system once again becomes homogeneous. However, this is inconvenient, because large amounts of vinyl acetate would be required and the end point, the onset of homogeneity, is difficult to detect. Instead, a second set of titrations is carried out, this time with water. The mixtures to be titrated are homogeneous mixtures of acetic acid and vinyl acetate of compositions v_1, v_2, v_3, and v_4. This set of titrations gives the data required to outline the left side of the two-phase curve from v_1' to v_4'.

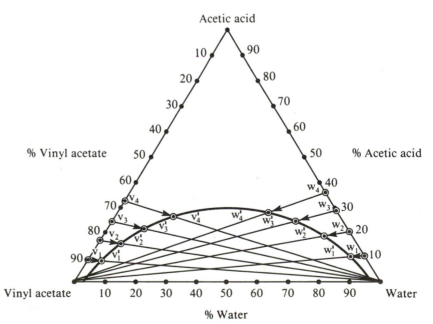

Figure 8–4 Determination of the two-phase boundary (acetic acid, vinyl acetate, and water).

Determination of the Tie Lines

The tie lines are determined by a third set of titrations after the two-phase boundary has been determined. Consider the four systems s_1, s_2, s_3, and s_4 shown in Figure 8–5. These three-component systems consist of two liquid layers, the composition of which is given by the ends of the tie lines. For example, the composition at s_1 is known, because a solution of such overall composition can be prepared gravimetrically. If the two layers s_1' and s_1'' are separated with a separatory funnel, their composition can be determined by titrating the acetic acid they contain with standard base. From the weight of the layer titrated and the weight of the acetic acid it contains, the weight fraction of acetic acid can be calculated.

Suppose that for layer s_1', the weight fraction of acetic acid is a_1'. The dotted line at a_1' gives the composition of *all* systems with a constant weight fraction of acetic acid, namely a_1'. It follows, then, that the intersection of this dotted line with the two-phase curve gives the overall composition of s_1'. Likewise, a titration of the second layer s_1'' gives its weight fraction of acetic acid, a_1''. A second dotted line at a_1'' must intersect the two-phase curve, indicating the composition of s_1''. If the work is carefully done, all three points s_1, s_1', and s_1'' should lie on a straight (tie) line. Three more such determinations on s_2, s_3, and s_4 suffice to determine the other tie lines.

Apparatus

The experiment requires eight 50-ml and four 100-ml glass-stoppered bottles or flasks; 50- and 10-ml burets with Teflon stopcocks; a 10-ml

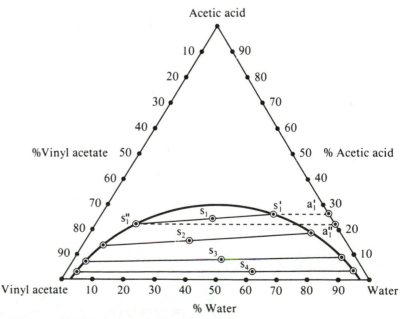

Figure 8–5 Determination of the tie lines.

measuring pipet; 10- and 25-ml volumetric pipets; a 100-ml separatory funnel with a Teflon stopcock; glacial acetic acid; vinyl acetate; 0.3 M sodium hydroxide (standardized); and a phenolphthalein indicator.

SAFETY CONSIDERATIONS

Glacial acetic acid smells bad and should be handled in a hood. The compound vinyl acetate is an analgesic, but is less toxic than carbon tetrachloride, 1,2-dichloroethane, or benzene, which are sometimes substituted in this experiment (Clark, 1974).

EXPERIMENTAL PROCEDURE

To determine the two-phase boundary, make up a series of solutions corresponding to the compositions w_1, w_2, w_3, w_4, v_1, v_2, v_3, and v_4 in Figure 8–4. These solutions should be made up gravimetrically, since the percent composition must be known accurately. It is convenient to deliver approximate volumes into a tared weighing bottle. Some suggested volumes are listed in Table 8–1.

Weigh each component to ±0.1 mg. Shake the solutions vigorously as you titrate them. The end point is the first occurrence of permanent cloudiness, caused by the appearance of a second liquid layer. The end point is somewhat easier to see if the background is black and the flask is well illuminated. Record the volume of titrant used, and calculate its mass from its density.

To determine the tie lines, prepare four solutions corresponding to the points s_1, s_2, s_3, and s_4 on Figure 8–5. Some suggested volumes are listed in Table 8–2. Weigh each component (±0.1 mg) into a glass-stoppered Erlenmeyer flask. To avoid loss by evaporation, add the vinyl acetate last, since it is the most volatile. Shake well and transfer to a separatory funnel with a Teflon stopcock. Separate the two phases into glass-stoppered flasks and titrate a 5-ml aliquot of

TABLE 8–1

Component Volumes (ml) for Phase Boundary Determination

For Titration with H_2O		For Titration with $CH_3COOCH=CH_2$	
CH_3COOH	$CH_3COOCH=CH_2$	CH_3COOH	H_2O
3	15	3	17
5	10	5	15
8	10	7	13
10	5	10	9

TABLE 8–2 ▬

Component Volumes (ml) for Tie Line Determination

$CH_3COOCH=CH_2$	H_2O	CH_3COOH
20	25	5
25	25	10
30	25	15
35	25	20

each phase with standard sodium hydroxide (approximately 0.3 M) with phenolphthalein as an indicator. The mass of the 5-ml aliquot must be known and can be determined either by delivering it into a tared flask and weighing, or by measuring the density of the separated phases with a Westphal balance.

This procedure lends itself to the investigation of numerous three-component systems consisting of water, acetic acid, and a third component that is relatively insoluble in water but miscible with acetic acid. Benzene and chloroform have been used (Heric, 1960), but both are more toxic than vinyl acetate. Toluene or cyclohexane would serve as a less toxic substitute for benzene. The water–1,2-dichloroethane–acetic acid system can be analyzed refractometrically (Clark, 1974) instead of titrimetrically.

Results and Calculations

At the end point, calculate the composition of each solution used to determine the two-phase boundary. Plot these points on triangular graph paper and draw a smooth curve through the point.

From the volume and molarity of the sodium hydroxide used to titrate the separated phases for the tie line determination, calculate the weight of acetic acid in the 5-ml aliquot and the weight fraction of acetic acid. Plot a point where the corresponding line of constant–weight-fraction acetic acid intersects the two-phase boundary curve. Repeat for the second phase, and draw a tie line through the three points. Determine the other three tie lines similarly.

REFERENCES

1. Clark, J. R., "Tie Lines in Phase Diagrams for Ternary Systems," *J. Chem. Educ.* 51:255, 1974.

2. Findlay, A. N., A. N. Campbell, and N. O. Smith, *The Phase Rule and Its Applications,* Dover, New York, 1951.

3. Francis, A. W., *Liquid–Liquid Equilibriums,* Interscience, New York, 1963.

4. Heric, E. L., "Tie Line Correlation and Plait Point Determination," *J. Chem. Educ.* 37:144, 1960.

5. Masing, G., *Ternary Systems,* Dover, New York, 1960.

Experiment 9

The Three-Component System: Water, *n*-Propanol, and Sodium Chloride (two liquids and one solid)

In this experiment, we investigate a three-component system consisting of two liquids and one solid. An approximate phase diagram is given in Figure 9–1. The two liquids, water and *n*-propanol, are miscible in all proportions. The solid, sodium chloride, is completely insoluble in *n*-propanol, but substantially soluble in water. At 25°C, 36 g NaCl dissolves in 100 g water, corresponding to point R in Figure 9–1.

In Figure 9–1, S = NaCl, A = *n*-propanol, and W = water. Point R represents a saturated solution of salt in water, and the line QR, saturated solutions of salt in dilute aqueous solutions of alcohol. The line AP represents saturated solutions of salt in dilute solutions of water in alcohol. The lines in the ruled portions of the diagram connect phases in equilibrium with each other. The addition of salt to water–alcohol mixtures makes them less soluble in each other, so they separate into two liquid layers in area PQT. The tie lines in PQT give the composition of the liquid layers in equilibrium with each other. The phase diagram may be divided into five different areas, three of which contain two phases. Area APS contains one solid phase and a liquid phase of compositions

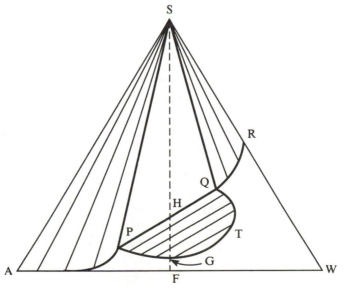

Figure 9–1 Water, *n*-propanol, sodium chloride.

along AP. Area QRS contains one solid phase and a liquid phase of compositions along QR. For these areas, $F = 3 - P = 1$.

The area APTQRW is homogeneous; it contains only one phase, which is a solution of salt in aqueous alcohol. To define the system in this area, two compositions much be given, since $F = 3 - P = 2$.

In the area PQS, three phases are present: solid salt and two liquid phases. One liquid phase is alcohol rich (P) and the other water rich (Q). Both contain dissolved salt. Here the system is invariant since $F = 3 - P = 0$.

A system in the PQT area consists of two liquid phases: an alcohol-rich liquid phase and a water-rich liquid phase. Both contain dissolved salt, but are not saturated with it.

Consider a two-component system F that is 50% water and 50% alcohol. When salt is added to this system, all compositions lie along the line FS. At G the mutual solubility of alcohol and water in each other has decreased so much that a second liquid layer begins to separate out. As more and more salt is added along FS, more and more of the second layer separates. When H is reached, both liquid layers are saturated with salt and it becomes a third phase. The compositions of the two layers are P and Q and their relative amounts are P:Q = HQ:PH. The less dense layer, P, is alcohol rich and floats over the water-rich layer, from which it may be separated. This is called the "salting out effect"; this effect can be taken advantage of in the organic synthesis laboratory to increase the yield of an organic compound produced in the presence of water.

Apparatus

The equipment required is a buret, 250-ml Erlenmeyer flasks, two 10-ml measuring pipets, and a 1-ml measuring pipet. In addition, 200 ml saturated aqueous sodium chloride solution, n-propanol, and iodine are needed.

SAFETY CONSIDERATIONS

None; n-propanol is not particularly toxic. It is somewhat inflammable when pure.

EXPERIMENTAL PROCEDURE

Consider the addition of n-propanol to aqueous solutions of NaCl in water represented by a, b, c, and d in Figure 9–2. The titration of the binary mixture of composition a with n-propanol results in systems whose compositions lie along aA. The initial composition is a; as alcohol is added, the composition gradually changes along the line aA. The compositions of all systems resulting from the titration of b, c, and d with alcohol lie along the lines bA, cA, and dA.

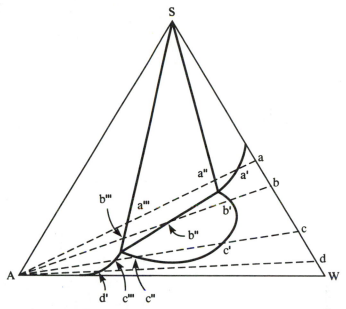

Figure 9–2 Titrations of aqueous NaCl with *n*-propanol.

Points a, b, c, and d represent binary mixtures of water and *n*-propanol. The lines Aa, Ab, Ac, and Ad represent all compositions that result from mixing pure *n*-propanol with each of these binary mixtures. For example, if mixture a is titrated with alcohol, the composition changes from a to a′, where solid NaCl precipitates out of solution. From a to a′ more and more NaCl (*s*) comes out, and when a″ is reached, a second liquid layer begins to separate. As the titration continues from a″ to a‴, more and more of the second layer forms as the first layer diminishes, until at point a‴, the first layer is gone entirely. The titration behavior is summarized in Table 9–1.

TABLE 9–1

Phase Diagram Interpretation

Line	Point	Observation
aA	a′	Solid separates
aA	a″	Second liquid layer appears
aA	a‴	First liquid layer disappears
bA	b′	Second liquid layer appears
bA	b″	Solid separates
bA	b‴	First liquid layer disappears
cA	c′	Second liquid layer appears
cA	c″	First liquid layer disappears
cA	c‴	Solid separates
dA	d′	Solid separates

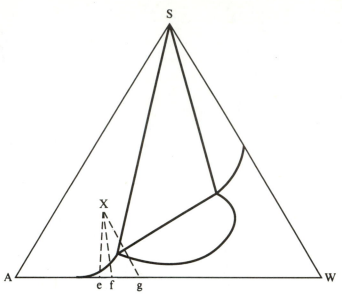

Figure 9–3 Titrations of composition X with *n*-propanol (*aq*).

In principle, a series of titrations of aqueous salt solutions with pure *n*-propanol should suffice to reveal the phase boundaries. However, because the phase boundaries are a bit crowded in the region around point P (Fig. 9–1), the strategy calls for a supporting attack from the south flank, illustrated in Figure 9–3. If a mixture of alcohol, salt, and water of composition X is titrated with three different binary mixtures of alcohol and water (e, f, and g), the compositions lie on the lines Xe, Xf, and Xg. The starting mixtures e, f, and g are homogeneous, but mixture X contains some solid NaCl. The compositions change from X to e and so on. Using arguments like those in the previous paragraph, you should be able

TABLE 9–2

**Aqueous Sodium Chloride Starting Mixtures
(to be titrated with pure *n*-propanol)**

NaCl (saturated) (ml)	Water (ml)
10.0	0.0
10.0	0.1
10.0	0.2
10.0	0.5
9.0	1.0
8.0	2.0
6.0	4.0
5.0	5.0
3.0	7.0
1.0	4.0

TABLE 9–3

Aqueous n-Propanol Titrants (for titrating samples of composition X)

n-Propanol (ml)	Water (ml)
30.0	10.0
25.0	5.0
35.0	5.0

to predict the expected observable changes as the titrations proceed. Tables 9–2 and 9–3 list the recommended starting mixtures for the titrations.

Several suitable starting mixtures of sodium chloride and water should be prepared for titration with n-propanol. The composition lines should intersect the various phase boundaries so that the appearance or disappearance of phases can be observed and the compositions where these observations occur can be calculated from the titration data. Starting mixtures other than those listed in Table 9–2 can be prepared if necessary.

For each titration with the solutions listed in Table 9–2, prepare a solution of composition X (Fig. 9–3) by adding 1.0 ml saturated sodium chloride solution to 10.0 ml pure n-propanol. A solution of composition X contains solid NaCl since it lies in area ASP of Figure 9–1.

Titrate the samples in Tables 9–2 and 9–3. Shake vigorously while titrating. Good lighting is necessary to detect the various end points—separations and phase changes.

Results and Calculations

From the recorded volumes used in the titrations, calculate the weight fraction of n-propanol, water, and sodium chloride with the data in Table 9–4. Plot the weight fractions of your observations on a triangular paper and draw smooth curves to represent the phase boundaries.

TABLE 9–4

Densities of the Liquids at 25°C

Liquid	Density (g/ml)
NaCl (sat'd)[a]	1.20
n-Propanol	0.80
Water	1.00

[a] Solubility is 36.2 g/100 g H_2O.

A three-component system consisting of one liquid and two solid phases may be determined by Schreinmakers's method of wet residues. The system sodium nitrate–lead nitrate–water is suitable for student work (Heric, 1958).

REFERENCES

1. Findlay, A. N., A. N. Campbell, and N. O. Smith, *The Phase Rule and Its Applications,* Dover, New York, 1951.
2. Francis, A. W., *Liquid–Liquid Equilibriums,* Interscience, New York, 1963.
3. Heric, E. L., "A Phase Rule Experiment: The System Lead Nitrate–Sodium Nitrate–Water," *J. Chem. Educ.* 35:510, 1958.
4. Masing, G., *Ternary Systems,* Dover, New York, 1960.

Experiment 10

Nonideal Solutions of Nonelectrolytes

The general relationship between the free energy per mole of a substance in standard state (noted by a superscript zero) and an arbitrary state is given by

$$G_A = G_A^0 + RT \ln a_A \qquad (10\text{-}1)$$

In a solution, Equation 10–1 becomes

$$G_A(\text{soln}) = G_A^0(\text{pure}) + RT \ln \frac{P_A}{P_A^0} \qquad (10\text{-}2)$$

where P_A^0 is the vapor pressure of pure A and P_A is the vapor pressure of A in solution. The activity $a_A = P_A/P_A^0$. These relationships are valid whether the solution is ideal or nonideal.

Ideal Solutions

In an ideal solution, both components follow Raoult's Law, so that

$$\frac{P_A}{P_A^0} = X_A \qquad \text{and} \qquad \frac{P_B}{P_B^0} = X_B \qquad (10\text{-}3)$$

The activity equals the mole fraction, and the standard state of the solvent is the state of the mole fraction equal to 1. The partial pressure of each component is a linear function of the mole fraction:

$$P_A = X_A P_A^0 \qquad \text{and} \qquad P_B = X_B P_B^0 \qquad (10\text{--}4)$$

This relationship is shown in the lower solid lines of Figure 10–1. The total pressure is the sum of the partial pressures:

$$P_T = P_A + P_B = X_A P_A^0 + X_B P_B^0 \qquad (10\text{--}5)$$

$$= (P_B^0 - P_A^0)X_B + P_A^0$$

Thus, the upper solid line in Figure 10–1 is a straight line of slope equal to $(P_B^0 - P_A^0)$ and intercept equal to P_A^0.

If the solution is ideal, it is also possible to calculate the composition of the gas phase:

$$Y_A = \frac{P_A}{P_T} \qquad \text{and} \qquad Y_B = \frac{P_B}{P_T} \qquad (10\text{--}6)$$

where the Y_i are the mole fractions in the gas phase. If the solution is nonideal, then the composition cannot be calculated but must be measured experimentally, which is what is done in this experiment.

Activities and Activity Coefficients in Nonideal Solutions

In an ideal solution, the activities are simply equal to the mole fractions of the liquid phase:

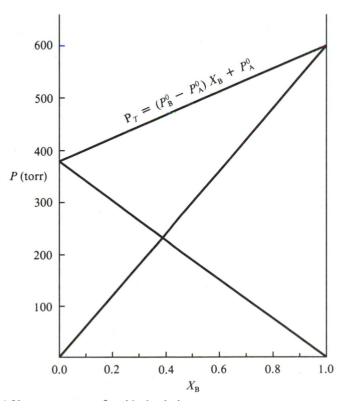

Figure 10–1 Vapor pressures of an ideal solution.

$$a_i = X_i \qquad\qquad (10\text{--}7)$$

In a nonideal solution, however, the activities are given by

$$a_i = \gamma_i X_i \qquad\qquad (10\text{--}8)$$

where γ_i are the activity coefficients of the ith components. Experimentally, it is necessary to determine the activity coefficient first and then calculate activities. Thus, from Equation 10–8,

$$\gamma_i = \frac{a_i}{X_i} \qquad\qquad (10\text{--}9)$$

or

$$\gamma_i = \frac{P_i}{P_i^0 X_i} \qquad\qquad (10\text{--}10)$$

where P_i is the partial pressure and P_i^0 is the vapor pressure of the pure component at the experimental temperature.

Dalton's Law gives the partial pressures in terms of the mole fraction in the gas phase and the total pressure:

$$P_i = Y_i P_T \qquad\qquad (10\text{--}11)$$

Substitution of Equation 10–11 in Equation 10–10 gives

$$\gamma_i = \frac{Y_i P_T}{X_i P_i^0} \qquad\qquad (10\text{--}12)$$

Equation 10–12 permits the experimental determination of the activity coefficient and thus forms the working equation for the experiment. Manometric measurements give P_T, the total pressure. The vapor pressures of the pure components P_i^0 are determined from literature values calculated from vapor-pressure equations. The compositions of the liquid phase X_i and of the gas phase Y_i are determined refractometrically. Once the activity coefficients have been determined, it is a simple matter to calculate the activities from Equation 10–8. The typical behavior of the partial pressure, activity coefficient, and activity is shown in Figure 10–2a, b, and c, where the broken lines represent ideal solutions.

Free Energy of Mixing

For an ideal solution, the free energy of mixing is given by

$$\Delta G_{\text{mix}} = X_A RT \ln X_A + X_B RT \ln X_B \qquad\qquad (10\text{--}13)$$

Because the X_i vary from 0 to 1, the $\ln X_i$ are always negative. Consequently, the free energy of mixing is also always negative. A plot of ΔG_{mix} versus X_A is shown as the broken line for an ideal solution in Figure 10–2d.

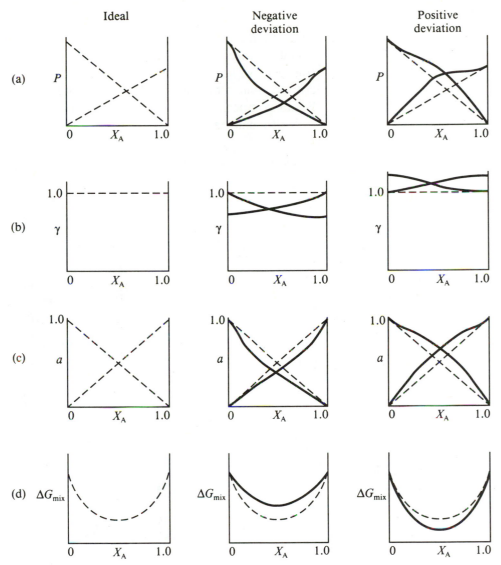

Figure 10-2 The comparative behavior of ideal and nonideal solutions.

For a nonideal solution, the free energy of mixing is given by

$$\Delta G_{\text{mix}} = X_A RT \ln \frac{P_A}{P_A^0} + X_B RT \ln \frac{P_B}{P_B^0} \tag{10-14}$$

Calculation of P_A and P_B with Equation 10–11 from Y_i measured refractometrically permit the calculation of ΔG_{mix} at each liquid composition X_i. The vapor pressures of the pure components P_i^0 are calculated from known vapor-pressure equations at the experimental temperature.

The broken lines in Figure 10–2d show the typical behavior of ideal solutions in comparison with a nonideal solution (solid line).

Apparatus

The apparatus, illustrated in Figures 10–3 and 10–4, consists of an electrically heated flask to which is sealed a reflux condenser and U-tube trap. Analysis of a sample of liquid from the U-tube gives the vapor phase composition, and analysis of a sample of liquid from the boiler gives the liquid composition. An Abbé refractometer is used with a calibration curve to analyze the liquid mixtures. The thermometer is graduated in

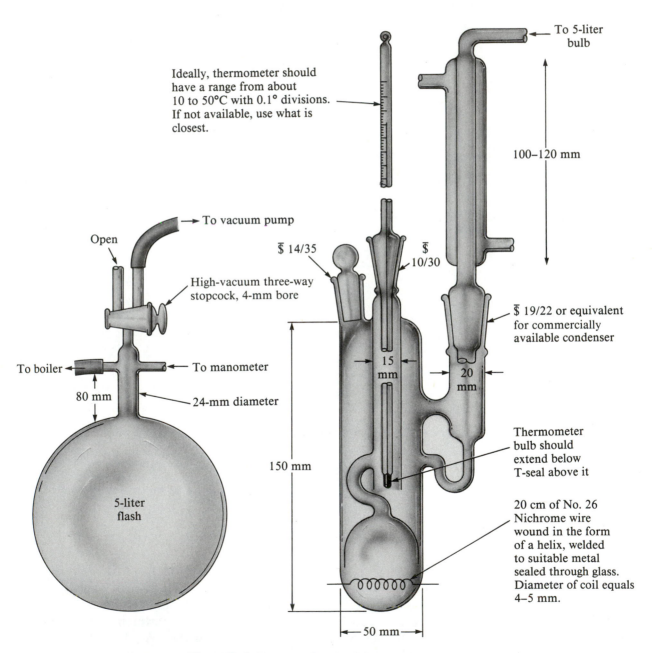

To 5-liter bulb

Ideally, thermometer should have a range from about 10 to 50°C with 0.1° divisions. If not available, use what is closest.

100–120 mm

To vacuum pump

Open

High-vacuum three-way stopcock, 4-mm bore

$\overline{\underline{S}}$ 14/35

$\overline{\underline{S}}$ 10/30

$\overline{\underline{S}}$ 19/22 or equivalent for commercially available condenser

To boiler

To manometer

80 mm

24-mm diameter

15 mm

20 mm

150 mm

Thermometer bulb should extend below T-seal above it

5-liter flask

20 cm of No. 26 Nichrome wire wound in the form of a helix, welded to suitable metal sealed through glass. Diameter of coil equals 4–5 mm.

50 mm

Figure 10–3 Apparatus for measuring equilibrium compositions and pressures at constant temperature.

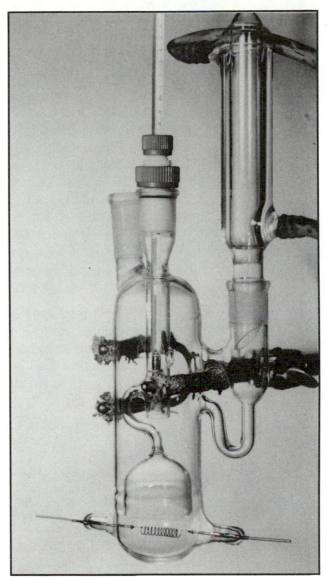

Figure 10–4 Apparatus for measuring equilibrium compositions and pressures at constant temperature, ready for use. (Karen R. Sime)

0.1°C intervals. Between the mercury manometer and the boiler is a 5-L flask connected to a mechanical vacuum pump through a three-way stopcock.

SAFETY CONSIDERATIONS

Both acetone and chloroform are known to be toxic. Avoid inhaling their vapors. Acetone is volatile and inflammable.

Be sure not to allow current to flow through the boiler heater unless it is completely covered with a liquid. Without a liquid to absorb the heat, damage to the wire heater may result.

EXPERIMENTAL PROCEDURE

Fill the boiler with 100 ml pure chloroform. Start the vacuum pump and evacuate to approximately 300 torr. Turn on the current to the boiler heater and adjust it until the pure chloroform begins to boil. Finally, adjust the pressure by either pumping or letting air in until the equilibrium temperature is exactly 35°C. Repeat this procedure for each run so that the equilibrium temperature for each run is 35°C. Record the pressure, turn off the current, and open the apparatus to the atmosphere. When the system has cooled, open it and remove a 1-ml sample from the boiler *and* the U-tube for subsequent analysis by refractometry.

Repeat this sequence for a total of ten runs of various compositions. Vary the composition by removing part of the liquid from the previous run and replacing it with the second component, according to the scheme set forth in Table 10–1.

Analysis of the Liquid Mixtures. Prepare three mixtures of acetone and chloroform with mole fractions of approximately 0.25, 0.5, and 0.75 acetone. Weigh the liquid samples sufficiently accurately so that the mole fraction can be calculated to less than 0.5%.

Measure the refractive index of the three mixtures and of pure acetone and chloroform. Plot refractive index versus mole fraction. Use this calibration curve to determine the compositions of the

TABLE 10–1

Initial Volumes of Chloroform and Acetone

A. **Chloroform-Rich Sequence (100 ml chloroform initially)**

Run Number	Remove (ml)	Add C_3H_8O (ml)	Approx. Mole Fraction X_{CHCl_3}
1	—	—	1.0
2	9	9	0.9
3	10	10	0.8
4	14	14	0.7
5	11	11	0.6

B. **Acetone-Rich Sequence (100 ml acetone initially)**

Run Number	Remove (ml)	Add $CHCl_3$ (ml)	Approx. Mole Fraction X_{CHCl_3}
6	—	—	0.0
7	16	16	0.15
8	20	20	0.3
9	20	20	0.45
10	20	20	0.55

boiling liquid X_{CHCl_3} (sampled from the boiler) and its equilibrium vapor Y_{CHCl_3} (sampled from the condensed vapor in the U-tube).

Prepare a table of the calibration data and a graph of refractive index versus composition.

Results and Calculations

Calculate the vapor pressure $P_i^0(torr)$ for pure acetone and pure chloroform from the following equations (Boublík et al., 1973):

$$\log_{10} P_{acetone}^0 = 7.11714 - \frac{1210.595}{t + 229.664} \tag{10–15}$$

$$\log_{10} P_{chloroform}^0 = 6.95465 - \frac{1170.966}{t + 226.252} \tag{10–16}$$

Calculate P_a, P_c, γ_a, γ_c, a_a, a_c, and ΔG_{mix} for each run. Prepare appropriate tables of the run experimental data and the calculated parameters. Prepare graphs of P_T and P_i versus X_a, γ_i versus X_a, a_i versus X_a, and ΔG_{mix} versus X_a. Compare your results with ideal solution behavior and with data from the literature. Calculate the propagated error in the activity coefficient.

Alternatively, the ethyl acetate–cyclohexane system at 40°C may be investigated. These substances are less toxic than acetone and chloroform. Analysis by refractive index measurement is also suitable. Vapor pressure data are listed in Table 6–1.

REFERENCES

1. Barrow, G., *Physical Chemistry,* 4th ed., McGraw-Hill, New York, 1979.

2. Boublík, T., V. Fried, and E. Hála, *The Vapor Pressures of Pure Substances,* Elsevier, New York, 1973.

3. Frigaria, N. A., "Vapor Pressure Measurements," *J. Chem. Educ.,* 39:35, 1962.

4. McGlashan, M. L., "Deviations from Raoult's Law," *J. Chem. Educ.* 40:516 1963.

5. Mueller, C. R., and E. Kearns, "Thermodynamic Studies of the System Acetone–Chloroform," *J. Phys. Chem.* 62:1441, 1958.

6. Ozog, J. Z., and J. A. Morrison, "Activity Coefficients of Acetone–Chloroform Solutions," *J. Chem. Educ.* 60:72, 1983.

7. Shearer, E. C., "Liquid–Vapor Equilibrium at Constant Temperature," *J. Chem. Educ.* 50:446, 1973.

8. Smyth, C. P., and E. W. Engel, "Molecular Orientation and the Partial Vapor Pressures of Binary Mixtures," *J. Am. Chem. Soc.* 51:2646, 1929.

9. Tobey, S. W., "Ideal Solution Laws," *J. Chem. Educ.* 39:258, 1962.

Properties of Solutions

Experiment 11

Partial Molal Volume

The phase rule in general is written $F = C - P + 2$. For a single phase and a single component, $F = 1 - 1 + 2 = 2$. This means that in order to specify the state of a system consisting of one component and one phase completely and unambiguously, a minimum of two properties of the system must be given. These are usually pressure and temperature, since it is experimentally convenient to control these variables. For any extensive thermodynamic property Y (such as V, G, H, S, and A), we can write, for a one-component system,

$$Y = f(P, T) \qquad (11\text{--}1)$$

The total differential is

$$dY = \left(\frac{\partial Y}{\partial P}\right)_T dP + \left(\frac{\partial Y}{\partial T}\right)_P dT \qquad (11\text{--}2)$$

In a two-component system, another variable is introduced, so we write

$$Y = f(P, T, n_1, n_2) \qquad (11\text{--}3)$$

and the total differential becomes

$$dY = \left(\frac{\partial Y}{\partial P}\right)_{T,n_1,n_2} dP + \left(\frac{\partial Y}{\partial T}\right)_{P,n_1,n_2} dT \qquad (11\text{--}4)$$

$$+ \left(\frac{\partial Y}{\partial n_2}\right)_{P,T,n_1} dn_1 + \left(\frac{\partial Y}{\partial n_1}\right)_{P,T,n_2} dn_2$$

The partial molal Y is defined as

$$\bar{Y}_i = \left(\frac{\partial Y}{\partial n_i}\right)_{P,T,n_j} \tag{11-5}$$

At constant temperature and pressure, Equation 11–3 becomes

$$Y = f(n_1, n_2) \tag{11-6}$$

and the total differential becomes

$$dY = \left(\frac{\partial Y}{\partial n_2}\right)_{P,T,n_1} dn_1 + \left(\frac{\partial Y}{\partial n_1}\right)_{P,T,n_2} dn_2 \tag{11-7}$$

which can be written more compactly as

$$dY = \bar{Y}_1 \, dn_1 + \bar{Y}_2 \, dn_2 \tag{11-8}$$

Of all the extensive thermodynamic properties, the volume is the easiest to visualize; this also holds true for the partial molar volume, defined as

$$\bar{V}_i = \left(\frac{\partial V}{\partial n_i}\right)_{P,T,n_j} \tag{11-9}$$

so that Equation 11–8 can be written

$$dV = \bar{V}_1 \, dn_1 + \bar{V}_2 \, dn_2 \tag{11-10}$$

The integrated form of this equation (Klotz, 1950) is

$$V = n_1 \bar{V}_1 + n_2 \bar{V}_2 \tag{11-11}$$

which is an interesting and surprising result. Let us see why.

Partial Molal Volume of Some Real Solutions

What is so surprising about this rather innocuous-appearing equation? Well, if a solution is ideal, then its volume is just the sum of the volumes of the solute and solvent:

$$V = n_1 \bar{V}_1^0 + n_2 \bar{V}_2^0 \tag{11-12}$$

Benzene and toluene form an ideal solution. The volume of 1 mole pure benzene is 88.9 ml; the volume of 1 mole pure toluene is 106.4 ml. Equation 11–12 states that 88.9 ml benzene mixed with 106.4 ml toluene results in 88.9 ml + 106.4 ml, or 195.3 ml of solution. Common sense suggests that the volumes add up, since the volumes of substances in solution are extensive properties.

On the other hand, water and ethanol do not form an ideal solution. The volume of 1 mole pure ethanol is 58.0 ml and the volume of 1 mole pure water is 18.0 ml. However, 1 mole water mixed with 1 mole ethanol does not result in 58.0 ml + 18.0 ml, or 76.0 ml, but rather 74.3 ml!

According to Equation 11–11, it is the partial molal volumes that are additive or extensive properties. When the mole fraction is 0.5, the partial molal volume of ethanol is 57.4 ml and the partial molal volume of water is 16.9 ml. With Equation 11–11, we can now calculate the volume of the solution:

$$1 \text{ mole} \times 57.4 \text{ ml/mol} + 1 \text{ mole} \times 16.9 \text{ ml/mol} = 74.3 \text{ ml} \quad \textbf{(11–13)}$$

which is exactly what is observed. Note that the values just cited for the partial molal volumes of ethanol and water are only for a particular concentration; in this case, the mole fraction equals 0.5 and applies only to the water–ethanol system.

Apparent Molal Volume

The experimental determination of partial molal volumes is, in principle, quite simple and involves the careful measurement of the densities of solutions of known concentrations. The calculation is simplified by the use of a related quantity, the apparent molal volume of the solute, ϕV. Consider the volume of a solution as n_2 moles of a solute are added to a fixed n_1 moles of solvent. The volume of the solution might change as shown in Figure 11–1. The volume due to the added solute (per mole) is called the apparent molal volume ϕV. Figure 11–1 shows that

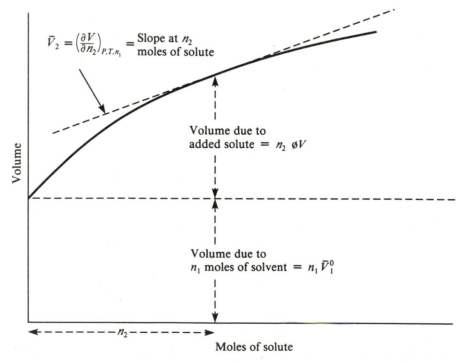

$$\bar{V}_2 = \left(\frac{\partial V}{\partial n_2}\right)_{P,T,n_1} = \frac{\text{Slope at } n_2}{\text{moles of solute}}$$

Volume due to added solute $= n_2\ \phi V$

Volume due to n_1 moles of solvent $= n_1\ \bar{V}_1^0$

Volume

n_2

Moles of solute

Figure 11–1 How the total volume of a solution V depends on the volume of the pure solvent and the apparent molal volume of the solute ϕV.

$$\phi V = \frac{V_{\text{solution}} - V_{\text{solvent}}}{\text{moles of solute}} \qquad (11-14)$$

or

$$\phi V = \frac{V - n_1 \bar{V}_1^0}{n_2} \qquad (11-15)$$

Thus, in Figure 11–1, the volume V of the solution at any particular added n_2 moles of solute is given by a rearrangement of Equation 11–15:

$$V = n_1 \bar{V}_1^0 + n_2 \phi V \qquad (11-16)$$

It should be clear from an examination of Figure 11–1 that it does not make sense to define an apparent molal volume for the solvent, since the number of moles of both solvent and solute cannot be held constant simultaneously.

Partial and Apparent Molal Volumes from Density Measurements

The Solute. The apparent molal volume is useful because it can be measured easily (as Fig. 11–1 suggests) and because the partial molal volumes of both the solute and solvent can be calculated from measurements of the apparent molal volumes. Relating the partial molal volume of the solute to the apparent molal volume is straightforward: partial differentiation of Equation 11–16 with respect to n_2 at constant n_1 gives

$$\bar{V}_2 = \left(\frac{\partial V}{\partial n_2} \right)_{n_1} = \phi V + n_2 \left(\frac{\partial \phi V}{\partial n_2} \right)_{n_1} \qquad (11-17)$$

Next we wish to relate the apparent molal volume to the usual experimental parameters, namely the molality m and the density d of the solution. Equation 11–15 defines apparent molal volume. The volume V of solution can be expressed as

$$V = \frac{\text{Wt. solvent} + \text{Wt. solute}}{d} \qquad (11-18)$$

$$= \frac{n_1 M_1 + n_2 M_2}{d} = \frac{1000 + m M_2}{d}$$

where M_1 and M_2 are the molecular weights of the solvent and solute, respectively. The term $n_1 \bar{V}_1^0$ in Equation 11–15 is just the volume of the solvent; when the concentration is expressed in molality, the mass of solvent is 1000 g, so the volume of the solvent is

$$n_1 \bar{V}_1^0 = 1000 / d_1 \qquad (11-19)$$

Substitution of V and $n_1 \bar{V}_1^0$ from Equations 11–18 and 11–19 into Equation 11–15 gives us our working equation for experimentally measuring ϕV, the apparent molal volume, with n_2 equal to m:

$$\phi V = \frac{1}{m}\left(\frac{1000 + mM_2}{d} - \frac{1000}{d_1}\right) \tag{11-20}$$

where d is the density of the solution and d_1 is the density of the solvent.

One last point: we should substitute m for n_2 on the right side of Equation 11–17, and then make the following substitution, since ϕV_2 is a linear function of $\sqrt{m}$, but not of m (Lewis and Randall, 1961):

$$\left(\frac{\partial \phi V}{\partial m}\right)_{n_1} = \left(\frac{\partial \phi V}{\partial \sqrt{m}}\right)_{n_1}\left(\frac{\partial \sqrt{m}}{\partial m}\right)_{n_1} = \frac{1}{2\sqrt{m}}\left(\frac{\partial \phi V}{\partial \sqrt{m}}\right)_{n_1} \tag{11-21}$$

When Equation 11–21 is combined with Equation 11–17, we obtain our working equation for $\bar{V}_2$:

$$\bar{V}_2 = \left(\frac{\partial V}{\partial n_2}\right)_{n_1} = \phi V + \frac{\sqrt{m}}{2}\left(\frac{\partial \phi V}{\partial \sqrt{m}}\right)_{n_1} \tag{11-22}$$

The Solvent. In order to calculate $\bar{V}_1$, let us first rewrite Equation 11–11 as

$$\bar{V}_1 = \frac{V - n_2 \bar{V}_2}{n_1} \tag{11-23}$$

Substitution of the right sides of Equations 11–16 and 11–17 for V and $\bar{V}_2$, respectively, results in

$$\bar{V}_1 = \left(\frac{\partial V}{\partial n_1}\right)_{n_2} = \frac{1}{n_1}\left[n_1 \bar{V}_1^0 - n_2^2\left(\frac{\partial \phi V}{\partial n_2}\right)_{n_1}\right] \tag{11-24}$$

Since we wish to use molality for concentration, $n_2 = m$ and $n_1 = 55.51$. The volume of 1 mole of solvent $\bar{V}_1^0$ equals M_1/d_1. When these values and the right side of Equation 11–21 are substituted into Equation 11–24, the result is our working equation for $\bar{V}_1$:

$$\bar{V}_1 = \left(\frac{\partial V}{\partial n_1}\right)_{n_2} = \frac{M_1}{d_1} - \frac{m^{3/2}}{111.02}\left(\frac{\partial \phi V}{\partial \sqrt{m}}\right)_{n_1} \tag{11-25}$$

SAFETY CONSIDERATIONS

None.

Apparatus

In addition to sodium chloride, the experiment requires two pycnometers of the Ostwald-Sprengel (Fig. 11–2) or Weld type with a volume of 25

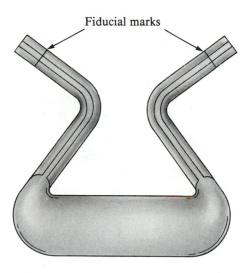

Fiducial marks

Figure 11–2 Ostwald-Sprengel pycnometer.

to 50 ml; five 100-ml volumetric flasks; a rubber bulb for filling the Ost-wald-Sprengel pycnometer; and clean, dry cotton cloths or tissue paper to wipe the pycnometers dry.

EXPERIMENTAL PROCEDURE

Prepare about 100 ml of five solutions of sodium chloride in distilled water ranging in molal concentration from about 0.3 to 3.0. Weigh the solute and solvent with an analytical balance into dry glass-stoppered flasks.

The densities of the solutions are measured with an Ostwald-Sprengel pycnometer that has been calibrated by calculating its volume from the mass of water it contains and the density of water (Table 14–2, Chapter 14) at the temperature of the measurement. First, rinse the pycnometer with a few portions of distilled water, followed with a few milliliters of acetone. Fill the pycnometer by dipping one arm into the liquid and pulling the liquid into the vessel with a rubber suction bulb attached to the other arm. With an aspirator, suck air through it to remove most of the acetone and then dry at about 110°C for 10 to 15 minutes. Let the pycnometer hang in the balance room for 30 minutes before determining the weight of the empty, dry pycnometer.

Exercise extreme care in weighing, filling, rinsing, cleaning, handling, and drying the pycnometer. After the pycnometer is clean, do not touch it with your fingers. Allow it to equilibrate with the humidity and temperature of the balance room for the same time period before each weighing.

After weighing the empty pycnometer, fill it with distilled water and hang it in a 25°C thermostat bath. Be sure that no air bubbles

```pascal
Program Apparent;

uses crt;

Const M2 = 58.443;   (* formula weight of NaCl *)
      d1 = 0.99707; (* density of water at 25 deg C *)
      n1 = 55.51;    (* moles of water in 1000g water *)
      ArrayLength = 20; (* maximum number of data pairs *)
Type Ary = Array[1..ArrayLength] of Real;
Var m,d: Ary;    (* lists of input data; molalities & densities *)
    N:  Integer;                    (* requested number of data pairs *)

(*************************************************************)

Function Phi (m,d: real): Real; (* Phi is apparent molal volume *)
Var V, Vsol: real;                        (* V is volume of solution *)
                                (* Vsol is volume of 1000g of water *)
Begin
   V:= (1000 + m*M2)/d;                    (* volume of solution    *)
   Vsol:= 1000/d1;                         (* volume of solvent     *)
   Phi:= (V - Vsol)/m                      (* apparent molal volume *)
End; (* Phi *)

(*************************************************************)

Procedure GetData;
Var I,
    X,Y: Integer;   (* Screen Coordinates *)
Begin
   CLRSCR;
   Write('How many molality-density pairs do you wish to input?  '  );
   eReadln(N);
   Write('Please enter ', N:1,' molality-density pairs:  ');
   Writeln;
   For I:= 1 to N do
      Begin
         Write('molality # ',I:1,': ');
         X:= WhereX; Y:= WhereY;
         Read(m[I]);
         GoToXY(X+10,Y);
         Write('   density  # ',I:1,': ');
         Readln(d[I]);
      End; (* For *)
   Writeln;
End; (* GetData *)

(*************************************************************)
```

Figure 11–3 *Program Apparent.*

```
Procedure CalcApparentMolalVolumes;
Var I: integer;
Begin
   Writeln;Writeln('molality  density  Apparent Molal Volume');
   Writeln('    m        d              NaCl');
   For I:= 1 to N do
        Writeln(m[I]:6:3, d[I]:10:3,Phi(m[I],d[I]):16:3);
End;

(*****************************************************************)

Begin (* M A I N   P R O G R A M *)
   GetData;
   CalcApparentMolalVolumes;
End.
```

Figure 11–3 (continued)

are pulled in. After 10 to 15 minutes for temperature equilibration, adjust the amount of liquid so that it is full exactly to the mark at 25°C. Do this by holding a piece of filter paper against the end of one arm of the pycnometer and tilting the pycnometer so that a small amount of water is drawn into the filter paper. Keep the main body of the pycnometer immersed in the thermostat bath while adjusting the water level.

It is important to eliminate errors due to variation in the weight of the invisible layer of moisture on the surface of the pycnometer and its supporting hook. Always treat the pycnometer surface in exactly the same manner before weighing. After removing it from the thermostat, wash the outside of the pycnometer with distilled water, dry it carefully with a clean, dry cloth, and let it hang in the balance room for 10 minutes before weighing.

To check your experimental technique, determine the mass of water contained in your pycnometer with two independent measurements. They should agree to within ±0.01%. If they do not, see your instructor before proceeding.

Measure the density of the sodium chloride solutions, taking the same care used in calibrating the pycnometer.

Results and Calculations

The weight of the liquid in the pycnometer should be corrected for buoyancy (Lewis and Woolf, 1971). The corrected weight is given by

$$W_{\text{liq}} = W_{\text{wts}} + W_{\text{wts}}\left(\frac{d_{\text{air}}}{d_{\text{liq}}} - \frac{d_{\text{air}}}{d_{\text{wts}}}\right) \qquad \textbf{(11–26)}$$

where W_{liq} is the corrected weight of the liquid, W_{wts} is the total mass of the balance weights, d_{air} equals 0.0012 g/ml, and d_{wts} is the density of

```pascal
Program Partial;

uses crt;

Const M1 = 18.01;          (* molecular weight of solvent, water *)
      M2 = 58.443;         (* formular weight of solute, NaCl *)
      d1 = 0.99707;        (* density of water at 25.0 deg C *)
      n1 = 55.51;          (* moles of water in 1000g *)
      ArrayLength = 20;(* maximum allowed data pairs *)
Type Ary = Array[1..ArrayLength] of Real;
Var  Slope:  Real;  (* slope of phi vs. sq.rt. of molality *)
     m,phi:  Ary;   (* arrays of input molalities and *)
                    (* apparent molal volumes *)
     N: Integer;       (* requested number of input pairs *)

(***************************************************************)

Function Vbar1(m,slope:  real): real;
Const n =1.5;
Var M32:  real;  (* molality to the 3/2 power *)
Begin
   M32:= exp(n*ln(m));  (* molality to the 3/2 power *)
   Vbar1:= ((M1/d1) - (m32*slope)/(2*n1));  (* equation (25) *)
End; (* Vbar1 *)

(***************************************************************)

Function Vbar2(m,slope,phi: real): real;
Begin
   Vbar2:= Phi + SQRT(m)*slope/2;
End; (* Vbar2 *)

(***************************************************************)

Procedure GetData;
Var I,
    X,Y: Integer;
Begin
   CLRSCR;
   Write('How many molality, apparent molal vol pairs do you wish to input? ');
   Readln(n);
   Write('What is the slope of your plot of ');
   Writeln('apparent molal volume vs. SQRT(molality)?');
   Writeln;Readln(slope);Writeln;
   Writeln('Please enter your ',N:1,' data pairs.');
   For I:= 1 to N do
      Begin
         Writeln;
         Write('molality # ',I:1,': ');
         X:= WhereX;Y:= WhereY;
         Read(m[I]);
         GoToXY(X+10,Y);
         Write('     Phi # ',I:1,': ');
         Readln(Phi[I]);
      End;  (* For *)
End; (* GetData *)
```

Figure 11–4 *Program Partial.*

```
(***************************************************************)

Procedure CalcApparentMolalVolumes;
Var I: Integer;
   Begin
   CLRSCR;
   Writeln;
   Writeln('  molality        apparent molal  Partial molal    Partial molal');
   Writeln('                        volume       volume water    volume NaCl');
   Writeln;
   For I:= 1 to N do
      Begin
         Write(m[I]:8:3,Phi[I]:17:3,Vbar1(m[I],slope):15:3);
         Writeln(Vbar2(m[I],slope,phi[I]):15:3)
      End; (* For *)
End;  (* CalcPartialMolalVolumes *)

(***************************************************************)

Begin (* M A I N   P R O G R A M *)
   GetData;
   CalcApparentMolalVolumes;
End.
```

Figure 11–4 (continued)

the balance weights. The density of brass is 8.4 g/ml and the density of stainless steel is 7.8 g/ml. The weights of single-pan balances are usually stainless steel, while the weights used with chain balances are normally brass.

With Equation 11–20 calculate the apparent molal volumes of the solute for each solution, and plot ϕV against $\sqrt{m}$. Determine the slope of the line $\partial \phi V / \partial \sqrt{m}$ by least squares, and note the standard deviation. For each solution, calculate $\bar{V}_1$ and $\bar{V}_2$ with Equations 11–25 and 11–22, respectively.

To ease the tedium of the calculations, two computer programs are given in Figures 11–3 and 11–4. They could easily be linked together with a least-squares program into a single program that would calculate all the required quantities by simply furnishing the molalities and densities. It is preferable, however, to plot the apparent molal volumes and critically examine the data before proceeding. If you use a computer to carry out the calculations, furnish a sample calculation for one solution verifying the output for the apparent molal volume, and the partial molal volume for the solute and solvent for the given molality.

REFERENCES

1. Bauer, N., and S. Z. Levin, in A. Weissberger and B. W. Rossiter, eds., *Techniques of Chemistry,* vol. I, *Physical Methods of Chemistry,* pt. IV, chap. II, Wiley-Interscience, New York, 1972.

2. Klotz, I. M., *Chemical Thermodynamics,* chap. 13, Prentice-Hall, Englewood Cliffs, New Jersey, 1950.

3. Lewis, G. N., and M. Randall, revised by K. Pitzer and L. Brewer, *Thermodynamics,* 2nd ed., McGraw-Hill, New York, 1961.

4. Lewis, J. E., and L. A. Woolf, "Air Buoyancy Corrections for Single-Pan Balances," *J. Chem. Educ.* 48:639, 1971.

5. Timmermans, J., *The Physico-Chemical Constants of Binary Systems in Concentrated Solutions,* Interscience, New York, 1960.

Experiment 12

Activities of a Solvent from Freezing Point Measurements

In this experiment we shall investigate the activities of water containing various concentrations of HCl. When water is pure, its activity is unity. In ideal solutions, the solvent activity equals its mole fraction:

$$a_1 = X_1 \tag{12-1}$$

and in real solutions, the activity of the solvent is given by

$$a_1 = \gamma_1 X_1 \tag{12-2}$$

where γ_1 is the activity coefficient of the solvent. When a solution of any solute in water freezes, the following equilibrium is established:

$$\text{water(solid)} = \text{water(soln)} \tag{12-3}$$

Since the system is at equilibrium, $\Delta G = 0$, so we can write

$$G_1(\text{solid}) = G_1(\text{soln}) \tag{12-4}$$

The relationship between the free energy per mole of the solvent $G_1(\text{soln})$ at any mole fraction of the solvent X_1 is given by

$$G_1(\text{soln}) = G_1^0(\text{liq}) + RT \ln a_1 \tag{12-5}$$

where $G_1^0(\text{liq})$ is the free energy per mole of pure solvent water and $a_1 = \gamma_1 X_1$. Substitution of Equation 12–4 into Equation 12–5 gives

$$G_1(\text{solid}) = G_1^0(\text{pure liq}) + RT \ln a_1 \tag{12-6}$$

Since $\partial(G/T)/\partial T = -H/T^2$ at constant pressure, differentiation of Equation 12–6 gives

$$H_1(\text{solid}) = H_1(\text{pure liq}) - RT^2 \left(\frac{\partial \ln a_1}{\partial T} \right)_P \tag{12-7}$$

where $H_1(\text{pure liq}) - H_1(\text{solid})$ is just the molar heat of fusion, ΔH, of the solvent water. Equation 12–7 may be rewritten:

$$\Delta H = RT^2 \left(\frac{\partial \ln a_1}{\partial T} \right)_P \qquad (12\text{–}8)$$

For precise work we must take into account the fact that the heat of fusion is a function of temperature and depends on the difference in heat capacities of the $H_2O(s)$ and $H_2O(\ell)$:

$$\Delta H = \Delta H_m - \Delta C_p (T_m - T) \qquad (12\text{–}9)$$

where ΔH_m is the heat of fusion at the melting point of pure solvent T_m. Substitution of Equation 12–9 into Equation 12–8 gives

$$\left(\frac{\partial \ln a_1}{\partial T} \right)_P = \frac{\Delta H_m}{RT^2} - \frac{\Delta C_p (T_m - T)}{RT^2} \qquad (12\text{–}10)$$

When Equation 12–10 is integrated, the result is

$$\ln a_1 = -\frac{\Delta H_m}{RT} + \frac{\Delta C_p T_m}{RT} + \frac{\Delta C_p \ln T}{R} + \text{constant} \qquad (12\text{–}11)$$

The integration constant can be evaluated, since $a_1 = 1$ when $T = T_m$. Subsequent rearrangement gives

$$\ln a_1 = -\left(\frac{\Delta H_m - \Delta C_p T_m}{RT_m} \right) \left(\frac{T_m}{T} - 1 \right) - \frac{\Delta C_p \ln (T_m/T)}{R} \qquad (12\text{–}12)$$

For water, $\Delta H_m = 6008 \pm 4$ J/mol, $T_m = 273.16$ K, and $\Delta C_p = 38.1 \pm 0.2$ J K^{-1} mol^{-1} (Lewis and Randall, 1961, p. 407). Substitution of these values for ΔH_m and ΔC_p gives

$$\ln a_1 = 1.9372 \left(\frac{273.16}{T} - 1 \right) - 4.583 \ln \left(\frac{273.16}{T} \right) \qquad (12\text{–}13)$$

With Equation 12–12, one can calculate the activity a_1 of the solvent water for a solution that freezes at temperature T. Then from the mole fraction of water in the solution, one can calculate the activity coefficient of the solvent:

$$\gamma_1 = \frac{a_1}{X_1} \qquad (12\text{–}14)$$

Apparatus

This experiment requires concentrated hydrochloric acid and cracked ice. A 500-ml Dewar flask must have a cork stopper drilled for a stirrer, thermometer, and pipet. Four 6-inch test tubes, six 500-ml glass-stoppered bottles, 10- and 25-ml pipets, a 50-ml buret, 0.1 M standard NaOH, and a phenolphthalein indicator are needed. For the concentrations suggested in this experiment, the thermometers should cover the range from about −8 to 0°C. The ASTM 12C thermometer with a range from −20 to +102°C in 0.2° divisions or the ASTM 63C with a range from −8 to +32°C in 0.1° divisions is suitable. A Beckman thermometer with its upper limit set to approximately 0°C is also satisfactory.

SAFETY CONSIDERATIONS

The Dewar flask should be taped and safety glasses worn at all times. Concentrated hydrochloric acid is dangerous.

EXPERIMENTAL PROCEDURE

From concentrated HCl (approximately 12 M), prepare 500 ml of 1, 1.5, 2, 2.5, and 3 M HCl. Store in a glass-stoppered bottle. Fill a sixth bottle with distilled water and place all six bottles in a tray of cracked ice. Pack the ice around the bottles and cool to 0°C.

Place about 250 ml of cracked ice in a 500-ml Dewar flask and just cover with distilled water that has already been cooled to 0°C. Stir the ice-water mixture vigorously until a constant temperature reading is observed. Tap the thermometer gently with a pencil to prevent the mercury from sticking.

The Beckman thermometer has a range of about 5°C, and its zero can be set anywhere between about −10 and +120°C. For this experiment, the reading of the ice water (273.16 K) should be near the top end of the Beckman scale, since the freezing points of the solutions will be slightly lower. See your instructor if this is not the case.

Drain off the water, replenish the ice if necessary, and add enough 3 M HCl to just cover the ice. Stir vigorously and record the new temperature. Withdraw about 25 ml of the solution and transfer to a suitable test tube. Allow the solution to warm to room temperature and measure the density of the solution with a Westphal balance. Dilute a 10-ml aliquot of the acid solution to 100 ml and titrate a 10-ml aliquot of the diluted solution with 0.1 M standard NaOH. Calculate the mole fraction of water in the original acid sample.

Repeat with the other solutions of HCl. Adjust the size of the aliquots so that sensible amounts of standard NaOH are required.

Results and Calculations

From the measured freezing points, calculate the activity of water in each solution with Equation 12–13. Calculate the activity coefficients with Equation 12–14. Present the experimental and calculated quantities in tabular form.

Calculate the number of moles of water in one liter of your most concentrated solution. Calculate the number of moles in one liter of pure water. Compare and comment.

Calculate and discuss the errors in the results.

REFERENCES

1. Klotz, I., *Chemical Thermodynamics,* Prentice-Hall, Englewood Cliffs, New Jersey, 1950, p. 318ff.
2. Lewis, G. N., and M. Randall, revised by K. S. Pitzer and L. Brewer, *Thermodynamics,* 2nd ed., chap. 26, McGraw-Hill, New York, 1961.

Surface Effects

Experiment 13

Surface Tension

As shown in Figure 13–1, molecules situated in the bulk liquid are in a quite different environment from molecules at the surface of the liquid. In the bulk liquid, attractive forces act symmetrically between molecules, but at the liquid–air interface, molecules are attracted more strongly to the molecules below them in the bulk liquid than they are toward molecules in the vapor phase above them. Consequently, the surface tends to contract to a minimum, giving rise to a force f in the surface plane (Fig. 13–2).

The force f required to generate a new surface by pulling the movable wire in the x direction is proportional to the length ℓ of the surface:

$$f = 2\gamma\ell \qquad \text{(13–1)}$$

The proportionality constant γ is the surface tension, and the factor 2 is required in this example because two surfaces are simultaneously generated above and below the fixed wire. The surface tension may be regarded as the force per unit length:

$$\gamma = \frac{f}{\ell} = \frac{\text{force}}{\text{length}} = \text{N/m} = \text{kg s}^{-2} \qquad \text{(13–2)}$$

Alternatively, the surface tension can be considered to be the work required to generate a unit area of surface:

$$\gamma = \frac{f\ell}{\ell^2} = \frac{w}{A} = \text{J/m}^2 = \text{kg s}^{-2} \qquad \text{(13–3)}$$

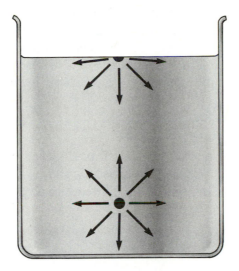

Figure 13–1 Intermolecular forces in the surface and in the bulk liquid.

The usual unit for surface tension is the newton per meter (N/m), although the dyne per centimeter prevails in the older literature. Since a dyne/cm is a g s^{-2}, while an N/m is a kg s^{-2}, the cgs and SI units differ by a factor of 1000. The surface tension of water is 71.97×10^{-3} N/m or 71.97 dyne/cm.

From Equation 13–3, $dw = \gamma \, dA$. When this is introduced into the usual expression for the combined First and Second Laws for the Gibbs free energy, we obtain

$$dG = -S \, dT + V \, dP + \gamma \, dA$$

which reduces at constant temperature and pressure to

$$dG = \gamma \, dA \qquad \qquad \textbf{(13–4)}$$

or

$$\gamma = \left(\frac{\partial G}{\partial A} \right)_{P,T}$$

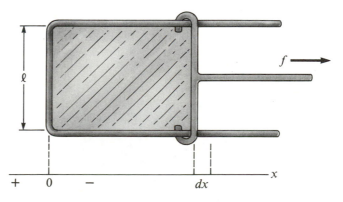

Figure 13–2 The force required to generate a new surface is proportional to the length ℓ of the surface.

Consequently, the surface tension γ is called the Gibbs free energy per unit area. Since γ is a constant at a given T and P, and dG tends to a minimum as a system approaches equilibrium, it follows that dA tends to decrease. This explains why soap bubbles and water droplets are spherical, since the sphere has the smallest surface for a given volume.

The Laplace Equation

In a soap bubble at equilibrium, the tendency for the surface area to decrease is balanced by a rise in internal pressure, $\Delta P = P_{in} - P_{out}$. For the bubble in Figure 13–3, the surface area A and the volume V are given by

$$A = 4\pi r^2 \tag{13-5}$$

$$V = \frac{4\pi r^3}{3} \tag{13-6}$$

If the bubble expands slightly by dr, then the change in area is $dA = 8\pi r dr$, so the work done in generating this new surface area w_s is

$$dw_s = 8\pi r\gamma dr \tag{13-7}$$

To maintain equilibrium, an equivalent amount of PV work of expansion w_{PV} is done by the system. The change in volume dV is $4\pi r^2 dr$, so the PV work of expansion is

$$w_{PV} = 4\pi r^2 \, dr \, \Delta P \tag{13-8}$$

Since $w_{PV} = w_s$,

$$4\pi r^2 \, dr \, \Delta P = 8\pi r\gamma dr$$

or

$$\Delta P = \frac{2\gamma}{r} \tag{13-9}$$

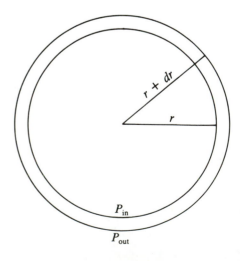

Figure 13–3 The expansion of a bubble of radius r.

Equation 13–9, the Laplace equation, is one of the fundamental equations of surface chemistry.

The surface tension depends on both temperature and concentration. The excess surface energy E^S with units of J/m^2 (Adamson, 1973, p. 306) is

$$E^S = \gamma - T\frac{d\gamma}{dT} \tag{13–10}$$

For surface-active solutes (surfactants), the concentration of the solute is greater on the surface than in the bulk solution. The excess concentration Γ_2 with units of mol/m^2 (Adamson, 1973, p. 387) is

$$\Gamma_2 = -\frac{1}{RT}\frac{d\gamma}{d\ln C} \tag{13–11}$$

Equation 13–11 is known as the Gibbs isotherm after J. Willard Gibbs, who first derived it. In SI units, Γ has units of mol m^{-2} since γ has units of J m^{-2}, R has units of J K^{-1} mol^{-1}, T is in kelvins, and C is in mol m^{-3}. Note that $d\ln C = dC/C$, which is dimensionless. When $d\gamma/d\ln C$ is negative, Γ_2 is positive and the solute is surface active.

In aqueous solutions, surfactants are usually polar organic molecules having an aliphatic end and a polar hydroxyl or carboxyl group.

Capillary Rise Method of Measuring Surface Tension

The rise of a liquid in a capillary is readily described in terms of the Laplace equation. Consider Figure 13–4, which shows a hemispherical meniscus at the liquid–air interface. Since the pressure P_2 under the meniscus is less than the pressure over the meniscus, the water rises until

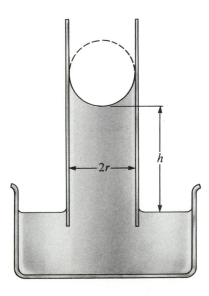

Figure 13–4 Capillary rise due to pressure difference.

the hydrostatic pressure of the water column equals the pressure difference:

$$\Delta P = dgh \tag{13–12}$$

where d is the density of water (kg/m^3), $g = 9.8\ ms^{-2}$, and h is the height of the column. When Equation 13–12 is combined with the Laplace equation, the result is

$$dgh = \frac{2\gamma}{r} \tag{13–13}$$

$$\gamma = \frac{dghr}{2} \tag{13–14}$$

Assuming that the meniscus is hemispherical is the same as assuming that the contact angle θ (Fig. 13–5) is zero. Instead of equalizing interfacial pressures, we may equalize the interfacial forces. The hydrostatic force down f_h is

$$f_h = \pi r^2\, dgh \tag{13–15}$$

while the force up f_s arising from the surface tension acting along the circumference of the meniscus is

$$f_s = 2\pi r\gamma \cos\theta \tag{13–16}$$

At equilibrium the two forces are equal, so that

$$\gamma = \frac{dghr}{2\cos\theta} \tag{13–17}$$

which is identical to Equation 13–14 when $\theta = 0$.

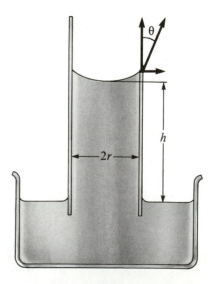

Figure 13–5 Contact angle in the capillary rise method.

Bubble Pressure Method

The bubble pressure method is a direct application and demonstration of the Laplace equation. Consider blowing an air bubble from a tube immersed to a depth h in a liquid of surface tension γ as illustrated in Figure 13–6. Initially, the force required to begin blowing the bubble (Fig. 13–6, A) is just the pressure required to overcome the hydrostatic pressure head dgh. As the bubble expands, an additional Laplacian pressure $2\gamma/r$ is needed, which is quite small as the bubble radius initially is large. The bubble radius gradually decreases as the bubble grows until it equals the radius of the tube (Fig. 13–6, C), after which the radius increases further. The Laplacian pressure increases to a maximum when the bubble radius equals the tube radius, and measurement of the maximum bubble pressure permits calculation of the surface tension from Equation 13–18:

$$P_{\max} = dgh + \frac{2\gamma}{r} \qquad (13\text{–}18)$$

Since it is not necessary to observe the bubble, and the pressure determination can be remote, this method offers some advantages. It is necessary only to predetermine the immersion depth of the tube. Thus, the method is adaptable to the high-temperature measurement of the surface tension of molten salts or even metals.

Apparatus

The apparatus includes a thermostat bath from 20 to 50°C; a calibrated capillary or a plastic scale fastened to a capillary about 150 mm long with

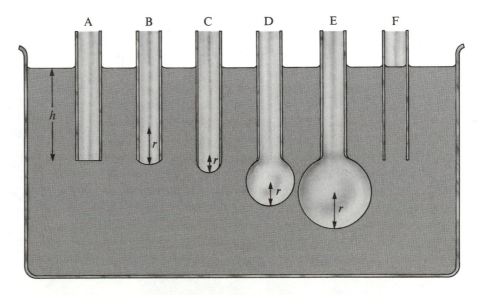

Figure 13–6 The formation of a bubble at the end of a capillary tube.

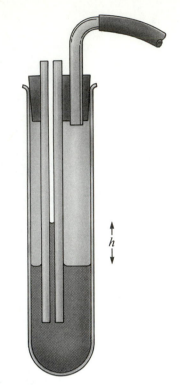

Figure 13–7 Capillary rise apparatus.

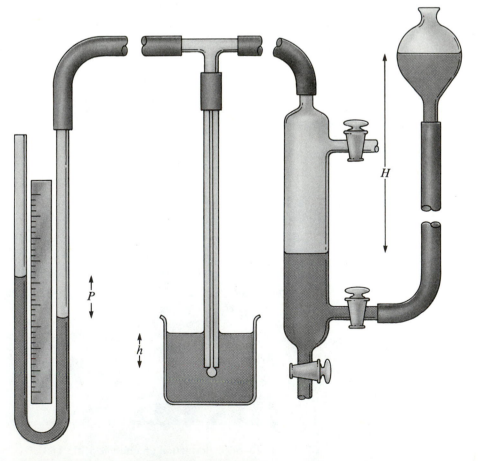

Figure 13–8 Maximum bubble pressure apparatus.

a bore of about 0.2 mm; a test tube; a glass angle tube; rubber tubing; and a stopper (Fig. 13–7). For the bubble pressure method, a similar capillary, a water-filled manometer, a reservoir filled with three small stopcocks, and a leveling bulb connected with rubber tubing are needed (Fig. 13–8). For measurements, many organic liquids are suitable: acetone, dioxane, dimethylformamide, methanol, and n-butanol.

SAFETY CONSIDERATIONS

Organic liquids vary in toxicity and flammability. Use caution. Wear safety goggles when using cleaning solution.

EXPERIMENTAL PROCEDURE

Clean the capillary tube in warm cleaning solution (dichromate and concentrated sulfuric acid) and rinse it well with water and the liquid to be measured. Do the cleaning shortly before making the measurements or store the clean capillary under distilled water.

Capillary Rise Method. Determine the radius of the capillary by measuring the rise of pure water at 25°C; at this temperature the surface tension equals 71.97×10^{-3} N/m. The height to which the water rises should be measured several times, both rising and falling to the equilibrium position. A small rubber bulb can be used to pull the liquid slightly above the equilibrium position. If the tube is absolutely clean, the same position should be achieved, indicating that the contact angle $\theta = 0$.

Measure the surface tension of one pure liquid from 0 to about 50°C at four or five temperatures. Measure one of the liquids at 25°C with both the capillary rise and bubble pressure methods. Prepare aqueous solutions of n-butanol ranging from about 0.8 to 0.1 M. Measure the surface tension of each solution at 25°C.

Bubble Pressure Method. Set up the apparatus shown in Figure 13–8 (Försterling and Kuhn, 1985). The rate at which bubbles are formed is determined by the difference in the levels of water in the reservoir and the leveling bulb. The sample should be immersed in a thermostat for all measurements. Read the maximum bubble pressure from the water-filled manometer. Read the depth of immersion from the calibrated capillary or from an attached scale.

Results and Calculations

Calculate the surface tension with Equation 13–14 for the capillary rise method and with Equation 13–18 for the bubble pressure method. To obtain γ in J m^{-2} with Equations 13–14 and 13–18, all measured quantities must be expressed in SI units: ΔP_{max} in Pa, and so forth. Plot γ

versus T, determine $d\gamma/dT$, and calculate E^S, the excess surface energy. Plot γ versus $\ln C$, determine $d\gamma/d\ln C$ and calculate Γ_2, the excess surface concentration. From Γ_2, calculate the cross-sectional area of the surfactant molecule.

Calculate the error in measurements with the capillary rise and bubble pressure methods. Compare those results with literature values.

REFERENCES

1. Adamson, A. W., *Physical Chemistry of Surfaces,* 2nd ed., Wiley-Interscience, New York, 1967.

2. Adamson, A. W., *A Textbook of Physical Chemistry,* Academic Press, New York, 1973.

3. Alberty, R. A., *Physical Chemistry,* 7th ed., Wiley, New York, 1987.

4. Alexander, A. E., and J. B. Hayter, "Determination of Surface and Interfacial Tension," in A. Weissberger and B. W. Rossiter, eds., *Techniques of Chemistry,* vol. I, *Physical Methods of Chemistry,* pt. V, Wiley-Interscience, New York, 1971.

5. Butler, J. A. V., and A. Wighman, "Adsorption at the Surface of Solutions," *J. Chem. Soc.* 1932:2089.

6. Försterling, H.-D., and H. Kuhn, *Praxis der Physikalischen Chemie,* VCH, Deerfield Beach, Florida, 1985.

7. Harkins, W. D., and R. W. Wampler, "Activity Coefficients and the Adsorption of Organic Solvents," *J. Am. Chem. Soc.* 53:850, 1931.

Experiment 14

Viscosity of Solutions of Macromolecules

When a liquid flows, whether through a tube or as the result of pouring from a vessel, layers of liquid slide over each other. The force f required is directly proportional to the area A and velocity v of the layers, and inversely proportional to the distance d between them, as shown in Figure 14–1. The proportionality constant η is the coefficient of viscosity, or more commonly, the viscosity:

$$f = \eta \frac{Av}{d} \tag{14-1}$$

Units of Viscosity

The traditional unit of viscosity is the **poise,** originally a cgs unit that has the symbol P. The dimensions based on Equation 14–1 are the g cm^{-1} sec^{-1}:

Figure 14–1 Origin of viscosity.

$$\frac{fd}{Av} = \frac{\text{g cm sec}^{-2}\ \text{cm}}{\text{cm}^2\ \text{cm sec}^{-1}} = \text{g cm}^{-1}\ \text{sec}^{-1} = 1\ \text{poise} = 1\ \text{P} \quad (14\text{–}2)$$

In SI units the poise has no special symbol and has units of kg m^{-1} s^{-1}. Thus, 1 P = 0.1 kg m^{-1} s^{-1}. Viscosities of gases and liquids are often reported in μP and cP, respectively. For example, at 25°C the viscosity of N$_2$ is 178 μP and the viscosity of water is 0.8937 cP.

Measurement

The rate of flow R (cm^3/sec) of a liquid through a cylindrical tube of radius r and length ℓ under a pressure head P is given by the Pousille equation:

$$R = \frac{V}{t} = \frac{\pi P r^4}{8 \eta \ell} \quad (14\text{–}3)$$

Measurement of $P, r, t, V,$ and ℓ permits the calculation of the viscosity:

$$\eta = \frac{\pi P r^4 t}{8 V \ell} \quad (14\text{–}4)$$

It is easier to measure the viscosity of a liquid by comparing it with another liquid of known viscosity (presumably already measured with Eq. 14–4). Since $P = \rho g h$,

$$\frac{\eta_1}{\eta_2} = \frac{\rho_1 t_1}{\rho_2 t_2} \quad (14\text{–}5)$$

The Ostwald viscometer (Fig. 14–2) is a simple device for comparing the flow times of two liquids of known density. If the viscosity of one liquid is known, the other can be calculated. After the reservoir is filled with a liquid, it is pulled by suction above the upper mark. The time required for the liquid to fall from mark 1 to mark 2 is recorded. Then the time required for the same volume of a liquid of unknown viscosity to flow under identical conditions is recorded, and the viscosity is calculated with Equation 14–5.

Viscosity of Solutions of High Polymers

While the viscosity of water and molasses arises from hydrogen bonding between flowing liquid layers, the viscosity of solutions of macromolecules

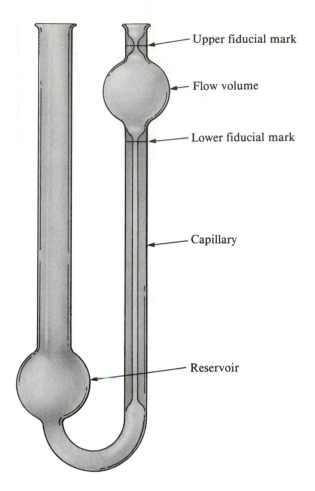

— Upper fiducial mark

— Flow volume

— Lower fiducial mark

— Capillary

— Reservoir

Figure 14–2 Ostwald viscometer.

arises from the entanglement of large molecules flowing along with the solvent. One would expect this effect to depend on the size, shape, and molecular weight of the macromolecules.

Indeed, the problem attracted the attention of Einstein, who showed that for large spherical molecules,

$$\eta_{sp} = 2.5(V/V_0) = 2.5\phi \qquad (14\text{–}6)$$

where ϕ is the fractional volume occupied by the macromolecule and η_{sp}, the specific viscosity, is defined as

$$\eta_{sp} = \frac{\eta - \eta_0}{\eta_0} \qquad (14\text{–}7)$$

where η and η_0 are the viscosities of the solution and solvent, respectively. Note that the specific viscosity is dimensionless.

Intrinsic Viscosity. The properties of individual macromolecules separated from their interaction with each other are found by extrapolating to infinite dilution:

$$[\eta] = \lim_{c \to 0} \frac{\eta_{sp}}{c} = \lim_{c \to 0} \frac{1}{c} \ln \frac{\eta}{\eta_0} \qquad (14\text{-}8)$$

The quantity $[\eta]$, the intrinsic viscosity, is important because it is most directly related to properties such as molecular weight and shape. Its dimensions are that of reciprocal concentration, usually expressed as $1/(g/100 \text{ ml})$. In dilute solution, both η_{sp}/c and $(1/c) \ln (\eta/\eta_0)$ are linearly dependent upon concentration c and can be extrapolated to infinite dilution to determine $[\eta]$.

Molecular Weight. It is found empirically that the intrinsic viscosity of a solution depends on the molecular weight of the solute:

$$[\eta] = KM^a \qquad (14\text{-}9)$$

where a falls between $1/2$ and 2 and is a parameter that is related to the shape of the molecule. When the solute molecule is nearly spherical, a is about 0.5, but when the solute molecule is long, extended, or rod shaped, its value may reach 1.7 to 1.8.

Both K and a must be determined from viscosity measurements with solutions of macromolecules of known molecular weight, determined by other methods, such as osmotic pressure measurements. Some typical values of K and a are listed in Table 14–1.

Because chain termination in polymerization reactions occurs randomly, polymers have a distribution of molecular masses about an average value. Experimental measurement of viscosity or colligative properties like osmotic pressure leads to M_n, the **number average molecular weight.** If the polymer sample is considered to be made up of a large number of fractions consisting of n_1 moles of mass M_1, n_2 moles of mass M_2, and so on, the number average molecular weight is defined as

$$M_n = \frac{n_1 M_1 + n_2 M_2 + \cdots}{n_1 + n_2 + \cdots} = \frac{\sum_i n_i M_i}{\sum_i n_i}$$

Colligative properties such as osmotic pressure depend on the number of particles in solution but not on the individual mass. Light scattering of polymer solutions, however, depends on the mass and number of particles and leads to the **mass average molecular weight** M_w (Matthews, 1984) where

TABLE 14–1

Parameters for Molecular Weight Determination (units of $[\eta]$ and K are $1/(gm/100 \text{ ml})$) $[\eta] = KM^a$

Macromolecule	Solvent	$K \times 10^4$	a
Cellulose acetate	Acetone	1.49	0.82
Methyl methacrylate	Benzene	0.94	0.76
Polystyrene	Toluene	3.7	0.62
Polyvinyl alcohol	Water	2.0	0.76

$$M_w = \frac{\sum_i n_i M_i^2}{\sum_i n_i M_i}$$

If all the molecules have the same molecular mass, the polymer is said to be monodisperse and $M_w/M_n = 1.0$. Synthetic polymers with M_w/M_n as low as 1.04 have been prepared. Vinyl polymers have M_w/M_n ratios from 2 to about 10 and low-density polyethylene has M_w/M_n as high as 20 (Rudin, 1969).

Apparatus

This experiment requires an Ostwald viscometer with a water flow time of about 100 sec; a 25°C thermostat; 10- and 25-ml pipets; a rubber bulb and tube; two or more 50-ml volumetric flasks; cleaning solution; polystyrene (the peanut-sized pieces used as protective packing material are satisfactory); and 300 ml toluene.

SAFETY CONSIDERATIONS

None.

EXPERIMENTAL PROCEDURE

Dissolve about 1 g polystyrene in toluene in a 50-ml volumetric flask. While the polymer is dissolving, clean the viscometer with cleaning solution. If it is not available, prepare the cleaning solution by dissolving 12 g sodium dichromate ($Na_2Cr_2O_7$) in 12–15 ml hot water. Cool. Cautiously, slowly, and with stirring, add 225 ml concentrated sulfuric acid. Store in a 250-ml glass-stoppered bottle. Use carefully on glassware that has previously been cleaned with detergent and rinsed with water.

Calibration. After cleaning the viscometer with cleaning solution, rinse it with distilled water; then add a 10-ml aliquot of water and let it come to thermal equilibrium with the thermostat. Both fiducial marks should be below the water level of the thermostat. With a rubber bulb, pull the water in the viscometer above the upper fiducial mark and measure the time it requires to flow from the upper to the lower mark. Repeat to determine the uncertainty in measuring time. The flow time with water, along with its density at 25°C (0.99777 g/cm^3) and viscosity (0.8937 cP) are used with Equation 14–5 to determine the viscosities of toluene η_0 and the polystyrene solutions η.

Drain the water from the viscometer; rinse it first with a few milliliters of acetone, then with toluene. Add a 10-ml aliquot of toluene, and after the viscometer and toluene are at thermal equilibrium, measure the flow time as before.

Measurements. Take a 25-ml aliquot of the polystyrene solution and dilute to 50 ml with toluene in a volumetric flask. Again measure the flow time. Dilute repeatedly until the final solution is 1/8 the original solution concentration; measure the flow time of each solution.

Results and Calculations

The density of toluene at 25°C is 0.866 g/cm^3. The density of the solutions may be assumed to be that of toluene. Tabulate c, η_0, η_{sp}, η_{sp}/c, and $(1/c) \ln (\eta/\eta_0)$. On a single sheet of graph paper, plot both η_{sp}/c and $(1/c) \ln (\eta/\eta_0)$ versus c. Extrapolate to $c = 0$ to obtain $[\eta]$. If your extrapolation is linear, you may run a least-squares analysis of the data. Calculate the molecular weight of your polystyrene sample with Equation 14–9 and the parameters listed in Table 14–1. The parameter K in Table 14–1 requires the concentration c to be expressed in g/100 ml.

REFERENCES

1. Allcock, H. R., and F. W. Lampe, *Contemporary Polymer Chemistry,* Prentice-Hall, Englewood Cliffs, New Jersey, 1981.

2. Bradbury, J. H., "Polymerization Kinetics and Viscometric Characterization of Polystyrene," *J. Chem. Educ.* 40:465–468, 1963.

3. Carraher, Jr., C. E., "Generation of Poly(vinyl alcohol) and Arrangement of Structural Units," *J. Chem. Educ.* 55:473–475, 1978.

4. Flory, P. J., *Principles of Polymer Chemistry,* Cornell University Press, Ithaca, New York, 1953.

5. Flory, P. J., and F. S. Leutner, "Occurrence of Head-to-Head Arrangements of Structural Units in Polyvinyl Alcohol," *J. Poly. Sci.* 3:880, 1948.

6. Goldberg, A. I., W. P. Hohenstein, and H. Mark, "Intrinsic Viscosity–Molecular Weight Relationships for Polystyrene," *J. Poly. Sci.* 2:502, 1947.

7. Johnson, J. F., J. R. Martin, and R. S. Porter, in A. Weissberger and B. W. Rossiter, eds., *Techniques of Chemistry,* vol. I, *Physical Methods of Chemistry,* pt. VI, chap. 2, Wiley-Interscience, New York, 1977.

8. Mathias, L. J., "Evaluation of Viscosity–Molecular Weight Relationship," *J. Chem. Educ.* 60:422, 1983.

9. Matthews, G. P., "Light Scattering by Polymers," *J. Chem. Educ.* 61:552–554, 1984.

10. Rabek, J. F., *Experimental Methods in Polymer Chemistry,* Wiley-Interscience, Chichester, England, 1980.

11. Rudin, A., "Molecular Weight Distributions of Polymers," *J. Chem. Educ.* 46:595, 1969.

12. Swindells, J. F., R. Ullman, and H. Mark, in A. Weissberger, ed., *Technique of Organic Chemistry,* vol. 1, pt. 1, chap. XII, Interscience, New York, 1959.

13. Tanford, C., *Physical Chemistry of Macromolecules,* Wiley, New York, 1961.

Experiment 15

Adsorption of Acetic Acid by a Solid

Since solids exist because of intermolecular forces between the repeating units that make up the lattice, those intermolecular forces are unsaturated or unsatisfied at the surface of the solid. In the interior of a solid, each molecule (if it is a molecular solid) is surrounded on all sides by identical molecules. At the surface of such a solid, however, each molecule is only partially surrounded by identical molecules; where it is not surrounded by identical molecules, any available molecule or ion is adsorbed to its surface. The solid phase is called the **adsorbent.** The molecules that are adsorbed on the adsorbent are collectively called the adsorbed phase or **adsorbate.** The adsorbate is either a gas (molecules) or a solute (molecules or ions) in solution. In this experiment, we will investigate the adsorption of acetic acid in aqueous solution on activated charcoal.

Adsorption by a solid is not a very important process unless the solid has a very large surface compared to its mass. Consequently, charcoal made from bone, blood, or coconut shells is especially effective because it has a highly porous structure. Charcoal is activated by being heated to quite high temperatures in a vacuum or in a stream of dry air. This treatment probably desorbs the hydrocarbons that are adsorbed when the charcoal is first produced. Charcoal, a covalently bonded solid, is more effective at adsorbing molecules than ions. Silver chloride, on the other hand, forms as a precipitate of nearly colloidal dimensions if precipitated rapidly; in colloidal form it has a very high surface-to-mass ratio and readily adsorbs ions from solutions, often to the dismay of analytical chemists.

The amount of adsorption, given the symbol Y, has units of moles adsorbate per mass adsorbent. If the adsorbate is a gas, Y may have units of volume adsorbate per mass adsorbent. The amount of adsorption Y increases with the concentration c of the adsorbate. The increase is very rapid at first, when the surface of the adsorbent is relatively free. As the surface fills with the adsorbate, the rate of adsorption dY/dc decreases. Eventually, the surface of the adsorbent becomes full, and further increases in the concentration cause no further increase in the amount adsorbed, as shown in Figure 15–1. The amount adsorbed when the surface is just covered with a monomolecular layer of the adsorbate is called Y_{max}. At a given concentration the amount adsorbed decreases with increasing temperature.

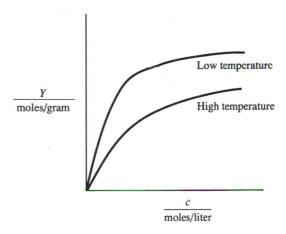

Figure 15–1 Typical adsorption isotherms.

The Freundlich Equation

One of the earliest attempts to describe mathematically the adsorption isotherms shown in Figure 15–1 was the Freundlich equation (Freundlich, 1909):

$$Y = kc^{1/n} \tag{15-1}$$

In this purely empirical equation, the units of Y are moles adsorbate per gram adsorbent, c is the concentration (mol/L), and k and n are experimentally determined constants. Since Equation 15–1 is valid only for a given adsorbed phase and adsorbent at a constant temperature, it is sometimes called the Freundlich isotherm. To test the validity of the Freundlich isotherm, take the logarithms of both sides of Equation 15–1:

$$\log_{10} Y = \log_{10} k + \frac{1}{n} \log_{10} c \tag{15-2}$$

If $\log_{10} Y$ is plotted against $\log_{10} c$, a straight line results with the slope equal to $1/n$ and the intercept equal to $\log_{10} k$.

The Langmuir Adsorption Isotherm

The Langmuir adsorption isotherm is based on theoretical considerations. The postulates of the theory are as follows:

1. The adsorbed phase forms a layer of molecules one molecule deep.
2. The system is in a state of equilibrium such that the rate of adsorption equals the rate of desorption.
3. The rate of adsorption is proportional to the concentration and the fraction of the surface that is vacant.

4. The rate of desorption is proportional to the fraction of the surface that is already covered.

If θ is the fraction of the surface that is covered, then $1 - \theta$ is the fraction of the surface that is vacant. Consequently,

$$k_1(1 - \theta)c = k_2 \tag{15-3}$$

$$\theta = \frac{k_1 c}{k_2 + k_1 c} \tag{15-4}$$

If k is defined to be k_1/k_2, Equation 15–4 becomes

$$\theta = \frac{kc}{1 + kc} \tag{15-5}$$

Since the fraction of the surface covered is also equal to Y/Y_{max}, we may write

$$\frac{Y}{Y_{max}} = \frac{kc}{1 + kc} \tag{15-6}$$

With rearrangement, Equation 15–6 becomes

$$\frac{c}{Y} = \frac{1}{kY_{max}} + \frac{c}{Y_{max}} \tag{15-7}$$

This is the equation of the Langmuir adsorption isotherm. If a system follows the Langmuir equation, then a plot of c/Y versus c is a straight line with a slope equal to $1/Y_{max}$ and an intercept equal to $1/kY_{max}$.

Determination of the Specific Area of the Adsorbent

If the adsorption of the adsorbate leads to a maximum of a single molecular layer when the adsorption is complete, it is possible to calculate the area of the adsorbent. When a monomolecular layer is adsorbed, it may be assumed that the area of the adsorbent equals the total area of the adsorbed molecules. The determination of the area of an adsorbed molecule is simple in the case of small adsorbed gas molecules such as helium, hydrogen, and nitrogen. When the adsorbed molecule is a carboxylic acid, however, the structure is clearly more complex. Nevertheless, studies of the adsorption of straight-chain aliphatic monocarboxylic acids indicate that the number of moles of acid adsorbed per gram adsorbent is independent of chain length (Hansen and Craig, 1954). This suggests that the acid molecules are adsorbed vertically, with the aliphatic chain up and the carboxyl group down and attached to the adsorbent. The cross-sectional area A_a of a straight-chain acid may be taken to be about

21×10^{-20} m^2. The specific area of the adsorbent S (m^2/g) is then given by

$$S = A_a N_A Y_{max} \qquad (15\text{--}8)$$

where N_A is Avogadro's number.

Apparatus

This experiment requires about 20 g activated charcoal (from blood); 12 glass-stoppered 125-ml Erlenmeyer flasks; 5-, 10-, 25-, and 50-ml pipets; medium-fine filter paper; 600 ml 0.4 M acetic acid; 500 ml 0.1 M sodium hydroxide (standardized); a phenolphthalein indicator; and a 50-ml buret.

SAFETY CONSIDERATIONS

If it is necessary to prepare the 0.4 M acetic acid from glacial acetic acid, take care not to breathe the fumes, and carry out the dilution in a ventilated hood.

EXPERIMENTAL PROCEDURE

Weigh about 1.5 g charcoal into each of the dry glass-stoppered Erlenmeyer flasks. Record the weight to ± 1 mg. Prepare a series of acetic acid solutions of various concentrations according to Table 15–1. Add 100 ml of each solution to each charcoal sample. Swirl the flasks vigorously and let them stand overnight. Filter the solutions and titrate a suitable size aliquot (Table 15–1) of each filtrate with standard 0.1 M sodium hydroxide. Use progressively larger aliquots for the more dilute solutions. Run each concentration in duplicate.

TABLE 15–1

Suggested Volumes of 0.4 M Acetic Acid to Dilute to 100.00 ml

Sample[a]	0.4 M Acetic Acid (ml)	Aliquot for Analysis (ml)
1	100	10
2	75	10
3	50	10
4	25	25
5	10	25
6	5	50

[a] Prepare samples in duplicate.

Results and Calculations

From the titration data determine the concentrations of the original acetic acid solution and of the acid solutions in equilibrium with the adsorbent. From the volume of the solutions, their equilibrium concentrations, and the original acid concentrations, calculate Y, the number of moles of acid adsorbed per gram of adsorbent.

Prepare suitable tables of the quantities needed to test the validity of the Freundlich and Langmuir isotherms for the aqueous acetic acid–charcoal system. Plot the duplicate runs independently (do not average before plotting). Calculate Freundlich and Langmuir parameters and discuss the observed results.

REFERENCES

1. Brunauer, S., "Adsorption of Gases in Multimolecular Layers," *J. Am. Chem. Soc.* 60:309, 1938.

2. Brunauer, S., *Physical Adsorption,* Princeton University Press, Princeton, New Jersey, 1971.

3. Duff, D. G., S. M. C. Ross, and D. H. Vaughn, "Adsorption from Solution," *J. Chem. Educ.* 65:815–816, 1988.

4. Dunicz, B. L., "Surface Area of Activated Charcoal by Langmuir Adsorption Isotherm," *J. Chem. Educ.* 38:357, 1961.

5. Freundlich, H. M. F., *Kappilarchemie,* Leipzig, 1909.

6. Hansen, R. S., and R. P. Craig, "The Adsorption of Aliphatic Alcohols from Aqueous Solutions by Non-Porous Carbons," *J. Phys. Chem.* 58:211, 1954.

7. Langmuir, I., "The Constitution and Fundamental Properties of Solids and Liquids," *J. Am. Chem. Soc.* 38:2221, 1916.

8. Popiel, W. J., "Adsorption by Solids from Binary Solutions," *J. Chem. Educ.* 43:415–418, 1966.

Equilibrium Constants

Dissociation of Dinitrogen Tetroxide

At room temperature, dinitrogen tetroxide, a colorless gas, is appreciably dissociated to nitrogen dioxide, a reddish gas. The equation for the equilibrium is

$$N_2O_4 = 2NO_2 \qquad (16\text{–}1)$$

The equilibrium constant for this reaction may be written

$$K = \frac{P^2_{NO_2}}{P_{N_2O_4}} \qquad (16\text{–}2)$$

If the total pressure at equilibrium is P, then the partial pressures of the two oxides are given by

$$P_{NO_2} = X_{NO_2} P \qquad (16\text{–}3)$$

$$P_{N_2O_4} = X_{N_2O_4} P \qquad (16\text{–}4)$$

where X_i are the mole fractions of NO_2 and N_2O_4. If n is the initial number of moles of N_2O_4 and α is the degree of dissociation, then

$$n_{NO_2} = 2n\alpha \qquad (16\text{–}5)$$

$$n_{N_2O_4} = n(1 - \alpha) \qquad (16\text{–}6)$$

$$n_{total} = n(1 + \alpha) \qquad (16\text{–}7)$$

and the mole fractions are

$$X_{NO_2} = \frac{2\alpha}{1 + \alpha} \qquad (16\text{--}8)$$

$$X_{N_2O_4} = \frac{1 - \alpha}{1 + \alpha} \qquad (16\text{--}9)$$

Combining Equations 16–3, 16–4, 16–8, and 16–9 with Equation 16–2 gives

$$K = \frac{4\alpha^2 P}{1 - \alpha^2} \qquad (16\text{--}10)$$

The degree of dissociation α may be measured experimentally by measuring the molecular weight of the gas mixture. If the subscripts 1 and 2 represent the monomer and dimer, respectively, then the average molecular weight M of the gas mixture is given by

$$M = X_1 M_1 + X_2 M_2 \qquad (16\text{--}11)$$

where X_1 and X_2 are mole fractions of monomer and dimer and $M_2 = 2M_1$. When mole fractions are expressed in terms of Equations 16–3 and 16–9 and substituted in Equation 16–11, the result after slight rearrangement is

$$\alpha = \frac{M_2 - M}{M} \qquad (16\text{--}12)$$

where M_2 is 92.06, the molecular weight of N_2O_4, and M is the average molecular weight of the mixture of monomer and dimer. Experimentally, the problem is reduced to determining M, the average molecular weight of the mixture of N_2O_4 and NO_2. This is easily accomplished by measuring the density of a sample of the equilibrium mixture:

$$PV = nRT = gRT/M \qquad (16\text{--}13)$$

$$d = g/V \qquad (16\text{--}14)$$

$$M = dRT/P \qquad (16\text{--}15)$$

Thus, measurements of density lead to the determination of M, α, and K.

Calculating the Second Law Heat of Reaction

If such measurements are carried out over a range of temperatures, then a plot of $\ln K$ versus $1/T$ permits evaluation of the heat and entropy of the dissociation reaction from

$$\ln K = \frac{-\Delta H^0}{RT} + \frac{\Delta S^0}{R} \qquad (16\text{--}16)$$

The slope equals $-\Delta H^0/R$ and the intercept equals $\Delta S^0/R$. The plot of $\ln K$ versus $1/T$ is linear if ΔC_P for the reaction is zero or nearly so. The heat of reaction obtained by this treatment of the data is sometimes called the **Second Law heat of reaction.** Another method of treating the data is described below, resulting in the so-called **Third Law heat of reaction.**

When you calculate K with partial pressures in atmospheres, you are choosing as the reference state the gases at 1 atm; hence, the heat and entropy are for the following process:

$$N_2O_4(1 \text{ atm}) = 2NO_2(1 \text{ atm}) \tag{16-17}$$

The units (joules or calories) depend only on your choice of R. If you calculate K with bar instead of atmospheres, then you are choosing as the reference state the gas at 1 bar; in this case the heat and entropy are for the following process:

$$N_2O_4(1 \text{ bar}) = 2NO_2(1 \text{ bar}) \tag{16-18}$$

The enthalpy change for Equations 16–17 and 16–18 is the same, since the enthalpy is essentially independent of pressure. The entropy (and free energy) changes, however, are slightly different. It is an international convention to use 1 bar as the reference state.

Calculating the Third Law Heat of Reaction

This method of calculating heats of reaction from equilibrium data is equivalent to calculating absolute entropies of the products and reactants by statistical thermodynamics and combining the resulting ΔS^0 of the reaction with ΔG^0 (from measured equilibrium constants) to obtain the heat of reaction. In practice it is more convenient to calculate free energy functions by statistical thermodynamics and to combine them with measured equilibrium constants to obtain the heat of reaction (Margrave, 1955). The free energy function is defined as

$$\frac{G^0 - H^0_{298}}{T}$$

so that the difference in the free energy functions of the products and reactants is given by

$$\Delta\left(\frac{G^0 - H^0_{298}}{T}\right)_{\text{reaction}} \tag{16-19}$$

$$= \Sigma \left(\frac{G^0 - H^0_{298}}{T}\right)_{\text{products}} - \Sigma \left(\frac{G^0 - H^0_{298}}{T}\right)_{\text{reactants}}$$

The free energy function can be calculated from the usual equations of statistical thermodynamics (Lewis and Randall, 1961) or may be found in extensive tabulations (Chase et al., 1986). Some values of the free energy functions for NO_2 and N_2O_4 are listed in Table 16–1. Furthermore,

$$\Delta\left(\frac{G^0 - H^0_{298}}{T}\right)_{\text{reaction}} = \frac{\Delta G^0}{T} - \frac{\Delta H^0_{298}}{T} \tag{16-20}$$

TABLE 16–1

Free-Energy Functions (J/K) for Two Oxides of Nitrogen (standard state pressure = 1 bar)

T (K)	$-\left(\dfrac{G^0 - H^0_{298}}{T}\right)_{N_2O_4}$	$-\left(\dfrac{G^0 - H^0_{298}}{T}\right)_{NO_2}$
200	311.015	243.325
250	305.613	240.634
300	304.377	240.034
350	305.343	240.491
400	307.557	241.524

Source: Chase, M. W. et al., *JANAF Thermochemical Tables,* 3rd ed., ACS and AIP, New York, 1986, pp. 1535, 1560.

and

$$\Delta\left(\frac{G^0 - H^0_{298}}{T}\right)_{\text{reaction}} = -R \ln K - \frac{\Delta H^0_{298}}{T} \qquad (16\text{–}21)$$

Solving for the heat of reaction, we find

$$\Delta H^0_{298} = -T\,\Delta\left(\frac{G^0 - H^0_{298}}{T}\right)_{\text{reaction}} - RT \ln K \qquad (16\text{–}22)$$

Notice that a *single* measurement of K combined with the change in free energy function gives a value of the heat of reaction, the so-called Third Law heat of reaction. The Second Law heat of reaction, discussed above, requires the measurement of equilibrium constants at two or more temperatures.

The free energy function is especially useful because not only can it be calculated easily by statistical thermodynamics, but—because it is a slowly changing function of temperature—it can also be interpolated easily.

Apparatus

The apparatus consists of a vacuum line in a hood or well-ventilated surroundings; liquid nitrogen or dry ice trap; a balance (0.1 mg); a top-loading balance (0.1 g); two sample bulbs of 200- to 250-ml capacity with a stopcock and O-ring connector; a small cylinder of N_2O_4 with suitable coupling to a glass vacuum line, for example, a Cajon clamp[1]; and a variable-temperature thermostat (25 to 60°C).

SAFETY CONSIDERATIONS

Nitrogen dioxide and dinitrogen tetroxide are poisonous and should be confined to a hood. Dewar flasks and vacuum lines can implode if

[1] Cajon Company, 9760 Shepard Road, Macedonia, Ohio 44056.

broken; safety glasses should be worn when in the vicinity of high-vacuum apparatus. In the compressed gas cylinder, N_2O_4 is in the liquid state. The gas in equilibrium with the liquid at room temperature is mostly N_2O_4 in equilibrium with a small amount of NO_2. The total equilibrium vapor pressure at 25°C is about 1.1 atm. This is sufficiently low that the full tank pressure may safely be applied to the vacuum line without danger of blowing out the stopcocks. Be sure that the tank is not warmed above room temperature by exposure to the sun, a hot vacuum pump, or other incidental source of heat.

EXPERIMENTAL PROCEDURE

Calibration. To measure the density of the equilibrium gas mixture, the two bulbs must be calibrated; that is, their volume must be determined. Evacuate and weigh the empty bulb to ±0.1 g on a top-loading balance. Then invert the bulb over water near room temperature and open the stopcock so that the bulb fills with water. Use a thin glass tube or hypodermic needle to finish filling the bulb up to the stopcock. Remove any excess with a pipe cleaner. Record the water temperature.

Weigh the filled bulb, look up the density of water (Table 14–2 on page 394) at the recorded temperature, and calculate the volume of the bulb. Remove the water from the bulb by inserting a thin plastic tube or a long hypodermic needle to let air in. After the water has been removed from the bulbs, attach them to the vacuum line and pump until they are absolutely dry. Do not grease the O-rings used to connect the bulbs. Handle the bulbs with a tissue paper now to prevent oil or perspiration from the skin from changing the weight of the bulb. Determine the weight of the dry evacuated bulb to ±0.1 mg.

Filling the Bulbs. A tank of N_2O_4 at room temperature consists of a condensed liquid in equilibrium with a mixture of NO_2 and N_2O_4. The total pressure just exceeds 1 atm at temperatures above about 22°C. Consequently, the full tank pressure may be exerted in an ordinary vacuum line without a pressure-reducing valve. The room (and tank) temperature should be above 22°C so that there is a *positive* pressure in the bulbs. *Safety goggles must be worn!*

Since oxides of nitrogen are corrosive, the pump is protected by stopcock D and a liquid nitrogen cold trap, as shown in Figure 16–1. The gas tank is connected to the vacuum line with a Cajon clamp below stopcock B. The clamp is opened by turning the knurled ring counterclockwise. It then slips easily over the glass tube below stopcock B; turning the ring clockwise tightens an O-ring vacuum-tight seal around the glass tube.

With the tank needle valve A closed, open the tank's main valve. Next, set the stopcocks so that when the pump is turned on,

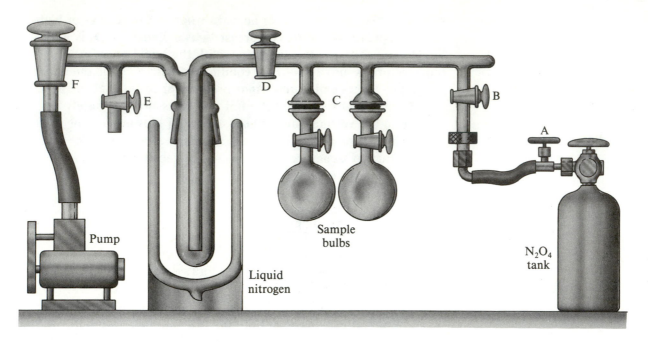

Figure 16–1 Gas-handling apparatus for nitrogen oxides.

the bulbs and the vacuum line are evacuated all the way to valve A. Continue pumping for 5 minutes or so, or until you hear a "hard" vacuum. You are now ready to fill the bulbs.

Next, close stopcock D. You now have a static system, so don't waste time. Open needle valve A. The line fills instantly with a red gas from A to D. Close the stopcocks on the two bulbs. Don't touch them with bare hands; use tissue paper. Close needle valve A and the main valve of the tank.

Now relax. You have filled the two bulbs with dinitrogen tetroxide to a pressure of a little over 1 atm. Carefully unscrew the knurled ring connecting the metal Cajon clamp to the glass tube. Be prepared for a puff of red gas to escape into the hood. The two bulbs may now be disconnected from the line and placed in a secure, clean, and dry place. The dinitrogen tetroxide left over in the line is most easily removed by flushing it out with a gentle flow of compressed air (or nitrogen) entering the system through a rubber tube attached at E. Be sure stopcock B is open and the Cajon clamp is removed before beginning the flushing.

Equilibrium Measurements. Set the controller on the thermostat to about 30°C and let it come to thermal equilibrium. Place the bulbs in the thermostat so that the water level is about half-way between the bulb and the stopcock; clamp securely in place. After about 10 minutes, rotate the stopcocks to bleed out the excess gas. As the system comes to equilibrium, N_2O_4 dissociates into $2NO_2$, so that the pressure increases. Periodically opening the stopcocks lets the system come to equilibrium at the prevailing atmospheric pressure. This should be carefully measured with a mercury barometer. When a puff of red gas no longer comes out upon turning the stopcock past the open position, it may be assumed that the system is at thermodynamic equilibrium.

After equilibrium is attained, remove the bulbs, dry them carefully, and let them equilibrate with room air before weighing. Be sure no moisture has condensed inside the inlet tube to the bulb. Dry it with a pipe cleaner if it has. Don't hurry the weighings. Try to treat the bulbs the same way after each weighing and in the same way as when the bulb was previously weighed evacuated. The increase in weight of the evacuated bulb should arise only from the mixture of oxides of nitrogen, *not* from spurious indeterminant amounts of moisture picked up from the thermostat. This is the largest source of error in the experiment. Careful, uniform weighing is necessary.

Make measurements at approximately 30, 40, 50, and 60°C. Temperature measurement and control should be to at least ±0.1°C.

Emptying the Bulbs of Nitrogen Oxides. Attach the bulbs to the vacuum line with their stopcocks closed and evacuate the line. Stopcocks F and D are open; all other stopcocks are closed. Fill the cold trap with liquid nitrogen. Slowly open the stopcocks on the two bulbs and allow the nitrogen oxides to be pumped into the trap, where they freeze out. Close stopcock F. Open stopcocks E and B to let air into the line. Attach a rubber tube to stopcock E and sweep out the nitrogen oxides through stopcock B with compressed air or nitrogen. Remove the Dewar flask of liquid nitrogen. It takes several minutes before the trap warms up. Check the state of the stopcocks carefully before sweeping with compressed air or nitrogen.

Results and Calculations

For each bulb at each temperature, calculate and tabulate the gas mixture density d, molecular weight of mixture M, degree of dissociation α, and equilibrium constants. Do not average the duplicate values at each temperature, but treat them as independent evaluations. Record P. As the standard state, use $P^0 = 1$ bar.

Plot $\ln K$ versus $1/T$ and calculate the equation of the line $\ln K = -A/T + B$, ΔH^0, and ΔS^0. These are the Second Law values.

Calculate the Third Law heats of reaction with the free energy functions listed in Table 16–1. Compare the Second Law heat of reaction, the average of the Third Law heats of reaction, and the literature values. Calculate ΔG^0 at 298 K and compare with the literature value.

Discuss the errors in the calculated values and, in particular, compare the errors in the Second and Third Law treatments of the data. The Third Law treatment is especially useful in identifying systematic errors in the heats of reaction. Spurious results are easily identified and examined for possible rejection.

REFERENCES

1. Chase, M. W., et al., *JANAF Thermochemical Tables,* 3rd ed., published by the American Chemical Society and the American Institute of Physics for the National Bureau of Standards, New York, 1986.

2. Giauque, W. F., and J. D. Kemp, "The Entropies of Nitrogen Tetroxide and Nitrogen Dioxide. The Heat Capacities from 15°K to the Boiling Point. The Heat of Vaporization and Vapor Pressure. The Equilibrium $N_2O_4 = 2\ NO_2 = 2\ NO + O_2$," *J. Chem. Phys.* 6:40, 1938.

3. Harris, L., and K. Churney, "Evaluation of the Equilibrium for the $N_2O_4(g) = 2NO_2(g)$ Reaction at 298.16°K from Light-Transmission Measurements," *J. Chem. Phys.* 47:1703, 1967.

4. Lewis, G. N., and M. Randall, revised by K. S. Pitzer and L. Brewer, *Thermodynamics,* McGraw-Hill, New York, 1961.

5. Margrave, J. L., "Thermodynamic Calculations," *J. Chem. Educ.* 32:520, 1955.

6. Verhoek, F. H., and F. Daniels, "The Dissociation Constants of Nitrogen Tetroxide and Nitrogen Trioxide," *J. Am. Chem. Soc.* 53:1250, 1931.

7. Wettach, F. S., "A Photometric Study of the N_2O_4–NO_2 Equilibrium," *J. Chem. Educ.* 49:557, 1972.

Experiment 17

Measurement of the Solution Equilibrium: $I^- + I_2 = I_3^-$

Carbon tetrachloride and water are essentially insoluble in each other. Placed together in a container, the denser carbon tetrachloride sinks to the bottom and the lighter water layer floats above. If a small amount of iodine is added to this two-phase, two-component system, a little iodine dissolves in each layer. The concentration of iodine in the organic layer (ct) is somewhat greater than in the aqueous layer (aq). If a little more iodine is added to the system, more dissolves in each layer, but the *relative* concentrations in the two phases remain constant. Consequently, for the equilibrium

$$I_2(aq) = I_2(ct) \tag{17-1}$$

an equilibrium constant can be written:

$$K_d = \frac{[I_2(ct)]}{[I_2(aq)]} \tag{17-2}$$

The equilibrium constant is called the **distribution** or **partition coefficient.** The distribution coefficient is easily determined by allowing iodine to equilibrate between the two phases and subsequently titrating the iodine contained in aliquots of each phase. From the analytically determined iodine concentrations, the distribution coefficient is calculated with Equation 17–2.

If a little iodide (as KI) is introduced into the system, a new equilibrium plays a role:

$$I_2(aq) + I^-(aq) = I_3^-(aq) \qquad (17\text{–}3)$$

$$K = \frac{[I_3^-(aq)]}{[I_2(aq)][I^-(aq)]} \qquad (17\text{–}4)$$

A complex ion, the triiodide ion, is formed from the reaction between I^- and I_2. In the two-phase system two equilibria occur simultaneously, each competing for I_2:

water layer $\qquad I_2(aq) + I^-(aq) \overset{K}{\rightleftarrows} I_3^-(aq)$

$\qquad\qquad\qquad \updownarrow K_d$

organic layer $\qquad I_2(ct)$

Again, the concentrations of all species at equilibrium can be determined titrimetrically, and the equilibrium constant for the formation of the complex ion (Eq. 17–4) can be calculated. Titration of the aqueous layer gives the *total* iodine present in the aqueous layer, that is,

$$[I_2(\text{total})] = [I_2(aq)] + [I_3^-(aq)] \qquad (17\text{–}5)$$

However, when K_d is known, titration of the organic phase permits calculation of the equilibrium concentration of iodine in the aqueous phase $[I_2(aq)]$ since

$$[I_2(aq)] = [I_2(ct)]/K_d \qquad (17\text{–}6)$$

Then the triiodide concentration can be calculated:

$$[I_3^-(aq)] = [I_2(\text{total})] - [I_2(aq)] \qquad (17\text{–}7)$$

The iodide at equilibrium is found from the iodide initially present in the system and the stoichiometry of Equation 17–3:

$$[I^-(aq)] = [I^-(aq)_{\text{initial}}] - [I_3^-(aq)] \qquad (17\text{–}8)$$

The equilibrium constant for the formation of the complex ion is then calculated with Equation 17–4. Actually the equilibrium constant should be written in terms of the activities and the activity coefficients:

$$K = \frac{[I_3^-(aq)]}{[I_2(aq)][I^-(aq)]} \times \frac{\gamma_3}{\gamma_2 \gamma^-} \qquad (17\text{–}9)$$

However, I^- and I_3^-, with the same charge and in the same ionic environment, have very nearly the same activity coefficients, which then tend to cancel out in Equation 17–9. Molecular iodine, in both water and carbon tetrachloride, behaves quite ideally, so its activity coefficient is very nearly unity. Consequently, both Equation 17–2 and Equation 17–4 may be used to calculate equilibrium constants.

Apparatus

The apparatus consists of six 250-ml glass-stoppered Erlenmeyer flasks; a thermostat (25°C); 10- and 50-ml burets; 10-, 25-, and 50-ml pipets; a rubber bulb for pipets; and a 10-ml graduated cylinder. The other requirements are 250 ml 0.06 M I_2 solution in carbon tetrachloride, which is approximately the saturation solubility at room temperature; 500 ml 0.1 M KI; 500 ml 0.1 M (standardized) sodium thiosulfate ($Na_2S_2O_3$); and 25 ml starch indicator (fresh soluble starch). The KI solution should be a standard solution; KI may be treated as a primary standard. The I_2 solution need not be standardized.

SAFETY CONSIDERATIONS

Solid iodine is corrosive to the skin; its solutions stain badly. Avoid inhaling carbon tetrachloride fumes.

EXPERIMENTAL PROCEDURE

In order to determine the equilibrium constant for the formation of the triiodide ion, it is necessary to measure the distribution coefficient for iodine with water and carbon tetrachloride. Some suggested reagent volumes are listed in Table 17–1. The first three contain no iodide and allow measurement of the distribution coefficient K_d. The last three contain various concentrations of iodine and iodide for the measurement of the equilibrium constant K_c.

Pipet the required aliquots into six 250-ml glass-stoppered Erlenmeyer flasks. Place the flasks in a 25°C thermostat for about 1 hour, and shake them vigorously every 5 minutes. It is important to shake the flasks periodically in order to attain equilibrium.

Analysis. Prepare 250 ml 0.01 M sodium thiosulfate by quantitatively diluting a 50-ml aliquot of 0.05 M sodium thiosulfate to 250 ml. Because the iodine concentration in the carbon tetrachloride

TABLE 17–1

Initial Volumes (ml) of Reactants and Solvents

No.	Aqueous Phase		Carbon Tetrachloride Phase	
	H_2O	0.1 M KI	0.06 M I_2 (aq)	CCl_4
1	200	0	50	0
2	200	0	25	25
3	200	0	15	35
4	0	200	50	0
5	0	200	25	25
6	150	50	50	0

layer is considerably higher than in the water layer, 0.05 M sodium thiosulfate is used to titrate the carbon tetrachloride aliquots and the 0.01 M solution is used for the water layer aliquots. Since the iodine concentrations vary widely, it is useful to vary both the volume of the aliquot of the iodine solution and the buret containing the thiosulfate titrant. Table 17–2 lists some suggested volumes.

After the samples have stood quietly and have separated into two homogeneous layers, pipet an aliquot of an aqueous layer into a 250-ml glass-stoppered Erlenmeyer flask containing about 10 ml 0.1 M KI. The KI addition need not be quantitative; it is present to suppress loss of iodine by vaporization.

Take care that the carbon tetrachloride solution does not contaminate the pipet used for the water layer aliquot; the iodine concentration in the carbon tetrachloride layer is much higher than in the water layer, so that a drop or two would cause a considerable error in the water layer concentration. With a 10-ml buret containing 0.01 M thiosulfate, titrate the iodine solution until the yellow color of the iodine has faded, but is still visible. Add 1 ml starch indicator solution. A deep blue color should develop. Continue titrating until the blue color disappears sharply. Titrate the remaining sample according to the scheme outlined in Table 17–2.

The stoichiometry of the titration reaction is

$$2S_2O_3^{2-} + I_3^- = S_4O_6^{2-} + 3I^- \qquad (17\text{--}10)$$

The half-reaction for the oxidation of the thiosulfate ion is

$$2S_2O_3^{2-} = \tfrac{1}{2}S_4O_6^{2-} + e^- \qquad (17\text{--}11)$$

Since 1 mole of thiosulfate donates 1 mole of electrons, the equivalent weight and the molecular weight of thiosulfate are identical. Thus the molarity and the normality are the same.

Standardization. If the sodium thiosulfate is not standardized, it may be easily standardized against primary standard potassium

TABLE 17–2

Volumes for Iodine Analysis

Run	Aqueous Phase (Titrate with 0.01 M $S_2O_3^{2-}$)		Carbon Tetrachloride Phase (Titrate with 0.05 M $S_2O_3^{2-}$)	
	$V_{aliquot}$	V_{buret}	$V_{aliquot}$	V_{buret}
1	50	10	10	50
2	50	10	25	50
3	50	10	25	50
4	10	50	25	50
5	10	50	25	50
6	10	50	10	50

iodate, KIO_3. Sodium thiosulfate itself is not a primary standard and neither is iodine. However, a quantitative amount of iodine may be liberated from a weighed sample of potassium iodate; the liberated iodine may then be titrated as above:

$$IO_3^- + 8I^- + 6H^+ = 3I_3^- + 3H_2O \qquad \text{(17–12)}$$

The iodine, liberated as the triiodide, is then titrated as before:

$$2S_2O_3^{2-} + I_3^- = S_4O_6^{2-} + 3I_3^- \qquad \text{(17–13)}$$

Procedure. Weigh about 0.45 to 0.55 g KIO_3 (± 0.1 mg) into a 1-L volumetric flask and dilute to the mark. The solution is about 0.08 N (0.48 M). Calculate the exact normality.

To about 1 L freshly boiled and cooled water, add about 12.5 g $Na_2S_2O_3$ $5H_2O$ and 0.1 g Na_2CO_3. The solution is about 0.05 N (0.05 M). Store in an amber bottle in the dark.

Stir together 2 g soluble starch, 10 mg HgI_2, and 30 ml water. Add to 1 L boiling water with constant stirring. The iodide acts as a preservative.

Pipet three 25-ml aliquots of the standard KIO_3 into 250-ml flasks. To the first, add 50 ml water, 1 g reagent-grade KI, and 10 ml 1 N HCl. Titrate immediately until the solution is a pale yellow color. Add 3 to 5 ml starch indicator and continue titrating until the blue color suddenly becomes colorless. Repeat with the second and third aliquots.

Results and Calculations

From the volume of the aliquots, the concentration of thiosulfate, the volume of thiosulfate, and the stoichiometry of Equation 17–10, calculate the total iodine concentrations in the aqueous and organic layers of the six systems.

Calculate the distribution coefficient for systems number 1, 2, and 3. Calculate the propagated error and compare your K_d with the value in the literature.

With equations 17–6, 17–7, and 17–8, calculate the equilibrium concentrations of $I_2(aq)$, $I_3^-(aq)$, and $I^-(aq)$. Calculate K_c and compare with the value in the literature. Assume that activity coefficients are unity. Discuss any systematic trends in K and K_d with concentration.

REFERENCES

1. Acherman, M. N., "Determination of the Equilibrium Constant for Triiode Formation: Use of a Less Toxic Solvent," *J. Chem. Educ.* 55:795, 1978.

2. Benoit, R. L., and M. Guay, "Stabilite d'ions Trihalogenures En Milieu Polaire Aprotique," *Inorg. Nucl. Chem. Lett.* 4:215–217, 1968.

3. Daniele, G., "Spectrophotometric Measurement of the Equilibrium Constant of Potassium Triiode," *Gazz. Chim. Ital.* 90:1068, 1960.

4. Dawson, H. M., "On the Nature of Polyiodides and Their Dissociation in Aqueous Solution," *J. Chem. Soc.* 79:238, 1901.

5. Glasstone, S., and D. Lewis, *Elements of Physical Chemistry,* 2nd ed., Van Nostrand, Princeton, New Jersey, 1960, p. 382.

6. Guggenheim, E. A., and J. E. Prue, *Physico-Chemical Calculations,* Interscience, New York, 1955, p. 348.

7. Kahwa, I. A., "A Graphical Procedure for the Simultaneous Determination of the Distribution Constant of Iodine and the Stability Constants of Trihalide Anions," *J. Chem. Educ.* 61:823–825, 1984.

8. Ramamurti, G., K. Renganathan, and L. R. Ganesan, "The Instability Constant of the $I_3^- \rightleftarrows I^- + I_2$ System," *J. Chem. Educ.* 53:326, 1976.

9. Ramette, R. W., "Equilibrium Constants from Spectrophotometric Data," *J. Chem. Educ.* 44:647, 1967.

10. Skoog, D. A., and D. M. West, *Analytical Chemistry,* 4th ed., chap. 13, Saunders College Publishing, Philadelphia, 1986.

Experiment 18

Monomer–Dimer Equilibrium

Hydrogen bonding is important in organic, inorganic, and biochemical compounds. A hydrogen bond can form when hydrogen is bonded to a highly electronegative atom, most frequently oxygen. It occurs in compounds as complex as proteins and as simple as H_2O and HF. It frequently plays a role in organic compounds when a hydrogen atom is bonded to an oxygen in a hydroxyl or carboxyl group.

It is well known that carboxylic acids tend to dimerize in nonpolar solvents, as illustrated in Figure 18–1a. This is an example of bonding between molecules or **intermolecular** hydrogen bonding. Such dimerization occurs in solution with nonpolar solvents, in the pure solid or liquid state, and even in the gas phase. When R is a simple hydrocarbon moiety, the dimerization is essentially complete at room temperature.

In this experiment, we study the dimerization of salicylic acid. In salicylic acid, hydrogen bonding can also take place within the molecule because of its particular structure. This is an example of **intramolecular** hydrogen bonding, bonding within a molecule (Fig. 18–1b).

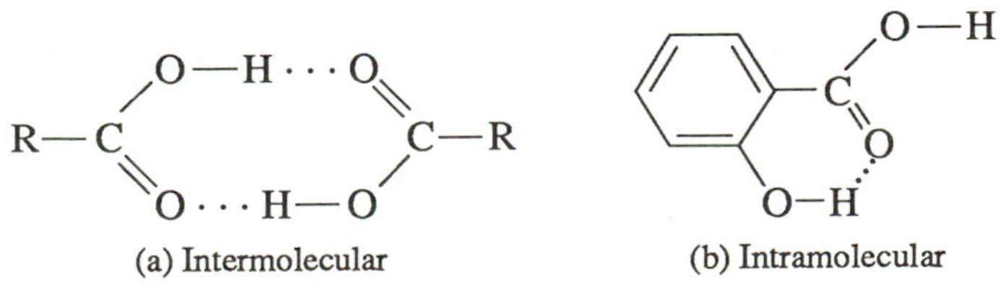

(a) Intermolecular   (b) Intramolecular

Figure 18–1 Hydrogen bonding.

In nonpolar solvents or dilute solutions, where molecules are quite isolated, intramolecular hydrogen bonding predominates. As the concentration increases, the chance of molecular encounters increases, and so does dimerization due to intermolecular hydrogen bonding. If some salicylic acid is added to a two-phase system consisting of water and an organic phase, the acid dissolves in both phases and distributes itself between the two phases. It exists as a monomer in water; but in the organic phase, there is an equilibrium between the monomer and dimer.

The distribution of salicylic acid between a water phase and an organic phase, where there is a monomer–dimer equilibrium, can be illustrated as follows:

$$\text{organic layer} \qquad \text{S}(or) \overset{K_2}{\rightleftharpoons} \tfrac{1}{2}\text{S}_2(or)$$

$$\updownarrow K_1$$

$$\text{water layer} \qquad \text{S}(aq)$$

The equilibrium constant for the distribution of the acid between the two phases is

$$K_1 = \frac{[\text{S}(or)]}{[\text{S}(aq)]} \tag{18-1}$$

For the equilibrium between the monomer and dimer, the equilibrium constant is written:

$$K_2 = \frac{[\text{S}_2(or)]^{1/2}}{[\text{S}(or)]} \tag{18-2}$$

The total concentration of salicylic acid present in the organic phase $C(or)$ is given by

$$C(or) = 2[\text{S}_2(or)] + [\text{S}(or)] \tag{18-3}$$

Similarly, in the aqueous phase, the total concentration is

$$C(aq) = [\text{S}(aq)] \tag{18-4}$$

When $[\text{S}(or)]$ and $[\text{S}_2(or)]$ are eliminated, the result is

$$\frac{C(or)}{C(aq)} = K_1 + K_1^2 K_2^2 C(aq) \tag{18-5}$$

When $C(or)/C(aq)$ is plotted against $C(aq)$, a straight line results with the slope equal to $2K_1^2 K_2^2$ and intercept equal to K_1.

Apparatus

The experiment requires toluene; saturated solutions of salicylic acid in toluene and water; a thermostat (25°C); 0.1 M sodium hydroxide; phenolphthalein; ten 125-ml Erlenmeyer flasks; 10-, 25-, and 50-ml pipets; a 250-ml volumetric flask; and six 250-ml separatory funnels with Teflon stopcocks.

SAFETY CONSIDERATIONS

Use the organic solvent in a well-ventilated hood and avoid contact with the skin. Dispose of the organic solutions in the organic waste container.

EXPERIMENTAL PROCEDURE

Add 25 ml of the saturated solution of salicylic acid in water to each of five 250-ml separatory funnels. Dilute each with 8, 15, 25, 50, and 75 ml water. Add about 25 ml of the saturated solution of salicylic acid in toluene to each of the five separatory funnels, and dilute the corresponding solutions with 8, 15, 25, 50, and 75 ml toluene. A graduated cylinder suffices for these measurements.

Place the separatory funnels in a 25°C thermostat. Wait about 10 minutes for the solutions to get close to thermal equilibrium. For the next 30 minutes, remove the separatory funnels every five minutes and shake vigorously for at least 15 seconds before replacing them in the thermostat. Use a timer for the 15 seconds or you will probably shake for a shorter period. To achieve equilibrium between the two phases, it is necessary to disperse the toluene and water phases into fine drops to increase the surface area, and this can be achieved only by vigorous and prolonged shaking.

After the system reaches equilibrium, let the separatory funnels stand quietly for at least 30 minutes so that the phases separate completely and no fine droplets remain dispersed. Remove 10-ml aliquots of each phase from each separatory funnel and drain into ten labeled 125-ml Erlenmeyer flasks. Add 25 ml water to the five flasks containing the toluene aliquots. Titrate the solutions with standard 0.05 M sodium hydroxide with a phenolphthalein indicator. The toluene–water mixture must be stirred vigorously during the titration as the salicylic acid must be extracted from the toluene phase into the water during the course of the titration. A magnetic stirrer and stirring bar are helpful.

Results and Calculations

Calculate the equilibrium molar concentrations of the salicylic acid in the toluene and water phases. Tabulate the experimental and calculated data. Plot $C(or)/C(aq)$ versus $C(aq)$ according to Equation 18–5.

Calculate K_1 and K_2 with a least-squares analysis of the data. Analyze the errors in K_1 and K_2.

REFERENCES

1. Ellison, H. R., "Simultaneous Equilibria in the Benzoic Acid–Benzene–Water System," *J. Chem. Educ.* 48:124, 1971.

2. Hendrixon, W. S., "Beiträge zur Kenntnis der Dissoziation in Lösungen," *Z. Anorg. Chem.* 13:73–80, 1897.

3. Huq, A. K. M. S., and S. A. K. Lodhi, "Distribution of Benzoic Acid Between Benzene and Water and Dimerization of Benzoic Acid in Benzene," *J. Phys. Chem.* 70:1354, 1966.

4. Pinmentel, G. C., and A. L. McClellan, *The Hydrogen Bond,* Reinhold, New York, 1960.

5. Vinogradov, S. N., and R. H. Linnell, *Hydrogen Bonding,* Van Nostrand, New York, 1970.

Experiment 19

The pK_a of a Weak Acid

A Brønsted acid is defined as a proton donor and a Brønsted base as a proton acceptor. Some familiar examples of weak Brønsted acids are

$$HA + H_2O = H_3O^+ + A^- \tag{19-1}$$

$$NH_4^+ + H_2O = H_3O^+ + NH_3 \tag{19-2}$$

$$HCN + H_2O = H_3O^+ + CN^- \tag{19-3}$$

$$HOC_6H_4NO_2 + H_2O = H_3O^+ + OC_6H_4NO^- \tag{19-4}$$

$$HIn + H_2O = H_3O^+ + In^- \tag{19-5}$$

In Equation 19–5 HIn represents an acid–base indicator—an organic compound in which HIn, a weak acid, has one color and its conjugate base, In$^-$, has another color. By Le Châtelier's principle, a solution containing the indicator has the color of HIn in acidic solutions and the color of In$^-$ in basic solutions. In general,

$$acid + base = conjugate\ acid + conjugate\ base \tag{19-6}$$

An equilibrium constant may in general be written

$$K_a = \frac{[H_3O^+][A^-]}{[HA]} \tag{19-7}$$

Since the pH can usually be measured easily, the equilibrium constant can be determined if the equilibrium concentration of HA and A$^-$ can be measured. In some favorable cases, the acid and its conjugate base have different absorption spectra, which allows the equilibrium concentrations of HA and A⁻ to be measured spectrophotometrically. This is certainly true for many acid–base indicators (Eq. 19–5) and some other colored compounds such as p-nitrophenol (Eq. 19–4).

The absorption of light is governed by the Beer-Lambert Law:

$$A = \log \frac{I_0}{I} = \epsilon[X]\ell \qquad \text{(19–8)}$$

where

A = the absorbance
I = the intensity of light transmitted by the solution
I_0 = the intensity of light transmitted by the pure solvent
ϵ = the molar absorptivity (or extinction coefficient, sometimes given the symbols a or k)
$[X]$ = the concentration in moles per liter of absorbing species X
ℓ = the path length in centimeters

In Figure 19–1a and b absorbance is plotted against wavelength; in Figure 19–1a, the absorption maxima are so far apart that the absorption at each maximum is independent of the other:

$$A_1 = \epsilon_{1,\text{HIn}}[\text{HIn}] \qquad \text{(19–9)}$$

$$A_2 = \epsilon_{2,\text{In}^-}[\text{In}^-] \qquad \text{(19–10)}$$

Normally the situation is not so favorable (Fig. 19–1b) and both species absorb appreciably at both wavelength 1 and wavelength 2:

$$A_1 = \epsilon_{1,\text{HIn}}[\text{HIn}] + \epsilon_{1,\text{In}^-}[\text{In}^-] \qquad \text{(19–11)}$$

$$A_2 = \epsilon_{2,\text{HIn}}[\text{HIn}] + \epsilon_{2,\text{In}^-}[\text{In}^-] \qquad \text{(19–12)}$$

When the molar absorptivity ϵ_i of each species $[X_i]$ is known at wavelength λ_1 and wavelength λ_2, the concentrations may be calculated from a measurement of absorption A_i at each wavelength, since Equations 19–11 and 19–12 are then two equations in two unknowns.

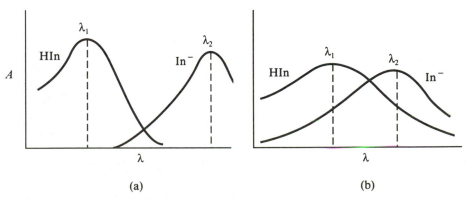

(a)

(b)

Figure 19–1 Absorption spectra.

Equation 19–7 may be rewritten as

$$pH = pK_a + \log \frac{[In^-]}{[HIn]} \qquad (19\text{--}13)$$

If the concentrations of the weak acid HIn and its conjugate base In^- are measured spectrophotometrically in solutions of various pH, then a plot of pH against $\log([In^-]/[HIn])$ gives a straight line with slope equal to unity and an intercept equal to pK_a.

Apparatus

A laboratory stock solution—prepared by dissolving 1 g crystalline methyl red in 300 ml 95% ethanol and diluting to 500 ml in a volumetric flask with distilled water—is convenient. In addition, the following solutions are required: 500 ml 0.04 M sodium acetate, 100 ml 0.01 M sodium acetate, 100 ml 0.02 M acetic acid, 25 ml 0.1 M hydrochloric acid, 100 0.01 M hydrochloric acid, and 500 ml 95% ethanol. The equipment needed includes a spectrophotometer, pH meter, two 250-ml volumetric flasks, nine 100-ml volumetric flasks, and 1-, 5-, 10-, 25-, and 50-ml volumetric pipets.

SAFETY CONSIDERATIONS

Methyl red, as well as many other indicators, is a derivative of dimethylaminoazobenzene, a well-known carcinogen. Ethanol is inflammable.

EXPERIMENTAL PROCEDURE

From the stock solution (see above) prepare a *working solution* by adding 10 ml of the stock solution to 125 ml 95% ethanol and diluting to 250 ml with water.

The Spectra. The wavelengths at which HIn and In^- show a maximum absorption must be determined. This is done by measuring the absorption spectra from 350 to 600 nm of an acid solution (solution A) and a basic solution (solution B) of the indicator. These are prepared from the working solution of the indicator.

For solution A, add a 10-ml aliquot of the indicator working solution to a 100-ml volumetric flask containing about 10 ml 0.1 M HCl and dilute to the mark. The pH of this acidic solution is about 2, so that the indicator is present entirely as HIn according to Equation 19–5.

For solution B, add a 10-ml aliquot of the indicator working solution to a 100-ml volumetric flask containing about 25 ml 0.04 M sodium acetate and dilute to the mark. In this basic solution the

pH is about 8 and the indicator is present entirely as In⁻ according
to Equation 19–5.

With matched cells and distilled water in the reference cell,
measure the spectra of HIn and In⁻ solutions, and determine λ_1
and λ_2, the absorption maxima.

The Molar Absorptivities. Dilute 15-, 20-, and 25-ml aliquots
of solution A to 50 ml with 0.01 M HCl. Measure the absorbance
of each solution (including undiluted solution A) at λ_1 and λ_2. Plot
A against [HIn], that is, plot A against the relative concentration
of HIn. For the dilutions suggested above, the relative concentrations are 0.3, 0.4, 0.5, and 1.0. Calculate the (relative) molar absorptivity of HIn at the two wavelengths from the slope of the plots
(Eq. 19–9).

In a similar manner, dilute 15-, 20-, and 25-ml aliquots of
solution B to 50 ml with 0.01 M sodium acetate. Plot A against the
relative [In⁻] and determine the relative molar absorptivity of In⁻
at each wavelength (Eq. 19–10). Place all four Beer's Law plots on
a single sheet of graph paper. Run a least-squares analysis of the
data to get the slopes and their standard deviations.

The Ionization Constant. Determine the pK_a of the indicator
with Equation 19–13. Make up solutions of differing pH from the
indicator working solution and various ratios of the acetic acid and
sodium acetate buffer (Table 19–1). Calibrate the pH meter with a
standard buffer at pH 6.00. For each solution in Table 19–1, measure
the pH and the absorbance at λ_1 and λ_2.

Results and Calculations

To determine the molar absorptivities ϵ_1 and ϵ_2 with Equations 19–9 and
19–10, plot absorbance against relative concentration of [HIn] and [In⁻]
at the two wavelength maxima. Graph the four Beer's Law plots on a
single sheet of graph paper, with an appropriate legend. Run a least-

TABLE 19–1

Dilution Volumes (ml) for Various Buffer Concentrations

Mixture	Working Solution	0.02 M HAc	0.04 M NaAc	Water	Total
1	10	50	25	15	100
2	10	25	25	40	100
3	10	20	25	45	100
4	10	15	25	50	100
5	10	10	25	55	100
6	10	5	25	60	100
7	10	1	25	64	100

squares analysis of the data to determine the slopes and their standard deviations. The equilibrium species in Equation 19–5 are $[H_3O^+]$, $[HIn]$, and $[In^-]$. The $[H_3O^+]$ concentration is determined directly with a pH meter.

The $[HIn]$ and $[In^-]$ concentrations are determined by solving two equations in two unknowns, Equations 19–11 and 19–12, from the measurements of absorbance at two wavelengths for each of the solutions in Table 19–1. These are actually the relative concentrations of $[HIn]$ and $[In^-]$ since relative concentrations are used in the Beer's Law plot. It is sufficient to determine relative concentration because only the raio $[In^-]/[HIn]$ is used in Equation 19–13. When Equations 19–11 and 19–12 are solved simultaneously, the results are

$$[HIn] = \frac{(A_1\epsilon_{2,In^-}) - (A_2\epsilon_{1,In^-})}{(\epsilon_{2,In^-})(\epsilon_{1,HIn}) - (\epsilon_{1,In^-})(\epsilon_{2,HIn})} \qquad (19\text{–}14)$$

$$[In^-] = \frac{(A_1\epsilon_{2,HIn}) - (A_2\epsilon_{1,HIn})}{(\epsilon_{1,In^-})(\epsilon_{2,HIn}) - (\epsilon_{2,In^-})(\epsilon_{1,HIn})} \qquad (19\text{–}15)$$

Plot pH against $\log([In^-]/[HIn])$. According to Equation 19–13, this should result in a straight line of slope equal to unity and an intercept equal to pK_a. Run a least-squares analysis of these data to obtain pK_a and its standard deviation. Analyze the errors, and compare your results with selected values from the literature.

REFERENCES

1. Ramett, R. W., "The Dissociation Quotient of Bromcresol Green," *J. Chem. Educ.* 40:252–254, 1963.

2. Ramett, R. W., "Equilibrium Constants from Spectrophotometric Data," *J. Chem. Educ.* 44:647, 1967.

3. Skoog, D. A., and D. M. West, *Analytical Chemistry*, 4th ed., Saunders College Publishing, Philadelphia, 1986.

4. Tobey, S. W., "The Acid Dissociation Constant of Methyl Red," *J. Chem. Educ.* 35:514, 1958.

Electrochemistry

Experiment 20

Experiment 20

Transference Numbers by the Moving Boundary Method

The conduction of electricity by solid and liquid metals involves the transfer of electrons under the influence of a potential difference. The electrons move, but the metal atoms do not. On the other hand, conduction of electricity by fused salts and solutions of weak or strong electrolytes is caused by the migration of ions, both positive and negative. Under the influence of a potential difference between two electrodes, negative ions migrate to the positive electrode and positive ions migrate to the negative electrode. This means that the total current I is the sum of the current carried by the anions and cations:

$$I = I_+ + I_- \qquad (20\text{--}1)$$

For a given potential difference, the rate at which an ion migrates depends on its size and the magnitude of its charge. The higher the charge and the smaller the size, the faster an ion migrates through the solution. Since the anion and cation of a given salt nearly always differ in size and often in the magnitude of their charge, they migrate at different velocities through the solution. Consequently, the anions and cations carry different currents through the solution, that is, $I_+ \neq I_-$. The transference number of an ion is simply the fraction of the total current that it transports through the solution:

$$t_+ = I_+/I \qquad t_- = I_-/I \qquad (20\text{--}2)$$

$$t_+ + t_- = 1 \qquad (20\text{--}3)$$

Let us examine these ideas quantitatively. Consider two parallel electrodes separated by a distance d and immersed in an electrolytic solution (Fig. 20–1). Let the ion velocities be v_+ and v_-. In 1 sec, all the positive ions closer than v_+ cm from the negative electrode will migrate to it. Let this number equal n_+. The total current carried by the cations (C/sec) is given by

$$I_+ = \frac{n_+ v_+ Z_+ e}{d} \qquad (20\text{–}4)$$

Similarly, the current carried by the anions is given by

$$I_- = \frac{n_- v_- Z_- e}{d} \qquad (20\text{–}5)$$

Since the total current equals the sum of I_+ and I_-,

$$I = \frac{n_+ v_+ Z_+ e + n_- v_- Z_- e}{d} \qquad (20\text{–}6)$$

The condition for electrical neutrality in an electrolytic solution is

$$n_+ Z_+ = n_- Z_- \qquad (20\text{–}7)$$

so that Equation 20–6 may be written

$$I = \frac{n_+ Z_+ e(v_+ + v_-)}{d} \qquad (20\text{–}8)$$

$$t_+ = \frac{I_+}{I} = \frac{v_+}{v_+ + v_-} \qquad (20\text{–}9)$$

$$t_- = \frac{I_-}{I} = \frac{v_-}{v_+ + v_-} \qquad (20\text{–}10)$$

Finally, from Equations 20–9 and 20–10,

$$\frac{t_+}{t_-} = \frac{v_+}{v_+} \qquad (20\text{–}11)$$

Thus, the relative amount of current carried by the anions and cations as they migrate through the solution is just equal to their relative velocities. Because the velocities with which ions migrate through a solution depend on their charge, size, and degree of hydration, anions and cations ordi-

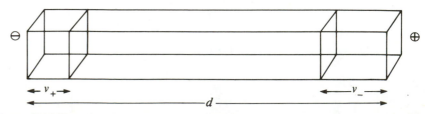

Figure 20–1 Transfer of charge by ion migration.

narily travel with quite different velocities and have different transference numbers. In some special cases, the transference numbers are very nearly 0.5. For example, in KCl, the K^+ and the Cl^- are isoelectronic and have very nearly the same size. Measurements show that $t_+ = 0.490$.

A variety of methods are available for the measurement of transference numbers. In the Hittorf method, the changes in concentration of anions and cations in the middle and end sections of a cell are measured directly, and from these data the transference numbers can be calculated. Electromotive force measurements on voltaic cells with transference permits calculation of transference numbers. When applicable, however, the most accurate measurements of transference numbers are made by the moving boundary method.

The moving boundary apparatus (Fig. 20–2) consists of a vertically arranged electrolytic cell. With HCl as an electrolyte, an Ag, AgCl cathode is placed at the top and a cadmium anode at the bottom. The cell itself is made from a 1-ml measuring pipet. The cell is filled with HCl solution containing a little methyl violet indicator. Before a current is passed through the cell, there are no Cd^{2+} ions in the solution. The indicator is nearly colorless in the presence of HCl.

As soon as a current is passed through the solution, cadmium is oxidized to Cd^{2+}. Under the influence of the applied potential, both the H^+ and the Cd^{2+} migrate upward. The Cd^{2+}, however, migrate more slowly. The formation of Cd^{2+} at the anode has the effect of replacing HCl (soln) with $CdCl_2$ (soln). The pH of the $CdCl_2$ is higher than that of HCl, so the indicator gives a violet color in the region containing the $CdCl_2$, with a sharp boundary at the interface between the two solutions. As more and more Cd^{2+} is generated and migrates upward, so does the visible boundary.

The total charge passed through the solution is Q. The charge transferred by the hydrogen ion is Q_+ and the charge transferred by the chloride ion is Q_-. The transference number of the hydrogen ion is

$$t_+ = \frac{Q_+}{Q} \tag{20–12}$$

When a current of I amperes flows for t seconds, $Q = It$. If the moving boundary sweeps out a volume of V liters during this time, and the HCl concentration is N equiv/L, then the charge carried by the hydrogen ions is

$$Q_+ = FNV \tag{20–13}$$

where $F = 96,485$ C/equiv. The transference number of the hydrogen ion is then calculated from

$$t_+ = \frac{FNV}{It} \tag{20–14}$$

or

$$V = \frac{t_+ It}{FN} \tag{20–15}$$

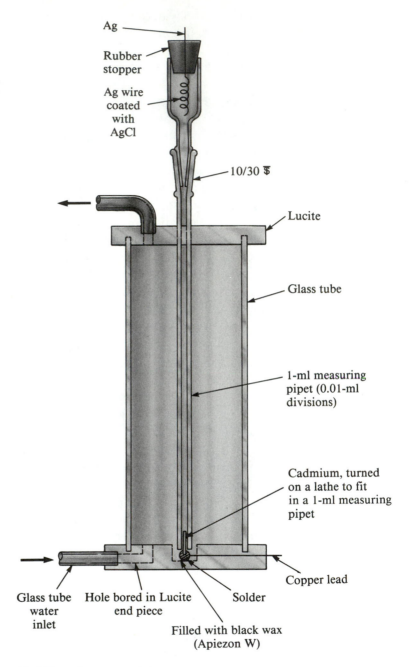

Figure 20–2 Transference number apparatus.

Apparatus

In the transference number apparatus (Fig. 20–2) the central tube is an ordinary 1-ml measuring pipet with 0.01-ml divisions. After the ends are trimmed, a 10/30 ₷ female joint is sealed to the top end to receive an Ag, AgCl electrode shown to the left. The Ag, AgCl electrode consists of a silver wire coated electrolytically with silver chloride after being wound into a spiral and inserted into a rubber stopper. Cadmium, usually supplied as short rods, can be turned down on a lathe until a 4- to 5-mm stub slip fits into the bore of the measuring pipet. A copper wire lead is soldered to the cadmium. The tube is surrounded by a glass tube 60 to 70 mm in diameter in order to thermostat the apparatus. The round end plates are constructed of 12-mm-thick Lucite plastic bored to receive the pipet and water outlet at one end, and the cadmium anode copper connecting lead and water inlet at the other.

The cathode consists of a silver wire coated with silver chloride. If the wire is not well coated, place it in a little dilute HCl and pass a small current between it and a platinum wire immersed in the HCl. Use two or three 1.5-volt dry cells. Since the silver wire is the anode for the coating process, it is attached to the plus terminal of a dry cell. The electrode reactions are

$$\text{Anode:} \qquad Ag + Cl^- = AgCl + e^-$$

$$\text{Cathode:} \qquad 2e^- + 2H_2O = H_2 + 2OH^-$$

Also required are a power supply with an output adjustable up to 400 volts DC and 5 mA; 0.1 N HCl; methyl violet indicator; an electric timer; a milliammeter, or for higher precision, a coulometer.

SAFETY CONSIDERATIONS

Since the constant current power supply may deliver well above 100 volts, care should be taken to use well-insulated leads and not to touch any bare wire. The area around the power supply should be kept dry.

EXPERIMENTAL PROCEDURE

Add a few drops of methyl violet to 10 ml 0.1 N HCl. Rinse and then fill the transference tube with this solution. Be sure that no bubbles are present in the tube. Connect the thermostat jacket tubes to a circulating pump. It is necessary to thermostat the transference tube to prevent the temperature rising due to electrical heating. Turn on the power supply and adjust the current to 3.00 mA. After a few minutes, a distinct violet color should develop at the bottom of the tube near the cadmium anode. Start the time when the boundary reaches the first convenient volume mark on the tube. Record the time every 0.1 ml. Repeat the procedure with 0.05 N HCl.

Results and Calculations

For both solutions, tabulate time and volume. Plot the volume swept out against the coulombs of charge passed through the solution. The plot should be linear. Run a least-squares analysis; the slope equals $t_+ I/FN$ according to Equation 20–15. Calculate the transference number from the least-squares slope. Compare with the literature values and analyze the errors.

REFERENCES

1. Bender, P., and D. A. Lewis, "Transference Number by the Moving Boundary Method," *J. Chem. Educ.* 24:454, 1947.

2. Longsworth, L. G., "An Experiment to Illustrate the Mechanism of the Conductance Process in Ionic Solutions," *J. Chem. Educ.* 11:420, 1934.

3. Harned, H. S., and B. B. Owen, *The Physical Chemistry of Electrolytic Solutions,* Reinhold, New York, 1950.

4. Hoyt, C. H., "Factors Affecting the Accuracy of Transference Number Determination," *J. Chem. Educ.* 14:472, 1937.

5. Lonergan, G. A., and D. C. Pepper, "Transport Numbers and Ionic Mobilities by the Moving Boundary Method," *J. Chem. Educ.* 42:82, 1965.

6. MacInnes, D. A., *The Principles of Electrochemistry,* Reinhold, New York, 1939.

7. Spiro, M., in A. Weissberger, ed., *Technique of Organic Chemistry,* vol. I, pt. IV, Wiley-Interscience, New York, 1960.

8. Yager, B. J., and P. Y. Smith, "A Semi-Automatic Moving Boundary Apparatus," *J. Chem. Educ.* 49:363, 1972.

Experiment 21

Conductance of Electrolytic Solutions

Solutions of electrolytes, like any conductor of electricity, obey Ohm's Law:

$$I = \frac{E}{R} \tag{21–1}$$

where I is the current in amperes, E is the voltage in volts, and R is the resistance in ohms. The resistance of a conductor is directly proportional to its length and inversely proportional to its cross-sectional area:

$$R = \rho \frac{\ell}{A} \tag{21–2}$$

where ρ, the specific resistivity, has units of ohm cm when ℓ is expressed in centimeters and A in square centimeters. Conductance is defined as the reciprocal of resistance:

$$L = \frac{1}{R} = \frac{1}{\rho}\frac{A}{\ell} = L_s\frac{A}{\ell} \tag{21-3}$$

where $L_s = 1/\rho$ is the specific conductance and has units of ohm^{-1} cm^{-1} or mho cm^{-1}. One can think of the specific conductance as the conductance of a cube of material, 1 cm on each edge. For electrolytic solutions, the "cube" of material is the solution subtended by two parallel plate electrodes, each 1 cm^2, as shown in Figure 21–1. Since it is difficult to construct such a cell with electrodes that are exactly parallel and exactly 1 cm^2, the cell is calibrated with a solution of exactly known specific conductance. From Equation 21–3,

$$L_s = \frac{\ell}{A}\frac{1}{R} = \frac{K}{R} \tag{21-4}$$

The cell constant K, with units cm^{-1}, is determined by measuring the resistance of a cell filled with a solution of known specific conductance, which is invariably aqueous KCl (Table 21–1).

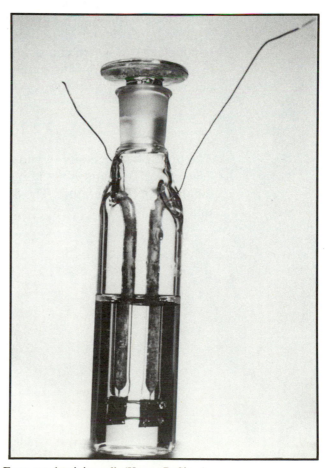

Figure 21–1 Freas conductivity cell. (Karen R. Sime)

TABLE 21–1

Specific Conductance of Aqueous KCl at 25°C

N (equiv/L)	L_s (ohm^{-1} cm^{-1})
0.01000	0.001412
0.02000	0.002768

The resistance of the cell is measured with a Wheatstone bridge (Fig. 21–2). An alternating current, usually about 1 kHz, is used to prevent polarization of the electrodes. The cell resistance is R_{cell}, R_s is a standard known resistance—several values of which can be selected so that a wide range of resistances can be measured—and R_1–R_2 is a slide wire. The galvanometer is usually replaced by an amplifier and an electric eye or oscilloscope. When the bridge is balanced, no current flows through the galvanometer. Under these conditions

$$R_{cell} = \frac{R_1}{R_2} R_s \tag{21-5}$$

The specific conductance increases as the concentration increases. A more fundamental unit of electrolytic conductance is the equivalent conductance Λ, which can be thought of as the value of L_s contributed by one equivalent of ions contained in 1000 cm^3 (1 L) of solvent. It is defined as

$$\Lambda = \frac{1000 L_s}{c} \tag{21-6}$$

The units of c are equiv/ℓ and the units of Λ are cm^2 equiv^{-1} ohm^{-1}. This is a cgs unit and is the unit of specific conductance most frequently tabulated in the literature. The SI unit is obtained by multiplying the cgs unit by 10^{-4}.

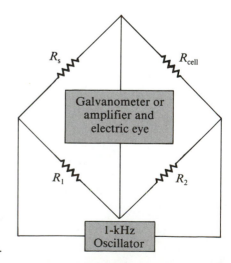

Figure 21–2 Wheatstone bridge.

Strong Electrolytes

While the specific conductance of a solution increases with concentration, the equivalent conductance decreases as the concentration increases. Early investigators of electrolytic conductivity noticed that the decrease in equivalent conductance is proportional to the square root of the concentration:

$$\Lambda = \Lambda_0 - b\sqrt{c} \qquad (21\text{-}7)$$

Below concentrations of about 0.1 M, a plot of Λ against $\sqrt{c}$ gives a straight line, the intercept of which equals Λ_0, the equivalent conductance at infinite dilution. It was also noticed by early investigators that Λ_0 values are additive. For example,

$$\Lambda_0(NaCl) = \Lambda_0(KCl) + \Lambda_0(NaNO_3) - \Lambda_0(KNO_3) \qquad (21\text{-}8)$$

Some representative values of Λ_0 are shown in Table 21-2.

This suggests that the equivalent conductance of a strong electrolyte is made up of independent contributions from the anions and cations of which it is composed. In other words, Equation 21-8 may be written:

$$\Lambda_0(NaCl) = \lambda_0(K^+) + \lambda_0(Cl^-) + \lambda_0(Na^+) \qquad (21\text{-}9)$$
$$+ \lambda_0(NO_3^-) - \lambda_0(K^+) - \lambda_0(NO_3^-)$$

or

$$\Lambda_0(NaCl) = \lambda_0(Na^+) + \lambda_0(Cl^-) \qquad (21\text{-}10)$$

Since it is possible by independent measurements to determine the fraction of current carried by the anions and cations—that is, the transference number (see Experiment 20 and Table 21-3)—the equivalent ionic conductances of individual ions are easily calculated:

$$\lambda_0(Na^+) = t_+\Lambda_0(NaCl) \qquad (21\text{-}11)$$

$$\lambda_0(Cl^-) = t_-\Lambda_0(NaCl) \qquad (21\text{-}12)$$

TABLE 21-2

Equivalent Conductances (cm^2 $equiv^{-1}$ ohm^{-1}) of Some Strong Electrolytes

Electrolyte	Λ_0	Electrolyte	Λ_0
HCl	426.16	$CaCl_2$	135.84
LiCl	115.03	$Ca(NO_3)_2$	130.94
NaCl	126.45	$BaCl_2$	139.98
KCl	149.86	$NaNO_3$	121.55
KBr	151.9	KNO_3	144.96
NH_4Cl	149.7	NaO_2CCH_3	91.0

Source: Harned, H. S., and B. B. Owen, *The Physical Chemistry of Electrolytic Solutions*, Reinhold, 1950, p. 537.

TABLE 21–3 ▬

Cation Transference Numbers at 25°C

Electrolyte	t_+
HCl	0.8209
NaO_2CCH_3	0.5507
KO_2CCH_3	0.6427
KNO_3	0.5027
NH_4Cl	0.4909
KCl	0.4906
KI	0.4892
KBr	0.4849
NaCl	0.3963
LiCl	0.3364

Source: Harned, H. S., and B. B. Owen, *The Physical Chemistry of Electrolytic Solutions,* Reinhold, 1950, p. 538.

From a table of equivalent conductances of individual ions (Table 21–4), it is then possible to calculate the equivalent conductances of a large number of strong electrolytes.

The Onsager Equation. The complete theoretical treatment of the conductance of electrolytes is very complicated because conductance depends upon numerous factors. Because there is an electrical attraction between positive and negative ions, each ion is surrounded by an **ionic atmosphere** of the opposite charge. As an ion moves through the solution, its ionic atmosphere is distorted in a manner that depends on its velocity, a phenomenon known as a **relaxation effect.** In addition, because the ionic atmosphere has the opposite charge of the ion it surrounds, it tends to move in a direction opposite to that of the ion itself, dragging solvent molecules along with it. This is called the **electrophoretic effect.**

TABLE 21–4 ▬

Equivalent Conductances (cm^2 $equiv^{-1}$ ohm^{-1}) of Ions at 25°C

Cation	λ_0	Anion	λ_0
H^+	349.8	OH^-	197.6
Li^+	38.69	Cl^-	76.34
Na^+	50.11	Br^-	78.3
K^+	73.52	I^-	76.8
NH_4^+	73.4	$C_2O_4^{2-}$	74.2
Mg^{2+}	53.06	NO_3^-	71.44
Ca^{2+}	59.50	$CH_3CO_2^-$	40.9
Zn^{2+}	53	SO_4^{2-}	80

Source: Harned, H. S., and B. B. Owen, *The Physical Chemistry of Electrolytic Solutions,* Reinhold, 1950, p. 172.

All of these factors depend on the charge of the ions, their radii, the dielectric constant and viscosity of the solvent, and the temperature of the solution. Based on such a model, P. Debye and E. Hückel derived an equation, later improved by L. Onsager, for the equivalent conductance of an electrolytic solution. For an aqueous solution of a 1:1 electrolyte at 25°C, the equation takes the form

$$\Lambda = \Lambda_0 - (60.32 + 0.2289\Lambda_0)\sqrt{c} \qquad \text{(21–13)}$$

The Onsager equation is in agreement with the observation (Eq. 21–7) that the equivalent conductance depends linearly on the square root of concentration and also predicts that the slope of that linear dependence is a simple function of the equivalent conductance at infinite dilution.

Weak Electrolytes

Because a weak electrolyte such as acetic acid is only slightly dissociated in aqueous solution, its specific conductance is considerably lower than that of a strong electrolyte at the same concentration. The equivalent conductance of a weak electrolyte is also lower than that of a strong electrolyte at ordinary concentrations, say, 0.1 to 1.0 M, but as the concentration of a weak electrolyte decreases the equivalent conductance becomes infinite. Consequently, it is not possible to extrapolate the equivalent conductance of a weak electrolyte to infinite dilution to obtain Λ_0. For a weak electrolyte Λ_0 is obtained from individual ionic equivalent conductances as already illustrated in Equation 21–8. For example, for the weak acid HAc, acetic acid:

$$\Lambda_0(\text{HAc}) = \Lambda_0(\text{HCl}) + \Lambda_0(\text{NaAc}) - \Lambda_0(\text{NaCl}) \qquad \text{(21–14)}$$

The Equilibrium Constant. For a weak electrolyte such as acetic acid, the dissociation is written:

$$\text{HAc} = \text{H}^+ + \text{Ac}^- \qquad \text{(21–15)}$$

for which the equilibrium constant is written:

$$K = \frac{[\text{H}^+][\text{Ac}^-]}{[\text{HAc}]}\frac{\gamma_+\gamma_-}{\gamma_a} \qquad \text{(21–16)}$$

If α is the degree of dissociation at equilibrium, then the equilibrium concentrations of HAc, Ac$^-$, and H$^+$ are $c(1 - \alpha)$, $c\alpha$, and $c\alpha$, respectively. The activity coefficient of the neutral molecule HAc is very likely near unity. Consequently, the concentration equilibrium constant K_c can be written in terms of the degree of dissociation:

$$K_c = \frac{[\text{H}^+][\text{Ac}^-]}{[\text{HAc}]} = \frac{c\alpha^2}{1 - \alpha} \qquad \text{(21–17)}$$

If α is small and the concentration low, the degree of dissociation α at a given concentration can be calculated from the equivalent conductance at that concentration and the equivalent conductance at infinite dilution, determined as described above in Equation 21–14:

$$\alpha = \frac{\Lambda}{\Lambda_0} \qquad (21\text{–}18)$$

Thus, the concentration equilibrium constant K_c can be determined from conductance measurements and, in dilute solutions, the activity coefficients can be calculated.

The Activity Coefficients. Since the HAc molecule is not charged, it may be assumed that γ_a equals 1. In a very dilute solution, the Debye-Hückel equation permits calculation of the ion activity coefficients:

$$\log \gamma_{\pm} = -0.5091 Z_+ Z_- \sqrt{\mu} \qquad (21\text{–}19)$$

The ionic strength μ is defined by

$$\mu = \frac{1}{2} \sum c_i Z_i^2 \qquad (21\text{–}20)$$

where c_i is the molar concentration of each ionic species and Z_i is the corresponding ionic charge. For the acetic acid dissociation,

$$\mu = \frac{1}{2} \left[c\alpha(1)^2 + c\alpha(1)^2 \right] = c\alpha \qquad (21\text{–}21)$$

The mean ionic activity coefficient is defined as follows:

$$\gamma_{\pm} = (\gamma_+ \gamma_-)^{1/2} \qquad (21\text{–}22)$$

Equation 21–16 may be rewritten as

$$\log K_c = \log K - \log \gamma_{\pm}^2 \qquad (21\text{–}23)$$

Combining Equations 21–19 and 21–23 with the mean ionic activity and the ionic strength gives

$$\log K_c = \log K + 1.018 \sqrt{c\alpha} \qquad (21\text{–}24)$$

Thus, a plot of $\log K_c$ against $\sqrt{c\alpha}$ is a straight line with a slope equal to 1.018 and an intercept equal to $\log K$, where K is the thermodynamic equilibrium constant for Equation 21–15 (Barrow, 1979, p. 598).

Apparatus

The experiment requires a self-contained alternating current Wheatstone bridge and detector;[1] a Freas conductivity cell; freshly boiled and cooled distilled water; 0.1 N HCl and 0.1 N acetic acid (standardized); 0.0200 N or 0.0100 N potassium chloride (see Table 21–1); a 25-ml pipet; and five 50-ml volumetric flasks.

[1] Yellow Springs Instrument Company, Box 279, Yellow Springs, Ohio 45387.

SAFETY CONSIDERATIONS

If you make up your dilute solutions from glacial acetic acid, work in the hood. Glacial acetic acid smells bad.

EXPERIMENTAL PROCEDURE

Standardize the acetic and hydrochloric acid solutions, if this has not already been done. The conductivity cell is stored with the electrodes under distilled water or dilute acid. Rinse the cell with several portions of distilled water. Do this over a large enameled photographic tray to catch any spilled mercury contained in the internal wells. Rinse with and then fill the cell to about 2 cm above the top of the electrodes with 0.0100 or 0.0200 N KCl; measure the cell resistance and calculate the cell constant. Let the cell and contents come to thermal equilibrium for about 10 minutes before making any measurements.

Hydrochloric Acid. Prepare solutions of concentrations equal to N/10, N/20, N/40, N/80, and N/160 by taking 25-ml aliquots and diluting to 50 ml in volumetric flasks. Calculate the exact concentration from the original normality. Clamp in the 25°C thermostat and let the solutions come to thermal equilibrium before measuring the conductances.

Acetic Acid. Prepare solutions of acetic acid as with the hydrochloric acid solutions: N/10, N/20, N/40, N/80, N/160, N/320, and N/640. Measure the conductances after the solutions have come to thermal equilibrium at 25°C.

Results and Calculations

Hydrochloric Acid. Calculate and tabulate concentrations, specific conductances, and equivalent conductances. Plot equivalent conductances against c and determine Λ_0 and the slope of the plot. Calculate the slope of the Onsager equation with your experimentally determined Λ_0 and compare with your measured slope. Analyze the propagated errors and compare your result with the literature value in Table 21–2.

Acetic Acid. Calculate and tabulate concentrations, specific conductances, and equivalent conductances. Plot equivalent conductances against $\sqrt{c}$. Calculate α's with your measured and equivalent ionic conductances from the literature (Table 21–4). Calculate K_c at each concentration. Note and discuss any trends. Plot log K_c versus $\sqrt{c\alpha}$ and compare your plot with Figure 7-6-1 in Harned and Owen (1950). Discuss whether or not the errors in your measurements warrant such a plot. Critically compare your K with the value in the literature.

REFERENCES

1. Barrow, G. M., *Physical Chemistry,* 4th ed., McGraw-Hill, New York, 1979.

2. Harned, H. S., and B. B. Owen, *The Physical Chemistry of Electrolytic Solutions,* Reinhold, New York, 1950.

3. MacInnes, D. A., *The Principles of Electrochemistry,* Reinhold, New York, 1939.

4. Spiro, M., in A. Weissberger, ed., *Technique of Organic Chemistry,* vol. I, pt. IV, Wiley-Interscience, New York, 1960.

5. Robinson, R. A., and R. H. Stokes, *Electrolytic Solutions,* 2nd ed., rev., Academic Press, New York, 1965.

6. Stock, J. T., "The Variation of Equivalent Conductance with Concentration," *J. Chem. Educ.* 31:410, 1954.

Experiment 22

Thermodynamic Data from Electromotive Force Measurements

For any reversible change of state, chemical or physical, occurring at constant temperature and pressure, the maximum work that can be obtained is equal to the decrease in the free energy:

$$w = -\Delta G \tag{22-1}$$

Electrical work equals the product of charge and voltage. When the balanced cell reaction involves the transfer of n moles of electrons, the total charge transferred equals nF coulombs, and the work (J) that can be obtained is given by

$$w = nFE \tag{22-2}$$

where the cell electromotive force is E (V). Consequently, the change in free energy in a reversible cell operating at constant pressure and temperature can be determined by simply measuring the cell potential

$$\Delta G = -nFE \tag{22-3}$$

and if all the products and reactants are present at unit activity, the standard free energy can be calculated from the standard cell potential:

$$\Delta G^0 = -nFE^0 \tag{22-4}$$

Since the free energy depends upon temperature according to

$$\left(\frac{\partial G}{\partial T}\right)_P = -\Delta S \tag{22-5}$$

measurement of the change in EMF with temperature permits calculation of the entropy change by combining Equations 22–3 or 22–4 and 22–5; for example,

$$\Delta S = nF\left(\frac{\partial E}{\partial T}\right)_P \tag{22-6}$$

Finally, the heat of a reaction can be determined from the electromotive force (EMF) measurements by combining Equations 22–3 and 22–6 with

$$\Delta G = \Delta H - T\,\Delta S \tag{22-7}$$

to give

$$\Delta H = -nFE + nFT\left(\frac{\partial E}{\partial T}\right)_P \tag{22-8}$$

If the cell measured is a standard cell, then superscript zeros should be written with ΔG, ΔH, ΔS, and E as indicated in Equation 22–4.

Since it is rather simple to measure cell EMFs with a great deal of accuracy, this is an attractive method of obtaining thermodynamic data when suitable cells can be constructed. The cell chosen for study (Gerke, 1922) in this experiment is

$$\text{Ag,AgCl,KCl(0.1 M),Hg}_2\text{Cl}_2\text{,Hg,Pt} \tag{22-9}$$

for which the overall cell reaction is

$$2\text{Ag}(s) + \text{Hg}_2\text{Cl}_2(s) = 2\text{AgCl}(s) + 2\text{Hg}(\ell) \tag{22-10}$$

Since all the products and reactants are present at unit activity, the measured cell potential equals E^0 and should be independent of the KCl concentration. Since all the products and reactants are pure substances, their absolute entropies can be determined from the Third Law from measurements of the heat capacities from 0 to 298.15 K and a Third Law ΔS^0 can be calculated for the reaction in Equation 22–10 above. From measurements of $(\partial E/\partial T)$, a Second Law ΔS^0 can be calculated with Equation 22–6, and the Second and Third Law entropies compared.

Apparatus

The experiment requires mercury; mercurous chloride (Hg_2Cl_2); 0.1 M potassium chloride; a strip of sheet silver, $0.1 \times 5 \times 50$ mm; an H-tube with fritted glass between compartments; a platinum electrode consisting of 1 cm of 26-gauge platinum wire sealed in a 6-mm soft glass tube; thermostats at 25 and 35°C; ice; a 2-L beaker; electrical leads; a potentiometer (± 0.1 mV) or a digital multimeter (DMM) such as the Keithley[1] Model 197 DMM; a 1.5-volt dry cell; and 1 M HCl.

[1] Keithley Instruments, Inc., 28775 Aurora Road, Cleveland, Ohio 44139.

SAFETY CONSIDERATIONS

Mercury is toxic and care should be taken to prevent spilling it on the desktop. It should be handled over a large enameled photographic developing tray.

EXPERIMENTAL PROCEDURE

The H-cell (Fig. 22–1) should be scrupulously clean. It is especially important that no traces of bromides are present. Prepare the cathode compartment first by adding about 5 ml mercury. Fill both compartments with 0.1 M KCl to the same level. Add about 100 mg Hg_2Cl_2 to the cathode compartment, and place the platinum electrode so that the platinum wire at the bottom is entirely submerged in the mercury. Clamp the H-cell into a 25°C thermostat, and while it is coming to thermal equilibrium prepare the Ag, AgCl electrode.

Remove any old AgCl from the silver electrode by scrubbing it in concentrated ammonia. Dry the silver strip and polish it clean with fine emery paper. Clean it with 1 M HCl; then connect it to the positive electrode of a 1.5-volt dry cell with an electrical lead and alligator clip and immerse it in a small clean beaker of 1 M HCl. Connect the negative electrode of the dry cell to a platinum electrode immersed in the HCl solution. Electrolyze the solution for about 5 minutes. Hydrogen is evolved at the platinum electrode and AgCl deposits as a dark brown to violet coating on the silver. Write electrode reactions for the cathode and anode reactions. Rinse the electrodes thoroughly with distilled water and store either in

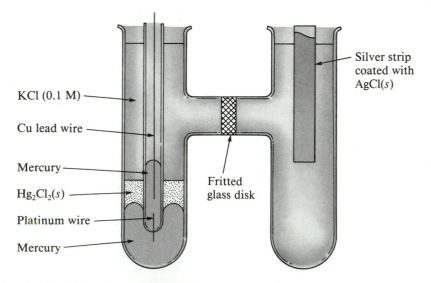

Figure 22–1 The cell: Ag,AgCl,KCl(0.1 M),Hg_2Cl_2,Hg,Pt.

distilled water or 0.1 M KCl. It is important that the electrode never be allowed to dry out.

Place the Ag, AgCl electrode in the anode compartment of the H-cell, connect it to the potentiometer, and allow the system to come to equilibrium. While the system is coming to equilibrium, determine whether the leads are correctly connected to the potentiometer by attempting to balance the potentiometer. If it is not possible to balance it, reverse the leads and try again.

Measure the cell EMF at 0°C in an ice water bath and at 25 and 35°C in a thermostat. Make several measurements at each temperature to be sure that the system is at equilibrium. If time permits measure at 0, 25, 35, and back down again at 25 and 0°C.

Results and Calculations

Tabulate the EMFs and temperatures. Since the EMF is a linear function of temperature over this temperature range, run a least-squares analysis of the data. With the slope of the line, calculate the entropy with Equation 22–6. Calculate E^0 at 25°C if your thermostat is not at exactly 25°C, and calculate ΔG^0 at 25°C with Equation 22–4. Calculate ΔH^0 with Equation 22–7.

Compare your measured E^0 at 25°C with data from tables of standard reduction potentials. Analyze the errors and discuss the difference. Calculate ΔS^0 with absolute entropies from the literature and compare this Third Law entropy with your Second Law ΔS^0 determined with Equation 22–6. Analyze the errors and discuss the difference in values. Finally, critically compare your value for ΔH^0 with a value calculated with heats of formation obtained from the literature (Gerke, 1922).

AN ALTERNATIVE CELL

The cell described above is interesting because it allows the experimentally determined Second Law entropy change of a chemical reaction to be compared with the direct calorimetrically determined Third Law entropy change. The reaction, however, does not lend itself to a direct calorimetric measurement of the heat of reaction. A cell that does (Hill and Moss, 1965) is

$$Zn\,|\,ZnSO_4(0.00625\ M)\,|\,|\,Ag_2SO_4(0.00625\ M)\,|\,Ag$$

The cell reaction is

$$Zn(s) + 2Ag^+(0.00625\ M) = Ag(s) + Zn^{2+}(0.00625\ M)$$

The calorimetric determination of the enthalpy change for this reaction is described in Experiment 4. A cell suitable for the potentiometric measurement of the heat of reaction is shown in Figure 22–2.

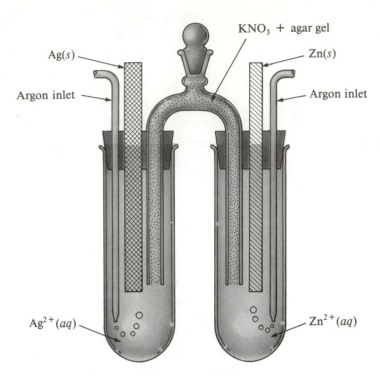

Figure 22–2 The cell: $Zn|ZnSO_4(M$ molar$)\|Ag_2(SO_4)(M$ molar$)|Ag$.

Apparatus

Construct the cell with two 2.5-by-10-cm test tubes connected by a U-shaped salt bridge (Fig. 22–2) inserted through two rubber stoppers through which three holes are drilled. One hole is for the salt bridge tube; the second hole allows for bubbling argon through the electrolytes; and the third hole admits the metal electrode and also allows the argon to escape.

Prepare the salt bridge by adding 1 g agar to 40 ml cold water. After heating the mixture to boiling while stirring, cool the solution somewhat, add 4 g potassium nitrate (KNO_3), and dissolve. Pour the warm solution into the U-tube and allow it to come to room temperature.

Preparing the Zinc Electrode. Prepare the zinc electrode from a strip of zinc about 4 to 5 mm wide, 100 mm long, and about 0.5 to 1 mm thick. Clean it briefly in 30% sulfuric acid, rinse it in distilled water, and immediately place it in a saturated solution of mercurous nitrate $Hg_2(NO_3)_2$ for 2 or 3 minutes. After rinsing it with distilled water and 0.00625 M zinc sulfate, place the electrode in the cell containing 0.00625 M zinc sulfate. The argon should have been flowing through the cell while you were preparing the electrode.

Preparing the Silver Electrode. The silver electrode can be made of sheet silver about the same dimensions as the zinc electrode, or it can be made of heavy-gauge silver wire. Anodize the length of the electrode dipping into the electrolyte in a 10% solution of potassium cyanide for about 10 minutes to obtain a clean surface. **Potassium cyanide is a dangerous poison!** After anodizing, electroplate fresh silver on the silver electrode by electrolyzing a solution that is 1% silver nitrate and 90% methanol. Use a piece of platinum wire as the second electrode. Let the current from two fresh 1.5-volt dry cells pass through the solution for about an hour. After rinsing it in distilled water and 0.00625 M silver sulfate, place the freshly plated silver electrode in the silver sulfate electrolyte through which argon is bubbling. The electrode should be freshly anodized and plated each day that it is used.

SAFETY CONSIDERATIONS

Potassium cyanide is exceedingly poisonous. Gloves should be worn at all times. When handling KCN, work over a large photographic tray so no material spills on the desktop. Do not throw KCN solutions into the sink; use the special disposal containers provided.

EXPERIMENTAL PROCEDURE

Measure the cell potential at 0°C in an ice water slush bath. Be sure that the ice extends to the bottom of the ice water mixture. Water is most dense at 4°C, so water of this temperature tends to settle to the bottom of the ice water mixture; this can result in large errors unless care is taken that ice and water are in equilibrium throughout the mixture.

In addition, measure at about 40°C and at one or two temperatures between 0 and 40°C. Take several readings at each temperature and be sure that the cell is at thermal equilibrium with the thermostat.

Results and Calculations

Tabulate the EMFs and temperatures. Since the EMF is a linear function of temperature over this temperature range, run a least-squares analysis of the data. With the slope of the line, calculate the entropy with Equation 22–6. Calculate E^0 at 25°C if your thermostat is not at exactly 25°C, and calculate ΔG^0 at 25°C with Equation 22–4. Calculate ΔH^0 with Equation 22–7.

Compare your measured E^0 at 25°C with data from tables of standard reduction potentials. Analyze the errors and discuss the difference. Calculate ΔS^0 with absolute entropies from the literature and compare this Third Law entropy with your Second Law ΔS^0 determined with Equation 22–6. Analyze the errors and discuss the difference in values. Finally, critically compare your value for ΔH^0 with a value calculated with heats of formation obtained from the literature.

If time permits, measure the heat of reaction calorimetrically (Experiment 4) and compare the calorimetric, potentiometric, and literature values. Calculate the propagated errors and discuss the comparisons in terms of the analysis of experimental errors.

REFERENCES

1. Gerke, R. H., "Temperature Coefficient of Electromotive Force of Galvanic Cells and Entropy of Reaction," *J. Am. Chem. Soc.* 44:1684, 1922.

2. Hill, D. L., and S. J. Moss, "Heat of Reaction in Aqueous Solution by Potentiometry and Calorimetry," *J. Chem. Educ.* 42:541, 1965.

3. Ives, D. J. G., and J. J. Janz, *Reference Electrodes,* Academic Press, New York, 1961.

4. LaMer, V. K., and W. G. Parks, "The Temperature Coefficients of the Electromotive Force of the Cell Cd (metal), $CdSO_4$, Cd (satd. amalgam)," *J. Am. Chem. Soc.* 56:90, 1934.

5. Latimer, W. M., *The Oxidation States of the Elements and Their Potentials in Aqueous Solutions,* Prentice-Hall, Englewood Cliffs, New Jersey, 1952.

6. MacInnes, D. A., *The Principles of Electrochemistry,* Dover, New York, 1961.

Experiment 23

Activity Coefficients from Electromotive Force Measurements

Since the potential E of a voltaic cell depends directly upon the activities of the participating reactants and products, measurement of the cell potential provides a straightforward method for evaluating activities and activity coefficients. A classic example of a suitable cell is the reversible concentration cell

$$Pt,H_2(p \text{ bar})|HCl(m \text{ molal});AgCl(s)|Ag \qquad (23\text{--}1)$$

At the anode, the half-cell reaction is

$$\tfrac{1}{2}H_2(p \text{ bar}) = H^+(m \text{ molal}) + e^- \qquad (23\text{--}2)$$

while at the cathode

$$AgCl(s) + e^- = Ag(s) + Cl^-(m \text{ molal}) \qquad (23\text{--}3)$$

so that the overall cell reaction is

$$\tfrac{1}{2}H_2(p \text{ bar}) + AgCl(s) \qquad (23\text{--}4)$$

$$= H^+(m \text{ molal}) + Cl^-(m \text{ molal}) + Ag(s)$$

Such a cell can physically take on a variety of forms, but essentially all of them consist of a hydrogen electrode and an Ag, AgCl electrode dipping into the same solution of HCl of molality m, as shown in Figure 23–1. The hydrogen electrode is reversible with respect to the H^+ ions, and the Ag, AgCl electrode is reversible with respect to the Cl^- ion. The electromotive force (EMF) of the cell is given by

$$E = E^0 - \frac{RT}{nF} \ln Q \qquad (23\text{–}5)$$

where $R = 8.31451$ J K^{-1} mol^{-1}, $n = 1$ equiv/mol, $F = 96485.31$ C/equiv

$$Q = \frac{a_{H^+} a_{Cl^-} a_{Ag}}{a_{AgCl} a_{H_2}^{1/2}} \qquad (23\text{–}6)$$

The activities of solid silver and silver chloride are unity. The activity of the hydrogen gas is its actual measured pressure in atmospheres, corrected for the vapor pressure and hydrostatic head of the water through which it bubbles:

$$P_{H_2} = P_{\text{barometric}} - P_{H_2O} + P_{\text{hydrostatic}} \qquad (23\text{–}7)$$

In terms of the molality m and the activity coefficients, the activity quotient then becomes

$$Q = \frac{\gamma_+ m_+ \gamma_- m_-}{P_{H_2}^{1/2}} = \frac{\gamma_\pm^2 m^2}{P_{H_2}^{1/2}} \qquad (23\text{–}8)$$

and the EMF of the cell is then given by

$$E = E^0 - \frac{RT}{F} \ln \frac{m^2}{P_{H_2}^{1/2}} - \frac{RT}{F} \ln \gamma_\pm^2 \qquad (23\text{–}9)$$

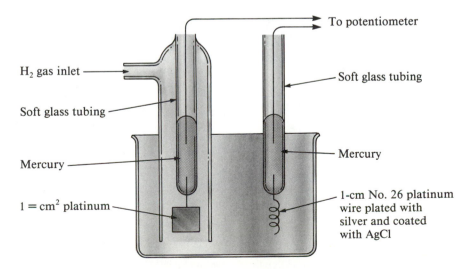

Figure 23–1 The cell: Pt,H_2(p atm)$|$HCl(m molal); AgCl(s)$|$Ag.

At 25.00°C, Equation 23–9 becomes

$$E + 0.05916 \log \frac{m^2}{P_{H_2}^{1/2}} = E^0 - 0.1183 \log \gamma_{\pm} \qquad (23\text{–}10)$$

To obtain E^0, it is necessary to measure E at progressively more dilute solutions and extrapolate to infinite dilution. It is best to plot E against $\sqrt{m}$ instead of against m since, according to the Debye-Hückel Theory,

$$\log \gamma_{\pm} = 0.5091 Z_+ Z_- \sqrt{\mu} = -0.5091 \sqrt{m} \qquad (23\text{–}11)$$

For a 1:1 electrolyte like HCl, the ionic strength μ is just equal to the molality m in a dilute solution. When Equations 23–10 and 23–11 are combined, the result is

$$E + 0.05916 \log \frac{m^2}{P_{H_2}^{1/2}} = E^0 + 0.06023 \sqrt{m} \qquad (23\text{–}12)$$

A plot of the left side of Equation 23–12 against $\sqrt{m}$ gives a straight line, the intercept of which equals E^0 and the slope of which equals 0.0623.

Once E^0 has been determined by extrapolation, Equation 23–10 is used to calculate activity coefficients for solutions of any molality m.

Apparatus

The apparatus consists of a potentiometer (± 0.1 mV) or a digital multimeter (DMM) such as the Keithley[1] Model 197 DMM; 0.1 M HCl (standardized); a hydrogen electrode; an Ag, AgCl electrode; a 200-ml flask; a 100-ml pipet; a tank of hydrogen with a pressure-reducing valve; clean rubber tubing, free from talc; and miscellaneous beakers and flasks.

SAFETY CONSIDERATIONS

The cell should be used in a well-ventilated hood and kept away from any open flames. The electrodes should be stored under distilled water when not in use and not allowed to dry out.

EXPERIMENTAL PROCEDURE

By successive dilutions of 0.1 M HCl (standardized) with a 100-ml pipet and 200-ml volumetric flask, prepare solutions of M/10, M/20, M/40, M/80 and M/160. Rinse the electrodes and cell with the HCl solution to be used. Bubble hydrogen gas through the hydrogen electrode while the cell comes to thermal equilibrium in a 25°C thermostat—about 10 minutes. Make three or four readings every 5 minutes or so until no drift is observed.

[1] Keithley Instruments, Inc., 28775 Aurora Road, Cleveland, Ohio 44139.

Results and Calculations

Convert molarities to molalities:

$$m = \frac{M}{d - M \times \text{HCl}} \tag{23-13}$$

where m is the molality (mol/kg), M is the molarity (mol/L), d is the density (kg/L), and HCl = 0.03646 kg/mol. For $M = 0$ to 0.1, $d = 1.0000$ to 1.0020, and d may be linearly interpolated. For experimental convenience we have used molarities in preparing solutions of various concentrations; however, most of the work in the literature is reported in terms of molalities. Calculate the H_2 pressure in bar, using SI units, in Equation 23–7. Note that 1 bar = 10^5 Pa and 1 atm = 101,325 Pa. By international agreement, the standard-state pressure = 1 bar has replaced 1 atm.

Tabulate m, $\sqrt{m}$, E, and the left side of Equation 23–12. Plot the left side of Equation 23–12 against $\sqrt{m}$, and determine E^0. Compare your measured E^0 with the value from the literature, which is just E^0 for the Ag, AgCl half-cell. Compare your plot with the data of Harned and Ehlers (1933). With Equation 23–10, calculate the activity coefficient at each experimental concentration. Compare your experimentally determined activity coefficients with those you calculate according to the extended Debye-Hückel Theory for a 1:1 electrolyte:

$$\log \gamma_\pm = \frac{-0.5091\sqrt{m}}{1 + aB\sqrt{m}} \tag{23-14}$$

where B equals 0.3286 at 25°C and a equals 4.0 (Harned and Owen, 1950, chap. 11).

Notes on the Construction of Electrodes

The Hydrogen Electrode. Cut a piece of thin platinum sheet to about 1 cm × 1 cm. Weld a 2-cm piece of 26-gauge platinum wire to the platinum sheet by heating to red heat and striking with a small hammer. Alternatively, look in the yellow pages of the phone directory under "Jewelers, Manufacturing" and you can have the welding operation done by an expert for a nominal charge.

Platinum seals very nicely to soft glass, but not to Pyrex glass. Obtain a Pyrex-to-soft glass graded seal and determine the soft glass end. (This end gives a yellow color to a torch flame more quickly than does the Pyrex end.) Heat the soft glass end until it begins to soften and close. When the opening is only slightly larger than the platinum wire diameter, cool the glass a bit, insert the platinum, and heat again until the glass shrinks down onto the wire and wets it. Continue heating with gradually decreasing intensity to anneal the seal. The remainder of the electrode is constructed of Pyrex glass. A few small holes are placed in the outer tube to allow the hydrogen to escape. These should be placed so that the water level rises about halfway up the platinum electrode.

Before using the hydrogen electrode, the platinum must be "platinized," that is, covered with a fine, even layer of platinum black. Prepare a solution containing 3 g platinic chloride per 100 ml water. Clean the metal to be platinized and a second electrode consisting of a piece of platinum wire by dipping them in warm aqua regia for a few minutes and then placing them in the solution. Connect the two electrodes to two 1.5-volt dry cells and pass a current through the solution for about 5 minutes. The result should be an even, nearly black coating of finely divided platinum. The electrode must never be allowed to dry out and should be stored under distilled water. The plating solution should be stored in a glass-stoppered bottle for further use.

The Silver, Silver Chloride Electrode. Form about 10 mm of 26-gauge platinum wire into a spiral (Fig. 23–1) and seal it into the soft glass end of a graded seal, as described above. Plate silver on the platinum by making it the cathode and a piece of pure silver the anode. The electrolyte consists of 10 g silver cyanide, 10 g potassium cyanide, 3 g potassium hydroxide, 15 g potassium carbonate, and 250 ml water. Connect the two electrodes and three 1.5-volt dry cells in series with a 1000-ohm rheostat and a milliammeter. Plate 3 or 4 hours at a current density of about 5 mA/cm². Wash the electrodes thoroughly. Save the plating solution for further use and label it POISON—CYANIDE!

Coat the silver with silver chloride by making it the anode and a piece of platinum wire the cathode in a 1 M HCl solution. As above, pass a current through the solution for about 5 minutes at a current density of 5 mA/cm². A visible layer of AgCl will form on the silver. The electrode should not be allowed to dry out. Store it in distilled water.

REFERENCES

1. Harned, H. S., and R. W. Ehlers, "The Thermodynamics of Aqueous Hydrochloric Acid Solutions from Electromotive Force Measurements," *J. Am. Chem. Soc.* 55:2179, 1933.

2. Harned, H. S., and B. B. Owen, *The Physical Chemistry of Electrolytic Solutions,* Reinhold, New York, 1950.

3. Ives, D. J. G., and G. J. Janz, *Reference Electrodes,* Academic Press, New York, 1961.

4. MacInnes, D. A., *The Principles of Electrochemistry,* Reinhold, New York, 1939.

Kinetics

Experiment 24

Stopped-Flow Kinetics of Fe(H$_2$O)$_5$(SCN)$^{2+}$ Formation

If a reaction is slow, it is possible to mix the reactants, periodically sample the mixture, and analyze the mixture for one of the reactants or products. If the reaction is fast, however, the time of mixing may be comparable to the half-life of the reaction, so that the determination of the initial time is uncertain. Even more important, the concentration and time are uncertain if the time required for the analysis of a sample is comparable to the half-life of the sample. The stopped-flow method overcomes the first of these difficulties by mixing the reactants together very rapidly.

The second difficulty can be overcome by using an analytical method that responds instantaneously to changes in concentration. The methods most commonly used with stopped flow are the measurement of absorbance and the measurement of conductance. The output of both these analytical methods is a changing electrical signal that is the electrical analog of the changing concentration. The changing electrical signal can then be displayed on a pen-and-ink recorder if the half-life is more than 2 or 3 seconds and the chart recorder is run at a comparably fast rate, say, 4 or 5 cm/sec.

An even better method of displaying the data is the use of a microcomputer such as an Apple IIe or IBM-PC. Equipped with an analog-to-digital converter, the computer can digitize the electrical analog signal from a spectrophotometer and store the digitized data for processing. With appropriate software, the experimental data can be displayed in numerical tabular or graphical form along with the resulting specific rate constants.

Stopped-Flow Technique

In the stopped-flow technique (Figs. 24–1 and 24–2), the two reactants are contained in two syringes about 2 ml in volume. The pistons of the two syringes are mechanically fixed together so that they may be driven simultaneously. The solutions flow together almost immediately into a mixing chamber, where they begin to react, and from the mixing chamber through a spectrophotometer cell, and finally into a stopping syringe, also about 2 ml in volume. The stopping syringe is backward so that the reacting mixture drives its piston out against a stopping block. As the stopping piston reaches the stopping block, it triggers an electrical switch that activates the data acquisition system. The data acquisition system can be a rapidly running chart recorder, or for fast reactions, an oscilloscope or analog-to-digital converter and a computer.

In addition, the stopped-flow apparatus has some kind of reservoirs containing the two reactants and suitable stopcocks and tubing so that the driving syringes can be conveniently refilled. The reservoirs are often constructed of larger syringes about ten times the volume of the driving syringes, or about 20 ml.

At the instant that the stopping syringe strikes the stopping block, the flow stops. However, the solution continues to react and the progress of the reaction can be followed by monitoring the change in absorbance of one of the reactants or products contained in the cell. The time required

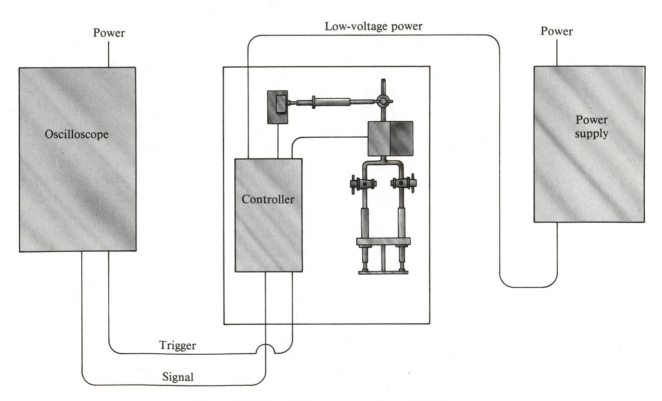

Figure 24–1 Electrical connections for the SF-1B.

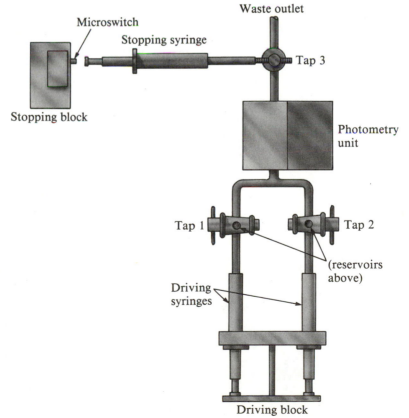

Figure 24–2 The SF-1B stopped-flow apparatus. (Karen R. Sime)

for the reacting mixture to travel from the mixing chamber to the spectrophotometer cell is called the **dead time**. Because the mixing time plus the dead time is on the order of a few milliseconds, it is possible to measure reactions with half-lives in the millisecond range with a stopped-flow apparatus.

Kinetics of Ferric Thiocyanate Formation

The reaction between iron(III) and the thiocyanate ion (SCN^-) forms an intensely red-colored complex and its rate may be followed spectrophotometrically with the stopped-flow apparatus. The stoichiometry is given by

$$Fe(H_2O)_6^{3+} + SCN^- = Fe(H_2O)_5SCN^{2+} + H_2O \qquad (24\text{--}1)$$

The Reaction Mechanism. Although the reaction mechanism is quite complex, it is well understood. The following sequence of elementary steps is consistent with the dependence of the observed first-order rate constant upon hydrogen ion concentration:

$$Fe(H_2O)_6^{3+} \underset{}{\overset{K_{h1}}{\rightleftharpoons}} Fe(H_2O)_5(OH)^{2+} + H^+ \qquad (24\text{--}2)$$

$$Fe(H_2O)_6^{3+} + SCN^- \underset{k_{-1}}{\overset{k_1}{\rightleftharpoons}} Fe(H_2O)_5(SCN)^{2+} + H_2O \qquad (24\text{--}3)$$

$$Fe(H_2O)_5(OH)^{2+} + SCN^- \underset{k_{-2}}{\overset{k_2}{\rightleftharpoons}} Fe(H_2O)_4(OH)(SCN)^+ + H_2O \qquad (24\text{--}4)$$

$$Fe(H_2O)_5(SCN)^{2+} \underset{}{\overset{K_{h2}}{\rightleftharpoons}} Fe(H_2O)_4(OH)(SCN)^+ + H^+ \qquad (24\text{--}5)$$

where K_{h1} and K_{h2} are the equilibrium constants for the hydrolysis reactions and k_i are the rate constants for the steps shown. It is assumed that the equilibria defined by K_{h1} and K_{h2} are reached rapidly compared to the rate of the reaction of SCN^- with $Fe(H_2O)_6^{3+}$ and $Fe(H_2O)_5(OH)^{2+}$.

According to the above mechanism, the rate of disappearance of SCN^- is given by

$$\frac{-d[SCN^-]}{dt} = k_1[Fe(H_2O)_6^{3+}][SCN^-] \qquad (24\text{--}6)$$

$$- k_{-1}[Fe(H_2O)_5(SCN)^{2+}] + k_2[Fe(H_2O)_5(OH)^{2+}][SCN^-]$$

$$- k_{-2}[Fe(H_2O)_4(OH)(SCN)^+]$$

According to Equation 24–2,

$$[Fe(H_2O)_5(OH)^{2+}] = \frac{K_{h1}[Fe(H_2O)_6^{3+}]}{[H^+]} \qquad (24\text{--}7)$$

Similarly, according to Equation 24–5,

$$[Fe(H_2O)_4(OH)(SCN)^+] = \frac{K_{h2}[Fe(H_2O)_5(SCN)^{2+}]}{[H^+]} \qquad (24\text{--}8)$$

Substitution of Equation 24–7 and Equation 24–8 into Equation 24–6 yields

$$\frac{-d[\text{SCN}^-]}{dt} = k_1[\text{Fe}(\text{H}_2\text{O})_6^{3+}][\text{SCN}^-] \tag{24–9}$$

$$- k_{-1}[\text{Fe}(\text{H}_2\text{O})_5(\text{SCN})^{2+}] + \frac{k_2 K_{h1}[\text{Fe}(\text{H}_2\text{O})_6^{3+}][\text{SCN}^-]}{[\text{H}^+]}$$

$$- \frac{k_{-2} K_{h2}[\text{Fe}(\text{H}_2\text{O})_5(\text{SCN})^{2+}]}{[\text{H}^+]}$$

which on rearrangement gives the differential form of the overall rate equation:

$$\frac{-d[\text{SCN}^-]}{dt} = (k_1 + k_2 K_{h1}/[\text{H}^+])[\text{Fe}(\text{H}_2\text{O})_6^{3+}][\text{SCN}^-] \tag{24–10}$$

$$- (k_{-1} + k_{-2} K_{h2}/[\text{H}^+])[\text{Fe}(\text{H}_2\text{O})_5(\text{SCN})^{2+}]$$

To simplify this expression, one can define forward and reverse rate constants, k_f and k_r, according to

$$k_f = k_1 + k_2 K_{h1}/[\text{H}^+] \tag{24–11}$$

$$k_r = k_{-1} + k_{-2} K_{h2}/[\text{H}^+] \tag{24–12}$$

Substitution of Equation 24–11 and Equation 24–12 into Equation 24–10 yields

$$\frac{-d[\text{SCN}^-]}{dt} = k_f[\text{Fe}(\text{H}_2\text{O})_6^{3+}][\text{SCN}^-] \tag{24–13}$$

$$- k_r[\text{Fe}(\text{H}_2\text{O})_5(\text{SCN})^{2+}]$$

If the initial conditions are such that $[\text{Fe}(\text{H}_2\text{O})_6^{3+}]$ is much greater than $[\text{SCN}^-]$—which is true in this experiment—then $[\text{Fe}(\text{H}_2\text{O})_6^{3+}]$ is essentially constant throughout the reaction and Equation 24–13 can be integrated (Deuel, 1988) to give

$$\ln \frac{[\text{Fe}(\text{H}_2\text{O})_5(\text{SCN})^{2+}]_\infty}{[\text{Fe}(\text{H}_2\text{O})_5(\text{SCN})^{2+}]_\infty - [\text{Fe}(\text{H}_2\text{O})_5(\text{SCN})^{2+}]_t} = k_{\text{obs}}t \tag{24–14}$$

and $k_{\text{obs}} = (k_f[\text{Fe}(\text{H}_2\text{O})_6^{3+}] + k_r)t$, which is the experimentally observed first-order rate constant for the reaction at a particular hydrogen ion concentration. In terms of the specific rate constants defined by Equations 24–3 and 24–4,

$$k_{\text{obs}} = (k_1[\text{Fe}(\text{H}_2\text{O})_6^{3+}] + k_{-1}) \tag{24–15}$$

$$+ (k_2 K_{h1}[\text{Fe}(\text{H}_2\text{O})_6^{3+}] + k_{-2} K_{h2})/[\text{H}^+]$$

Monitoring the Changing Concentration of $\text{Fe}(\text{H}_2\text{O})_5\text{SCN}^-$.

With a Spectrophotometer. For a fixed hydrogen ion concentration, the absorbance A of the reaction mixture is proportional to the concentration of iron thiocyanate ion, as this is the principal species that absorbs in the 450-nm range. Hence,

$$[Fe(H_2O)_5(SCN)^{2+}]_\infty \propto A_\infty - A_0 \qquad (24\text{--}16)$$

and

$$[Fe(H_2O)_5(SCN)^{2+}]_\infty - [Fe(H_2O)_5(SCN)^{2+}]_t \propto A_\infty - A_t \quad (24\text{--}17)$$

Substitution of Equations 24–16 and 24–17 into Equation 24–15 gives the overall rate equation in terms of absorbance:

$$\ln (A_\infty - A_t) = -k_{obs}t + \ln (A_\infty - A_0) \qquad (24\text{--}18)$$

With a normal spectrophotometer this would be the working equation, as spectrophotometers often generate absorbance units directly; that is, I and I_0 are measured in a spectrophotometer and an output signal from a spectrophotometer is the electrical analog of the absorbance A, defined as

$$A \equiv \log (I_0/I) \qquad (24\text{--}19)$$

The spectrophotometer makes this conversion from I and I_0 to A electronically, internally, and automatically.

With a Photomultiplier. In the present case, however, the output of the stopped-flow photomultiplier generates a voltage V that is proportional to the light intensity I, so that Equation 24–19 may be rewritten in terms of voltage. Thus,

$$I_0/I = V_0/V \qquad (24\text{--}20)$$

$$A = \log (V_0/V) \qquad (24\text{--}21)$$

Thus, a working equation for using the voltage output of a photomultiplier detector is given by Equation 24–22

$$\ln [\ln (V_t/V_\infty)] = -k_{obs}t + \ln [\ln (V_0/V_\infty)] \qquad (24\text{--}22)$$

The voltages in Equation 24–22 are directly measurable experimental parameters. Equation 24–22 is of the form $y = mx + b$, and hence a plot of $\ln [\ln (V_t/V_\infty)]$ versus t gives a straight line with slope $-k_{obs}$ and intercept $\ln [\ln (V_0/V_\infty)]$.

The change in voltage with time displayed on an oscilloscope screen is shown in Figure 24–3. Trace A at the top of the screen corresponds to 0 volts (no light falling on the photomultiplier). Trace B corresponds to the voltage V_L generated when the cell is filled with pure water and light is passing through the cell and falling on the photomultiplier. Trace C is generated as a colored product is produced after the flow is stopped. The voltage when the reaction is complete is V_∞.

It takes a finite time for the reaction mixture to travel from the mixing chamber to the photocell; this dead time equals about 15 msec for the SF-1B. If the reaction is very fast, some reaction takes place during the dead time, so that the voltage at time equals zero, V_0, is not equal to

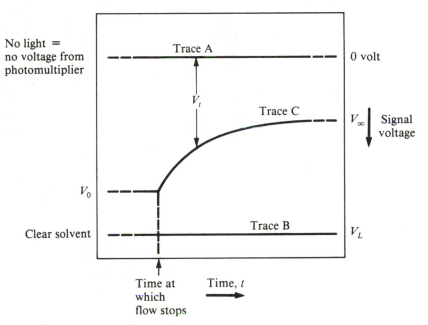

Figure 24–3 Hypothetical oscilloscope trace.

V_L, the voltage for pure water. If the reaction half-life is appreciably less than the dead time, then V_0 may be too close to V_∞ to permit accurate measurements. However, the half-life of the iron(III)–thiocyanate reaction is about 50 msec (depending on pH), so the dead time is not a problem.

Actually, the oscilloscope trace shows only the changes taking place after being triggered, which is taken as time equals zero and is the time at which the flow stops.

Apparatus

The requirements for the experiment are a spectrophotometer; a storage oscilloscope; the stopped-flow apparatus;[1] a microcomputer with analog-to-digital converter (optional); perchloric acid ($HClO_4$); ferric ammonium sulfate ($Fe(NH_4)_3(SO_4)_3 \cdot 4H_2O$); potassium thiocyanate (KSCN); sodium chlorate ($NaClO_4$); six 250-ml flasks; four 1-L glass-stoppered bottles; and 5-, 10-, and 25-ml pipets.

SAFETY CONSIDERATIONS

Sodium chlorate and 60% perchlorate acid are dangerously explosive oxidizing agents. They should be kept away from sun, heat, and all oxidizing agents, including all organic compounds. Their solutions should be diluted with water before disposal.

[1] Construction details are given by Gray and Workman (1980). An inexpensive teaching stopped-flow apparatus, as well as more sophisticated models, are available from HI-Tech Scientific, Ltd., Brunel Road, Salisbury Wiltshire, England, SP2 7PU, phone (0722)23643.

EXPERIMENTAL PROCEDURE

The rates of ionic reaction are dependent not only on the concentrations of reactants, but also on ionic strength. With this in mind, prepare a series of solutions of various hydrogen ion concentrations at constant ionic strength. It is convenient to use prepared stock solutions (Table 24–1) to make the working solutions (Table 24–2). Each of the working solutions is diluted with water to 250 ml total volume so that the final ionic strength is constant and equal to 0.4. The concentration of iron(III) ion in all the Fe_i solutions (Table 24–2) is 0.010 M; the concentration of thiocyanate ion in the Y solution is 0.0010 M. Since the iron is in great excess over the thiocyanate, its concentration is essentially constant and the reaction is pseudo–first order with respect to the thiocyanate. The hydrogen ion concentration for each of the Fe_1 through Fe_5 solutions is 0.020 M, 0.040 M, 0.060 M, 0.120 M, and 0.220 M, respectively. Note that these are the concentrations in the working solutions before mixing in the stopped-flow apparatus; after mixing in the stopped-flow apparatus the concentrations are decreased by one half.

Carry out a series of runs with constant concentrations of Fe^{3+} and SCN^-. The run compositions are listed in Table 24–2.

Using the Stopped-Flow Apparatus with an Oscilloscope. Set the wavelength to 425 nm. The settings of the two driving stopcocks (D) and the stopping stopcock (T for trigger) are used in a variety of combinations. It is convenient to label some standard settings X, Y, and Z as shown in Figure 24–4. The setting label X/XX, for example, represents the setting in Figure 24–4a, in which the upper stopcock T and the two lower stopcocks D are all in their X settings. The barrels of each of the stopcocks should bear a mark to indicate the direction in which the bottom of the T-bore points.

1. In the X/XX setting, fill the reservoirs with distilled water. Pull the driving pistons back and forth to repeatedly rinse them with water.

TABLE 24–I

Preparation of Stock Solutions

Solution	Concentration	Solutes
A	1.00 M $HClO_4$	105.0 ml 61.4% w/w $HClO_4$ diluted to 1000 ml with H_2O
B	1.00 M $NaClO_4$	122.46 g $NaClO_4$ diluted to 1000 ml with H_2O
C	0.100 M Fe^{3+}, 0.200 M $HClO_4$	48.2 g $FeNH_4(SO_4)_2$ and 200 ml 1.00 M $HClO_4$ diluted to 1000 ml with H_2O
D	0.0100 M KCNS	0.972 g KCNS diluted to 1000 ml with H_2O

TABLE 24–2

Preparation of Working Solutions[a]

Solution	Stock Solution Aliquot (volumes/ml)			
	A	B	C	D
Fe_1	0	95	25	0
Fe_2	5	90	25	0
Fe_3	10	85	25	0
Fe_4	25	70	25	0
Fe_5	50	45	25	0
SCN^-	0	100	0	25

[a] Dilute with water to 250.0 ml total volume.

2. Pull the driving pistons fully back and set to X/YY. Push vigorously on the driving block. As the driving pistons move forward, water is forced through the system into the stopping syringe, which moves until it strikes the stopping block.

3. Change the setting to Y/YY (Fig. 24–4b). Now push the stopping syringe T to empty its contents into the sink S. Repeat steps 1, 2, and 3 until the system is clean and free of bubbles.

4. Change the setting to Z/ZZ (Fig. 24–4c) and permit the water in the reservoirs to drain. When each reservoir is almost empty, change the drive stopcock to setting X. When both reservoirs are almost empty, change the stopping syringe T to setting X also. The setting is now X/XX.

5. Fill one reservoir about half-full of working solution Fe_i and the other about half-full of reagent SCN^- (Table 24–2). Repeat

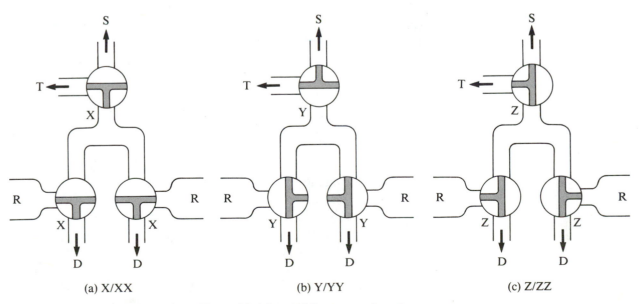

(a) X/XX   (b) Y/YY   (c) Z/ZZ

Figure 24–4 Stopped-flow stopcock settings.

steps 1, 2, and 3 above. Notice that the reaction mixture shows a slight color due to the $Fe(H_2O)SCN^{2+}$ formed.

Next prepare the oscilloscope. Check that the two cables terminating in BNC connectors are connected to the proper terminals: (1) from the *trigger* on the stopped-flow apparatus to *external trigger* on the oscilloscope; and (2) from the *signal* on the stopped-flow apparatus to *signal* or *vertical input* on the oscilloscope.

On the stopped-flow apparatus turn the lamp on and the bias off. The bias adjustment has essentially the same effect as the vertical position control of the oscilloscope and so is redundant. Make the following oscilloscope settings:

1. The sensitivity (vertical gain) to 0.050 V/division.
2. The sweep time to 50 msec/division.
3. The coupling to DC.
4. Switch to *Free Run*.
5. Adjust the *position* knob so the sweep is in the middle of the screen.

Check that the trigger signal is operating properly. On the oscilloscope set the *trigger slope* to +, which corresponds to a trigger going from off to a plus voltage triggering the oscilloscope. Switch from *Free Run* to *Single Sweep*. Push *Erase* to clear the storage screen. Push the stopping block trigger on the stopped-flow apparatus manually. This should result in a trace sweeping horizontally across the screen. If nothing happens adjust the oscilloscope *trigger level* knob in the less positive direction and try manually triggering again, after pushing the *trigger reset* button. *The trigger reset button must be reset before every use of the external trigger.*

Since the reaction mixture is already in the syringes, the system is ready for a run. The stopcock setting should be X/YY. Check that the stopping block syringe is in and the drive syringes are out, full, and free of bubbles. Push the drive syringes vigorously forward. They will move the stopping block syringe and stop when it stops triggering the signal to the oscilloscope. The trace on the screen will appear approximately as trace C in Figure 24–3. Traces A and B are not seen and not needed. Take the quantity $V_t - V_\infty$ and the number of divisions between the trace and the level on the screen at which V_∞ is observed. This is equivalent to setting V_∞ equal to zero and measuring the division downward from it. It is not necessary to convert to voltage. Tabulate $V_t - V_\infty$, $\ln(V_t - V_\infty)$, and t at 0.05-sec intervals until $V_t - V_\infty$ equals zero. Plot $\ln(V_t - V_\infty)$ versus t. The slope equals $-k_{obs}$.

This simplified method of reading the oscilloscope trace is valid if the absorption is small; that is, if I/I_0 is less than about 0.85. In such a case $A \propto \Delta V \propto \Delta I$, where $\Delta I = I_0 - I$. Then $I_0/I = x + 1$, where $x = \Delta I/I$. When the absorbance is expanded as a power series,

TABLE 24–3

A/D Conversion Arithmetic

V_{analog} Input to A/D (from an experiment)		12-Bit A/D Converter (transparent to user)	Integer Output (stored in a program array)
0.0000	0.0000	0000 0000 0000	0
0.0002	0.0012	0000 0000 0001	1
0.0005	0.0024	0000 0000 0010	2
0.0007	0.0037	0000 0000 0011	3
$\vdots$	$\vdots$	$\vdots$	$\vdots$
0.7598	3.7988	1100 0010 1000	3112
$\vdots$	$\vdots$	$\vdots$	$\vdots$
0.9995	4.9976	1111 1111 1101	4094
0.9998	4.9988	1111 1111 1110	4095
1.0000	5.0000	1111 1111 1111	4096

$$A = \ln \frac{I_0}{I} = \ln (x + 1) = \frac{x}{1} - \frac{x^2}{2} + \frac{x^3}{3} - \frac{x^4}{4} + \cdots \quad \textbf{(24–23)}$$

If x is small, then I is approximately equal to I_0, so A is directly proportional to ΔI. Thus ΔI can be plotted against t instead of plotting $\ln A$ versus t.

Using the Stopped-Flow Apparatus with an Analog-to-Digital Converter. The output voltage of a photomultiplier is the electrical analog of the light intensity. The output voltage of a spectrophotometer is the electrical analog of the absorbance. The output voltage of most measuring instruments is the electrical analog of some physical quantity that varies in a continuous manner. An analog-to-digital (A/D) converter is an accessory to a computer that converts a continuous analog signal into a binary number, which in turn is translated into an integer number. The Metrabyte Dash-16[2] is a typical A/D converter card that fits into an accessory slot of an IBM-PC or compatible computer. It has 12 registers, each of which can be 0 or 1. Coupled together, each register represents a placeholder to the left of a decimal point, so that the range of digits, 2^{12}, or 4096 values, are equivalent to the input voltage range, say, 1.000 volt. In this case the resolution of the A/D converter is 1/4096 or 0.00024 volt.

Like most A/D converters, the Dash-16 has several input analog voltage ranges: 0 to 1.0 volt, 0 to 5.0 volts, and so on. The value of the input voltage can be calculated from the integer output from the A/D converter:

$$V_{analog} = V_{range} \times \frac{\text{integer output}}{4096} \quad \textbf{(24–24)}$$

[2] MetraByte Corporation, 440 Myles Standish Boulevard, Tauton, Massachusetts 02780, phone (617)880-3000.

Thus, if the card is calibrated for the 5.0-volt range, and from the A/D card an analog signal produces an integer of value 3112, then V_{analog} equals 3.799 volts, as shown in Table 24–3. In practice, a series of integers are output from the cards for successive measurement over a period of time; these are stored in the elements of an array called *DigitalData* (Fig. 24–5).

The A/D card also has a timer and the collection of data can be initiated by sending a trigger signal to the card. The trigger signal is a small electrical voltage that changes the value from 0 to 1 volt. When the trigger voltage is received by the card, the timer begins and the first analog signal is sampled, converted to a digit, and stored as the first element in the array *DigitalData, DigitalData[0]*. The total number of measurements to be made (the value of a variable called *Count*) is previously input by the user. The rate at which measurements are made by the A/D converter is also previously input by the user by assigning a value to the variable *Rate*. Since the units of the variable *Count* are samples and the units of *Rate* are samples per second, the time *t* between measurements is

$$t = \frac{Count \, (\text{samples})}{Rate \, (\text{samples}/\text{sec})} \tag{24–25}$$

With the Dash-16, the *Rate* can be set as high as 50,000 samples/ sec, permitting the measurement of very fast reactions. If a reaction has a half-life of 10 msec and we wish to sample it over three half-lives, then in 30 msec 1500 samples could be measured at a rate of 50,000 samples/sec! In practice there is little point in collecting more than 300 to 600 points, which corresponds to the number of pixels across a computer monitor on which the output may eventually be graphically displayed. An example of such a display is shown in Figure 24–6.

The A/D converter cards usually come with a BASIC software driver, which permits simple interfacing between the user and the computer. Several vendors support the MetraByte D-16; the program shown in Figure 24–5 was written with a Turbo Pascal driver furnished by Quinn-Curtis.[3] Such a driver allows full use of Turbo Pascal's graphics capability. *Program StopFlow* is a brief program that illustrates the use of the A/D converter. The pseudocode may be written as follows:

Program StopFlow

Procedure GetData. This consists of three short procedures:

> *Procedure Initialize.* After assigning values to the card parameters according to the Quinn-Curtis manual, the call to the procedure *d16_init* initializes the card.

(Text continues on page 593.)

---

[3] Quinn-Curtis, 49 Highland Avenue, Needham, Massachusetts 02194, phone (617)444–7721.

```pascal
Program StopFlow;
(* A Program to demonstrate A/D conversion with the        *)
(* MetraByte Dash-16 Card                                  *)

uses
  crt,
  tp4d16;    (* This is tp4d16.tpu furnished by MetraByte  *)
             (* as tp4d16.pas and compiled with TP5.0      *)

Const
  MaxPoints = 400;

Type
  IntAry = array[0..MaxPoints] of integer;
  RealAry = array[0..MaxPoints] of Real;

Var
  DigitalArray:  IntAry;
  TimeArray, VoltageArray: RealAry;
  Count: Integer;
  Rate: Real;

(*****************************************************************)

Procedure GetData(Var DigitalData: IntAry;
                  Var TimeData, VoltageData: RealAry);

Var i, mode, cycle, trigger,
    base_adr, err_code, int_level, dma_level,
    board_num, dataval, chanhi, chanlo:        integer;
    {The above variable identifiers are all from the
     Quinn-Curtis Pascal Driver user's manual}
{-----------------------------------------------------------------}
Procedure Initialize;
Var i:  integer;
Begin {Initialize}
  For i:= 0 to MaxPoints do
    Begin
      DigitalData[i]:= 0;
      TimeData[i]:= 0;
    End;
  ClrScr;
  board_num:= 0; base_adr:= $330; int_level:= 5; dma_level:= 3;
  d16_init(board_num,base_adr,int_level,dma_level,err_code);
End; {Initialize}
{-----------------------------------------------------------------}
```

Figure 24–5 Simplified stopped-flow data acquisition program.

```
Procedure SetParameters;
Begin {SetParameters}
  chanlo:= 0;   mode:= 0; cycle:= 0; trigger:= 1;
  gotoxy(10,3);
  Write('What is "COUNT"? (How many data points?) Maximum is 400:  ');
  ReadLn(Count);
  GotoXY(10,4);
  Write('What is the sampling RATE?:  ');
  ReadLn(Rate);
  GotoXY(1,5);
  Write('Push plunger on stop flow apparatus to trigger');
  Writeln(' data collection.');
  Write('Then data will be collected for the ');
  Writeln('next ',(count/rate):6:2,' seconds');
End;{SetParameters}
{----------------------------------------------------------         }
Procedure CollectandConvert;
Var i:  integer;
Begin {CollectandConvert}
  {The following call awaits the trigger voltage going from 0 to +}
  d16_ainm(board_num,chanlo,mode,cycle,trigger,
          count,rate,DigitalData[0],err_code);
  if err_code <> 0 then
    Begin                                          {if something      }
      writeln('Execution error ',err_code:4); {goes wrong        }
    End
  else
      For i:= 1 to Count-1 do {Build arrays of time and voltage   }
        Begin {For }
          TimeData[i]:= TimeData[i-1] + 1/rate; {units are seconds}
          VoltageData[i]:= DigitalData[i] / 4095; {units are volts}
        End { For }
End; { CollectandConvert }
{-------------------------------------------------------------------}
Begin {GetData}
  Initialize;
  SetParameters;
  CollectandConvert;
End;   {GetData}

(*******************************************************************++*)

Procedure WriteToScreen(VoltageData, TimeData:  RealAry);
Var i:  integer;
Begin
  Writeln;
  Writeln('  Time/s    Voltage/V');
  Writeln;
  For i:= 0 to Count - 1 do
    Writeln(TimeData[i]:5:3,'         ',VoltageData[i]:6:4);
End;

(*****************************************************************)

Begin { StopFlow }
  GetData( DigitalArray, TimeArray, VoltageArray);
  WriteToScreen(VoltageArray, TimeArray);
End. { StopFlow }
```

Figure 24–5 (continued)

Real Time Data Acquisition....   14:39 Wednesday, 23 November 1988
Saved as : hans\5-6h.dat

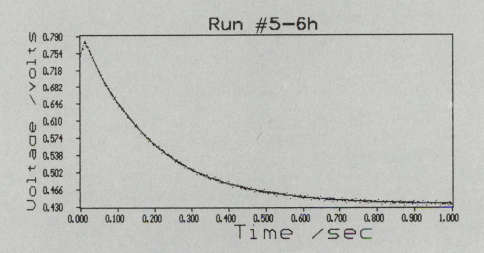

Count = 1000   Rate = 1000.0

Run #5-6h : Stopped Flow TIME vrs VOLTAGE DATA

TIME/s	V /v	TIME/s	V /v	TIME/s	V /v	TIME/s	V /v
0.000	0.7512	0.240	0.5385	0.480	0.4654	0.720	0.4444
0.015	0.7761	0.255	0.5289	0.495	0.4628	0.735	0.4452
0.030	0.7526	0.270	0.5258	0.510	0.4608	0.750	0.4427
0.045	0.7258	0.285	0.5158	0.525	0.4601	0.765	0.4422
0.060	0.7013	0.300	0.5092	0.540	0.4567	0.780	0.4425
0.075	0.6799	0.315	0.5043	0.555	0.4596	0.795	0.4410
0.090	0.6576	0.330	0.4979	0.570	0.4547	0.810	0.4413
0.105	0.6447	0.345	0.4928	0.585	0.4525	0.825	0.4405
0.120	0.6286	0.360	0.4901	0.600	0.4515	0.840	0.4435
0.135	0.6139	0.375	0.4808	0.615	0.4501	0.855	0.4408
0.150	0.6012	0.390	0.4818	0.630	0.4484	0.870	0.4393
0.165	0.5888	0.405	0.4786	0.645	0.4486	0.885	0.4393
0.180	0.5758	0.420	0.4745	0.660	0.4427	0.900	0.4396
0.195	0.5653	0.435	0.4730	0.675	0.4459	0.915	0.4383
0.210	0.5551	0.450	0.4711	0.690	0.4459	0.930	0.4386
0.225	0.5468	0.465	0.4667	0.705	0.4447	0.945	0.4369

Figure 24–6 Output of a complete stopped-flow data acquisition program (Deuel, 1988).

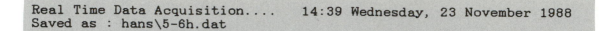

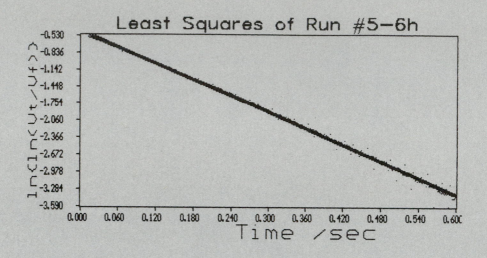

Run #5-6h : Least Squares Calculation Results

───

Number of points = 587
V(inf)   = 0.4379
t[initial]   = 0.0110
t[final]   = 0.5990

Slope   =   -4.915040   s =   0.000420
Intercept   =   -0.419900   s =   0.000147
Correlation Coefficient   =   -0.998748

        s = standard deviation of the mean

    [Error during calculation of V(inf)]

───

Figure 24–6 (continued)

Procedure SetParameters. This procedure allows the user to input interactively the values for *Count* and *Rate*. When the variable *trigger* is assigned the value 1, the card awaits a trigger voltage changing from 0 to 1 before acquiring data.

Procedure CollectAndConvert. The call to the Quinn-Curtis procedure *d16_ainm* initiates data acquisition as soon as a positive trigger voltage is sent to the card from the stopped-flow apparatus. The elements of the array *DigitalData* are filled with integers corresponding to the analog input voltage every *Count/Rate* seconds until *Count* measurements have been made. This is done automatically by the Quinn-Curtis driver unit. Then the integers are converted to volts and assigned to the elements of the array *VoltageData*. A parallel array of time, *TimeData,* is also built.

Procedure WriteToScreen. The values of time and voltage are listed to the screen.

This sample program illustrates the essential feature of timed data acquisition. The A/D converter can also be used to measure V_∞. The remainder of the data reduction, least-squares analysis, and graphical display can be carried out with appropriate procedures. In this way the output illustrated in Figure 24–6 was generated with a Turbo Pascal program.

Results and Discussion

The observed rate constant k_{obs} is found to vary with $[H^+]$, because both $FeSCN^{2+}$ and $Fe(OH)SCN^+$ are formed. The concentration of $FeOH^{2+}$, which is much more reactive than Fe^{3+}, decreases as $[H^+]$ increases. According to Equation 24–15, k_{obs} is only constant at a particular hydrogen ion concentration and k_{obs} is a linear function of $1/[H^+]$. Thus Equation 24–15 is an equation of the form $y = mx + b$, where

$$b = k_1[Fe(H_2O)_6^{3+}] + k_{-1} \tag{24–26}$$

and

$$m = (k_2 K_{h1}[Fe(H_2O)_6^{3+}] + k_{-2} K_{h2}) \tag{24–27}$$

Hence, by performing a series of experimental rate determinations at different hydrogen ion concentrations, one can obtain the necessary data for a least-squares analysis according to Equation 24–15. A plot of the different k_{obs}'s versus $1/[H^+]$ should yield a straight line with slope m and intercept b.

Obtaining k_1, k_{-1}, k_2, and k_{-2} from k_{obs}. If one takes the hydrolysis equilibrium constant values for K_{h1} and K_{h2} from the literature, then

Equations 24–26 and 24–27 represent two equations in four unknowns, k_1, k_{-1}, k_2, and k_{-2}. One still needs two more equations.

By definition, at equilibrium the concentration of each of the species in Equation 24–3 is constant. Hence,

$$k_1[\mathrm{Fe(H_2O)_6^{3+}}][\mathrm{SCN^-}] = k_{-1}[\mathrm{Fe(H_2O)_5(SCN)^{2+}}] \quad \text{(24–28)}$$

or

$$\frac{k_1}{k_{-1}} = \frac{[\mathrm{Fe(H_2O)_5(SCN)^{2+}}]}{[\mathrm{Fe(H_2O)_6^{3+}}][\mathrm{SCN^-}]} = K_{eq} \quad \text{(24–29)}$$

which is the expression for the equilibrium constant of Equation 24–3, and can be taken from the literature. Similarly, from Equation 24–4 at equilibrium one can write

$$\frac{k_2}{k_{-2}} = \frac{[\mathrm{Fe(H_2O)_4(OH)(SCN)^+}]}{[\mathrm{Fe(H_2O)_5(OH)^{2+}}][\mathrm{SCN^-}]} \quad \text{(24–30)}$$

Substituting Equations 24–7 and 24–8 into Equation 24–30 and combining with Equation 24–29 yields

$$\frac{k_2}{k_{-2}} = \frac{K_{h2}k_1}{K_{h1}k_{-1}} \quad \text{(24–31)}$$

Equations 24–26, 24–27, 24–29, and 24–31 allow one to determine experimentally the four rate constants of Equations 24–3 and 24–4. First, plot k_{obs} versus $1/[\mathrm{H^+}]$. From a linear regression determine the slope m and the intercept b. Second, calculate k_1 and k_{-1} with the intercept b and Equations 24–26 and 24–29. Third, calculate k_2 and k_{-2} with the slope m and Equations 24–27 and 24–31.

Compare your results with values from the literature (Below et al., 1958; Funahashi et al., 1973; Goodall et al., 1972; Meiling and Pardue, 1978). Take note of the differences in ionic strength between the conditions of this experiment and those in the literature.

Some suitable values for the equilibrium constants at 25°C needed for these calculations are: $K_{h1} = 1.89 \times 10^{-3}$ (Goodall et al., 1972), $K_{h2} = 6.5 \times 10^{-5}$ (Lister et al., 1955), and $K_{eq} = 139$ (Goodall et al., 1972).

REFERENCES

1. Below, J. F., R. E. Connick, and C. P. Coppel, "Kinetics of the Formation of Ferric Thiocyanate Complex," *J. Am. Chem. Soc.* 80:2961, 1958.

2. Caldin, E. F., *Fast Reactions in Solution,* Wiley, New York, 1964.

3. Deuel, H., B.S. Thesis, California State University, Sacramento, December 1988.

4. Funahashi, S., S. Adachi, and M. Tanaka, "Kinetics and Mechanism of Formation of Monothiocyanato and Monotrilotriacetato Complexes of Iron(III)," *Bull. Chem. Soc. Japan* 46:479, 1973.

5. Goodall, D. M., P. W. Harrison, M. J. Hardy, and C. J. Kirk, "Relaxation Kinetics of Ferric Thiocyanate: A Temperature Jump and Flash Photolysis Experiment," *J. Chem. Educ.* 49:675–678, 1972.

6. Gray, Jr., E. T., and H. J. Workman, "An Easily Constructed and Inexpensive Stopped-Flow System for Observing Rapid Reactions," *J. Chem. Educ.* 57:752, 1980.

7. Försterling, H.-D., and H. Kuhn, *Praxis der physikalischen Chemie,* VCH, Weinheim, 1985.

8. Gibson, Q. H., "Rapid Mixing: Stopped Flow," in *Methods in Enzymology,* vol. XVI, *Fast Reactions,* Academic Press, New York, 1969.

9. Lister, M. W., and D. E. Rivington, "Some Measurements on the Iron(III) Thiocyanate System in Aqueous Solution," *Can. J. Chem.* 33:1572, 1955.

10. Marcotte, R. E., "A Functional Fast Flow Kinetics Apparatus," *J. Chem. Educ.* 57:388, 1980.

11. Meiling, G. E., and H. L. Pardue, "Evaluation of a Computer-Controlled Stopped-Flow System for Fundamental Kinetic Studies," *Anal. Chem.* 50:1333, 1978.

12. Morelli, B., "A Kinetic Experiment Using a Spring Powered, Stopped-Flow Apparatus," *J. Chem. Educ.* 53:119, 1976.

13. Patel, R. C., G. Atkison, and R. J. Boe, "Fast Reactions: Rapid Mixing and Concentration Jump Experiments," *J. Chem. Educ.* 47:800, 1970.

14. Reimsborough, V. C., and B. H. Robinson, "A Convenient Stopped-Flow Experiment," *J. Chem. Educ.* 58:586, 1981.

15. Wilkins, R. G., *The Study of Kinetics and Mechanisms of Reactions of Transition Metal Complexes,* Allyn and Bacon, Boston, 1974.

Experiment 25

Kinetics of Nitric Oxide Oxidation

Not many reactions are of third order, and it appears that most known homogeneous gas reactions of the third order involve nitric oxide (NO) as one of the reactants. In this experiment we study the homogeneous gas phase reaction between nitric oxide and oxygen. Some of the characteristics of this and a few other reactions of nitric oxide are shown in Table 25–1.

Reaction Order and Mechanism

The mechanism for these reactions of nitric oxide has not been completely established, but it is possible that the reaction is not only third order but also trimolecular. Remember the difference between order and molecularity. The order is just the sum of the exponents in the experimentally observed rate law:

$$\text{Rate} = k[\text{A}]^a[\text{B}]^b[\text{C}]^c \dots \qquad \textbf{(25–1)}$$

TABLE 25–1

Some Third-Order Reactions of Nitric Oxide at 25°C

Reaction	k ($L^2\ mol^{-2}\ sec^{-1}$)	E_a (kJ/mol)	Ref.
$2NO + O_2 = 2NO_2$	7.1×10^3	-4.6	a
$2NO + Cl_2 = 2ClNO$	21.0	$+17.0$	b
$2NO + Br_2 = 2BrNO$	3.1×10^3	-5.6	c
$2NO + H_2 = N_2O + H_2O$	3.1×10^3	$+197.0$	d

a. Bodenstein, M., "Velocity of Reaction Between Nitric Oxide and Oxygen," *Z. Electrochem.* 24:183, 1918.
b. Trautz, M., "Temperature Coefficient of Nitrosyl Chloride from Nitric Oxide and Oxygen," *Z. anorg. Chem.* 88:285, 1914.
c. Trautz, M., and V. P. Dalal, "Velocity of the Formation of Nitrosyl Bromide," *Z. anorg. Chem.* 102:149, 1918.
d. Hinshelwood, C. N., and J. W. Mitchell, "The Reaction of Nitric Oxide with Hydrogen and with Deuterium," *J. Chem. Soc.* 378–384, 1936. (k is at 800°C.)

The order equals $a + b + c + \ldots$ and can be determined experimentally by measuring rates and concentrations. The order with respect to individual reactants is often a small integer, or nearly so because of experimental error. The reaction between nitric oxide and oxygen is third order (Table 25–1):

$$Rate = [NO]^2[O_2]$$

The molecularity, on the other hand, is the number of molecules entering into the formation of the activated complex. This must be an integer and usually equals two, since activation is believed to occur because of collisions between two molecules that then form an activated complex. Such an elementary reaction is called **bimolecular**. Some reactions are called **unimolecular**, but even those are believed to form an activated complex by collision, either with themselves, another molecule, or the walls of the containing vessel. The collision theory for unimolecular reactions is well established (Robinson and Holbrook, 1972). Table 25–1 reveals some unusual features of the reactions of nitric oxide. Notice that the activation energy E_a is small or even negative. For example, the reaction between nitric oxide and oxygen is about 10% slower at 100°C compared with 25°C. This behavior is highly unusual: rates of nearly all reactions increase as the temperature increases. In addition, the rate constants indicate that these reactions of nitric oxide are unusually slow compared to typical bimolecular gas phase reactions.

Trimolecularity. The slow rate of the reaction and the stoichiometry suggest that the reaction is trimolecular. Defining a three-body collision, however, poses some difficulties. A collision between hard spheres takes place in zero time; that is, the contact time between colliding hard spheres is zero. Consequently, a three-body collision between hard spheres is essentially impossible. Collisions between soft spheres, however, lead to nonzero contact time; these are sometimes called sticky collisions. One can visualize a collision between a third body and two molecules stuck together. But when two molecules are "stuck" together, are they not bonded?

One way of getting out of this semantic difficulty is to define a collision as an event occurring whenever two molecules get within a molecular (or collision) diameter of each other. On the basis of such a model it has been suggested that the probability of a three-body collision compared to the probability of a two-body collision equals the ratio of the collision diameter σ to the mean free path λ, illustrated in Figure 25–1. Since the mean free path of ordinary molecules at 1 atm is about 10^{-7} m and collision diameters are typically around 10^{-10} m, this suggests that one trimolecular collision occurs for every 1000 bimolecular collisions. Thus, other things being equal, we would expect trimolecular reactions to be relatively slow.

Two-Step Mechanisms. A two-step mechanism for the reaction between NO and O_2 offers two choices for the first step: either two NO molecules collide to form an intermediate (N_2O_2), or an NO molecule and an O_2 molecule collide to form an intermediate (NO_3). Let us consider the formation of NO_3 first.

NO_3 as an Intermediate.

$$NO + O_2 \underset{k_2}{\overset{k_1}{\rightleftarrows}} NO_3 \qquad \text{fast} \qquad \qquad (25\text{–}2)$$

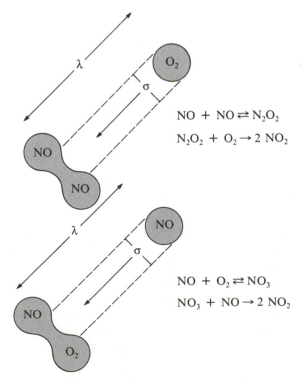

$$NO + NO \rightleftarrows N_2O_2$$
$$N_2O_2 + O_2 \rightarrow 2\,NO_2$$

$$NO + O_2 \rightleftarrows NO_3$$
$$NO_3 + NO \rightarrow 2\,NO_2$$

Figure 25–1 Mean free path and collision diameter.

$$NO_3 + NO \xrightarrow{k_3} 2NO_2 \qquad \text{slow} \qquad \text{(25–3)}$$

If it is assumed that the steady-state approximation holds for the intermediate species NO_3, then one may write

$$\frac{d[NO_3]}{dt} = 0 = k_1[NO][O_2] - k_2[NO_3] - k_3[NO_3][NO] \qquad \text{(25–4)}$$

Solving for the small, constant steady-state concentration of the intermediate NO_3, we find

$$[NO_3] = \frac{k_1[NO][O_2]}{k_2 + k_3[NO]} \qquad \text{(25–5)}$$

The rate of the reaction is just equal to half the rate of production of NO_2:

$$\text{Rate} = \frac{d[NO_2]}{2\,dt} \qquad \text{(25–6)}$$

By writing the rate law for this elementary step and then substituting the steady-state concentration of the intermediate NO_3, we obtain an expression for the rate in terms of the concentrations of the reactants:

$$\text{Rate} = k_3[NO_3][NO] = \frac{2k_3k_1[NO]^2[O_2]}{k_2 + k_3[NO]} \qquad \text{(25–7)}$$

If the first step is fast and the second step is much slower, then $k_2 \gg k_3[NO]$, and so

$$\text{Rate} = \frac{2k_3k_1}{k_2}[NO]^2[O_2] \qquad \text{(25–8)}$$

Since the forward and the reverse reactions of the first step are elementary reactions, the equilibrium constant K for this reaction equals k_1/k_2 and then we may write

$$\text{Rate} = 2k_3K[NO]^2[O_2] \qquad \text{(25–9)}$$

Thus, $k_{obs} = k_3K$. The observed rate constant equals the product of k_3, a second-order bimolecular rate constant, and K, an equilibrium constant for an association. Since the equilibrium constant for an association (an exothermic process) must decrease as the temperature rises, k_{obs} decreases slightly as the temperature rises. In other words, the observed activation energy equals the sum of the enthalpy change for the fast association reaction and the activation energy of the slow bimolecular second step. Since these energies are opposite in sign, their sum could be small or even slightly negative. This is consistent with the observed activation energies listed in Table 25–1.

N_2O_2 as an Intermediate. The two-step reaction that produces N_2O_2 is similar:

$$NO + NO \underset{k_2}{\overset{k_1}{\rightleftarrows}} N_2O_2 \qquad \text{fast} \qquad (25\text{--}10)$$

$$N_2O_2 + O_2 \overset{k_3}{\rightarrow} 2NO_2 \qquad \text{slow} \qquad (25\text{--}11)$$

$$\frac{d[N_2O_2]}{dt} = 0 = k_1[NO]^2 - k_2[N_2O_2] - k_3[N_2O_2][O_2] \quad (25\text{--}12)$$

$$[N_2O_2] = \frac{k_1[NO]^2}{k_2 + k_3[O_2]} \qquad (25\text{--}13)$$

$$\text{Rate} = k_3[N_2O_2][O_2] = \frac{k_3k_1[NO]^2[O_2]}{k_2 + k_3[O_2]} \qquad (25\text{--}14)$$

If $k_2 \gg k_3[O_2]$, then

$$\text{Rate} = \frac{k_3k_1}{k_2}[NO]^2[O_2] \qquad (25\text{--}15)$$

$$\text{Rate} = k_3K[NO]^2[O_2] \qquad (25\text{--}16)$$

which is formally the same rate law as Equation 25–9.

Unfortunately, none of the aforementioned arguments permits us to distinguish between these two-step mechanisms. This discussion illustrates how one can deduce the same rate law with two different mechanisms.

Transition State Theory. If the reaction between nitric oxide and oxygen is truly trimolecular, then two NO molecules and one O_2 molecule form an activated complex containing six atoms altogether: two nitrogen atoms and four oxygen atoms. With the Transition State Theory, we do not concern ourselves about the nature of the collision process that accomplishes all this; we just assume that the reactants are in equilibrium with the activated complex. The rate constant according to the Transition State Theory is then given by

$$k = \frac{kT}{h} \frac{Q^{\#}}{Q_{NO}^2 Q_{O_2}} e^{-\Delta E^{\#}/RT} \qquad (25\text{--}17)$$

When the expressions for the translational, rotational, and vibrational partition functions for the reactants and the activated complex are substituted into Equation 25–17, the result (Gershinowitz and Eyring, 1935) is

$$k = T^{-7/2}e^{-\Delta E^{\#}/RT} \qquad (25\text{--}18)$$

It is reasonable to assume that $\Delta E^{\#}$ is zero, since NO has one unpaired electron and O_2 has two unpaired electrons. This is almost like the reaction between two radicals, for which it is assumed that the activation energy is zero. It follows that

$$k \approx T^{-7/2} \qquad (25\text{--}19)$$

a result consistent with the observed negative temperature coefficient.

Although none of the mechanisms for the reaction between nitric oxide and oxygen gives a complete picture, together they give us considerable insight into the reaction.

Experimental Considerations for Gas Phase Rate Measurements

When nitrogen oxide reacts with oxygen at room temperature, two reactions must be taken into consideration.

$$NO + O_2 \rightarrow 2NO_2$$

$$2NO_2 \rightarrow N_2O_4$$

At 100°C the dimerization of nitric oxide to form dinitrogen tetroxide is negligible and may be neglected. The rate of the reaction between NO and O_2 is slow enough that it may be measured directly. Because the stoichiometry of the reaction involves a decrease in the number of moles as reactants change into products, the pressure decreases as the reaction progresses. This allows one to follow the course of the reaction manometrically. So that the reaction occurs at a conveniently measurable rate, it is necessary to use initial partial pressures of NO and O_2 in the 10- to 30-torr range. For this reason, α-bromonaphthalene ($d = 1.48$ g/ml) is used as a manometric fluid instead of mercury.

Stoichiometric Amounts of NO and O_2. The use of stoichiometric initial partial pressures of NO and O_2 simplifies the calculations, but complicates establishing the initial conditions: the initial partial pressure of NO must be exactly twice that of O_2:

$$
\begin{array}{lccc}
 & 2NO & + O_2 \rightarrow & 2NO_2 \\
\text{at } t = 0 & 2P^0 & P^0 & 0 \\
\text{at } t & 2P & P & 2P^0 - 2P \\
\text{at } t = \infty & 0 & 0 & 2P^0
\end{array}
$$

If the reaction is third order, then the rate law is

$$\frac{dP}{dt} = k[2P]^2[P] = 4kP^3 \qquad \textbf{(25–20)}$$

Separation of variables and integration leads to

$$\frac{1}{P^2} = 8kt + \left(\frac{1}{P^0}\right)^2 \qquad \textbf{(25–21)}$$

The reaction may be tested to determine whether it is third order by plotting $1/P^2$ against time. If the reaction is third order, the plot is linear and the slope equals $8k$. The partial pressure of O_2 is P, which equals $P_T - 2P^0$, where P_T is the total observed pressure of reactants and P^0 is the initial pressure of O_2. (The initial pressure of NO is $2P^0$.)

Nonstoichiometric Reactants. The use of nonstoichiometric initial pressures of nitric oxide and oxygen is experimentally simpler, but the integrated rate equation is more complex. For a third-order reaction of the stoichiometry $2A + B = X + \cdots$ the integrated rate equation (Laidler, 1965) is as follows:

$$\frac{1}{2b_0 - a_0}\left(\frac{1}{a} - \frac{1}{a_0}\right) + \frac{1}{(2b_0 - a_0)^2}\ln\frac{b_0 a}{a_0 b} = kt \qquad (25\text{--}22)$$

Since pressure is proportional to moles at constant volume and constant temperature, we may write

$$\frac{1}{2P_{O_2}^0 - P_{NO}^0}\left(\frac{1}{P_{NO}} - \frac{1}{P_{NO}^0}\right) + \frac{1}{(2P_{O_2}^0 - P_{NO}^0)^2}\ln\frac{P_{O_2}^0 P_{NO}}{P_{NO}^0 P_{O_2}} = kt \qquad (25\text{--}23)$$

When the left-hand side of this equation is plotted against t, the result is a straight line passing through the origin with a slope equal to the third-order rate constant. Determination of the partial pressures of NO and O_2 at any time t is based on the total pressure, the initial partial pressures, and the stoichiometry of the reaction:

$$2NO + \qquad O_2 \qquad = \qquad 2NO_2 \qquad (25\text{--}24)$$

at $t = 0$ $\quad P_{NO}^0 \qquad\qquad P_{O_2}^0$

at t $\qquad P_{NO} \quad P_{O_2}^0 - \tfrac{1}{2}(P_{NO}^0 - P_{NO}) \quad P_{NO}^0 - P_{NO}$

The total pressure P_T is the sum of the partial pressures:

$$P_T = P_{O_2}^0 + \tfrac{1}{2}P_{NO}^0 + \tfrac{1}{2}P_{NO} \qquad (25\text{--}25)$$

From this relationship, the partial pressure of NO can be expressed in terms of the measured total pressure at time t and the initial pressures:

$$P_{NO} = 2P_T - 2P_{O_2}^0 - P_{NO}^0 \qquad (25\text{--}26)$$

Similarly, the partial pressure of oxygen at any time t can be calculated from the measured total pressure at time t and the initial pressures:

$$P_{O_2} = 2P_{O_2}^0 - P_T \qquad (25\text{--}27)$$

Apparatus

In addition to tanks of oxygen and nitrogen oxide, a corrosion-resistant pressure transducer or manometer filled with α-bromonaphthalene is needed. The apparatus consists of a thermostat bath capable of maintaining the reaction vessel at 90 to 100°C and a vacuum rack with bulbs for storing, manipulating, and reacting NO and O_2. The exact configuration will vary from laboratory to laboratory, but will incorporate most of the features shown in Figure 25–2.

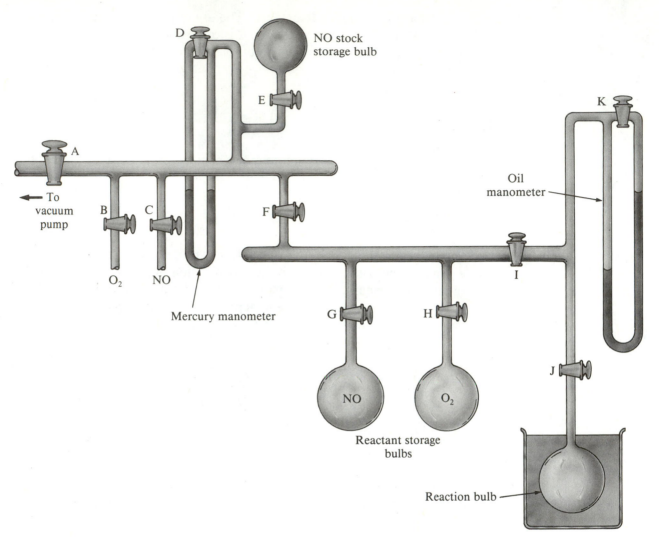

Figure 25–2 Gas-handling apparatus for 2 NO + O$_2$ = 2 NO$_2$.

SAFETY CONSIDERATIONS

The oxides of nitrogen are toxic, so provision should be made for venting the gases into a hood. Safety glasses should be worn at all times when working on a vacuum rack.

EXPERIMENTAL PROCEDURE

There are two direct measurements to be made experimentally; the initial pressure (P^0) of one of the reactants and the total reaction pressure (P_T) against time. To find the rate argument using Equation 25–23 one needs the total reaction pressure against time and the initial pressure of each reactant. The second initial reactant pressure, not measured directly, comes by extrapolation back to time zero of a plot of reaction pressure versus time, since

$$P_T^0 = P_{O_2}^0 + P_{NO_2}^0$$

Before you attempt to measure these pressures it is a good idea to become familiar with the vacuum rack assembly and its dynamics.

Vacuum Rack Assembly. Figure 25–2 is a schematic diagram of the apparatus. It includes the following features: a vacuum pump assembly for evacuation ($\sim 10^{-2}$ torr), a primary line consisting of two gas inlets, a stock NO bulb (2000 ml), a mercury manometer, and stopcocks labeled A through E. Stopcock F separates the primary line from a secondary line, to which are attached the two reactant storage bulbs (250 ml) and a reaction line connecting the α-bromonaphthalene oil manometer, the reaction bulb (250 ml), and stopcocks labeled G through K.

Filling the Stock NO Bulb. To eliminate the need for a high-pressure gas tank of NO for each run, it is convenient to use a large bulb stocked with NO. If a 2000-ml bulb is initially filled with NO at a pressure of 500 to 600 torr, then a large number of runs can be made by transferring NO from the stock bulb to the reactant storage bulb at 20 to 50 torr and 250 ml volume.

Attach a tank of NO to the gas inlet (Fig. 25–2). Close stopcock F to evacuate the primary line, with stopcocks A, C, D, and E open and B closed. When a "hard" vacuum exists, close stopcock A.

When it has been established that the line is evacuated up to the closed needle valve of the NO tank, fill the stock NO bulb to a pressure of approximately 450 to 500 torr. First close stopcock D to activate the mercury manometer. Next, check that the needle valve on the NO tank is closed, and crack the gas tank main valve to admit the full tank pressure to the needle valve. The needle valve now controls the admission of the NO to the stock bulb. Carefully crack the needle open and slowly fill the stock bulb to 450 to 500 torr, reading the pressure from the mercury manometer. When the stock bulb pressure reaches 500 to 600 torr, close the needle valve, stopcock C, stopcock E, and then the NO tank valve.

Because the large amount of residual NO may not be safely removed using the vacuum pump, one must aspirate this waste NO from the primary line into a hood. Open stopcock D to zero the mercury manometer. Attach the aspirator line (equipped with a safety trap) to gas inlet C. Open stopcock C. The aspirator functions only at fairly high pressures, so it may be necessary to let air into the system and repeatedly aspirate. Repeat the above procedure until you see a significant dilution in the NO_2 color. At low enough pressures the residual gas may be removed with the vacuum pump by closing C first and then opening A.

Filling the Reactant Storage Bulbs. From this point on, all pressure readings refer to pressures measured manometrically with an α-bromonaphthalene oil manometer (9.137 mm oil = 1 mm Hg). Some suggested storage bulb pressures are listed in Table 25–2.

The reactant storage bulbs hold the two gases just before a run is made. It is in the filling of these two bulbs that the experimenter has the ability to control the relative initial pressures of the reactants. If the storage bulbs are filled to approximately the same pressures, then, during a run, one gas is first admitted to the reaction bulb and its initial pressure is accurately measured with the oil manometer. This initial pressure is slightly less than 1/2 of the pressure of the gas in its storage bulb. The reaction is initiated by admitting the second gas from its storage bulb to the reaction bulb. The resultant initial pressure of this second gas is roughly 1/4 of its storage bulb pressure. Keep this in mind when deciding to what pressure to fill the storage bulbs; also, do not exceed storage bulb pressures of more than 500 mm (oil) as this strains the capacity of the oil manometer.

Evacuate the primary and secondary lines—both storage bulbs—up to closed stopcock J. Also closed are stopcocks B, C, and E. Stopcocks A, D, F, G, H, I, and K are open. When a "hard" vacuum is attained, close A and either G or H, depending on which gas you want to fill first. To register the pressures, close K and note that the two levels should not move while under vacuum. If they do, open K again and check for leaks or reevacuate the line. Keep in mind that the lightweight oil is sensitive to sudden pressure changes. Inattentive use of the oil manometer can result in serious complications should the oil shoot up into the vacuum line.

With the oil manometer active and holding at zero, fill the NO storage bulb (stopcock H closed). Open F and I, then slowly crack E to admit NO into its storage bulb, carefully monitoring the pressure

TABLE 25–2

Some Suggested Storage Bulb Pressures

Desired Initial Reactant Pressure Ratio	Storage Pressure (mm oil)	
$P_{NO}^0/P_{O_2}^0$	P_{NO}	P_{O_2}
~1:1[a]	500	300
~2:1[a]	570	200
~1:2[b]	200	500

[a] Reaction method: nitric oxide added to oxygen.
[b] Reaction method: oxygen added to nitric oxide.
Note: The pressure of the second gas added to the first must be high enough to allow it to flow into the reaction flask against the opposing first gas pressure of $P^0(X)$.

on the oil manometer. Once the desired amount of NO is added, close E and read the storage bulb pressure accurately. Close G, open K, and reevacuate the line in preparation for filling the O_2 bulb.

When equipped with a regulator, the O_2 tank provides a controllable source for filling the O_2 storage bulb. Evacuate the line with H and B open up to the needle valve of the O_2 tank. Set the regulator to 2 to 3 psi (10 to 20 KPa) of pressure with the needle valve closed. Again, activate the manometer by closing K. Note that here the levels will not remain zeroed due to a leakage of O_2 through the needle valve. While observing the oil manometer closely, close the needle valve, and open and fill the O_2 storage bulb to the desired pressure. Close B and then the needle valve before reading the pressure. With the O_2 storage bulb filled to an accurate pressure, close H, open K, and reevacuate the line to prepare for making a run.

Making a Run. With the entire line now evacuated—including the reaction bulb—and the oil bath operating at the desired temperature, shut stopcock F to isolate the secondary and reaction lines from the pump assembly. Stopcocks G and H are closed; J and I are open. Close stopcock K to activate the oil manometer, making certain that the levels hold at zero. Depending on the reaction method being used, NO or O_2 enters the reaction bulb first. Open the first storage bulb to admit the gas. Read the initial pressure of the first reactant gas accurately. Close J and the storage bulb stopcock and open K. Evacuate the line again by opening F.

Addition of the second gas initiates the reaction and indicates time zero, so have the timer ready. As before, close F and close K to activate the manometer. With J still closed, open the second storage bulb to admit the gas to the line. Just like the first gas, the second gas must squirt through the capillary of the reaction flask and the manometer level will drop. However, the total reaction pressure also decreases with time ($2NO + O_2 \rightarrow 2NO_2$; three molecules react to form two). Open J, count 1 second to allow the first traces of the second gas to contact the other, and start the timer. Note the dropping level of the manometer due to the introduction of the second gas to the reaction flask. After 30 to 35 seconds have elapsed to ensure that as much of the second reactant as possible has entered the reaction flask, close stopcock I, which now isolates the reaction line. Record the total reaction pressures at 60-second intervals for the first 500 to 600 seconds when the pressure is changing rapidly, and at 80- to 100-second intervals thereafter. Sufficient data accumulate after the reaction has run for about 2000 seconds—the pressure begins to level out.

To clean out the line in preparation for another run, close the second storage bulb if you have not done so already so as not to waste the remaining gas inside. Open K, I, and F to reevacuate the entire line, including the reaction bulb.

(Text continues on page 608.)

```
Program RateNO2Oxidation;
(* program to process NO2 + O2 rate data *)

uses crt;

Type Ary = Array[1..50] of Real;

Var PzNO,           (* Pressure of NO at t = zero *)
    PzO2:  Real;    (* Pressure of O2 at t = zero *)
    PNO,            (* Array of NO pressures      *)
    PO2,            (* Array of O2 pressures      *)
    PNO2,           (* Array of NO2 pressures     *)
    P,              (* Array of total pressures   *)
    t,              (* Array of time              *)
    Arg:   Ary;     (* Array of arguments of integrated rate law *)
    n: Integer;     (* Number of P and t pairs to be input       *)
(*********************************************************************)

Function A(PtNO,PtO2:  Real):  Real;
(* This function calculates the argument of the integrated rate    *)
(* equation for the general case of unequal initial pressures of   *)
(* reactants. PtNO and PtO2 are partial pressures at time = t      *)

VAR Zterm,          (* Depends only on pressure of reactants at t = 0 *)
    B,C:  Real;

Begin
  Zterm:= 1/(2*PzO2 - PzNO);
  B:= Zterm*((1/PtNO) - (1/PzNO));
  C:= Sqr(Zterm)*Ln(PzO2*PtNO/PzNO/PtO2);
  A:= B + C;
End;

(*********************************************************************)

Procedure GetTimeAndPressure;
Var I,
    X,Y:  Integer;  (* Screen Coordinates *)
Begin (* GetTimeAndPressure *)
   CLRSCR;
   Write('What is the initial pressure of NO?  ');
   Read(PzNO);Writeln;
   Write('What is the initial pressure of O2?  ');
   Read(PzO2);Writeln;
   Write('What is the number of P and t pairs you wish to enter?  ');
   Readln(N);
   Writeln;
   For I:= 1 to N Do
     Begin (* For *)
       Write(     'P(',I,'):  ');
       X:= WhereX; Y:= WhereY;
       Read(P[I]);
       GoToXY(X+10,Y);
       Write('       t(',I,'):  ');
```

Figure 25–3 *Program RateNO2Oxidation.*

```
      Readln(t[I]);
    End;  (* For *)
End;  (* GetTimeAndPressure *)

(****************************************************************)

Procedure CalcResults;
(* This procedure cacluates the pressure of NO and O2 at time t   *)
(* and then calculates an array of arguments of the integrated   *)
(* rate law by calling the function A, which does this calculation*)

VAR  I:  Integer;

Begin  (* CalculateResults *)
   For I:= 1 to N do
     Begin (* For *)
       PO2[I]:= P[I] - PzNO;              (* These three lines follow *)
       PNO[I]:= 2*P[I] -PzNO - 2*PzO2;  (* from the reaction        *)
       PNO2[I]:=2*(PzO2 - PO2[I]);       (* stoichiometry            *)
       Arg[I]:= A(PNO[I],PO2[I]);            (* a call to function A *)
     End;  (* For *)
   Readln;
End;    (* CalculateResults *)

(****************************************************************)

Procedure WriteResults;
CONST Sp = '        ';
VAR I:  Integer;
Begin (* WriteResults *)
   CLRSCR;
   Write(' time       PNO         PO2         PNO2');
   Writeln(Sp,'   Pt     Rate Law Argument');
   Writeln;
   For I:= 1 to N do
     Begin (* For *)
       Write(t[I]:5:0,Sp,PNO[I]:5:1,Sp,PO2[I]:5:1,Sp,PNO2[I]:5:1);
       Writeln(P[I]:13:1,Sp,Arg[I]:10);
     End; (* For *)
End;  (* WriteResults *)

(****************************************************************)

Begin (* M A I N   P R O G R A M *)
   GetTimeAndPressure;
   CalcResults;
   WriteResults
End.  (* M A I N   P R O G R A M *)
```

Figure 25–3 (continued)

Since the reaction rate increases approximately with the cube of the reaction concentrations, it is desirable to use initial pressures of about 6 to 20 torr. At these pressures the reaction progresses slowly enough to ensure accurate readings. The low pressures require the use of α-bromonaphthalene as the manometric fluid. It is 1/9.137 as dense as mercury (9.137 mm α-bromonaphthalene corresponds to 1 mm Hg), affording more than a ninefold increase in the degree to which one can read these dilute gas pressures.

Results and Calculations

Record the initial pressures of oxygen and nitric oxide. Tabulate the time and total pressure. Calculate and tabulate the partial pressures of oxygen, nitric oxide, and nitrogen dioxide with time.

If stoichiometric amounts of the reactants were used, plot the left side of Equation 25–21 against time and calculate the rate constant. If nonstoichiometric amounts of the reactants were used, plot the left side of Equation 25–23 to calculate the rate constant. A Pascal program to calculate the rate equation argument (Eq. 25–23) is given in Figure 25–3. For each run, the interactive input consists of the initial pressures of nitric oxide and oxygen, followed by a list of total pressures and their corresponding times.

Compare the experimental rate constant with the values reported in the literature. Rate constants in the literature are reported in both $mm^{-2}\,sec^{-1}$ and $cm^3\,mol^{-2}\,sec^{-1}$. Report the experimental value in both units at the reaction temperature.

REFERENCES

1. Desrochers, P., B.S. Thesis, California State University, Sacramento, June 1987.

2. Gershinowitz, H., and H. Eyring, "The Theory of Termolecular Reactions," *J. Am. Chem. Soc.* 57:985, 1935.

3. Hasche, R., and W. A. Patrick, "Studies on the Rate of Oxidation of Nitric Oxide," *J. Am. Chem. Soc.* 47:1207, 1925.

4. Johnston, H. S., and L. W. Slentz, "Oxidation of Nitric Oxide at High Pressures of Reactants," *J. Am. Chem. Soc.* 73:2948, 1951.

5. Laidler, K. J., *Chemical Kinetics,* 2nd ed., McGraw-Hill, New York, 1965.

6. Ritchie, M., *Chemical Kinetics in Homogeneous Systems,* Wiley, New York, 1966.

7. Robinson, P. J., and K. A. Holbrook, *Unimolecular Reactions,* Wiley-Interscience, London, 1972.

8. Sykes, A. G., *Kinetics of Inorganic Reactions,* Pergamon Press, London, 1966.

Experiment 26

Hydrolysis of Ethyl Acetate

Ethyl acetate ($CH_3COOC_2H_5$) hydrolyzes in alkaline solution according to the reaction

$$EtAc + OH^- = EtOH + Ac^- \qquad \textbf{(26–1)}$$

Even at room temperature the rate of hydrolysis is appreciable. Since this aqueous solution contains ions, it conducts electricity. Initially, only Na^+ and OH^- are present in the solution, but as the reaction progresses Ac^- is produced. Since the conductivities of the hydroxyl and acetate ions are very different, the conductivity of the solution changes with time as the hydrolysis proceeds.

The conductance of an ion depends on its ionic mobility, which in turn depends on the size of the ion. Since the conductance of the large acetate ion is less than that of the smaller, more mobile hydroxyl ion, the conductivity of the solution decreases as the hydrolysis proceeds. In this experiment, the progress of the reaction is followed by measuring the conductance of the reaction mixture as a function of time.

Since the reaction is found to be second order and goes essentially to completion (Walker, 1906), the rate law is

$$\frac{-d(a - x)}{dt} = k(a - x)(b - x) \qquad \textbf{(26–2)}$$

where

a = initial molarity of ethyl acetate
b = initial molarity of sodium hydroxide
x = moles of reactants/L reacting in time t
k = specific rate constant ($L\ mol^{-1}\ sec^{-1}$)

The integrated form of Equation 26–2 depends on whether the initial concentrations of the reactants are equal or not equal. If a does not equal b, then integration of Equation 26–2 (Moore and Pearson, 1981) leads to

$$\ln \frac{b(a - x)}{a(b - x)} = k(a - b)t \qquad \textbf{(26–3)}$$

If a equals b, then the integrated form of Equation 26–2 is

$$\frac{x}{a(a - x)} = kat \qquad \textbf{(26–4)}$$

In either case, it is possible to follow the course of the reaction by measuring the conductance as the reaction progresses with time.

Conductance, Equivalent Conductance, and Ionic Concentrations

Conductance L (mhos) is just the reciprocal of resistance R (ohms):

$$L = \frac{1}{R} \tag{26-5}$$

The resistance of any conductor is directly proportional to its length and inversely proportional to its cross-sectional area,

$$R = \rho \frac{\ell}{A} \tag{26-6}$$

where R is the resistance in ohms, ℓ is the length in centimeters, and A is the cross-sectional area in square centimeters. The proportionality constant ρ is called the specific resistance. Equation 26–6 may be written in terms of the conductance L,

$$L = L_s \frac{A}{\ell} \tag{26-7}$$

where L_s is the specific conductance with units of mho per centimeter.

The usual laboratory conductance cell is constructed with two approximately parallel platinum electrodes, each with an area A of approximately 1 cm^2 separated by a distance ℓ of about 1 cm. Since A and ℓ are difficult to measure directly with such a cell, a cell constant is defined:

$$k_{\text{cell}} = \frac{\ell}{A} \tag{26-8}$$

Equation 26–7 is then written

$$L = \frac{L_s}{k_{\text{cell}}} \tag{26-9}$$

The cell constant is determined by filling the cell with a substance of exactly known specific conductance and measuring the cell resistance very precisely with a Wheatstone bridge, as described in Experiment 21. In this experiment, however, it is not necessary to know the cell constant; it suffices to measure the resistance of the cell as a function of time.

The conductance L of the reacting solution depends on the equivalent conductances of the conducting species in the solutions and their concentrations. The equivalent conductance (mho cm^2 equiv^{-1}) is defined as

$$\Lambda = \frac{1000 L_s}{c} \tag{26-10}$$

where L_s is the specific conductance of the solution, c is the concentration in equiv/L, and the constant is 1000 cm^3/L. In this experiment, 1 equiv

= 1 mole. Thus, the conductance of the solution in terms of the individual ionic conductances λ_i is given by

$$\frac{1}{R} = L = \frac{L_s}{k_{\text{cell}}} = \frac{c\Lambda}{1000k_{\text{cell}}} = \frac{\sum_i c_i^+ \lambda_i^+ + \sum_i c_i^- \lambda_i^-}{1000k_{\text{cell}}} \qquad (26\text{--}11)$$

Consequently, the course of the reaction can be followed by monitoring the conductance of the solution as time passes. This is possible because the acetate ion produced by the reaction and the hydroxyl ion consumed have substantially different equivalent ionic conductances (at 25°C, $\lambda_{\text{Ac}^-} = 40.9$ and $\lambda_{\text{OH}^-} = 197.6$ mho cm^{-2} equiv^{-1}).

It is convenient to treat separately the two cases, equal and unequal initial concentrations of reactants.

Reactant Concentrations Are Equal

Since the conductance of the solution is proportional to the equivalent conductance of all the ions in the solution, we can write expressions for the conductance at times $t = 0$, t, and t_c, the time for the reaction to go to completion. The expressions for the concentrations at various times are summarized in Table 26–1. According to Equation 26–11 and Table 26–1, the conductances are given by

$$L_0 = (a\lambda_{\text{Na}^+} + a\lambda_{\text{OH}^-})/1000k_{\text{cell}} \qquad (26\text{--}12)$$

$$L_t = (a\lambda_{\text{Na}^+} + a\lambda_{\text{OH}^-} - x\lambda_{\text{OH}^-} + x\lambda_{\text{Ac}^-})/1000k_{\text{cell}} \qquad (26\text{--}13)$$

$$L_c = (a\lambda_{\text{Na}^+} + a\lambda_{\text{Ac}^-})/1000k_{\text{cell}} \qquad (26\text{--}14)$$

It follows by substitution that

$$\frac{x}{a - x} = \frac{L_0 - L_t}{L_t - L_c} \qquad (26\text{--}15)$$

Thus, Equation 26–4 may be written in terms of experimentally measured conductances:

$$\frac{L_0 - L_t}{L_t - L_c} = kat \qquad (26\text{--}16)$$

TABLE 26–1

Concentrations at Various Times: Equal Reactant Concentrations

Time[a]	[EtAc]	[OH$^-$]	[EtOH]	[Ac$^-$]
$t = 0$	a	a	0	0
t	$a - x$	$a - x$	x	x
t_c	0	0	a	a

[a] It is assumed that the reaction goes essentially to completion after a long period of time t_c.

Determining L_0. The value of L_0 can be obtained by two methods. First, extrapolation of a graph of the measured resistances back to $t = 0$ gives a value for R_0 from which $L_0 = 1/R_0$. Second, direct measurement of a solution of sodium hydroxide of the same concentration as the initial concentration in the run can be measured. Remember that the prepared stock standard solution of sodium hydroxide must be diluted with water to the same concentration as it would have initially in a run.

Determining L_c. With Equation 26–11 and Table 26–1 we can determine the value of L_c,

$$L_c = (a\lambda_{Ac^-} + a\lambda_{Na^+})/1000k_{cell} \qquad (26\text{--}17)$$

and similarly for L_0:

$$L_0 = (a\lambda_{OH^-} + a\lambda_{Na^+})/1000k_{cell} \qquad (26\text{--}18)$$

While the conductance at $t = 0$ is just the conductance of a solution of sodium hydroxide, the conductance when the reaction has gone to completion is just the conductance of a solution of sodium acetate. To calculate L_c, the conductance at completion, divide Equation 26–17 by Equation 26–18 and solve for L_c:

$$L_c = \left(\frac{\lambda_{Ac^-} + \lambda_{Na^+}}{\lambda_{OH^-} + \lambda_{Na^+}}\right)L_0 \qquad (26\text{--}19)$$

If the left side of Equation 26–16 is plotted against t, a straight line results, the slope of which equals ka.

Reactant Concentrations Are Unequal

When the reactant concentrations are unequal, one of the reactants is a limiting reagent and its concentration equals zero when the reaction is at completion. The changes in concentration for this condition are summarized in Table 26–2. The concentration of the spectator ion Na^+ remains constant, while the concentrations of OH^- and Ac^- change with

TABLE 26–2

**Concentrations at Various Times:
Unequal Reactant Concentrations**

Time	[EtAc]	[OH⁻]	[EtOH]	[Ac⁻]
$t = 0$	a	b	0	0
t	$a - x$	$b - x$	x	x
$t_c{}^a$	0	$b - c$	c	c
$t_c{}^b$	$a - c$	0	c	c

[a] Let c equal the concentration of the limiting reagent, a.
[b] Let c equal the concentration of the limiting reagent, b.

the course of the reaction. With the data in Table 26–2 and Equation 26–11 one may show that

$$L_0 - L_t = x(\lambda_{OH^-} - \lambda_{Ac^-})/1000k_{cell} \qquad (26\text{–}20)$$

$$L_0 - L_c = c(\lambda_{OH^-} - \lambda_{Ac^-})/1000k_{cell} \qquad (26\text{–}21)$$

If Equations 26–20 and 26–21 are solved for x, the result is

$$x = \frac{c(L_0 - L_t)}{(L_0 - L_c)} \qquad (26\text{–}22)$$

Determination of L_0 and L_c. The determination of L_0 is exactly the same as discussed above for the case of equal reactant concentrations. The calculation of L_c is similar. For ethyl acetate as the limiting reagent, we can determine L_c with Equation 26–11 and Table 26–2:

$$L_c = [(b - c)\lambda_{OH^-} + c\lambda_{Ac^-} + b\lambda_{Na^+}]/1000k_{cell} \qquad (26\text{–}23)$$

The same expression is valid if sodium hydroxide is the limiting reagent, since the first term drops out when $b = c$. At $t = 0$ Equation 26–23 is still valid since the solution is simply a sodium hydroxide solution. If Equation 26–18 is combined with Equation 26–23, L_c can be calculated for the case of unequal reactant concentrations:

$$L_c = L_0\left[\frac{(b - c)\lambda_{OH^-} + c\lambda_{Ac^-} + b\lambda_{Na^+}}{b(\lambda_{OH^-} + \lambda_{Na^+})}\right] \qquad (26\text{–}24)$$

Now, with values of x calculated from Equation 26–22, it is possible to plot $\ln[b(a - x)]/[a(b - x)]$ against t. According to Equation 25–3, a straight line results, the slope of which equals $k(a - b)$, so that the specific rate constant may be evaluated.

Apparatus

The experiment requires miscellaneous glass and volumetric ware, a conductance cell, a conductance bridge, a buret, 25 and 35°C thermostats, and a timer.

SAFETY CONSIDERATIONS

Many conductance cells make electrical contact by a bare lead inserted into a well in the cell filled with mercury. Mercury is very toxic. Take care to avoid spilling it. Handle the cell over a large enameled photographic tray.

The electrodes of the conductance cell should always be covered with water and not allowed to dry out. If they dry out, they may need a fresh coat of platinum black.

EXPERIMENTAL PROCEDURE

In a 250-ml volumetric flask, prepare a 0.04 M solution of ethyl acetate. Weigh the required amount of ethyl acetate into a tared weighing bottle containing water. Adding the ethyl acetate (±0.001 g) to the water prevents loss of the ester from evaporation. Keep the weighing bottle tightly closed before washing the solution of the ester into the volumetric flask and diluting to the mark. Calculate the exact molarity. This solution should be prepared on the same day it is used, since appreciable hydrolysis can take place in a few days even in a neutral solution.

Calculate the number of milliliters of standard 0.1 M NaOH required to make 250.0 ml NaOH with exactly the same molarity as the ethyl acetate solution. With a buret, deliver the required NaOH solution into a 250-ml volumetric flask and dilute to the mark.

Prior to making a run, immerse the solutions of the ester and the base, and the conductance cell, in the 25°C thermostat and allow them to come to thermal equilibrium. A flask for mixing should also be immersed.

A special flask (Daniels et al., 1970) containing the ester and base in two compartments is especially convenient (Fig. 26–1). Pipet a 25-ml aliquot of base into the inner cylinder and 50 ml of ester solution into the region outside the cylinder. Replace in the thermostat until it reaches thermal equilibrium. Clamp the empty conductance cell in the thermostat.

Figure 26–1 Mixing flask. (Source: Daniels et al., 1970.)

Alternatively, clamp in place in the thermostat a glass-stoppered test tube containing a 25-ml aliquot of the ethyl acetate solution and a stoppered 250-ml Erlenmeyer flask containing a 50-ml aliquot of the sodium hydroxide solution. After the solutions have attained thermal equilibrium, mix them quickly in the flask and transfer a portion to a thermostatted conductance cell.

After mixing the solutions, start the timer and begin measuring resistance every minute for the first 10 minutes, then every 2 minutes, and finally every 5 or 10 minutes until a total run time of about 45 minutes is achieved.

Although R_0 can be determined by extrapolating a plot of R versus t back to zero time, it can also be measured with the conductance bridge directly. Mix an aliquot of the NaOH solution with an aliquot of water equal in size to the ester aliquot used. The resistance of this solution equals R_0. This is a good exercise to do if you are unfamiliar with the conductance bridge and need a practice solution. Having measured R_0 directly, you can preset your bridge settings to facilitate initiating a run.

Make three more runs, reversing the ratios of ester and base in one, and with equal concentrations of ester and base in the other two. Of the four runs, make two at 25°C and two at 35°C.

Remember to calculate the initial concentrations of the reactions *after mixing!*

Results and Calculations

Equal Reactant Concentrations. For each run, identify R_0, R_c, T, and a. Tabulate and graph R and t. Determine R_0 graphically and by direct measurement. Use Equation 26–16 to treat the data when $a = b$. Tabulate and graph the left side of Equation 26–16 against time t in seconds. Calculate k. Report the error in k determined from a least-squares analysis of the data.

Unequal Reactant Concentrations. For each run, identify R_0, R_c, T, a, b, and c. Tabulate and graph R and t. Determine R_0 graphically and by direct measurement. Compare and comment on the direct and graphical measurements. When a and b are unequal, tabulate and plot the left side of Equation 26–3 against time t in seconds. List R_0, R_c, a, b, c, and T. From a least-squares analysis of the data, determine the specific rate constant k and its standard deviation.

Activation Energy. Prepare a tabular summary of T, k, $\ln k$, and $1/T$. From an Arrhenius plot of $\ln k$ versus $1/T$, calculate the activation energy E_a and the preexponential factor A:

$$\ln k = -\frac{E_a}{RT} + \ln A \qquad (26\text{–}25)$$

(Text continues on page 620.)

```
Program EtAcRate;
(* A program to calculate rate constants for               *)
(*    EtAc  +  OH-  =  Ac-  +  EtOH                         *)
(*     a       b                   molarities at t = 0      *)
(*   a - x   b - x     x       x   molarities at t          *)
(*                                                          *)
(* if a equals b within 5%, they will be averaged and treated *)
(* as if equal.  Otherwise, they will be treated as unequal. *)

uses crt,printer;

TYPE Ary = Array[1..30] of Real;

VAR a,            (* initial molarity of ethyl acetate        *)
    b,            (* initial molarity of sodium hydroxide     *)
    c,            (* molarity of limiting reagent             *)
    Ac,           (* ionic conductance of Acetate (mho cm2/mole) *)
    OH,           (* ionic conductance of hydroxyl(mho cm2/mole) *)
    Na,           (* ionic conductance of sodium   (mho cm2/mole) *)
    Ro,           (* resistance at t = 0                      *)
    Rc,           (* resistance at t = infinity               *)
    k:  Real;     (* specfic rate constant (liter/mole/sec)   *)
    N:  Integer;  (* number of time(sec),resistance(ohm) pairs *)
    t,            (* array of time(sec)                       *)
    R,            (* array of resistances(ohm)                *)
    LnAB,         (* array of log((AR+1)/(BR+1))              *)
    L:  Ary;      (* array of ((Lo-Lt)/(Lt-Linf))            *)
    aEQUALb:  Boolean;
    Answer:   Char;

(**************************************************************)

Procedure LeastSquares(X,Y:  Ary;        (* Input array of RAeal   *)
                       Var int,s:  Real; (* output intercept, slope *)
                             N  :  Integer);     (* also input      *)
(* Fits a straight line to a set of N  X-Y pairs   *)
(* The line has a slope equal to A and intercept B *)

Var I:  Integer;
    SumX,SumY,SumXY,SumX2,SumY2,
    SEE,XI,YI,SXY,SXX,SYY:  Real;

Begin (* LeastSquares *)
   SumX:= 0;
   SumY:= 0;
   SumXY:=0;
   SumX2:=0;
   SumY2:=0;
   For I:= 1 to N do
      Begin (* For *)
      XI:= X[I];
      YI:= Y[I];
      SumX:= SumX + XI;
      SumY:= SumY + YI;
      SumXY:= SumXY + XI*YI;
      SumX2:= SumX2 + XI*XI;
      SumY2:= SumY2 + YI*YI;
      End;  (* For *)
   SXX:= SumX2 - SumX*SumX/N;
   SXY:= SumXY - SumX*SumY/N;
```

Figure 26–2 *Program EtAcRate.*

```
    SYY:= SumY2 - SumY*SumY/N;
    s:= SXY/SXX;
    int:= ((SumX2*SumY - SumX*SumXY)/N)/SXX;
End;  (* LeastSquares *)

(**************************************************************)

Procedure Hardcopy;

    Procedure PrintAA;
    VAR I: Integer;
    Begin (* WriteAA *)
      Clrscr;
      Writeln(Lst,'Ro,Rc,a,b =  ',Ro:8:1,Rc:8:1,a:8:4,b:8:4);
      Writeln(Lst);
      Writeln(Lst,'        R/ohm    t/sec            L[I]');
      Writeln(Lst);
      For I:= 1 to N Do
        Writeln(Lst,R[I]:10:1,t[I]:10:1,'          ',L[I]:6:3);
      Writeln(Lst);
      Writeln(Lst,'The Rate constant k equals ',k:10, ' liter/mole/sec');
    End; (* PrintAA *)

   Procedure PrintAB;
   VAR I:  Integer;
     Begin (* WriteAB *)
       Clrscr;
       Writeln(Lst,'Ro,Rc,a,b,c =  ',Ro:8:1,Rc:8:1,a:8:4,b:8:4,c:8:4);
       Writeln(Lst);
       Writeln(Lst,'       R/ohm     t/sec     ln(b(a-x))/(a(b-x))');
       Writeln(Lst);
       For I:= 1 to N Do
          Writeln(Lst,R[I]:10:1,t[I]:10:1,LnAB[I]:17:3);
       Writeln(Lst);
       Writeln(Lst,'The rate constant k equals ',k:10,' liter/mole/sec');
     End;  (* PrintAB *)

  Begin (* HardCopy *)
    Clrscr;
    If aEQUALb then
        PrintAA
    Else PrintAB;
  End;  (* HardCopy *)

(**************************************************************)

Procedure DoAAcalc;
VAR Lo,Lc,Lt,slope,intercept:  Real;

    Procedure CalcAA;
    VAR I:  Integer;
    Begin (* CalcAA *)
      Lo:= 1/Ro;
      Lc:= 1/Rc;
      For I:= 1 to N Do
        Begin
          Lt:= 1/R[I];
          L[I]:= (Lo - Lt)/(Lt - Lc)
        End;
```

Figure 26–2 (continued)

```pascal
        LeastSquares(t,L,intercept,slope,N);
        k:= slope/a;
     End; (* CalcAA *)

   Procedure WriteAA;
   VAR I: Integer;
   Begin (* WriteAA *)
     Clrscr;
     Writeln('Ro,Rc,a,b =  ',Ro:8:1,Rc:8:1,a:8:4,b:8:4);
     Writeln;
     Writeln('        R/ohm    t/sec             L[I]');
     Writeln;
     For I:= 1 to N Do
       Writeln(R[I]:10:1,t[I]:10:1,'           ',L[I]:6:3);
     Writeln;
     Writeln('The Rate constant k equals ',k:10, ' liter/mole/sec');
   End; (* WriteAA *)

Begin (* DoAAcalc *)
   CalcAA;
   WriteAA;
   Writeln;
   Writeln('Do you want a Hard Copy printed?(Y or N?)');
   Writeln('If Yes be sure your printer is ready; then enter a Y.');
   Read(answer);
   If (answer = 'Y') OR (answer = 'y') then Hardcopy;
End; (* DoAAcalc *)

(*********************************************************************)

Procedure DoABcalc;
(* Calculates argument of integrated rate equation for the case  *)
(* of a not equal to b                                           *)

VAR slope,intercept:  Real;
    x: Ary;

  Procedure CalcAB;
  VAR I:  Integer;
    Begin (* CalcAB *)
      For I:= 1 to N Do
        Begin (* For *)
          x[I]:= c*(1/Ro -1/R[I])/(1/Ro - 1/Rc);
          lnAB[I]:= ln(b*(a - x[I])/(a*(b - x[I])));
        End;  (* For *)
      LeastSquares(t,LnAB,intercept,slope,N);
      k:= slope/(a - b);
    End; (* Calc AB *)

  Procedure WriteAB;
  VAR I:  Integer;
    Begin (* WriteAB *)
      Clrscr;
      Writeln('Ro,Rc,a,b,c =  ',Ro:8:1,Rc:8:1,a:8:4,b:8:4,c:8:4);
      Writeln;
      Writeln('       R/ohm      t/sec       ln(b(a-x))/(a(b-x))');
      Writeln;
      For I:= 1 to N Do
         Writeln(R[I]:10:1,t[I]:10:1,LnAB[I]:17:3);
      Writeln;
```

Figure 26–2 (continued)

```
        Writeln('The rate constant k equals ',k:10,' liter/mole/sec');
      End;  (* WriteAB *)

Begin (* DoABcalc *)
   CalcAB;
   WriteAB;
   Writeln;
   Writeln('Do you want a Hard Copy printed?(Y or N?)');
   Writeln('If Yes be sure your printer is ready; then enter a Y.');
   Read(answer);
   If (answer = 'Y') OR (answer = 'y') then Hardcopy;
End;  (* DoABcalc *)

(****************************************************************)

Procedure GetIonConductivities;
Begin (* GetIonConductivities *)
   CLRSCR;
   Writeln('Please enter the individual ionic conductivities.  Some');
   Writeln('suitable values (ohm cm2) are:');
   Writeln('at 25 deg C:  40.8 for Ac-, 192 for OH- and 50.9 for Na+');
   Writeln('at 35 deg C:  48.0 for Ac-, 229 for OH- and 62.3 for Na+');
   Writeln('enter your values');
   Writeln;
   Write('for acetate ion:  ');
   Readln(AC);
   Write('for hydroxyl ion:  ');
   Readln(OH);
   Write('for sodium ion:  ');
   Readln(Na);
End;  (* GetIonConductivities *)

(****************************************************************)

Procedure GetInitialMolarities;
Var Ave:  Real;
Begin (* GetInitialMolarities *)
   Writeln;
   Write('What is the value in ohms of Ro?  ');
   Readln(Ro);Writeln;
   Write('What is the initial molarity of ethyl acetate?  ');
   Readln(a);
   Write('What is the initial molarity of sodium hydroxide?  ');
   Readln(b);
   aEQUALb:= False;      (* initialize the boolean variable *)
   Ave:= (a + b)/2;
   If (Abs(a - b))/Ave <0.05 Then (* check if a is about equal to b *)
     Begin (* If Then *)
       aEQUALb:= True;
       Rc:= Ro*(OH + Na)/(Ac + Na);
       a:= Ave;
       DoAAcalc
     End    (* If Then *)
   Else
     Begin  (* Else *)
       If (a - b)<0 Then
           c:= a           (* molarity of limiting reagent is a *)
       Else c:= b;         (* molarity of limiting reagent is b *)
       Rc:= (Ro*b*(OH + Na))/((b - c)*OH + c*Ac + b*Na);
       DoABcalc
```

Figure 26–2 (continued)

```
     End; (* Else *)
End; (* GetInitialMolarities *)

(*************************************************************)

Procedure GetTimeAndResistance;
Var I,
    X,Y:   Integer;    (* Screen Coordinates *)
Begin (* GetTimeAndResistance *)
   CLRSCR;
   Write('What is the number of R and t pairs you wish to enter?  ');
   Readln(N);
   Writeln;
   For I:= 1 to N Do
     Begin (* For *)
       Write(      'R(',I,'):  ');
       X:= WhereX; Y:= WhereY;
       Read(R[I]);
       GoToXY(X+10,Y);
       Write('      t(',I,'):   ');
       Readln(t[I]);
     End;  (* For *)
End;  (* GetTimeAndResistance *)


(*************************************************************)

Begin (*  M A I N   P R O G R A M  *)
   GetIonConductivities;
   GetTimeAndResistance;
   GetInitialMolarities;
End.  (*  M A I N   P R O G R A M  *)
```

Figure 26–2 (continued)

Critically examine the errors in your experiment. In light of the error evaluation, compare your results with those in the literature.

REFERENCES

1. Daniels, F., R. A. Alberty, J. W. Williams, C. D. Cornwell, P. Bender, and J. E. Harriman, *Experimental Physical Chemistry,* 7th ed., McGraw-Hill, New York, 1970.

2. Guggenheim, E. A., and J. E. Prue, *Physico-Chemical Calculations,* Interscience, New York, 1955, p. 445.

3. Moelwyn-Hughs, E. A., *Kinetics of Reaction in Solution,* 2nd ed., Oxford Press, London, 1947.

4. Moore, J. W., and R. G. Pearson, *Kinetics and Mechanism,* 3rd ed., Wiley, New York, 1981.

5. Saldick, J., and L. P. Hammett, "Rate Measurements by Continuous Titration in a Stirred Flow Reactor," *J. Am. Chem. Soc.* 72:286, 1950.

6. Stieglitz, J., and E. M. Terry, "The Coefficient of Saponification of Ethyl Acetate by Sodium Hydroxide," *J. Am. Chem. Soc.* 49:2216, 1927.

7. Walker, J., "A Method for Determining Velocities of Saponification," *Proc. R. Soc. London, Ser. A,* 78:157–160, 1906.

Experiment 27

The Rate of Reaction Between Acetone and Bromine

The rate of a chemical reaction is one of its fundamental properties. Generally, ionic reactions encountered in inorganic chemistry are much faster than reactions between molecules studied in the organic chemistry laboratory. Precipitation reactions, acid–base reactions, and oxidation–reduction reactions seem to take place virtually instantaneously. In the organic laboratory, on the other hand, it is often necessary to heat reaction mixtures for many minutes, even hours, in order that the reaction mixture move appreciably toward its equilibrium state. In this experiment, we investigate a relatively slow reaction, primarily involving molecules: the reaction between acetone and bromine.

The Rate Law and the Rate of a Reaction

The rate of a reaction always has units of change in concentration per unit time, usually mol L^{-1} sec^{-1}. For the generalized reaction

$$aA + bB = dD + eE \qquad (27\text{--}1)$$

the rate of reaction can be expressed in terms of rates of change of concentration of any of the products or reactants:

$$rate = \frac{-1}{a}\frac{d[A]}{dt} = \frac{-1}{b}\frac{d[B]}{dt} = \frac{1}{d}\frac{d[D]}{dt} = \frac{1}{e}\frac{d[E]}{dt} \qquad (27\text{--}2)$$

Thus, the *rate* is the same whether the course of the reaction is actually followed by determining the change in concentration of A, B, D, or E. Any convenient analytical procedure that permits measuring the change in concentration of a component without disturbing the reaction can be used.

The rate of a chemical reaction depends on the concentrations of the reactants and products:

$$rate = k[A]^p[B]^q[D]^r[E]^s \qquad (27\text{--}3)$$

This equation is called the rate law for the reaction. The proportionality constant k is the specific rate constant. It is temperature dependent and always increases with temperature. The exponents are usually small integers (-1, 0, 1, 2), are temperature independent, and must always be determined experimentally for real solutions.

Determination of the Order

The exponent p in Equation 27–3 is the order of the reaction with respect to the component A, etc. If p were 0, the rate of the reaction would be independent of the concentration of A; the reaction would be zero order with respect to A. The sum of the exponents (if they are small, integral, and positive) is the overall order of the reaction. Otherwise the order is said to be complex.

The rate law for a reaction is determined experimentally by either integral or differential methods.

The Integrated Rate Equation. The rate equation can be integrated for integral orders, such as 0, 1, and 2. These are shown in Table 27–1. The rate law for a chemical reaction is characterized by its order and its specific rate constant. These parameters may be determined either by studying the differential form of the rate law directly, or by application of the integrated rate law.

In the integral method, the applicable rate law is determined by plotting [A], ln [A], and 1/[A] against t. One of these is linear and the corresponding order is the correct one. The other plots would be nonlinear. The slope of the linear plot equals $-k$ or k as the case may be. This method is illustrated in Experiment 26.

The Method of Initial Rates. In this experiment the differential method is used. Integral methods have a number of drawbacks. The reaction may not be of an exactly integral order, side reactions may interfere, and since the reaction must be allowed to go appreciably toward completion, the reverse reaction may invalidate the integrated rate equation.

Many of these difficulties can be avoided by studying the rate of the reaction in its initial phase only, before the reaction advances appreciably. When one of the components is colored, the change in concentration of that component can be measured in situ spectrophotometrically; this is the case with the bromination of acetone:

$$CH_3COCH_3 + Br_2 = CH_3COCH_2Br + Br^- + H^+ \qquad (27\text{–}4)$$

TABLE 27–1

Integrated Rate Equations

Zero Order	First Order	Second Order
$\dfrac{-dA}{dt} = k[A]^0$	$\dfrac{-dA}{dt} = k[A]$	$\dfrac{-dA}{dt} = k[A]^2$
$[A] = [A_0] - kt$	$\ln [A] = \ln [A_0] - kt$	$\dfrac{1}{[A]} = \dfrac{1}{[A_0]} + kt$

The rate equation for the acid-catalyzed reaction is

$$R = \frac{-d[Br_2]}{dt} = k[Ac]^p[Br_2]^q[H^+]^r \tag{27-5}$$

where $Ac = CH_3COCH_3$. The rate of the reaction is conveniently studied by measuring the decrease in absorption of visible light by bromine spectrophotometrically.

In the differential method or method of initial rates, the dependence of the rate on the concentration of each component is studied while the concentrations of the others are held constant. For example, if the initial rate of the reaction of acetone with bromine is studied at several concentrations of acetone, while the concentrations of bromine and hydrogen ions are held constant, the order with respect to acetone can be determined. If the log of both sides of Equation 27–5 is taken, the result is

$$\log R = \log k + p \log [Ac] + q \log [Br_2] + r \log [H^+] \tag{27-6}$$

With the concentrations of bromine and hydrogen ion held constant, Equation 27–6 reduces to

$$\log R = p \log [Ac] + \text{constant} \tag{27-7}$$

A plot of $\log R$ against $\log [Ac]$ gives a straight line with a slope equal to p, the order with respect to acetone. Similarly, the slope of a plot of $\log R$ against $\log [Br_2]$ at constant concentrations of acetone and hydrogen ion gives q, the order with respect to bromine. The order with respect to hydrogen ion is determined by a third plot.

Once the order with respect to each component is determined, the rate constant k can be calculated for each run with Equation 27–5:

$$k = \frac{-d[Br_2]/dt}{[CH_3COCH_3]^p[Br_2]^q[H^+]^r} \tag{27-8}$$

The Reaction Mechanism

The mechanism of a chemical reaction is a list of the elementary reaction steps that lead to the observed stoichiometric reaction. The sum of the elementary reaction steps equals the overall stoichiometry of the overall reaction. If the elementary steps are known, then the rate law can be deduced from them.

The difficulty lies in establishing the elementary steps that make up the mechanism: they *cannot* be deduced, but instead must be discovered by intuitive means. Thus, reaction mechanisms are always *hypothetical* and cannot be proven. Evidence to support a given mechanism can be shown, however. For example, if the rate law deduced from the mechanism is consistent with the experimentally observed rate law, then that is taken as evidence in support of the proposed mechanism. It does not prove the mechanism correct, however, because it is possible to find more than one mechanism from which the same rate law may be deduced!

Other factors may be used to choose among alternative mechanisms. For example, thermodynamic arguments may eliminate certain elementary steps from consideration, or a mechanism may require unusual or implausible species, and consequently must be rejected. Thus, the construction of a mechanism, like the construction of any hypothesis, is tentative.

In acid solution, the rate of the reaction between bromine and acetone is found to be first order with respect to the hydrogen ion concentration and independent of the bromine concentration. In fact, the rate is the same for chlorine, bromine, and iodine. Any mechanism for the reaction between bromine and acetone must be consistent with these observations. For example (Ac = acetone, Int = intermediate),

$$CH_3COCH_3 + H^+ \rightleftarrows \left[CH_3 - \overset{\overset{\displaystyle OH}{\|}}{C} - CH_3 \right]^+ \qquad (27\text{–}9)$$
$$\text{(Ac)} \qquad\qquad\qquad \text{(Int)}$$

$$\left[CH_3 - \overset{\overset{\displaystyle OH}{\|}}{C} - CH_3 \right]^+ + H_2O \underset{k_{-1}}{\overset{k_1}{\rightleftarrows}} CH_3 - \overset{\overset{\displaystyle OH}{|}}{C} = CH_2 + H_3O^+ \quad (27\text{–}10)$$
$$\text{(Int)} \qquad\qquad\qquad\qquad\qquad \text{(Enol)}$$

$$CH_3 - \overset{\overset{\displaystyle OH}{|}}{C} = CH_2 + Br_2 \overset{k_2}{\rightarrow} CH_3COCH_2Br + H^+ + Br^- \quad (27\text{–}11)$$
$$\text{(Enol)} \qquad\qquad\qquad \text{(AcBr)}$$

With the steady-state approximation, the rate of formation of the enol is given by

$$\frac{d[\text{Enol}]}{dt} = 0 = k_1[\text{Int}] - (k_{-1}[H^+] + k_2[Br_2])[\text{Enol}] \quad (27\text{–}12)$$

and the steady-state approximation for the rate of formation of brominated acetone is given by

$$\frac{d[\text{AcBr}]}{dt} = 0 = k_2[Br_2][\text{Enol}] \qquad (27\text{–}13)$$

With the equilibrium constant for Equation 27–9 given by

$$K = \frac{[\text{Int}]}{[\text{Ac}][H^+]} \qquad (27\text{–}14)$$

Equations 27–12, 27–13, and 27–14 can be solved to give the rate law from this mechanism, that is, the rate of formation of the product of the reaction, AcBr, in terms of reactants whose concentrations can be measured:

$$\frac{d[\text{AcBr}]}{dt} = \frac{k_1 k_2 K[\text{Ac}][H^+][Br_2]}{k_{-1}[H^+] + k_2[Br_2]} \qquad (27\text{–}15)$$

If the reaction between the enol and the hydrogen ion is slow (Eq. 27–10) and the reaction between the enol and the bromine is fast (Eq. 27–11), then in the denominator of Equation 29–15,

$$k_{-1}[\text{H}^+] \ll k_2[\text{Br}_2] \qquad \qquad (27\text{–}16)$$

and the rate law (Eq. 27–15) simplifies to

$$\frac{d[\text{AcBr}]}{dt} = k_1 K[\text{Ac}][\text{H}^+]$$

or

$$\frac{d[\text{AcBr}]}{dt} = k_{\text{obs}}[\text{Ac}][\text{H}^+]$$

The observed rate constant equals Kk_1, and the observed rate depends upon the concentrations of acetone and hydrogen ion but is independent of the bromine concentration. The rate is independent not only of the concentration of bromine, but also of halogen, and for this reason the rate of the reaction is the same for chlorine, bromine, and iodine.

Apparatus

Acetone, saturated bromine solution, volumetric ware, a spectrophotometer, and cells are needed for this experiment.

SAFETY CONSIDERATIONS

Bromine causes painful burns on the skin; its vapors are irritating and dangerous. Handle the solution in the hood and wear protective gloves. Acetone is flammable; avoid inhaling acetone vapor.

EXPERIMENTAL PROCEDURE

The experiment should be designed so that the concentrations of acetone, bromine, and hydrogen ion are each varied at least three times as the others are held constant. A range of concentrations must be selected so that the slowest and fastest rates are convenient to measure. If the bromine color disappears within seconds upon mixing, the concentrations of one or more constituents is too high. On the other hand, if many minutes pass before the absorbance changes noticeably, the concentrations are too dilute. Some suggested concentration ranges are listed in Table 27–2.

Start with some intermediate combination and experiment until you have a feel for the rates that result. If the maximum concentrations are used for all the constituents, the rate is probably too

TABLE 27–2

Suggested Minimum and Maximum Molarities After Mixing

	Acetone	Bromine	HCl
Max	4.0	0.02	1.0
Min	0.1	0.001	0.02

fast to measure. Similarly, if the minimum concentrations are used for all the components, the measured rate is probably far too slow to measure conveniently.

Notice that these concentrations are *after* mixing. The use of a constant final volume simplifies the design of the experiment. For example, if aliquots of some stock solutions of acetone, bromine, and HCl are taken and always mixed and diluted in a 100-ml volumetric flask, then the final concentrations may be easily varied or held constant, and calculated. In Table 27–3 some possibilities are listed. Here the concentrations of bromine and HCl are held constant as the concentration of acetone after mixing is varied from 1.0 to 0.2 M.

The initial rates are found by measuring the decrease in bromine absorbency at a selected wavelength. The wavelength should be selected so that the absorbencies range from about 0.1 to 0.9; the initial absorbency should lie near the upper end of this range. The absorbency, defined as

$$A = -\log \frac{I}{I_0} \qquad (27\text{–}17)$$

is directly proportional to the concentration of the absorbing species, in this case bromine, when the solution is sufficiently dilute to obey Beer's Law:

$$A = \epsilon bc \qquad (27\text{–}18)$$

where ϵ is the molar absorptivity (L mol^{-1} cm^{-1}), b is the path length (cm), and c is the concentration (mol/L). The molar absorptivities at various wavelengths are given in Table 27–4.

TABLE 27–3

Aliquots (ml) to Be Mixed and Diluted to 100.0 ml

Run	4 M Acetone	0.02 M Br$_2$	1.0 M HCl
1	25	25	10
2	10	25	10
3	5	25	10
⋮	⋮	⋮	⋮

TABLE 27–4

Molar Absorptivities of Bromine in Aqueous Solution

Wavelength (nm)	Molar Absorptivity ε (L mol^{-1} cm^{-1})
400	160
450	100
500	30
550	8

Source: Daniels, F., et al., *Experimental Physical Chemistry,* 7th ed., McGraw-Hill, New York, 1970.

A 5:1 dilution of a bromine saturated solution gives an approximately 0.03 M bromine solution. A more exact value can be found by measuring the absorbance at a wavelength, which results in an absorptivity between about 0.2 and 0.7. The concentration can then be calculated with Equation 27–18 and the data in Table 27–4.

Results and Calculations

Logically organize your data into appropriate tables. Prepare graphs of bromine concentration against time for variable concentrations of acetone, of bromine, and of HCl, and from the slopes of these graphs, calculate the rates for the various runs. Use more than one graph per page.

Prepare graphs of the log of rate against the log of concentration of acetone, bromine, and HCl. From the slopes of these graphs, calculate the order of the reaction with respect to acetone, bromine, and H^+ according to Equation 27–6. After determining the order, calculate the specific rate constant k for each run with Equation 27–8. Critically evaluate the errors in the orders and in k. Compare the average k with the literature value.

REFERENCES

1. Barrow, G. M., *Physical Chemistry,* 4th ed., chap. 19, McGraw-Hill, New York, 1979.

2. Bartlett, P. D., "Enolization as Directed by Acid and Base Catalysts, II. Enolic Mechanism of the Haloform Reaction," *J. Am. Chem. Soc.* 56:967–969, 1934.

3. Bell, R. P., and P. Jones, "Binary and Ternary Mechanisms in the Iodination of Acetone," *J. Chem. Soc.* 1930:2186.

4. Daniels, F., et al., *Experimental Physical Chemistry,* 7th ed., McGraw-Hill, New York, 1970.

5. Ingold, C. K., *Structure and Mechanism in Organic Chemistry,* G. Bell & Sons, London, 1964.

6. Lapworth, A., "The Action of Halogens on Compounds Containing the Carbonyl Group," *J. Chem. Soc.* 85:30, 1904.

7. Moore, J. W., and R. G. Pearson, *Kinetics and Mechanisms,* 3rd ed., Wiley, New York, 1981.

8. Zucker, L., and L. P. Hammett, "Kinetics of the Iodination of Acetophenone in Sulfuric and Perchloric Acid Solutions," *J. Am. Chem. Soc.* 61:2791, 1939.

Experiment 28

Computerized Data Acquisition of a Second-Order Reaction

In acid solution, the hexacyanoferrate(III) ion is reduced to hexacyanoferrate(II):

$$[Fe(CN)_6]^{3-} + e^- = [Fe(CN)_6]^{4-} \qquad E^0 = 0.36 \text{ volt} \quad \textbf{(28-1)}$$

Because the 3+ state of iron is stabilized by the complexing ligands, it is a weaker oxidizing agent than the iron(III) ion:

$$[Fe(H_2O)_6]^{3-} + e^- = [Fe(H_2O)_6]^{4-} \qquad E^0 = 0.77 \text{ volt} \quad \textbf{(28-2)}$$

At room temperature the hexacyanoferrate(III) ion reacts at a moderate rate with ascorbic acid ($C_6H_8O_6$), which is oxidized to dehydroascorbic acid ($C_6H_6O_6$) (Mehrotra et al., 1969):

$$2[Fe(CN)_6]^{3-} + C_6H_8O_6 = 2[Fe(CN)_6]^{4-} + C_6H_6O_6 + 2H^+ \quad \textbf{(28-3)}$$

The structures of ascorbic acid and dehydroascorbic acid are shown in Figure 28–1. The reaction is first order with respect to hexacyanoferrate(III) and first order with respect to ascorbic acid. The reaction stoichiometry is of the type

$$a\text{A} + b\text{B} \rightarrow \text{products} \quad \textbf{(28-4)}$$

and the rate law is

$$\frac{-d[\text{A}]}{dt} = k[\text{A}][\text{B}] \quad \textbf{(28-5)}$$

where [A] and [B] represent the concentrations of A and B at time t. Integration of Equation 28–5 (Watkins and Olson, 1980) leads to

$$\ln\frac{[\text{A}]}{[\text{B}]} = \frac{b[\text{A}]_0 - a[\text{B}]_0}{a} kt + \ln\frac{[\text{A}]_0}{[\text{B}]_0} \quad \textbf{(28-6)}$$

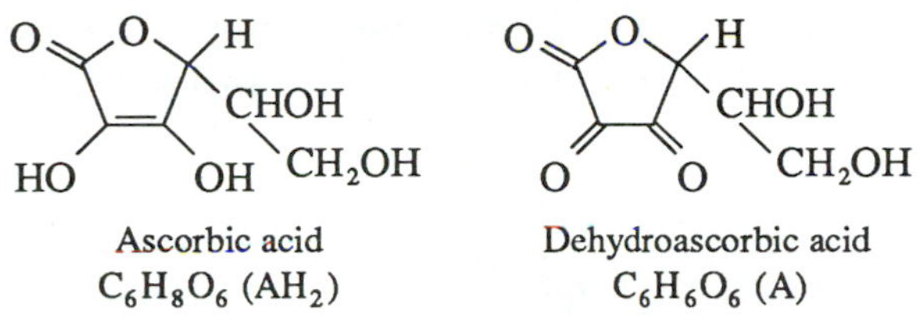

Ascorbic acid
$C_6H_8O_6$ (AH$_2$)

Dehydroascorbic acid
$C_6H_6O_6$ (A)

Figure 28–1 Structural formulas of ascorbic acid and dehydroascorbic acid.

In Equation 28–6, $[A]_0$ and $[B]_0$ are the initial molar concentrations of A and B. If $[A]$ represents the molar concentration of ascorbic acid and $[B]$ represents the molar concentration of hexacyanoferrate(III) at time t, then $a = 1$ and $b = 2$. According to Equation 28–6, a plot of $\ln [A]/[B]$ versus time gives a straight line of slope equal to $(b[A]_0 - a[B]_0)k/a$, permitting the calculation of the second-order specific rate constant k. In this experiment, $b = 2$ and $a = 1$, so the slope equals $(2[A]_0 - [B]_0)k$.

Analysis

Because none of the products, intermediates, or reactants is colored except the hexacyanoferrate(III) ion, it is convenient to follow the course of the reaction spectrophotometrically. Since the absorption coefficient of the yellow aqueous hexacyanoferrate(III) ion equals 1012 M^{-1} cm^{-1} (Aziz and Mirza, 1964), the absorbance of hexacyanoferrate(III) at time t is given by

$$\text{absorbance} = 1012[\text{Fe(CN)}_6]^{3-} \qquad (28\text{–}7)$$

It follows from the stoichiometry of Equation 28–3 that the concentration of ascorbic acid at time t is given by

$$[C_6H_8O_6] = [C_6H_8O_6]_0 - \tfrac{1}{2}[\text{Fe(CN)}_6]_0^{3-} - [\text{Fe(CN)}_6]^{3-} \qquad (28\text{–}8)$$

Mechanism

The mechanism for the reaction (Mehrotra et al., 1969) consists of the formation of an ascorbate anion (AH$^-$) in a rapid equilibrium ionization of ascorbic acid (AH$_2$):

$$AH_2 \underset{k_{-1}}{\overset{k_1}{\rightleftharpoons}} AH^- + H^+ \qquad \text{fast} \qquad (28\text{–}9)$$

The ionization is followed by the slow rate-determining step, the oxidation of the ascorbate ion to the short-lived ascorbate free radical (AH·):

$$[\text{Fe(CN)}_6]^{3-} + AH^- \overset{k_2}{\rightarrow} [\text{Fe(CN)}_6]^{4-} + AH\cdot \qquad \text{slow} \qquad (28\text{–}10)$$

In the final step, an electron is rapidly transferred from the ascorbate free radical to the hexacyanoferrate(III), producing dehydroascorbic acid (A):

$$[\text{Fe(CN)}_6]^{3-} + AH\cdot \overset{k_3}{\rightarrow} [\text{Fe(CN)}_6]^{4-} + A + H^+ \qquad \text{fast} \qquad (28\text{–}11)$$

Application of the steady-state approximation to the elementary steps of the reaction mechanism leads to the theoretical rate law (Mehrotra et al., 1969):

$$\frac{-d}{dt}[Fe(CN)_6]^{3-} = \frac{2k_1k_2[AH_2][Fe(CN)_6]^{3-}}{k_{-1}[H^+] + k_2[Fe(CN)_6]^{3-}}$$

Since it is postulated in the mechanism that $k_{-1} \gg k_2$, it follows that $k_{-1}[H^+] \gg k_2[Fe(CN)_6]^{3-}$ and that

$$\frac{-d}{dt}[Fe(CN)_6]^{3-} = \frac{2k_1k_2[AH_2][Fe(CN)_6]^{3-}}{k_{-1}[H^+]}$$

Since this is also the form of the observed rate law, the postulated mechanism is supported.

Effect of Ionic Strength

According to the Transition State Theory for ionic reactions in aqueous solution (Moore and Pearson, 1981) the specific rate constant depends upon the ionic strength I of the solution and on the charges Z_A and Z_B of the ionic species reacting to form the activated complex:

$$\log k = \log k_0 + 1.02 Z_A Z_B \frac{I^{1/2}}{1 + I^{1/2}} \qquad \text{(28–12)}$$

The constant 1.02 arises from the Debye-Hückel Theory for aqueous solutions at 25°C. The specific rate constant for an ideal solution (ionic strength equals zero, activity coefficients equal unity) is given by k_0.

Equation 28–12 suggests that a plot of $\log k$ versus $I^{1/2}/(1 + I^{1/2})$ has a slope equal to $1.02 Z_A Z_B$ and an intercept equal to $\log k_0$. Depending on the charges Z_A and Z_B of the species entering into the formation of the activated complex, the slope may be positive, zero, or negative. Since 1.02 is approximately equal to unity, slopes of 0, ±1, ±2, ±3, and so on should be observed (Fig. 28–2).

If the reacting ions are of the same charge, the rate of the reaction should increase with ionic strength, but if the ions are of different charges, the rate of the reaction should decrease with ionic strength. If one of the moieties forming the activated complex is a neutral molecule, then the rate of the reaction should be independent of the ionic strength, corresponding to zero slope in Equation 28–12.

For the reaction described in this experiment, you will measure the kinetic salt effect and determine whether the observed dependence of the reaction rate on ionic strength is consistent with the proposed reaction mechanism.

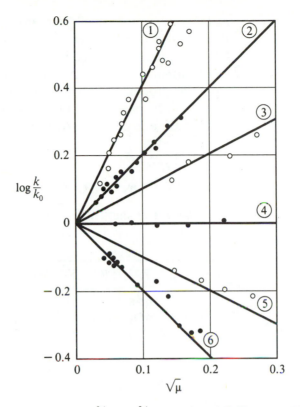

1. $2[Co(NH_3)_5Br]^{2+} + Hg^{2+} + 2H_2O \rightarrow 2Co[(NH_3)_5H_2O]^{3+} + HgBr_2$

2. $S_2O_8^{2-} + I^- \rightarrow I_2 + 2SO_4^{2-}$

3. $[NO_2{=}N{-}COOC_2H_5]^- + OH^- \rightarrow N_2O + CO_3^{2-} + C_2H_5OH$

4. $C_{12}H_{22}O_{11} + OH^- \rightarrow$ invert sugar

5. $H_2O_2 + 2H^+ - 2Br^- \rightarrow 2H_2O + Br_2$

6. $[Co(NH_3)_5Br]^{2+} + OH^- \rightarrow [Co(NH_3)_5OH]^{2+} + Br^-$

Figure 28–2 Kinetic salt effect. (Source: LaMer, V. K., *Chem. Rev.* 10:179, 1932.)

Apparatus

The experiment requires potassium hexacyanoferrate(III), $K_3Fe(CN)_6$; ascorbic acid, $C_6H_8O_6$; 0.010 M HNO_3 (standardized); disodium EDTA dihydrate; four 100-ml volumetric flasks; a 500-ml volumetric flask; a 250-ml Erlenmeyer flask; a 100-ml beaker; a 25-ml pipet; a second timer; and a spectrophotometer (visible wavelength range).

In addition, it is convenient to use a microcomputer and analog-to-digital (A/D) converter. An inexpensive configuration consists of an Apple computer (II+ or IIe) equipped with an AdaLab card[1] and

[1] Available from Interactive Microware, Inc., P.O. Box 139, College Station, Pennsylvania 16804, phone (814)238-8294.

accompanying software. Alternatively, an IBM-PC computer (or clone) fitted with a Dash-16 A/D card[2] may be used. Other computers and A/D converters may also be used. Although the description of the data acquisition in this experiment is for the Apple computer with an AdaLab card, the program can be used with other configurations with minor changes to the specific software. Some features of data acquisition with an IBM-PC and Dash-16 A/D card are described in Experiment 24.

SAFETY CONSIDERATIONS

Potassium hexacyanoferrate (III) is toxic.

EXPERIMENTAL PROCEDURE

Prepare four solutions that are 1×10^{-3} M in $K_3Fe(CN)_6$ and of varied $NaNO_3$ concentrations: 0.025, 0.05, 0.1, and 0.2 M. Weigh the salts to ± 0.1 mg into 100-ml volumetric flasks and dilute to the mark.

Prepare 500 ml of 2.5×10^{-4} M ascorbic acid. As a solvent use 0.01 M HNO_3 containing 0.001% disodium EDTA dihydrate. Weigh the ascorbic acid to ± 0.1 mg. Because ascorbic acid solutions are slowly oxidized by atmospheric oxygen, they are made up fresh daily. The nitric acid is a standard solution of exactly the same concentration for all the kinetic runs.

If the spectrophotometer cell is thermostatted, allow the solutions to come to thermal equilibrium in a 25°C thermostat. Otherwise, let the solutions, glassware, spectrophotometer, and its cuvettes stand together overnight to come to thermal equilibrium. In either case, record the reaction temperature. Pipet a 25-ml aliquot of one solution into a 250-ml Erlenmeyer flask and a 25-ml aliquot of the other solution into a 100-ml beaker. The glassware and solutions should be at the reaction temperature. Set the spectrophotometer to 418 nm and zero the absorbance with two matched 1-cm cells in the cell compartment.

Manual Data Collection. Pour the solution from the beaker to the Erlenmeyer and start the second timer *immediately*. Swirl for 2 or 3 seconds, pour the reaction mixture into a 1-cm cuvette, and place the cuvette in the sample holder of the spectrophotometer. Read the absorbance when the timer reaches 30 seconds and every 30 seconds thereafter. Measure the specific rate constant at three or four ionic strengths ranging from about 0.025 to 0.2 M.

[2] Supplied by MetraByte Corp., 440 Myles Standish Boulevard, Taunton, Massachusetts 02780, phone (617) 880-3000.

Computerized Data Acquisition. Boot up the data acquisition program shown in Figure 28–3 by inserting a DOS disk in the Apple computer equipped with an AdaLab card and turning it on. Replace the DOS disk with one with *Program GetData* (Fig. 28–2) and enter *Load GetData.* When the program has been loaded, enter *Run* and answer the interactive questions that follow.

1. WHAT IS THE RUN NUMBER? Enter an integer corresponding to your run number. This is just for convenience in organizing your output hard copy.
2. WHAT IS THE INITIAL CONCENTRATION (M) OF ASCORBIC ACID IN MOLES/LITER? Enter the concentration *after mixing.*
3. WHAT IS THE INITIAL CONCENTRATION (M) OF POTASSIUM HEXACYANOFERRATE(III) IN MOLES/LITER? Enter the value *after mixing.*
4. WHAT IS THE DELAY TIME IN SECONDS? If the time between measurements is 30 seconds, enter 30.
5. WHAT IS THE SAMPLE SIZE? Enter the number of points at which the sample absorbance is to be measured. If you enter 40, and 30 was entered in item 4 above, 40 points will be measured every 30 seconds and the total run time will be 1200 seconds, or 20 minutes.

At this point the computer responds with the message HIT RETURN TO BEGIN.

To initiate the run, pour the solution from the beaker to the Erlenmeyer flask and hit the return key of the computer *immediately.* The computer will clear the screen and write to the screen the flashing message: RUN IN PROGRESS: DO NOT DISTURB.

Swirl for 2 or 3 seconds, pour the reaction mixture into a 1-cm cuvette, and place the cuvette in the sample holder of the spectrophotometer. Be sure that the printer is turned on.

The program in Figure 28–3 records the first absorbance 30 seconds after the return key is pressed at the time the solutions are mixed. Consequently, the mixing, swirling, and pouring of the solutions and their placement into the spectrophotometer must be done in *less than 30 seconds.* The remainder of the data acquisition and reduction to the specific rate constant is automatic. At the end of the run, the run data will be printed on the screen and on the printer. Measure the specific rate constant at three or four ionic strengths ranging from about 0.025 to 0.2 M.

Results and Calculations

Manual Data Acquisition. From the molar absorbance of $K_3Fe(CN)_6$, which equals 1020 L mol^{-1} cm^{-1}, calculate the molar concentration of $K_3Fe(CN)_6$ at each time at which the absorbance was

```
10 REM   *********************
20 REM   *                   *
30 REM   *   PROGRAM GETDATA  *
40 REM   *                   *
50 REM   *********************
60 REM
70 REM  LINES 100 & 110 INITIALIZE ADALAB CARD
80 REM   THE ADALAB CARD
90 REM
100  HIMEM: 36095:DZ = 0: DIM CZ(5),QZ(5),DZ(100)
110  PRINT  CHR$ (4)"BRUN QUICKI/O"
120 REM   *********************
130 REM  LINE 170 BEGINS
140 REM   INTERACTIVE INPUT
150 REM   *********************
160  HOME : DIM TZ(100)
170  INPUT "WHAT IS THE RUN NUMBER? ";RN
180  PRINT
190  INPUT "WHAT IS THE INIT. CONC. OF ASCORBIC ACID (MOLES/LITER)";AO
200  PRINT
210  INPUT "WHAT IS THE INIT. CONC. OF HEXACYANOFERRATE)MOLES/LITER)";FO

220  PRINT
230  INPUT "WHAT IS THE DELAY TIME IN SECONDS? ";DT
240 DT = DT * 10
250  PRINT
260  INPUT "WHAT IS THE SAMPLE SIZE? ";X
270  PRINT
280  INPUT "HIT RETURN TO BEGIN RUN";A$
290 REM   *********************
300 REM  LINE 360 INITIALIZES
310 REM      THE CLOCK
320 REM   *********************
330  HOME
340  FLASH
350  VTAB 12: PRINT "RUN IN PROGRESS: DO NOT DISTURB"
360  & TI1:T = DZ: & AIO
370 REM   *********************
380 REM  L#'S 500-70 = DATA
390 REM   AQUISITION LOOP
400 REM   *********************
410 REM  LINES 430-440 DISCARD FIRST VALUE
420  FOR SAMPLE = 1 TO X
430  & AIO
440 DV = DZ * FO
450 T = T - DT
460 TZ(SAMPLE) = T
470  & TI1: IF DZ > T GOTO 470
480  & AIO,SAMPLE: NEXT SAMPLE
490 REM   *********************
500  REM  BEGIN DATA PROCESSING
510  REM  *********************
520 AS = 0
530  PR# 1
540  HOME
550  NORMAL
560  PRINT "               RUN#";RN
570  PRINT "----------------------------------"
580  PRINT "T(SEC)"; TAB( 12)"LOG(AS/FE)"; TAB( 30)"ABS"
590  PRINT
600  FOR SAMPLE = 1 TO X
610  REM  *********************
620  REM  LINE 670 CONVERTS DIGIT
630  REM       TO ABSORBANCE
640  REM  *********************
650 M = 4.994E - 3
660 B = 0.0493
670 A = M * DZ(SAMPLE) + B
680  REM  *********************
690  REM  LINE 730 CONVERTS ABS,
700  REM   TO CONC. BASED ON LIT.
710  REM   VALUE FOR M. ABS. COEF.
720  REM  *********************
730 C = A / 1012
740 AS = AO - ((FO - C) / 2)
750  IF AS < = 0 THEN 790
760 L =  LOG (AS / C)
770  PRINT ( ABS (TZ(SAMPLE) - TZ(1)) / 10); TAB( 12)L; TAB( 30)A
780  GOTO 800
790  PRINT "NEGATIVE ARGUMENT FOR LOG"
800  NEXT SAMPLE
810  PRINT
820  PRINT "INIT CONC OF ASCORB ACID = ";AO;"(MOLES/LITER)"
830  PRINT "INIT CONC OF HEXAFERRATE = ";FO;"(MOLES/LITER)"
840  PR# 0
850  END
```

Figure 28–3 *Program GetData.*

measured. From the stoichiometry of the reaction (Eq. 28–3) and the initial concentrations of the reactants, calculate the concentrations of ascorbic acid (Eq. 28–8).

Prepare appropriate tables of experimental and calculated data and plot $\log [C_6H_8O_6]/[K_3Fe(CN)_6]$ versus t (sec). Run a least-squares analysis of the plot, and from the slope calculate the specific rate constant k. According to the reaction stoichiometry (Eq. 28–3) and the rate law (Eq. 28–6), the slope is given by

$$\text{slope} = (2[A]_0 - [B]_0)k \qquad (28\text{–}13)$$

To investigate the kinetic salt effect, plot $\log k$ versus $I^{1/2}/(1 + I^{1/2})$. Compare the slope of this plot with the slope expected from Equation 28–13. Discuss the kinetic salt effect in terms of the proposed mechanism for the reaction. Compare the experimental specific rate constant with the value reported in the literature and discuss the difference in terms of an analysis of the errors affecting the experiment.

Program Documentation. *Program GetData* (Fig. 28–3) consists of four distinct sections. The first section initializes the system, the second provides for user input, the third does the data sampling and A/D conversion, and the fourth writes the output to the printer.

Line 100 is essentially unchanged as provided by Interactive Microware (IM), the designer of the AdaLab data acquisition card. The statements in line 100 initialize some integer arrays, and the last statement runs IM's associated software, which is on the disk in machine language and is called *Quick I/O*. Our program, called *GetData*, communicates with the A/D converter through *Quick I/O*.

The rest of the program is written in AppleSoft BASIC, which includes a few extensions for controlling input and output between the AdaLab card and the computer memory. Note that *Quick I/O* recognizes the identifiers of only two variables in the computer memory. The first is the integer variable with the identifier $D\%$. The second is the integer array with the exact same identifier $D\%$.[3] A legal assignment to the first is, for example,

50 D% = 48

An example of a legal assignment to the integer array is

60 D%(3) = 57

Remember, these are the only two variables in memory with which the AdaLab card directly communicates.

In line 160, we conclude the initialization section of the program by dimensioning the integer array $T\%$, which will hold a list of sampling times.

---

[3] In BASIC, an integer variable and an integer array variable may be given the same identifier. They are still independent variables and really have nothing to do with each other. Interactive Microware has chosen to use the same identifier, $D\%$, for an integer variable and an integer array variable.

The second section, the input section, includes statements 170 through 280. This section allows the user to vary the parameters of the experiment, such as the initial concentrations of the reactants, the delay time between samples, and the number of samples. Here, a sample is a reading of the analog voltage from the spectrophotometer by the A/D converter. This section is pure AppleSoft BASIC.

The third section is the heart of the data collection program, statements 290 through 480. Only two statements are used that are not standard BASIC, but part of IM's *Quick I/O: &TI1* and *&AI0*.

Actually, the ampersand (&) is used in any AppleSoft program that must communicate with a user-supplied (that's us) machine language program (that's *Quick I/O*). When we make a kinetics run, what do we do? We measure time and concentration. In *Program GetData* we measure time with the statements using the *&TI1* expression and we measure concentration (actually analog voltage) with the expression *&AI0*. Let us analyze these *Quick I/O* statements to see how they are used in a program and what they do.

First, consider *&TI1*. The & tells AppleSoft to go to the AdaLab card. The *T* stands for Timer, which is a count-down timer in the card. It ticks away every 0.1 seconds as soon as the program is run. The *1* stands for timer number 1 (there are two others in the card, but they are not needed here). The *I* stands for input *to the computer memory*. Where? If no parameters are supplied, the *current* reading of the timer is input to *D%*, where it is stored until you do something with it. In other words, the program segment

50 &TI1

60 PRINT D%

prints the current reading of timer number 1. The value is some arbitrary integer number, output by the timer. If the timer is read exactly 10 seconds later, the reading is again an arbitrary integer. The difference between the two integers, in this example, would be exactly 100, since 100 increments would have been ticked off by the timer.

The reading of the timer can also be stored in the integer array *D%* by specifying the number of the array element where the value is to be stored. For example,

50 &TI1,23

60 PRINT D%(23)

Line 50 says, "Read timer number 1, and input the value of the time into the 23rd element of the integer array *D%*." Line 60 prints the value in *D%(23)*, which would be the current timer reading.

The instruction *&AI0* is used with the A/D converter in the AdaLab card. The & has the same meaning as before, the *A* stands for A/D converter, and the *0* means that the AdaLab card has been installed in slot 0 of the Apple computer. Once again *I* stands for input to the *computer* *memory.* Where? You guessed it. With *Quick I/O,* there are only two possibilities: $D\%$ the integer variable or $D\%$ the integer array variable.

First, an example with the integer variable $D\%$:

50 &AI0

60 PRINT D%

When line 50 is executed, the & tells AppleSoft to go to *Quick I/O* and AdaLab card. The *A* stands for A/D converter and the *0* tells *Quick I/O* that the card is in slot 0 of the Apple computer. When line 50 is executed, the current value of the analog voltage is sampled (or read), converted to a digit, and stored in the computer memory in the integer variable $D\%$. Line 60 prints the current value in $D\%$.

Now an example with the integer array variable $D\%$. If the *&AI0* instruction is followed by a comma and an integer value, then *Quick I/O* understands that the integer is the number of an element in the integer array $D\%$:

50 &AI0,23

60 PRINT D%(23)

In this case, the analog voltage is converted to a digit, which is stored in the 23rd element of the integer array, that is, in $D\%(23)$.

The fourth and final section of the program (lines 490 to 850) is written in pure AppleSoft BASIC (no *Quick I/O*). The concentration of hexacyanoferrate(III) is calculated from the absorbance and the molar absorption coefficient in line 730. Then the concentration of ascorbic acid is calculated in line 740, which is a BASIC version of Equation 28–8. The left side of Equation 28–6 is calculated in line 760 and, along with time and absorbance, is tabulated in line 770.

These examples serve to illustrate the actual logic of *Program GetData.* Line-by-line documentation is shown in Figure 28–3.

Calibrating the Data Acquisition Card with *Program Calibrate.* Unlike a first-order reaction, it is necessary to know the actual molar concentration of the reactants as a function of time when analyzing a second-order reaction. In a first-order reaction, only the ratios of concentration appear in the integrated rate law, e.g., (C/C_0), so the units of concentration are unimportant. When applying A/D conversion to first-order kinetics, the ratio of the digital output can be used, since it is proportional to concentration (Experiments 24 and 29).

Consequently, for a second-order reaction, it is necessary to calibrate the A/D converter to obtain the relationship between the concentration [of hexacyanoferrate(III)] and the digital output of the converter. This is done by placing several concentrations of any colored solution in the spectrophotometer, noting the absorbance, running the program, and recording the digital output for each concentration. In effect, this

```
10    REM    **********************
20    REM    *                    *
30    REM    *      PROGRAM        *
40    REM    *      CALIBRATE      *
50    REM    *                    *
60    REM    **********************
70    REM
80    REM   L#100 INITIALIZES ADALAB
90    HIMEM: 36095:D% = 0: DIM C%(5),Q%(5),D%(100)
100   PRINT  CHR$ (4)"BRUN QUICKI/O"
110   DIM A(100)
120   HOME
130   PRINT "THIS PROGRAM CALCULATES A"
140   PRINT "CALIBRATION CURVE FOR USE"
150   PRINT "WITH PROGRAM GETDATA"
160   PRINT
170   INPUT "INPUT THE NUMBER OF POINTS";N
180   & TI1:T = D%: & AIO:DLAY = 1
190   REM    ***********************
200   REM   BEGIN DATA AQUISITION
210   REM    ***********************
220   FOR SAMPLE = 1 TO N
230   INPUT "INPUT THE ABSORBANCE";A
240 A(SAMPLE) = A
250   T = T - DLAY
260   REM   LINES 270-280 DISCARD FIRST SAMPLE
270   & AIO
280 DV = D% * A
290   & TI1: IF D% > T GOTO 290
300   & AIO,SAMPLE
310   NEXT SAMPLE
320   REM    ***********************
330   REM     BEGIN LEAST SQUARES
340   REM    ***********************
350 X = 0
360 Y = 0
370 XY = 0
380 X2 = 0
390 Y2 = 0
400 M = 0
410 B = 0
420 R = 0
430   FOR J = 1 TO N
440 X = X + D%(J)
450 Y = Y + A(J)
460 XY = XY + D%(J) * A(J)
470 X2 = X2 + D%(J) ^ 2
480 Y2 = Y2 + A(J) ^ 2
490   NEXT J
500 M = (N * XY - (X * Y)) / (N * X2 - (X ^ 2))
510 B = (X2 * Y - (X * XY)) / (N * X2 - (X ^ 2))
520 R1 = (N * XY - X * Y)
530 R2 =  SQR (N * X2 - X ^ 2) *  SQR (N * Y2 - Y ^ 2)
540 R = R1 / R2
550   PR# 1
560   PRINT
570   PRINT
580   PRINT "   CALIBRATION  CURVE"
590   PRINT "---------------------------"
600   PRINT "Y (ABS)","X (DIGIT)"
610   PRINT "---------------------------"
620   FOR I = 1 TO N
630   PRINT A(I),D%(I)
640   NEXT I
650   PRINT
660   PRINT "SLOPE = M = ";M
670   PRINT "INTERCEPT = B = ";B
680   PRINT "CORR. COEF. = ";R
690   PRINT
700   PRINT "TRANSFER M & B VALUES TO LINES 650 & 660 OF GETDATA"
710   PR# 0
720   END
```

Figure 28–4 *Program Calibrate.*

procedure gives the data for a Beer's Law plot of absorbance against digits (instead of against concentration),

$$A = mD + b \qquad (28\text{--}14)$$

where A is absorbance, D is digital output, and m and b are the slope and intercept of the linear plot of A versus D determined by a least-squares analysis of the data. As with a Beer's Law plot, b should be zero, but with an A/D converter there may be a small voltage that is converted to a (relatively) small digit. The determination of Equation 28–14 must be done prior to performing the experiment, since the actual values depend on the particular A/D card and spectrophotometer used. Equation 28–14 appears as line 670 in *Program GetData;* the values are for an AdaLab card and a Beckman 25 spectrophotometer. Once these parameters have been determined for a given A/D card and spectrophotometer, they are quite constant and need to be checked only occasionally.

Program Calibrate (Fig. 28–4) provides a convenient method for calibrating any spectrophotometer to an AdaLab data acquisition card. To use the program, the spectrophotometer is turned on and set to the desired wavelength, the program is booted, and several solutions with absorbencies between about 0.1 and 0.9 are inserted into the spectrophotometer sample compartment. The values of the absorbencies are read from the spectrophotometer and entered interactively to *Program Calibrate.* The data are input with the **for . . . next** loop from line 220 to line 310. After the last sample absorbency has been entered, the program runs a least-squares fit on the absorbencies and the corresponding digits generated by the AdaLab card (lines 350 to 540). At line 550 the printer is activated, and the remainder of the program prints out the data and prompts the user to transfer the slope m and intercept b calculated by the least-squares routine to lines 650 and 660 of *Program GetData.*

Program Calibrate must be run at least once to calibrate a given AdaLab card and spectrophotometer. The amount of drift that occurs is a matter of experience in each laboratory.

REFERENCES

1. Aziz, F., and G. A. Mirza, "Spectrophotometric Determination of Hydrogen Peroxide in Alkaline Solution," *Talanta* 11:889, 1964.

2. Barnaal, D., *Digital and Microprocessor Electronics for Scientific Applications,* Breton, North Scituate, Massachusetts, 1982.

3. Benson, S. W., *The Foundations of Chemical Kinetics,* McGraw-Hill, New York, 1960.

4. Diefendoerfer, A. J., *Principles of Electronic Instrumentation,* Saunders College Publishing, Philadelphia, 1979.

5. Higgins, R. J., *Electronics with Digital and Analog Integrated Circuits,* Prentice-Hall, Englewood Cliffs, New Jersey, 1983.

6. Malmstadt, H. V., and C. G. Enke, *Digital Electronics for Scientists,* Benjamin/Cummings, Menlo Park, California, 1969.

7. Martins, L. J., and J. B. daCosta, "The Oxidation of Ascorbic Acid by

Hexacyanoferrate(III) Ion in Acidic Aqueous Media," *J. Chem. Educ.* 65:176–178, 1988.

8. Mehrotra, U. S., M. C. Agrawal, and S. P. Mushran, "Kinetics of the Reduction of Hexacyanoferrate by Ascorbic Acid," *J. Phys. Chem.* 73:1996, 1969.

9. Moore, J. W., and R. G. Pearson, *Kinetics and Mechanism,* 3rd ed., Wiley, New York, 1981.

10. Watkins, K. W., and J. A. Olson, "Ionic Strength Effect on the Rate of Reduction of Hexacyanoferrate(III) by Ascorbic Acid," *J. Chem. Educ.* 57: 158–159, 1980.

Experiment 29

First-Order Decay of the Triplet State

Consider a molecule in its ground electronic state. Absorption of a photon of electromagnetic radiation, usually in the visible or ultraviolet region of the spectrum, may lead to the molecule reaching an excited electronic state. When the absorption spectrum of a molecule is measured, the absorption of photons is observed, but not the fate of the molecules in their excited states. Usually the excitation energy is simply converted to the translational, rotational, and vibrational energies of neighboring molecules through collision. Since this energy ultimately ends up slightly heating the sample and its surroundings, this process is called **thermal degradation.**

Another way an electronically excited molecule can lose its energy is through reaction with itself (dimerization, polymerization) or with a different molecule, if present, to form new chemical substances. The study of these reactions is called **photochemistry.**

A third path by which an electronically excited molecule can rid itself of its excess energy is by **spontaneous emission** of photons, the study of which is called **photophysics.** Such radiative decay is called luminescence, which is further classified into fluorescence and phosphorescence. If light (usually ultraviolet) falling upon a sample results in the emission of light (always of a longer wavelength), then the phenomenon is called **luminescence.** If the emission of light continues for a noticeable time after the incident radiation is removed, then it is called **phosphorescence.** If the emission of light ceases immediately after the incident radiation is removed, it is called **fluorescence.**

A more exact spectroscopic definition requires that fluorescence involve singlet-to-singlet transitions and that phosphorescence involve triplet-to-singlet transitions.

Ordinary molecules have no unpaired electrons in their ground state. Consequently, their total spin angular momentum $S = 0$ with a multiplicity of $2S + 1$, corresponding to a singlet state. If all the electrons in an excited electronic state are paired, that state is also a singlet state. If, however, the spins of a pair of electrons become unpaired in an excited electronic state, then the total spin angular momentum equals $2(1/2)$, so that $S = 1$. When $S = 1$, the multiplicity of the state is $2S + 1 = 3$, so the state is called a triplet state or simply a triplet. This corresponds to the three allowed orientations that the total angular momentum vector can have with respect to an applied external magnetic field ($M_S = +1$, 0, -1). In Figure 29–1, the singlet states are labeled with an S and the triplet states with a T. The subscripts 0, 1, and 2 refer to the ground, first excited, and second excited electronic states, respectively.

Initially (Fig. 29–1a), all molecules are in the ground vibrational level of the ground electronic state S_0. Absorption of radiation takes the molecule to one of the upper vibrational levels of the first excited singlet state S_1. The time required for the absorbance transition is 10^{-15} to 10^{-18} sec, which is much faster than a typical molecular vibration (10^{-13} sec). Thus, the interatomic separation r appears constant during the time of the transition from the S_0 to the S_1 state, in accord with the Franck-Condon Principle. As molecules in the S_1 state collide with their neighboring molecules, they lose vibrational energy and trickle down to the ground vibrational state of the S_1 state, a process called **vibrational relaxation.** Trapped momentarily, the molecules hesitate, but not for long

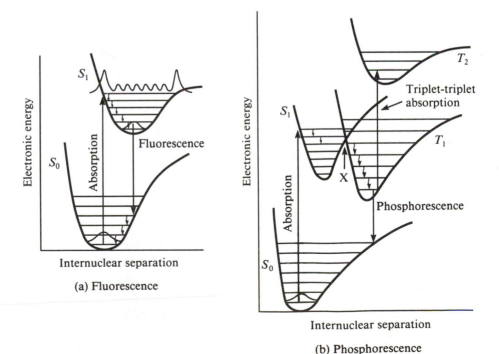

(a) Fluorescence

(b) Phosphorescence

Figure 29–1 Fluorescence and phosphorescence in a diatomic molecule.

(about 10^{-6} to 10^{-9} sec), before dropping to the S_0 state and emitting a photon of energy corresponding to the energy difference. Since the energy difference labeled "fluorescence" in Figure 29–1a is less than the energy difference labeled "absorption" in Figure 29–1a, the fluorescence light is of longer wavelength than the incident radiation.

The initial sequence of transitions leading to phosphorescence is similar to that leading to fluorescence, as shown in Figure 29–1b, but in addition to the S_0 and S_1 states a triplet state T_1 is required. Ordinarily, singlet-to-triplet transitions are forbidden, but if T_1 is positioned so that its potential-energy curve crosses the S_1 potential-energy curve, intersystem crossing is possible. **Intersystem crossing** is the radiationless transition between states of different multiplicities. At point X in Figure 29–1b, a molecule is at the extreme of a vibration in both the S_1 and T_1 states and has the same interatomic separation r. Thus, molecules excited from S_0 to S_1 trickle down by means of vibrational relaxation until they reach the vibrational level at X, cross over from S_1 to T_1, and then again trickle down by means of vibrational relaxation to the lowest vibrational level in T_1. Here the molecules pile up because the transitions from T_1 to S_0 are forbidden. Nevertheless, because of spin orbit coupling, molecules do make the T_1 to S_0 transition, emitting photons of energy corresponding to the energy difference. Since the energy difference labeled "phosphorescence" in Figure 29–1b is less than the energy difference labeled "absorption" in Figure 29–1b, the phosphorescence light is of longer wavelength than the incident radiation. Because of the forbidden nature of the T_1 to S_0 transition, phosphorescence is much slower than fluorescence.

The object of this experiment is the measurement of the lifetimes of triplet states in anthracene, phenazine, and 1,2,5,6-dibenzanthracene. It is assumed that the rate-determining step in phosphorescence is the T_1 to S_0 transition, which originates in the lowest vibrational level in the T_1 state, and that the radiative decay is first order, so that

$$-dN/dt = kN \qquad (29–1)$$

where N is the number of molecules populating the T_1, $v = 0$ state, and the lifetime τ of the triplet state is defined as

$$\tau = 1/k \qquad (29–2)$$

Although the three molecules chosen for study are all colorless in their ground state (S_0), they are colored in the T_1 state. Consequently, the change in "concentration" of molecules in the state T_1 can be followed spectrophotometrically by following the change in intensity of the triplet–triplet absorption indicated in Figure 29–1b. The wavelengths λ for maximum absorption are listed in Table 29–1.

The experiment consists of generating a batch of molecules in the T_1 state and then watching them decay by observing the decrease in intensity of the triplet–triplet absorption. The apparatus for accomplishing

TABLE 29–1

Triplet–Triplet Absorption Maxima

Molecule	λ (nm)
Anthracene	424
Phenazine	440
1,2,5,6-Dibenzanthracene	538

this is a flash lamp and spectrophotometer. Absorption of incident radiation from a powerful (about 40-J) flash lamp of short duration (<100 μsec) generates a batch of molecules in the T_1 state. Subsequently, the continuously supplied monochromatic radiation incident on the sample is absorbed with an absorbance that decreases as the population of the T_1, $v = 0$ state decreases according to first-order kinetics (Fig. 29–2).

In principle, determining the first-order rate constant for the triplet-state decay should be similar to determining the first-order rate constant for a chemical reaction. In practice, two differences arise. First, the lifetimes are very short, ranging between 0.005 and 1.5 sec for the chosen samples. Second, the output of the spectrophotometer detector is not the absorbance but the voltage supplied by a photocell, which is proportional to the intensity I of radiation.

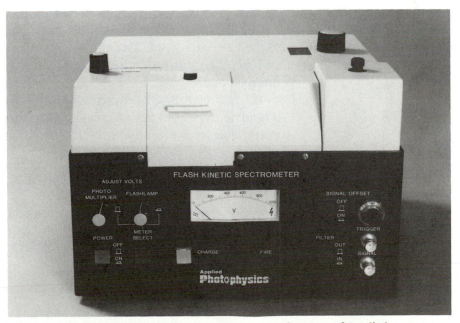

Figure 29–2 The FKS 858 flash kinetic spectrometer. (Courtesy of Applied Photophysics, Leatherhead, Surrey, England.)

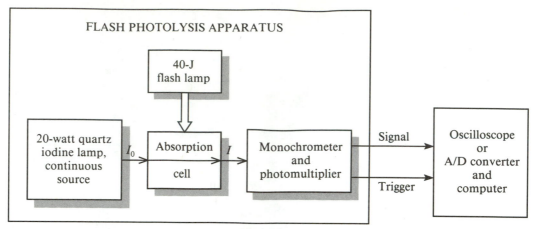

Figure 29–3 Schematic of apparatus.

Apparatus

Applied Photophysics[1] manufactures a flash kinetic spectrometer (Fig. 29–3), the FKS 858, which is reasonably priced and designed for the undergraduate laboratory. It is self-contained, consisting of a flash lamp (40 J), continuous source, cell holder, monochromator (400 to 800 nm), photomultiplier, and power supplies. Its overall design and performance appear similar to the apparatus described in the literature (Goodall et al., 1972; Porter and West, 1970; Yamanashi and Nowak, 1968).

The samples are solid solutions of anthracene, phenazine, and 1,2,5,6-dibenzanthracene in polymethylmethacrylate (PMM) formed into polished cylinders 1 cm in diameter and 5 cm long. These may be prepared (Ashmore et al., 1967) or purchased from Applied Photophysics.

The output of the photomultiplier is a voltage that is proportional to the light intensity I. The change in voltage with time can be measured and recorded with either a storage oscilloscope or an A/D converter and microcomputer. We use a PC clone with a MetraByte Dash-16F A/D card.[2] Although BASIC drivers are furnished with the MetraByte cards, it is advantageous to use a Pascal language driver (an inexpensive version is available from Quinn-Curtis[3]).

SAFETY CONSIDERATIONS

Because a flash photolysis apparatus generates intensely bright light and dangerously high voltages, it is important to exercise considerable care in using the instrument. The cell compartment

[1] Applied Photophysics, 203/205 Kingston Road, Leatherhead, Surrey KT22 7PB, England, phone 0372 386537.
[2] MetraByte Corporation, 440 Myles Standish Boulevard, Taunton, Massachusetts 02780, phone (617)880–3000.
[3] Quinn-Curtis, 49 Highland Avenue, Needham, Massachusetts 02194, phone (617)444–7721.

should be not only shielded but equipped with an interlock switch. Remember that large capacitors charged to high voltages retain their charge long after the instrument is turned off. If maintenance work is performed, the capacitors should be discharged by qualified personnel. Although relatively inexpensive, the FKS 858 appears to be a well-built and safe instrument.

EXPERIMENTAL PROCEDURE

If the samples (solid solutions of aromatic hydrocarbons in PMM) have been exposed to air for several months they should be outgassed in a vacuum desiccator at 90°C for 12 hours before use because the PMM absorbs oxygen surprisingly readily. Oxygen, a triplet in its ground state, acts as an effective quencher of the triplet state of the hydrocarbons, since energy transfer from the hydrocarbon triplet to the oxygen triplet occurs readily. The observed effect is a marked shortening of the measured triplet lifetimes of the hydrocarbons.

Oscilloscope. Before turning the power on to the FKS 858, turn its three potentiometers fully counterclockwise. They are labeled "Flash Volts," "Photomultiplier Volts," and "Signal Offset." Place a sample in the sample compartment of the FKS 858 and set the wavelength according to the sample used (Table 29–1).

Turn the power on to both the oscilloscope and the FKS 858 and permit them to warm up for 15 minutes. Two cables terminating in BNC connectors are furnished. One cable runs from the FKS

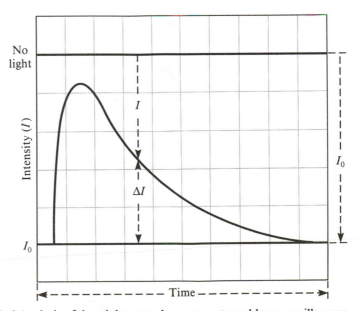

Figure 29–4 Analysis of the triplet-state decay curve traced by an oscilloscope.

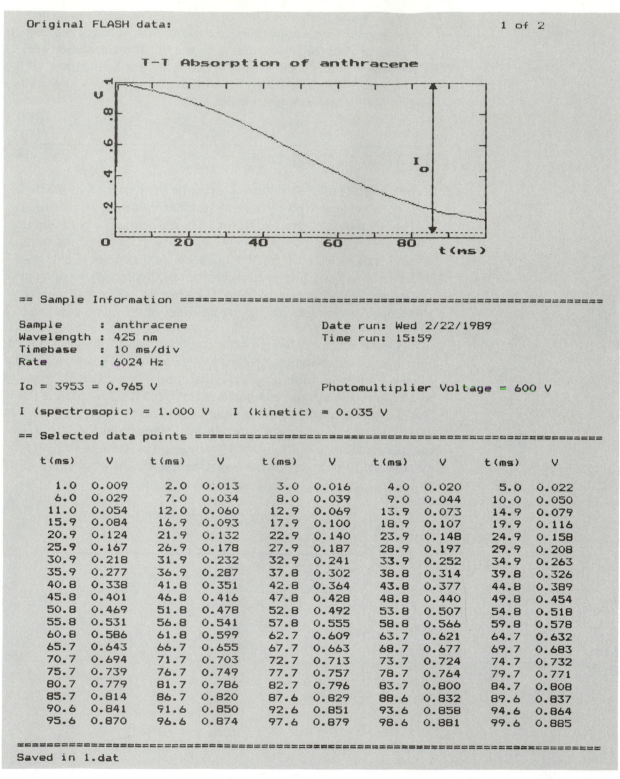

T-T Absorption of anthracene

== Sample Information ===

Sample   : anthracene   Date run: Wed 2/22/1989
Wavelength : 425 nm   Time run: 15:59
Timebase   : 10 ms/div
Rate   : 6024 Hz

Io = 3953 = 0.965 V   Photomultiplier Voltage = 600 V

I (spectrosopic) = 1.000 V   I (kinetic) = 0.035 V

== Selected data points ==

t(ms)	V	t(ms)	V	t(ms)	V	t(ms)	V	t(ms)	V
1.0	0.009	2.0	0.013	3.0	0.016	4.0	0.020	5.0	0.022
6.0	0.029	7.0	0.034	8.0	0.039	9.0	0.044	10.0	0.050
11.0	0.054	12.0	0.060	12.9	0.069	13.9	0.073	14.9	0.079
15.9	0.084	16.9	0.093	17.9	0.100	18.9	0.107	19.9	0.116
20.9	0.124	21.9	0.132	22.9	0.140	23.9	0.148	24.9	0.158
25.9	0.167	26.9	0.178	27.9	0.187	28.9	0.197	29.9	0.208
30.9	0.218	31.9	0.232	32.9	0.241	33.9	0.252	34.9	0.263
35.9	0.277	36.9	0.287	37.8	0.302	38.8	0.314	39.8	0.326
40.8	0.338	41.8	0.351	42.8	0.364	43.8	0.377	44.8	0.389
45.8	0.401	46.8	0.416	47.8	0.428	48.8	0.440	49.8	0.454
50.8	0.469	51.8	0.478	52.8	0.492	53.8	0.507	54.8	0.518
55.8	0.531	56.8	0.541	57.8	0.555	58.8	0.566	59.8	0.578
60.8	0.586	61.8	0.599	62.7	0.609	63.7	0.621	64.7	0.632
65.7	0.643	66.7	0.655	67.7	0.663	68.7	0.677	69.7	0.683
70.7	0.694	71.7	0.703	72.7	0.713	73.7	0.724	74.7	0.732
75.7	0.739	76.7	0.749	77.7	0.757	78.7	0.764	79.7	0.771
80.7	0.779	81.7	0.786	82.7	0.796	83.7	0.800	84.7	0.808
85.7	0.814	86.7	0.820	87.6	0.829	88.6	0.832	89.6	0.837
90.6	0.841	91.6	0.850	92.6	0.851	93.6	0.858	94.6	0.864
95.6	0.870	96.6	0.874	97.6	0.879	98.6	0.881	99.6	0.885

==
Saved in 1.dat

Figure 29–5 Output of a program to acquire data from the flash kinetic spectrometer. (Andrei Tokmakoff, B.S. thesis, California State University, Sacramento, 1988.)

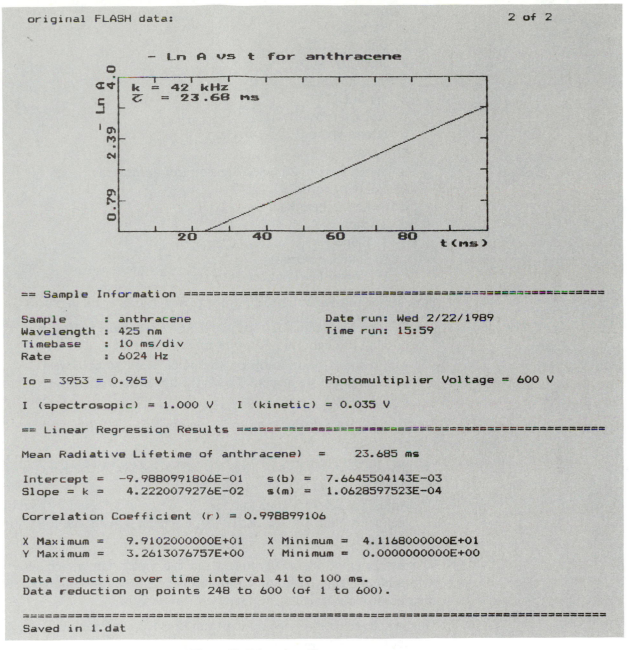

Figure 29-5 (continued)

858 terminal marked "Synk" to the oscilloscope terminal marked "Trigger." The other runs from the FKS 858 terminal marked "Signal" to the oscilloscope terminal marked "Vertical (or Y) Amplifier."

Make the following oscilloscope settings:

1. General
   a. Storage: On
   b. Enhance: Off
   c. Intensity: Max
2. Vertical amplifier
   a. Set to DC
   b. Normal (not invert)
   c. Sensitivity to 0.20 V/division
3. Horizontal sweep
   a. Sweep time to 20 msec/division (for anthracene)
   b. Coupling to DC
   c. Switch to Free Run
   d. Trigger
      1. Slope +
      2. Source external
      3. Trigger level (adjust until it works)

Now set the trace to indicate 0% transmission (sample still in place).

4. Set the movable mirror (top left of FKS 858) to the Spectroscopic position (no light on photomultiplier).
5. Adjust the position knob on the oscilloscope so the sweep is near the top of the screen. This is the upper horizontal trace in Figure 29–4.

Next set the trace to indicate 100% transmission (sample still in place):

6. Set the movable mirror (top left of FKS 858) to the Kinetic position (no light on photomultiplier).
7. Adjust the Set PM Volts knob so the sweep is deflected by five major divisions on the oscilloscope screen. This is the lower horizontal trace labeled I_0 in Figure 29–4.

The instrument is now ready to generate the triplet-state molecules with a photolyzing flash and to monitor their decay by following their absorption of light at the selected wavelength.

8. Switch the oscilloscope setting from Free Run to External Trigger (or Single Sweep).
9. Press the oscilloscope Trigger Reset button to ready the trigger mode.
10. On the FKS 858, push the Meter Select switch in to indicate the flash lamp voltage.
11. Press the Charge switch in. Then slowly turn the Set Flash Volts potentiometer to 600 V. (The sample compartment cover must be in place to close a safety interlock switch.)
12. Press the Flash button. Simultaneously with the beginning of the flash, the duration of which is less than 100 μsec, the trigger

signal causes a single horizontal sweep at the preset time base. The result is a first-order decay trace on the oscilloscope similar to the drawing in Figure 29–4, the screen coordinates of which can be recorded in your notebook.

13. Repeat steps 9 through 13 for repeat runs. *The Trigger Reset button on the oscilloscope must be reset before each run!*

A/D Converter. See the discussion in Experiment 24. The analog voltage corresponding to I_0 can be measured with an A/D converter and stored as the value of a variable in a computer program. Just as the oscilloscope can be triggered by a signal from the FKS 858, so can the A/D converter after the user inputs the value of *Count* (the number of sample voltages to be measured) and *Rate* (the number of samples per second to be measured). The time interval between sample voltage measurements is *Count/Rate*. The data stored in the parallel array of time and voltage can be treated with a linear regression analysis to give the rate constant for the first-order decay of the triplet state. A skeleton program for the data acquisition is shown earlier in Figure 24–5. A typical printout of the data reduction (Fig. 29–5) shows the graphics capability of Turbo Pascal (Tokmakoff, 1988).

Results and Calculations

From a tracing of the oscilloscope screen (Fig. 29–4), prepare a table of time (in seconds), I (in divisions), I/I_0, and A. Record the number of divisions corresponding to I_0, normally 5.0 divisions ($A = \log_{10} I_0/I$).

Prepare a plot of ln A against t. Determine the slope m of the line with a linear regression. The first-order rate constant $k = -m$. The lifetime of the triplet state $\tau = 1/k$.

If computerized data acquisition is available, make at least one run with both the computer and the oscilloscope. Critically compare the results with literature values.

REFERENCES

1. Ashmore, P. G., F. S. Dainton, and T. M. Sugden, eds., *Photochemistry and Reaction Kinetics,* Cambridge University Press, Cambridge, England, 1967.

2. Birks, J. B., *Photophysics of Aromatic Molecules,* Wiley, New York, 1970.

3. Boyer, R., et al., "The Photophysical Properties of 2-Naphthol," *J. Chem. Educ.* 62:630–632, 1985.

4. Chambers, K. W., and I. M. Smith, "An Inexpensive Flash Photolysis Apparatus and Demonstration Experiment," *J. Chem. Educ.* 51:354–356, 1974.

5. Cundall, R. B., and A. Gilbert, *Photochemistry,* Nelson, London, 1970.

6. Goodall, D. M., et al., "Flash Photolysis Experiments for Teaching Kinetics and Photochemistry," *J. Chem. Educ.* 49:669–674, 1972.

7. Legenza, M. W., and C. J. Marzzacco, "The Rate Constant for Fluorescence Quenching," *J. Chem. Educ.* 54:183–184, 1977.

8. Levanon, H., et al., "Triplet-State Formation of Porphycenes," *J. Phys. Chem.* 92:2429–2433, 1988.

9. Moore, J. W., and R. G. Pearson, *Kinetics and Mechanisms,* 3rd ed., Wiley-Interscience, New York, 1981, p. 273.

10. Porter, G., and M. A. West, "An Inexpensive Flash Kinetic Spectrophotometer," *Educ. Chem.* 7:230–231, 1970.

11. Rabek, R. F., *Experimental Methods in Photochemistry and Photophysics,* Wiley, Chichester, 1982.

12. Siebrand, W., "Mechanism of Radiationless Triplet Decay in Aromatic Hydrocarbons and the Magnitude of the Franck-Condon Factors," *J. Chem. Phys.* 44:4055–4056, 1966.

13. Siebrand, W., "Radiationless Transitions in Polyatomic Molecules, II. Triplet–Ground State Transitions in Aromatic Hydrocarbons," *J. Chem. Phys.* 47:2411–2422, 1967.

14. Tokmakoff, A., B.S. Thesis, California State University, Sacramento, December, 1988.

15. Turro, N. J., *Molecular Photochemistry,* Benjamin, New York, 1967.

16. Van Stam, J., and J.-E. Lofroth, "The Photolysis of Singlet Excited β-Naphthol," *J. Chem. Educ.* 63:181–184, 1986.

17. West, M. A., K. J. McCallum, and R. J. Woods, "Absorption and Emission Spectra and Triplet Decay of some Aromatic and N-Heterocyclic Compounds in Polymethylmethacrylate," *Trans. Faraday Soc.* 66:2135–2147, 1970.

18. Yamanashi, R., and A. V. Nowak, "Recombination of Iodine via Flash Photolysis," *J. Chem. Educ.* 45:705–710, 1968.

Spectroscopy

Experiment 30

Visible Spectrum of the Hydrogen Atom

Compared to the separation of rotational and vibrational energy levels, the separation of electronic energy levels in atoms and molecules is large. Consequently, transitions between rotational levels are observed in the microwave, between vibrational levels in the infrared, and between electronic levels in the ultraviolet–visible region of the spectrum.

Light—electromagnetic radiation in the ultraviolet–visible region—can be separated into its spectral components either by diffraction with a grating or by refraction with a prism. Diffraction is greater with longer wavelengths and refraction is greater at shorter wavelengths. For visible radiation, photographic film is convenient for detecting the diffracted spectrum.

To produce transitions between electronic energy levels in common gases, it is convenient to subject the gas at a few millimeters' pressure to a very high voltage in a Geissler discharge tube. The resulting electroluminescence arises from the emission of photons of electromagnetic radiation in the ultraviolet–visible region.

The more electrons in an atom, the more complex the resulting spectrum. The spectrum appears as a series of vertical lines, which are just the image of the spectrographic slit falling on the grating. The number of lines present is large and the spectrum complex for heavier atoms. If molecules are present, the lines characteristically crowd together in groups called bands (e.g., the band spectrum of nitrogen). The simplest spectrum is that of hydrogen atoms, studied first by Balmer in 1885.

Experimental Background

The spectrum observed by Balmer lies in the visible region. The regularity of the lines prompted Balmer to fit their wavelengths to a simple empirical equation,

$$\lambda = 3646 n^2 / (n^2 - 4) \tag{30-1}$$

where $n = 3, 4, 5, \ldots$ and λ is the wavelength in angstroms. Nowadays the same equation is generally written

$$\tilde{\nu} = R\left(\frac{1}{n_2^2} - \frac{1}{n_1^2}\right) \tag{30-2}$$

where $\tilde{\nu}$ is the wave number of the line in cm^{-1} and R, the Rydberg constant, equals 109,677.8 cm^{-1}. When $n_2 = 2$, Equation 30–2 is equivalent to Equation 30–1.

A variety of units of length are commonly used in spectroscopic measurements,

$$1 \text{ m} = 10^9 \text{ nm} = 10^8 \text{ Å} = 10^6 \text{ } \mu\text{m} = 10^3 \text{ mm} = 10^2 \text{ cm} \tag{30-3}$$

where nm = nanometer, Å = angstrom, μm = micron,[1] mm = millimeter, and cm = centimeter. In addition, the wave number $\tilde{\nu}$ is defined as $1/\lambda$, where λ has units of cm and $\tilde{\nu}$ has units of cm^{-1}.

Theory

Both quantum mechanics and the Bohr Theory of the hydrogen atom give the same expression for the allowed electronic energies of an H atom,

$$E = -\frac{2\pi^2 \mu e^4 Z^2}{n^2 h^2} \tag{30-4}$$

where h is Planck's constant, e is the charge on an electron, μ is the reduced mass of the electron and the nucleus, and Z is the nuclear charge ($=1$ for an H atom). The reduced mass is given by

$$\mu = \frac{m_e m_p}{m_e + m_p} \tag{30-5}$$

where m_e is the mass of an electron and m_p is the mass of a proton.

A photon of electromagnetic radiation is emitted when an atom changes from a state of higher electronic energy E_2 to a lower one E_1,

$$E_2 - E_1 = h\nu = hc/\lambda = hc\tilde{\nu} \tag{30-6}$$

where h is Planck's constant, ν is the frequency, and $\tilde{\nu}$ is the frequency in wave numbers (cm^{-1}). Equations 30–4 and 30–6 may be combined to give

$$\tilde{\nu} = R\left(\frac{1}{n_2^2} - \frac{1}{n_1^2}\right) \tag{30-7}$$

[1] Rarely used.

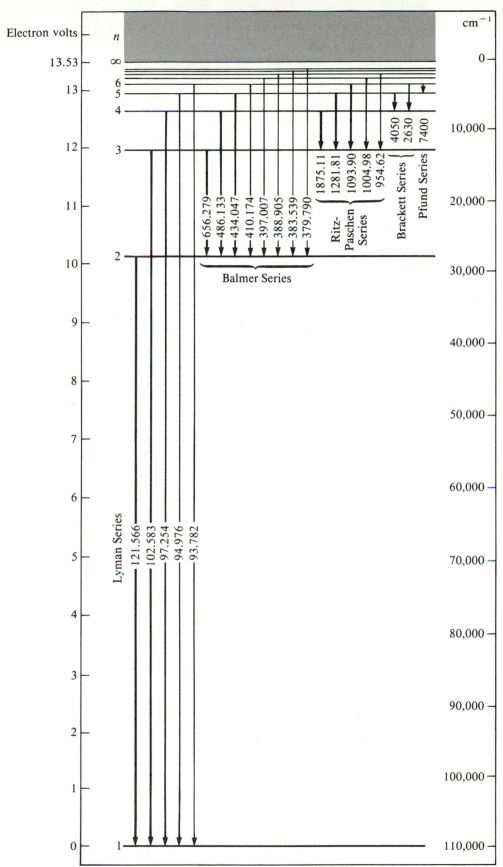

Figure 30–1 Energy-level diagram for the H atom. Wavelengths of transitions are given in nanometers.

where R (cm^{-1}), the Rydberg constant, is given by

$$R = \frac{2\pi^2 \mu e^4}{c h^3} \qquad (30\text{--}8)$$

In SI units R (m^{-1}) is given by

$$R = \frac{2\pi^2 \mu e^4}{(4\pi\epsilon_0)^2 c h^3} \qquad (30\text{--}9)$$

where the permittivity $4\pi\epsilon_0 = 1.11264 \ 10^{-10} \ \text{C}^2 \ \text{N}^{-1} \ \text{m}^2$. Much early spectroscopic data is reported in cgs units, for example, cm^{-1}.

The energy-level diagram for a hydrogen atom is given in Figure 30–1 along with some electronic transitions. All the transitions terminating at the level n_2 equal to 2 lie in the visible region of the spectrum and are called the Balmer series. A number of other series have been observed, one of which, the Lyman series, lies in the ultraviolet region. The other series lie in the infrared region of the spectrum.

Apparatus

The apparatus consists of a spectrograph (e.g., Baird-Atomic 1.5-m grating spectrograph, model SB-1, covering 3700 to 7400 Å, with a dispersion of 1.5 nm/mm); gas discharge tubes (H_2, N_2, Ar, He, and Ne); a high-voltage transformer for the tubes; a film developing darkroom; Plus-X film or equivalent;[2] D-76 developer or equivalent; glacial acetic acid; Kodak Acid Fix; and three 200-ml graduated cylinders for developing tanks.

SAFETY CONSIDERATIONS

When changing discharge tubes, be sure to both turn off and unplug the high-voltage transformer to avoid electric shock. Let the tubes cool before touching them, as they become quite hot after a few minutes, especially the thin middle section. Be careful not to exert any severe leverage on the tubes when pulling them from their holder, as the thin section is susceptible to breaking.

Avoid skin contact with developer or fixer. Developers contain strong organic reducing agents that cause an allergic reaction in many people; fixing agents are essentially sodium thiosulfate, $Na_2S_2O_3$. If you are known to be allergic to sulfur-containing drugs, you should take extra precautions to avoid contact with fixing solutions.

---

[2] Kodachrome color slide film may be substituted for black and white film. See the note at the end of Experimental Procedure for directions.

EXPERIMENTAL PROCEDURE

Although the following procedure conforms closely to operation of the Baird-Atomic 1.5-m grating spectrograph, the procedures for other spectrographs of comparable resolution are similar.

Loading the Film. The film, Eastman Kodak Plus-X, is stored in a refrigerator on a 100-foot spool. Open only in a darkroom in total darkness (no red safelight). Before using unexposed film, practice loading the spectrographic film holder in daylight with a practice piece of film precut to the proper length, furnished by your instructor. Notice that the film holder has two slots: the upper one for the film holder slide and the lower one for the film. Without looking at the film and holder, practice slipping the film into the proper slot and sliding the film holder slide into place. It is easy to get the film into the wrong slot. The slide should slip in easily; if it does not, it is probably because the film is in the wrong slot. Do not force the slide in, as it can be bent easily. When you feel comfortable with the loading procedure, set the paper cutter in the darkroom to 30.0 cm, place the unopened film container and film holder conveniently near the film cutter, and turn out the lights. Cut and load the film into the holder. Replace the slide, put the felt pad on the film holder in the closed position, and turn the lights on.

Adjusting the Source. The source, lens, slit, and grating must be collinear, with an image of the source made by the lens falling sharply on the slit. Turn the high voltage off *and* unplug it (for safety), insert a mercury discharge tube, and turn it on. Remove the slit, Hartman slide, and film holder from the spectrograph (Fig. 30–2). Look straight down the spectrograph at the grating and you should see bright, wide vertical "lines," mostly green in color. As you move your head slowly from side to side, you should see yellow lines to the right, then green, and finally blue lines to the far left. Replace the slit holder and look again. The lines should be fainter but sharper. Set the slit for 32 μm (middle position). The yellow doublet you see is at 579.06 and 576.96 nm; the brightest green line is at 546.07 nm.

Insert the film holder into the spectrograph, but leave the slide in. With the Hartman slide out, adjust the source and/or lens until a sharp image of the mercury discharge tube falls on the slit. The image should be centered sharply and exactly on the slit. Close the shutter and carefully pull the slide from the film holder until a line scratched on it is visible. Leave the slide in position at this point. Replace the Hartman slide and set to position 1.

Be very careful to avoid the slightest change in position of the source or lens during the exposure and especially while changing tubes. A small change in the position of the source or lens causes a

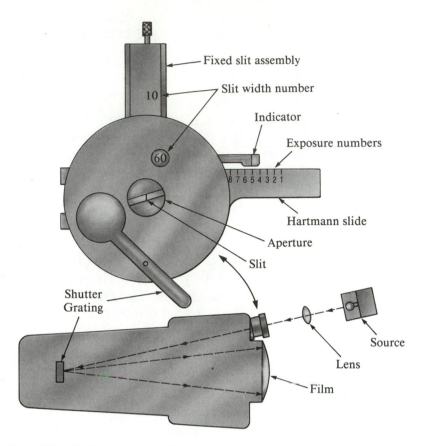

Figure 30–2 Grating spectrograph.

large change in the position of the image, which may cause the image to no longer fall on the slit. Since the shutter must be open to adjust the image onto the slit, the run is ruined and must be repeated.

Taking the Exposures. First check the following:

1. Shutter closed.
2. Source, lens, and slit lined up.
3. Image of source falls sharply on slit.
4. Slide out to scratched line.
5. Hartman slide at position 1.
6. Slit at 32 μm.

To take the exposures, have a stopwatch or timer at hand. Open the shutter, time the exposure, close the shutter, and move the Hartman slide to the next position. Change source tube *without moving the tube holder.* Continue until the exposures indicated in Table 30–1 have been taken.

The exposures may be taken in any sequence, but should be in the positions indicated in Table 30–1.

TABLE 30–1

Recommended Exposure Times[a]

Position	Tube	Time (sec)
1	Hg	100
2	Hg	10
3	Ar	50
4	He	50
5	N_2	10
6	N_2	300
7	H_2	1
8	H_2	300
9	Ne	10
10	Hg	10

[a] Plus-X film, normal development, 32-μm slit, and Geissler tube source.

Developing. After all the exposures have been taken, push the slide into the film holder as far as it will go, and develop the film in total darkness. Three 200-ml graduated cylinders are convenient as developing "tanks." The steps are

1. Develop: 5 min in Kodak D-76.
2. Short stop: 5 sec in short stop (about 1 ml glacial acetic acid per 100 ml water).
3. Fix: 5 min in Kodak Acid Fix.
4. Wash: 20 min in running water.

Hang the film overnight before measuring.

Using the Comparator. The comparator is a traveling microscope that can measure the line positions with respect to the calibration lines with great precision. Before placing the film in the comparator, tape it to a light box and identify three or four Hg lines, such as the doublet at 576.96 and 579.06 nm and the intense line at 546.07 nm (Fig. 30–3). On the light box, measure the distance between these lines with a transparent millimeter ruler and make a rough calculation of the dispersion in nanometers per millimeter. Next, again with the ruler, measure the distance between one of the prominent Hg lines you have identified and an Hg line you are unsure of. Calculate its wavelength with your crude dispersion and the distance from a known line. Compare the value with a table of Hg lines; you will probably find its value listed, and you have identified another line. Mark it with a dot or two with a fine felt-tip pen so it will be easy to find under the comparator lens. Repeat this procedure to identify a few more lines in the He and Ar spectra. Mark them so they can be found easily under the microscope lens. Without some preliminary identification of this type, it is difficult

Figure 30–3 Mercury calibration spectrum.

to identify lines in the comparator because of its narrow field of view.

The comparator is nothing but a fancy millimeter ruler; it gives the same information as above, but with greater precision. Position your film, emulsion side up, between the two glass plates and place in the traveling microscope. Adjust the film so that the shortest wavelength lines of the Balmer series can be seen at one extreme end of the traveling microscope bed. A travel distance of 10.0 cm permits all of the Balmer lines except the one of longest wavelength to be measured without moving the film. After all are measured except this last line at the red end, it can be measured by moving the film, measuring the distance between it and a conveniently located calibration line, and using the dispersion calculated from the film's first setting.

The film must be lined up parallel to the direction of travel of the microscope traveling bed. At each end of the film, adjust the film manually so that corresponding Hg lines in film positions 2 and 10 remain centered in the eyepiece cross hair when the film bed is translated perpendicular to the spectra. Check at each end of the film to be sure that the film is lined up correctly.

Notice that the perpendicular motion also has a scale in millimeters. This is useful for identifying the position of each spectrum on the film. For your own convenience, for each spectrum in Table 30–1 you should note the perpendicular scale position in millimeters. Then you can go to any spectrum by adjusting the perpendicular travel to the proper scale position without repeatedly searching.

Record the scale position of at least 10 to 12 calibration lines (Hg, He, and Ar). Without any movement of the film on the traveling bed, record the scale readings of all the Balmer series lines located on the short-wavelength half of the film. Be extremely careful not to move the film on the bed, as this makes it impossible to calibrate the film. All readings, calibration lines, and hydrogen lines must be measured at one sitting, or at least without disturbing the position of the film on the traveling bed.

If the first line of the Balmer series, lying at the red end of the spectrum, is out of reach, move the film until it and a convenient calibration line are measurable. Measure the distance between them and record the scale reading.

Note. Kodachrome color slide film may be substituted for black and white film. Cut the leader off the film. In total darkness, feed the film from the spool into the film holder. Replace the slide and let the spool remain attached to the film, dangling at the end of the film holder. Double the exposure times in Table 30–1. In the darkroom, wind the film from the film holder back into the spool. Take the film to a local film processor and request that it be developed without cutting into the slides. The lines on the resulting film strip can be measured as described above.

Results and Calculations

Prepare a table of calibration lines, their wavelengths λ, and comparator scale readings X. Since the dispersion of a grating spectrograph is linear across the film, these data can be fit (by least squares) to an equation of the form

$$\lambda = mX + b \qquad (30\text{–}10)$$

where m is the dispersion in nanometers per millimeter and b is the intercept. Record the standard deviations in m and b for analyzing the propagated errors in the wavelengths of the Balmer series lines. Tabulate the comparator scale readings for the observed Balmer lines and calculate their wavelengths with Equation 30–10. Use this value of the dispersion to calculate the wavelength of the first (red) line of the Balmer series if it was out of reach with your comparator. Index your lines by calculating the Rydberg constant with each line. If you choose n correctly, R is a constant. Report the average value of R and the standard deviation. Compare it and the measured wavelengths with literature values and a theoretical value calculated with Equation 30–9.

Alternatively, plot $\tilde{\nu}$ against n_1^2 with $n_2 = 2$. According to Equation 30–2, the result should be a straight line of slope equal to $-R$. Fit the data by least squares and report the standard deviation in R. Compare the standard deviation in R with the propagated error.

REFERENCES

1. Herzberg, G., *Atomic Spectra and Atomic Structure,* 2nd ed., Dover, New York, 1944.

2. Kuhn, H. G., *Atomic Spectra,* Academic Press, New York, 1962.

3. Marquis, J. A., "Color Photography of Spectra," *J. Chem. Educ.* 37:580, 1960.

4. Moore, C., *Atomic Energy Levels*, National Bureau of Standards Circular 467, vol. I, U.S. Government Printing Office, Washington, D.C., 1949.

5. Pearse, R. W. B., and A. G. Gaydon, *The Identification of Molecular Spectra*, Wiley, New York, 1963.

Experiment 31

Absorption Spectrum and Dissociation Energy of Iodine and Bromine

The beautiful violet vapor visible over iodine even at room temperature arises because gaseous iodine molecules absorb electromagnetic radiation that lies in the visible region of the spectrum. Since the color of the absorbed light lies in the yellow region of the visible spectrum, the vapor appears violet. The transitions that give rise to the absorption spectrum take place between the ground electronic state of the iodine molecule and an excited electronic state. At room temperature, all iodine molecules are in the ground electronic state, and nearly all are in the ground vibrational state ($v'' = 0$) as well. For this reason, transitions in the absorption spectrum of iodine all originate in the state $v'' = 0$. Figure 31–1 illustrates how all the absorption transitions begin in the state $v'' = 0$ and terminate in the upper electronic state in all possible vibrational states; that is, v' equals anything from 0 to infinity. Since only a single vibrational state is involved in the lower electronic state, the spectrum is simpler than the emission spectrum that arises from transitions between all possible vibrational levels in both electronic states.

In Figure 31–1 you can see some of the transitions that give rise to the absorption spectrum of molecular iodine. In practice, not all the transitions are observed because electronic transitions take place much more quickly than the period of a molecular vibration (Franck-Condon Principle). Furthermore, one would expect that absorption transitions would originate in the state $v'' = 0$, where the probability is a maximum; that is, where ψ_0^2 is a maximum. This is shown in Figure 31–1 for the $v'' = 0$ state (D'alterio et al., 1974). In addition, the most intense transitions terminate in upper-level states, where there also is a region of high probability; that is, where $\psi_{v'}^2$ is a maximum. The B electronic state of the iodine molecule is positioned so that vertical lines (transitions) drawn

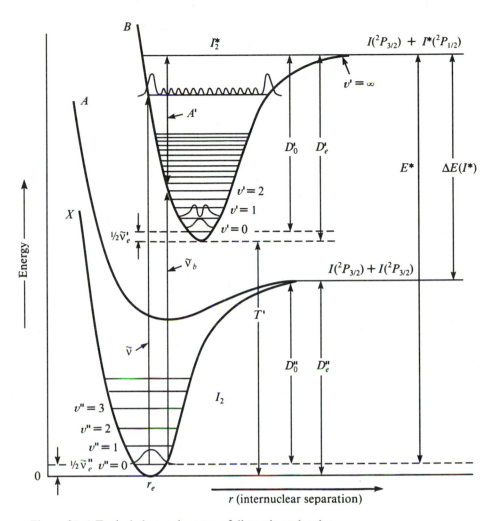

Figure 31–1 Typical electronic states of diatomic molecules.

from the vibrational state $v'' = 0$ to v' vibrational states intersect the B state in a way that the most intense transitions are, for the upper-level vibrational numbers, from about $v' = 20$ to $v' = 50$, or nearly to the convergence limit. Lines drawn from the $v'' = 0$ state tend to miss the upper-level vibrational state of low values of v'. For this reason, these transitions are weak or not observed, as shown in Figure 31–1.

The nomenclature used in Figure 31–1 follows Herzberg (1950), except that $\tilde{\nu}$ is used instead of ω. If you are new to this nomenclature, it might be helpful to notice that primed ($'$) parameters are properties of the upper state only, while the double-primed ($''$) parameters are lower-state properties. Most values with no prime at all correspond to energy differences between states. It is customary to use the wave number (cm^{-1}) as a unit of "energy" for both states and transitions between them. The wave number is a quasi-cgs unit and is also called the kayser. With this nomenclature in mind, the upper $G'(v')$ and lower $G''(v'')$ state energies are written:

$$G'(v') = T' + \tilde{\nu}'_e(v' + \tfrac{1}{2}) \qquad - \tilde{\nu}'_e x'_e(v' + \tfrac{1}{2})^2 \qquad \text{(31–1)}$$

$$+ \tilde{\nu}'_e y'_e(v' + \tfrac{1}{2})^3 + \cdots$$

$$G''(v'') = T'' + \tilde{\nu}''_e(v'' + \tfrac{1}{2}) \qquad - \tilde{\nu}''_e x''_e(v'' + \tfrac{1}{2})^2 \qquad \text{(31–2)}$$

$$+ \tilde{\nu}''_e y''_e(v'' + \tfrac{1}{2})^3 + \cdots$$

Note that $T'' = 0$ for the ground state. The absorption transitions between these sets of energy levels are given by

$$\tilde{\nu} = G'(v') - G''(v'') \qquad \text{(31–3)}$$

or

$$\tilde{\nu} = T' + \tilde{\nu}'_e(v' + \tfrac{1}{2}) - \tilde{\nu}'_e x'_e(v' + \tfrac{1}{2})^2 + \tilde{\nu}'_e y'_e(v' + \tfrac{1}{2})^3 \qquad \text{(31–4)}$$

$$- \tilde{\nu}''_e(v'' + \tfrac{1}{2}) + \tilde{\nu}''_e x''_e(v'' + \tfrac{1}{2})^2 - \tilde{\nu}''_e y''_e(v'' + \tfrac{1}{2})^2$$

When v'' is set equal to zero, Equation 31–4 becomes an expression for the absorption spectrum, that is, for all possible absorption transitions between $v'' = 0$ and $v' = $ anything. As already pointed out, not all of these transitions may be sufficiently intense to be observable.

The Birge-Sponer Extrapolation

Of special interest are the spacings between vibrational energy levels, and you can see by examining the transitions drawn in Figure 31–1 that the absorption spectrum can give us some information about these spacings. For example, in the B state, the spacing $\Delta\tilde{\nu}$ between the $v' = 0$ state and the $v' = 1$ is given by

$$\Delta\tilde{\nu} = [T' + G'(1) - G''(0)] - [T' - G'(0) - G''(0)] \qquad \text{(31–5)}$$

$$\Delta\tilde{\nu} = G'(1) - G'(0) \qquad \text{(31–6)}$$

In general, the spacing between upper levels is given by

$$\Delta\tilde{\nu} = G'(v' + 1) - G'(v') \qquad \text{(31–7)}$$

If the second-order anharmonicity constants $\tilde{\nu}_e y_e$ are ignored, substitution of Equations 31–1 and 31–2 into Equation 31–7 gives

$$\Delta\tilde{\nu} = \tilde{\nu}'_e - 2\tilde{\nu}'_e x'_e(v' + 1) \qquad \text{(31–8)}$$

Equation 31–8 is the equation of a linear Birge-Sponer extrapolation. The values of $\Delta\tilde{\nu}$ are experimentally accessible, since they are just the differences in energy (cm^{-1}) between adjacent pairs of absorption bands. Also, v' is the upper-state vibrational quantum number of the lower energy member of the pair. Remember that the $\Delta\tilde{\nu}$ are just the vibrational energy level spacings in the B electronic state. Equation 31–8 shows that the spacings are a linear function of the vibrational quantum number. If $\Delta\tilde{\nu}$ is plotted against $(v' + 1)$, the result is a straight line, the slope of which equals $-2\tilde{\nu}'_e x'_e$ and the intercept of which equals $\tilde{\nu}'_e$. The plot is called a Birge-Sponer extrapolation. It is linear if second-order anharmonicity constants are sufficiently small, and they often are. The plot is shown in Figure 31–2a for the case where $\tilde{\nu}_e y_e$ and higher anharmonicity constants are negligible, and in Figure 31–2b where they are not negligible.

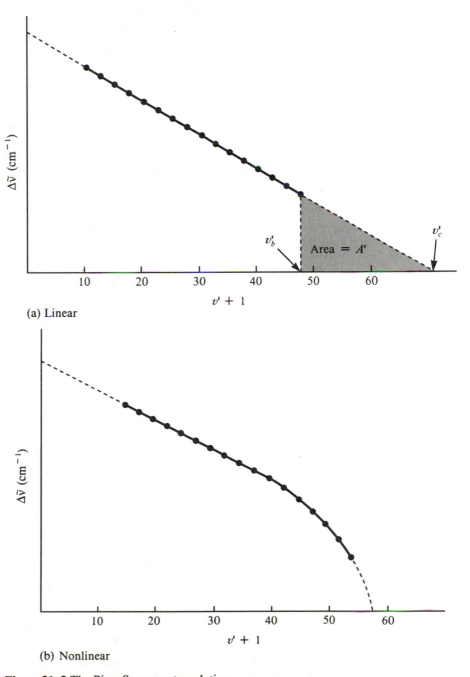

Figure 31–2 The Birge-Sponer extrapolation.

In addition to $\tilde{\nu}'_e$ and $\tilde{\nu}'_e x'_e$, the dissociation energies D'_0 and D'_e can be determined from the Birge-Sponer extrapolation. In Figure 31–1, compare D'_0 with the spacings between the upper-state vibrational energy levels. The sum of all the spacings is exactly equal to D'_0. The sum of the spacings can be expressed as

$$D'_0 = \sum \Delta\tilde{\nu} \tag{31–9}$$

$$= (\tilde{\nu}'_e - 2\tilde{\nu}_e x'_e) + (\tilde{\nu}'_e - 4\tilde{\nu}_e x'_e) + (\tilde{\nu}_e - 6\tilde{\nu}_e x_e) + \cdots$$

Since many spacings fall between the state $v' = 0$ and the convergence limit at $v' =$ infinity, you can replace the summation sign with an integral:

$$D'_0 = \sum_{v'=0}^{v'_c} \Delta\tilde{\nu} = \int_0^\infty \Delta\tilde{\nu} \, dv \tag{31–10}$$

$$= \text{area under the Birge-Sponer curve}$$

This area is the total area of the triangle in Figure 31–2a ($= 1/2$ base $\times$ height). Even if the Birge-Sponer extrapolation is nonlinear, as it is when the higher anharmonicity constants are not negligible, the area under the curve can still be measured—for example, by Simpson's method. (See the section on numerical integration in Chapter 6, Part Two). Thus, even in this case, the dissociation energy D'_0 can be determined.

So far, we have determined three important upper-state parameters: $\tilde{\nu}'_e$, $\tilde{\nu}'_e x'_e$, and D'_0. Examination of Figure 31–1 reveals that the difference between D'_0 and D'_e is just $\tilde{\nu}'_e/2$, the zero point energy. Consequently, you can easily determine D'_e:

$$D'_e = D'_0 + \tilde{\nu}'_e/2 \tag{31–11}$$

The Convergence Limit

Because the intensities of the absorption transitions fall off at high upper-state quantum numbers, it is not possible to observe the convergence limit directly for either the bromine or the iodine spectrum. However, examination of Figure 31–1 reveals that

$$E^* = \tilde{\nu}_b + A' \tag{31–12}$$

where $\tilde{\nu}_b$ is the energy in cm^{-1} of the absorption band at which the extrapolation is begun and A' is the area under the Birge-Sponer curve from v'_b to v'_c. The vibrational quantum number at convergence is v'_c, and v'_b is the vibrational quantum number at $\tilde{\nu}_b$. The shaded area in Figure 31–2a equals A'. The total area under the curve (shaded and unshaded areas) equals D'_0, the upper-state dissociation energy. The convergence limit E^* is given by Equation 31–12.

Lower-State Parameters

Up to this point, we have calculated only upper-state parameters. From the absorption spectrum alone, we cannot calculate any of the lower-state parameters shown in Figure 31–1. However, with data from the literature of atomic energy levels, we can get a value for $\Delta E(I^*)$, which is the energy of an iodine atom in its first excited state ($^2P_{1/2}$) compared to its ground state ($^2P_{3/2}$). When an iodine molecule in its ground electronic state (X) dissociates, it dissociates into two iodine atoms, both of which are in their ground state ($^2P_{3/2}$). On the other hand, an iodine molecule in an excited electronic state (B) dissociates into two iodine atoms, one of which is in the ground state ($^2P_{3/2}$) while the other ends up in a higher atomic energy level ($^2P_{1/2}$). The value of $\Delta E(I^*)$, obtained from the atomic spectra, is 7603.15 cm^{-1} for iodine atoms (Moore, 1958) and 3685 cm^{-1} for bromine atoms (Gaydon, 1953). Although E^* is sketched off to the right side of Figure 31–1, it is actually an absorption transition, namely the convergence limit, which corresponds to the transition from the state $v'' = 0$ to $v' = $ infinity. Examination of Figure 31–1 now reveals that

$$D_0'' = E^* - \Delta E(I^*) \tag{31–13}$$

Unfortunately, the experimental data we have at hand so far do not permit determination of D_e''. By analogy with Equation 31–11, we need $\tilde{\nu}_e''$, the lower-state fundamental. If you could observe just one absorption transition from the state $v'' = 1$ to an upper state of known v', then you could get $\tilde{\nu}_e''$ from the difference in energy between two transitions terminating at the same upper state. Since you may not have observed such transitions, you may use the literature values of $\tilde{\nu}_e$ for iodine and bromine, 215 cm^{-1} and 323 cm^{-1}, respectively (Herzberg, 1950). Then, as with Equation 31–11 for the upper state, you may calculate D_e'' from

$$D_e'' = D_0'' + \tilde{\nu}_e''/2 \tag{31–14}$$

Naturally, these dissociation energies of iodine in its ground state are of more importance or utility than the excited-state values. Usually, when we speak loosely of the (molecular) dissociation energy of a molecule, we mean D_0''. This is related to the thermodynamic (molar) dissociation energy by

$$N \times D_0'' = \Delta E_0^0 = \Delta H_0^0 \tag{31–15}$$

where N is Avogadro's number. The usual units for D_0'' are cm^{-1}/molecule or eV/molecule, while the customary units for ΔE_0^0 and ΔH_0^0 are kJ/mol or kcal/mol. You can think of D_0'' as the energy required to dissociate 1 molecule of iodine at 0 K, and ΔE_0^0 or ΔH_0^0 as the energy required to dissociate 1 mole of iodine at 0 K. At 0 K, $\Delta E_0^0 = \Delta H_0^0$ because

$$\Delta H_0^0 = \Delta E_0^0 + PV = \Delta E_0^0 + RT = \Delta E_0^0 \tag{31–16}$$

since $RT = 0$ at 0 K. The molecular dissociation energy is most often seen in the Morse equation (Alberty, 1987):

$$U = D_e\{1 - exp\,[-a(r - r_e)]\}^2 \tag{31–17}$$

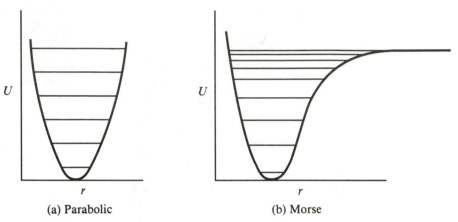

$$U = \tfrac{1}{2}k(r - r_e)^2 \qquad\qquad U = D_e\{1 - exp[-\beta(r - r_e)]\}^2$$

U		U	
	r		r

(a) Parabolic   (b) Morse

Figure 31–3 Potential energy curves.

This is the potential energy of an anharmonic oscillator, as compared with a harmonic oscillator, for which

$$U = k(r - r_e)^2 \qquad\qquad (31\text{–}18)$$

Since a vibrating molecule is a one-dimensional system, the Schrödinger wave equation may be written

$$-\frac{h^2}{8\pi^2 m}\frac{\partial^2\psi}{\partial x^2} + U = \epsilon\psi \qquad\qquad (31\text{–}19)$$

When Equation 31–18 is substituted for the potential energy U, the eigenvalues that result are $\tilde{\nu}_v = (v + \tfrac{1}{2})\tilde{\nu}_e$; but when Equation 31–17 is substituted for U, the resulting eigenvalues are a power series in $(v + \tfrac{1}{2})$,

$$\tilde{\nu}_v = \tilde{\nu}_e(v + \tfrac{1}{2}) - \tilde{\nu}_e x_e(v + \tfrac{1}{2})^2 + \tilde{\nu}_e y_e(v + \tfrac{1}{2})^3 - \cdots \qquad (31\text{–}20)$$

See also Equations 31–1 and 31–2. In Figure 31–3, Equations 31–17 and 31–18 are plotted to compare the familiar parabolic potential-energy curve of a harmonic oscillator with the Morse curve of an anharmonic oscillator.

Apparatus

The experiment requires a spectrophotometer, such as a Cary model 11, Beckman DK-2, or Perkin-Elmer 3840; bromine and/or iodine; and cells with 1.0-cm and 10.0-cm path lengths.

SAFETY CONSIDERATIONS

Bromine and iodine are highly corrosive to the skin, especially bromine. Use rubber gloves when handling bromine and work in a well-ventilated hood.

EXPERIMENTAL PROCEDURE

Since the vapor pressure of bromine is around 200 torr at room temperature, a few drops in the bottom of a 1.0-cm spectrophotometer cell gives sufficient absorption. Run the spectrum from about 500 to 650 nm. With a 10.0-cm cell, the vapor pressure of iodine is sufficient to give a usable spectrum at room temperature, although warming to 35 to 40°C is advisable. A Perkin-Elmer 3840 requires only about 1 or 2 seconds to generate a spectrum. Consequently, it is possible to place a few crystals of iodine in a 1.0-cm cell, warm the cell by dipping it in a Dewar flask of hot water, wipe the cell dry, place it in the cell compartment, and get a spectrum before the cell cools appreciably.

Results and Calculations

Tabulate the experimental values of the absorption maxima in nm and cm^{-1}. Between 500 and 550 nm, all the observed absorption transitions originate in the state $v'' = 0$. At wavelengths longer than about 550 nm, some "shoulders" appear on the absorption bands arising from the $v'' = 0$ to v' transitions. These shoulders are "hot" bands arising from the $v'' = 1$ to v' transitions and are not used in this experiment. They can, however, cause some confusion in deciding which bands arise from the ground state. Figure 31–4 shows how the changing intensities can be used to determine which peaks arise from the ground state where $v'' = 0$.

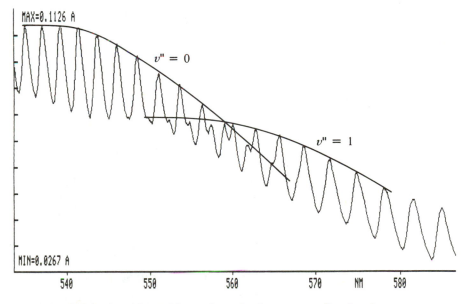

Figure 31–4 Using band intensities to determine lower-state vibrational quantum numbers in molecular iodine.

TABLE 31-1

Bromine and Iodine Transitions:
$v'' = 0$ to v'

Molecule	v'	nm
Br_2	15	558.47
Br_2	16	555.12
Br_2	17	551.93
I_2	27	541.2
I_2	28	539.0
I_2	29	536.9

Sources: Bromine data are from Rosen (1951); iodine data are from McNaught (1980).

Assign upper-state vibrational quantum number v' with the assistance of a few values taken from the literature (Table 31-1 and Fig. 31-4).

Tabulate the difference in energy (cm^{-1}) of adjacent transitions according to Equation 31-7 and index according to the vibrational quantum number of the higher member of the pair. Prepare a Birge-Sponer extrapolation. Calculate, tabulate, and compare your values with literature values of $\tilde{v}'_e$, $\tilde{v}'_e x'_e$, D'_e, D'_0, D''_e, and D''_0. Discuss the errors.

REFERENCES

1. Alberty, R. A., *Physical Chemistry,* 7th ed., Wiley, New York, 1987.

2. Armanious, M., and M. Shoja, "Analysis of the Band Spectrum of I_2 Using Apple II," *J. Chem. Educ.* 63:627, 1986.

3. Davies, M. J., "Simple Molecular Spectra Experiments," *J. Chem. Educ.* 28:474, 1951.

4. Barrow, G., *Physical Chemistry,* McGraw-Hill, New York, 1961.

5. D'alterio, R., R. Mattson, and R. Harris, "Potential Curves for the I_2 Molecule," *J. Chem. Educ.* 51:282, 1974.

6. Gaydon, A. G., *Dissociation Energies,* 2nd ed., rev., Chapman and Hall, London, 1953.

7. Glasstone, S., *Theoretical Chemistry,* Van Nostrand, New York, 1944.

8. Herzberg, G., *Molecular Spectra and Molecular Structure,* vol. 1, *Spectra of Diatomic Molecules,* 2nd ed., Van Nostrand, Princeton, New Jersey, 1950.

9. McNaught, I. J., "The Electronic Spectrum of Iodine Revisited," *J. Chem. Educ.* 57:101, 1980.

10. Moore, C. E., "Atomic Energy Levels," vol. III, Circular of the National Bureau of Standards, U.S. Government Printing Office, Washington, D.C., 1958.

11. Rosen, B., ed., *Tables de Constantes et Données Numériques, 4, Données Spectroscopiques,* Herman and Co., Paris, 1951.

12. Snadden, R. B., "The Iodine Spectrum Revisited," *J. Chem. Educ.* 64:919, 1987.

13. Stafford, F. E., "Band Spectra and Dissociation Energies," *J. Chem. Educ.* 39:626, 1962.

Experiment 32

The Electronic Spectrum of Nitrogen: The Second Positive Series

The electronic spectra of molecules generally lie in the visible or ultraviolet region. Transitions between two electronic states give the appearance of a large number of bands rather than lines. Each band arises from the closely spaced rotational transitions that are superimposed upon an electronic transition between two vibrational levels in the upper and lower electronic states. With high resolution, each band is seen to consist of a number of lines, arising from rotational transitions accompanying the vibrational and electronic changes.

The total energy of a molecule in an excited electronic state is the sum of its electronic, vibrational, and rotational energies:

$$E_{\text{total}} = E_{\text{el}} + E_v + E_r \qquad (32\text{--}1)$$

Division of energy by hc gives wave numbers (cm^{-1})

$$T_{\text{total}} = T_{\text{el}} + G(v) + F(J) \qquad (32\text{--}2)$$

where T_{el} is the energy of an electronic state (expressed in wave numbers), $G(v)$ is the vibrational energy, and $F(J)$ is the rotational energy. For the ground electronic state, $T_{\text{el}} = 0$. For an anharmonically vibrating molecule, the vibrational energies (expressed in wave numbers) are given by

$$G(v) = \tilde{\nu}_e(v + \tfrac{1}{2}) - \tilde{\nu}_e x_e(v + \tfrac{1}{2})^2 + \tilde{\nu}_e y_e(v + \tfrac{1}{2})^3 + \cdots \qquad (32\text{--}3)$$

and the rotational energies (expressed in wave numbers) of a nonrigid rotating molecule are given by

$$F(J) = B_e J(J + 1) - D_e J^2(J + 1)^2 \qquad (32\text{--}4)$$

The energy $\tilde{\nu}$ of a transition (in wave numbers) between an upper state, designated by a single prime, and a lower state, designated by a double prime, is given by

$$\tilde{\nu} = (T'_{\text{el}} - T''_{\text{el}}) + (G' - G'') + (F' - F'') \qquad (32\text{--}5)$$

Since the rotational energies are much smaller than the vibrational energies, $(F' - F'')$ is set equal to zero. Replacing $(T'_{\text{el}} - T''_{\text{el}})$ with $\tilde{\nu}_{\text{el}}$, and substituting Equation 32–3 with a single ' to indicate upper states and a double " to indicate lower states, we can write Equation 32–5 as

$$\tilde{\nu} = \tilde{\nu}_{\text{el}} + \tilde{\nu}'_e(v' + \tfrac{1}{2}) - \tilde{\nu}'_e x'_e(v' + \tfrac{1}{2})^2 + \tilde{\nu}'_e y'_e(v' + \tfrac{1}{2})^3 \qquad (32\text{--}6)$$

$$- \tilde{\nu}''_e(v'' + \tfrac{1}{2}) + \tilde{\nu}''_e x''_e(v'' + \tfrac{1}{2})^2 - \tilde{\nu}''_e y''_e(v'' + \tfrac{1}{2})^3$$

The relationship among these parameters is shown in Figure 32–1.

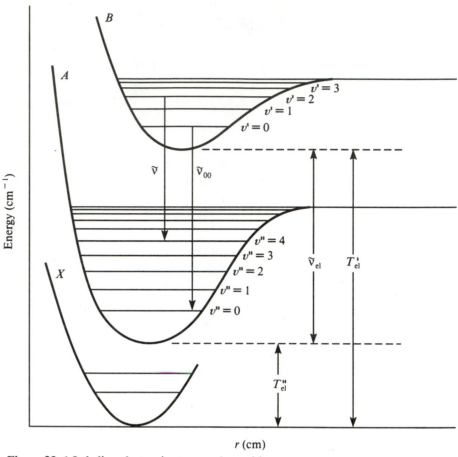

Figure 32–1 Labeling electronic states and transitions.

Notice that $\tilde{\nu}_{el}$ is not an observable transition, but is just the difference in energy between the two electronic states. An actual transition, close but not equal in energy to $\tilde{\nu}_{el}$, is the transition from $v' = 0$ in the upper electronic state to $v'' = 0$ in the lower electronic state. This characteristic transition is labeled $\tilde{\nu}_{00}$. The relationship between $\tilde{\nu}_{00}$ and $\tilde{\nu}_{el}$, derived from Equation 32–6, is

$$\tilde{\nu}_{00} = \tilde{\nu}_{el} + (\tilde{\nu}'_e - \tilde{\nu}''_e)/2 - (\tilde{\nu}'_e x'_e - \tilde{\nu}''_e x''_e)/4 + (\tilde{\nu}'_e y'_e - \tilde{\nu}''_e y''_e)/8 \quad \textbf{(32–7)}$$

Since Δv can be anything (there is no selection rule), all possible transitions between vibrational levels in the upper and lower electronic states are permitted. Organizing the observed transitions into progressions and especially sequences simplifies finding the relationship between the positions of the bands on the film and their origin in the energy-level diagram (Fig. 32–2).

In a Deslandres table, the observed transitions are displayed in an array; the rows correspond to the upper-state vibrational quantum number v' and the columns to v''. As an example, Table 32–1 is the Deslandres table for carbon monoxide. The values in brackets are differences between

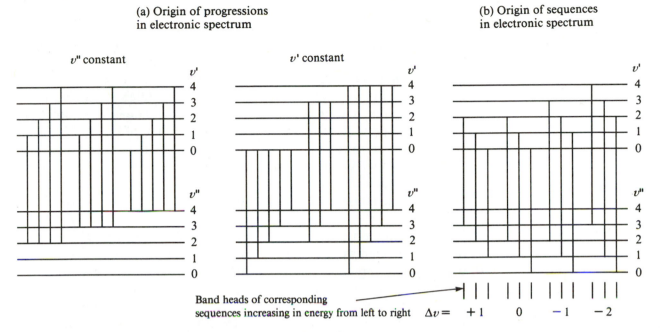

(a) Origin of progressions in electronic spectrum

(b) Origin of sequences in electronic spectrum

v'' constant

v' constant

Band heads of corresponding sequences increasing in energy from left to right $\Delta v = \quad +1 \qquad 0 \qquad -1 \qquad -2$

Figure 32–2 Progression and sequences.

adjacent values in the same row. The values in parentheses are differences between adjacent values in the same column. The [differences] correspond to vibrational energy-level spacings in the lower electronic state, while the (differences) correspond to vibrational energy-level spacings in the upper electronic state.

Thus, the spacing between the $v'' = 0$ and $v'' = 1$ levels (taking average values) is [2166], the spacing between the $v'' = 1$ and $v'' = 2$ levels is [2119], and so on for lower-level spacings. Similarly, differences

TABLE 32–1

Deslandres Table for CO Bands ($\tilde{\nu}$ in cm^{-1})

v'	v''				
	0	1	2	3	4
0	—	62602 [2117] (1486)	60485 [2092] (1488)	58393 [2064] (1489)	56329
1	66231 [2233] (1444)	64088 (1445)	—	59882 [2064] (1443)	57818
2	67675 [2142] (1413)	65533 [2117] (1411)	63416 [2091] (1412)	61325	—
3	69088 [2144] (1381)	66944 [2116] (1379)	64828 (1371)	—	60675 (1380)
4	70469 [2146]	68323 [2124]	66199 [2082]	64117 [2062]	62055
Average	[2166]	[2119]	[2088]	[2063]	

between rows gives spacings in the upper level. Thus, the spacing between $v' = 0$ and $v' = 1$ is (1487), between $v' = 1$ and $v' = 2$ it is (1444), and so on for lower levels. In general, all the different bands in the same horizontal row in a Deslandres table have the same v' (upper-state vibrational quantum number). These correspond to the progressions in the right half of Figure 32–2a. Similarly, all the different bands in the same vertical column have the same v'' (lower-state quantum number). These correspond to the progressions in the left half of Figure 32–2a.

Examination of the transitions lying on the main diagonal of the Deslandres table reveals that these correspond to $\Delta v = 0$. These are seen to be the sequence in Figure 32–2b labeled $\Delta v = 0$. The diagonal parallel to the main diagonal corresponds to the sequence in Figure 32–2b labeled $\Delta v = +1$.

Agreement between row and column differences for a given entry into the Deslandres table is good evidence that the entry is correct. In this way the vibrational quantum numbers v' and v'' can be assigned to the bands observed on the film.

The emission spectrum is more complex than the absorption spectrum because any upper-state vibrational level may be populated with molecules, and transitions to any lower-state level are allowed, since Δv can be anything. The Deslandres table serves to organize the observed transitions in a manner that facilitates indexing the observed spectrum.

A large number of electronic states of the nitrogen molecule are known, and many spectral series arising from transitions between them have been thoroughly studied (Lofthus and Krupenie, 1977). In this experiment we study a band spectrum, called the second positive series, lying in the blue region of the visible spectrum and arising from transitions between the C and B states of the nitrogen molecule. A few transitions are shown in Figure 32–3. The first positive series arises from transitions between the B and A states and lies in the red end of the spectrum. Both series should be clearly visible on your film. The ground state (X) is not involved in transitions of the first or second positive series.

Apparatus

The apparatus consists of a grating spectrograph (e.g., Baird-Atomic 1.5-m grating spectrograph, model SB-1, covering 370 to 740 nm, with a dispersion of 1.5 nm/mm); gas discharge tubes (H_2, N_2, Ar, He, and Ne); a high-voltage transformer for the tubes; film developing darkroom; Plus-X film or equivalent; D-76 developer or equivalent; glacial acetic acid; Kodak Acid Fix; and three 200-ml graduated cylinders for developing tanks.

SAFETY CONSIDERATIONS

When changing discharge tubes, be sure to both turn off and unplug the high-voltage transformer to avoid electric shock. Let the tubes

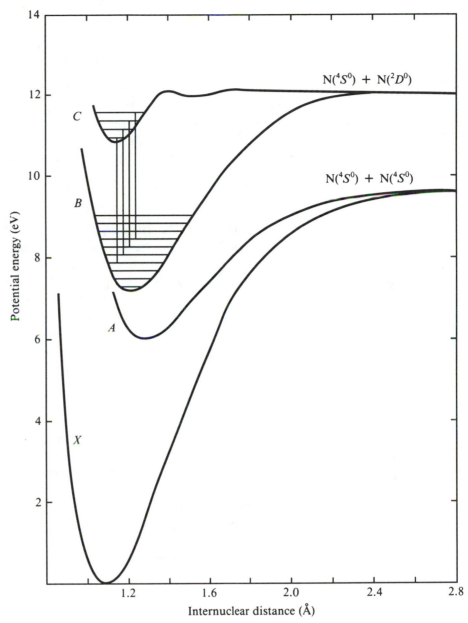

Figure 32–3 Some prominent electronic energy levels for the nitrogen molecule. The vertical lines between the C and B states represent a few transitions from the second positive series. (Source: Lofthus, 1977.)

cool before touching them, as they become quite hot after a few minutes, especially the thin middle section. Be careful not to exert any severe leverage on the tubes when pulling them from their holder, as the thin section is susceptible to breaking.

Avoid skin contact with developer or fixer. Developers contain strong organic reducing agents that cause an allergic reaction in many people; fixing agents are essentially sodium thiosulfate, $Na_2S_2O_3$.

EXPERIMENTAL PROCEDURE

See the Experimental Procedure for Experiment 30.

Results and Calculations

Prepare a table of the calibration lines, their wavelengths λ, and comparator scale readings X. Since the dispersion of a grating spectrograph is linear across the film, these data can be fit (by least squares) to an equation of the form

$$\lambda = mX + b \tag{32-8}$$

where m is the dispersion in nanometers per millimeter and b is the intercept. Record the standard deviations in m and b for analyzing propagated errors. Tabulate the comparator scale readings for the observed band heads for the second positive series for N_2 and calculate their wavelengths with Equation 32–8. Determine the propagated error in wavelength. After indexing (determining v' and v'' for each observed band head), compare your wavelengths with the literature values.

As an aid in indexing, compare your spectrum with Figure 32–4. The wavelengths of the more intense transitions are shown in Figure 32–4. Notice how neighboring transitions tend to belong to the same sequence. Normally about 20 transitions are of measurable intensity when Plus-X film is exposed as recommended. Transform the observed wavelengths to wave numbers, and with the aid of Figure 32–4, construct a Deslandres table for your transitions. See also the figures in Herzberg (1950) and Pearse and Gaydon (1963). After indexing the observed transitions in a Deslandres table, calculate row and column differences as previously described, and use the consistency of these numbers to aid in indexing the band heads. Sketch an energy-level diagram indicating your observed transitions. Label the energy-level spacings with the values you have just determined from the row and column differences in your Deslandres table.

These energy-level spacings correspond to the differences between adjacent vibration energy levels, $G_{v+1} - G_v$, given by Equation 32–3. For the upper electronic state, $v = v'$, and from Equation 32–3 the upper-level spacings are given by

$$G_{v'+1} - G_{v'} = \tilde{\nu}'_e - 2\tilde{\nu}'_e x'_e(v' + 1) \tag{32-9}$$

$$+ 3\tilde{\nu}'_e y'_e(v'^2 + 2v' + 13/12) + \cdots$$

Similarly, for the lower electronic state, $v = v''$, and the lower-level spacings are given by

$$G_{v''+1} - G_{v''} = \tilde{\nu}''_e - 2\tilde{\nu}''_e x''_e(v'' + 1) \tag{32-10}$$

$$+ 3\tilde{\nu}''_e y''_e(v''^2 + 2v'' + 13/12) + \cdots$$

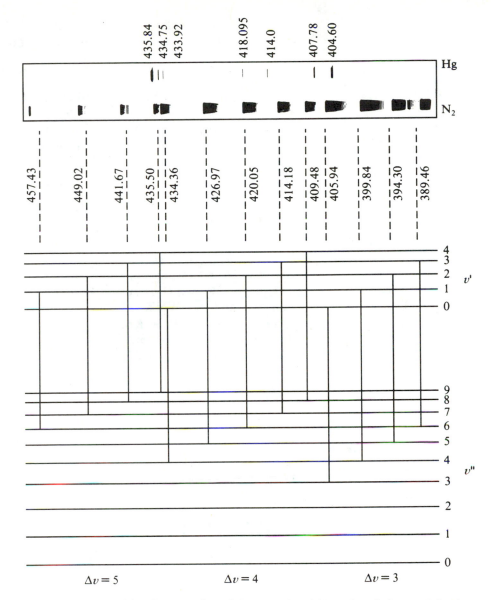

Figure 32–4 The origin of that portion of the second positive series of nitrogen lying in the visible region of the spectrum. The top spectrum shows some prominent mercury calibration lines; the lower spectrum shows three sequences of the nitrogen bands. Wavelengths are in nanometers.

For the lower-level state, $\tilde{\nu}_e'' y_e''$ is very nearly zero, so that when $G_{v''+1} - G_{v''}$ is plotted against $(v'' + 1)$, a straight line results, the slope of which equals $-2\tilde{\nu}_e'' x_e''$ and the intercept of which equals $\tilde{\nu}_e''$. This graphical approach to determining the vibrational parameters is called a Birge-Sponer extrapolation. Compare Equation 32–9 and Equation 31–8.

For the upper state, $\tilde{\nu}_e' y_e'$ is not negligible and must be determined. Since three variables ($\tilde{\nu}_e'$, $\tilde{\nu}_e' x_e'$, and $\tilde{\nu}_e' y_e'$) are unknown, at least three equations must be known. For example, if the measurement $G_{v'+1} - G_{v'} = 1922$ for $v' = 0$, then one equation would be

$$1922 = x - 2y + 39z/12 \qquad (32\text{--}11)$$

where $x = \tilde{\nu}'_e$, $y = \tilde{\nu}'_e x'_e$, and $z = \tilde{\nu}'_e y'_e$. With two more equations, this system of three equations in three unknowns can be solved by Cramer's Rule. Your instructor may have a computer program available online or in the library for solving systems of linear equations; or see Part Two.

Compare your values for $\tilde{\nu}'_e$, $\tilde{\nu}'_e x'_e$, $\tilde{\nu}''_e$, $\tilde{\nu}''_e x''_e$, and $\tilde{\nu}''_e y''_e$ with the literature values in terms of the errors calculated for your measurements.

REFERENCES

1. Barrow, G. M., *Molecular Spectroscopy,* McGraw-Hill, New York, 1962.

2. Herzberg, G., *Molecular Spectra and Molecular Structure,* vol. I, *Spectra of Diatomic Molecules,* 2nd ed., Van Nostrand, Princeton, New Jersey, 1950.

3. Huber, K. P., and G. Herzberg, *Molecular Spectra and Molecular Structure,* Van Nostrand, New York, 1979.

4. Lofthus, A., and P. H. Krupenie, "The Spectrum of Molecular Nitrogen," *J. Phys. Chem. Ref. Data* 6:113–290, 1977.

5. Pearse, R. W. B., and A. G. Gaydon, *The Identification of Molecular Spectra,* Wiley, New York, 1963.

Experiment 33

Vibration–Rotation Spectra of a Diatomic Molecule

Pure rotation spectra are observable in the microwave region, and vibration–rotation spectra in the infrared. In this experiment, the infrared spectra of HCl and DCl are measured. From an analysis of the rotational fine structure, some molecular properties are calculated for the molecules in their ground and first excited vibrational states.

Energies of a Rotating and Vibrating Diatomic Molecule

The energy levels of a nonrigid rotor are given by

$$F(J) = B_e J(J + 1) - D_e J^2(J + 1)^2 \qquad (33\text{--}1)$$

where B_e is the rotational constant and D_e is the centrifugal stretching constant. In a harmonic oscillator, D_e equals zero. The rotational constant and centrifugal stretching constants are not quite constant, but instead depend linearly on the vibrational state v:

$$B_v = B_e - \alpha_e(v + 1/2) \qquad (33\text{–}2)$$

$$D_v = D_e - \beta_e(v + 1/2) \qquad (33\text{–}3)$$

In this experiment, we shall assume that β_e is negligibly small, but we shall determine α_e, B_e, B_0, and B_1. The constant α_e reflects the degree of coupling between the rotational and vibrational modes of the molecule. You can regard B_e as a vibration-free rotational constant. In other words, it is what the rotational constant would be if the molecule were not vibrating.

The energy levels of an anharmonic oscillator are given by a power series in $(v + 1/2)$, where the vibrational quantum number v has values of 0, 1, 2, 3, . . . :

$$G(v) = \tilde{\nu}_e(v + 1/2) - \tilde{\nu}_e x_e(v + 1/2)^2 \qquad (33\text{–}4)$$

$$- \tilde{\nu}_e y_e(v + 1/2)^3 + \cdots$$

The fundamental vibration frequency (wave numbers) of the molecule is $\tilde{\nu}_e$, and the anharmonicity constants are $\tilde{\nu}_e x_e$, $\tilde{\nu}_e y_e$, ... In this experiment we shall determine only the first-order anharmonicity constants of the form $\tilde{\nu}_e x_e$ for the states $v = 0$ and the state $v = 1$. In shorthand notation, then, the energies of a rotating and vibrating diatomic molecule are given by

$$T_{(v,J)} = F(J) + G(v) \qquad (33\text{–}5)$$

The Vibration–Rotation Spectrum of a Diatomic Molecule

The infrared spectrum of a diatomic molecule arises from transitions between the ground vibration state, for which $v'' = 0$, and the first excited state, for which $v' = 1$:

$$\Delta T = T_{(v'=1,J)} - T_{(v''=0,J)} \qquad (33\text{–}6)$$

The values of ΔT are the energies of the transitions in wave numbers. The selection rules for changes in the rotational quantum number J are, in general, $\Delta J = 0, +1$, or -1. However, in diatomic molecules such as hydrogen chloride, which have no net angular momentum in a direction perpendicular to the molecular axis, the selection rules are restricted to $\Delta J = +1$.

Those transitions that correspond to $\Delta J = +1$ form the R branch of the spectrum, and those corresponding to $J = -1$ form the P branch. In cases where transitions for which $J = 0$ are observed (e.g., in the corresponding transitions in nitric oxide, which has an unpaired electron with net angular momentum about the molecular axis), the additional absorption lines appear close together in the center of the spectrum and are termed the Q branch. Figure 33–1 is an energy-level diagram with

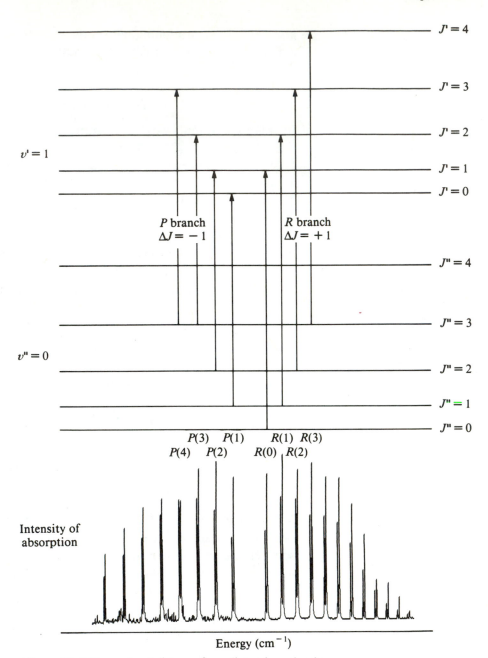

Figure 33–1 Energy-level diagram for a diatomic molecule.

some absorption transitions from the ground state $v'' = 0$ and the first excited state $v' = 1$. Note that the J in $P(J)$ and $R(J)$ refers to the value of J in the lower state from which the transition originates.

It simplifies matters if transitions arising from $\Delta J = +1$ (the R branch) are considered separately from $\Delta J = -1$ (the P branch). Let the lower-state rotational quantum number take on the values: $J'' = 0, 1, 2, 3, \ldots$. Then the rotational quantum numbers in the upper state are given by $J' = J'' + 1$.

The energies in wave numbers of the R branch are given by

$$\tilde{\nu}_R = (v' - v'')\tilde{\nu}_e - x_e\tilde{\nu}_e[(v' - v'')(v' + v'' + 1)] \tag{33-7}$$

$$+ (B_{v'} + B_{v''})(J'' + 1) + (B_{v'} - B_{v''})(J'' + 1)^2$$

$$- 4D_e(J'' + 1)^3$$

The energies in wave numbers of the P branch are given by

$$\tilde{\nu}_P = (v' - v'')\tilde{\nu}_e - x_e\tilde{\nu}_e[(v' - v'')(v' + v'' + 1)] \tag{33-8}$$

$$- (B_{v'} + B_{v''})J'' + (B_{v'} - B_{v''})J''^2 + 4D_eJ''^3$$

Calculation of Molecular Parameters from the Spectrum

The calculation of $B_{v'}$, $B_{v''}$, D_e, $\tilde{\nu}_e$, and $x_e\tilde{\nu}_e$ are facilitated by some systematic comparisons of the lines in the P branch and R branch of the experimental infrared spectrum. Let $R(J)$ symbolize the energy in wave numbers of the lines in the R branch that originate in the state $J'' = J$. Similarly, $P(J)$ are the P branch transitions originating in the state $J'' = J$. Thus, the two lines on each side of the band center are $P(1)$ and $R(0)$. Three useful arrangements of Equations 33–7 and 33–8 are

$$\frac{R(J) - P(J)}{4(J + 1/2)} = B_{v'} - 2D_e(J^2 + J + 1) \tag{33-9}$$

$$\frac{R(J-1) - P(J+1)}{4(J + 1/2)} = B_{v''} - 2D_e(J^2 + J + 1) \tag{33-10}$$

$$\frac{R(J) + P(J+1)}{2} = (v' - v'')\tilde{\nu}_e[1 - x_e(v' + v'' + 1)] \tag{33-11}$$

$$+ (B_{v'} - B_{v''})(J + 1)^2$$

Note that the left side of Equation 33–9 plotted against $J^2 + J + 1$ gives a straight line with slope equal to $-2D_e$ and intercept equal to $B_{v'}$. A similar plot with Equation 33–10 gives $B_{v''}$ and D_e.

When the left side of Equation 33–11 is plotted against $(J + 1)^2$, the slope equals $B_{v'} - B_{v''}$, while the intercept equals the band origin (BO).

$$\tilde{\nu}_{\text{BO}} = (v' - v'')\tilde{\nu}_e[1 - x_e(v' + v'' + 1)] \tag{33-12}$$

Under ordinary conditions, the change in vibrational quantum number accompanying the infrared absorption spectrum is a change from $v'' = 0$ to $v' = 1$. For this case Equation 33–12 reduces to

$$\tilde{\nu}_{\text{BO}} = \tilde{\nu}_e - 2\tilde{\nu}_e x_e \tag{33-13}$$

Although our measurement of the $v'' = 0$ to $v' = 1$ infrared absorption spectrum permits evaluation of the band origin $\tilde{\nu}_{\text{BO}}$, not enough information is available from this measurement alone to evaluate the fundamental vibration frequency $\tilde{\nu}_e$ (cm^{-1}) and the anharmonicity constant $\tilde{\nu}_e x_e$ (cm^{-1}).

Determination of the fundamental vibration frequency and the anharmonicity constant requires knowledge of overtones, that is, transitions from $v'' = 0$ to $v' = 2, 3, 4, \ldots$. When $v'' = 0$, Equation 33–12 reduces to

$$\frac{\tilde{\nu}_{BO}}{v'} = \tilde{\nu}_e - x_e \tilde{\nu}_e (v' + 1) \qquad \textbf{(33–14)}$$

If two overtones are known, we have two equations in two unknowns ($\tilde{\nu}_e$ and $x_e \tilde{\nu}_e$), which may be calculated. If several overtone bands are available, the left side of Equation 33–13 may be plotted against ($v' + 1$); the slope equals $x_e \tilde{\nu}_e$ and the intercept equals $\tilde{\nu}_e$. If not experimentally observed, overtones are available in Table 33–1 or the literature (e.g., Herzberg, 1950, Table 8, p. 55).

The rotational constant B_e and the coupling constant α_e may be determined with Equation 33–2 and the values of B_0 and B_1 calculated with Equations 33–9 and 33–10. A better value might be determined by using the calculated values of B_0 and B_1 along with additional values of B_v from Table 33–1 or the literature (e.g., Herzberg, 1950, Table 13, p.

TABLE 33–1

Selected Molecular Parameters for HCl (all units in cm^{-1})

Molecular Parameter	H^{35}Cl	H^{37}Cl	D^{35}Cl	D^{37}Cl
$\tilde{\nu}_0(1\text{–}0)$	2885.9775[a]	2883.8705[b]	2091.0613[a]	2088.073[b]
$\tilde{\nu}_0(2\text{–}0)$	5667.9841[a]	5663.926[b]	4128.4330[a]	4122.68[d]
$\tilde{\nu}_0(3\text{–}0)$	8346.782[a]		6112.79[d]	6104.28[d]
$\tilde{\nu}_0(4\text{–}0)$	10922.81[a]			
$\tilde{\nu}_0(5\text{–}0)$	13396.19[a]			
$\tilde{\nu}_e$	2990.97424[a]		2145.1630[a]	2141.82[d]
B_0	10.440254[a]	10.4243[b]	5.392261[a]	5.3757[b]
B_1	10.136223[a]	10.1209[b]	5.279816[a]	5.2673[b]
B_2	9.834663[a]	9.8213[b]	5.168106[a]	5.1537[b]
B_3	9.534909[a]			
B_4	9.2360[a]			
B_5	8.942[a]			
B_e	10.593404[a]	10.578[c]	5.448794[a]	5.432[d]
$\tilde{\nu}_e x_e$	52.84579[a]		27.18252[a]	26.99[d]
α_e	−0.307139[a]	−0.3035[c]	−0.1132911[a]	−0.111[d]
$D_e/10^{-4}$	−5.32019[a]	−5.30[b]	−1.40[a]	−1.36[d]

[a] Rank, D. H., D. P. Eastman, B. S. Rao, and T. A. Wiggins, "Rotational and Vibrational Constants of the HCl^{35} and DCl^{35} Molecules," *J. Opt. Soc. Am.* 52:1–7, 1962.
[b] Webb, D. U., and K. N. Rao, "Vibration Rotation Bands of Heated Hydrogen Halides," *J. Mol. Spectrosc.*, 28:121, 1968.
[c] Pickworth, J., and H. W. Thompson, "The Fundamental Vibration-Rotation Band of Deuterium Chloride," *Proc. R. Soc. London, Ser. A*, 218:37, 1953.
[d] Van Horne, B. H., and C. D. House, "Near Infrared Spectrum of DCl," *J. Chem. Phys.* 25:56–59, 1956.

114). A plot of B_v against $(v + 1/2)$ gives α_e as the slope and B_e as the intercept.

The equilibrium internuclear separation r_e can be calculated from B_e. The dependence of the rotational constant on v is discussed by Herzberg (1950, p. 106). Compare

$$B_e = \frac{h}{8\pi^2 c \mu r_e^2} \quad \text{and} \quad B_v = \frac{h}{8\pi^2 c \mu \bar{r}_v^2} \qquad (33\text{–}15)$$

where r_e is the equilibrium internuclear separation—that is, the true interatomic distance in a "vibration-free" molecule—while $\sqrt{\bar{r}_v^2}$ is the root-mean-square internuclear separation in a vibrating molecule. Since $\sqrt{\bar{r}_v^2} > r_e$, $B_e > B_v$ because of anharmonicity.

Apparatus

The apparatus consists of an infrared spectrophotometer with a resolution of about 0.25 wave number or less, and a 10-cm gas cell with KBr windows, filled with approximately 100 torr each of HCl and DCl.

SAFETY CONSIDERATIONS

Hydrogen chloride is a toxic, corrosive gas. The KBr cell windows must not be touched with the fingers. The cell should be stored in a desiccator when not in use.

EXPERIMENTAL PROCEDURE

A prefilled gas cell will be provided. Because many excellent infrared spectrophotometers are currently available, it is not practical to give specific operating instructions. Detailed guidelines for the operation of your instrument will be provided. Some general features of infrared spectrophotometers are discussed in Part Two, Chapter 14.

Measure the values in wave numbers of the $P(J)$ and $R(J)$ transition in HCl and DCl. If the resolution of your spectrophotometer permits, measure the peaks arising from the two isotopes of chlorine so that the molecular parameters may be calculated for four isotopes of HCl: $H^{35}Cl$, $H^{37}Cl$, $D^{35}Cl$, and $D^{37}Cl$.

Results and Calculations

Tabulate J, $P(J)$, $R(J)$, $[R(J) + P(J + 1)]/2$, $J^2 + J + 1$, $(J + 1)^2$, $[R(J) - P(J)]/[4(J + 1/2)]$, and $[R(J - 1) - P(J + 1)]/[4(J + 1/2)]$ for each isotopic species and prepare plots of Equations 33–9, 33–10, and 33–11 to determine $B_{v'}$, $B_{v''}$, D_e, and the band origin $\tilde{\nu}_{BO}$.

(Text continues on page 686.)

```
Program IRSpec;

(*****************************************************************
*                                                                *
* This is a program to process the observed energies in the      *
* infra-red absorption spectrum of the HCl molecule.             *
*                                                                *
* Input consists of P(J) value from J = 1 up to the observed     *
* maximum J followed by R(J) values from J = 0 up to the         *
* observed maximum J.  Here, J equals the value of the rota-     *
* tional quantum number of the rotational state from which the   *
* transition originates, i.e, the value of J".                   *
*                                                                *
* Output consists of tables of input data, and the tables        *
* necessary to calculate the usual molecular parameters of a     *
* non-rigid rotor and anharmonic oscillator.                     *
*                                                                *
*  June 6, 1987                                                  *
*                                                                *
*****************************************************************)
uses crt;

Const h = 6.66186E-34;        (* joule sec                 *)
      c = 2.9979025E8;        (* meters/sec                *)
      Jmax = 50;              (* max allowed J             *)

Type Ary  = Array [0..Jmax] of Real;
     AryI = Array [0..Jmax] of Integer;

Var R,P,                 (* arrays of input data from R and P branch *)
    BorAry,              (* array to calculate band origin           *)
    BupAry,              (* array to calculate Bv'  in upper state   *)
    BloAry:  Ary;        (* array to calculate Bv'' in lower state   *)
    Ycalc:   Ary;        (* array of calc val of dependent variable  *)
    J1Ary,               (* array of (J + 1)2                        *)
    JJ1Ary:  AryI;       (* array of J2 + J + 1                      *)
    N:   Integer;        (* The maximum value of J for  R(J)         *)
    De,                  (* Centrifugal-distortion constant          *)
    Sigmam,              (* Standard deviation in slope              *)
    Sigmab,              (* Standard deviation in intercept          *)
    CorrelCoef:  Real; (* Correlation Coefficient for Lst. Squares   *)
    Name:  String[35]; (* HCl-35, DCl-37, etc.                      *)

Procedure GetSpectralData;
   (* Accepts values of R(J) and P(J) interactively              *)
   Var J,                            (* rotational quantum number     *)
      X,Y:  Integer;                 (* screen coordinates            *)
   Begin (* GetSpectralData *)
      CLRSCR;
      Writeln;Writeln('Please prepare to enter values for');
      Writeln('R(J) ranging from J = 0 to N and for');
      Writeln('P(J) ranging from J = 0 to N ');
      Writeln('Since P(0) is not defined, enter a 0 for P(0).');
      Writeln; Writeln('Press the enter key when ready');
      Readln;
```

Figure 33–2 *Program IRSpec.*

```
        CLRSCR;Write('What is your maximum value of J?   ');
        Readln(N); Writeln;
        For J:= 0 to N Do
            Begin
            Write('         R(',J,'):   ');
            X:= WhereX; Y:= WhereY;
            Read(R[J]);
            GoToXY(X+15,Y);
            Write('  P(',(J),'):   ');
            Readln(P[J]);
            End;  (* For *)
    End;   (* GetSpectralData *)

Procedure CalculateArrays;
    Var J:  Integer;
    Begin (* Calculate Arrays *)
    For J:= 0 to (N - 1) Do
        Begin
        BorAry[J]:= ((R[J] +P[J + 1])/2);
        J1Ary[J]:= Sqr(J + 1)
        End;
    For J:= 1 to (N - 1) Do
        Begin (* For *)
        BupAry[J]:= (R[J] - P[J])/(4*(J + 0.5));
        BloAry[J]:= (R[J - 1] - P[J + 1])/(4*(J + 0.5));
        JJ1Ary[J]:= (J*J + J +1)
        End;  (* For *)
    End;   (* Calculate Arrays *)

Procedure EchoInput;
Const Sp = '                  ';
Var J:  Integer;
    Begin (* WriteData *)
    CLRSCR;
    Writeln(Sp,'      Table 1.  Input Data.');
    Writeln;
    Writeln(Sp,'   J       P(J)              R(J)');
    Writeln;
    For J:= 0 to N Do
        Writeln(Sp,J:5,P[J]:10:2,R[J]:16:2);
    End;  (* WriteData *)



(**************************************************************)

(   PROCEDURE ReadFile(VAR X,Y:  Ary;
                       VAR N:  Integer);

    Var I:  Integer;
        F:  Text;

Begin (* ReadFile *)
    Assign(F,'LS.DAT');
    Reset(F);
    Readln(F,N);
```

Figure 33-2 (continued)

```
   For I:= 1 to N Do
      Readln(F,X[I],Y[I]);
   Close(F)
End;  (* ReadFile *)  }


 (*************************************************************)

 Procedure LeastSquares(X:   AryI;
                        Y:   Ary;
                        Var Ycalc:  Ary;
                        Var b,m  :  Real;
                            N   :  Integer);
 (* Fits a straight line to a set of N  X-Y pairs   *)
 (* The line has a slope equal to m and intercept b *)

 Var I:  Integer;
     SumX,SumY,SumXY,SumX2,SumY2,
     XI,YI,SXY,SXX,SYY:  Real;

 Begin (* LeastSquares *)
    SumX:= 0;
    SumY:= 0;
    SumXY:=0;
    SumX2:=0;
    SumY2:=0;
    For I:= 1 to N do
       Begin (* For *)
       XI:= X[I];
       YI:= Y[I];
       SumX:= SumX + XI;
       SumY:= SumY + YI;
       SumXY:= SumXY + XI*YI;
       SumX2:= SumX2 + XI*XI;
       SumY2:= SumY2 + YI*YI;
       End;  (* For *)
    SXX:= SumX2 - SumX*SumX/N;
    SXY:= SumXY - SumX*SumY/N;
    SYY:= SumY2 - SumY*SumY/N;
    m:= SXY/SXX;
    b:= ((SumX2*SumY - SumX*SumXY)/N)/SXX;
    CorrelCoef:= SXY/SQRT(SXX*SYY);
    Sigmam:= (SQRT((SumY2 - b*SumY -m*SumXY)/(N - 2)))/SQRT(SXX);
    Sigmab:= Sigmam*SQRT(SumX2/N);
    For I:= 1 to N do
       Ycalc[I]:= b + m*X[I];
 End;  (* LeastSquares *)


 (*************************************************************)

 Procedure WriteTable2;
 (* Writes data for calculating band origin *)
 Const Sp = '                       ';
 Var J:  Integer;
```

Figure 33–2 (continued)

```
      Slope, Intercept, BandOrigin, DeltaB:  Real;
Begin (* WritelTable1 *)
   Readln;
   CLRSCR;
   Writeln(Sp,'Table 2. Data for Calculation of De and Band Origin.');
   Writeln;
   Writeln(Sp,'    J        Sqr(J + 1)    R(J) + P(J + 1)');
   Writeln(Sp,'                          ---------------');
   Writeln(Sp,'                                 2        ');
   Writeln;
   For J:= 0 to (N - 1) Do
      Writeln(Sp,J:5,J1Ary[J]:10,BorAry[J]:20:2);
   LeastSquares(J1Ary,BorAry,Ycalc,intercept,slope,J);
   DeltaB:= slope;
   BandOrigin:= intercept;
   Writeln;
   Write(Sp,'Bv(v=1)-Bv(v=0) equals:  ',DeltaB:5:4);
   Writeln(' Sigma = ',Sigmam:6);
   Writeln;
   Write(Sp,'The Band Origin equals:  ',BandOrigin:6:2);
   Writeln('  Sigma = ', Sigmab:6);
End;

Procedure WriteTable3;
(* Writes data for calculating Bv' *)
Const Sp = '                         ';
Var J:  Integer;
    Slope, Intercept, De,Bv0:  Real;
Begin (* WritelTable3 *)
   Readln;
   CLRSCR;
   Writeln(Sp,'Table 3.  Data for Calculation of De and Bv(v = 0).');
   Writeln;
   Writeln(Sp,'    J    J*J + J + 1     (R(J - 1) - P(J + 1)');
   Writeln(Sp,'                        --------------------');
   Writeln(Sp,'                              4*(J + 0.5)     ');
   Writeln;
   For J:= 1 to (N - 1) Do
      Writeln(Sp,J:5,JJ1Ary[J]:10,BloAry[J]:20:3);
   LeastSquares(JJ1Ary,BloAry,Ycalc,intercept,slope,J);
   De:= slope/(-2);
   Bv0:= intercept;
   Writeln;
   Writeln(Sp,'       De equals:  ',De:7:5,' Sigma =  ',Sigmam/2:6);
   Writeln(Sp,'  Bv(v=0) equals:  ',Bv0:5:2,' Sigma =  ',Sigmab:6);

End;  (* WritelTable3 *)

Procedure WriteTable4;
(* Writes data for calculating Bv' *)
Const Sp = '                         ';
Var J:  Integer;
    Slope,Intercept,De,Bv1:  Real;
Begin (* WritelTable4 *)
   Readln;
```

Figure 33–2 (continued)

```
    CLRSCR;
    Writeln(Sp,'Table 4.   Data for Calculation of De and Bv(v=1).');
    Writeln;
    Writeln(Sp,'    J      J*J + J + 1     (R(J) - P(J)');
    Writeln(Sp,'                          ------------');
    Writeln(Sp,'                           4*(J + 0.5)');
    Writeln;
    For J:= 1 to (N - 1) Do
        Writeln(Sp,J:5,JJ1Ary[J]:10,BupAry[J]:20:3);
    LeastSquares(JJ1Ary,BupAry,Ycalc,intercept,slope,J);
    De:= slope/(-2);
    Bv1:= intercept;
    Writeln;
    Writeln(Sp,'          De equals:  ',De:7:5,'  Sigma =  ',Sigmam/2:6);
    Writeln(Sp,'    B(v=1) equals:  ',Bv1:5:2,' Sigma =  ', Sigmab:6);
End;  (* WriteTable4 *)

Begin (* M A I N   P R O G R A M*)
    GetSpectralData;
    EchoInput;
    CalculateArrays;
    WriteTable2;
    WriteTable3;
    WriteTable4;
End.  (* M A I N   P R O G R A M*)
```

Figure 33–2 (continued)

A computer program for calculating these parameters is listed in Figure 33–2. Input of J, $P(J)$, and $R(J)$ to the program is interactive. If you use a computer program to assist in the calculations, you should include a detailed longhand sample calculation for each of the equations (Eqs. 33–9, 33–10, and 33–11) to demonstrate (1) that you understand the calculation, and (2) that the program is functioning properly. Include appropriate plots even if the data are processed by computer.

Prepare a table of your experimentally determined B_0 and B_1 and a few literature values with higher v. Graphically determine B_e and α_e according to Equation 33–2.

With your experimentally determined band origin and overtones from Table 33–1 or the literature (Herzberg, 1950, p. 55), determine $\tilde{\nu}_e$ and $x_e\tilde{\nu}_e$ from the slope and intercept of a plot of Equation 33–14.

Determine the slopes and intercepts of all plots by means of a least-squares computer program. Take careful note of the standard deviations in slopes and intercepts when reporting the uncertainties in your calculated values. Compare the observed scatter of points, especially in plots of Equations 33–9, 33–10, and 33–11, with various measures of precision output by your computer program (correlation coefficient, $Y - Y_{calc}$, etc.)

Finally, calculate the equilibrium internuclear distances in HCl and DCl with Equation 33–15 and compare with the literature values. How does r_e compare in HCl and DCl? What does this suggest about the nature of the chemical bond in the two molecules? Predict the values of $\tilde{\nu}_e$ and r_e in tritium chloride.

REFERENCES

1. Barrow, G. M., *Introduction to Molecular Spectroscopy*, McGraw-Hill, New York, 1962.

2. Hardy, J. D., E. F. Barker, and D. M. Dennison, "The Infrared Spectrum of H^2Cl," *Phys. Rev.* 42:279, 1932.

3. Huber, K. P., and G. Herzberg, *Molecular Spectra and Molecular Structure*, Van Nostrand, Princeton, New Jersey, 1979.

4. Herzberg, G., *Spectra of Diatomic Molecules*, 2nd ed., Van Nostrand, Princeton, New Jersey, 1950.

5. King, G. W., *Spectroscopy and Molecular Structure*, Holt, Rinehart and Winston, New York, 1964.

6. Pauling, L., and E. B. Wilson, *Introduction to Quantum Mechanics*, McGraw-Hill, New York, 1935.

7. Pickworth, J., and H. W. Thompson, "The Fundamental Vibration-Rotation Band of Deuterium Chloride," *Proc. R. Soc. London, Ser. A*, 218:37, 1953.

8. Prais, M. G., "Analysis of the Vibrational-Rotational Spectrum of Diatomic Molecules," *J. Chem. Educ.* 63:747, 1986.

9. Rank, D. H., D. P. Eastman, B. S. Rao, and T. A. Wiggins, "Rotational and Vibrational Constants of the HCl^{35} and DCl^{35} Molecules," *J. Opt. Soc. Am.* 52:1, 1962.

10. Richards, L. W., "The Infrared Spectra of Four Isotopes in HCl," *J. Chem. Educ.* 43:552–554, 1966.

11. Roberts, B., "The HCl Vibrational Rotational Spectrum," *J. Chem. Educ.* 43:357, 1966.

12. Stafford, F. E., C. W. Holt, and G. L. Paulson, "Vibration-Rotation Spectrum of HCl," *J. Chem. Educ.* 40:245, 1963.

13. "Tables of Wavenumbers for the Calibration of Infra-red Spectrometers," *Pure Appl. Chem.* 1:537–699, 1960.

Experiment 34

Spectrum of a Particle in a Box

The Schrödinger wave equation is written

$$H\psi = E\psi \tag{34-1}$$

In three dimensions, E is a function of x, y, and z and is the allowed energy of a particle. The solution to the wave equation is ψ, a wave function or eigenfunction, which is also a function of x, y, and z. It is a solution to Equation 34–1 for certain values (eigenvalues) of E. For a particle in a three-dimensional system, the Hamiltonian operator H is given by

$$H = -\frac{h^2}{8\pi m^2}\left(\frac{\partial^2}{\partial x^2} + \frac{\partial^2}{\partial y^2} + \frac{\partial^2}{\partial z^2}\right) + U \qquad (34\text{–}2)$$

The potential energy is given by U, which in general is also a function of x, y, and z and varies from system to system. The mass of the particle is m, and h is the Planck constant. In SI units, m is in kilograms, U in joules, and $h = 6.6262 \times 10^{-34}$ J s.

When a particle is constrained to move along a one-dimensional path, the problem becomes one dimensional. If the particle is permitted to move only on a path of fixed length L, where x ranges from 0 to L, then the potential energy is infinite for $x < 0$ and $x > L$. This is what is meant by *constrained* to a one-dimensional path. If the potential energy is zero in the permitted region, $0 < x < L$, then the operator simplifies to

$$H = -\frac{h^2}{8\pi^2 m}\frac{\partial^2}{\partial x^2} \qquad (34\text{–}3)$$

and the Schrödinger wave equation is written

$$-\frac{h^2}{8\pi^2 m}\frac{\partial^2 \psi}{\partial x^2} = E\psi \qquad (34\text{–}4)$$

where U and E are functions of x only.

The solution of this differential equation is straightforward and is described in detail in most contemporary physical chemistry textbooks (for example, Barrow, 1979, p. 82). Solutions (eigenfunctions) of the form

$$\psi_n = A \sin \frac{n\pi x}{L} \qquad (34\text{–}5)$$

satisfy the postulates of quantum theory when $n = 1, 2, 3, \ldots$ and $A = (2/L)^{1/2}$, where L is the maximum value of the coordinate x. The constant A is a normalization factor and is found by setting equal to unity the integral from $x = 0$ to $x = 1$ of ψ^2. Certain energies (eigenvalues) associated with the set of wave functions (eigenfunctions) are given by

$$E_n = \frac{n^2 h^2}{8mL^2} \qquad (34\text{–}6)$$

If the particle is an electron, then its mass m equals 9.1095×10^{-31} kg and the problem is sometimes described as an electron on a wire, a particle in a box, or a particle in a one-dimensional well.

An energy-level diagram corresponding to Equation 34–6 is shown in Figure 34–1. In Figure 34–1a, an electron is shown occupying the

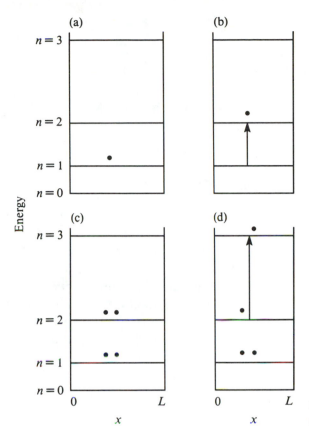

Figure 34–1 An electron in a one-dimensional box.

ground state, for which the quantum number $n = 1$. When $n > 1$, the electron is in an excited state. In Figure 34–1b the electron is in the first excited state. If a system contains more than one electron, then two electrons can occupy each level, since the Pauli Exclusion Principle requires that no two electrons in a given energy level can have the same quantum numbers. In this system, the quantum numbers are n and s, which can be $\pm 1/2$. In Figure 34–1c and d, the system contains four electrons. Figure 34–1c corresponds to the ground state and Figure 34–1d to the first excited state. The energy change for a transition from one state to another is given by

$$E_n = h^2(n_2^2 - n_1^2)/(8mL^2) \qquad (34–7)$$

What kind of real world systems are described by the particle in a box? One of the most common types of molecules that fit this model is conjugated systems. An example with four electrons, corresponding to Figure 34–1c and d, is butadiene (Fig. 34–2). The framework of the molecule, connecting the carbon and hydrogen atoms, is held together by σ bonds formed from the overlap of three sp^2 hybrid orbitals surrounding each carbon atom. The fourth bonding electron from each carbon is in an unhybridized p_z orbital perpendicular to the carbon–hydrogen

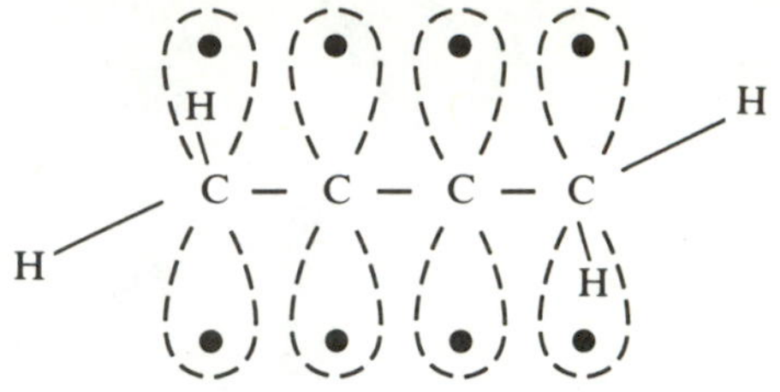

Figure 34–2 Butadiene, $CH_2=C-C=CH_2$.

skeleton. These overlap to form a single π orbital extending from the first carbon atom to the fourth. These four electrons are constrained to this π orbital; that is, the potential energy is essentially infinite outside the π orbital and essentially zero in it. The transition shown in Figure 34–1c and d lies in the ultraviolet region at a wavelength of 217.0 nm (Kuhn, 1949).

Another group of organic compounds that absorb visible light are the symmetrical polymethine dyes (Kuhn, 1948, 1949). The structures of some typical members of this class of compounds are shown in Figure 34–3. In these cyanine dyes the polymethine chain forms a conjugated chain extending from the nitrogen atom on one end of the molecule to the opposite nitrogen atom plus one carbon (on the ethyl group). As in the butadiene example, each carbon atom contributes one π electron to the molecular orbital that extends from one end of the chain to the other.

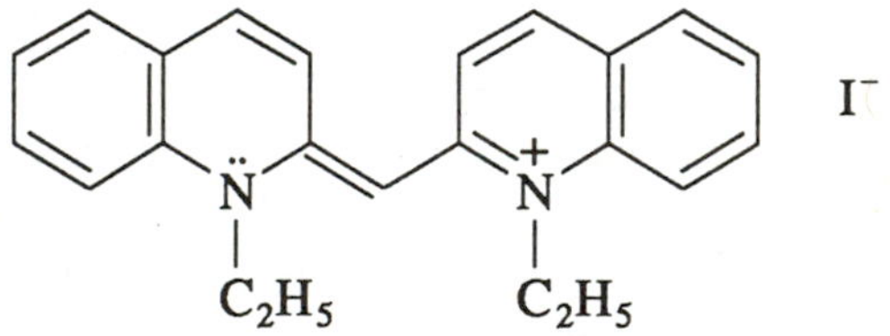

1,1′-diethyl-2,2′-cyanine iodide

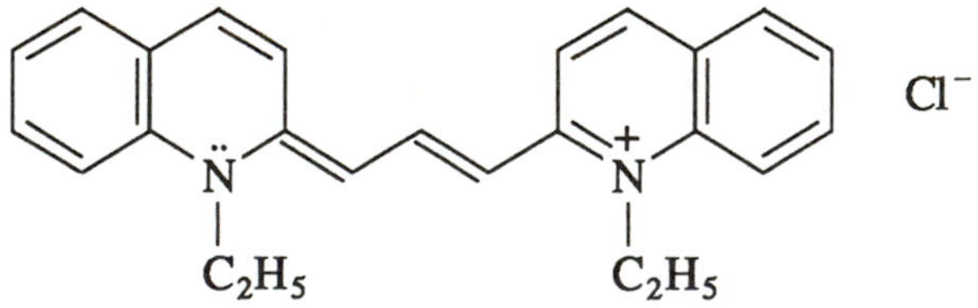

1,1′-diethyl-2,2′-carbocyanine chloride

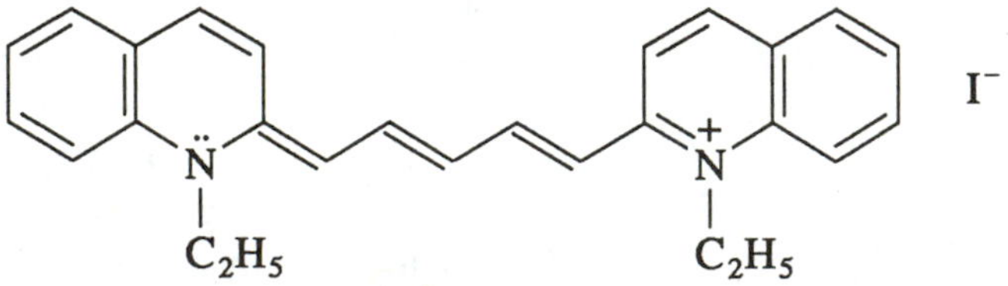

1,1′-diethyl-2,2′-dicarbocyanine iodide

Figure 34–3 Symmetrical polymethine dyes.

As shown in Figure 34–3, the nitrogen atom on the left side of the molecule is drawn with two dots to indicate the two electrons it contributes to the extended π orbital. The nitrogen on the right side is shown with a plus charge so it has only one electron to contribute. Thus, the total number N of π electrons equals the number p of carbon atoms in the polymethine chain between the nitrogen atoms plus the 3 electrons from the two nitrogen atoms:

$$N = p + 3$$

Thus, $N = 6$, 8, and 10 for the molecules shown in Figure 34–3.

Equation 34–7 relates the energy of a transition to the length of the "box," that is, the length L of the π orbital. It is of interest to obtain from Equation 34–7 a relationship between the structure of the molecule (i.e., the length L) and the measured wavelength at which absorption of electromagnetic radiation occurs for a given transition.

If N is the number of π electrons in the orbital, then for the ground state, each of the $N/2$ levels contains two electrons, as indicated in Figure 34–4 (the electrons are represented by dots). Thus, the quantum number n in Equation 34–6 equals $N/2$. The remaining levels are empty. The absorption band is associated with the excitation of an electron from the highest occupied level, labeled $N/2$, to the lowest unoccupied level, labeled $[(N/2) + 1]$. The energy difference between these two levels is

$$\Delta E = \frac{h^2}{8mL^2}[(N/2 + 1)^2 - (N/2)^2] \qquad (34\text{–}8)$$

where $\Delta E = h\nu$. Upon substituting $\nu = c/\lambda$ and simplifying, we obtain

$$L^2 = \frac{(N + 1)h\lambda}{8mc} \qquad (34\text{–}9)$$

Upon substituting values for the natural constants h, m, and c and converting L and λ to nanometers:

$$L = 0.01742[(N + 1)\lambda]^{1/2} \qquad (34\text{–}10)$$

The lowest energy transition (longest wavelength) corresponds to a transition from the state labeled $(N/2)$ to the state $(N/2 + 1)$. Thus, a measurement of the wavelength λ at which absorption occurs permits calculation of the length L of the "box," or in this case the length of the conjugated chain of atoms.

Apparatus

The apparatus consists of a spectrophotometer, preferably recording, such as a Beckman 25, and miscellaneous pipets and volumetric flasks. Also required are methanol and stock solutions (about 0.001 M) of a series of related cyanine dyes,[1] for example,

---

[1] Available from Aldrich Chemical Co., Inc., 940 West St. Paul Avenue, Milwaukee, Wisconsin 53201; Eastman Kodak Co., 343 State Street, Rochester, New York 14650; and other suppliers of fine chemicals.

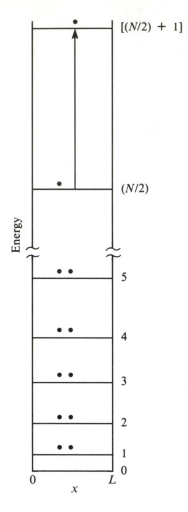

Figure 34–4 Energy levels.

1,1'-diethyl-n,n'-cyanine iodide
1,1'-diethyl-n,n'-carbocyanine chloride
1,1'-diethyl-n,n'-dicarbocyanine iodide

where n equals 2, 3, or 4.

SAFETY CONSIDERATIONS

Methanol is toxic. When the toxicity of a compound is unknown, as is the case with these dyes, the compound should be treated as if it were toxic and carcinogenic.

EXPERIMENTAL PROCEDURE

Review the operating instructions for the spectrophotometer before the day of the experiment. Turn the spectrophotometer on before assembling your equipment and let it warm up for at least half an hour. Stock solutions of the dye in concentrations of about 0.001 M should be available.

Choose a series of dyes of related structure, that is, dyes with a common value of n (see above). If a recording spectrophotometer is not available, measure the absorbance about every 10 nm over a 100-nm range centered about the maximum. Use methanol in the reference cell. The concentrations of the dyes should be adjusted by dilution if necessary so that the maximum absorption is about 1.0 (percent transmission equals about 10).

Results and Calculations

From the measured absorption maxima for your series of dyes, calculate the length L of the π orbital. Compare your values with the length calculated from the known structure of the molecule.

According to Kuhn (1949), the length L of the chain is "measured by the length of the polymethine zig-zag chain between the nitrogen atoms plus one bond distance to either side." However, since resonance structures can be written for the dyes, the bonds between *each* nitrogen and the first ethyl carbon to which it is bonded are counted. Thus the number of bonds N_b in the 1,1'-diethyl-2,2'-cyanine iodide chain in Figure 34–3 is 6:

$$C_{Et}-N-C-C-C-N-C_{Et}$$

and the length L (nm) of the chain should be

$$L = 0.139 N_b \qquad (34\text{–}11)$$

where 0.139 nm equals the length of a $C-C$ bond with bond order equal to 1.5 as found in, for example, benzene (Kuhn, 1949).

Compare your absorption maxima with values from the literature. Critically evaluate your data, calculate the errors, and compare your values in terms of your analysis of the errors with the literature values. Calculate the value of the absorption maximum in butadiene and compare with that in the literature (Kuhn, 1949).

REFERENCES

1. Barrow, G. M., *Physical Chemistry,* 4th ed., chap. 14, McGraw-Hill, New York, 1979.

2. Brooker, L. G. S., "Absorption and Resonance in Dyes," *Rev. Mod. Phys.* 14:275–293, 1942.

3. Chang, R., *Basic Principles of Spectroscopy,* McGraw-Hill, New York, 1971.

4. El-Issa, B. D., "The Particle in a Box Revisited," *J. Chem. Educ.* 63:761, 1986.

5. Fisher, N. I., and F. M. Hamer, "A Comparison of the Absorption Spectra

of Some Typical Symmetrical Cyanine Dyes," *Proc. R. Soc. London, Ser. A,* 154:703–723, 1936.

6. Gerkin, R. E., "A Molecular Spectral Corroboration of Elementary Operator Quantum Mechanics," *J. Chem. Educ.* 42:400–491, 1965.

7. Jaffé, H. H., and M. Orchin, *Theory and Application of Ultraviolet Spectroscopy,* Wiley, New York, 1962.

8. Kuhn, H., "Elektronengasmodell zur quantitativen Deutung der Lichtabsorption von organischen Farbstoffen," *Helv. Chim. Acta* 31:1441, 1948.

9. Kuhn, H., "A Quantum-Mechanical Theory of Light Absorption of Organic Dyes and Similar Compounds," *J. Chem. Phys.* 17:1198–1212, 1949.

10. Olsson, L., "Band Breadth of Electronic Transitions and the Particle-in-a-Box Model," *J. Chem. Educ.* 63:756, 1986.

Electric and Magnetic Properties

Experiment 35

Dipole Moment of Polar Molecules in Solution

The dipole moment μ, an electrical property of molecules, is defined as

$$\mu = qr \tag{35-1}$$

where q is the magnitude of two charges ($-q$ and $+q$) separated by a distance r. In molecules, the dipole moment is a measure of the electrical asymmetry present, and thus gives information on the nature of molecular bonding.

In cgs units, if the magnitude of q is that of an electron (4.803×10^{-10} esu) and the two charges are separated by 1 Å (10^{-8} cm), then

$$\mu = 4.803 \times 10^{10} \times 10^{-8} = 4.803 \times 10^{-18} \text{ esu cm} \tag{35-2}$$

Dipole moments are usually reported in debyes (D), defined as

$$1 \text{ D} = 10^{-18} \text{ esu cm} \tag{35-3}$$

The debye is named in honor of Peter Debye, Nobel laureate and early worker in the study of dipole moments.

In SI units, if the magnitude of q is that of an electron (1.6022×10^{-19} C) and the two charges are separated by 1 Å (10^{-10} m), then

$$\mu = 1.6022 \times 10^{-19} \text{ C} \times 10^{-8} \text{ m} = 1.6022 \times 10^{-29} \text{ C m} \quad \text{(35–4)}$$

Thus, 1 D = 3.336×10^{-30} C m.

Dipole Moment and Molar Polarization

The dipole moment is not measured directly, but rather calculated from a measurement of the dielectric constant ϵ, a bulk property of material substances. The dielectric constant is related to another bulk property, the total molar polarization P_T, which may be interpreted as the dipole moment per unit volume of the bulk substance:

$$\frac{\epsilon - 1}{\epsilon + 2} \frac{M}{d} = P_T \quad \text{(35–5)}$$

Since M is the molecular weight and d the density, P_T has units of cubic centimeters per mole.

In polar substances—that is, substances having a permanent dipole moment—the total molar polarization arises from two contributions, P_d and P_μ. The distortion polarization P_d can be thought of as the induced dipole moment per unit volume, and P_μ as the permanent dipole moment per unit volume. To measure the dielectric constant, the substance is placed in an electric field, which induces a dipole moment in the molecules of the substance whether it has a permanent dipole or not. Since we are primarily interested in the permanent dipole moment, we must subtract from the total molar polarization the contribution from the induced dipole moment, namely the distortion polarization P_d. Thus,

$$P_\mu = P_T - P_d \quad \text{(35–6)}$$

Fortunately, the distortion molar polarization is easily measured, since it is exactly equal to the molar refraction M_r, which can easily be calculated from a measurement of the refractive index n:

$$P_d = M_r = \frac{n^2 - 1}{n^2 + 2} \frac{M}{d} \quad \text{(35–7)}$$

Finally, having isolated P_μ, we can calculate the dipole moment μ, since it is related to P_μ by an equation first derived by Debye:

$$P_\mu = \frac{4\pi N \mu^2}{9kT} \quad \text{(35–8)}$$

A convenient form of this equation results when the cgs values for k (1.3807×10^{-16} erg deg^{-1} mol^{-1}) and N (6.022×10^{23}/mol) are substituted in Equation 35–8,

$$\mu = 0.01281 \sqrt{P_\mu T} \quad \text{(35–9)}$$

where μ is in debyes.

Dipole Moment Measurement in Solution

While these relationships are valid for polar molecules in the gas phase, there is a complication in calculating dipole moments from measurements of dielectric constants of a polar solute in a nonpolar solvent: the molecular dipoles interact strongly with each other and with the solvent, and the solvent itself contributes to the molar polarization by virtue of its distortion polarization even though it is nonpolar. Consequently, some means of carrying out the measurements in dilute solution must be used and the measurements must be extrapolated to infinite dilution.

A variety of methods are available for separating the effect of the solvent from that of the solute whose dipole moment is sought. The method of Guggenheim (1949) requires measurement of the dielectric constant and refractive indices for a range of concentrations, but unlike most other methods, it does not require precise measurements of the density—often a tedious task. According to Guggenheim, the orientation polarization P_μ of the solute is related to the dielectric constant ϵ_{12} of the solution, the refractive index n_{12} of the solution, and the concentration c (mol/ml) by

$$\frac{\epsilon_{12} - 1}{\epsilon_{12} + 2} - \frac{n_{12}^2 - 1}{n_{12}^2 + 2} = P_\mu c + \text{constant} \qquad \textbf{(35–10)}$$

A plot of the left side of Equation 35–10 against the concentration c (mol/ml) gives a straight line with slope P_μ.

Measuring the Dielectric Constant of a Solution

Between about 1960 and 1980, a large number of laboratories imported from Germany the WTW DM01 dipolemeter. Since this instrument is still in widespread use, its operation will be described here. Some alternative methods for measuring dielectric constants are described by Bonilla (1977), Kurtz (1977), and Chao (1988).

With the WTW DM01, the liquid sample is held in a fixed capacitor consisting of two concentric gold-plated cylinders surrounded by a water jacket and connected electrically in parallel with a variable-frequency LC oscillator operating at approximately 2.0 MHz (Fig. 35–1). The output of the LC oscillator is then mixed with the output of a fixed-frequency oscillator operating at approximately the same frequency to form a low beat frequency, which is then compared with a low fixed frequency, namely the 60-cycle line frequency. By applying the 60-cycle line frequency to the horizontal scan of a built-in oscilloscope and the low beat frequency to the vertical plates, a Lissajous figure is formed depending on the ratio of the two low frequencies. At exact resonance the Lissajous figure is a straight horizontal line that moves up and down once per second for each beat per second deviation from resonance. Consequently the determination of the resonance point is very precise: $\Delta C/C = 7 \times 10^{-6}$.

TABLE 35–I

Dielectric Constants at 20°C

Solvent	ϵ	$d\epsilon/dT$ (deg^{-1})
Benzene	2.2825	−0.00196
Carbon tetrachloride	2.2363	−0.0020
Cyclohexane	2.0228	−0.0016
Dibutyl ether	3.0830	−0.00709

Cells are available for the instrument to cover a wide range of dielectric constants. For the solvents and solutes suggested in this experiment, the DFL 1, covering the range of dielectric constants from 1.0 to 3.4, is suitable. The companion cell, the DFL 2, covers the range 2.0 to 6.9.

The variable capacitor in the variable-frequency LC oscillator is mechanically coupled to a double-geared precision scale with 4500 divisions (equivalent to a scale length of 10 meters), which must be calibrated with two or more liquids of known dielectric constant (Table 35–1). The cell is filled with a calibration liquid and brought to resonance, and the dial division reading D and the dielectric constant ϵ are recorded. The process is repeated for the liquids in Table 35–1. Since the relationship is linear, the calibration line ($\epsilon = mD + b$) can be determined by least squares. If the thermostat bath is held constant for subsequent use, it is optional to recalibrate for each use, since the calibration line remains quite constant for a given cell and temperature. The apparatus described by Kurtz (1977) is also calibrated in this manner.

Once the dial is calibrated in terms of dielectric constant, the dielectric constants of the solutions to be investigated are placed in the cell, the resonance point is established, the dial is read, and the dielectric constant is read from the already determined calibration curve: $\epsilon = mD + b$.

While the cells for the WTW DM01 are ruggedly built and gold-plated, and should last forever, the electronic components of many WTWs are aging badly. As a replacement for these components, the very inexpensive (less than $10) oscillator described by Kurtz (1977) is completely satisfactory. When it is used with an ordinary-frequency counter,[1] no reduction in precision is observed (Vanicek, 1989). Because the dielectric constant ϵ and the oscillator frequency f are inversely related, the calibration equation is $\epsilon = m/f + b$, where m and b are the slope and the intercept, respectively. The calibration is carried out as described above, except that ϵ is plotted against $1/f$.

---

[1] Heathkit Model IM-2401, Heath Co., Benton Harbor, Michigan 49022; or B&K Model 1803, B&K-Precision/Maxtec International Corp., 6460 West Cortland, Chicago, Illinois 60635.

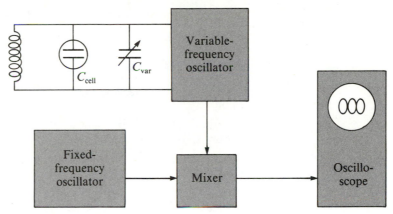

Figure 35–1 Dipolemeter.

Apparatus

The experiment requires eight 100-ml volumetric flasks, assorted pipets, a thermostat, chlorobenzene, 1,2-dichlorobenzene, 1 L dry cyclohexane, and an apparatus for measuring dielectric constants (Fig. 35–1), such as the WTW dipolemeter, model DM01.

SAFETY CONSIDERATIONS

All the organic solvents should be considered toxic and should be handled with protective rubber gloves. Avoid breathing fumes. Dispose of organic solvents in the container provided.

EXPERIMENTAL PROCEDURE

Prepare 100 ml each of four solutions of chlorobenzene in cyclohexane with concentrations ranging from 0.05 to 0.20 M and four solutions of 1,2-dichlorobenzene in cyclohexane with concentrations in the same range. Calibrate the dipolemeter with solutions of known dielectric constant (Table 35–1). Measure the dielectric constant ϵ_{12} and refractive index n_{12} of each solution. Record the temperature.

Results and Calculations

Tabulate the data for the chlorobenzene and dichlorobenzene solutions. Prepare a calibration curve by plotting ϵ against D (dial divisions) or $1/f$ (frequency). Plot the left side of Equation 35–10 against c (mol/ml). From a least-squares fit of the data, calculate P_μ and μ. With the experimental value of the C—Cl bond moment determined for chlorobenzene, calculate the dipole moment of 1,2-dichlorobenzene, assuming a 60° angle between the C—Cl bond moments in 1,2-dichlorobenzene. Compare the results of this calculation with the experimental value. Evaluate the errors and compare your results critically with values found in the literature.

REFERENCES

1. Bender, P., "Measurement of Dipole Moments," *J. Chem. Educ.* 23:179, 1946.

2. Bonilla, A., and B. Vassos, "A Novel Approach for Dipole Moment Laboratory Experiments," *J. Chem. Educ.* 54:130–131, 1977.

3. Braun, C. L., W. H. Stockmayer, and R. A. Orwell, "Dipole Moments of 1,2-Disubstituted Ethanes and Their Homologs," *J. Chem. Educ.* 47:287, 1970.

4. Chao, T. H., "A Modified Resonance Apparatus for the Determination of Dielectric Constants," *J. Chem. Educ.* 65:837–838, 1988.

5. Chien, J.-Y., "Dielectric Constants by the Heterodyne-Beat Method," *J. Chem. Educ.* 24:494, 1947.

6. Guggenheim, E. A., "A Proposed Simplification in the Procedure for Computing Electric Dipole Moments," *Trans. Faraday Soc.* 45:714–720, 1949.

7. Halverstadt, I. F., and W. D. Kumler, "Solvent Polarization Error and Its Elimination in Calculating Dipole Moments," *J. Am. Chem. Soc.* 64:2988, 1942.

8. Janini, G. M., and A. H. Katrib, "Determination of the Dipole Moment of Polar Compounds in Non-Polar Solvents," *J. Chem. Educ.* 60:1087, 1983.

9. Kurtz, S. R., O. T. Anderson, and B. R. Willeford, Jr., "A New Simple Apparatus for the Measurement of Dipole Moments," *J. Chem. Educ.* 54:181–182, 1977.

10. LeFevre, R. J. W., *Dipole Moments, Their Measurement and Application in Chemistry,* Methuen, London, 1953.

11. Maryott, A. A., and E. R. Smith, *Tables of Dielectric Constants of Pure Liquids,* National Bureau of Standards Circular No. 514, U.S. Government Printing Office, Washington, D.C., 1951.

12. McClellan, A. L., *Tables of Experimental Dipole Moments,* Freeman, San Francisco, 1963.

13. Moffatt, J. B., "Thermodynamics from Dipole Moments," *J. Chem. Educ.* 43:74, 1966.

14. Smith, J. W., *Electric Dipole Moments,* Butterworths, London, 1955.

15. Smith, J. W., "Some Developments of Guggenheim's Simplified Procedure for Computing Electric Dipole Moments," *Trans. Faraday Soc.* 46:394, 1950.

16. Smyth, C. P., *Dielectric Behavior and Structure,* McGraw-Hill, New York, 1955.

17. Smyth, C. P., "Determination of Dipole Moments," in A. Weissberger and B. W. Rossiter, eds., *Techniques of Chemistry,* vol. I, *Physical Methods of Chemistry,* 4th ed., pt. IV, chap. 6, Wiley-Interscience, New York, 1972.

18. Thompson, H. B., "The Determination of Dipole Moments in Solution," *J. Chem. Educ.* 43:66, 1966.

19. Vanicek, J. A., California State University, Sacramento, private communication, August 1989.

20. Wesson, L. G., *Tables of Electric Dipole Moments,* Technology Press, M.I.T., Cambridge, Massachusetts, 1948.

Experiment 36

Magnetic Moments of Paramagnetic Salts

Knowledge of the number of paired and unpaired electrons in metal complex ions is especially useful in understanding their bonding, magnetic, and spectral characteristics. Measurement of the molar magnetic susceptibility leads quite directly to the number of paired and unpaired electrons in a complex ion. Knowledge of the configuration of outer paired and unpaired electrons leads directly to understanding the magnetic properties of complex ions.

In a free isolated transition element ion, the d orbital consists of a set of five degenerate (having the same energy) d orbitals. When the central metal ion forms a complex ion, the ligands, such as H_2O, NH_3, and Cl^-, form an electrostatic crystal field of high symmetry, usually tetrahedral or octahedral, about the central metal ion and its five d orbitals. The d orbitals are named d_{xy}, d_{xz}, d_{yz}, $d_{x^2-y^2}$, and d_{z^2} according to their orientation to a coordinate system the origin of which is at the central metal ion (Fig. 36–1). In the absence of an external field, these d orbitals have exactly the same energy and are said to be degenerate.

Octahedral ligand fields are very common; all of the compounds suggested for study in this experiment are of octahedral symmetry. The ligands are considered to lie at the corners of a regular octahedron, and in turn the corners of the regular octahedron lie on the axes of the coordinate system. Examination of the d orbitals shows that the $d_{x^2-y^2}$ and d_{z^2} orbitals, which lie on the axis, interact more strongly than the d_{xy}, d_{xz}, and d_{yz} orbitals, which lie between the axes. Consequently, in the presence of an octahedral ligand field, the degeneracy of the orbitals is removed so that the $d_{x^2-y^2}$ and d_{z^2} orbitals fall into one set (by virtue of their identical relative axial orientation) called the e_g, while the d_{xy}, d_{xz}, and d_{yz} orbitals fall into another set called the t_{2g} orbitals, as shown in Figure 36–2. The crystal-field splitting energy is called Δ_0. Notice that the e_g orbitals, which approach the ligands more closely, lie at a higher energy than the t_{2g} orbitals.

The splitting Δ_0 between the e_g and t_{2g} levels depends on the magnitude of the electrostatic field provided by the ligand. Because the magnitude of Δ_0 also affects the spectral properties of the complex ion, the ordered list of ligands from weak field to strong field is called the Spectrochemical Series:

$$I^- < Br^- < Cl^- < F^- < OH^- < H_2O < NH_3 < en < phen < CN^-$$

Weak ligand field	Strong ligand field
Small Δ_0	Large Δ_0
High-spin complexes	Low-spin complexes

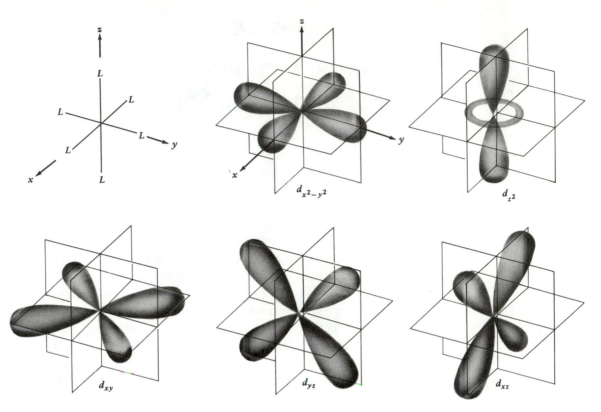

Figure 36–1 The five *d* orbitals and their relation to the ligands *L* on the *x*, *y*, and *z* axes. The ligands lie at the corners of a regular octahedron (not shown).

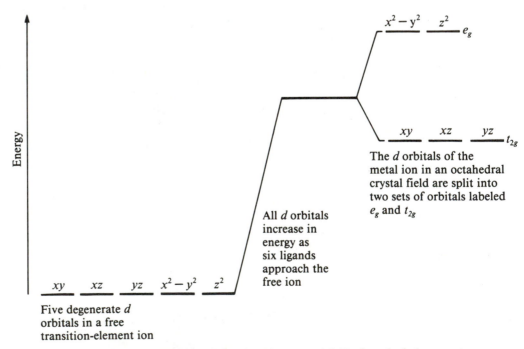

Figure 36–2 The splitting of *d* orbitals in a crystal field of octahedral symmetry.

How the d electrons occupy the e_{2g} and t_{2g} orbitals depends on the magnitude of Δ_0 and Hund's Rule. According to Hund's Rule, the electrons occupy the orbitals of the lowest-energy degenerate quantum state singly (spin unpaired) as much as possible.

Since the ten d orbitals can hold one to ten electrons, it is useful to look at each situation individually, as shown in Figure 36–3. When the d electrons are subjected to a weak field, as shown in the bottom half of Figure 36–3, the d orbital electrons keep their spins unpaired and their energy low, filling the available d orbitals as shown. In the presence of a high field, the crystal-field splitting Δ_0 is sufficiently large that the need to keep the energy as low as possible overcomes the tendency to keep the spins unpaired. The result, for example, for a d^4 set of electrons is that two electrons pair up in the t_{2g} orbital in a high-field environment, while all four spins remain unpaired in a low-field environment, three unpaired electrons occupying the t_{2g} oribtal and one unpaired electron occupying the e_g orbital. For this case the ion has four unpaired electrons in the presence of a weak field, but only three unpaired electrons in the presence of a strong field. As indicated in Figure 36–3, high- and low-spin behavior is also shown by d^5, d^6, and d^7 systems.

Macroscopic magnetic properties of paramagnetic substances depend on the number of unpaired electrons in a molecule or ion. The

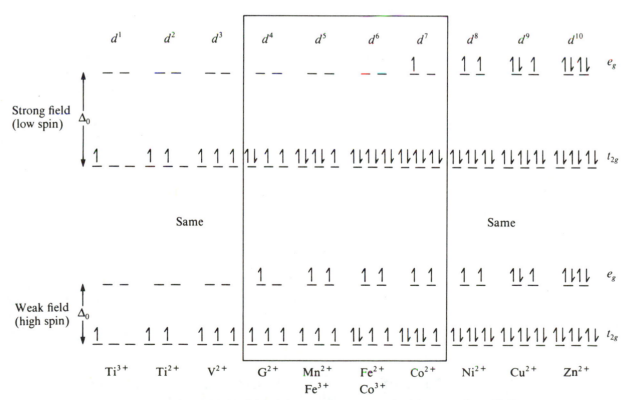

Figure 36–3 High-spin and low-spin states for ions with d^1 to d^{10} electron configurations.

measurement of magnetic susceptibility, a macroscopic magnetic property, permits determination of the magnetic moment, a molecular property from which the number of unpaired electrons can be calculated.

Some Definitions

When *any* bulk substance is placed in a magnetic field $\mathbf{H}$, a magnetic moment $\mathbf{M}$ is induced in it. The magnitude of the induced magnetic moment depends upon the strength of the magnetic field:

$$\mathbf{M} = \chi\mathbf{H} \qquad \text{(36–1)}$$

Since $\mathbf{M}$ is the magnetic moment per unit volume, the proportionality constant χ is called the magnetic susceptibility per unit volume, or just the volume susceptibility. The volume susceptibility is dimensionless.

Two more definitions are important. The susceptibility per unit mass, or weight susceptibility, χ_w is defined as

$$\chi_w = \frac{\chi}{d} \qquad \text{(36–2)}$$

where d is the density of the substance (g/cm^3). The weight susceptibility has units of cubic centimeters per gram. Finally, the molar susceptibility χ_m is defined as

$$\chi_m = M\chi_w = \frac{M\chi}{d} \qquad \text{(36–3)}$$

where M is the molecular weight of the substance. The molar susceptibility χ_m has units of $cm^{-3}\,mol^{-1}$.

If a substance is placed in a magnetic field of a certain intensity, then the intensity of the field within the substance may be either smaller ($-\chi$) or larger ($+\chi$) than the intensity in the surrounding space. In the first case the substance is called **diamagnetic** and in the second, **paramagnetic.**

Early experimental measurements on magnetic susceptibility showed that it is temperature dependent:

$$\chi_m = \frac{C}{T} - A \qquad \text{(36–4)}$$

where the constant C is known as the Curie constant. The first term is found to be much larger than the second term, A.

So far our discussion has been limited to the magnetic properties of *bulk* substances. Two tasks remain. First, how are bulk magnetic properties such as χ_m related to magnetic properties on a molecular scale? Second, how are bulk magnetic properties measured in the laboratory?

Connecting Bulk Magnetism and Molecular Magnetism

In 1905 P. Langevin developed a simple theory for paramagnetism. Each molecule or ion with an unbalanced electron orbit behaves as a tiny magnet with a definite moment. The axis of each tiny magnet tends to align itself parallel to the direction of the applied field. A Boltzmann distribution of these aligned magnets indicates that the temperature tends to oppose the alignment, giving rise to a $1/T$ term in Curie's Law (Eq. 36–4). The desired result, connecting the bulk measurable χ_m to the magnetic moment μ_m of an individual molecule or ion, is

$$\chi_m = N_A \alpha_m + \frac{N_A \mu_m^2}{3kT} \tag{36–5}$$

where N_A is Avogadro's number, k is the Boltzmann constant, and T is the absolute temperature. The magnetic susceptibility is seen to arise from a diamagnetic term and a paramagnetic term. The comparison with Curie's Law is obvious. The diamagnetism per molecule α_m is negative and very small compared to the paramagnetic term. Diamagnetism is a universal property of all matter; in diamagnetic materials, the paramagnetic second term equals zero. The first term is never zero, but often is so small in comparison that it is negligible. In this case, the magnetic moment is given by

$$\mu_m = \sqrt{\frac{3kT\chi_m}{N_A}} \tag{36–6}$$

Thus, a measurement of χ_m at temperature T permits calculation of the magnetic moment μ_m. Equation 36–6 is not generally used directly, since a special unit for the magnetic moment is usually used, the Bohr magneton μ_B, the magnitude of which is given by

$$\mu_B = \frac{eh}{4\pi mc} = 9.2741 \times 10^{-21} \text{ erg/oersted} \tag{36–7}$$

Verify this value by substituting the cgs values of e, h, m, and c. (A straightforward derivation is given in Barrow (1979) and other physical chemistry textbooks.

If Equation 36–6 is squared and multiplied top and bottom by the square of the Bohr magneton, the result is

$$\mu_m^2 = \frac{3kT\chi_m}{N} \frac{\mu_B^2}{\mu_B^2} \tag{36–8}$$

$$\mu_m = \mu_B \sqrt{\frac{3kT\chi_m}{N\mu_B^2}} \tag{36–9}$$

$$\mu_m = 2.8278\mu_B\sqrt{\chi_m T} \tag{36–10}$$

Verify the constant 2.8278 by substituting the cgs values for $3k$, N, and μ_B. Equation 36–10 gives the magnetic moment in Bohr magnetons μ_B when the molar magnetic susceptibility is in cubic centimeters per mole and the temperature is in kelvins. If χ_m is expressed in SI units (m^3

mol^{-1}), then the constant becomes 2827.8, but the magnetic moment still has units of Bohr magnetons.

Measuring Magnetic Susceptibility

If a tube of cross-sectional area A containing a substance with volume susceptibility χ is placed in a magnetic field of strength $\mathbf{H}$, there is a magnetic force of attraction (if *para*magnetic) given by

$$f = \frac{\chi A \mathbf{H}^2}{2} \tag{36-11}$$

In the Gouy (pronounced "gwee") method, this force is measured directly by suspending the sample from the weighing arm of an analytical balance to the midpoint between the poles of a strong magnet, as shown in Figure 36–4. The force is determined by measuring the weight of the suspended sample with and without the magnetic field present:

$$m_{\text{w}}g - m_{\text{wo}}g = \frac{\chi A \mathbf{H}^2}{2} \tag{36-12}$$

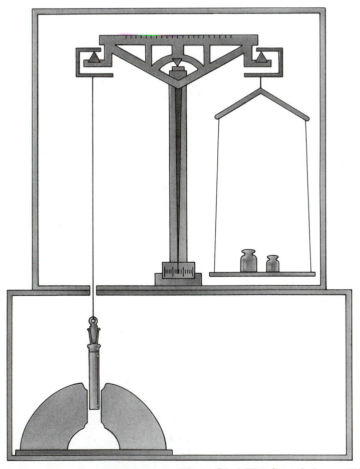

Figure 36–4 The Gouy balance. (Photo by Karen R. Sime.)

In Equation 36–12, m_w and m_{wo} are the masses (grams) with and without the magnetic field required to balance the beam, and g is the acceleration due to gravity (cm/sec^2). The Gouy balance is calibrated by using a substance of known volume susceptibility χ and calculating the factor $A\mathbf{H}^2/2$, which is the apparatus constant. With the apparatus constant known, measurement of the change in weight with and without the field permits calculation of μ_m for an unknown substance.

Apparatus

The apparatus includes a Gouy balance, as illustrated in Figure 36–4, consisting of an analytical balance (± 0.1 mg) modified to suspend a sample through its base; a permanent magnet, about 5000 gauss with a polar gap of about 2 cm and pole faces about 2 cm in diameter; a sample tube with a diameter slightly less than the magnet's polar gap; and six 100-ml volumetric flasks. A variety of transition-element salts are needed; for example, zinc sulfate, copper sulfate, nickel chloride, cobalt sulfate, manganous sulfate, ferric ammonium sulfate, ferrous ammonium sulfate, potassium ferrocyanide, potassium ferricyanide, and potassium permanganate.

SAFETY CONSIDERATIONS

Soluble salts of nearly all transition elements are toxic. Take care not to get your watch near the powerful magnet; remove it and keep it at a safe distance. Keep all magnetic materials and tools away from the magnet.

EXPERIMENTAL PROCEDURE

Calibration. With $NiCl_2 \cdot 6H_2O$, prepare about 100 ml of a solution about 30% by weight $NiCl_2$. Determine the percent to three significant figures. The weight susceptibility of aqueous nickel chloride (Selwood, 1956) is given by

$$\chi_w = \left[\frac{10{,}030p}{T} - 0.72(1 - p) \right] \times 10^{-6} \ cm^3/g \quad \textbf{(36–13)}$$

where T is the absolute temperature and p is the *weight* fraction of $NiCl_2$ (*not* $NiCl_2 \cdot 6H_2O$) in the aqueous solution. Measure the density of the solution d and calculate the volume susceptibility χ:

$$\chi = \chi_w d \quad \textbf{(36–14)}$$

The apparatus constant is then given by

$$\frac{A\mathbf{H}^2}{2} = \frac{(m_w - m_{wo})g}{\chi} \quad \textbf{(36–15)}$$

where $g = 980$ cm/sec^2 and m_w and m_{wo} are the weights in grams with and without the magnetic field.

Measuring the Molar Susceptibility. With a Gouy balance, measure the weight changes with and without the magnetic field for the samples listed in Table 36–1. Also measure the density of each solution to three significant figures. The density of a powdered solid is the actual density of the powdered solid in the sample tube. Since it depends on the tightness of packing, the density of the solid should be measured in the sample tube by filling to a line of known volume and weighing the empty and filled sample tube. The density of the packing must be very uniform. Nonhomogeneity in packing is a serious source of error in measuring the magnetic properties of solid powdered samples. You might consider repacking and remeasuring some solid samples to get an idea of the reproducibility of the measurements.

Results and Calculations

For both the solutions and the solid samples the volume susceptibility is calculated with

$$\chi = \frac{\Delta mg}{A\mathbf{H}^2/2} \qquad (36-16)$$

where Δm is the weight change of the sample in the Gouy balance with and without the magnetic field present, the gravitational constant equals 980 cm/sec^2, and $A\mathbf{H}_2/2$ is the apparatus constant determined in the calibration procedure. The weight susceptibility of the sample (solid or solution) is then calculated from

$$\chi_w = \chi/d \qquad (36-17)$$

Now we wish to calculate the weight susceptibility of the anhydrous salt, which means we must correct for the presence of water present as

TABLE 36–1

Useful Data for the Experimental Samples

Compound	Suggested Sample Mass (g) per 100 ml Solution	Formula Weight (g/m)	
		Anhydrous	**Hydrated**
$CoSO_4 \cdot 7H_2O$	30	155.01	281.2
$CoSO_4 \cdot 7H_2O$	Repeat as a solid	155.01	281.12
$CuSO_4 \cdot 5H_2O$	Run as a solid	159.60	249.69
$MnSO_4 \cdot H_2O$	10	151.00	169.01
$NiCl_2 \cdot 6H_2O$	40	129.62	237.72
$K_3Fe(CN)_6$	30	329.36	—
$K_4Fe(CN)_6 \cdot 3H_2O$	30	368.39	422.39
$Fe(NH_4)_2(SO_4)_2$	30	284.16	392.16

solvent and/or water of hydration. It is assumed that we can treat a solid compound as a solution of 1 mole of anhydrous salt in *n* moles of water of hydration. In that case, we can use the same equation for calculating the weight susceptibilities of the solutions and the solids:

$$\chi_w(\text{anhy. salt}) = [\chi_w(\text{soln. or hydrate}) \qquad \textbf{(36–18)}$$
$$+ 0.720 \times 10^{-6}(1 - p)]/p$$

In Equation 36–18, *p* is the weight fraction of the *anhydrous* salt in either the solution or the hydrate. We are assuming that the weight susceptibility

```
Program Paramagnetism;

Procedure CalibrateApparatus;

    Procedure EnterNiCl2Data;
        Begin
        End;   (* EnterNiCl2Data *)

    Procedure Calibrate;
        Begin
        End;   (* WriteOutCalibrationData *)

    Begin (* CalibrateApparatus *)
        EnterNiCl2Data;
        Calibrate;
    End;   (* Calibrate Apparatus *)


Procedure CalculateSusceptibility;

    Procedure EnterCompoundData;
        Begin
        End;   (* EnterCompoundData *)

    Procedure CalculateData;
        Begin  (* Calculate Data *)
        End;   (* CalculateData *)


Begin (* CalculateSusceptibility *)
    EnterCompoundData;
    CalculateData;
End;  (* CalculateSusceptibility *)


Begin (* M A I N   P R O G R A M   *)
    CalibrateApparatus;
    CalculateSusceptibility;
End.  (* M A I N   P R O G R A M   *)
```

Figure 36–5 Summary of Pascal program for calculating magnetic susceptibility.

```
Program Paramagnetism;
(* This program calculates molar magnetic susceptibility  *)
(* and number of unpaired electrons in selected           *)
(* transition element ions                                *)

uses crt;

CONST Xwater = 0.72E-6; (* mass magnetic suscep. of water *)
          g = 980;      (* cm/sec/sec gravity constant    *)

VAR AppConst:  Real;      (* Apparatus Constant           *)
    DeltaWt,    (* weight change w & w/o magnetic field   *)
    Rho:  Real;(* density solution in grams              *)
    Finished:  Boolean;
    Answer:  Char;

(***********************************************************************)

Procedure CalibrateApparatus;
(*  This procedure calculates the apparatus constant      *)
(*  of a Gouy balance                                     *)

Const NiCl2 = 129.62;

Var p,          (* weight fraction of anhydrous salt      *)
    T,          (* temperature in kelvins                 *)
    WtNiCl2,    (* mass anhydrous salt in grams           *)
    WtSoln,     (* mass NiCl2 solution in grams           *)
    X,          (* volume susceptibility  solution        *)
    Xg:  Real; (* Wt  susceptibility solution            *)

  (*---------------------------------------------------------*)

  Procedure EnterNiCl2Data;

  Begin  (* EnterNiCl2Data *)
    ClrScr;
    Write('Please enter the temperature (in Kelvins) of the NiCl2 soln:  ');
    Readln(T);
    Write('Please enter mass of anhydrous NiCl2:  ');
    Readln(WtNiCl2);
    Writeln('Please enter the mass of the solution.');
    Write('This is the mass of the hydrate plus the');
    Write(' mass of the water:  ');
    Readln(WtSoln);
    Write('Please enter the weight change (in grams)');
    Write(' w & w/o the magnetic field:  ');
    Readln(DeltaWt);
    Write('Please enter the density of the solution(gm/ml):  ');
    Readln(Rho);
  End;  (* EnterNiCl2Data *)

  (*---------------------------------------------------------*)
```

Figure 36–6 *Program Paramagnetism.*

of a mixture of a salt and water is the weighted average of the two components.

Figure 36–5 is a summary of a Pascal computer program for carrying out the calculations described up to this point. The entire program is listed in Figure 36–6. The summary, which is just a list of the Pascal

```
Procedure Calibrate;

    Begin (* Calibrate *)
       p:= WtNiCl2/WtSoln;
       Xg:= ((10030*p/T) - 0.72*(1 - p))*1E-6;
       X:= Xg*Rho;
       AppConst:= DeltaWt*g/X;
       Writeln;
       Writeln('At ',T:7:2,' K and a weight fraction NiCl2 = ',p:6:4);
       Writeln('The apparatus constant = ',AppConst:10);
       Writeln;writeln;writeln;
       Writeln('Please press return to continue');
       Readln;
    End;  (* WriteOutCalibrationData *)

  (*--------------------------------------------------------*)

Begin (* CalibrateApparatus *)
    EnterNiCl2Data;
    Calibrate;
End;  (* Calibrate Apparatus *)

(*****************************************************************)

Procedure CalculateSusceptibility;

VAR T,
    M,                  (* formula weight of ANHYDROUS salt      *)
    WtAnhyCmpd,         (* weight of ANHYDROUS salt in grams     *)
    WtSoln,             (* wt water + wt HYDRATE in grams        *)
    Rho,                (* density of solution in grams/ml       *)
    p,                  (* weight fraction ANHYDROUS salt in grams   *)
    DeltaWt,            (* wt change w & w/o magnetic field in grams *)
    XgSoln,             (* Wt susceptibility of solution         *)
    XgSolute,           (* Wt susceptibility of solute           *)
    Xmolar,             (* Molar susceptibility of solute        *)
    MagMoment:  Real;   (* Magnetic moment of solute             *)

  (*--------------------------------------------------------*)
```

Figure 36–6 (continued)

procedures declared in the program, is actually a legal Pascal program. The program in Figure 36–5 will compile and run, but unlike the program in Figure 36–6, it has no output. Nevertheless, Figure 36–5 illustrates how self-documenting Pascal is and how programs written in Pascal closely mimic the procedures we use in scientific calculations.

The molar magnetic susceptibility (uncorrected for the diamagnetism of ligands and so denoted with a prime) is given by

$$\chi'_m = M\chi_w \qquad\qquad (36\text{–}19)$$

Correction for the Diamagnetism of Ligands

The molar magnetic susceptibilities are additive quantities whether they arise from paramagnetic or from diamagnetic atoms and ions. Usually

```pascal
   Procedure EnterCompoundData;
   Begin  (* EnterCompoundData *)
      ClrScr;
      Write('Please enter the temperature (in Kelvins) of the salt:  ');
      Readln(T);
      Write('Please enter the formula weight of the ANHYDROUS compound:  ');
      Readln(M);
      Write('Please enter mass of anhydrous Compound:  ');
      Readln(WtAnhyCmpd);
      Writeln('Please enter the mass of the solution.');
      Write('This is the mass of the hydrate plus the');
      Write(' mass of the water:  ');
      Readln(WtSoln);
      Write('Please enter the weight change (in grams)');
      Write(' w & w/o the magnetic field:  ');
      Readln(DeltaWt);
      Write('Please enter the density of the solution(gm/ml):  ');
      Readln(Rho);
   End;  (* EnterCompoundData *)

   (*---------------------------------------------------------*)

   Procedure CalculateData;

   VAR n:  Real;  (* number of unpaired electrons *)

      Begin  (* CalculateData *)
         p:= WtAnhyCmpd/WtSoln;
         XgSoln:= DeltaWt*g/AppConst/Rho;
         XgSolute:= (XgSoln + 0.72E-6*(1 - p))/p;
         Xmolar:= M*XgSolute;
         MagMoment:= 2.824*SQRT(Xmolar*T);
         Writeln;
         Writeln('The Molar magnetic susceptibility equals ',Xmolar:9);
         Writeln;
         Writeln('The magnetic moment equals ',MagMoment:5:3,' Bohr magnetons');
         n:= (-2 + SQRT(4 + 4*SQR(MagMoment)))/2;
         Writeln;
         Writeln('The number of unpaired electrons equals ',n:3:1);
         Writeln;Writeln;
      End;     (* CalculateData *)

   (*---------------------------------------------------------*)

Begin (* CalculateSusceptibility *)
   EnterCompoundData;
   CalculateData;
End;  (* CalculateSusceptibility *)


(*****************************************************************************)

Begin (*  M A I N   P R O G R A M  *)
   CalibrateApparatus;
   REPEAT
      Finished:= False;
      CalculateSusceptibility;
      Write('Run Again?  Enter a Y or an N, please.  ');
      Readln(Answer);
      If (Answer = 'N') OR (Answer = 'n') Then Finished:= True;
   UNTIL Finished;
End.  (*  M A I N   P R O G R A M  *)
```

Figure 36–6 (continued)

TABLE 36-2

Molar Magnetic Susceptibilities: Ionic, Atomic, and Group Contributions

	Ions in Solution		
	$\chi_m \times 10^6$ cgs Units		$\chi_m \times 10^6$ cgs Units
F^-	−11	Ca^{2+}	−8
Cl^-	−26	Ba^{2+}	−24
Br^-	−36	Na^+	−5
I^-	−52	Cs^+	−31
Cr^{2+}	−14	Mn^{2+}	−14
Fe^{2+}	−13	Co^{2+}	−12
Ni^{2+}	−11	Cu^{2+}	−11
NO_3^-	−20	SO_4^{2+}	−40
CN^-	−18	OH^-	−12
	Atomic and Group Contributions		
F	−6.3	H	−2.93
Cl	−20.1	C	−6.0
Br	−30.6	O (ether)	−4.61
I	−44.6	O (keto)	+1.73
S	−15.0	O (ester)	−3.36
N	−5.6	Benzene ring	−1.44
C=C	+5.5	C≡C	+0.8
N=N	+1.8	C=N	+8.2

Source: Landolt-Börnstein, New Series, *Group II, vol 2,* Magnetic Properties of Coordination and Organo-Metallic Transition Metal Complexes, *Springer-Verlag, Berlin, 1966.*

an anhydrous salt contains a single paramagnetic atom or ion, but it may be bonded to a number of diamagnetic ligands. A short list of molar magnetic susceptibilities of some paramagnetic substances is given in Table 36–2. The corrected magnetic susceptibility is χ_m:

$$\chi_m = \chi'_m + \sum \chi_m(\text{dia}) \tag{36-20}$$

Note that the first two terms are positive, but the third term, the summation, is negative. Some sources use the opposite sign convention (Selwood, 1956) and subtract *positive* diamagnetic susceptibilities. In either case χ_m is greater than χ'_m.

The Magnetic Moment

In general the magnetic moment of an atom or ion consists of two parts: the orbital contribution and the electron spin contribution. In most cases of paramagnetism, the orbital contribution appears to be quenched, so that only the unpaired electrons contribute to the paramagnetism. The theoretical magnetic moment in Bohr magnetons may be calculated from

$$\mu_m = 2\mu_B\sqrt{S(S+1)} = \mu_B\sqrt{n(n+2)} \tag{36-21}$$

where S is the total spin quantum number and n is the number of unpaired electrons. Since each unpaired electron has a spin $s = 1/2$, $S = ns = n(1/2)$.

Tabulate the weight changes for the various salts; the susceptibilities; the magnetic moments, calculated from Equation 36–10; and the number of unpaired electrons, calculated with Equation 36–21. Include the calibration data. Discuss the errors. Discuss the number of unpaired electrons in the metal ions in terms of their electron configuration. For those ions that can form both high-spin and low-spin complexes, discuss the electron configuration in terms of the crystal field strength and the observed number of unpaired electrons. Compare with literature values for χ_m and μ_m.

REFERENCES

1. Barrow, G., *Physical Chemistry,* 4th ed., McGraw-Hill, New York, 1979.

2. Bates, L. F., *Modern Magnetism,* Cambridge University Press, Cambridge, England, 1951.

3. Brubacher, L. J., and F. E. Stafford, "Magnetic Susceptibility," *J. Chem. Educ.* 39:574, 1962.

4. Crooks, J. E., "B or H? A Chemist's Guide to Modern Teaching on Magnetism," *J. Chem. Educ.* 56:301, 1979.

5. Figgis, B. N., and J. Lewis, in H. B. Jonassen and A. Weissberger, eds., *Technique of Inorganic Chemistry,* vol. 4, Interscience, New York, 1965.

6. Figgis, B. N., and J. Wilkins, in J. Lewis and R. G. Wilkins, eds., *Modern Coordination Chemistry,* Interscience, New York, 1960.

7. Föex, G., in *Tables de Constantes et Données Numériques,* No. 7, "Diamagnetisme et Paramagnetisme," Masson et Cie, Paris, 1957.

8. Gerlock, M., *Magnetism and Ligand Field Analysis,* Cambridge University Press, Cambridge, England, 1983.

9. Kirschner, S., M. J. Albinak, and J. G. Bergman, "Determination of Magnetic Moments of Solids, Liquids and Solutions," *J. Chem. Educ.* 39:576, 1962.

10. Mulay, L. N., in A. Weissberger and B. W. Rossiter, eds., *Techniques of Chemistry,* vol. I, *Physical Methods of Chemistry,* 4th ed., Wiley, New York, 1972.

11. Mulay, L. N., *Magnetic Susceptibility,* Interscience, New York, 1972.

12. Pass, G., and H. Sutcliffe, "Measurement of Magnetic Susceptibility and the Adoption of SI Units," *J. Chem. Educ.* 48:180, 1971.

13. Schuler, R. H., "Laboratory Experiments in Magnetochemistry," *J. Chem. Educ.* 27:591, 1950.

14. Selwood, P. W., *Magnetochemistry,* 2nd ed., Interscience, New York, 1956.

X-Ray Diffraction

Experiment 37

Diffraction Studies of Powdered Crystals

In the solid state most pure substances (elements and compounds) are crystalline as opposed to amorphous. The units of which the crystal is composed repeat themselves periodically in space in rather simple geometric patterns. The number of patterns is surprisingly few. In two dimensions, only five translational patterns or lattices are possible (Fig. 37–1). If the two translations in the lattices shown in Figure 37–1 are combined with a third, a set of 14 three-dimensional (space) lattices can be identified. These are the 14 Bravais lattices (Fig. 37–2).

The 14 Bravais lattices are further classified among the six crystal systems given in Table 37–1. The lattices are labeled according to their crystal type, i.e., presence or absence of centering. A simple lattice is given the symbol P (from the German *Primitiv*), a face-centered lattice the symbol F (from the German *Flächenzentriert*), and a body-centered lattice the symbol I (from the German *Innenzentriert*). In addition, the symbols A, B, and C label the centering on the three possible sides. The symbol R stands for a primitive rhombohedron, a rarely used alternative representation of a hexagonal lattice.

Points, Planes, and Cells

The unit cell of a crystal is described in terms of a set of x, y, and z axes. The repeat distance along the x axis is a, which is located at the fractional cell coordinates (1,0,0). Similarly, the repeat distance along the y axis is b, and the repeat distance along the z axis is c.

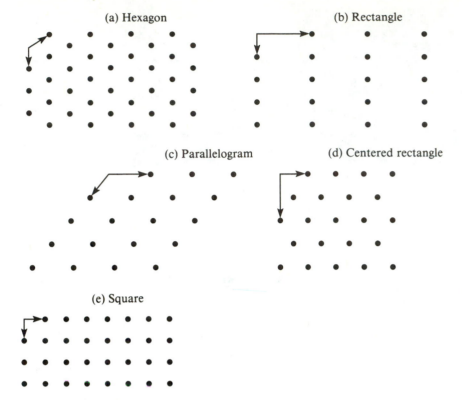

Figure 37-1 The five plane lattices.

Lines drawn through lattice points facilitate visualizing the unit cell. It is then easy to imagine that the unit cell is bounded by sets of parallel planes drawn through the lattice points. Other planes not passing through lattice points can also be drawn. It is useful to be able to describe sets of planes and their orientations.

Miller Indices. Consider the general oblique plane near the origin in Figure 37-3. This plane intercepts the x axis at $a/3$ or, more simply, at $1/3$. The same plane intercepts the y axis at $3/4$ and the z axis at $1/2$. The Miller indices of a plane are a set of integers that are the reciprocals of the axial intercepts. Because the Miller indices of parallel planes are identical, the Miller indices provide a practical method of identifying the members of a set of parallel planes and their orientations.

The axial intercepts, their reciprocals, and the cleared fractions, which are the Miller indices, are shown in Table 37-2 for the three oblique planes illustrated in Figure 37-3. The Miller indices of some special planes, namely those that bound the unit cell, are of interest. The intercepts of a plane parallel to an axis is at infinity, so the corresponding Miller index is 1/infinity, which equals zero. Thus, every unit cell in a crystal is bounded by the (100) planes, the (010) planes, and the (001) planes. In a cubic crystal, the distances between these planes equal the repeat distances a, b, and c of the unit cell. The distance between parallel

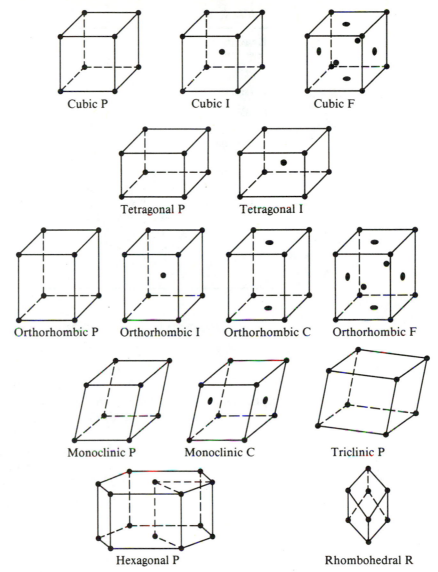

Figure 37–2 The fourteen Bravais lattices.

planes is denoted by $d_{hk\ell}$, where h, k, and ℓ are the Miller indices of the planes. If the crystal is cubic, the following relationships hold:

$$a = d_{100} = d_{010} = d_{001} \qquad (37\text{–}1)$$

If the crystal axes are orthogonal (cubic, tetragonal, or orthorhombic), then

$$a = d_{100} \qquad (37\text{–}2)$$

$$b = d_{010} \qquad (37\text{–}3)$$

$$c = d_{001} \qquad (37\text{–}4)$$

TABLE 37–1

Bravais Lattices and the Six Crystal Systems

Crystal Class	Lattice Parameters	Translational Symmetry				
		P	A (B, C)	F	I	R
Cubic	$a = b = c$ $\alpha = \beta = \gamma = 90°$	x	. . .	x	x	. . .
Hexagonal	$a = b \neq c$ $\alpha = \beta = 90°; \gamma = 120°$	x	. . .	. . .	. . .	x
Tetragonal	$a = b \neq c$ $\alpha = \beta = \gamma = 90°$	x	. . .	. . .	x	. . .
Orthorhombic	$a \neq b \neq c$ $\alpha = \beta = \gamma = 90°$	x	x	x	x	. . .
Monoclinic	$a \neq b \neq c$ $\alpha = \beta = 90° \neq \gamma$	x	x	. . .	. . .	. . .
Triclinic	$a \neq b \neq c$ $\alpha \neq \beta \neq \gamma$	x	. . .	. . .	. . .	. . .

The Distance Between Planes

In Figure 37–4, the plane defined by points A, B, and C intercepts the axes at a/h, b/k, and c/ℓ.

The perpendicular distance from the origin O to a point P on the plane equals the spacing between the planes, $d_{hk\ell}$.

The direction cosines of the line OP are

$$\cos POA = \frac{d}{a/h} \tag{37-5}$$

$$\cos POB = \frac{d}{b/k} \tag{37-6}$$

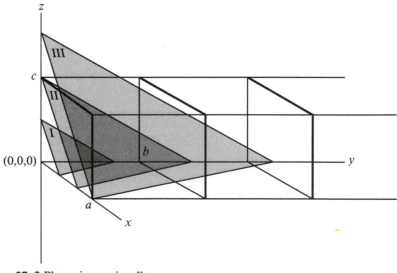

Figure 37–3 Planes in a unit cell.

TABLE 37–2

Miller Indices from Axial Intercepts of the Oblique Planes in Figure 37–3

Axial Intercepts on a, b, and c	Reciprocal of Intercept	Cleared Fractions
Plane I		
1/3	3	9
3/4	4/3	4
1/2	2	6
Plane II		
2/3	3/2	9
6/4	4/6	4
1	1	6
Plane III		
1	1	9
9/4	4/9	4
3/2	2/3	6

$$\cos \text{POC} = \frac{d}{c/\ell} \tag{37–7}$$

According to the Law of Direction Cosines,

$$\cos^2 \text{POA} + \cos^2 \text{POB} + \cos^2 \text{POC} = 1 \tag{37–8}$$

so

$$\frac{d^2 h^2}{a^2} + \frac{d^2 k^2}{b^2} + \frac{d^2 \ell^2}{c^2} = 1 \tag{37–9}$$

or

$$d = \frac{1}{\sqrt{\dfrac{h^2}{a^2} + \dfrac{k^2}{b^2} + \dfrac{\ell^2}{c^2}}} \tag{37–10}$$

It is customary to label d with the Miller indices of the planes for which it is the spacing: $d_{hk\ell}$.

Diffraction of X Rays by Crystal Planes

If you pick up a crystal and examine it, you can often see that the external morphology of the crystal reflects the translational symmetry of the crystal. You have probably seen the cubic symmetry of salt or the hexagonal symmetry of quartz. Such a simple optical examination, however, reveals nothing of the lattice centering, not to mention the positions of the atoms or ions in the crystal. In this experiment, x rays are used to determine the lattice type for crystals in the cubic system.

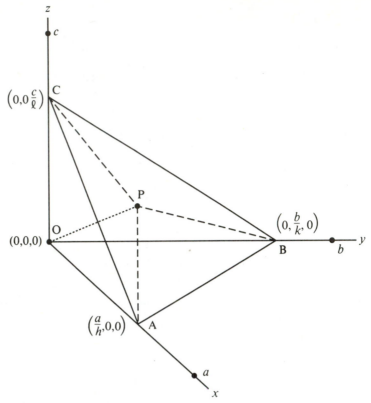

Figure 37–4 The spacing between planes.

X rays are a form of electromagnetic radiation of wavelength ranging from about 0.1 to 10 angstroms. In 1912 Max von Laue suggested that the wavelength of x rays is about the same order of magnitude as the translational repeat distances in crystals and consequently might be expected to be diffracted just as ordinary light was known to be diffracted by gratings ruled with spacings of the same approximate wavelength as light. Such x-ray diffraction effects were first observed by directing a beam of x rays at a single crystal of copper sulfate. Subsequently, many crystals were investigated by x-ray diffraction, and the nature of the diffraction was explained by William Bragg.

The Bragg Equation. A crystalline substance may be considered a three-dimensional stack of two-dimensional planes of scattering points, as shown in Figure 37–5. If a beam of x rays is directed at the crystal, the wave fronts of the incoming beam are reflected from sets of parallel planes. The relative phase of the reflected beam depends upon the wavelength λ, the angle of reflection θ, and the spacing between planes.

The extra distance traveled by wavelet I compared to wavelet II equals BC + CD. From the geometry of reflection it is also true that BC = CD = $d \sin \theta$. As long as the path lengths of the reflected wavelengths differ by an integral number of wavelengths ($n\lambda$), the reflected waves are

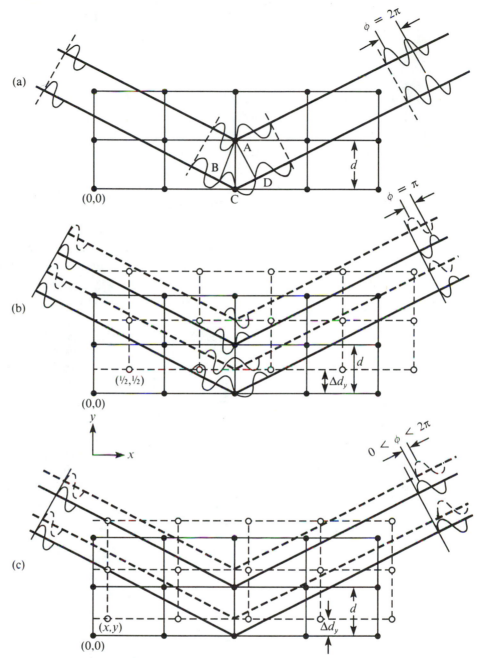

Figure 37–5 Bragg reflection of x rays. (a) One atom per unit cell. (b) Two atoms per unit cell. The cell is centered with one atom located at $(0,0)$ and the other at $(\frac{1}{2}, \frac{1}{2})$. (c) Two atoms per unit cell. One atom is at $(0,0)$ and the other at a general position (x, y).

in phase, and constructive interference takes place, i.e., the condition for reinforcement is that $BC + CD = n\lambda$. Thus, the Bragg condition is met when

$$n\lambda = 2d \sin \theta \qquad (37\text{–}11)$$

Equation 37–11 is the Bragg equation, d is the interplanar spacing, and θ is the Bragg diffraction angle. Since the sets of diffracting planes are identified by their Miller indices, Equation 37–11 is written

$$\lambda = 2 d_{hk\ell} \sin \theta \qquad (37\text{–}12)$$

The angle θ is known from the geometry of the apparatus. The wavelength λ is known and constant. Thus, $d_{hk\ell}$, calculated from Equation 37–12, corresponds to the spacing between the diffracting planes. If many reflections are observed, it is possible to identify the Miller indices $hk\ell$ of each of the sets of diffracting planes. The process of identifying which Miller indices belong to which reflecting planes is called indexing the reflections, and will be described later.

The Debye-Scherrer Camera. Many experimental arrangements are in use for establishing a known and controllable geometry between the crystal, the incident beam, the diffracted beam, and detecting the diffracted beam. In this experiment we shall use a Debye-Scherrer camera, shown in Figure 37–6. For the Debye-Scherrer method the sample consists of a

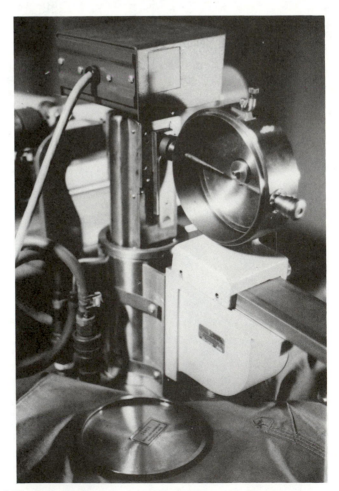

Figure 37–6 The Debye-Scherrer x-ray camera. (Karen R. Sime)

powder, which is just a ground-up crystal, contained in a thin-walled glass tube. The sample may also be a wire, which may be looked upon as a tightly packed powder. In either case, the powder consists of a very large number of tiny crystals close together in random orientations with respect to the incident x-ray beam. The diffracted beam is detected by the blackening of a strip of photographic film surrounding the sample. The broad x-ray beam from an x-ray tube (Figs. 37–6 and 37–7) is collimated into a thin, well-defined beam about 0.5 mm in diameter and directed at the sample. Much of the beam is not diffracted but passes directly through the sample and is caught in a beam catcher to prevent random scattering and fogging of the film. The physical arrangement is shown in Figure 37–8.

The many tiny crystals in the sample tube present essentially all the possible angles θ to the incident x-ray beam. Consequently, every plane satisfies the Bragg condition (Eq. 37–12) and may diffract the beam onto the film. If only one crystal were present and satisfied the Bragg condition, only a tiny spot corresponding to the intersection of the diffracted beam with the film would be observed. However, in a powder many tiny crystals present a great number of possible angles with respect to the beam axis. Consequently, rotation about the beam axis of one of these tiny crystals that happens to satisfy the Bragg condition generates a cone of reflection that intersects the nearly planar film in a circle. The film is too narrow to contain the whole circle; instead one observes sections of the reflection circle symmetrically placed about the beam axis, as shown in Figure 37–8.

From the geometry of the camera, it is seen that the reflected beam makes an angle 2θ with the incident beam diffracted from an arbitrary plane in a particular crystal satisfying the Bragg condition. Thus, the two halves of the reflection circle subtend 4θ of angle, which results in the proportion

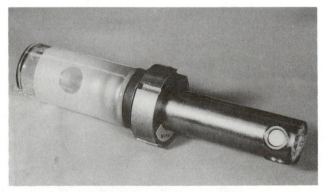

Figure 37–7 An x-ray tube. The overall length is about 32 cm. The high-voltage connector is inserted into the glass end. The anode is located at the upper end. Two of the three ports are visible. The port facing outward, about 1 cm in diameter, is covered with thin beryllium window, opaque to light but transparent to x rays because of its low atomic number. (Karen R. Sime)

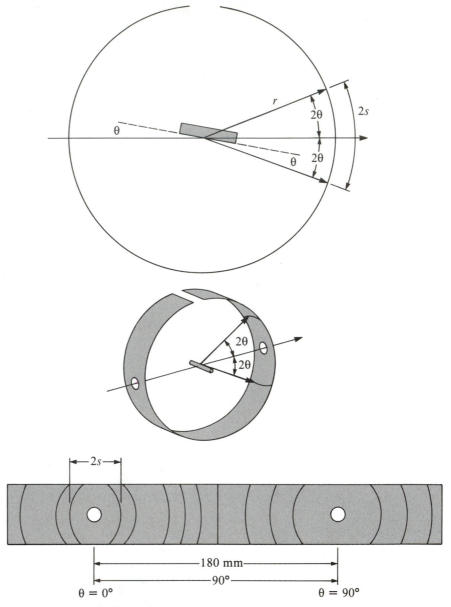

Figure 37–8 Geometry of the Debye-Scherrer x-ray camera. Dimensions correspond to a camera radius $r = 57.296$ mm.

$$\frac{2s}{4\theta} = \frac{2\pi r}{360} \qquad (37\text{–}13)$$

where s is the distance measured along the film, θ is the Bragg reflection angle, and r is the radius of the camera. In order that 1 degree of angle (θ) corresponds to 2 mm of film (s), $\theta/s = 1$ degree/2 mm and $r = 180/\pi = 57.296$ mm. Smaller cameras with $r = 90/\pi = 28.648$ mm are also available; for these, 1 degree corresponds to 1 mm.

The unrolled film is shown at the bottom of Figure 37–8. The Bragg angle $\theta = 0°$ lies at the center of the beam stop hole in the film, while $\theta = 90°$ lies at the center of the beam collimator hole. The distance between the holes is 90 mm in the small camera and 180 mm in the larger one. Thus, the Bragg angle can be measured with a ruler or a more precise film measuring instrument, and the spacing d of the set of planes giving rise to this line on the film can be calculated with Equation 37–12.

Indexing Reflections. According to Equation 37–10, the relationship between the d spacing of the planes and the unit translation of a cubic cell for which $a = b = c$ is given by

$$d_{hk\ell} = \frac{a}{\sqrt{h^2 + k^2 + \ell^2}} \tag{37–14}$$

Thus, planes with the smallest Miller indices are the most widely spaced. If Equations 37–12 and 37–14 are combined to eliminate $d_{hk\ell}$, the result is

$$\frac{\lambda^2}{4a^2} = \frac{\sin^2 \theta}{M} \tag{37–15}$$

where M equals $h^2 + k^2 + \ell^2$. For a given cubic unit cell, the left side of Equation 37–15 is a constant. Since M takes on values of 1, 2, 3, 4, 5, 6, 8, etc., it is possible to index the reflections (find $hk\ell$) by finding a series of M for which the right side of Equation 37–15 is constant for the measured values of θ. Notice that the sum of the squares of three integers cannot equal 7 and some other higher values. See Table 37–3.

In practice, indexing is not always that simple, because some of the lines may be too weak to observe and some lines are extinguished because of the translational symmetry of the lattice, namely body- or face-centering. The possible reflections for primitive, face-centered, and body-centered cubic lattices are listed in Table 37–3.

TABLE 37–3

Extinctions for Cubic Crystals[a]

M	P	F	I	M	P	F	I
1	100			10	310		310
2	110		110	11	311	311	
3	111	111		12	222	222	222
4	200	200	200	13	320		
5	210			14	321		321
6	211		211	15			
7				16	400	400	400
8	220	220	220	17	410,322		
9	300,221			18	411,330		411,330

[a] Extinguished reflections are omitted.

Rearrangement of Equation 37–14 leads to a graphical method for indexing the reflections from a cubic crystal (Azaroff and Buerger, 1958):

$$a = \sqrt{M}d_{hk\ell} \qquad\qquad \textbf{(37–16)}$$

In this form, a is seen to be a linear function of $d_{hk\ell}$. This is the equation of a family of lines passing through the origin and with slope equal to $\sqrt{M}$, as illustrated in Figure 37–9. Shown in the inset on the graph are the d values calculated from the measured Bragg angles of a powder diffraction photograph. The d scale is the same as the main graph. The $d = 0$ point of the overlay is lined up with the $d = 0$ line (the ordinate) of the main graph. When the overlay is moved up and down, a position is found where the d values on the overlay coincide with the lines on the main graph. The value of a is then read off the main graph ordinate and the proper Miller indices are assigned from the labels on the main graph lines.

Systematic Extinctions and the Phase Angle. Although a good number of the lines may line up, some may be missing entirely. These systematic extinctions can tell us whether the cubic unit cell is primitive, body centered, or face centered. In Figure 37–5a, the unit cell is assumed

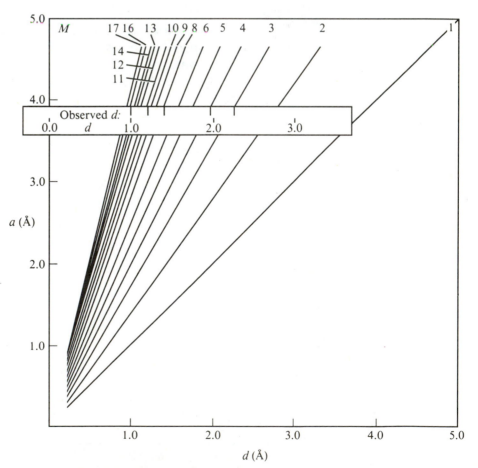

Figure 37–9 Graphical indexing of cubic crystals.

to contain only one kind of atom, and the Bragg angle is drawn so that constructive interference occurs. But what if the cell contains two or more atoms of the same or different kinds? Compare parts (a) and (b) of Figure 37–5. In (a), the unit cell contains *one* atom, which can be considered to be located at the origin (0,0) and lying on the solid horizontal planes (lines in this two-dimensional representation). The angle θ has been adjusted so that the Bragg condition is satisfied, and the wave fronts after reflection from the horizontal planes of atoms are exactly in phase; that is, the phase angle ϕ equals 2π, corresponding to constructive interference.

In part (b), the unit cell contains *two* atoms, one at the origin (0,0) and the other at the cell center $(\frac{1}{2},\frac{1}{2})$, shown as open circles lying on dashed horizontal planes (dashed lines in this two-dimensional representation). Since the distance Δd between the dashed planes is half the distance d between the solid planes, the phase angle ϕ between the solid and dashed planes is π, corresponding to destructive interference. If the centering atom and the origin atom are *identical*, the interference is complete, the wavelets cancel in pairs, and no reflection is observed.

In general, the following proportion holds (Buerger, 1942, p. 50):

$$\frac{\phi}{2\pi} = \frac{\Delta d}{d} \tag{37–17}$$

In part (c), the unit cell again contains two atoms, the first at the origin (0,0) and the second in a general position (x,y), where x and y are the fractional coordinates of the atoms along the cell repeat distances (a, b, and c in three dimensions). Thus, in part (c), in the y direction, $\Delta d_y = yb$. Recalling that the Miller indices are the fractional intercepts along the axes, $d = b/k$. Substitution of Δd_y and d in Equation 37–17 gives

$$\frac{\phi_y}{2\pi} = \frac{yb}{b/k} = ky$$

or

$$\phi_y = (2\pi)ky$$

For a general plane $(hk\ell)$ in three dimensions, the phase angle ϕ for the one atom located at (x,y,z) is

$$\phi = 2\pi(hx + ky + \ell z)$$

If there are j atoms in the unit cell, then all j atoms reflect from the $hk\ell$ plane with phase angles ϕ_j where

$$\phi_j = 2\pi(hx_j + ky_j + \ell z_j) \tag{37–18}$$

Suppose a unit cell has two atoms. Since the first atom can always be placed at the origin ($x = y = z = 0$ or 1), the phase angle ϕ for the first atom is always 0 or 2π, i.e., constructive interference. If the second atom is at some arbitrary general position, ϕ is probably not equal to 2π. But suppose that the second atom is located at the exact center of the

TABLE 37–4

Extinctions for Face-Centering

h	k	ℓ	All Even or All Odd?	Reflection Observed? (See Fig. 37–10)
1	0	0	No	No
1	1	0	No	No
1	1	1	Yes	Yes
2	0	0	Yes	Yes
2	1	0	No	No
2	1	1	No	No
2	2	0	Yes	Yes
2	2	1	No	No
3	0	1	No	No

cell, a special position for which $x = 1/2$, $y = 1/2$, and $z = 1/2$. Then the phase angle ϕ would be

$$\phi = 2\pi[h(1/2) + k(1/2) + \ell(1/2)] = \pi[h + k + \ell] = n\pi \quad (37\text{–}19)$$

If n is odd, destructive interference occurs; if n is even, constructive interference occurs. If the atoms are identical, as they are in a centered lattice, this means that reflections with $h + k + \ell$ equal to an uneven number are extinct, extinguished, and not observed.

For a face-centered lattice, systematic extinctions of reflections are observed when h, k, and ℓ are either not all even or not all odd. For the first few $hk\ell$ sets, ask yourself if h, k, and ℓ are all odd or all even. The answers are in Table 37–4. In Figure 37–10, a comparison is made of the appearance of diffraction photographs of primitive and centered cubic cells, assuming that the cell dimensions are the same. In actual practice, about 30 reflections are observed for a primitive cubic cell with a cell dimension of about 4 Å.

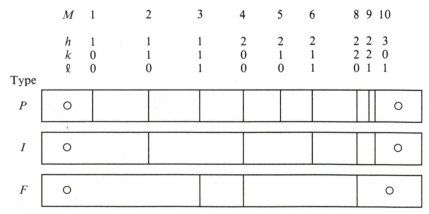

Figure 37–10 Systematic extinctions for P, I, and F cubic lattices.

Systematic Extinction and the Structure Factor. Let us look in more detail at the nature of the intensities of diffracted x rays in order to understand the origin of such systematic extinctions as are illustrated in Figure 37–10. The intensities of waves diffracted by a crystal lattice depend on the number of atoms j, their scattering factors f_j, and the phase angle ϕ. It is easy to visualize how two waves add up constructively (Fig. 37–5a) when ϕ equals 2π and how they cancel (Fig. 37–5b) when ϕ equals π. Even when the unit cell contains just two atoms, but one atom is at a general position (x, y, z), it is awkward to visualize the net amplitude of the two reflected waves (shown in Fig. 37–5c). Because unit cells in real crystals contain not just one or two but dozens, even hundreds, of atoms, it is necessary to find a better way to add up the wavelets from all the atoms. A simple solution is the vector addition of waves, called **superposition**. A wave can be represented by a rotating vector (Fig. 37–11), and the vectors can easily be added (Stout and Jensen, 1968, p. 213).

In Figure 37–11, the amplitudes of the dotted and dashed waves to be added are represented by the vectors of amplitude f_1 and f_2; their phase angles are ϕ_1 and ϕ_2. The vector sum of f_1 and f_2 is F, which is called the **structure factor** of the reflection. Because of the ability of an atom to scatter x rays is proportional to its number of electrons, the f_i are called the **scattering factors**. In fact, the scattering factor f of an atom equals the atomic number Z at small Bragg angles θ. To see how the waves on the right side of Figure 37–11 are generated, consider the three vectors f_1, f_2, and F locked together and rotating with a frequency equal to that of the incident x-ray radiation. The projection of the rotating vectors on the vertical axis equals the amplitude of the reflected wave.

It is convenient, as we shall see, to express the structure factor F as a complex number. The two-atom vector diagram in Figure 37–11 is shown in Figure 37–12a on the complex plane; below it (Fig. 37–12b)

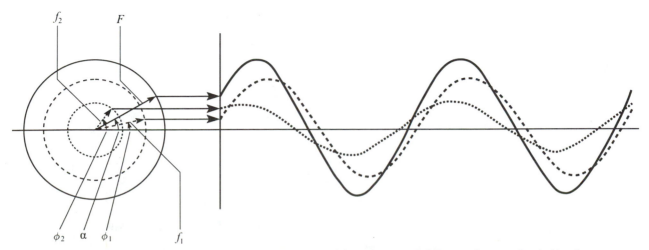

Figure 37–11 The vector addition of waves of different phase angles (ϕ_i) and amplitudes (f_i).

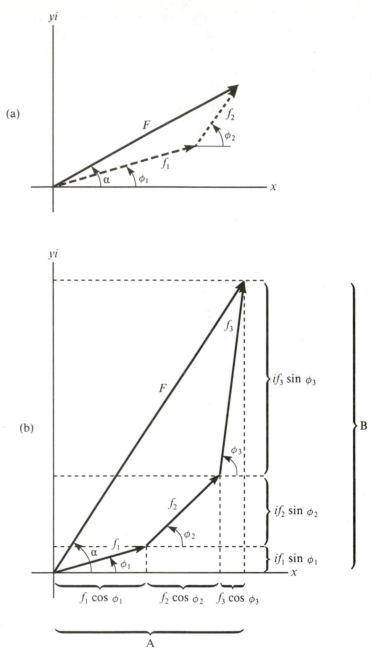

Figure 37–12 The superposition of waves. (a) Two waves. (b) Three waves. Components on the coordinate axes of a sum of vectors are equal to the sums of the components of the individual vectors.

is shown a three-atom structure factor diagram, where F is the vector sum of three waves of amplitudes (scattering factors) f_1, f_2, and f_3. From the sums of the component vectors along the x and yi axes, the structure factor F is given by Equation 37–19:

$$F = f_1 \cos \phi_1 + f_2 \cos \phi_2 \cdots + if_1 \sin \phi_1 + if_2 \sin \phi_2 + \cdots \quad \textbf{(37–19)}$$

$$F = \sum f_\alpha \cos \phi_\alpha + i \sum f_\alpha \sin \phi_\alpha \qquad \text{(37–20)}$$

$$F = A + iB \qquad \text{(37–21)}$$

where A and B are the two summations over all the j atoms in the cell. The magnitude of the structure factor is given by Equation 37–22:

$$|F| = \sqrt{A^2 + B^2} \qquad \text{(37–22)}$$

This is the form that is convenient for actually calculating the magnitude of the structure factors from the known positions of the atoms.

Two different forms of the structure factor F are useful, the trigonometric form (Eq. 37–19) and an exponential form. These can be converted to each other with Euler's Theorem, Equation 37–23:

$$\cos \phi + i \sin \phi = e^{\phi i} \qquad \text{(37–23)}$$

Although the exponential form is not convenient for actually calculating structure factors, it is useful for deriving the extinction conditions for $h, k,$ and ℓ, which arise from lattice-centering. If an exponential term is substituted for each pair of sine and cosine terms in Equation 37–19 according to Equation 37–23, the result is

$$F = f_1 e^{i\phi_1} + f_2 e^{i\phi_2} + f_3 e^{i\phi_3} + \cdots \qquad \text{(37–24)}$$

where the phase angles are given by Equation 37–18. As an example, suppose that a lattice is body centered, i.e., that the unit cell contains two atoms, one at $x = 0$, $y = 0$, and $z = 0$, and the other at $x = 1/2$, $y = 1/2$, and $z = 1/2$. According to Equation 37–24, the structure depends upon the contribution from each of the two atoms:

$$F = f_1 e^{2\pi i (hx_1 + kx_1 + \ell x_1)} + f_2 e^{2\pi i (hx_2 + ky_2 + \ell z_2)} \qquad \text{(37–25)}$$

If the lattice is centered, the two atoms are identical, so their scattering factors are identical; thus, $f_1 = f_2 = f$. If their positions are substituted in Equation 37–25, the result is

$$F = fe^{(0)} + fe^{2\pi i (h/2 + k/2 + \ell/2)} \qquad \text{(37–26)}$$

$$F = f(1 + e^{n\pi i}) \qquad \text{(37–27)}$$

where $n = h + k + \ell$. If you substitute $n\pi$ for ϕ in Equation 37–23, you will discover that

$$e^{n\pi i} = +1 \text{ for even } n \qquad \text{(37–28)}$$

$$e^{n\pi i} = -1 \text{ for odd } n \qquad \text{(37–29)}$$

Thus, when $h + k + \ell = n =$ odd, the reflection is extinguished, since

$$F = f(1 - 1) = 0 \qquad \text{(37–30)}$$

but when $h + k + \ell = n =$ even, the reflection is permitted, since

$$F = f(1 + 1) = 2f \qquad \text{(37–31)}$$

These conditions for systematic extinctions hold for all the body-centered lattices (type I) shown in Figure 37–2.

Now notice the difference between molybdenum and cesium chloride illustrated in Figure 37–13. For molybdenum, the centering atom is *identical* to the origin atom ($f_1 = f_2 = f_{Mo}$), so Equations 37–26 through 37–31 apply and the systematic extinctions characteristic of a body-centered lattice are observed. The cesium chloride lattice is not centered because the ion located at $x = 1/2$, $y = 1/2$, $z = 1/2$ is not *identical* to the ion located at $x = 0$, $y = 0$, $z = 0$. For cesium chloride, $f_1 = f_{Cs}$ and $f_2 = f_{Cl}$. Consequently the structure factor for cesium chloride is given by

$$F = f_{Cs^+}e^{(0)} + f_{Cl^-}e^{2\pi i(h/2+k/2+\ell/2)} \qquad (37\text{–}32)$$

$$F = f_{Cs^+} + f_{Cl^-}e^{n\pi i} \qquad (37\text{–}33)$$

When $h + k + \ell = n$ is even, then

$$F = f_{Cs^+} + f_{Cl^-} \qquad (37\text{–}34)$$

and a reflection is observed; but when $h + k + \ell = n = $ odd,

$$F = f_{Cs^+} - f_{Cl^-} \qquad (37\text{–}35)$$

and although the reflection is not extinguished, it is weak because the structure factor equals the *difference* between the scattering factors for Cs^+ and Cl^-. If you examine your CsCl photograph, you will notice that there are many lines, since none is extinguished, and that the lines alternate in intensity, as $n = h + k + \ell$ alternates from even to odd.

What do you suppose the powder diffraction photographs for cesium bromide and cesium iodide look like? Remember that the scattering factor depends on the number of electrons in an atom or ion. In fact, for small Bragg angles, f equals Z, the atomic number. The more electrons in an atom or ion, the more strongly it scatters x rays. Think of the scattering factor as the "scattering power," which is probably a better name.

For cesium bromide, the structure factor equations are similar. When $h + k + \ell = n$ is even, then

$$F = f_{Cs^+} + f_{Br^-} \qquad (37\text{–}36)$$

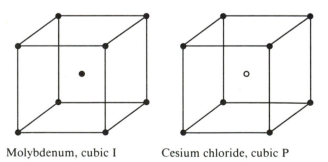

Molybdenum, cubic I   Cesium chloride, cubic P

Figure 37–13 Molybdenum (I) and cesium chloride (P) lattices.

and a reflection is observed; but when $h + k + \ell = n = $ odd,

$$F = f_{Cs^+} - f_{Br^-}$$ (37–37)

The same alternating intensities are observed for cesium bromide. In fact, the weak lines are even weaker than for cesium chloride because the difference given by Equation 37–37 is even smaller than the difference given by Equation 37–35.

For cesium iodide, the structure factor equations are similar. When $h + k + \ell = n$ is even, then

$$F = f_{Cs^+} + f_{I^-}$$ (37–38)

and a reflection is observed; but when $h + k + \ell = n = $ odd,

$$F = f_{Cs^+} - f_{I^-}$$ (37–39)

and not only are the odd reflections weak, but their intensities are virtually zero because Cs^+ and I^- are isoelectronic. Because these two ions have exactly the same number of electrons and the same electron configuration, their ability to scatter electrons is virtually identical. Consequently, when $h + k + \ell = n$ is odd, $F = 0$, and no reflection is observed.

Although its pattern appears the same as that of molybdenum, cesium iodide is no more a centered lattice than is cesium bromide or cesium chloride. A comparison of the cesium halide diffraction patterns dramatically displays the expected alternating intensities (Boer and Jordan, 1965).

Apparatus

This experiment requires an XRD-5 x-ray diffraction apparatus or equivalent; a Debye-Scherrer powder diffraction camera; a film punch and cutter; Kodak x-ray film; a darkroom; an x-ray developer and fixer; a powder diffraction film-measuring instrument; molybdenum or tungsten wire; copper wire; cesium chloride; mortar and pestle; and thin-walled capillary tubes for powder samples.

SAFETY CONSIDERATIONS

The x rays generated from this apparatus are exceedingly dangerous. When the apparatus is properly used, stray radiation is virtually undetectable and the apparatus is safe. Be sure you know the location of all beam ports on the x-ray tube or tubes and understand how to check that the ports are closed or collimated into a proper beam stop or camera. Do not adjust the camera, beam stops, or port windows when the apparatus is turned on.

EXPERIMENTAL PROCEDURE

Your instructor will guide you through each part of the procedure for the first sample, and you will work independently with the last two samples.

Loading the Sample. The three samples selected for this experiment are copper wire, molybdenum wire, and powdered cesium chloride, $CsCl$, contained in a 0.2-mm thin-walled capillary tube. The metal wires diffract as if they were tightly packed powders and have the advantage that no fragile capillary tube is necessary for support. These three samples demonstrate the three different cubic lattice types: primitive, body centered, and face centered.

Lay the camera on its back on the table. Remove the cover. Remove the collimator and the beam stop and place them where they will not roll off the table onto the floor, since the tips are easily damaged. With long-nosed pliers, place the brass sample holder into the receptacle in the camera, where it is held by friction. Replace the collimator and place on it the special magnifying glass. Place the camera on a light box so that you can look through the magnifying glass at the sample, silhouetted against the lighted background. If the sample is not visible, rotate the pulley on the side of the camera; the sample should come into view and appear to go up and down as the pulley is turned, rotating the sample. Center the sample by rotating until the sample is as high as possible; if it goes out of view, estimate where it would be as high as possible. Then screw in the small knurled knob at the top of the camera. It moves up against the sample holder, which is magnetically held to the camera; continue screwing until the sample is pushed into the center of the illuminated field. Screw the knob back out of the way and rotate the sample again. It may go up and down, but not as much as before. Repeat the adjustment until the sample shows no up-and-down motion.

Loading the Film. Before cutting and loading the camera, obtain a previously cut and punched practice piece of film. Insert it in the film punch and cutter to familiarize yourself with its operation. With this same film strip, practice loading the camera in daylight, first watching, then with your eyes averted, until you are convinced you can load the camera in the dark. When you can load the film in the dark, obtain a roll of film in its light-tight container, which should be stored in a refrigerator. Take the film and camera to the darkroom. *Do not open the film except in total darkness.* Remove the cover, collimator, and beam stop from the camera and place them in a secure place where you can find and identify them in total darkness. In total darkness, open the film container, feel for the film, and slip the end of the roll into the film punch and cutter until the end of the film touches the end of the track. Replace the film in its light-tight container before proceeding further. Punch the two holes, and cut the film to the preset length. Remove the film, curl it into a loop, and slip it into the camera, taking care not to come into the slightest contact with the sample tube or wire, which have already been carefully aligned. Bring one end of the film to the fixed stop and the other to the moveable stop. Then tighten the stop against the film and secure the stop. Insert the collimator and beam stop. When the cover is again in place, the lights may be turned on.

Making an Exposure. Place the camera on its track by the x-ray tube port. Power to the x-ray generator should be off. Insert the brass radiation shield over the end of the collimator. Slide the camera toward the tube; pull the spring-loaded window cover down and slide the camera toward the tube until the shield is against the nickel filter in the window of the copper x-ray tube. Follow in detail the instructions given by your instructor for completing the alignment and turning on the power, the x rays, and the timer. A step-by-step checklist should be posted by the x-ray generator. The metal wire samples require about a 30-minute exposure in a camera with a 118-mm diameter. These can be completed in one laboratory period. The cesium chloride requires about 3 hours. This can be started toward the end of the period, with the timer set to turn the generator off. The film can be removed and stored in an envelope of photographic black paper and developed when convenient.

Developing the Film. The developer, shortstop, and fixer are contained in three 200-ml graduated cylinders, which serve as developing tanks. When the exposure is complete, open the camera in *total darkness* and remove the film. Attach a stainless-steel clip to each end and develop for 5 min at 20°C. See the appropriate charts for other times and temperatures. Remove the film from the developer, drain it a few seconds, dip it in the shortstop, drain, and fix for about 5 minutes. At this time the lights may be turned on. Wash the film in an 8-by-10-inch enlarging tray under a slow stream of running water for about 30 minutes. Then remove the film, wipe it with a clean, moist film sponge, and hang it to dry. To distinguish your films from others, identify them with a small pattern of notches cut with scissors.

Measuring the Film. Tape the dry film to the illuminated surface of the film-measuring device (Fig. 37–14) with masking tape. The film should be parallel to the graduated scale. Determine the scale position of the two holes, x_L and x_R, in the film by measuring the scale positions of pairs of lines symmetrically located about the centers (Fig. 37–15). The center of the left hole, for example, is equal to $(x'_1 + x_1)/2$. The average of similar calculations with x_2, x'_2, x_3, and x'_3 gives the best value for x_L. Take the averages of as many pairs as are measurable.

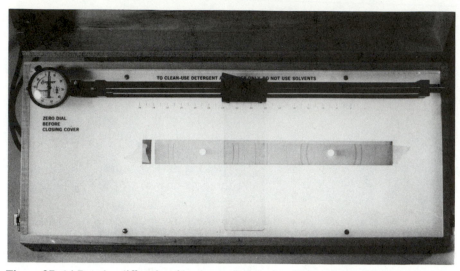

Figure 37–14 Powder diffraction film measuring device. When the film is taped to the light box, the distances between diffraction lines may be measured by sliding the indicator along the film and recording readings from the scale and micrometer. If the lines are sharp, the precision is about .005 mm. (Karen R. Sime)

If the distance between holes were exactly 180 mm (corresponding to 90°, when the camera diameter equals 114 mm), the values of $x_1, x_2 \ldots$ in mm would be equal to the angle 2θ. Because of film shrinkage, a correction factor K must be calculated:

$$K = \frac{180}{(x_R - x_L)} \tag{37-40}$$

Then for each reflection, the Bragg angle θ is calculated from

$$\theta = K(x_i - x_L)/2 \tag{37-41}$$

where x_i is the scale reading of the ith line.

For high-angle reflections, closely spaced double lines may be observed, both of which should be measured. These correspond to resolved reflection from the K_{α_1} and K_{α_2} wavelengths of the incident radiation. For copper, these wavelengths and their intensity-weighted average are given in Table 37–5. If the lines are not resolved, as is the case at low angles, use the average wavelength to calculate Bragg angles.

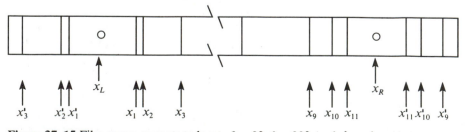

Figure 37–15 Film measurements to locate $\theta = 0°$, $\theta = 90°$, and d spacings.

TABLE 37–5

Wavelengths for Copper Radiation

Radiation	Wavelength (Å)
CuK_{α_1}	1.54050
CuK_{α_2}	1.54434
$CuK_{\alpha_{ave}}$	1.54180

Results and Calculations

For each of the three substances studied, prepare a table of the data used to calculate the film shrinkage correction factor with Equation 37–40. Calculate d for each reflection with Equation 37–12. Plot these values on a strip of graph paper with the same scale as in Figure 37–9, place the strip on Figure 37–9, and move it up and down until the observed lines match. Index each observed reflection and graphically obtain an approximate value for the cell constant a. Once the lines are indexed, deduce the lattice type (P, I, or F) and calculate a precise value for a with Equation 37–15 for each reflection.

Tabulate x_i, θ, $\sin \theta$, λ, d, $hk\ell$, and a_0 for each reflection. Calculate an average value of a for the high-angle reflections and compare your value with the value from the literature.

REFERENCES

1. Azaroff, L. V., *Elements of X-Ray Crystallography,* McGraw-Hill, New York, 1968.

2. Azaroff, L. V., and M. J. Buerger, *The Powder Method in X-Ray Crystallography,* McGraw-Hill, New York, 1958.

3. Boer, F. P., and T. H. Jordan, "X-ray Crystallography Experiment," *J. Chem. Educ.* 42:76, 1965.

4. Buerger, M. J., *X-Ray Crystallography,* Wiley, New York, 1942.

5. Bunn, C. W., *Chemical Crystallography,* 2nd ed., Clarendon Press, Oxford, 1961.

6. Henry, N. F. M., H. Lipson, and W. A. Wooster, *The Interpretation of X-Ray Diffraction Photographs,* Macmillan, London, 1960.

7. Nuffield, E. W., *X-Ray Diffraction Methods,* Wiley, New York, 1966.

8. Stout, G. H., and L. H. Jensen, *X-Ray Structure Determination,* Macmillan, London, 1968.

Experiment 38

Single-Crystal Rotation Photographs

In the real lattice of a crystal, the Miller indices provide a means of describing and naming a family of planes. Planes with the same Miller indices are characterized by a common orientation in space and a common spacing. The spacing (for orthogonal crystals) is given by

$$d_{hk\ell} = 1/\sqrt{a^2/h^2 + b^2/k^2 + c^2/\ell^2} \qquad (38-1)$$

where h, k, and ℓ are the Miller indices and a, b, and c are the repeat distances along the x, y, and z axes, respectively. A Miller index of a plane is best interpreted as the reciprocal of the plane's intercept on an axis.

For a given wavelength, the Bragg equation (Equation 38–2) gives the relationship between

$$\frac{2\sin\theta}{\lambda} = \frac{1}{d_{hk\ell}} \qquad (38-2)$$

the diffraction angle and the spacing between diffracting planes. In the form of Equation 38–2, you can see immediately that there is a reciprocal relationship between theta and $d_{hk\ell}$. More explicitly, there is a reciprocal relationship between what is measured (the diffraction angle in an instrument) and a fundamental property of a crystal (the interplanar spacing).

For most of us, visualizing planes in three-dimensional space is not particularly easy and is made more complex by the reciprocal relationship between angle and spacing.

The solution to simplifying the relationship between diffraction instrumentation and crystal planes is the **reciprocal lattice,** a construct introduced by P. Ewald[1] in 1921. Each *point* in a reciprocal lattice represents the spacing and direction of a set of planes with Miller indices $hk\ell$.

The Reciprocal Lattice

The relationship between the real lattice and the reciprocal lattice is given in Figure 38–1. The repeat distances in the crystal are a, b (not shown), and c. Each point in the reciprocal lattice represents a set of

[1] Ewald, P. P., "Das 'reziproke Gitter' in der Strukturtheorie," *Z. Krist.* (A) 56:148, 1921.

planes $hk\ell$ in the crystal lattice and lies perpendicular to those planes at a distance $1/d_{hk\ell}$ from the origin. The crystal lattice origin and the reciprocal lattice origin coincide at 000.

To more completely understand the reciprocal lattice, it is helpful to actually draw the one shown in Figure 38–1.

1. The real lattice and the reciprocal lattice share a common origin at 000.
2. From the common origin, construct a normal to each plane in the real lattice.
3. Set the length of the normal equal to the reciprocal of the interplanar spacing.

$$a^* = \frac{1}{d_{100}} \qquad\qquad b^* = \frac{1}{d_{010}} \qquad\qquad c^* = \frac{1}{d_{001}} \qquad (38\text{–}3)$$

4. Notice that the interplanar spacings $d_{hk\ell}$ are not equal to the unit cell translations (a, b, and c) unless the cell is orthogonal ($\alpha = \beta = \gamma = 90°$). From the geometry of Figure 38–1 (monoclinic; the b axis is normal to the ac plane),

$$d_{100} = a \sin \beta \qquad\qquad d_{010} = b \qquad\qquad d_{001} = c \sin \beta$$

$$\beta^* = 180° - \beta$$

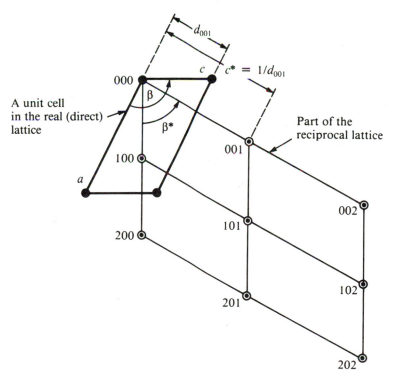

Figure 38–1 Relationship between the reciprocal and real (direct) lattices. Notice that a point in the reciprocal lattice represents an *entire set of planes* and their spacings in the real lattice.

As an example, let us sketch a couple of reciprocal lattice points for the cell illustrated in Figure 38–1. Suppose that the cell is monoclinic with $a = 5.00$ Å, $c = 3.00$ Å, and $\beta = 117°$. For a two-dimensional drawing in the ac plane, it is not necessary to consider b, which is normal to the ac plane.

With a ruler and protractor, sketch the unit cell (heavy lines in Figure 38–1). First, select an arbitrary but convenient scale, say, 1 cm on the paper equals 1 Å. Put in the origin, and draw the 3-cm-long line labeled c. Measure 117° with a protractor to get the direction for a, and draw the 5-cm-long line labeled a. Similarly, draw in the other two heavy lines outlining the unit cell. Two parallel 100 planes normal to the paper pass through the horizontal heavy lines, and two parallel 001 planes normal to the paper pass through the two slanted heavy lines. The 010 plane lies in the plane of the paper.

Now sketch some reciprocal lattice points. The spacing between the 100 and 001 planes is calculated as follows:

$$d_{100} = a \sin \beta = 5.00 \sin 117 = 4.45 \text{ Å}$$

$$d_{001} = c \sin \beta = 3.00 \sin 117 = 2.67 \text{ Å}$$

Next, calculate the $a*$, the distance in reciprocal space from the origin to the first reciprocal lattice points.

$$a* = \frac{1}{d_{100}} = \frac{1}{4.45 \text{ Å}} = 0.225 \text{ Å}^{-1}$$

$$c* = \frac{1}{d_{001}} = \frac{1}{2.67 \text{ Å}} = 0.3745 \text{ Å}^{-1}$$

Thus the 100 reciprocal lattice point lies on a line normal to the 100 planes of the real lattice and at a distance from the origin equal to 0.225 Å^{-1} in reciprocal space.

The 001 reciprocal lattice point lies on a line normal to the 001 planes of the real lattice and at a distance from the origin equal to 0.3745 Å^{-1} in reciprocal space.

To sketch the reciprocal lattice, we need to choose a convenient scale for reciprocal space, say, 1 cm on the paper equals 0.075 Å^{-1}. How far from the origin does the 100 point in reciprocal space lie?

$$\frac{0.225 \text{ Å}^{-1}}{0.075 \text{ Å}^{-1}/\text{cm}} = 3.00 \text{ cm}$$

Similarly for the 001 point:

$$\frac{0.3745 \text{ Å}^{-1}}{0.075 \text{ Å}^{-1}/\text{cm}} = 5.00 \text{ cm}$$

Now, along a normal from the origin to the 100 plane, draw a line 3.00 cm long, terminate it with a small open circle, and label it 100.

Next, draw a normal from the origin to the 001 plane 5.00 cm long, terminate it with a small open circle, and label it 001. These are the first two points of the reciprocal lattice in Figure 38–1; they are closest to the origin. The remaining points can be drawn similarly.

Notice that the scale factor was chosen to be 0.075 Å^{-1}/cm so that the reciprocal lattice and the real lattice would appear about the same size on the drawing. This was for artistic reasons, not crystallographic reason. Change the scale to 0.10 Å^{-1}/cm and redraw; maybe you will like this scale better.

Remember that a crystal has an infinite number of 001 planes, all parallel to each other and separated by a distance d_{001}. The 001 point in the reciprocal lattice represents that whole family of planes. It represents their two properties: their direction (by the normal) and their separation (by the distance of the reciprocal lattice point from the origin in reciprocal space). Thus the reciprocal lattice point lies on the tip of a vector.

In summary, the reciprocal lattice consists of all points located on perpendiculars drawn from the origin to all planes $hk\ell$ at a distance $1/d_{hk\ell}$ from the origin. Each reciprocal lattice point then represents a set of planes and their orientation with respect to the origin.

The Ewald Sphere

The **Ewald sphere,** also known as the sphere of reflection, is the geometric construct that relates the geometry of the diffraction apparatus (or camera) to the geometry of the crystal, i.e., to the reciprocal lattice (Fig. 38–2). The Ewald sphere is constructed as follows:

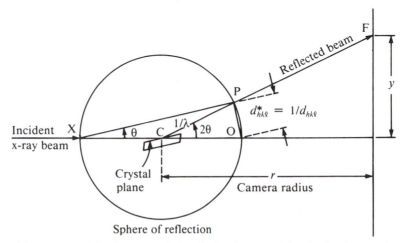

Figure 38–2 The Ewald sphere, or sphere of reflection, provides the fundamental relationship between reciprocal space and real space. The radius of the Ewald sphere equals $1/\lambda$.

1. In two dimensions, the sphere of reflection becomes a circle of reflection with a radius $1/\lambda$.
2. The incident x-ray beam lies on XCO.
3. The origin of the reciprocal lattice lies at point O, and for simplicity a single reciprocal lattice point is shown at P at a distance $1/d_{hk\ell}$ from O.

The beam, the sphere of reflection, and the camera (film) are considered to be locked together. Likewise, the crystal and its reciprocal lattice are considered to be locked together. In a diffraction experiment, the crystal and reciprocal lattice are rotated about the origin until a reciprocal lattice point P touches the sphere of reflection. Whenever a reciprocal lattice point intersects the sphere of reflection, the Bragg condition is satisfied and reflection occurs along the line CPF.

The geometry of the sphere of reflection reveals that the Bragg condition is satisfied when point P lies on the sphere.

$$\sin \theta = \frac{OP}{OX} = \frac{1/d_{hk\ell}}{2/\lambda} \tag{38-4}$$

or

$$\lambda = 2 d_{hk\ell} \sin \theta \tag{38-5}$$

which is Bragg's Law. The sphere of reflection simplifies understanding the geometrical relationships between the diffraction apparatus and the reciprocal lattice.

Diffractometers and x-ray diffraction cameras are devices (Table 38-1) for bringing points on the reciprocal lattice onto the sphere of reflection and detecting or recording the resulting reflection. A number of mechanical means in use for accomplishing this arrangement are shown in Table 38-1.

The Rotation Photograph

The mechanical features of a rotation photograph are shown in Figure 38-3. A crystal is mounted so that the rotation axis coincides with a principal crystal axis. The film, rolled into a cylinder with an opening to admit the x-ray beam, is stationary. The axis of the cylinder is collinear

TABLE 38-1

X-Ray Diffraction Apparatus

Name	Detection	State
Debye-Scherrer	Film	Powder
Weissenberg	Film	Single crystal
Buerger precession	Film	Single crystal
Four-circle diffractometer	Scintillation counter	Single crystal

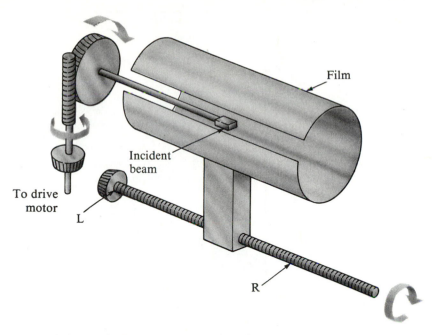

Figure 38–3 The mechanical arrangement of the film and camera for a rotation photograph. The film is stationary as the crystal rotates. The gears at the lower left are not engaged.

with the rotation axis, and the x-ray beam is perpendicular to the rotation axis.

When the film is unrolled and developed, the diffraction pattern consists of a series of "layer lines." The plane (in the rolled film) of the central layer line is parallel to the x-ray beam. On each side of the layer line lie pairs of layer lines, which are mirror images of each other (Fig. 38–4). Each line is displaced a distance y from the central line and the plane (in the rolled film) of each line.

Origin of Rotation Photograph Diffraction Pattern

Figure 38–5a shows the relationship between the crystal (reciprocal lattice), the sphere of reflection, the x-ray beam, and the film. In this figure think of the film, beam, and sphere as fixed, while the lattice slowly (about 1 rpm) rotates 360°.

The reciprocal lattice in Figure 38–5a is drawn to emphasize reciprocal lattice planes.[2] As the crystal rotates, the points on the reciprocal lattice intercept the sphere of reflection, Bragg's Law is satisfied, and a

[2] Not to be confused with planes in the real crystal. A plane in the real crystal has a unique set of Miller indices $hk\ell$. A plane in the reciprocal lattice is a *family* of planes having *one common* Miller index.

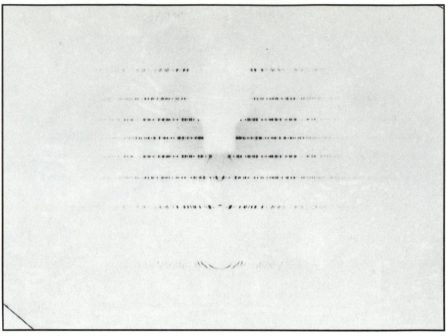

Figure 38–4 A rotation photograph of a crystal of *tris*-acetylacetonatoruthenium(III).
Because the rotation is along the *b* axis, the spacing between the layer lines permits
calculation of the repeat distance along the *b* axis.

reflection takes place. Consider point P in the reciprocal lattice. As the
lattice rotates from the position shown in Figure 38–5a, point P eventually
reaches point S on the Ewald sphere, and Bragg's Law is satisfied for this
reciprocal lattice point (set of planes in the real crystal). The direction
of the diffracted beam is given by a line passing through C, the center of
the Ewald sphere, and S, the reciprocal lattice point. When this line is
extended, it hits the film at F, resulting in a black spot on the developed film.

In Figure 38–5a, the crystal is shown mounted so that it rotates
along the b^* axis. All reciprocal lattice points $0k0$ lie on the rotation axis.
The rotation axis is tangent to the Ewald sphere at the origin (000). The
reciprocal lattice layer that includes the origin consists of the $h0\ell$ plane.
When the reciprocal lattice points in the $h0\ell$ reciprocal lattice plane (Fig.
38–5a) pass through the Ewald sphere, the resulting diffracted beams
generate the **central layer line,** also labeled $h0\ell$ in Figure 38–5b.

The reciprocal lattice planes in Figure 38–5b lying immediately
above and below the $h0\ell$ level are the $h1\ell$ and $h\bar{1}\ell$ levels, respectively.
When these planes of reciprocal lattice points pass through the Ewald
sphere, they generate the layer lines immediately above and below the
central layer line. Some higher-level layer lines are also labeled in Figure
38–5b; they can be recognized in Figure 38–4, an actual photograph of
the layer lines.

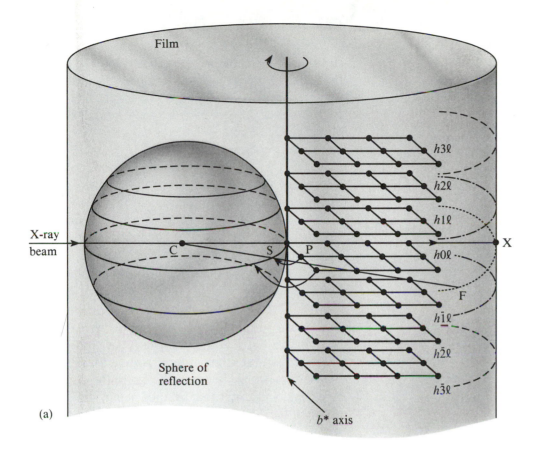

(a)

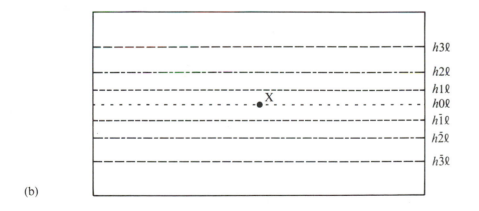

(b)

Figure 38–5 The rotation photograph and the sphere of reflection. The x-ray beam, the sphere of reflection, and the film are stationary. As the crystal rotates, its reciprocal lattices slice through the sphere of reflection. As each point P in the reciprocal lattice touches the sphere of reflection, the Bragg condition is satisfied and a reflected beam CP zaps the films on a layer line. When the film is unrolled, the layer lines appear as shown. The undiffracted beam would strike the film at X if it were not caught in a beam stop. The shadow of the beam stop is visible in Figure 38–3.

Calculation of the Unit Cell Translation Along the Rotation Axis

In Figure 38–5a, the rotation axis is labeled the $b*$ axis, and you can see that the reciprocal lattice planes are separated by $b*$. From this figure you can see that the separation between the observed layer lines on the film is directly related to the separation between the reciprocal lattice plane and indirectly related to b, the unit-cell translation of the direct (crystal) lattice.

The Ewald sphere simplifies the relationship between the reciprocal lattice and the direct lattice of the crystal. In Figure 38–2, y represents the distance on the film between a layer line and the central line in, say, millimeters. This is the only parameter measured in a rotation photograph. In both reciprocal space and real space, 2θ is the angle between the direct beam and the diffracted beam.

In the real space of the film, beam, and camera, for the nth layer line (Fig. 38–2),

$$\tan 2\theta_n = \frac{y_n}{r} = \frac{2y_n}{d} = \frac{2y_n}{57.296} \qquad \text{(38–6)}$$

where r and d are the camera radius and diameter, respectively, y is the measured distance between the central layer line on the film and the nth layer line, and $2y_n$ is the distance between two corresponding layers, e.g., between the 300 layer line and the $\bar{3}$00 layer line.

In reciprocal space for the nth layer line (Fig. 38–2),

$$\sin 2\theta_n = \frac{n d^*_{hk\ell}}{1/\lambda} = \frac{n\lambda}{d_{hk\ell}} \qquad \text{(38–7)}$$

or

$$d_{hk\ell} = \frac{n\lambda}{\sin\left(\tan^{-1}\left(2y_n/d\right)\right)} \qquad \text{(38–8)}$$

Equation 38–8 shows that the repeat distance along the rotation axis is a function of the layer line separation $2y_n$ only, for a given camera diameter d and radiation wavelength λ. Some sample data calculated with Equation 38–8 are shown in Table 38–2.

TABLE 38–2

Rotation Photograph Calculation (camera diameter equal to 57.3 mm, $\lambda = 1.5418\text{Å}$)

Layer Line	$2Y_n$ (mm)	$\tan 2\theta_n$	$\sin 2\theta_n$	d_{100} (Å)
1	12.0	0.2094	0.2050	7.521
2	25.8	0.4503	0.4106	7.510
3	44.6	0.7784	0.6143	7.530
			Average	7.520

Apparatus

Rotation photographs are taken with a Weissenberg camera with the film screen removed (see Figs. 38–3 and 38–6) and the camera bed locked motionless. Weissenberg cameras are available from the Charles E. Supper Co., Massachusetts, the Nonius Company, Delft, Holland, and other manufacturers in England and Germany. Used cameras are occasionally offered for sale in the monthly bulletin of the American Crystallographic Association, a division of the American Institute of Physics.

Camera Setup for Rotation Photographs

A. The worm-gear engaging lever at the right side of the camera bed should be disengaged so that the bed runs freely back and forth.

B. Center the bed with respect to the beam collimator and fix it in place by sliding the two integrating stops up against each side of the spoked wheel. The wheel is in front of the bed; the stops slide along the millimeter scale in front of the carriage. Tighten gently. Figure 38–6 shows the setup at this point.

C. Remove the collimator.

D. Place the film cassette on the four points of suspension of the bed. Take care that the small wheel of the right rear suspension point fits into the groove on the right side of the cassette. The "wing" of the cassette should be forward and resting on the black pin of the integrating arm. If gentle finger pressure is applied to the bottom of the wing, the cassette should rotate readily.

E. Replace the collimator. It should not touch the cassette slot.

Figure 38–6 Camera with film holder removed. (Karen R. Sime)

Figure 38–7 Setup for rotation photograph. (Karen R. Sime)

   F. Check that both set screws at the ends of the rotation axis are tight. Figure 38–7 shows the setup at this point.

   G. Plug the camera electric cord into the table outlet.

   H. Set the tube selector on power supply to "Cu K Alpha."

   I. Follow directions for turning on the power.

   J. Note for the oscillation photograph:

      1. Set the "switch stops" on the run-of-scale wheel at the right-hand end of the rotation axis so that oscillation is through 10 to 15 degrees.

      2. Exposure time is about 15 minutes for oscillation, 1 hour for rotation.

SAFETY CONSIDERATIONS

X rays for diffraction cameras are sufficiently intense to be dangerous. All x-ray tubes should be equipped with warning lights and an illuminated international radiation symbol (Fig. 38–6). The junction between the camera and the tube should be well shielded and periodically monitored. The camera and power generator should be housed in a room of their own. Even if it is believed that the radiation level is insignificant, you should not loiter in the x-ray room when the power is on.

    The cables to the x-ray tubes carry exceedingly dangerous levels of electrical power, typically 40,000 volts at 20 milliamperes.

    X-ray film developers contain organic reducing reagents to which many people are sensitive or even allergic.

EXPERIMENTAL PROCEDURE

Your instructor may choose to furnish you with a crystal already mounted in a goniometer head (Fig. 38–8) and aligned with a crystal axis along the rotation axis of the camera. The relationship between the collimator, goniometer, and beam stop may be seen in Figures 38–6 and 38–9. Disengage the camera so that the spindle can be turned by hand. While rotating the crystal slowly, examine it with the camera telescope. While being rotated, the crystal should not appear to translate up or down. If you must start from scratch, follow the steps outlined below.

Mounting the Crystal. A variable-power microscope (2 to 50×) is a useful tool for mounting crystals. Choose a substance that crystallizes in needles if possible. The presence of a heavy atom (Z > 21) will shorten the exposure time relative to that of a hydrocarbon. An orthorhombic or tetragonal crystal is easier to investigate than a monoclinic or triclinic one. Lonsdale (1964) has suggested urea (tetragonal) and hexamethylene tetramine (cubic). The short dimension of a needle should be about 0.3 to 0.6 mm.

Figure 38–8 Goniometer. (Karen R. Sime)

Figure 38–9 Collimator, beam stop, and goniometer. (Karen R. Sime)

The needle is glued to a short piece of glass fiber (drawn from a rod). The needle is touched to some fresh glue and then to the crystal. Some quick-drying glues dry from the outside to form a nonsticky film on the surface, so that the needle may not stick. Epoxy glues that polymerize tend not to form such a film and are handy for mounting crystals.

The glass fiber should be mounted in a brass pin that fits into the goniometer head. Hold the pin with tweezers, heat it until it is just too hot to touch, and then touch it to a piece of beeswax. The melted wax will fill the hole in the brass mounting pin, and if a glass fiber is inserted while the wax is hot, the wax will support the fiber when the pin cools. The pin is then inserted into the goniometer head.

Optical Alignment. If the needle is well formed, with sharp parallel edges, optical alignment may suffice to get a satisfactory rotation photograph. Methods for more precise alignment are described by Stout and Jensen (1968). The following steps for the optical alignment of a crystal are for a Nonius Weissenberg camera but are essentially the same for other Weissenberg cameras.

A. No screens or camera holder should be in place.
B. Rotate the beam stop 1/4 turn up out of the vision path.
C. Locate the two knurled set screws near the black knob at the far right end of the rotation axis.
   1. Tighten the larger set screw (parallel to the rotation axis) and loosen the smaller set screw (perpendicular to the rotation axis.)
   2. The black knob may now be turned to place the crystal in the center of the cross hairs; then retighten the small screw.
   3. Now loosen the large set screw; this permits the crystal and the goniometer to be rotated manually for optical alignment.
D. First with one, then the other, goniometer scale parallel to the rotation axis, adjust the angle setting to bring the needle parallel to the horizontal cross hair. Readjust the translation on the goniometer head after each angle adjustment.
E. Repeat this maneuver until manual rotations show no translatory movement of the crystal (up or down) with the needle axis remaining parallel to the horizontal cross hair through 360° rotation.
F. Turn the beam stop down to intercept the beam.

Loading the Camera with Film. The steps described below apply to a Nonius camera but are nearly identical for other common Weissenberg cameras.

A. In total darkness, trim a 5-by-7-inch sheet of x-ray film to 5 by 6 5/16 inches with a paper cutter and trim off one corner.
B. Place a film cylinder vertically before you with the open slot opposite; note the film stop along the slot on the inside.
C. Place the film in black opaque paper with the cut corner to the lower left.
D. Roll the film and paper so they are slightly smaller than the cylinder diameter and insert them; slide the film snugly up against the film stop in one direction and snugly down into the groove on the inside of the film cylinder on the end.
E. To make the film lie flat, install the "slot clamp" and press against the film edge before turning the wing nuts to hold it in place.
F. Finally, the circular end piece is clipped in place.
G. Before turning on the lights in the darkroom, run a fingernail along both inside edges to determine that the ends of the film are inside the grooves and protected from light.

Taking the Exposure. The details of the x-ray power generator, timer, etc., vary from lab to lab and will be provided by your instructor.

Instead of taking a full 360° rotation photograph, you may wish to take an oscillation photograph. For this the camera is set so that the crystal rotates 10° to 20° back and forth instead of a full 360°. This approach has the advantage that the exposure time is correspondingly shorter.

Results and Calculations

Measure the distance between layer lines and calculate the unit-cell translation along the rotation axis with Equation 38–8.

Compare your results with a value from the literature. Calculate the propagated error in the repeat distance.

REFERENCES

1. Azaroff, L. V., *Elements of X-Ray Crystallography*, McGraw-Hill, New York, 1968.
2. Buerger, M. J., *X-Ray Crystallography*, Wiley, New York, 1942.
3. Lonsdale, K., "Crystal Structure Analysis for Undergraduates," *J. Chem. Educ.* 41:240–244, 1964.
4. Nuffield, E. W., *X-Ray Diffraction Methods*, Wiley, New York, 1966.
5. Stout, G. H., and L. H. Jensen, *X-Ray Structure Determination*, Macmillan, London, 1968.

Experiment 39

The Weissenberg Method

Examination of Figure 38–5 in the previous experiment reveals that in a rotation photograph the two-dimensional information of a reciprocal lattice plane is compressed into a one-dimensional layer line. Consequently, it is difficult to index the reflections. Furthermore, it is necessary to remount the crystal to determine the other two cell constants, unless the crystal is cubic.

The **Weissenberg method** is a moving-film technique that spreads the points of a single layer line out over the film. Because the two-dimensional reciprocal lattice plane is now photographed on a two-dimensional region of film, it is a relatively simple matter to index the diffraction pattern. In addition, the remaining two cell constants can be calculated.

Camera Mechanics

The same camera is used for Weissenberg and rotation photographs. In order to isolate and photograph a single reciprocal lattice plane, diffraction spots from undesired reciprocal lattice layers are screened off by means of a slotted metal cylinder placed between film and crystal. Diffracted beams from the desired reciprocal lattice layer pass through the slot onto the film, while the cylindrical metal shield absorbs all other diffracted beams.

In order to spread the layer line over the film, the cylindrical film is moved parallel to the rotation axis as the crystal rotates. The layer line screen is stationary. Figures 39–1 and 39–2 show a view of the Weissenberg camera with the screen tubes in place, defining between them a slot that is parallel to the x-ray beam and perpendicular to the rotation axis. The cylindrical film holder is in place. The film holder sits on a bed that moves back and forth parallel to the rotation axis of the crystal. The rotary motion of the crystal is synchronized with the translatory motion of the film (Fig. 39–1) so that 1 mm of film translation (in the x direction) exactly corresponds to 2° of rotation. In addition, the camera's radius $(28.65° = 90°/\pi)$ is such that 1 mm of film (in the y direction) corresponds to 1° of diffraction angle. Compare Figure 39–1 with Figure 38–7 and Figure 39–2 with Figure 38–3.

Origin of Row Lines

Figure 39–3 shows a cross section through a Weissenberg camera, i.e., an end-on view of the cylindrical film, the sphere of reflection, and two central rows of a reciprocal lattice. The central rows are three in number and intersect at the origin of the reciprocal lattice. In Figure 39–3, only reciprocal lattice points lying on the $h00$ and 00ℓ rows are shown.

Figure 39–1 Weissenberg camera with screen tubes and film holder. (Karen R. Sime)

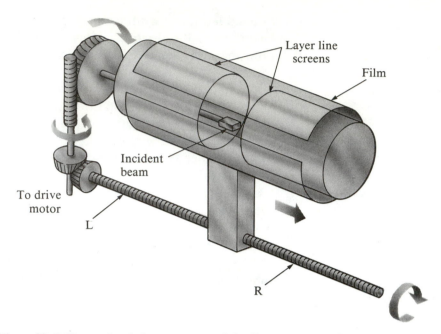

Figure 39-2 The mechanical arrangement of the film and camera for a Weissenberg photograph. The screen is stationary, but the film translates as the crystal rotates. As the long drive screw rotates, the screwing motion in the camera causes the camera to translate to the right, as the arrow indicates. The connecting worm gears also cause the crystal to rotate. When the camera reaches R, a switch reverses the drive motor, causing the camera to reverse its translatory motion back toward L, where the cycle repeats. The stationary cylindrical metal screens inside the film permit isolating the reflections belonging to a single selected layer line.

The $0k0$ row (not shown) passes through the origin out of the plane of the paper. The acute angle between the row lines is β^*, where $\beta = 180° - \beta^*$.

In Figure 39-3, consider the film, the sphere of reflection, and the x-ray beam to be fixed. Allow the reciprocal lattice to rotate counterclockwise. As the reciprocal lattice rotates successively through rotation angles ω_1, ω_2, ω_3, and ω_4, reciprocal lattice points 001, 002, 003, and 004 intercept the sphere of reflection, giving rise to diffracted beams CF_1, CF_2, CF_3, and CF_4. When the film is unrolled and developed, diffraction spots arising from the 001 row line lie on a line of slope equal to 2 on which the spots are labeled 001, 002, 003, and 004, as shown in Figure 39-3.

After the reciprocal lattice has rotated through β^*, a new row line begins to form on the upper half of the film, starting with the 100 reflection and followed by the 200, 300, 400, and 500 (Fig. 39-3). Since the film translates 1 mm for every 2° of reciprocal lattice rotation, the film will have translated $\beta^*/2$ mm from the point at which the 001 row line appears to the point at which the $h00$ row line appears. In other words, X_1 (measured in millimeters) equals $\beta^*/2$ (measured in degrees). After 180° of

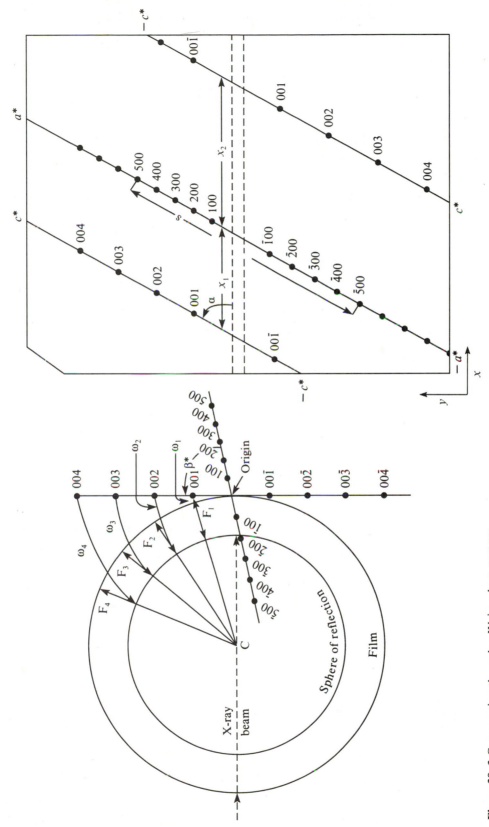

Figure 39-3 Cross section through a Weissenberg camera.

rotation, the 001 point is in the same position as the 001 point was at the start of the rotation. As a result,

$$X_1 + X_2 = 90 \text{ mm always}$$

$$X_1 = X_2 = 45 \text{ mm if } \beta^* = 90°$$

Calculating Cell Dimensions from a Weissenberg Photograph

Since 1 mm along the long, or y, direction of the film corresponds to 1° of Bragg angle θ, it is quite simple to measure the Bragg angle for any diffraction spot lying on a row line. The d spacing of the plane can then be calculated directly from the Bragg equation:

$$d_{hk\ell} = \lambda/2 \sin \theta \qquad (39\text{--}1)$$

For row lines in Figure 39–3, the calculated spacings are d_{h00} and $d_{00\ell}$.

Sample Calculation. In practice it may be convenient to measure the distance s between Friedel pairs[1] and calculate the distance y from the slope, which equals 2, of the row line as shown in Figure 39–3.

$$y = s \cos \alpha = s \cos (\tan^{-1} 2) = 0.89445s \qquad (39\text{--}2)$$

In Figure 39–4, the measured distance between the $(14\,0\,0)$ reflection and the $(\overline{14}\,0\,0)$ reflection is 127.0 mm, so that the Bragg angle θ can be calculated as follows:

$$\theta = y/2 = 0.8944 \times 127.0/2 = 56.79° \qquad (39\text{--}3)$$

The spacing between the $d_{14\,0\,0}$ planes is then calculated from Bragg's Law in the usual manner:

$$d_{1400} = \lambda/2 \sin \theta = 1.5418 \text{ Å}/2 \sin 56.79 = 0.9900 \text{ Å} \qquad (39\text{--}4)$$

Next we can calculate d_{100} from $d_{14\,0\,0}$:

$$d_{100} = hd_{14\,0\,0} = 14 \times 0.9900 = 13.86 \text{ Å} \qquad (39\text{--}5)$$

In an orthogonal crystal, $\alpha = \beta = \gamma = 90°$, so that

$$a = d_{100} \qquad (39\text{--}6)$$

$$b = d_{010} \qquad (39\text{--}7)$$

$$c = d_{001} \qquad (39\text{--}8)$$

In a monoclinic crystal, $\alpha = \gamma = 90°$ and $\beta \neq 90°$. Examination of Figure 38–1 shows that

$$a = \frac{d_{100}}{\sin \beta} \qquad (39\text{--}9)$$

---

[1] Reciprocal lattice points $hk\ell$ and $\overline{hk\ell}$ are Friedel pairs.

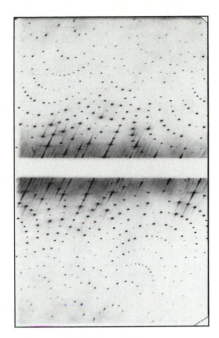

Figure 39–4 Weissenberg photograph of the $a*c*$ net.

$$b = d_{010} \qquad \text{(39–10)}$$

$$a = \frac{d_{100}}{\sin \beta}$$

Thus, in this example,

$$a = \frac{d_{100}}{\sin \beta} = \frac{13.86 \text{ Å}}{\sin 99.2} = 14.04 \text{ Å} \qquad \text{(39–11)}$$

The cell parameter c is calculated in the same manner, but $b = d_{010}$ since $\alpha = \gamma = 90°$ in a monoclinic crystal.

The angle β is calculated from the distance between row lines. In the sample Weissenberg photograph (Fig. 39–4), the measured distance X_1 equals 49.6 mm, so that

$$\beta = 2°/\text{mm} \times 49.6 \text{ mm} = 99.2° \qquad \text{(39–12)}$$

Apparatus

The apparatus is the same as that used for the rotation photograph in the previous experiment, but an additional accessory is used, the layer line screen.

SAFETY CONSIDERATIONS

See the previous experiment.

EXPERIMENTAL PROCEDURE

Mounting the Crystal. See the discussion of the rotation method in the previous experiment.

Aligning the Crystal. If the crystal is aligned for a rotation photograph, no further alignment is necessary. If the crystal is mounted on the b axis, i.e., the crystal is rotated about the b axis, then a rotation photograph gives the cell parameter b, while the Weissenberg photograph with the same setting gives the cell parameters a, c, and β.

Taking the Exposure. After setting up the camera, but before turning on the x rays, turn on the camera motor and let the camera run long enough that you can see the placement of the camera bed on the track. Allow the bed to translate as far as mechanically possible.

The exposure time for a full Weissenberg photograph (about 200° oscillation) may require several hours or even overnight, so it may be necessary to set an automatic timer on the power supply. Check the manual or ask your instructor for the details of setting the timer.

Results and Calculations

Measure the distance between several Friedel pairs on two principal row lines. Calculate the cell parameters associated with those row lines.

If your crystal is monoclinic and mounted on the b axis, measure the distance between row lines and calculate β.

From the uncertainties in the distances measured on the film, calculate the propagated errors in the cell parameters.

If you have previously made and interpreted a rotation photograph of the crystal with the same mounting, you now have a complete set of cell dimensions if the crystal is of monoclinic or higher symmetry. In that case, calculate the volume of the unit cell.

If time permits, measure the density of the crystal by the flotation method. Prepare a mixture of miscible liquids in which your crystal is totally insoluble. Drop two or three small crystals in a 10-ml graduated cylinder containing the liquid mixture. Adjust the density of the mixture by adding one of the liquids until the crystal neither floats nor sinks. The density of the liquid now equals the density of the crystal.

By comparing the x-ray density with the measured density, calculate the number (Z) of moieties of your compound contained in the unit cell, which usually equals 1, 2, or 4.

REFERENCES

1. Azaroff, L. V., *Elements of X-Ray Crystallography,* McGraw-Hill, New York, 1968.

2. Buerger, M. J., *X-Ray Crystallography,* Wiley, New York, 1942.

3. Lonsdale, K., "Crystal Structure Analysis for Undergraduates," *J. Chem. Educ.* 41:240–244, 1964.

4. Nuffield, E. W., *X-Ray Diffraction Methods,* Wiley, New York, 1966.

5. Stout, G. H., and L. H. Jensen, *X-Ray Structure Determination,* Macmillan, London, 1968.

Experiment 40

The Buerger Precession Photograph: The Precession Method

If you have been thinking that the reciprocal lattice is a weird and convoluted way of looking at a crystal, then you are in for a pleasant surprise. It is possible to take an undistorted photograph of the reciprocal lattice, one layer at a time. And this is almost as easy as taking a snapshot of your best friend. Figure 40–1 shows two zero-layer precession photographs of a monoclinic crystal, *tris*-ruthenium acetonate (Chao, 1973).

In a monoclinic crystal one axis is perpendicular to both the other axes. By convention this axis is always called the *b* axis and in the jargon of the trade is often referred to as the **unique axis.**

The crystal in Figure 40–1 was mounted on the b^* axis. After taking a photograph of the $hk0$ layer (a), the crystal was rotated by β (99.1° in this crystal) and a photograph of the $0k\ell$ layer (b) was taken. Consequently the b^* axis appears identically as the horizontal set of points in both photographs. Because b^* is the unique axis, both the a^*b^* and c^*b^* nets are orthogonal.

Because the interpretation of the reciprocal lattice as displayed in a precession photograph is so simple, the precession camera is one of the most commonly used tools in x-ray crystallography.

Theory

In the precession method, not only the crystal but the film moves during an exposure to x rays. In contrast to a rotation photograph, the crystal does not rotate, but instead the crystal and film *precess* about the x-ray beam (Fig. 40–2). The film and the crystal (i.e., a reciprocal lattice plane) are held parallel at all times while they precess about the x-ray beam.

The geometric relationship between the reciprocal lattice plane, the film, the beam, the sphere of reflection, and the precession angle μ is shown in Figure 40–2. In Figure 40–2, consider the Ewald sphere and the x-ray beam to be stationary while the film and lattice plane precess about the beam.

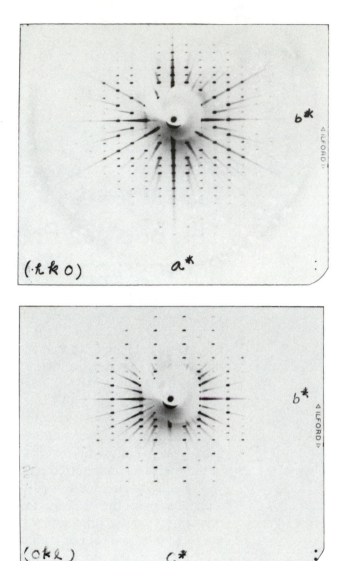

Figure 40–1 Precession photographs of a monoclinic crystal. Since the crystal is mounted along the b axis, the b^* row line is the central horizontal row line on both photographs. The crystal was rotated through the angle β between photographs.

The lines ON and O′N′ are normals to the lattice plane and film plane, respectively. The angle between these normals and the x-ray beam is the precession angle $\bar{\mu}$. Consider the normals ON and O′N′ coupled or locked together and always perpendicular to the lattice and film planes. As ON and O′N′ precess about the beam, so do the lattice and film planes.

The precession motion of the reciprocal lattice plane brings reciprocal lattice points onto the sphere of reflection, as shown by point P on Figures 40–2 and 40–3. As usual, when a reciprocal lattice point falls on the Ewald sphere, Bragg's Law is satisfied, and diffraction occurs. The direction of the diffracted beam lies along line CPF. The lattice point P shows up as a spot on the film at point F.

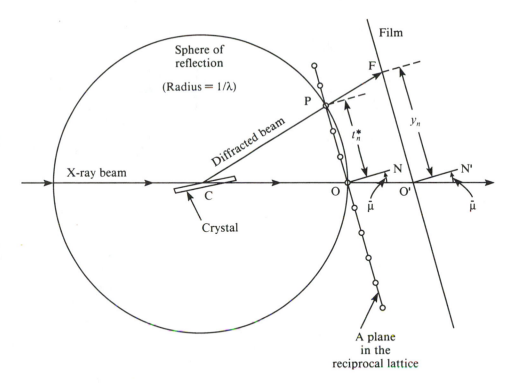

Figure 40–2 The precession motion. The beam and sphere of reflection are stationary. The reciprocal lattice plane and the film are always parallel as they precess about points O and O′. The precession angle is $\bar{\mu}$. The crystal is mounted so that the reciprocal lattice plane is a principal plane: $hk0$, $h0\ell$, or $0k\ell$. Whenever a reciprocal lattice point P slices the sphere of reflection, the Bragg condition is satisfied, and the diffracted beam (CP) zaps the film at F. See also Figure 40–3.

Inclination of the lattice layer by the precession angle $\bar{\mu}$ causes the lattice plane to intersect the sphere of reflection on a circle of diameter OP, as shown in Figure 40–3. As the lattice plane precesses about the x-ray beam, the circle of intersection (dashed line, Fig. 40–3) sweeps out a larger circle (dotted line, Fig. 40–3). Since all of the points within this dotted circle fall on the sphere of reflection during one precession revolution, they all cause a corresponding diffraction spot on the film, shown schematically in Figure 40–4. The intensities of the spots are different, depending on several physical factors as well as on the positions of the atoms in the unit cell. The intensities of Friedel pairs ($hk\ell$, $\bar{h}\bar{k}\bar{\ell}$) are, however, identical, as Figure 40–4 indicates.

The Buerger Precession Camera

The mechanical arrangement for allowing the crystal and film to remain parallel to each other while precessing about the x-ray beam is shown schematically in Figure 40–5 (Buerger, 1964). The beam and the sphere of reflection are rigid and unmovable. The crystal and the film are held in two forks, which are coupled together mechanically. The motion about the forks' vertical axes combined with the corresponding

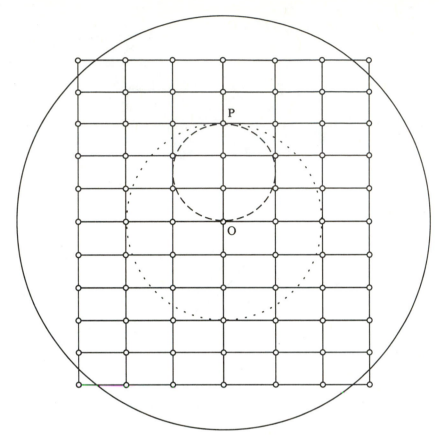

Figure 40–3 Diffraction in a precession camera. The dashed circle shows where the reciprocal lattice plane in Figure 40–2 slices the sphere of reflection. As the reciprocal lattice plane precesses about point O, the slicing extends out to those reciprocal lattice points included within the dotted circle.

horizontal motion of the axes between the tongs of the forks results in the two normals ON and O'N' precessing about the x-ray beam.

A typical precession camera is shown in Figure 40–6. The annular ring between the crystal and the film is necessary to screen off all reciprocal lattice layers except the one layer desired. The precession rate is about one revolution per minute, and exposure times range from a few minutes to several hours depending on crystal size, compositions, and precession angle. A small precession angle results in shorter exposure time, but fewer lattice points fall on the sphere of reflection (compare Figs. 40–2 and 40–3).

Calculation of Lattice Constants from a Precession Photograph

Point P in Figures 40–2 and 40–3 is a lattice point that lies a distance (in reciprocal space) t_n^* from the origin, where

$$d_n^* = t_n^* / n$$

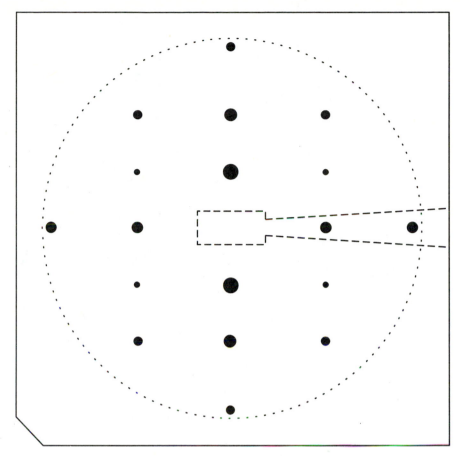

Figure 40–4 The diffraction pattern on the film. The shadow of the beam catcher blocks out two spots. Notice that the symmetry of the diffraction intensity follows Friedel's law: $I_{hk\ell} = I_{\bar{h}\bar{k}\bar{\ell}}$.

From corresponding parts of similar triangles

$$\frac{\text{OP}}{\text{OC}} = \frac{\text{O'F}}{\text{O'C}} \tag{40-1}$$

$$\frac{t_n^*}{1/\lambda} = \frac{y_n}{M}$$

$$d = \frac{nM\lambda}{y_n}$$

The film to crystal distance M is usually 60.00 mm and is a constant. When M is expressed in mm, y_n must be measured in mm; then d has whatever units are chosen for λ, usually nm or Å. When the radiation is MoK_α, λ equals 0.7107 Å, so

$$d = \frac{60.00 \times 0.7170 \times n}{y_n} \tag{40-2}$$

$$d = \frac{42.64n}{y_n}$$

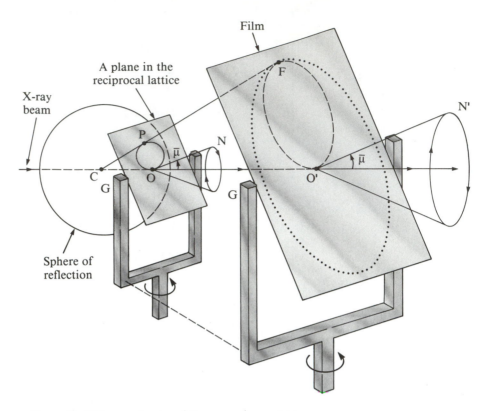

Figure 40–5 The mechanics of the precession camera.

Figure 40–6 The Nonius precession camera. (Karen R. Sime)

where n is the number of row lines between which the spacing is measured on the film. For accurate measurements, the row lines should be widely spaced.

As an example let us calculate d_{010} from the repeat distance along the $b*$ axis. The distance between the 050 and the $0\bar{5}0$ reflections is 56.7 mm. These lie at the extreme left and right sides of both photos in Figure 40-1.

$$d_{010} = \frac{42.64 \times 10}{56.7} = 7.52 \text{ Å}$$

which agrees well with the value 7.53 Å obtained by a rotation photograph (previous experiment). In a monoclinic crystal, b and $b*$ are collinear so $b = d_{010} = 7.52$ Å also.

It is important to remember that Equation 40-2 gives the interplanar spacing, either d_{100}, d_{010}, or d_{001}, but not necessarily a, b, or c, depending on the row line measured. Because b and $b*$ are collinear, $b = d_{010}$. However, $a \neq d_{100}$ and $c \neq d_{001}$ in a monoclinic crystal. The relationship between a or c and the corresponding interplanar spacing is shown in Figure 38-1.

$$a = \frac{d_{100}}{\sin \beta} \tag{40-3}$$

$$c = \frac{d_{001}}{\sin \beta} \tag{40-4}$$

and, for the sake of completeness, for a monoclinic crystal

$$b = d_{100}$$

In cubic, tetragonal, and orthorhombic cells, $\beta = 90$, $\sin \beta = 1$, and the cell d spacings correspond directly to the repeat distances a, b, and c.

Apparatus

The apparatus consists of a Buerger precession camera, a goniometer, a suitable crystal, mounting supplies, and a darkroom.

EXPERIMENTAL PROCEDURE

In order to measure β it is preferable to mount a monoclinic crystal along $a*$ or $c*$. When mounted along $a*$ and precessed about $b*$, the precession photograph shows the $a*c*$ net. The angle between $\alpha*$ and $\gamma*$ is $\beta*$, which can be readily measured directly from the film. A second photograph obtained after rotating the crystal through 90° gives the $a*b*$ net (when mounted about $a*$). From two such photographs, film measurements give a, b, c, and β, completely characterizing the monoclinic unit cell. The relationship between β and $\beta*$ is

$$\beta = 180 - \beta*$$

It is conventional to choose β to be greater than $90°$ and $\beta*$ less than $90°$.

Higher-Symmetry Crystals. If the crystal is orthorhombic or monoclinic, it is necessary to make two zero-level photographs. With an orthorhombic crystal, this is simple because the direct axes and reciprocal axes are collinear. If one photograph contains the $a*c*$ net, then rotation through $90°$ reveals the $a*b*$ net if the crystal is mounted along the a axis.

If the crystal is cubic or tetragonal, it is necessary to make only one zero-level precession photograph to obtain all the cell dimensions (if the tetragonal crystal is mounted along the nonunique axis).

Results and Calculations

If your crystal is monoclinic and mounted along $a*$ or $c*$, measure the angle between $a*$ and $c*$ on the $a*c*$ net, and report $\beta*$ and β. Measure the distance between row lines and calculate d_{100} and d_{001} with Equation 40–2. Calculate a and c with Equations 40–3 and 40–4. If you have a second photograph with the same mounting but with the crystal having been rotated through $90°$, measure the vertical row spacing and calculate d_{010} ($=b_0$). Verify that the angle between $b*$ and $a*$ is $90°$.

If your crystal is of higher symmetry, calculate a_0 if it is cubic, a and b if tetragonal, or a, b, and c if orthorhombic.

REFERENCES

1. Buerger, M. J., *The Precession Method in X-Ray Crystallography,* Wiley, New York, 1964.
2. Buerger, M. J., *X-Ray Crystallography,* Wiley, New York, 1942.
3. Stout, G. H. and L. H. Jensen, *X-Ray Structure Determination,* Macmillan, London, 1968.
4. Nuffield, E. W., *X-Ray Diffraction Methods,* Wiley, New York, 1966.
5. Azaroff, L. V., *Elements of X-Ray Crystallography,* McGraw-Hill, New York, 1968.
6. Chao, G. K., R. L. Sime, and R. J. Sime, *Acta. Cryst.* B29:2845, 1973.
7. Chao, G. K., Master's Thesis, California State University, Sacramento, 1973.

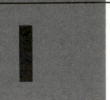

Bilingual Guide to Landolt-Börnstein

This guide permits the user who has no knowledge whatsoever of the German language to locate numerical data in the **sixth edition** of Landolt-Börnstein with ease. Landolt-Börnstein has no index, but instead has a multilevel table of contents, the first three levels of which are presented here, facing an English translation. No guide is necessary for the **New Series,** since the introduction, tables of contents, and other information are printed in English or in German and English.

Constants of Transistors and Boundary Layers

The International System of Units (SI)

The fifth edition of *Le Systeme International d'Unites (SI)* was published in French in 1985 by the Bureau International des Poids et Mesures (BIPM), known in English as the International Bureau of Weights and Measures and identified by the same acronym. The BIPM operates under the supervision of the Comité International des Poids et Mesures (CIPM), known in English as the International Committee for Weights and Measures, also CIPM. The CIPM edits the international journal *Metrologia,* in which articles relevant to SI units, symbols, and constants are published (in English). The CIPM, in turn, operates under the authority of the Conference General des Poids et Mesures (CGPM), which in English is the General Conference on Weights and Measures, also CGPM. The CGPM consists of delegates from all member countries, which numbered 45 as of December 31, 1980. The CGPM meets every four years to improve, confirm, and communicate the International System of Units (SI). The agency in the United States that cooperates with the CGPM on these matters is the National Institute of Standards and Technology (NIST), until recently known as the National Bureau of Standards (NBS). The NIST also publishes changes in the International System of Units in its organ, the *Journal of Research of the National Institute of Science and Technology.* Periodically the NIST summarizes the status and history of SI units in a publication called NBS Special Publication 330, "The International System of Units (SI)." It was published in 1981 and 1986 with the same title and number and will probably appear again in the early 1990s as NIST Special Publication 330.

Length

The **meter (m)** is the length of the path traveled by light in a vacuum in a measured time equal to 1/299,792,458 second. The SI unit of area is

TABLE I

Basic SI Units of Measure

Physical Quantity	Name of Unit	Symbol
Length	meter	m
Mass	kilogram	kg
Time	second	s
Electric current	ampere	A
Thermodynamic temperature	kelvin	K
Luminous intensity	candela	cd
Amount of substance	mole	mol
Plane angle	radian	rad
Solid angle	steradian	str

the square meter (m^2). Fluid volume is measured by the liter (0.001 m^3). However, the 12th CGPM declared that "the word 'liter' may be employed as a special name for the cubic decimeter" (dm^3). The 16th CGPM declared that "the name liter, although not included in the International System of Units, must be admitted for general use within the System," and elected, "as an exception, to adopt the two symbols l and

TABLE 2

Names and Symbols for Certain SI-Derived Units

Physical Quantity	Name	Symbol	Expression in Terms of Other Units	Expression in Terms of SI Base Units
Force	newton	N		$m \, kg \, s^{-2}$
Pressure	pascal	Pa	$N \, m^{-2}$	$m^{-1} \, kg \, s^{-2}$
Energy, work, heat	joule	J	$N \, m$	$m^2 \, kg \, s^{-2}$
Power, radiant flux	watt	W	$J \, s^{-1}$	$m^2 \, kg \, s^{-3}$
Electric charge	coulomb	C	$A \, s$	$s \, A$
Electric potential	volt	V	W/A	$m^2 \, kg \, s^{-3} \, A^{-1}$
Capacitance	farad	F	C/V	$m^{-2} \, kg^{-1} \, s^4 \, A^2$
Inductance	henry	H	$V \, A^{-1} \, s$	$m^2 \, kg \, s^{-2} \, A^{-2}$
Electric resistance	ohm	Ω	V/A	$m^2 \, kg \, s^{-3} \, A^{-2}$
Electric conductance	siemen	S	A/V	$m^{-2} \, kg^{-1} \, s^3 \, A^2$
Magnetic flux	weber	Wb	$V \, s$	$m^2 \, kg \, s^{-2} \, A^{-1}$
Magnetic flux density	tesla	T	$V \, s \, m^{-2}$	$kg \, s^{-2} \, A^{-1}$
Frequency	hertz	Hz		s^{-1}
Activity (of a radionuclide)	becquerel	Bq		s^{-1}
Adsorbed dose	gray	Gy	$J \, kg^{-1}$	$m^2 \, s^{-2}$
Dose-equivalent	sievert	Sv	J/kg	$m^2 \, s^{-2}$

Sources: National Bureau of Standards, "Policy for NBS Usage of SI Units," *J. Chem. Educ.* 48: 569–572, 1971; and National Bureau of Standards Special Publication 330, U.S. Government Printing Office, Washington, D.C., 1986.

TABLE 3

SI-Derived Units Without Special Symbols

Physical Quantity	SI Unit Name	Expression in Terms of SI Base Units
Area	square meter	m^2
Volume	cubic meter	m^3
Speed, velocity	meter per second	m/s
Acceleration	meter per second squared	m/s^2
Wave number	1/meter	m^{-1}
Density	kilogram per cubic meter	kg/m^3
Specific volume	cubic meter per kilogram	m^3/kg
Current density	ampere per square meter	A/m^2
Magnetic field strength	ampere per meter	A/m
Concentration	mole per cubic meter	mol/m^3
Luminance	candela per square meter	cd/m^2

Sources: National Bureau of Standards, "Policy for NBS Usage of SI Units," *J. Chem. Educ.* 48: 569–572, 1971; and National Bureau of Standards Special Publication 330, U.S. Government Printing Office, Washington, D.C., 1986.

L as symbols to be used for the unit liter." Consequently, any of the following may be freely used:

$$1 \text{ L} = 1 \text{ l} = 1 \text{ dm}^3 = 10^{-3} \text{ m}^3 = 1000 \text{ cm}^3 = 1000 \text{ ml} = 1000 \text{ mL}$$

Mass

The standard for the unit of mass, the **kilogram (kg),** is a cylinder of platinum-iridium alloy kept by the International Bureau of Weights and Measures at Paris, France. A duplicate in the custody of the National Bureau of Standards serves as the mass standard for the United States. This is the only SI base unit still defined by an artifact.

Time

The **second (s)** is defined as the duration of 9,192,631,770 cycles of the radiation associated with a specified transition of the cesium-133 atom.

Electric Current

The **ampere (A)** is that constant current which, if maintained in two straight parallel conductors of infinite length, of negligible circular cross section, and placed 1 meter apart in vacuum, would produce between these conductors a force equal to 2×10^{-7} newton per meter of length. The SI unit of voltage is the volt (V), which is defined as 1 V = 1 W/

TABLE 4

∎

Some SI-Derived Units Having Special Names

| Quantity | SI Unit | | |
	Name	Symbol	Expression in Terms of SI Base Units
Dynamic viscosity	Pascal second	Pa s	m^{-1} kg s^{-1}
Moment of force	newton meter	N m	m^2 kg s^{-2}
Surface tension	newton per meter	N/m	kg s^{-2}
Heat flux density	watt per square meter	W/m^2	kg s^{-3}
Heat capacity, entropy	joule per kelvin	J/K	m^2 kg s^{-2} K^{-1}
Specific heat capacity	joule per kilogram kelvin	J/(kg K)	m^2 s^{-2} K^{-1}
Specific energy	joule per kilogram	J/kg	m^2 s^{-2}
Thermal conductivity	watt per meter kelvin	W(m K)	m kg s^{-3} K^{-1}
Energy density	joule per cubic meter	J/m^3	m^{-1} kg s^{-2}
Electric field strength	volt per meter	V/m	m kg s^{-3} A^{-1}
Electric charge density	coulomb per cubic meter	C/m^3	m^{-3} s A
Electric flux density	coulomb per square meter	C/m^2	m^{-2} s A
Permittivity	farad per meter	F/m	m^{-3} kg^{-1} s^4 A^2
Permeability	henry per meter	H/m	m kg s^{-2} A^{-2}
Molar energy	energy per mole	J/mol	m^2 kg s^{-2} mol^{-1}
Molar entropy, heat capacity	joule per mole kelvin	J/(mol K)	m^2 kg s^{-2} K^{-1} mol^{-1}
Exposure (x and γ rays)	coulomb per kilogram	C/kg	kg^{-1} s A

Sources: National Bureau of Standards, "Policy for NBS Usage of SI Units," *J. Chem. Educ.* 48: 569–572, 1971; and National Bureau of Standards Special Publication 330, U.S. Government Printing Office, Washington, D.C., 1986.

1 A. The SI unit of resistance is the ohm (Ω), which is defined as 1 Ω = 1 V/1 A.

On January 1, 1990, new practical reference standards for the volt and ohm went into effect.[1,2]

Thermodynamic Temperature

The thermodynamic, or Kelvin, scale of temperature used in SI has its origin at absolute zero. The triple point of water is defined as 273.16 K. The **kelvin (K)** is the fraction 1/273.16 of the thermodynamic temperature of the triple point of water. The Celsius scale is derived from the

[1] Taylor, B. N., "New Measurement Standards for 1990," *Physics Today* 42:23–26, 1989.
[2] Quinn, T. J., "News from the BIPM," *Metrologia* 26:69–74, 1989.

Kelvin scale and is defined by the equation $t\,°C = T - T_0$, where $T_0 = 273.15$ K by definition. In other words, $0\,°C$ is defined to be exactly 0.01 K lower than the triple point of water.

A new temperature scale[3] was implemented on January 1, 1990: the International Temperature Scale of 1990 (ITS-90). It supersedes the International Practical Temperature Scale of 1968, Amended Edition of 1975 [IPTS-68(75)]. The difference between the old and new scales amounts to approximately $\pm 0.2\,°C$ in the range from $630\,°C$ to $1064\,°C$. Details of the implementation of ITS-90 should appear in an NIST publication, "Guidelines for Realizing the International Temperature Scale of 1990 (ITS-90)," NIST Technical Note 1265.

Amount of Substance

The **mole (mol)** is the amount of substance of a system that contains as many elementary entities as there are atoms in 0.012 kg of carbon-12.

Luminous Intensity

The **candela (cd)** is the luminous intensity, in the perpendicular direction, of a surface of $1/600,000$ square meter of a blackbody at the temperature of freezing platinum under a pressure of 101,325 newtons per square meter.

Plane Angle

The **radian (rad)** is the plane angle between two radii of a circle that cut off on the circumference an arc equal in length to the radian.

Solid Angle

The **steradian (str)** is the solid angle that, having its vertex in the center of a sphere, cuts off an area of the surface of the sphere equal to that of a square with sides of length equal to the radius of the sphere.

REFERENCES

Tables were adapted from the following:
1. National Bureau of Standards, *J. Chem. Educ.* 48:569, 1971.
2. National Bureau of Standards Special Publication 330, U.S. Government Printing Office, Washington, D.C., 1986.

[3] Magnum, B. W., National Institute of Standards and Technology, Gaithersburg, Maryland 20899 (private communication, August 1989).

Thermoelectric Voltage Versus Temperature

Thermoelectric Voltage as a Function of Temperature (°C) for Type K[a] Thermocouples, Reference Junctions at 0°C

°C	Absolute Millivolts											°C
	0	1	2	3	4	5	6	7	8	9	10	
−270	−6.458											−270
−260	−6.441	−6.444	−6.446	−6.448	−6.450	−6.452	−6.453	−6.455	−6.456	−6.457	−6.458	−260
−250	−6.404	−6.408	−6.413	−6.417	−6.421	−6.425	−6.429	−6.432	−6.435	−6.438	−6.441	−250
−240	−6.344	−6.351	−6.358	−6.364	−6.371	−6.377	−6.382	−6.388	−6.394	−6.399	−6.404	−240
−230	−6.262	−6.271	−6.280	−6.289	−6.297	−6.306	−6.314	−6.322	−6.329	−6.337	−6.344	−230
−220	−6.158	−6.170	−6.181	−6.192	−6.202	−6.213	−6.223	−6.233	−6.243	−6.253	−6.262	−220
−210	−6.035	−6.048	−6.061	−6.074	−6.087	−6.099	−6.111	−6.123	−6.135	−6.147	−6.158	−210
−200	−5.891	−5.907	−5.922	−5.936	−5.951	−5.965	−5.980	−5.994	−6.007	−6.021	−6.035	−200
−190	−5.730	−5.747	−5.763	−5.780	−5.796	−5.813	−5.829	−5.845	−5.860	−5.876	−5.891	−190
−180	−5.550	−5.569	−5.587	−5.606	−5.624	−5.642	−5.660	−5.678	−5.695	−5.712	−5.730	−180
−170	−5.354	−5.374	−5.394	−5.414	−5.434	−5.454	−5.474	−5.493	−5.512	−5.531	−5.550	−170
−160	−5.141	−5.163	−5.185	−5.207	−5.228	−5.249	−5.271	−5.292	−5.313	−5.333	−5.354	−160
−150	−4.912	−4.936	−4.959	−4.983	−5.006	−5.029	−5.051	−5.074	−5.097	−5.119	−5.141	−150
−140	−4.669	−4.694	−4.719	−4.743	−4.768	−4.792	−4.817	−4.841	−4.865	−4.889	−4.912	−140
−130	−4.410	−4.437	−4.463	−4.489	−4.515	−4.541	−4.567	−4.593	−4.618	−4.644	−4.669	−130
−120	−4.138	−4.166	−4.193	−4.221	−4.248	−4.276	−4.303	−4.330	−4.357	−4.384	−4.410	−120
−110	−3.852	−3.881	−3.910	−3.939	−3.968	−3.997	−4.025	−4.053	−4.082	−4.110	−4.138	−110
−100	−3.553	−3.584	−3.614	−3.644	−3.674	−3.704	−3.734	−3.764	−3.793	−3.823	−3.852	−100
−90	−3.242	−3.274	−3.305	−3.337	−3.368	−3.399	−3.430	−3.461	−3.492	−3.523	−3.553	−90
−80	−2.920	−2.953	−2.985	−3.018	−3.050	−3.082	−3.115	−3.147	−3.179	−3.211	−3.242	−80
−70	−2.586	−2.620	−2.654	−2.687	−2.721	−2.754	−2.788	−2.821	−2.854	−2.887	−2.920	−70
−60	−2.243	−2.277	−2.312	−2.347	−2.381	−2.416	−2.450	−2.484	−2.518	−2.552	−2.586	−60
−50	−1.889	−1.925	−1.961	−1.996	−2.032	−2.067	−2.102	−2.137	−2.173	−2.208	−2.243	−50
−40	−1.527	−1.563	−1.600	−1.636	−1.673	−1.709	−1.745	−1.781	−1.817	−1.853	−1.889	−40
−30	−1.156	−1.193	−1.231	−1.268	−1.305	−1.342	−1.379	−1.416	−1.453	−1.490	−1.527	−30
−20	−0.777	−0.816	−0.854	−0.892	−0.930	−0.968	−1.005	−1.043	−1.081	−1.118	−1.156	−20
−10	−0.392	−0.431	−0.469	−0.508	−0.547	−0.585	−0.624	−0.662	−0.701	−0.739	−0.777	−10
−0	0.000	−0.039	−0.079	−0.118	−0.157	−0.197	−0.236	−0.275	−0.314	−0.353	−0.392	−0

797

	Absolute Millivolts											
°C	0	1	2	3	4	5	6	7	8	9	10	°C
0	0.000	0.039	0.079	0.119	0.158	0.198	0.238	0.277	0.317	0.357	0.397	0
10	0.397	0.437	0.477	0.517	0.557	0.597	0.637	0.677	0.718	0.758	0.798	10
20	0.798	0.838	0.879	0.919	0.960	1.000	1.041	1.081	1.122	1.162	1.203	20
30	1.203	1.244	1.285	1.325	1.366	1.407	1.448	1.489	1.529	1.570	1.611	30
40	1.611	1.652	1.693	1.734	1.776	1.817	1.858	1.899	1.940	1.981	2.022	40
50	2.022	2.064	2.105	2.146	2.188	2.229	2.270	2.312	2.353	2.394	2.436	50
60	2.436	2.477	2.519	2.560	2.601	2.643	2.684	2.726	2.767	2.809	2.850	60
70	2.850	2.892	2.933	2.975	3.016	3.058	3.100	3.141	3.183	3.224	3.266	70
80	3.266	3.307	3.349	3.390	3.432	3.473	3.515	3.556	3.598	3.639	3.681	80
90	3.681	3.722	3.764	3.805	3.847	3.888	3.930	3.971	4.012	4.054	4.095	90
100	4.095	4.137	4.178	4.219	4.261	4.302	4.343	4.384	4.426	4.467	4.508	100
110	4.508	4.549	4.590	4.632	4.673	4.714	4.755	4.796	4.837	4.878	4.919	110
120	4.919	4.960	5.001	5.042	5.083	5.124	5.164	5.205	5.246	5.287	5.327	120
130	5.327	5.368	5.409	5.450	5.490	5.531	5.571	5.612	5.652	5.693	5.733	130
140	5.733	5.774	5.814	5.855	5.895	5.936	5.976	6.016	6.057	6.097	6.137	140
150	6.137	6.177	6.218	6.258	6.298	6.338	6.378	6.419	6.459	6.499	6.539	150
160	6.539	6.579	6.619	6.659	6.699	6.739	6.779	6.819	6.859	6.899	6.939	160
170	6.939	6.979	7.019	7.059	7.099	7.139	7.179	7.219	7.259	7.299	7.338	170
180	7.338	7.378	7.418	7.458	7.498	7.538	7.578	7.618	7.658	7.697	7.737	180
190	7.737	7.777	7.817	7.857	7.897	7.937	7.977	8.017	8.057	8.097	8.137	190
200	8.137	8.177	8.216	8.256	8.296	8.336	8.376	8.416	8.456	8.497	8.537	200
210	8.537	8.577	8.617	8.657	8.697	8.737	8.777	8.817	8.857	8.898	8.938	210
220	8.938	8.978	9.018	9.058	9.099	9.139	9.179	9.220	9.260	9.300	9.341	220
230	9.341	9.381	9.421	9.462	9.502	9.543	9.583	9.624	9.664	9.705	9.745	230
240	9.745	9.786	9.826	9.867	9.907	9.948	9.989	10.029	10.070	10.111	10.151	240
250	10.151	10.192	10.233	10.274	10.315	10.355	10.396	10.437	10.478	10.519	10.560	250
260	10.560	10.600	10.641	10.682	10.723	10.764	10.805	10.846	10.887	10.928	10.969	260
270	10.969	11.010	11.051	11.093	11.134	11.175	11.216	11.257	11.298	11.339	11.381	270
280	11.381	11.422	11.463	11.504	11.546	11.587	11.628	11.669	11.711	11.752	11.793	280
290	11.793	11.835	11.876	11.918	11.959	12.000	12.042	12.083	12.125	12.166	12.207	290
300	12.207	12.249	12.290	12.332	12.373	12.415	12.456	12.498	12.539	12.581	12.623	300
310	12.623	12.664	12.706	12.747	12.789	12.831	12.872	12.914	12.955	12.997	13.039	310
320	13.039	13.080	13.122	13.164	13.205	13.247	13.289	13.331	13.372	13.414	13.456	320
330	13.456	13.497	13.539	13.581	13.623	13.665	13.706	13.748	13.790	13.832	13.874	330
340	13.874	13.915	13.957	13.999	14.041	14.083	14.125	14.167	14.208	14.250	14.292	340
350	14.292	14.334	14.376	14.418	14.460	14.502	14.544	14.586	14.628	14.670	14.712	350
360	14.712	14.754	14.796	14.838	14.880	14.922	14.964	15.006	15.048	15.090	15.132	360
370	15.132	15.174	15.216	15.258	15.300	15.342	15.384	15.426	15.468	15.510	15.552	370
380	15.552	15.594	15.636	15.679	15.721	15.763	15.805	15.847	15.889	15.931	15.974	380
390	15.974	16.016	16.058	16.100	16.142	16.184	16.227	16.269	16.311	16.353	16.395	390
400	16.395	16.438	16.480	16.522	16.564	16.607	16.649	16.691	16.733	16.776	16.818	400
410	16.818	16.860	16.902	16.945	16.987	17.029	17.072	17.114	17.156	17.199	17.241	410
420	17.241	17.283	17.326	17.368	17.410	17.453	17.495	17.537	17.580	17.622	17.664	420
430	17.664	17.707	17.749	17.792	17.834	17.876	17.919	17.961	18.004	18.046	18.088	430
440	18.088	18.131	18.173	18.216	18.258	18.301	18.343	18.385	18.428	18.470	18.513	440
450	18.513	18.555	18.598	18.640	18.683	18.725	18.768	18.810	18.853	18.895	18.938	450
460	18.938	18.980	19.023	19.065	19.108	19.150	19.193	19.235	19.278	19.320	19.363	460
470	19.363	19.405	19.448	19.490	19.533	19.576	19.618	19.661	19.703	19.746	19.788	470

°C	0	1	2	3	4	5	6	7	8	9	10	°C
					Absolute Millivolts							
480	19.788	19.831	19.873	19.916	19.959	20.001	20.044	20.086	20.129	20.172	20.214	480
490	20.214	20.257	20.299	20.342	20.385	20.427	20.470	20.512	20.555	20.598	20.640	490
500	20.640	20.683	20.725	20.768	20.811	20.853	20.896	20.938	20.981	21.024	21.066	500
510	21.066	21.109	21.152	21.194	21.237	21.280	21.322	21.365	21.407	21.450	21.493	510
520	21.493	21.535	21.578	21.621	21.663	21.706	21.749	21.791	21.834	21.876	21.919	520
530	21.919	21.962	22.004	22.047	22.090	22.132	22.175	22.218	22.260	22.303	22.346	530
540	22.346	22.388	22.431	22.473	22.516	22.559	22.601	22.644	22.687	22.729	22.772	540
550	22.772	22.815	22.857	22.900	22.942	22.985	23.028	23.070	23.113	23.156	23.198	550
560	23.198	23.241	23.284	23.326	23.369	23.411	23.454	23.497	23.539	23.582	23.624	560
570	23.624	23.667	23.710	23.752	23.795	23.837	23.880	23.923	23.965	24.008	24.050	570
580	24.050	24.093	24.136	24.178	24.221	24.263	24.306	24.348	24.391	24.434	24.476	580
590	24.476	24.519	24.561	24.604	24.646	24.689	24.731	24.774	24.817	24.859	24.902	590
600	24.902	24.944	24.987	25.029	25.072	25.114	25.157	25.199	25.242	25.284	25.327	600
610	25.327	25.369	25.412	25.454	25.497	25.539	25.582	25.624	25.666	25.709	25.751	610
620	25.751	25.794	25.836	25.879	25.921	25.964	26.006	26.048	26.091	26.133	26.176	620
630	26.176	26.218	26.260	26.303	26.345	26.387	26.430	26.472	26.515	26.557	26.599	630
640	26.599	26.642	26.684	26.726	26.769	26.811	26.853	26.896	26.938	26.980	27.022	640
650	27.022	27.065	27.107	27.149	27.192	27.234	27.276	27.318	27.361	27.403	27.445	650
660	27.445	27.487	27.529	27.572	27.614	27.656	27.698	27.740	27.783	27.825	27.867	660
670	27.867	27.909	27.951	27.993	28.035	28.078	28.120	28.162	28.204	28.246	28.288	670
680	28.288	28.330	28.372	28.414	28.456	28.498	28.540	28.583	28.625	28.667	28.709	680
690	28.709	28.751	28.793	28.835	28.877	28.919	28.961	29.002	29.044	29.086	29.128	690
700	29.128	29.170	29.212	29.254	29.296	29.338	29.380	29.422	29.464	29.505	29.547	700
710	29.547	29.589	29.631	29.673	29.715	29.756	29.798	29.840	29.882	29.924	29.965	710
720	29.965	30.007	30.049	30.091	30.132	30.174	30.216	30.257	30.299	30.341	30.383	720
730	30.383	30.424	30.466	30.508	30.549	30.591	30.632	30.674	30.716	30.757	30.799	730
740	30.799	30.840	30.882	30.924	30.965	31.007	31.048	31.090	31.131	31.173	31.214	740
750	31.214	31.256	31.297	31.339	31.380	31.422	31.463	31.504	31.546	31.587	31.629	750
760	31.629	31.670	31.712	31.753	31.794	31.836	31.877	31.918	31.960	32.001	32.042	760
770	32.042	32.084	32.125	32.166	32.207	32.249	32.290	32.331	32.372	32.414	32.455	770
780	32.455	32.496	32.537	32.578	32.619	32.661	32.702	32.743	32.784	32.825	32.866	780
790	32.866	32.907	32.948	32.990	33.031	33.072	33.113	33.154	33.195	33.236	33.277	790
800	33.277	33.318	33.359	33.400	33.441	33.482	33.523	33.564	33.604	33.645	33.686	800
810	33.686	33.727	33.768	33.809	33.850	33.891	33.931	33.972	34.013	34.054	34.095	810
820	34.095	34.136	34.176	34.217	34.258	34.299	34.339	34.380	34.421	34.461	34.502	820
830	34.502	34.543	34.583	34.624	34.665	34.705	34.746	34.787	34.827	34.868	34.909	830
840	34.909	34.949	34.990	35.030	35.071	35.111	35.152	35.192	35.233	35.273	35.314	840
850	35.314	35.354	35.395	35.435	35.476	35.516	35.557	35.597	35.637	35.678	35.718	850
860	35.718	35.758	35.799	35.839	35.880	35.920	35.960	36.000	36.041	36.081	36.121	860
870	36.121	36.162	36.202	36.242	36.282	36.323	36.363	36.403	36.443	36.483	36.524	870
880	36.524	36.564	36.604	36.644	36.684	36.724	36.764	36.804	36.844	36.885	36.925	880
890	36.925	36.965	37.005	37.045	37.085	37.125	37.165	37.205	37.245	37.285	37.325	890
900	37.325	37.365	37.405	37.445	37.484	37.524	37.564	37.604	37.644	37.684	37.724	900
910	37.724	37.764	37.803	37.843	37.883	37.923	37.963	38.002	38.042	38.082	38.122	910
920	38.122	38.162	38.201	38.241	38.281	38.320	38.360	38.400	38.439	38.479	38.519	920
930	38.519	38.558	38.598	38.638	38.677	38.717	38.756	38.796	38.836	38.875	38.915	930
940	38.915	38.954	38.994	39.033	39.073	39.112	39.152	39.191	39.231	39.270	39.310	940

					Absolute Millivolts							
°C	0	1	2	3	4	5	6	7	8	9	10	°C
950	39.310	39.349	39.388	39.428	39.467	39.507	39.546	39.585	39.625	39.664	39.703	950
960	39.703	39.743	39.782	39.821	39.861	39.900	39.939	39.979	40.018	40.057	40.096	960
970	40.096	40.136	40.175	40.214	40.253	40.292	40.332	40.371	40.410	40.449	40.488	970
980	40.488	40.527	40.566	40.605	40.645	40.684	40.723	40.762	40.801	40.840	40.879	980
990	40.879	40.918	40.957	40.996	41.035	41.074	41.113	41.152	41.191	41.230	41.269	990
1,000	41.269	41.308	41.347	41.385	41.424	41.463	41.502	41.541	41.580	41.619	41.657	1,000
1,010	41.657	41.696	41.735	41.774	41.813	41.851	41.890	41.929	41.968	42.006	42.045	1,010
1,020	42.045	42.084	42.123	42.161	42.200	42.239	42.277	42.316	42.355	42.393	42.432	1,020
1,030	42.432	42.470	42.509	42.548	42.586	42.625	42.663	42.702	42.740	42.779	42.817	1,030
1,040	42.817	42.856	42.894	42.933	42.971	43.010	43.048	43.087	43.125	43.164	43.202	1,040
1,050	43.202	43.240	43.279	43.317	43.356	43.394	43.432	43.471	43.509	43.547	43.585	1,050
1,060	43.585	43.624	43.662	43.700	43.739	43.777	43.815	43.853	43.891	43.930	43.968	1,060
1,070	43.968	44.006	44.044	44.082	44.121	44.159	44.197	44.235	44.273	44.311	44.349	1,070
1,080	44.349	44.387	44.425	44.463	44.501	44.539	44.577	44.615	44.653	44.691	44.729	1,080
1,090	44.729	44.767	44.805	44.843	44.881	44.919	44.957	44.995	45.033	45.070	45.108	1,090
1,100	45.108	45.146	45.184	45.222	45.260	45.297	45.335	45.373	45.411	45.448	45.486	1,100
1,110	45.486	45.524	45.561	45.599	45.637	45.675	45.712	45.750	45.787	45.825	45.863	1,110
1,120	45.863	45.900	45.938	45.975	46.013	46.051	46.088	46.126	46.163	46.201	45.238	1,120
1,130	46.238	46.275	46.313	46.350	46.388	46.425	46.463	46.500	46.537	46.575	46.612	1,130
1,140	46.612	46.649	46.687	46.724	46.761	46.799	46.836	46.873	46.910	46.948	46.985	1,140
1,150	46.985	47.022	47.059	47.096	47.134	47.171	47.208	47.245	47.282	47.319	47.356	1,150
1,160	47.356	47.393	47.430	47.468	47.505	47.542	47.579	47.616	47.653	47.689	47.726	1,160
1,170	47.726	47.763	47.800	47.837	47.874	47.911	47.948	47.985	48.021	48.058	48.095	1,170
1,180	48.095	48.132	48.169	48.205	48.242	48.279	48.316	48.352	48.389	48.426	48.462	1,180
1,190	48.462	48.499	48.536	48.572	48.609	48.645	48.682	48.718	48.755	48.792	48.828	1,190
1,200	48.828	48.865	48.901	48.937	48.974	49.010	49.047	49.083	49.120	49.156	49.192	1,200
1,210	49.192	49.229	49.265	49.301	49.338	49.374	49.410	49.446	49.483	49.519	49.555	1,210
1,220	49.555	49.591	49.627	49.663	49.700	49.736	49.772	49.808	49.844	49.880	49.916	1,220
1,230	49.916	49.952	49.988	50.024	50.060	50.096	50.132	50.168	50.204	50.240	50.276	1,230
1,240	50.276	50.311	50.347	50.383	50.419	50.455	50.491	50.526	50.562	50.598	50.633	1,240
1,250	50.633	50.669	50.705	50.741	50.776	50.812	50.847	50.883	50.919	50.954	50.990	1,250
1,260	50.990	51.025	51.061	51.096	51.132	51.167	51.203	51.238	51.274	51.309	51.344	1,260
1,270	51.344	51.380	51.415	51.450	51.486	51.521	51.556	51.592	51.627	51.662	51.697	1,270
1,280	51.697	51.733	51.768	51.803	51.838	51.873	51.908	51.943	51.979	52.014	52.049	1,280
1,290	52.049	52.084	52.119	52.154	52.189	52.224	52.259	52.294	52.329	52.364	52.398	1,290
1,300	52.398	52.433	52.468	52.503	52.538	52.573	52.608	52.642	52.677	52.712	52.747	1,300
1,310	52.747	52.781	52.816	52.851	52.886	52.920	52.955	52.989	53.024	53.059	53.093	1,310
1,320	53.093	53.128	53.162	53.197	53.232	53.266	53.301	53.335	53.370	53.404	53.439	1,320
1,330	53.439	53.473	53.507	53.542	53.576	53.611	53.645	53.679	53.714	53.748	53.782	1,330
1,340	53.782	53.817	53.851	53.885	53.920	53.954	53.988	54.022	54.057	54.091	54.125	1,340
1,350	54.125	54.159	54.193	54.228	54.262	54.296	54.330	54.364	54.398	54.432	54.466	1,350
1,360	54.466	54.501	54.535	54.569	54.603	54.637	54.671	54.705	54.739	54.773	54.807	1,360
1,370	54.807	54.841	54.875									1,370

[a] Nickel-chromium alloy versus nickel-aluminum alloy.
Source: National Bureau of Standards Monograph 125, "Thermocouple Reference Tables Based on IPTS-68," U.S. Government Printing Office, Washington, D.C., 1974.

INDEX